T0406312

Engineering Geology for Society and Territory – Volume 6

Giorgio Lollino • Daniele Giordan
Kurosch Thuro • Carlos Carranza-Torres
Faquan Wu • Paul Marinos
Carlos Delgado
Editors

Engineering Geology for Society and Territory – Volume 6

Applied Geology for Major Engineering Projects

Part II

Editors
Giorgio Lollino
Daniele Giordan
Institute for Geo-Hydrological Protection
National Research Council (CNR)
Turin
Italy

Kurosch Thuro
Department of Engineering Geology
Technical University of Munich
Munich
Germany

Carlos Carranza-Torres
Department of Civil Engineering
University of Minnesota Duluth
Duluth, MN
USA

Faquan Wu
Institute of Geology and Geophysics
Chinese Academy of Sciences
Beijing
China

Paul Marinos
Geotechnical Engineering Department
National Technical University of Athens
Athens
Greece

Carlos Delgado
Escuela Universitaria de Ingeniería Técnica
de Obras Públicas
Universidad Politécnica de Madrid
Madrid
Spain

ISBN 978-3-319-09059-7 ISBN 978-3-319-09060-3 (eBook)
DOI 10.1007/978-3-319-09060-3
Springer Cham Heidelberg New York Dordrecht London

Library of Congress Control Number: 2014946956

Cover Illustration: Pont Ventoux, Val di Susa, north western Italy. Tunnel Boring Machine(TBM) used during the construction of the gallery used as deviation channel for a hydroelectric power plant. The TBM was used to drill a gallery of 4.3 km long and with circular diameter of 4.05 meters. *Photo*: Giorgio Lollino.

Printed on acid-free paper

Springer is part of Springer Science+Business Media (www.springer.com)

Foreword

It is our pleasure to present this volume as part of the book series on the Proceedings of the XII International IAEG Congress, Torino 2014.

For the 50th anniversary, the Congress collected contributions relevant to all themes where the IAEG members were involved, both in the research field and in professional activities.

Each volume is related to a specific topic, including:

1. Climate Change and Engineering Geology;
2. Landslide Processes;
3. River Basins, Reservoir Sedimentation and Water Resources;
4. Marine and Coastal Processes;
5. Urban Geology, Sustainable Planning and Landscape Exploitation;
6. Applied Geology for Major Engineering Projects;
7. Education, Professional Ethics and Public Recognition of Engineering Geology;
8. Preservation of Cultural Heritage.

The book series aims at constituting a milestone for our association, and a bridge for the development and challenges of Engineering Geology towards the future.

This ambition stimulated numerous conveners, who committed themselves to collect a large number of contributions from all parts of the world, and to select the best papers through two review stages. To highlight the work done by the conveners, the table of contents of the volumes maintains the structure of the sessions of the Congress.

The lectures delivered by prominent scientists, as well as the contributions of authors, have explored several questions ranging from scientific to economic aspects, from professional applications to ethical issues, which all have a possible impact on society and territory.

This volume testifies the evolution of engineering geology during the last 50 years, and summarizes the recent results. We hope that you will be able to find stimulating contributions, which will support your research or professional activities.

Giorgio Lollino

Carlos Delgado

Preface

Engineering geology, a relatively young field, emerged through recognition of the need for geologic input into engineering projects. Today, this primary field has expanded as the statutes of its learned society, the IAEG, define: "Engineering geology is the science devoted to the investigation, study and solution of the engineering and environmental problems which may arise as the result of the interaction between geology and the works and activities of man as well as to the development of measures for prevention or remediation of geological hazards."

The role of engineering geology for major engineering projects and infrastructure construction is well represented in the papers included in this volume of the proceeding of the 12th IAEG congress, devoted to major engineering projects. The geologic input is not only confined to the initial stage of such projects but the contribution of engineering geology includes all stages for their completion, reflecting the present standing of engineering geology in geotechnical engineering.

A retrospective review of the development of engineering geology shows that in the early days, up to the 1950s or even the 1960s, what was understood as engineering geology was restricted to assessments, with general and qualitative engineering descriptions. Then this is followed by a second period of development until about the 1980s. The demands of the development of society required more knowledge for the behaviour of the ground. Now meaningful geological models could be provided. However, the quantitative component was weak, and contributions to the design of structures were limited. Although improved, the understanding of geology in the engineering milieu is not satisfactory. A third period starts from the 1980s but mainly from the 1990s. Engineering geology, keeping the core values so far developed, is now evolving towards geoengineering.

Indeed, today engineering geology not only offers services but is also a substantial and an integral component of geotechnical engineering in construction. It is present in all phases of investigation, design and construction:

1. Engineering geology defines the geological conditions, provides the geological model (formations, tectonics and structure), and translates it into engineering terms, providing suitable ground profiles at the appropriate scale. Its role is decisive for detecting the presence of geological hazards, in the selection of the site or the alignment of the engineering structure and for the basic principles of the construction method. It makes no sense to proceed without a sound knowledge of the geological model. Let us be a little dogmatic here: in the absence or misinterpretation of the geological model the construction or operation will almost certainly be associated with problems either small or large, as accidents, delays, cost over-runs or even failures may occur. On the contrary, if this model is known from the very beginning of the design, half the game has already been won ... *if at the very start the geological structure of the site is misinterpreted, then any subsequent ... calculation may be so much labour in vain.* (Glossop 1968, 8th Rankine Lecture). Therefore: start from the forest and then look at the trees.

2. After having understood the behaviour of the ground, engineering geology contributes to the definition of the properties of the geometrical, the selection of suitable design parametres and of the appropriate criteria. This a stage with a close synergy with engineering. An understanding of in situ stresses and groundwater conditions complete this stage.
3. Engineering geology is and should also be present at the design phase to ensure that calculations and simulations do not misinterpret the geological reality. John Knill in his first Hans Cloos lecture, in 2002, expressed strong concern that the *effectiveness of the integration of engineering geology within the geotechnical engineering remains to be improved*. This integration is a field of development in today's engineering geology, and papers in this volume contribute towards such advance.
4. Engineering geology is involved in construction in order to validate the assumptions of the design, to contribute in the application of measures in unforeseen or unforeseeable circumstances and to secure the implementation of the contract.

And, undoubtedly, geological and engineering judgement should never be neglected in this whole process of creating an engineering project. Next to knowledge, experience is needed for this judgment. Mark Twain said *Good judgment comes from experience. But where does experience come from? Experience comes from bad judgment*. However, the correct application of geological and engineering principles means that experience can also come from good judgement.

It is very satisfactory that this volume of proceedings of the 12th congress of IAEG embraces all the above mentioned, and a large variety of cases of engineering works is presented. Dams and tunnels are the majority of these cases but also foundations, offshore structures, roads, railroads, slope design, construction material, tailings, repositories are dealt with. Papers on engineering properties and geotechnical classifications, site investigation issues and influence of groundwater are present together with contributions on the behaviour of soft rocks and weak rock masses. Active tectonics also attract special attention.

The volume is expected to constitute a valuable and lasting source of reference in the field of engineering geology, in particular, and in geotechnical engineering, in general.

Contents

Part I Keynote

1 **Problems in Buildings and Public Works Derived from Soils with Unesteable Structure and Soils with Large Volume Instability** 3
Carlos Delgado Alonso-Martirena

2 **Translating Geotechnical Risk in Financial Terms** 11
Alessandro Palmieri

3 **Large Deformation of Tunnel in Slate-Schistose Rock** 17
Faquan Wu, Jinli Miao, Han Bao, and Jie Wu

Part II Addressing Geological Uncertainties in Major Engineering Projects

4 **Effect of Petrogenesis on the Suitability of Some Pelitic Rocks as Construction Aggregates in the Tropics** 27
Tochukwu A.S. Ugwoke and Celestine O. Okogbue

5 **Geological Society of London Engineering Group Working Party on Periglacial and Glacial Engineering Geology** 31
David Giles, Martin Culshaw, Laurance Donnelly, David Evans, Mike de Freitas, James Griffiths, Sven Lukas, Christopher Martin, Anna Morley, Julian Murton, David Norbury, and Mike Winter

6 **New Methods of Determining Rock Properties for Geothermal Reservoir Characterization** 37
Mathias Nehler, Philipp Mielke, Greg Bignall, and Ingo Sass

7 **Application of Reliability Methods to Tunnel Lining Design in Weak Heterogeneous Rockmasses** 41
John C. Langford, N. Vlachopoulos, M.S. Diederichs, and D.J. Hutchinson

8 **Geological and Geotechnical Difference on Both Sides of the Same Tunnel** 47
Pedro Olivença and Vítor Santos

9 **Development of Probabilistic Geotechnical Ground Models for Offshore Engineering** 53
Konstantinos Symeonidis and Clark Fenton

10 **Baixo Sabor (Portugal) Upstream Dam Foundation: From Design Geological Predictions to Construction Geological Facts and Geotechnical Solutions** 59
Jorge Neves, Celso Lima, Fernando Ferreira, and João Machado

11 **The Foundations of Constructionsin Dobrogea—Romania, on WaterSensitive Soils, Loess** 65
Gabriela Brîndusa Cazacu, Nicolae Botu, and Daniela Grigore

12 **Influence of Micro-texture on the Geo-engineering Properties of Low Porosity Volcanic Rocks** 69
Ündül Ömer and Amann Florian

13 **Conceptual Geological Models, Its Importance in Interpreting Vadose Zone Hydrology and the Implications of Being Excluded** 73
Matthys A. Dippenaar and J. Louis van Rooy

14 **Treatment of Fossil Valley in Dam Area: A Case Study** 79
A.K. Singh and Bhatnagar Sharad

Part III Applied and Active Tectonics

15 **A Case Study of Three-dimensional Determination of Stress Orientation to Crystalline Rock Samples in Wenchuan Earthquake Fault Scientific Drilling Project Hole-2** 87
Weiren Lin, Lianjie Wang, Junwen Cui, Dongsheng Sun, and Manabu Takahashi

16 **Mathematical-Numerical Modeling of Tectonic Fault Zone (Tadzhikistan)** 91
Ernest V. Kalinin, Olga S. Barykina, and Leili L. Panasyan

17 **Neotectonic and Mass Movements on the New Fez-Taza Highway (Northern Morocco)** 95
Tabyaoui Hassan, Deffontaines Benoît, Chaouni Abdel-Ali, El Hammichi Fatima, Lahsaini Meriem, Mounadel Ahlam, Magalhaes Samuel, and Fortunato Gérardo

18 **Importance of Geological Map Updates in Engineering Geology, Application to the Rif-Chain and Its Foreland (Northern Morocco)** 101
Deffontaines Benoît, Tabyaoui Hassan, El Hammichi Fatima, Chaouni Abdel-Ali, Mounadel Ahlam, Lahsaini Meriam, Magalhaes Samuel, and Fortunato Gérardo

19 **Disaster Awareness Education for Children in Schools Around Geological Hazard Prone Areas in Indonesia** 107
Muslim Dicky, Evi Haerani, Motohiko Shibayama, Masaaki Ueshima, Naoko Kagawa, and Febri Hirnawan

20 **Analysis of Recent Deformation in the Southern Atlas of Tunisia Using Geomorphometry** 113
Mehdi Ben Hassen, Benoît Deffontaines, and Mohamed Moncef Turki

21 **Spatial Analysis of Remote Sensing Data in Early Stage of a Seismo-tectonic Research** 119
Novakova Lucie

22 **Geomorphic Evidence of Active Tectonics: The Case of Djemila Fault (Eastern Algeria)** 125
Youcef Bouhadad

23 **Seismic Cycle of the Southern Apennine Deformation Front: The Taranto Gulf Marine Terraces Inputs and Implications** 129
Benoît Deffontaines, Gérardo Fortunato, and Samuel Magalhaes

24 **New Structural and Geodynamic Coastal Jeffara Model (Southern Tunisia) and Engineering Implications** 139
Rim Ghedhoui, Benoît Deffontaines, and Mohamed Chedly Rabia

25 **Tunisia: A Mature Case Example of Structural Extrusion** 147
Benoît Deffontaines, Mehdi Ben Hassen, and Rim Ghedhoui

26 **The Extrusion of South-West Taiwan: An Offshore-Onshore Synthesis** 153
Benoît Deffontaines, Liu Char-Shine, and Chen Rou-Fei

27 **Active Tectonic Risk Assessment—Problems with Soil and Soft Sediment Deformation Structures** 161
Philip E.F. Collins

28 **Formation of Earthquake Faults by the Fukushima Hamadori Earthquake and an Estimation of Displacement Distribution Around the Faults Using Airborne LiDAR Data** 167
Shunsuke Shinagawa, Shuji Anan, Yasuhito Sasaki, Sakae Mukoyama, Shin-ichi Homma, and Yoko Kobayashi

Part IV Applied Geology for Infrastructure Projects

29 **Field Monitoring of the Behavior of Pile-Net Composite Foundation in Oversize-Deep-Soft Soil** 175
Yu-feng Wang, Qian-gong Cheng, and Jiu-jiang Wu

30 **Deformation Behavior of Excavated High Loess Slope Reinforced with Soil Nails and Pre-reinforced-Stabilizing Piles** 185
Qian-gong Cheng, Yu-feng Wang, and Jiu-jiang Wu

31 **Effects of Alkali Silica/Aggregate Reaction on Concrete Structures in Bundelkhand Region, Central India** 195
Suresh Chandra Bhatt and Bhuwan Chandra Joshi

32 **Applied Engineering Geology Methods for Exemplar Infrastructure Projects in Malopolskie and Podkarpackie Provinces** 203
Zbigniew Bednarczyk and Adam Szynkiewicz

33 **A Brief Overview of the Typical Engineering Characteristics of Tropical Red Soils** 211
George Brink

34 **Remote Analysis of Rock Slopes with Terrestrial Laser Scanning for Engineering Geological Tasks in Reservoir Planning** 215
Hieu Trung Nguyen, Tomás M. Fernandez-Steeger, Hans-Joachim Köhler, and Rafig Azzam

35 **Dynamically Loaded Anchorages** 219
Santoro Federica, Monia Calista, Antonio Pasculli, and Nicola Sciarra

36 **LiDAR and Discrete Fracture Network Modeling for Rockslide Characterization and Analysis** 223
Matthieu Sturzenegger, Tim Keegan, Ann Wen, David Willms, Doug Stead, and Tom Edwards

37 **Experimental Study on Water Sensitivity of the Red Sand Foundation in Angola** 229
Wei Zhang, Zhenghong Liu, Jianguo Zheng, Sumin Zhang, and Yongtang Yu

38 **Geological Characterization and Stability Conditions of the Motorway Tunnels of Arrangement Project of the NR43, Melbou (W. Béjaïa)** 237
Nassim Hallal and Rachid Bougdal

39 **A Modified Freeze-Thaw Laboratory Test for Pavement Sub Soils Affected by De-icing Chemicals** 243
Assel Sarsembayeva and Philip Collins

40 **A Geotechnical and Geochemical Characterisation of Oil Fire Contaminated Soils in Kuwait** 249
Humoud Al-Dahanii, Paul Watson, and David Giles

41 **Comparison Between Neural Network and Finite Element Models for the Prediction of Groundwater Temperatures in Heat Pump (GWHP) Systems** 255
Glenda Taddia, Stefano Lo Russo, and Vittorio Verda

42 **Ground Stiffness Evaluation Using the Soil Stiffness Gauge (SSG)** 259
Mário Quinta-Ferreira

43 **Influence of Fracture Systems and Weathering on the Sustainability of Rock Excavation Made for the Purpose of Infrastructure Construction** 263
Radoslav Varbanov, Miroslav Krastanov, and Rosen Nankin

44 **Route Alignment and Optimization of Railway Based on Geological Condition** 269
Weihua Zhao, Nengpan Ju, and Jianjun Zhao

45 **Engineering Properties of Permian Clay Tuffs** 273
John Johnston, Stephen Fityus, Olivier Buzzi, Chris Rodgers, and Robert Kingsland

46 **Vertical Harbour Quay Rehabilitation Using Ground Anchors** 279
Liliana Ribeiro and Alexandre Santos-Ferreira

47 **The Importance of the Existing Engineering Geological Conditions During the Building Construction on the Terrain Affected by Sliding** 285
Dragoslav Rakić, Zoran Berisavljević, Irena Basarić, and Uroš Đurić

48 **GIS Based, Heuristic Approach for Pipeline Route Corridor Selection** 291
Ludwig Schwarz, Klaus Robl, Walter Wakolbinger, Harry Mühling, and Pawel Zaradkiewicz

49 **A Study of Ground Natural Temperature Along Tabriz Metro Line 2, Iran** 295
Ebrahim Asghari–Kaljahi, Karim Yousefi-bavil, and Mahyar Babazadeh

50 **The Challenges of Site Investigations, Dredging, and Land Reclamation: A Port Hedland (Western Australia) Project Perspective** 299
P. Baker, J. Woods, M. Page, and F. Schlack

51 **The Role of Geological Analysis in the Design of Interventions for the Safety of the Road Asset. Some Examples** 303
Serena Scarano, Roberto Laureti, and Stefano Serangeli

52 **Groundwater Level Variation and Deformation in Clays Characteristic to the Helsinki Metropolitan Area** 309
Tiina-Liisa Toivanen and Jussi Leveinen

53 **Determine of Tunnel Face Stability Pressure in EPB Machine with Use Analytical Methods (Case Study: Mashhad Metro Line2)** 313
Mehdi Abbasi and Mohsen Abbasi

54 **Numerical Modeling of Interrelationships Between Linear Transportation Infrastructures and Hydro-geological Hazard in Floodplains** 317
Rosamaria Trizzino

55 **Convergence Predictions and Primary Support Optimization of the Tunnel Progon** 323
Zoran Berisavljevic, Svetozar Milenkovic, Dusan Berisavljevic, and Nenad Susic

56 **Quarry Site Selection and Geotechnical Characterization of Ballast Aggregate for Ambo-Ijaji Railway Project in Central Ethiopia: An Integrated GIS and Geotechnical Approach** 329
Regessa Bayisa, Raghuvanshi Tarun Kumar, and Kebede Seifu

57 **Radon Emanation Techniques as an Added Dimension in Site Investigation of Water Storage Facilities** 337
Gary Neil Davis and Mannie Levin

Part V Capturing and Communicating Geologic Variability and Uncertainty

58 **Improving Geotechnical Uncertainty Evaluation in Reliability-Based Design** 343
Fred H. Kulhawy

59 **Communicating Geological Uncertainty: The Use of the Conceptual Engineering Geological Model** 347
Christopher Jack and Steve Parry

60 **Evaluating the Effects of Input Cost Surface Uncertainty on Deep-Water Petroleum Pipeline Route Optimization** 351
William C. Haneberg

61 **Scanline Sampling Techniques for Rock Engineering Surveys: Insights from Intrinsic Geologic Variability and Uncertainty** 357
Helder I. Chaminé, Maria José Afonso, Luís Ramos, and Rogério Pinheiro

62 **A Suggested Geologic Model Complexity Rating System** 363
Jeffrey R. Keaton

63 **Managing Uncertainty in Geological Engineering Models for Open-Pit Feasibility** 367
Rosalind Munro and Jeffrey R. Keaton

Part VI Construction in Complex Geological Settings—The Problematic of Predicting the Nature of the Ground

64 **Engineering Geological and Geotechnical Cartographic Modeling as a Methodological Basis for Engineering Surveys and Design in Complex Geological Environment** 373
Felix Rivkin, I. Kuznetsova, A. Popova, I. Parmuzin, and I. Chehina

65 **Experiences Learned from Engineering Geological Investigation of Headrace Tunnel on Sedimentary Rock—Xekaman3 Hydropower Project—Lao PDR** 377
Nguyen Song Thanh and Dao Dang Minh

66 **Engineering Properties of Badlands in the Canadian Prairies** 381
Khan Fawad and Azam Shahid

67 **Case Studies of Post Investigation Geological Assessments: Hunter Expressway** ... 387
David J. Och, Robert Kingsland, Sudar Aryal, Henry Zhang, and Geoff Russell

68 **Contribution to the Behavior Study and Collapse Risk of Underground Cavities in Highly Saline Geological Formations** ... 393
Mohamed Chikhaoui, Ammar Nechnech, Dashnor Hoxha, and Kacem Moussa

69 **Seabed Properties for Anchoring Floating Structures in the Portuguese Offshore** ... 399
Joaquim Pombo, Aurora Rodrigues, and A. Paula F. da Silva

70 **Model of Permafrost Thaw Halo Formation Around a Pipeline** ... 405
Pavel Novikov, Elizaveta Makarycheva, and Valery Larionov

71 **Assessing Rock Mass Properties for Tunnelling in a Challenging Environment. The Case of Pefka Tunnel in Northern Greece** ... 409
Vassilis Marinos, George Prountzopoulos, Petros Fortsakis, Fragkiskos Chrysochoidis, Konstantinos Seferoglou, Vassilis Perleros, and Dimitrios Sarigiannis

72 **Prediction of RMR Ahead Excavation Front in D&B Tunnelling** ... 415
Vítor Santos, A. Paula F. da Silva, and M. Graça Brito

73 **The Medium- to Long-Term Effects of Soil Liquefaction in the Po Plain (Italy)** ... 421
Elio Bianchi, Lisa Borgatti, and Luca Vittuari

Part VII Engineering Geological Problems in Deep Seated Tunnels

74 **Leaching Characteristics of Heavy Metals from Mineralized Rocks Located Along Tunnel Construction Sites** ... 429
Nohara Yokobori, Toshifumi Igarashi, and Tetsuro Yoneda

75 **Hydrogeological Controls on the Swelling of Clay-Sulfate Rocks in Tunneling** ... 435
Christoph Butscher

76 **Geomechanical Characterisation of Hard Rocks for Disc Cutting in Deep Tunnels** ... 439
Marlène C. Villeneuve

77 **Geotechnical Design of an Underground Mine Dam in Gyöngyösoroszi, Hungary** ... 443
Vendel Józsa, Zoltán Czap, and Balázs Vásárhelyi

78 **An Approach on the Types and Mechanisms of Water Inrush in Traffic Tunnel Constructions in China** ... 449
Li Tianbin, Zuo Qiankun, Meng Lubo, and Xue Demin

79 **Numerical Analysis of the Influence of Tunnel Dimensions on Stress and Deformation Around Tunnels in Rocks** 453
G.E. Ene, C.T. Davie, and C.O. Okogbue

80 **Analysis of Stress Conditions at Deep Seated Tunnels—A Case Study at Brenner Base Tunnel** 459
Johanna Patzelt and Kurosch Thuro

81 **Acoustic Emission Technique to Detect Micro Cracking During Uniaxial Compression of Brittle Rocks** 465
Carola Wieser, Heiko Käsling, Manuel Raith, Ronald Richter, Dorothee Moser, Franziska Gemander, Christian Grosse, and Kurosch Thuro

82 **Towards a Uniform Definition of Rock Toughness for Penetration Prediction in TBM Tunneling** 469
Lisa Wilfing, Heiko Käsling, and Kurosch Thuro

83 **Stability Analysis of Accidental Blocks in the Surrounding Rockmass of Tunnels in Zipingpu Hydroelectric Project** 475
Yanna Yang, Mo Xu, Shuqiang Lu, and Hong Liu

Part VIII Engineering Geological Problems Related to Geological Disposal of High-level Nuclear Waste

84 **Feasible Study of the Siting of China's High-Level Radioactive Waste Repository in an Area of Northwest China** 483
Yuan Gexin, Zhao Zhenhua, Chen Jianjie, Jia Mingyan, Han Jimin, and Gao Weichao

85 **A New Apparatus for the Measurement of Swelling Pressure Under Constant Volume Condition** 489
C.S. Tang, A.M. Tang, Y.J. Cui, P. Delage, and E. De Laure

86 **2D and 3D Thermo-Hydraulic-Mechanical Analysis of Deep Geologic Disposal in Soft Sedimentary Rock** 493
Feng Zhang and Yonglin Xiong

87 **Anisotropy in Oedometer Test on Natural Boom Clay** 499
Linh-Quyen Dao, Yu-Jun Cui, Anh-Minh Tang, Pierre Delage, Xiang-Ling Li, and Xavier Sillen

88 **The OECD/NEA Report on Self-sealing of Fractures in Argillaceous Formations in the Context of Geological Disposal of Radioactive Waste** 503
Helmut Bock

89 **Permeability and Migration of Eu(III) in Compacted GMZ Bentonite-Sand Mixtures as HLW Buffer/Backfill Material** 507
Zhang Huyuan, Yan Ming, Zhou Lang, and Chen Hang

90 **Correlative Research on Permeability and Microstructure of Life Source Contaminated Clay** 511
Liwen Cao, Yong Wang, Pan Huo, Zhao Sun, and Xuezhe Zhang

91 **Diffusion of La^{3+} in Compacted GMZ Bentonite Used as Buffer Material in HLW Disposal** 515
Yonggui Chen, Lihui Niu, Yong He, Weimin Ye, and Chunming Zhu

92 **Soil Mechanics of Unsaturated Soils with Fractal-Texture** 519
Yongfu Xu and Ling Cao

93 **Thermal Effects on Chemical Diffusion in Multicomponent Ionic Systems** 525
Hywel R. Thomas and Majid Sedighi

94 **Unsaturated Hydraulic Conductivity of Highly Compacted Sand-GMZ01 Bentonite Mixtures Under Confined Conditions** 529
W.M. Ye, Wei Su, Miao Shen, Y.G. Chen, and Y.J. Cui

95 **Adsorption, Desorption and Competitive Adsorption of Heavy Metal Ions from Aqueous Solution onto GMZ01 Bentonite** 533
W.M. Ye, Yong He, Y.G. Chen, Bao Chen, and Y.J. Cui

96 **Enhanced Isothermal Effect on Swelling Pressure of Compacted MX80 Bentonite** 537
Snehasis Tripathy, Ramakrishna Bag, and Hywel R. Thomas

97 **Effects of Stress and Suction on the Volume Change Behaviour of GMZ Bentonite During Heating** 541
Wei-Min Ye, Qiong Wang, Ya-Wei Zhang, Bao Chen, and Yong-Gui Chen

98 **Preliminary Assessment of Tunnel Stability for a Radioactive Waste Repository in Boom Clay** 545
P. Arnold, P.J. Vardon, and M.A. Hicks

99 **Measurements of Acoustic Emission and Deformation in a Repository of Nuclear Waste in Salt Rock** 551
Jürgen Hesser, Diethelm Kaiser, Heinz Schmitz, and Thomas Spies

Part IX Engineering Geology and Design of Hydroelectric Power Plants

100 **Construction of the Underground Powerhouse at Dagachhu Hydropower Project, Bhutan** 557
Reinhold Steinacher and Gyeltshen Kuenga

101 **The Influence of Microbiological Processes on Subsurface Waters and Grounds in River Dam Basement** 563
N.G. Maksimovich and V.T. Khmurchik

102 Factors Controlling the Occurrence of Reservoir-Induced Seismicity 567
Xin Qiu and Clark Fenton

103 Condition of Boguchany Concrete Dam Foundation According to Instrumenal Observations 571
E.S. Kalustyan and V.K. Vavilova

104 Study on Reservoir and Water Inrush Characteristic in Nibashan Tunnel, Sichuan Province, China 577
Sixiang Ling, Yong Ren, Xiyong Wu, Siyuan Zhao, and Limao Qin

105 Differential Settlement Control Technologies of the Long Submarine Tunnel Covered by Municipal Road 583
Cuiying Zhou and Zhen Liu

106 Chontal HPP, Geological Features on Site Location and Dam Type 591
Cristina Accotto, Giuseppe Favata, Enrico Fornari, Nikolaos Kazilis, Marco Rolando, and Attilio Eusebio

Part X Geological Model in Major Engineering Projects

107 Evaluation of Geological Model in Construction Process of Sabzkuh Tunnel (Case Study in Iran) 599
Majid Taromi, Abbas Eftekhari and Jafar Khademi Hamidi

108 Geological Design for Complex Geological-Structural Contest: The Example of SS 125 "Nuova Orientale Sarda" 611
Serena Scarano, Roberto Laureti and Stefano Serangeli

109 The "A12—Tor dè Cenci" Motorway: Geological Reference Model and Design Solutions in Presence of Soft Soils 617
Stefano Serangeli, Roberto Laureti, and Serena Scarano

110 The Geological Reference Model for the Feasibility Study of the Corredor Bioceanico Aconcagua Base Tunnel (Argentina-Chile Trans-Andean Railway) 623
Marini Mattia, Mancari Giuseppe, Damiano Antonio, Alzate Marta, and Stra Michel

111 UHE Belo Monte: Geological and Geomechanical Model of Intake Foundation of Belo Monte Site 627
Jose Henrique Pereira and Nicole Borchardt

112 Geological Reference Model in the Design of the SS 182 "Trasversale Delle Serre": Ionian Calabria 633
Roberto Laureti, Serena Scarano, and Stefano Serangeli

Part XI Impacts of Environmental Hazards to Critical Infrastructures

113 The Vulnerability Shadow Cast by Debris Flow Events 641
M.G. Winter

114 Active Faults at Critical Infrastructure Sites: Definition, Hazard Assessment and Mitigation Measures 645
Alexander Strom

115 Foundation Damage Responses of Concrete Infrastructure in Railway Tunnels Under Fatigue Load 649
L.C. Huang and G. Li

116 Assessment and Risk Management for Integrated Water Services 653
Loretta Gnavi, Glenda Taddia, and Stefano Lo Russo

117 Multi-risk Assessment of Cuneo Province Road Network 657
Murgese Davide, Giraudo Giorgio, Testa Daniela, Airoldi Giulia, Cagna Roberto, Bugnano Mauro, and Castagna Sara

118 Superficial Hollows and Rockhead Anomalies in the London Basin, UK: Origins, Distribution and Risk Implications for Subsurface Infrastructure and Water Resources 663
Philip E.F. Collins, Vanessa J. Banks, Katherine R. Royse, and Stephanie H. Bricker

119 Analysis of the Interaction Between Buried Pipelines and Slope Instability Phenomena 667
Lisa Borgatti, Alessandro Marzani, Cecilia Spreafico Margherita, Gilberto Bonaga, Luca Vittuari, and Francesco Ubertini

120 The Importance of Rockfall and Landslide Risks on Swiss National Roads 671
Philippe Arnold and Luuk Dorren

Part XII Innovative Methods in Characterization and Monitoring of Geotechnical Structures

121 TLS Based Determination of the Orientation of Discontinuities in Karstic Rock Masses 679
Th. Mutschler, D. Groeger, and E. Richter

122 Characterization of the Dagorda Claystone in Leiria, Portugal, Based on Laboratory Tests 685
A. Veiga and M. Quinta-Ferreira

123 Acoustic Monitoring of Underground Instabilities in an Old Limestone Quarry 689
Cristina Occhiena, Charles-Edouard Nadim, Arianna Astolfi, Giuseppina Emma Puglisi, Louena Shtrepi, Christian Bouffier, Marina Pirulli, Julien de Rosny, Pascal Bigarré, and Claudio Scavia

124 **Investigative Procedures for Assessing Subsidence and Earth Fissure Risk for Dams and Levees** 695
Kenneth C. Fergason, Michael L. Rucker, Bibhuti B. Panda, and Michael D. Greenslade

125 **An Integrated Approach for Monitoring Slow Deformations Preceding Dynamic Failure in Rock Slopes: A Preliminary Study** 699
Chiara Colombero, Cesare Comina, Anna Maria Ferrero, Giuseppe Mandrone, Gessica Umili, and Sergio Vinciguerra

126 **Combining Finite-Discrete Numerical Modelling and Radar Interferometry for Rock Landslide Early Warning Systems** 705
Francesco Antolini and Marco Barla

127 **A Tool for Semi-automatic Geostructural Survey Based on DTM** 709
Sabrina Bonetto, Anna Facello, Anna Maria Ferrero, and Gessica Umili

128 **New Perspectives in Long Range Laser Scanner Survey Focus on Structural Data Treatment to Define Rockfall Susceptibility** 715
Andrea Filipello, Leandro Bornaz, and Giuseppe Mandrone

129 **Structural Data Treatment to Define Rockfall Susceptibility Using Long Range Laser Scanner** 721
Andrea Filipello, Giuseppe Mandrone, and Leandro Bornaz

130 **Artificial Neural Networks in Evaluating Piezometric Levels at the Foundation of Itaipu Dam** 725
Bruno Medeiros, Lázaro Valentin Zuquette, and Josiele Patias

131 **Use of an Advanced SAR Monitoring Technique to Monitor Old Embankment Dams** 731
Giovanni Nico, Andrea Di Pasquale, Marco Corsetti, Giuseppe Di Nunzio, Alfredo Pitullo, and Piernicola Lollino

132 **Modelling and Optimization of the Biological Treatment in the Conception of Water-Treatment Plants Whith Activated Sludge** 739
Moncef Chabi and Yahia Hammar

Part XIII Large Projects Impact Assessment, Mitigation and Compensation

133 **Mining with Filling for Mitigating Overburden Failure and Water Inrush Due to Coalmining** 749
Wanghua Sui, Gailing Zhang, Zhaoyang Wu, and Dingyang Zhang

134 **Empirical Cutting Tool Wear Prognosis for Hydroshield TBM in Soft Ground** 753
Florian Köppl, Kurosch Thuro, and Markus Thewes

135 **Risk and Mitigation of the Large Landslide of Brindisi di Montagna** . . . 757
Giuseppe Spilotro, Filomena Canora, Roberta Pellicani, and Francesco Vitelli

136 Environmental Impact of a Motorway Tunnel Project on an Important Karst Aquifer in Southern Latium Region: The Case of Mazzoccolo Spring (Formia, Italy) 761
Giuseppe Sappa, Flavia Ferranti, and Sibel Ergul

Part XIV Properties and Behaviour of Weak and Complex Rock Masses in Major Engineering Projects

137 Rock Mass Quality Rating (RMQR) System and Its Application to the Estimation of Geomechanical Characteristics of Rock Masses 769
Ömer Aydan, Resat Ulusay, and N. Tokashiki

138 Investigation and Treatment of Problematic Foundations for Storage Dams: Some Experience 773
Wynfrith Riemer and rer nat

139 Dissolution Influences on Gypsum Rock Under Short and Long-term Loading: Implications for Dams 779
Nihad B. Salih, Philip E.F. Collins, and Stephen Kershaw

140 Underground Works in Weak and Complex Rock Mass and Urban Area 785
Serratrice Jean François

141 A Dam with Floating Foundations 789
Vinod Kumar Kasliwal

142 Classification of "Loosened Rock Mass" Based on Cases of Dam Construction 793
Takahiro Eguchi, Katsuhito Agui, and Yasuhito Sasaki

143 Numerical Analysis of a Crossover Cavern Excavated in a Complex Rock Mass as Part of the Hong Kong Express Rail Link Project 799
D.K. Koungelis and R. Lyall

144 The Influence of Geological History on Preferred Particle Orientation and the Observed Anisotropy of Over Consolidated UK Mudrocks 805
Stephen Wilkinson and Clark Fenton

145 Mechanical Characterization of Weathered Schists 809
Thomas Le Cor, Damien Rangeard, Véronique Merrien-Soukatchoff, and Jérôme Simon

146 Deformation of Soil and Rock Transition Belt Caused by the Mining Damage 813
Qinghong Dong, Fei Liu, and Qiang Zhang

147 Geomechanical Assessment on a Metasedimentary Rock Cut Slope (Trofa, NW Portugal): Geotechnical Stability Analysis 819
M.J. Afonso, R.S. Silva, P. Moreira, J. Teixeira, H. Almeida, J.F. Trigo, and H.I. Chaminé

148 **Geomechanical Characterization of a Weak Sedimentary Rock Mass in a Large Embankment Dam Design** 825
Gian Luca Morelli and Ezio Baldovin

149 **Experimental Study of Anisotropically Mechanical Features of Phyllite and Its Engineering Effect** 831
Meng Lubo and Li Tianbin

150 **Quantification of Rock Joint Roughness Using Terrestrial Laser Scanning** 835
Maja Bitenc, D. Scott Kieffer, Kourosh Khoshelham, and Rok Vezočnik

151 **Elaboration and Interpretation of Ground Investigation Data for the Heterogeneous 'Athens Schist' Formation; from the 'Lithological Type' to the 'Engineering Geological Formation'** 839
Georgios Stoumpos and Konstantinos Boronkay

152 **Incorporating Variability and/or Uncertainty of Rock Mass Properties into GSI and RMi Systems Using Monte Carlo Method** 843
Mehmet Sari

153 **Using of Multivariate Statistical Analysis in Engineering Geology at the Pest Side of the Metro Line 4 in Budapest, Hungary** 851
Nikolett Bodnár, József Kovács, and Ákos Török

154 **Evaluation of the Swelling Pressure of the Corumbatai Formation Materials** 855
R.F.C. Souza and O.J. Pejon

155 **Classification of Weak Rock Masses in Dam Foundation and Tunnel Excavation** 859
V. Marinos, P. Fortsakis, and G. Stoumpos

156 **Applicability of Weathering Classification to Quartzitic Materials and Relation Between Mechanical Properties and Assigned Weathering Grades: A Comparison with Investigations on Granitic Materials** 865
A. Basu

157 **Performance of Forepole Support Elements Used in Tunnelling Within Weak Rock Masses** 869
J. Oke and N. Vlachopoulos

158 **The Research of Shear Creep Behaviors of Saturated Sericite-Quartz Phyllite** 875
Guang Ming Ren, Xin lei Ma, Bo Wen Ren, and Min Xia

Part XV Radioactive Waste Disposal: An Engineering Geological and Rock Mechanical Approach

159 Investigation of Mineral Deformation and Dissolution Problems Under Various Temperature Conditions. 883
J.H. Choi, B.G. Chae, C.M. Jeon, and Y.S. Seo

160 Analysis of Permeability Coefficient Along a Rough Fractures Using a Homogenization Method. 887
Chae Byung-Gon, Choi Jung Hae, Seo Yong-Seok, and Woo Ik

161 In Situ Quantification of Hydrocarbon in an Underground Facility in Tight Salt Rock 893
Benjamin Paul, Hua Shao, Jürgen Hesser, and Christian Lege

162 Relationship Between the Fractal Dimension and the Rock Mass Classification Parameters in the Bátaapáti Radioactive Waste Repository 897
Rita Kamera, Balázs Vásárhelyi, László Kovács, and Tivadar M. Tóth

163 Direct Shear Strength Test on Opalinus Clay, a Possible Host Rock for Radioactive Waste. 901
Buocz Ildikó, Török Ákos, Zhao Jian, and Rozgonyi-Boissinot Nikoletta

164 Significance of Joint Pattern on Modelling of a Drill and Blast Tunnel in Crystalline Rock 905
Dániel Borbély, Tamás Megyeri, and Péter Görög

165 Special Requirements for Geotechnical Characterization of Host Rocks and Designing of a Radioactive Waste Repository 909
László Kovács and Balázs Vásárhelyi

166 Rock Mechanical and Geotechnical Characterization of a Granitic Formation Hosting the Hungarian National Radioactive Waste Repository at Bátaapáti. 915
László Kovács, Eszter Mészáros, and Gábor Somodi

Part XVI Subsurface Water in Tunnels: Prediction, Estimation, Management

167 Ground Water Management for Large Under-Ground Storage Caverns 921
Saikat Pal, G. Kannan, Vijay Shahri, and A. Nanda

168 Experience from Investigation of Tectonically Extremely Deteriorated Rock Mass for the Highway Tunnel Višňové, Slovakia 927
Rudolf Ondrášik, Antonín Matejček, and Tatiana Durmeková

169 **Verification and Validation of Hydraulic Packer Test Results in a Deep Lying Tunnel Project** 931
Ulrich Burger, Paolo Perello, Sacha Reinhardt, and Riccardo Torri

170 **Change in Hydraulic Properties of Rock Mass Due to Tunnelling** 937
Bernard Millen, Giorgio Höfer-Öllinger, and Johann Brandl

171 **Groundwater Ingress in Head Race Tunnel of Tapovan: Vishnugad Hydroelectric Project in Higher Himalaya, India** 941
P.C. Nawani

172 **Investigation Constraints in Subsurface Water Aspect of Hydropower Development in the Indian Himalaya** 947
Y.P. Sharda and Yogendra Deva

173 **Prediction and Management of Ground Water for Underground Works in Himalayas** 955
Akhila Nath Mishra and S. Kannan

Part XVII Sustainable Water Management in Tunnels

174 **Methodological Approach for the Valorisation of the Geothermal Energy Potential of Water Inflows Within Tunnels** 963
Riccardo Torri, Nathalie Monin, Laudo Glarey, Antonio Dematteis, Lorenzo Brino, and Elena Maria Parisi

175 **Impacts on Groundwater Flow Due to the Excavation of Artificial Railway Tunnels in Soils** 967
Gabriele Bernagozzi, Gianluca Benedetti, Francesca Continelli, Cristiano Guerra, Renato Briganti, Santo Polimeni, Giuseppe Riggi, and Fabio Romano

176 **Chemical and Isotope Composition of Waters from Firenzuola Railway Tunnel, Italy** 971
L. Ranfagni, F. Gherardi, and S. Rossi

177 **Hydrogeological Modeling Applications in Tunnel Excavations: Examples from Tunnel Excavations in Granitic Rocks** 975
Baietto Alessandro, Burger Ulrich, and Perello Paolo

178 **Effects on the Aquifer During the Realization of Underground Railway Works in Turin** 981
Stefano Ciufegni, Fabrizio Bianco, Adriano Fiorucci, Barbara Moitre, Massimiliano Oppizzio, and Francesco Sacchi

179 **Proposal for Guidelines on Sustainable Water Management in Tunnels** 985
Antonio Dematteis

Part XVIII Uncertainty and Risk in Engineering Geology

180 The Design Geological and Geotechnical Model (DGGM) for Long and Deep Tunnels 991
Alessandro Riella, Mirko Vendramini, Attilio Eusebio, and Luca Soldo

181 Research on Overall Risk Assessment and Its Application in High Slope Engineering Construction 995
Tao Lianjin, An Junhai, Li Jidong, and Cai Dongming

182 The Research of Geological Forecast Based on Muti-source Information Fusion 1001
S. Cui, B. Zhang, F. Feng, and L. Xie

183 Presentation of the Activity of the AFTES WG 32: Considerations Concerning the Characterization of Geotechnical Uncertainties and Risks for Underground Projects 1007
G.W. Bianchi, J. Piraud, A.A. Robert, E. Egal, and L. Brino

184 Development of 3D Models for Determining Geotechnical-Geological Risk Sharing in Contracts—Dores de Guanhães/MG/Brazil Hydroelectric Powerplant Case Study 1013
Isabella Figueira, Laurenn Castro, Luiz Alkimin de Lacerda, Amanda Jarek, Rodrigo Moraes da Silveira, and Priscila Capanema

185 Use of Rock Mass Fabric Index in Fuzzy Environment for TBM Performance Prediction 1019
Mansour Hedayatzadeh and Jafar Khademi Hamidi

186 The Risk Analysis Applied to Deep Tunnels Design—El Teniente New Mine Level Access Tunnels, Chile 1023
Lorenzo Paolo Verzani, Giordano Russo, Piergiorgio Grasso, and Agustín Cabañas

187 Development of Measurement System of Seismic Wave Generated by the Excavation Blasting for Evaluating Geological Condition Around Tunnel Face 1031
Masashi Nakaya, Kazuhiro Onuma, Hiroyuki Yamamoto, Shinji Utsuki, and Hiroaki Niitsuma

188 Multidisciplinary Methodology Used to Detect and Evaluate the Occurrence of Methane During Tunnel Design and Excavation: An Example from Calabria (Southern Italy) 1035
S. Lombardi, S. Bigi, S. Serangeli, M.C. Tartarello, L. Ruggiero, S.E. Beaubien, P. Sacco, and D. De Angelis

189 Combined Geophysical Survey at the A2 Tunnel Maastricht 1039
O. Brenner and D. Orlowsky

190 **The Combined Use of Different Near Surface Geophysics Techniques and Geotechnical Analysis in Two Case Histories for the Advanced Design of Underground Works in Urban Environment: Rome Metro B and Torino-Ceres Railway** 1045
Riccardo Enrione, Simone Cocchi, and Mario Naldi

191 **Landslides Induced by Intense Rainfall and Human Interventions—Case Studies in Algeria** 1049
Ramdane Bahar, Omar Sadaoui, and Samir Sadaoui

Part XIX Physical Impacts to the Environment of Infrastructure Development Projects – Engineering Geology Data for Environmental Management

192 **Using of Man-Made Massives in Russian Mining (Engineering: Geological Aspects)** 1057
Galperin Anatoly

193 **Modeling Optimized UCG Gas Qualities and Related Tar Pollutant Production Under Different Field Boundary Conditions** 1063
Stefan Klebingat, Rafig Azzam, Marc Schulten, Thomas Kempka, Ralph Schlüter, and Tomás M. Fernández-Steeger

194 **Considerations About the Integration of Geological and Geotechnical Studies Applied to Engineering Projects and to Environmental Impact Assessment in São Paulo State, Brazil** 1067
Bitar Omar Yazbek, Sofia Julia A.M. Campos, Amarilis Lucia C.F. Gallardo, Braga Tania de Oliveira, and Caio Pompeu Cavalhieri

195 **Integrated Geological, Geotechnical and Hydrogeological Model Applied to Environmental Impact Assessment of Road Projects in Brazil** 1071
Sofia Julia A.M. Campos, Adalberto Aurelio Azevedo, Amarilis Lucia F.C. Gallardo, Pedro Refinetti Martins, Lauro Kazumi Dehira, and Alessandra Gonçalves Siqueira

Author Index 1077

Consiglio Nazionale delle Ricerche
Istituto di Ricerca per la Protezione Idrogeologica

The Istituto di Ricerca per la Protezione Idrogeologica (IRPI), of the Italian Consiglio Nazionale delle Ricerche (CNR), designs and executes research, technical and development activities in the vast and variegated field of natural hazards, vulnerability assessment and geo-risk mitigation. We study all geo-hydrological hazards, including floods, landslides, erosion processes, subsidence, droughts, and hazards in coastal and mountain areas. We investigate the availability and quality of water, the exploitation of geo-resources, and the disposal of wastes. We research the expected impact of climatic and environmental changes on geo-hazards and geo-resources, and we contribute to the design of sustainable adaptation strategies. Our outreach activities contribute to educate and inform on geo-hazards and their consequences in Italy.

We conduct our research and technical activities at various geographical and temporal scales, and in different physiographic and climatic regions, in Italy, in Europe, and in the World. Our scientific objective is the production of new knowledge about potentially dangerous natural phenomena, and their interactions with the natural and the human environment. We develop products, services, technologies and tools for the advanced, timely and accurate detection and monitoring of geo-hazards, for the assessment of geo-risks, and for the design and the implementation of sustainable strategies for risk reduction and adaptation. We are 100 dedicated scientists, technicians and administrative staff operating in five centres located in Perugia (headquarter), Bari, Cosenza, Padova and Torino. Our network of labs and expertizes is a recognized Centre of Competence on geo-hydrological hazards and risks for the Italian Civil Protection Department, an Office of the Prime Minister.

Part IX

Engineering Geology and Design of Hydroelectric Power Plants

Convener Luca Soldo—*Co-convener* Eng Massimo Cadenelli

The inherent technical, economical and environmental benefits of hydroelectric power make it an important contributor to the future world energy mix, particularly in developing countries. Engineering Geology represents a wealth of expertise which can ensure that future projects will be planned, constructed and operated with full respect for the environment. Engineering geologists support the design of hydroelectric power plants from the very beginning phases of planning and interpretation of the investigations, understanding the design geotechnical parameters. In the following, they must recognize the geological, hydrogeological and geotechnical hazard factors that may arise as a result of the interaction between geology and the planned works studying the necessary measures for their prevention and mitigation. This session aims to be a round table for an open analysis of the engineering geology "best practices" along the design protocol of hydropower plants.

Construction of the Underground Powerhouse at Dagachhu Hydropower Project, Bhutan

100

Reinhold Steinacher and Gyeltshen Kuenga

Abstract

A large-span underground Powerhouse cavern and a medium-size Transformer cavern with an average overburden of 280 m have been excavated following the New Austrian Tunnelling Method (NATM). The two caverns are separated in between by 35 m rock ledge. The encountered rock mass of crystalline assemblages are meta-sediments consisting of an intercalation of Mica Schists, Quartzites and Gneisses. This intercalation of brittle and ductile acting rock mass with very close to close spacing of schistosity planes did not pose major problems during excavation. The support system designed has proved to be adequate during construction and was verified by geotechnical monitoring. For western means sparse site investigation was compensated by sound engineering geological mapping and strict site supervision.

Keywords

Cavern • Hydropower • Himalaya • Bhutan • Q system

100.1 Introduction

The Dagachhu Hydropower Project (DHPP) is a run-of-river scheme being developed on the Dagachhu River in the south-western Dzongkhag (district) of Dagana in Bhutan. The major components of the project consist of a diversion dam, intake structures, surface desilting chambers, a mainly underground water conductor system of 9,100 m lengths, a surge tank, a vertical pressure shaft (270 m) and penstocks. Electricity is produced in an underground Powerhouse (PH). The water is finally diverted back to the main river course through the 680 m long tailrace tunnel (TRT). With an installed capacity of 126 MW, the Powerhouse owns two vertical axis Pelton turbines of 63 MW each. The project is expected to generate around 515 GWh of clean energy annually and to contribute to the reduction of greenhouse gas emissions in the region.

Situated in the heart of the eastern part of the 2,500 km long Himalayan range, Bhutan has a complicated topography and geomorphology due to the geo-historical processes. Harnessing the potential hydropower resource of estimated 30,000 MW is a major challenge due to the complex Himalayan geological set-up, logistics and accessibility and seasonal severe weather conditions. Due to heavy vegetation, high overburden and rugged nature of the terrain, geological investigations are difficult to perform. Therefore answering questions about rock mass quality within the planning phase of the project becomes a major issue. This led to severe cost and time overrun in several earlier hydropower projects in Bhutan.

The Powerhouse Complex of Dagachhu Hydropower Project consists of two main caverns: the Powerhouse (PH), 62.5 m × 23.88 m × 37 m (l × b × h) and the Transformer Cavern (TF), 52 m × 14.5 m × 16 m (l × b × h). The two are separated in between by 35 m rock column connected by two short bus-duct tunnels. A Main Access Tunnel (ACT), two Penstock tunnels and an Emergency Exit tunnel (EET)

R. Steinacher (✉)
Engineering Geologist, Bernard Consulting Engineers SE, Hall, Austria
e-mail: reinhold.steinacher@bernard-ing.com

G. Kuenga
Dagachhu Hydro Power Corporation Limited, Dagana District, Kingdom of Bhutan

G. Lollino et al. (eds.), *Engineering Geology for Society and Territory – Volume 6*,
DOI: 10.1007/978-3-319-09060-3_100, © Springer International Publishing Switzerland 2015

at the eastern side of PH as well as other auxiliary tunnels for construction access had to be excavated. It is located in spur projection at the left bank of the river as shown in the layout Fig. 100.1. Both caverns are aligned in North-South direction. For a more detailed (geological) description of the project see Holzleitner et al. (2008), Holzleitner and Fish (2012), Steinacher (2013), Gyeltshen and Steinacher (2013) and Gyeltshen et al. (2013).

This paper presents engineering geological aspects of the rock mass as encountered during investigation and construction and the support systems implemented to ensure immediate stability and safety as well as long term service performance. Further, instrumentation for the geotechnical monitoring and its interpretation used for design verification are presented.

100.2 Site Investigation

During the feasibility level study, several geological missions by the consultant had taken place with the aim to gain understanding and to be able to adjust the project to the existing geological settings. Engineering geological mapping of the project area in the Scale 1:10,000 was executed. Based on these results the layout for a detailed subsurface investigation program was developed. A detailed geological mapping of selected areas such as adits and portals in the Scale 1:5,000 followed. Three core drillings >200 m depth were executed at the Powerhouse area to detect fault zones and probable slope instabilities, investigate rock type, get information on the groundwater conditions and to gain rock samples for laboratory testing. The in situ stress conditions at the Powerhouse site were measured by conducting hydrofracturing tests.

Following the detailed investigation program, geotechnical interpretation and assessment of the rock mass behaviour were prepared. Some direct consequences arose of the site investigations:

- The Powerhouse cavern was rotated to gain a north-south direction to reach a sub perpendicular position to the strike of schistosity.
- It was decided to go for full face excavation of top heading rather than execute subdivided excavation.

100.3 Encountered Geological Conditions

The encountered rocks are dominated by an alternation of phyllitic Biotite–Muscovite–Garnet–Schist (Micaceous Schist), micaceous Quartzite and Gneiss in layers reaching from centimetres to several metres. At some places a very narrow intercalation is present, where a distinction is barely possible (see Fig. 100.2).

Rock mass encountered was of fair to poor quality in the top heading and of fair to good quality in the benches.

Distribution of Rock Quality Designation values (RQD) and Q-value ranges are shown in Fig. 100.3. Shear zones were encountered only very subordinate, mostly as 5–10 m long 1–5 cm gouge filled foliation parallel shear zones. One major weakness zone with about 1–3 m thickness was encountered parallel foliation from top heading down to the 3rd bench.

The whole Powerhouse Complex lies below the groundwater table. During excavation of the top heading and benching about 1.5 l/s of dripping to slightly flowing water was observed. Two litres per second are present as slightly dripping seepage from top heading and anchors on side walls after full excavation.

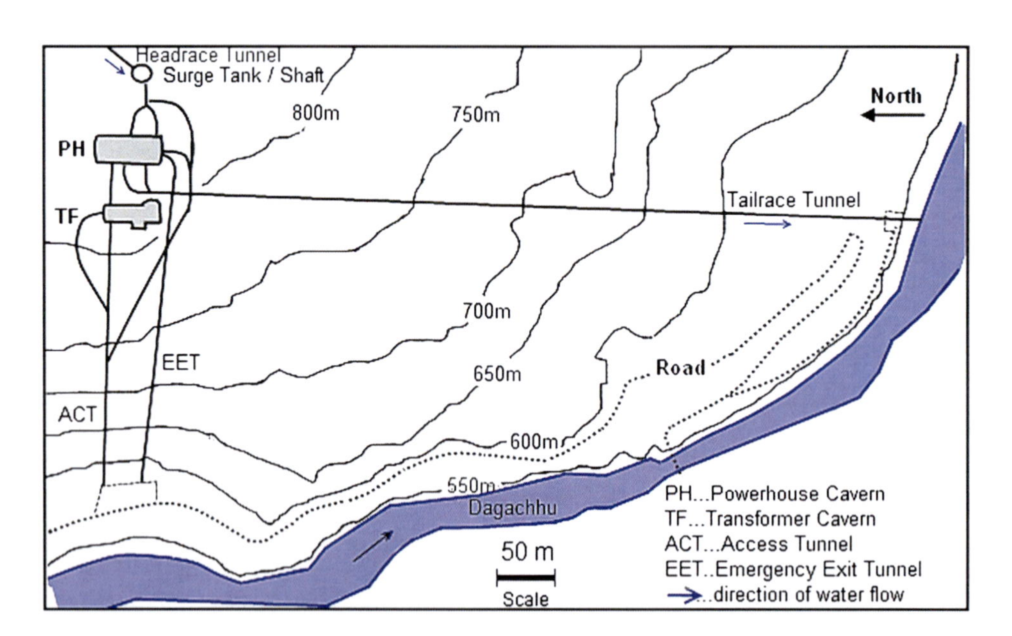

Fig. 100.1 Layout of the Powerhouse Complex at Dagachhu Project

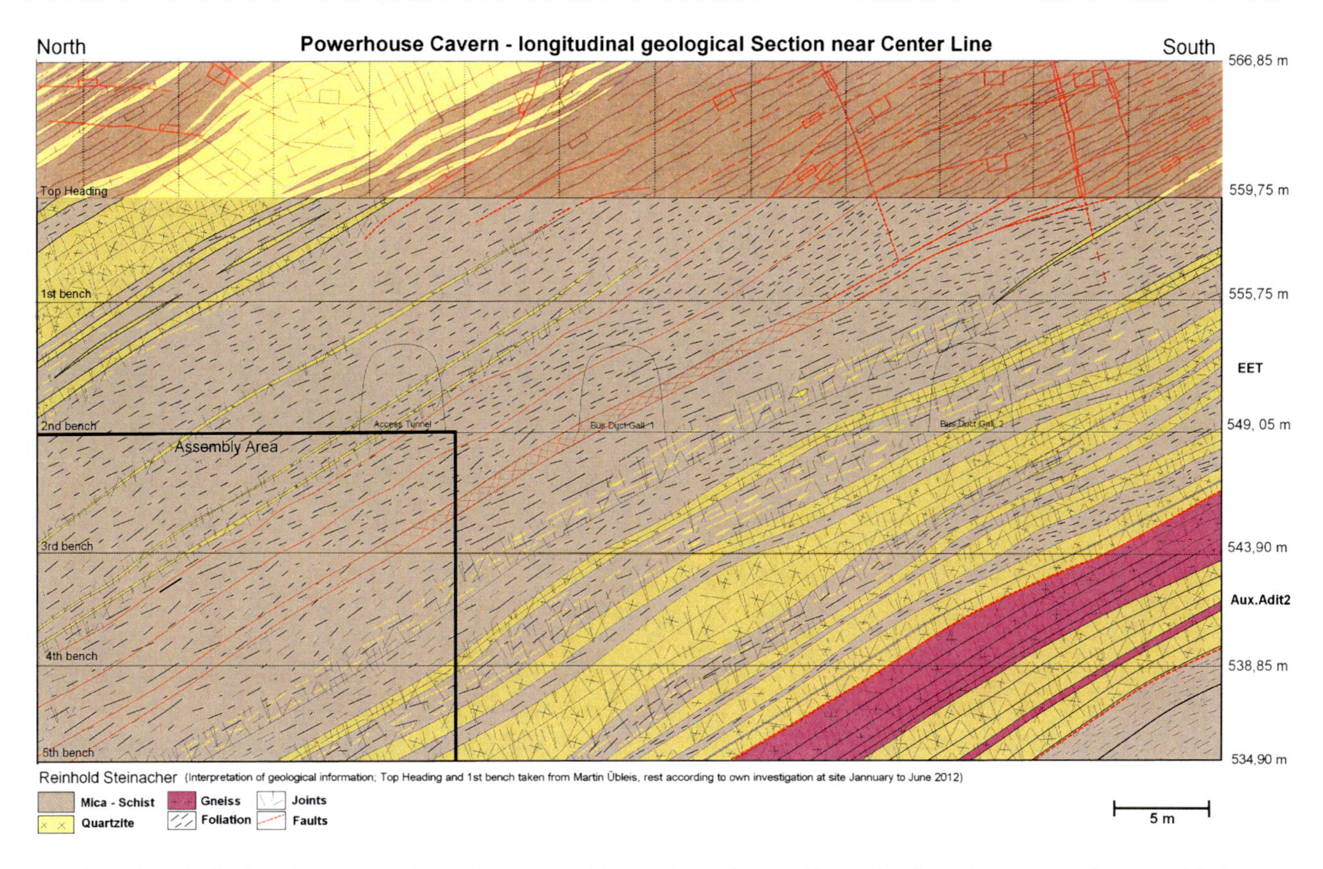

Fig. 100.2 Longitudinal section at *center line* of Powerhouse Cavern (*brown* Mica-Schist; *yellow* Quartzite; *magenta* Gneiss; *red* faults)

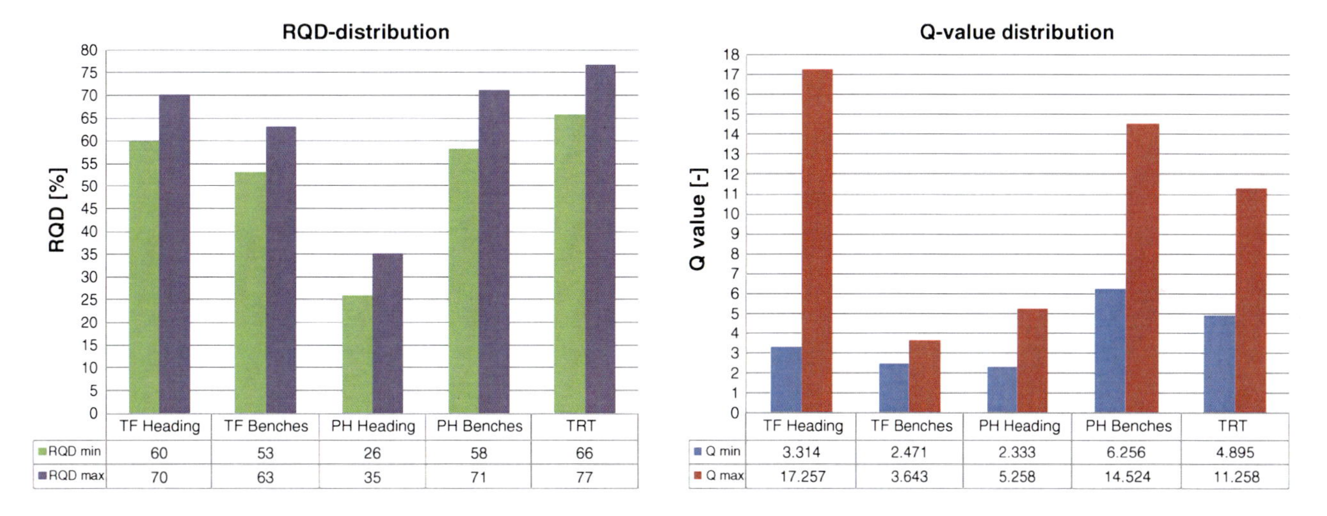

Fig. 100.3 Arithmetic mean of minimum and maximum values for RQD (*left*) and Q-values (*right*) for Transformer Cavern (*TF*), Powerhouse (*PH*) and Tailrace Tunnel (*TRT*)

100.4 Evaluation of Site Investigation

The drill cores were of bad quality and showed an unnaturally high degree of fracturing and core loss. The fracturing was assigned to the drilling process rather than the natural rock mass quality. An RQD could not been gained from the drill cores and had to be estimated from the field. Documentation of TCR (Total Core Recovery) brought at least some hints on rock mass quality but as well was not satisfactory. Performed laboratory tests (f.i. UCS) showed a high variety in results. Nevertheless, the performed site investigation proved sufficient to plan and construct the Powerhouse and Transformer Caverns. No unforeseen geological conditions were encountered. The rock mass at top heading was stable enough to be excavated full face and the chosen support measures were adequate. Geological mapping proved to be an efficient tool to gain information on rock mass conditions and general rock mass behaviour. This is especially true for this remote area in southern Bhutan where performance of core drilling as well as in situ and laboratory testing are not comparable with western standards. Rock mass characterization in the field hereby plays a crucial role to get at least approximate values for rock mass strength and an assessment of rock mass behaviour.

100.5 Support, Instrumentation and Monitoring

The concurrent support systems provided consisted of 32 mm diameter, 6 m long rock bolts at spacing of 3 m centre to centre (c/c) staggered both ways, lattice girder at spacing of 1.5 m c/c and three layers of welded wire mesh embedded in shotcrete having characteristic strength of 35 MPa. The minimum overall built-up thickness of shotcrete was 450 mm applied in four layers in the crown and 400 mm thick in three layers in the walls.

After completion of the heading excavation, installation of the primary support was started. Permanent Bar Anchor (PBA), 36 mm diameter, 15 m long ground anchor system with pre-stressing were installed. This was the main (primary) support system designed to support the rock mass deformation expected to occur during the subsequent bench excavation. The 1st bench excavation was started only after the complete installation and application of pre-stressing force to all PBAs in the top heading. Each bench excavation was performed with bench height of 4–5 m with provision of immediate support and installation of PBAs. The excavation of the entire Powerhouse cavern was completed on 21st June 2012.

As one of the fundamental principles of NATM bireflex targets (70), load cells (19) and multipoint borehole extensometers (19 MPBX) have been installed to observe rock mass behaviour during construction to ensure safety and adequacy of the support systems (construction control) and for monitoring of the long-term performance of the support system and the overall structure.

Against contractual agreements zero readings of the bireflex-targets were taken 1–2 days after excavation and support. Therefore monitored displacements do not represent absolute values. Due to the relative moderate excavation advance in Top Heading the magnitude of the "pre-displacement" may be of a magnitude of 30–50 mm. This would bring the total displacements into the expected and estimated range of about 60–80 mm. The crown targets showed settlements while the side wall targets tended to show a combination of settlement and movement into the excavated cavity. This might also be correlated with encountered geological features where a foliation parallel weakness zone could be identified during excavation dipping approximately with 30°–40° towards North-East. Many targets showing deformation greater than 5 mm lie near or above this weakness zone. Also load cells showed the influence of this zone. Both MPBX and Load Cells indicated maximum levels below 2D numerical modelling results of 60 mm and 780 kN respectively. The support system designed has proven to be adequate during construction.

100.6 Challenges and Conclusions

A close interaction between the *Dagachhu Hydropower Corporation*, *Bernard Engineers* as consultant and the contractor (*Hindustan Construction Company*) were crucial for the project success. NATM has proven an efficient and safe way to construct the Powerhouse Complex at Dagachhu, even though the Indian contractor had little experience with this kind of tunnelling concept. An independent and strict consultant was very efficient for all parties to ensure an up-to-date excavation and support methodology.

Geological field mapping was crucial for the assessment of engineering geological parameters in this remote area where in situ and laboratory testing does not reach western standards. The use of rock mass classification schemes though, should only be performed by an experienced geologist. These schemes are suitable to recognise trends in rock mass quality changes and to have a base of discussion with non-geologists. Q-values from different processors taken at the same outcrops rather show disillusioning results (Steinacher 2013).

Challenges encountered during construction should be mitigated beforehand by strict contract agreements and their enforcement on site. On the Indian subcontinent typical challenges may be:

- Seepage water management generally is not taken serious during construction stage. Soaked working areas lead to problems with accessibility and therefore time delay during mucking and supply with equipment. Strict guidance and enforcement of contractual agreements must prevent bad water management.
- Sub-contractors of the main contractor are employed for works like installation of Permanent Bar Anchors or Geotechnical measurement of bireflex targets. If conflicts arise or payment delays occur the sub-contractor stops work, which in most cases also hampers main construction works.
- Indian contractors are quite inflexible to changes in methodology. Financial stimulations in contract may help to change this attitude.

References

Gyeltshen K, Steinacher R (2013) Engineering geological aspects of the underground powerhouse at Dagachhu Hydropower Project, Bhutan. In: Druk Green Power Corporation Limited (eds) 25 years of learning, Proceedings of the 1st Druk Green technical conference, Chuckha, Dec 2013, Bhutan, pp 337–355

Gyeltshen K, Wangdi S, Dorji T (2013) Application of Permanent Bar Anchors (PBA) support system in underground powerhouse of Dagachhu Hydropower Project in Bhutan. In: Druk Green Power Corporation Limited (eds) 25 years of learning, Proceedings of the 1st Druk Green technical conference, Chuckha, Dec 2013, Bhutan, pp 1–19

Holzleitner W, Aichinger J, Fish M (2008) The Dagachhu Hydropower Project, Bhutan. Jahrbuch Tunnelbau 2008:1–3

Holzleitner W, Fish M (2012) NATM underground structures of the Dagachhu HPP. In: Stipek W, Galler R, Bauer M (eds) 50 years NATM—Experience reports. ITA Austria, Wien, pp 163–168

Steinacher R (2013) Engineering geological aspects of the 126 MW Hydropower Plant Dagachhu in Bhutan. In: Proceedings of the 19 conference for engineering geology, Technical University of Munich, pp 75–81

The Influence of Microbiological Processes on Subsurface Waters and Grounds in River Dam Basement

101

N.G. Maksimovich and V.T. Khmurchik

Abstract

Microbes are ubiquitous on the Earth and take an active part in the transformation of the geological environment. Their activity can change the geochemical parameters of ground and groundwater and lead to undesirable consequences after the building of hydrotechnical facilities, especially pressure ones. The geological and chemical survey of one of the ground dams of Kama–Volga rivers cascade (The Ural, Russian Federation) revealed the unusual deviations in chemical content of drain water and the presumable suffusion process at the dam basement. The aim of our study was investigation of the dam's ground and water to reveal another deviations in their characteristics, which in turn could help to assess the stability of the dam. We investigated the composition of water-dissolved organic matter (gas chromatography–mass spectrometry analysis), the composition of subsoil gases of the dam (gas analysis), and performed mineralogical analysis of sediments settled at the bottom of the dam's drain system (X-ray diffraction analysis); chemical analysis of water of various aquifers under the dam basement was performed too; also, the microbiological investigations of the dam's ground and water samples were made. We suppose, the results of investigation demonstrate the presence of an active microbiota in dam's ground and water, and microbiota metabolism could lead to hazardous changes in physical-mechanical properties of dam's ground and, eventually, the unstable state of the dam itself.

Keywords

Ground dam • Alluvial aquifer • Microbial processes • Dam's stability

101.1 Introduction

Microbes are ubiquitous on the Earth and take an active part in the transformation of the geological environment, including the impact on ground and groundwater (Bolotina and Sergeev 1987; Kuznetsov et al. 1962; Maksimovich and Hmurchik 2012; Radina 1973). Microbial processes can change the geochemical parameters of ground and groundwater and lead to undesirable consequences after the building of hydrotechnical facilities, especially pressure ones (Koff and Kozhevina 1981; Maksimovich et al. 2001).

The geological and chemical survey of one of the ground dams of Kama–Volga rivers cascade (The Ural, Russian Federation) revealed the unusual deviations in chemical content of drain water and the presumable suffusion process at the dam basement. We supposed, the main cause of deviations observed was the existence of microbiological processes in the dam's ground and water, and tried to explain the detected phenomena from this point of view.

The aim of our study was more proper investigation of the dam's ground and water to reveal another deviations in their characteristics, which in turn could help to assess the stability of the dam.

N.G. Maksimovich · V.T. Khmurchik (✉)
Natural Science Institute, Perm State National Research University, Perm, Russian Federation
e-mail: khmurchik.vadim@mail.ru

N.G. Maksimovich
e-mail: nmax54@gmail.com

DOI: 10.1007/978-3-319-09060-3_101, © Springer International Publishing Switzerland 2015

101.2 The Description of the Dam's Ground and Water

The studied dam is the part of Kama–Volga rivers cascade dams (The Ural, Russian Federation). The basement of the dam is presented of alluvial sediments up to 18 m in thickness. The sediments are consisted of clays, loams, sandy loams, and, in the upper part of geological column, of fine sands. Sand bands are observed in clays and loams. Sand and gravel deposits are in the lower part of geological column. The distribution and the composition of gravel-pebble strata are not uniform. Lenses of fine sand and interlayers of clays are observed in the gravel-pebble strata, peat is presented in the strata too (Mamenko 1967).

101.2.1 Hydrological and Hydrochemical Conditions Before the Dam Construction

Groundwater was widespread in alluvial deposits and located in sands and gravel-pebble rocks before the dam construction. Water table was 3–9 m below ground surface and had slope gradient to riverbed in 0.006–0.012 grad m^{-1}. Average filtration coefficient of alluvial horizon was 14, and 26–82 m day^{-1} in it gravel-pebble layer. Alluvial groundwater was weakly mineralized and HCO_3–Ca in chemical content.

101.2.2 Hydrological and Hydrochemical Conditions After the Dam Construction

Significant changes in hydrodynamical and hydrochemical conditions, which were caused by a number of factors, such as change in hydrostatic heads of groundwater, filtration of fresh water from the reservoir etc., have occurred since the dam construction. Alluvial aquifer became confined and has a local hydrostatic head at present. It water table is located at the depth of 1 m below the dam surface under it basement. Total salinity of alluvial aquifer's water is 100–200 mg dm^{-3} (up to 400 mg dm^{-3} sometimes) and it increases downstream in general; water of alluvial aquifer is HCO_3–Cl–Na–Ca in chemical content at present. The alluvial aquifer has a close hydraulic connection with the waters of the Kama River. Moreover, it seems to exist the local hydrogeological windows between alluvial aquifer and underlying aquifers, which could cause the increased content of chloride ions (up to 168 mg dm^{-3}), which was observed in individual observational wells of alluvial aquifer.

Geological and chemical survey of the dam revealed the elevated concentration of Fe^{2+} ions in drain water and the settlement of Fe–(hydr)oxides at the bottom of the dam's drain system. As well as this features resembled suffusion process, more proper investigation was done to assess the stability of the dam.

101.3 Methods

We investigated the composition of water-dissolved organic matter (gas chromatography–mass spectrometry analysis on "Agilent 6890/5973N"), the composition of subsoil gases of the dam (gas analysis on "Ecoprobe-5"), and performed mineralogical analysis of sediments settled at the bottom of the dam drain system (X-ray diffraction analysis on "D2 Phaser"). Chemical analysis of water of various aquifers under the dam basement was performed too. Also, the microbiological investigations of the dam's ground and water samples were made.

101.4 Results of Investigation

The chemical analysis of alluvial horizon's water under the dam basement revealed the presence of zone with elevated contents of Fe^{2+}, HCO_3^- and NH_4^+ ions. This water contains the high amount of water-dissolved organic matter—108–122 mg dm^{-3} (whereas it content did not exceeds 30–40 mg dm^{-3} in groundwater of the Ural region). The main features of water-dissolved organic matter were non-hydrocarbon character and technogenic origin. The content of chloroform-extracted bitumen was 1.1–1.6 mg dm^{-3} in range, and the oil-product content was less than 0.07 mg dm^{-3}. The hexane fraction of water-dissolved organic matter consisted of oxygen-containing compounds mainly, the presence of sulfide sulfur (up to 6 %) was found in the hexane fraction too.

The analysis of subsoil gases of the dam revealed the occurrence of regions with elevated contents of CH_4, C_2–C_5 hydrocarbons and volatile organic compounds.

The studies of the mineralogical composition of the sediments settled at the bottom of dam's drain system revealed a predominance of authigenic minerals' complex (calcite, amorphous iron hydroxides, goethite, hydrogoethite, and pyrite) over allotigenic one (quartz minerals). The newly-formed minerals—slices of calcite and pyrite—were detected.

Microbiological investigations of the dam's ground and water revealed the presence of an active metabolizing microbiota in them. Bacteria, isolated from the core and water samples, consumed organic matter, SO_4^{2-} and NO_3^- ions, produced gases and leached Fe ions from the dam's ground samples.

101.5 Conclusions

We suppose that the results of investigation, which were described in Sect. 101.4, demonstrate the presence of an active microbiota in dam's ground and water. Microbial activity could lead to the mobilization and removal of substances (for example, in the form of chemical elements' atoms) from the body and the basement of the dam due to the following processes and factors: bacterial formation of gases could increase tense state of the ground and cause the unconsolidation process; the removal of individual chemical elements from the ground could lead to destruction of it mineral skeleton and reduction of the mechanical firmness of ground; microbiological processes could change microaggregate and chemical composition of ground, dispergate clay minerals, increase ground's wetting ability, and decrease it filtration capacity; exometabolites, formed by microorganisms, could exhibit surface-active properties and reduce the strength of the structural bounds in the ground. So, the intensification of bacterial processes, which could be caused by supply of elevated concentrations of organic matter from anywhere, could lead to hazardous changes in physical-mechanical properties of dam's ground and, eventually, the unstable state of the dam itself.

References

Bolotina IN, Sergeev EM (1987) Microbiological studies in engineering geology. Eng Geol 5:3–17 (in Russian)
Koff GL, Kozhevina LS (1981) The role of microorganisms in the change of geological environment. Eng Geol 6:63–74 (in Russian)
Kuznetsov SI, Ivanov MV, Lyalikova NN (1962) The introduction to geological microbiology. The USSR Academy of Sciences, Moscow, 239 p (in Russian)
Maksimovich NG, Hmurchik VT (2012) The influence of microorganisms on the mineral composition and properties of grounds. Perm State Univ Bull, Ser Geol 3(16):47–54 (in Russian)
Maksimovich NG, Menshikova EA, Kazakevich SV (2001) Possibility of increasing the aggressiveness of groundwater during construction on pyrite-containing clay soils. In: Proceedings of the international symposium on "Engineering-geological problems of urban areas", vol 2. Aqua–Press, Ekaterinburg, pp 545–551
Mamenko GK (1967) The dam on the Kama river. In: Geology and dams, vol 5. Energiya, Moscow (in Russian)
Radina VV (1973) The role of microorganisms in the formation of ground properties and their tense state. Hydrotech Eng 9:22–24 (in Russian)

Factors Controlling the Occurrence of Reservoir-Induced Seismicity

102

Xin Qiu and Clark Fenton

Abstract

Reservoir-induced seismicity (RIS) is defined as the failure of a pre-existing fault due to reservoir impoundment or water level fluctuations. 127 RIS cases have been recorded around the world, with four events of $M > 6$. RIS is triggered by a complex interaction between a number of factors including reservoir size, stress regime, hydrogeological condition and reservoir-filling history. Using statistical evaluation of worldwide data the relationships among these factors are investigated. The occurrence of RIS shows a strong correlation with reservoir size, faulting regimes, rock types and background seismicity. However these factors alone are certainly not necessary or sufficient conditions for the triggering of RIS. As the interactions between water movement and geology can be significantly complex, a detailed study on the Xinfengjiang reservoir is presented. Examination of fault location, orientation and permeability structure indicate that the NNW Shijiao-Xingang-Baitian fault is responsible for the majority of seismic events, including the $M_s6.1$ mainshock. The hydrogeological conditions causing RIS are discussed.

Keywords

RIS • Hydrogeological regime

102.1 Introduction

Reservoir-induced seismicity (RIS) is defined as the failure of a pre-existing fault due to the presence of a reservoir impoundment or water level fluctuations. Of the 127 RIS cases reported around the world 4 cases triggered earthquakes $M \geq 6$, 15 cases involved earthquakes $5.9 \geq M \geq 5$ and 32 were $4.9 \geq M \geq 4$ (Qiu 2012). Through the examination of the observations recorded at different RIS sites around the world as well as the detail case study on Xinfengjiang reservoir, correlation between the occurrence of RIS and different inducing factors is assessed.

Although RIS occurs in reservoirs with varying dam heights, the likelihood of occurrence increases with increasing dam height; 17 % of the deep reservoirs with dam heights >150 m have triggered seismicity. A similar relationship is observed with reservoir capacity. Although there is a positive correlation between reservoir size (dam height and capacity) and the probability of RIS occurrence, reservoir size is neither a necessary nor a sufficient condition for RIS to occur since many RIS cases occur in small reservoirs and many large reservoirs do not trigger seismicity.

Examining tectonic setting it is noted that 79 % of the RIS reservoirs are located in normal or strike-slip faulting environments, while only 21 % are in reverse faulting environments (Qiu 2012). Simple Mohr-coulomb failure models show that the addition of vertical loading from reservoir impoundment will act to promote fault stability in a reverse faulting regime, hence the lack of RIS (Qiu 2012).

X. Qiu · C. Fenton (✉)
Department of Civil and Environmental Engineering, Imperial College London, London, UK
e-mail: c.fenton@imperial.ac.uk

G. Lollino et al. (eds.), *Engineering Geology for Society and Territory – Volume 6*,
DOI: 10.1007/978-3-319-09060-3_102, © Springer International Publishing Switzerland 2015

102.2 Xinfengjiang Reservoir, China

Xinfengjiang reservoir in southeast China has a capacity of 1.4×10^{10} m^3 and a dam height of 105 m. It is situated above a large E-W trending Late Mesozoic granite body. Soon after impoundment in October 1959, an increase in earthquake frequency was observed. On 19 March 1962, a M_s6.1 earthquake occurred 5 months after the first peak reservoir level was reached. Focal mechanisms indicated that the main shock was on a steep, NNW-striking left-lateral strike-slip fault. Seismic activity started to decrease after 1965 (Ding 1989).

Most of the earthquakes $M_s \geq 3$ are located within three regions: A, B and C (Fig. 102.1). Region A is characterised by the intersection several faults. Regions B and C are associated with the NE- to NNE-striking faults.

The location of a fault relative to the reservoir determines whether the oscillating reservoir loads have a stabilising or destabilising effect (Roeloffs 1988). The Heyuan fault (1, Fig. 102.1) is a shallow-dipping thrust fault inclined towards the SE with the reservoir located on its footwall. The oscillating reservoir load should not induce seismicity on the fault. Few earthquakes have occurred on the Heyuan fault, except for the middle segment where it is intersected by Shijiao-Xingang-Baitian fault (4, Fig. 102.1). The Renzishi fault (2, Fig. 102.1) is a steeply-dipping reverse fault; the northern segment of the fault dips SE while the southern segment dips towards the NW. The reservoir is located on the footwall of the northern segment, and on the hanging wall of the southern segment. Most RIS events are located along the northern segment of Renzishi fault in the reservoir

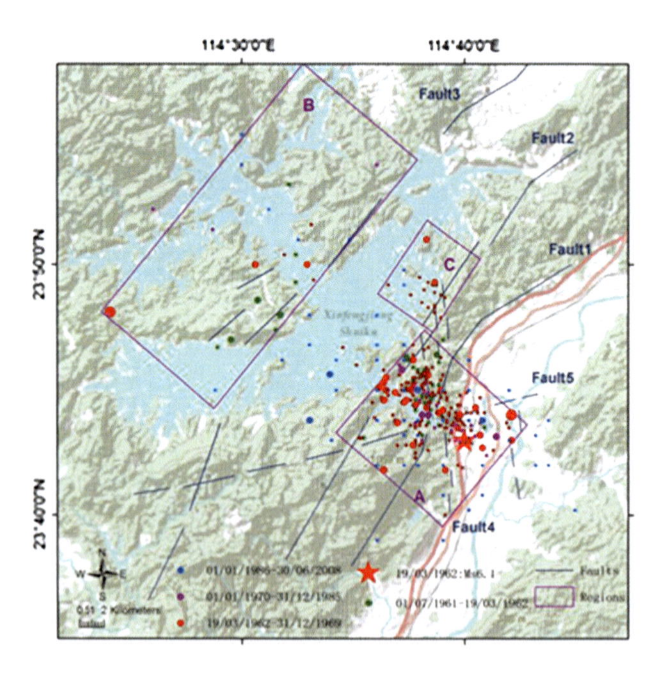

Fig. 102.1 Distribution of epicentres of $M_s \geq 3$ earthquakes (July 1961–June 2008)

and the middle segment is directly beneath the reservoir (Fig. 102.1). Earthquake focal mechanisms indicate that Shijiao-Xingang-Baitian fault (4) is the fault that is responsible for the M_s6.1 mainshock (Ding 1989). The Roeloffs (1988) fault location model explains the spatial distributions of epicentres around the reservoir.

Most RIS cases are caused by the reactivation of existing faults rather than the development of new ruptures (Morris et al. 1996). The possibility of reactivation for major faults in the Xinfengjiang area can be evaluated based on their orientations with respect to the regional stress field. This slip-tendency analysis is carried out using a MATLAB plug-in application (Neves et al. 2009). The results of this analysis indicate that the faults striking NNW have the highest slip tendency while faults striking NE/NNE direction are less likely to move (Qiu 2012).

Faults are structurally anisotropic and lithologically heterogeneous. In terms of permeability, they can either assist or impede water flow depending on their permeability structures (Caine et al. 1996). NE/NNE-striking faults around Xinfengjiang reservoir are reverse faults, usually ductile in nature. As there are no circulating paths for water to diffuse in this type of fault, RIS is unlikely to be triggered. The NNW-striking faults are strike-slip with distributed conduit type of permeability, leading to a greater possibility of RIS.

Fracture permeability around the reservoir is estimated to see if it lies within the seismogenic permeability range proposed by Talwani et al. (2007). Fractures within the seismogenic permeability range allow pore water to diffuse as Darcian flow, thus making it easier to induce seismicity. The estimated hydraulic diffusivity values (1–5 m^2/s) correspond to a permeability value range of 5×10^{-15}–2.5×10^{-14} m^2. This is within the seismogenic permeability range, indicating that RIS is likely to occur.

Analysis of fault location, slip tendency and permeability structure indicate that the NNW Shijiao-Xingang-Baitian fault (4, Fig. 102.1) is the structure responsible for the majority of earthquakes, including the M_s6.1 mainshock.

Although there is a reasonably good correlation between reservoir level and seismicity a period of delayed response is observed, when earthquakes occur sometime after peak reservoir level is attained (Qiu 2012). The M_s5.1 and M_s5.3 earthquakes occurred immediately after the exceedance of previous peak water level. This may be due to the Kaiser effect; if a material is experiencing cyclic loading with increasing stress, there is an increase in microseismicity (in the form of acoustic emission) if the highest stress level of the previous loading cycle (maximum water level) is exceeded (Lavarov 2003). It is notable that the M_s6.1 mainshock did not occur immediately after the first peak water level, but occurred when the water level was decreasing. One possible explanation could be although the

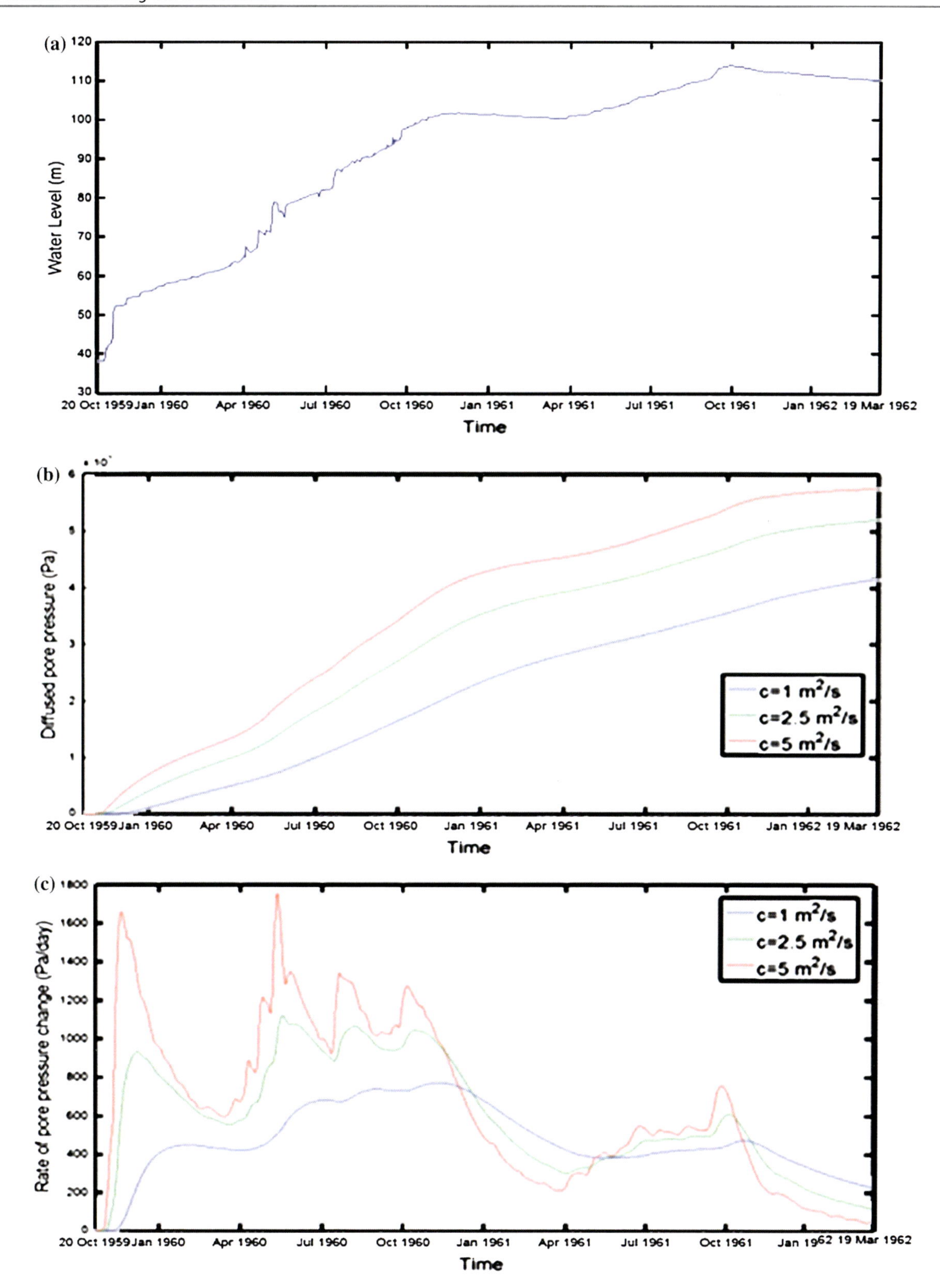

Fig. 102.2 Water level fluctuation (**a**), pore pressure diffusion history (**b**) and rate of pore pressure change (**c**), at the hypocentre of 19th Mar 1962 M_s6.1 main shock

reservoir level is decreasing, the pore pressure at the hypocentral location is still increasing. This hypothesis is only true if pore pressure diffusion is the dominant mechanism, since both the vertical elastic stress increase and undrained pore pressure increase due to compression are short term effects. The Mohr-Coulomb failure model used to calculate the stress field during the reservoir filling history indicates that the increase in vertical stress has negligible effect on triggering seismicity since the reservoir is located in a strike-slip faulting regime (Qiu 2012). This implies that RIS is only triggered by either the instantaneous undrained pore pressure increase due to elastic compression or pore pressure diffusion. To evaluate the relative importance of each of these effects the magnitudes of each are estimated. The maximum undrained pore pressure increase at the hypocentre of the main shock is estimated to be around 70 kPa using the Skempton effect (Talwani and Acree 1984). The diffused pore pressure is estimated using the approach of Talwani et al. (2007). The calculated pore pressures for 50 $M_s \geq 4$ earthquakes in the reservoir area during the period of 1961–2008 vary from 281 to 680 kPa, much larger than the undrained pore pressure increase, implying pore pressure diffusion should be the dominant mechanism in increasing pore pressure and reducing the effective stress. As the effect of undrained pore pressure increase due to elastic compression is instantaneous, by the time pore pressure front diffuses to the hypocentral location, this undrained effect may have already disappeared. Thus, the threshold of the diffused pore pressure can be interpreted as the threshold pore pressure for inducing seismic events. However, the threshold pore pressure values only indicate the pore pressure required to trigger an earthquake at a particular location, such as the epicentre of the M_s6.1 mainshock. It does not effectively explain why this mainshock occurred during a period of decreasing water level. To explain this phenomenon, a MATLAB routine is built to model the pore pressure diffusion history (Fig. 102.2) at the hypocentre of the M_s6.1 mainshock (Qiu 2012). Several important observations can be made. Firstly, the shape of pore pressure diffusion and water level fluctuations appear to be very similar, indicating a direct correlation. Secondly, the variations of the diffused pore pressure at hypocentral locations are dependent on the value of hydraulic diffusivity. When the hydraulic diffusivity is higher ($c = 5$ m/s^2), the diffused pore pressure will rise or diminish faster after experiencing a water level change, compared to low values of diffusivity. In addition, for the lower diffusivity case ($c = 1$ m/s^2), there is a delay between the reservoir impoundment and the onset of pore pressure increase. The initial increase in pore pressure for the lower diffusivity case is negligible. This delay of pore pressure could be due to the fact it takes longer for pore pressure front to arrive at the hypocentral locations for such low diffusivity values. Prior the M_s6.1 mainshock, although the water level was decreasing, the diffused pore pressure at the hypocentre was stilling increasing. A similar trend is observed for the M_s4.1 earthquake on 20th Feb 1962 foreshock. The rate of pore pressure increase is higher during the foreshock period than during the mainshock, indicating that in the long term, if the water fluctuations are small compared to the initial filling stage, the diffused pore pressure at a particular location may slowly diminish through time. However for the aftershock event of M_s4.3 on 6th December 1963, both the water level and the diffused pore pressure at the hypocentre were decreasing. This implies that failure of a fault does not always occur when the effective stress is at a minimum. Future approaches to explain this phenomenon should focus on earthquake-induced hydrological changes and deviatory effects of pore pressure drop (Qiu 2012).

102.3 Conclusions

Induced earthquakes are possible but not an inevitable consequence of the impoundment of a reservoir. Because of the complexity and variety of factors, any strategy for the limitation of RIS hazards should consider the overall complexity of the phenomenon. Analysis of global data shows that only when a combination of factors are present, can seismicity can be triggered. It is important therefore to maintain focus on the interrelation of diverse concurrent factors rather than to isolate any single one.

The evolution of the hydrogeological regime of the reservoir area plays the most important role in triggering and timing RIS. Concurrently, the stress regimes and the filling history can also be seen to contribute significantly, as highlighted by the case of Xinfengjiang reservoir. In order to develop a more comprehensive understanding of the triggering mechanisms, a series of RIS case histories should be subject to similar analysis. The modelling of pore pressure diffusion history is a particularly promising approach as it allows investigation of RIS sensitivity in relation to the hydrological conditions at hypocentral depths.

References

Caine JS, Evans JP, Forster CB (1996) Fault zone architecture and permeability structure. Geology 24:1025–1028

Ding YZ (1989) Reservoir-induced Seismicity. Seismological Press, Beijing (in Chinese)

Lavarov A (2003) The Kaiser effect in rocks: principles and stress estimation technique. Intl J Rock Mech Min Sci 40:151–171

Morris A, Ferril D, Henderson D (1996) Slip-tendency analysis and fault reactivation. Geology 24:275–278

Neves M, Paiva L, Luis J (2009) Software for slip-tendency analysis in 3D: a plug-in for Coulomb. Comput Geosci 35:2345–2352

Qiu X (2012) Reservoir-induced seismicity. Unpublished ME dissertation, Imperial College, London, 139 p

Roeloffs E (1988) Fault stability changes induced beneath a reservoir with cyclic variations in water level. J Geophys Res 93(B3): 2107–2124

Talwani P, Acree S (1984) Pore pressure diffusion and the mechanism of reservoir induced seismicity. PAGEOPH 122:947–964

Talwani P, Chen L, Gahalaut K (2007) Seismogenic permeability, ks. J Geophys Res 112:B07309. doi:10.1029/2006JB004665

Condition of Boguchany Concrete Dam Foundation According to Instrumenal Observations

103

E.S. Kalustyan and V.K. Vavilova

Abstract

Construction of 76 m high Boguchany concrete gravity dam on the Angara River in Siberia was finished in 2012 and reservoir filling started. The dam foundation is composed of trap intrusion presented by dolerites, heterogeneous by composition and properties. Since the start of construction in 1983 there have been conducted observations over the state of the rock foundation, which is subject to impacts of severe climatic conditions with annual temperature fluctuations up to 100 °C. During the period from 1998 to 2006 actually there was no construction work at the dam, however, instrumental observations continued. In the article the analysis is given of 30-year instrumental observations over the rock foundation deformations at various depths and over the temperature of rocks; relation was revealed between the foundation deformations with loads from the dam concrete, the water level of reservoir and temperature of ambient air.

Keywords

Dolerites • Concrete gravity dam • Monitoring deformations

103.1 Introduce

The Angara River originates from Lake Baikal. Due to this fact the river runoff under natural conditions is well regulated and in the high-water season it is only two times as much as that in the low-water season. The length of Angara is 1,850 km, the annual river runoff is 145 km^3. Taking into account its favorable hydrological conditions the Angara River was an ideal site for construction of hydro power plants already since the twenties of the XX-th century. The Angara water resources development in the interest of power generation was started only after the World War II with construction of the Irkutsk Hydro Power Plant (HPP). In 1955–1967 the second stage of the cascade—Bratsk HPP—was built. The Bratsk HPP reservoir with a full storage of 179.1 km^3 is one of the largest in the world. In 1980 the third stage of the Angara cascade was created on the Angara River—the Ust-Ilim HPP. At present the next stage of the cascade—the Boguchany HPP has been put into commercial operation the energy of which will make it possible to start developing the natural resources of the Angara region.

103.2 Monitoring Deformation of the Boguchany Concrete Dam

The Boguchany concrete gravity dam site is located in the wide Angara river valley. From the total length of the water retaining structures of 2,670.6 m and a height of 76 m the concrete gravity dam occupies 809.3 m in the left-bank part which are divided into 34 sections (Fig. 103.1).

The concrete structures fully rest on the trappean intrusion represented by dolerites varying in composition and properties. The peculiarity of the concrete dam foundation was determined by the composition and extent of the

E.S. Kalustyan · V.K. Vavilova (✉)
Branch of Institute Hydroproject, Geodynamic Research Center, Volokolamskoe Shosse 2, 125993, Moscow, Russia
e-mail: vergor@pochta.ru

E.S. Kalustyan
e-mail: office@geodyn.ru

G. Lollino et al. (eds.), *Engineering Geology for Society and Territory – Volume 6*,
DOI: 10.1007/978-3-319-09060-3_103, © Springer International Publishing Switzerland 2015

Fig. 103.1 General view of the Boguchany concrete gravity dam

mineralogical alteration of the dolerites. Excavation operations in these rocks caused significant de-stressing of the rock mass up to formation of dolerite detritus. Core logs of the boreholes arranged in the instrumentation installation areas before placing the concrete into the dam showed that more jointed rock with swelling minerals amounted to 1–5 % of the total volume in sections 6–19 of the left-bank abutment; 6–30 %—in river channel sections 20–24, 30–31, 33–34 and 30–70 %—in sections 25–29. It was considered that this fact should have an effect on properties of the rock mass in the concrete dam foundation, on its deformation indices and deformation uniformity as well as on the behavior during the construction and at the initial stage of operation—during the reservoir filling. Taking this into account the concrete placement into the concrete dam body was started from section 24 in 1983 year.

During construction of the Boguchany concrete dam the assessment of the foundation condition based on the results of in situ tests along with check measurements of deformations in the near-contact and active zones and temperature included special studies of the foundation. The data obtained during these special studies made it possible to assess the deformation irregularity of the foundation before the reservoir filling and to obtain the input data for determining deformation indices for different areas of the dam foundation (Smulskiy 1992).

The peculiarity of the Boguchany dam foundation consists in significant length of the water-retaining structures where measurements of deformations had to be conducted in the near-surface and active foundation zones. Heterogeneity of rocks was complemented here by harsh external conditions such as climatic effects with considerable variations of external air temperature which became the most important load in conditions of long-term development. In the 1990s of the last century the construction progress was impeded by the absence of financing.

Considerable temperature variations together with other factors defined the expediency of conducting special studies. Annual temperature variations during the whole period of observations at the Boguchany HPP site attained 100° (Fig. 103.2). In 2012 year the reservoir filling commenced under conditions of positive water temperatures in the Angara River introduced significant changes in distribution of foundation temperatures in the near-surface zone from the upstream side of the concrete dam.

Project of the Boguchany HPP concrete structures started in 1982 with laying the concrete in section 24 is practically completed at present. At that, stresses in the foundation of the first sections of the concrete dam due to concrete placement vary on the whole from 1.40 to 2.41 MPa. The crucial point is that stresses achieved by the time of reservoir

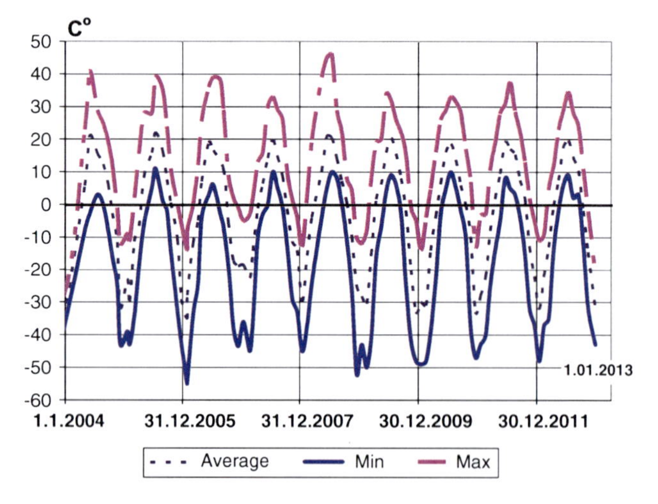

Fig. 103.2 Temperature of ambient air

filling at the foot of the concrete dam in all sections considerably exceeded the extent of de-stressing of the foundation as a result of rock excavation in the dam excavation pit (Kalustian et al. 2000).

Consolidation measures in the foundation of the concrete dam sections started in 1987 have been completed. The contact grouting, the 45 m deep single row grout curtain, and drainage measures have been perform.

For measure deformations in the surface zone of the concrete dam foundation to a depth of 2–5 m vibrating wire transducers of linear deformations LDT-10 with extension rods, 2,000 and 5,000 mm long, were used. Rock strain gauges were installed in the foundation of the majority of sections (from 6 to 34 sections) since commencement of concrete placement in 1983. Practically in all sections the instruments were installed by the beginning of the concrete placement into the foundation of the respective section. Due to this fact it was considered that the vibrating wire transducers measured relative deformations of the surface layer of the rock foundation from the moment of concrete placement into the concrete dam body of the Boguchany HPP. Based on the results of field observations it was possible to determine actual deformations of the surface and active areas of the dam foundation during all periods of the construction and at the initial stage of reservoir filling and the unique material was obtained for the analysis of foundation deformations in the severe climatic conditions of the building site. In the process of the field observations relative deformations in the surface zone of foundation were measured by the vibrating wire transducers on a monthly basis. The interval of taking readings at the initial stage was determined by concrete placement rate in the dam sections.

Out of 34 sections of the Boguchany HPP concrete dam, four sections—Nos. 12, 21, 28, 31 are measuring sections. These sections are most fully equipped with instruments for measuring relative deformations and temperature both in the body and in the foundation of the dam to depths of 2–7.5 m from the upstream and downstream faces side. The remaining 30 sections of the concrete dam are equipped with instruments for assessing deformations and temperature of the foundation from the concrete dam upstream face side.

The analysis of the results of the long-term field observations showed that vibrating wire instruments of domestic production—jointmeters LDT-10 with special extension rods continue to record deformations in the foundation surface part under the concrete dam sections both to depths of 2–16 m.

The civil work at the construction of the Boguchany concrete dam was conducted in the severe climatic conditions of Siberia with considerable drop of temperatures (Fig. 103.2). Therefore, along with recorded deformations it was essential to assess the values and distribution pattern of outer temperatures and the temperatures in the foundation both in the "rock—dam concrete" near-contact zone and in

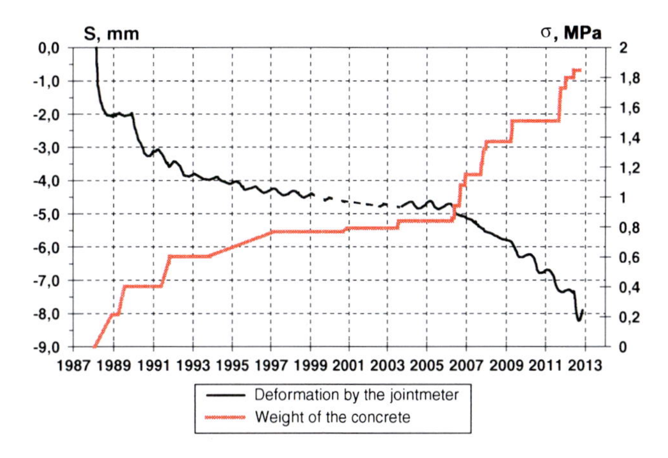

Fig. 103.3 Deformations near-contact 2 m zones foundation in the section 18 during the construction of the dam

depth within the active rock mass zone as well as water temperatures.

For the Boguchany HPP concrete dam foundation thermal actions represented additional loads causing de-stressing in the areas of uncovered rock mass especially in the hazardous HPP excavation pit area. In these conditions one of the possible scenarios of unfavourable condition development could form, i.e. cracks formation in the open areas of the foundation from the dam upstream face side. According to this scenario cracks could form from the upstream face side long before the onset of the reservoir filling, even at the building stage. The field observations showed that such developments had been avoided.

Special investigations included the assessment of stresses acting on the toe of the dam due to the weight of the concrete being placed (Fig. 103.3).

This made it possible to obtain data required for assessment of deformation indices for the foundation of different sections and to assess the deformation nonuniformity of the near-contact and active zones of the concrete dam foundation.

103.3 Results of Special Studies of the Modulus of Deformation of the Concrete Dam Foundation

The analysis of operation of damaged concrete dams showed that with different correlations of deformation indices of the foundation and dam the conditions of failure risk realization tended to develop. For estimation of the modulus of deformation value in separate sections of the Boguchany concrete dam and deformation non-uniformity of its foundation it was required to assess the stresses at the toe of the concrete dam foundation and respective settlements of the surfaces and active zones on the basis of the monitoring data.

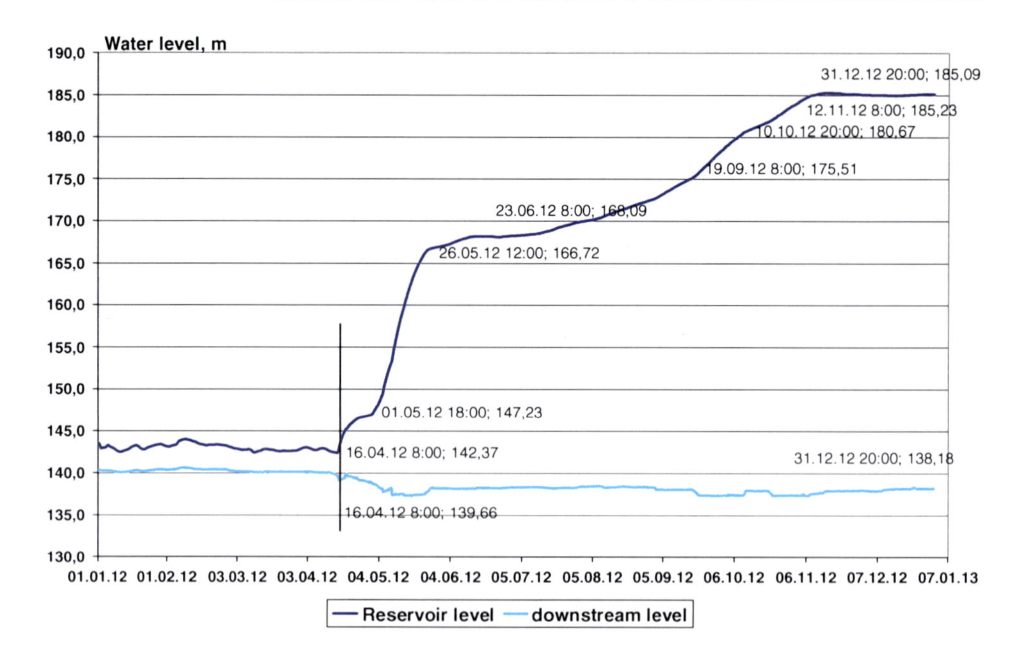

Fig. 103.4 Graphic of the reservoir filling

The deformation of the compressed foundation under the weight of the placed concrete is taken into account when calculating the values of the modulus of deformation. Characteristic of dolerite composing the Boguchany concrete dam foundation is a long time period when the rock mass comes to its natural condition that can be explained by the impact of environmental conditions and temperature variation in particular (Kalustian and Vavilova 2011).

The monitoring data on the foundation deformations for the entire period since 1983 as well as the results of special studies for determining the stresses at the base of dam concrete during concrete placement have been used as initial data for determination of the modulus of deformation of dolerites in the foundation of different sections of the Boguchany dam. Based on the obtained data a preliminary conclusion was drawn that the modulus of deformation in the foundation of sections No. 12–17 was higher ranging from 1,4984 to 11,166 Mpa than in sections No. 18–32. The modulus of deformation in sections No. 18–32 ranged from 6,013 to 10,644 Mpa with stresses in sections No. 18–24 ranging within 0.83–1.70 Mpa and in the foundation of sections No. 24–32 ranging within 1.48–1.76 Mpa. Generally the modulus of deformation of the concrete dam foundation may be considered lower than the modulus of deformation of the dam concrete.

The computations show that modulus of deformation in the near-contact and active zones of the foundation are different. In particular, a more deformable interlayer has formed in the concrete dam-rock foundation near-contact zone featuring the modulus of deformation equal to 1.6–3.3 GPa. The presence of this interlayer must be advantageous for stress distribution in the "dam-foundation" contact zone

In consideration of commencement of the Boguchany reservoir filling since 16.04.2012, new stage began in the dam operation (Fig. 103.4). Instrumental monitoring of the foundation in this case acquires a special importance as it allows for the areas of possible most unfavorable condition in the foundation to be identified even before the full supply level is attained and engineering protection measures to be taken. At the same time with water level rise in the reservoir there appeared new factors determining the deformations and temperature of the near-contact and active zones as well as general conditions of the foundation (Kalustian 2013).

Several typical features of deformations in the surface zone of the concrete dam foundation taking place with commencement of the reservoir filling are shown in Fig. 103.5.

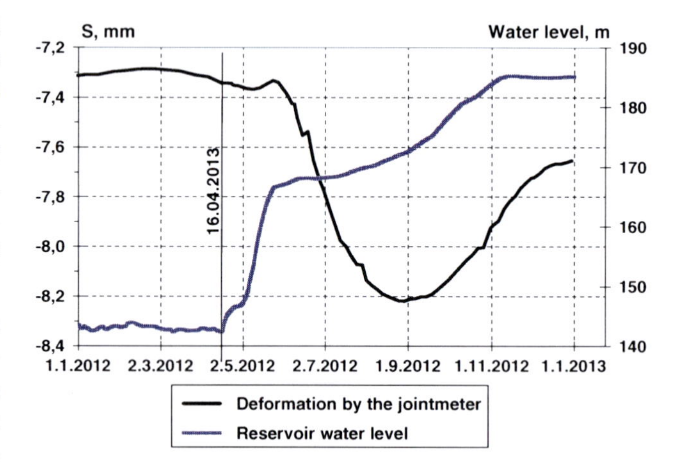

Fig. 103.5 Deformations near-contact 2 m zones foundation in the section 18 during of the reservoir filling

103.4 Conclusion

With start of reservoir filling the deformations of concrete dam foundation at Boguchany HPP both in the active and contact zones are depend by the reservoir water pressure. At that, foundation vertical deformations induced by the weight of water in the reservoir considerably exceeded horizontal displacements. On the whole, at reservoir filling to the intermediate elevation 185 m the condition of the structure is in the project state; the joint «rock—concrete» from the upstream side of the concrete dam is in the compressed state.

References

Kalustian E (2013) Dam construction and geomechanics. Moscow, Rads-RRL, 120 p

Kalustian ES, Gorbushina VK, Koryabin IA (2000) Condition of the foundation bed for the concrete dam of the Boguchany hydroelectric power plant as determined from field observations. Hydrotech Construct 34(4):163–167

Kalustian ES, Vavilova VK (2011) Boguchany HPP monitoring. Hydrotech Construct 12:36–37 (in Russian)

Smulskiy PY (1992) The Boguchany dam on the Angar River, vol 12. Geology and Dams, Moscow (in Russian)

Study on Reservoir and Water Inrush Characteristic in Nibashan Tunnel, Sichuan Province, China

104

Sixiang Ling, Yong Ren, Xiyong Wu, Siyuan Zhao, and Limao Qin

Abstract

The Nibashan tunnel is a major project mountain tunnel on Ya'an to Lugu lake expressway in Sichuan province. The tunnel relative elevation of about 2,100 m and the maximum depth of 1,701 m, belongs to the deep buried and extra-long tunnel. The tunnel's strata are mainly igneous rock of Suxiong Formation, Lower Sinian. The rock type belongs to rhyolite (69 %), andesite (21 %) and a small amount of pyroclastic rocks, dykes and terrigenous clastic rocks. Based on the field investigation and drilling data of Nibashan tunnel, analysis of igneous rock reservoir space and performance factor by the lithology, lithofacies, fractures, structure and diagenesis of igneous rock. It considered that the igneous rock reservoir water storability was mainly affected by tectonism, and the reservoir site is located in the tectonic fracture zone. Finally, through the high pressure water pressure test and pumping test of drilling on Nibashan tunnel, obtaining the water reservoir and water inrush characteristic of igneous rock and geological flaws in tunnel, and providing a theoretical basis to construct safely and smoothly the tunnel.

Keywords

Nibashan tunnel • Igneous rock • Reservoir • Water inrush characteristic

104.1 Introduction

The hydrodynamic system and dynamic equilibrium of surrounding rocks undergo drastic changes due to the excavation of underground projects, which causes instant release of the energy stored in underground water gush to the tunnel face (Lin and Song 2012). Water inrush in tunnels is characterized by suddenness, high speed, and high pressure, and it is highly destructive, which can do tremendous damage to construction and operation of tunnels and damage the environment, even serious safety accident (Wang et al. 2004; Li et al. 2011; Li and Li 2014). Therefore, the water inrush scourge was a very serious problem for the deep tunnel. In recent years, more than 100 cases of water inrush have been observed in China, causing serious losses of economics and deterioration of construction conditions (Zhao et al. 2013). The previous researchers have studied water critical pressure, water inrush prediction method, water inrush risk assessment, and treatment measures in the mining roadway, expressway tunnel and railway tunnel (Zhang and Peng 2005; Meng et al. 2012; Li et al. 2013; Zhao et al. 2013).

The Nibashan tunnel is located in the edge of Longmeshan fault, and has the igneous rocks. The maximum depth of the tunnel is 1,701 m, and about 5 km length tunnel deeper than 1,600 m. As a result, a large number of underground water is stored in the igneous rock fissures and joints in mountain. The formation of water inrush passages is mostly the geological

S. Ling · Y. Ren · X. Wu (✉) · S. Zhao · L. Qin
Department of Geological Engineering, Southwest Jiaotong University, Chengdu, Sichuan 610031, People's Republic of China
e-mail: wuxiyong@126.com

X. Wu
Key Laboratory of High-Speed Railway Engineering, Ministry of Education, Chengdu, Sichuan 610031, People's Republic of China

DOI: 10.1007/978-3-319-09060-3_104,

flaws, including fault, fracture, joints and unfavorable section in the surrounding rock over a tunnel (Shi and Singh 2001). Some scholars have researched the microscopic pore characteristics in acidic volcanic reservoir from the oil and gas exploration (Pang et al. 2007). The water reservoir space has characteristics of complexity, instability and heterogeneity, because of the igneous compositions are uneven, the condensation environmental differences and the uneven degree of porosity and fissures. The water inrush is likely to occur in the fault zone and joint fissures developing zone. Generally, the water inrush accident occurs unexpectedly, especially cause the serious accident at fault zones. Therefore, it is essential to accurately understand the reservoir characteristic of the igneous rocks and the water inrush features in Nibashan tunnel, and it can provide some effective technical countermeasures to assure the safety of tunnel construction.

104.2 Geological and Engineering Setting

The overall length of the Nibashan tunnel is about 10 km (Fig. 104.1). The tunnel is located on an anticlinal fold and the geological structure is controlled by geofracture, eg. Baohuang and Caodaping fault (Fig. 104.1a). From field investigation, the tunnel is covered with igneous rock and carbonates in Sinian Age (Fig. 104.1b). And also, we designed five drills to do the injection test and pumping test in the geological prospecting (Fig. 104.1b). The tunnel construction method is a drilling and blasting method. The tunnel will be established a twin tunnel, and each tunnel is designed two-lane single-way and the design speed is estimated to be 80 km/h. The lithology of tunnel is mainly composed of rhyolite, andesite and carbonatite, and few of granite porphyry, pyroclastic rock and dike. The Nibashan mountain ridge is the watershed of surface water, the mean annual rainfall is 742 mm in the north of watershed, and 1,300 mm in the south of watershed. Therefore, the atmospheric precipitation can recharge amply groundwater at the location of tunnel.

104.3 The Characteristic of Water Reservoir Space

104.3.1 The Features of Lithology and Petrography

The lithology of tunnel is mainly belongs to igneous rock, including rhyolite (69 %), andesite (21 %) and some other clastic rock. From the field investigation and drilling data, the numbers of fissures in rock is granite porphyry > rhyolite > andesite > diabasic dike. From the perspective of rock lithofacies, the flowing facies, erupting facies and volcanic sedimentary facies is equal to 59, 37 and 3 %, respectively,

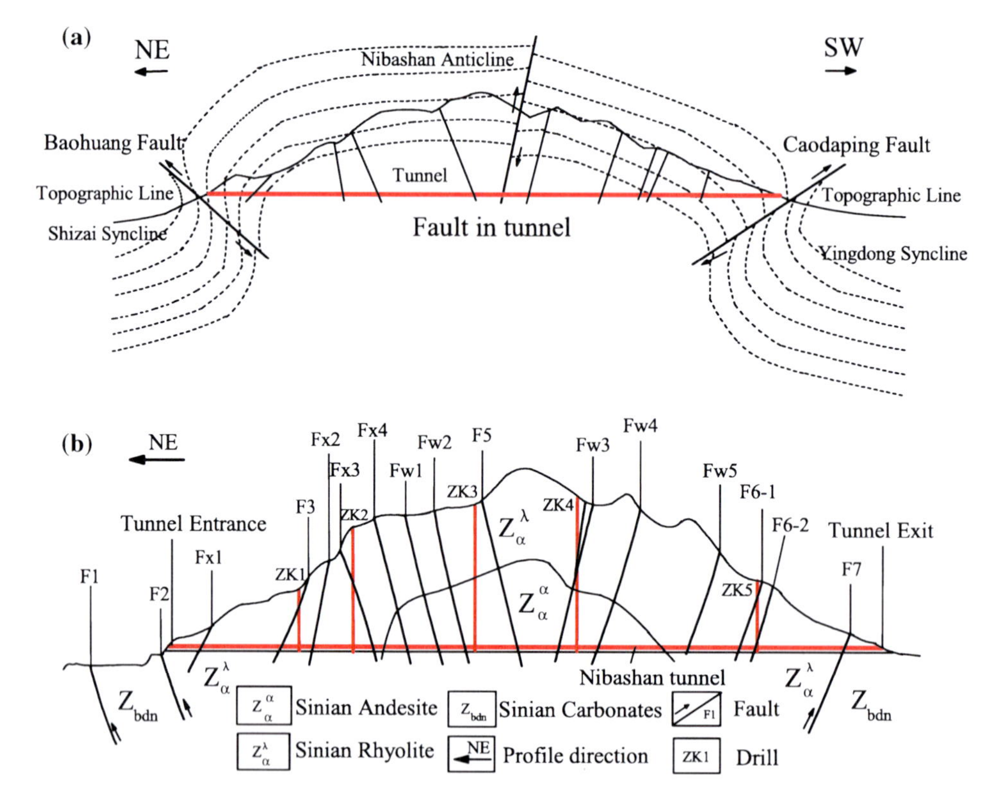

Fig. 104.1 The anticline profile and geological structure diagram of tunnel in Nibashan region

Table 104.1 Different type of pore in different rock at Nibashan tunnel

Petrography	Lithology	Pore type
Flowing facies	Stomatal and striated rhyolite	Stomata, almond stomata, dissolved pore
Erupting facies	Rhyolitic and crystal tuff, volcanic breccia	Stomata, intergranular pore, matrix pore
Volcanic sedimentary facies	Tuff, tuffaceous sandstone/breccia	Interparticle pore
Volcanic channel facies	Granite porphyry, diabasic dike	Columnar joints, contact zone fissures

from the drilling data. The porosity and permeability of rock, belongs to flowing facies and erupting facies, has a higher level, while the volcanic sedimentary facies rock has a lower level because of the sedimentary environment and diagenesis. The water storage performance statistics are shown in Table 104.1. From the combination of lithology and petrography, the most beneficial part for water reservoir is volcanic breccia and tufflava. However, the rhyolite proportion is 69 % in tunnel and the rhyolite is the main storage space at rock part.

104.3.2 The Effect of Fissures

From the field investigation, the joints and fissures of rock on tunnel surface are divided into four directions: N10°–50° W, N50°–80°E, N10°–20°E and N70°–90°W, respectively (Fig. 104.2). The dip of fissures and joints can be divided into three types: 0°–10° (horizontal fissure), 10°–60° (oblique or reticular fissure) and 60°–90° (high angle fissure), respectively. The proportion of high angle fissures, oblique fissures and horizontal fissures are 56, 42 and 2 %, respectively. The joints and fissures are one of types to water storage spaces. On the other hand, the fissures and joints are the seepage channel, which make the isolated primary pore connected with each other and promote significantly the development of secondary pores. Therefore, the joints and fissures can improve the water reservoir performance of igneous rock in Nibashan tunnel. From the drilling data, we found the deep fissures can reach up 1,000 m depth by the data of water color (incrustation), indicating that provide the water reservoir and seepage channel arrive to tunnel.

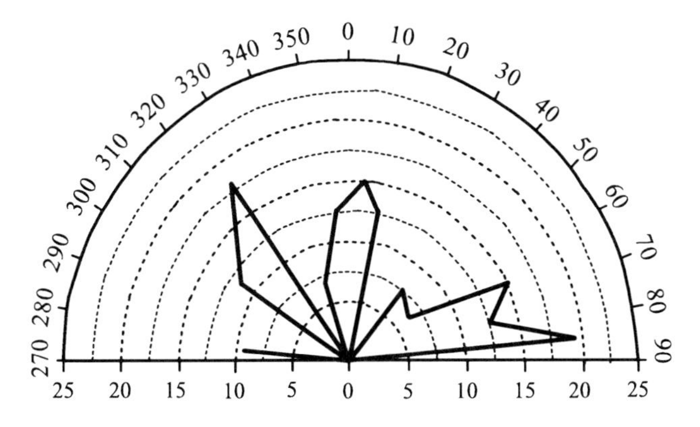

Fig. 104.2 The strike of joints rose diagram in tunnel overlay surfaces area

104.3.3 Reservoir of Structure and Dyke

The geological structure of Nibashan Mountain belongs to anticlinal fold and it displays the "Ω" shape (Fig. 104.1). The reverse fault (F2 and F7) located in the sides of the anticline, making the anticline into a symmetrical shape. The surface water flow to the anticlinal flanks, and then infiltrate into the underground affected by the terrain, fault and secondary folds. The faults of tunnel are mainly the compressor-shear fault, and this fault and the tensile fissures nearby the fault are the important part of water reservoir. Affected by the Nibashan anticline, the fault enriched the water at axis of anticline, such as F5 and Fw3 fault.

104.4 The Features of Water Inrush

In order to understand the extent of fissures and rock permeability, it should be tested by high-pressure water injection test. In the field, we choose the ZK1, ZK2 and ZK4 drill to do this test and the maximum pressure can reach up 4.5 MPa. Test objects including the complete bedrock and faults. The test diagrams (P–Q curves) are listed in Fig. 104.3 and the analytical results are shown as follows:

(1) In the ZK1 drilling, the location range of 128–136 m and 139–149 m is the fracture zone of fault F3, namely that the test results represent the water inrush model of fault F3. The curve of Fig. 104.3a, b belong to the "washout type" (D type) in the Water Resources and Hydropower Engineering borehole water pressure test procedures (SL 31-2003) (Ministry of Water Resources, P. R. China 2003), because the curve has obvious inflection point at 2 MPa in the Fig. 104.3a, b, and the seepage flux abruptly increases with decreasing the slope of line when the pressure is greater than 2 MPa. It shows that the fracture zone of fault F3 contains large number of broken rocks and the permeability is range of 0.012–0.0775 m/d from test in fault F3. Therefore, the blocking materials, such as clay materials, will be washed away when the pressure is greater than 2 MPa,

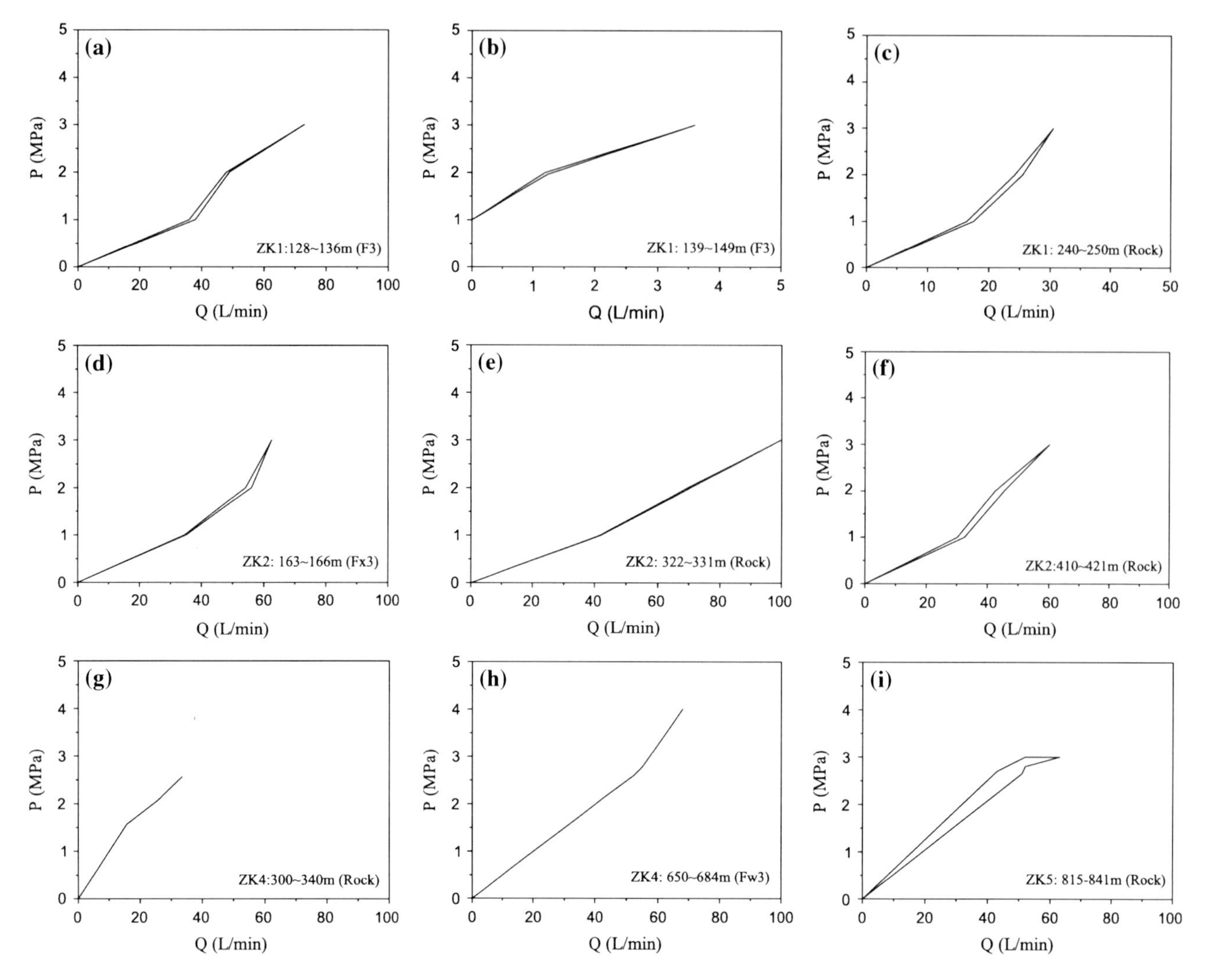

Fig. 104.3 The P–Q curves of high-pressure water injection test in tunnel site area

namely that the point of F3 in tunnel will be threatened under this pressure.

(2) The location range of 163–166 m represent the fault of Fx3 in the ZK2 drill. The P–Q curve belongs to the "turbulence model type" (B type) in the Water Resources and Hydropower Engineering borehole water pressure test procedures (SL 31-2003). From this curve (Fig. 104.3d) (Ministry of Water Resources, P. R. China 2003), the rock permeability and the flux are in higher level and the flux gradually decreases with increasing the pressure. Through test, the permeability of fault Fx3 is equal to 0.0768 m/d. Compared with F3, the blocking materials (e.g. clay) of fault F3 is washed easily from the fissures at the high pressure.

(3) The part of 300–340 m in ZK4 drill represent the complete rock-mass. The P–Q curve has obvious inflection point at 1.6 MPa, and then the seepage flux increases with the decreasing the slope of line when the pressure is greater than 1.6 MPa (Fig. 104.3g) (Ministry of Water Resources, P. R. China 2003). This curve belongs to the "expanding type"(C type) in the Water Resources and Hydropower Engineering borehole water pressure test procedures (SL 31-2003) (Ministry of Water Resources, P. R. China 2003). From the drilling data, most of the drill core RQD value is less than 30. Therefore, the fissure fillings are occurred the displacement and deformation when the water pressure greater than 1.6 MPa.

(4) The part of 650–684 m in ZK4 drill represent the fault Fw3. The P–Q curve belongs to "washout type" (D type) in the Water Resources and Hydropower Engineering borehole water pressure test procedures (SL 31-2003) (Ministry of Water Resources, P. R. China 2003), because this section has a higher seepage flux under the lower pressure and the curve has obvious inflection point at 2.5 MPa in the Fig. 104.3h and the seepage flux

abruptly increases above 2.5 MPa. It indicates that the rock of fault Fw3 has lots of fractures or joints and the permeability is in higher level under low pressure. However, the blocking materials are washed out at 2.5 MPa pressure, and the water channel is opened and lead to the seepage flux abruptly increase above 2.5 MPa. The P–Q curve convex to the P line as increasing pressure, it maybe consist with the rock hasn't reach up the saturation state. The seepage flux hasn't increase abruptly when the pressure is greater than 3 MPa (Fig. 104.3h). Therefore, the critical pressure of hydraulic fracturing fissure should be greater than 3 MPa as increasing depth under 684 m in ZK4 drill.

(5) In the ZK5 drill, the location range of 815–841 m represent the complete rock-mass. The P–Q curve belongs to the "laminar flow model type" (A type) from the specification of SL 31-2003 (Fig. 104.3i). The permeability is less than 10^{-3} m/d, consisting with the ROD is greater than 65 % from the drilling data.

(6) The part of 240–250 m in ZK1, 322–331 m and 410–421 m in ZK2 drill represent the complete rock-mass. The P–Q curves belong to the "turbulence model type" (B type) from the specification of SL 31-2003, indicating that the rock-mass has good permeability when the pressure is in lower condition (Fig. 104.3c, e, f, h).

From the high-pressure water injection test data, showing that the rock-mass are developed lots of fissures in tunnel site area. The rock permeability is in higher state when pressure in lower condition, while the flux gradually decreases with increasing pressure. The fillings of fissure and fault are washed away to make the fissure form the flow channel when the pressure is in 1.6–2.0 MPa, and lead to the flux gradually increase under this state. That is to say, the fissure and fault occurs the "washout type" water inrush when the piezometric head reach about 1.6 MPa. Expect for the location 815–841 m of ZK5 drill, we have choose the section of 820–825 m of ZK4 drill to do the high-pressure water injection test, while pressure gage was broken abruptly when the water pressure reached up 4 MPa. According to the test found that the deep rock fissure occurred hydraulic fracturing when the critical pressure is greater than 4 MPa when depth reach up 800 m, and this pressure increase with increasing depth. And also, we choose the 1,388 m point of ZK4 drill to do the pumping test, and the permeability is equal to 7×10^{-4} m/d. The rock core is the rhyolite, and the rock has lots of joints and fissures. The black iron manganese disseminated body and chloritization was found on the surface of joints and fissures. Therefore, the water reservoir mainly stored in rhyolite, and the water can arrive to deeper tunnel location.

Affected by water reservoir, the water inrush type in the tunnel can be categorized into two. One is the water flux decreases rapidly and short duration, and the other is the water flux keeps the stable state and longer duration. The former one is reasons for the water reservoir is debunked, but lack of the recharge into tunnel and lead to the flux decreases rapidly. The latter one is closely linked with the groundwater and water reservoir, the water pressure keep in higher level state in a certain period to make the water inrush last longer time. From the field and tunnel excavation monitor, the volcanic rock undoubtedly has a good water storage space, and the maximum water flux can reach 25,000 m^3/d.

104.5 Conclusion

We studied the features of water reservoir and water inrush in Nibashan tunnel. Some conclusions can be drawn from the above discussion:

(1) The main water reservoir is the pores and fissures of rhyolite and andesite in tunnel, and some groundwater can be as deep as 1,000 m. The fault fracture zone and dike are often contain large number of groundwater and closely contact with the movement of groundwater, where is the most possibility to occur the serious water inrush accident in tunnel.

(2) The shallow of entrance are enriched phreatic water, and the part near the anticlinal axis enriched the groundwater as increasing depth. The deep water reservoir affected by the oblique and reticular fissure.

Acknowledgements This work was supported by research funds awarded by the National Natural Science Foundation of China (Grant No. 41172261) and the Key Technology Research and Development Program of Sichuan Province, China (Grant No. 2012SZ0051).

References

Li S, Zhou Z, Li L et al (2013) Risk assessment of water inrush in karst tunnels based on attribute synthetic evaluation system. Tunn Undergr Space Technol 38:50–58

Li X, Li Y, Zhou S (2011) Study and application of forecasting system for water inrush under high pressure in Xiamen Submarine Tunnel construction based on GIS. Procedia Environ Sci 10:999–1005

Li X, Li Y (2014) Research on risk assessment system for water inrush in the karst tunnel construction based on GIS: case study on the diversion tunnel groups of the Jinping II Hydropower Station. Tunn Undergr Space Technol 40:182–191

Lin G, Song R (2012) Research on the mechanism and treatment technology of mud gushing in karst tunnel. Tunnel Construct 32 (2):169–174 (in Chinese)

Meng Z, Li G, Xie X (2012) A geological assessment method of floor water inrush risk and its application. Eng Geol 143–144:51–60

Ministry of Water Resources, P. R. China (2003) Water Resources and hydropower engineering borehole water pressure test procedures (SL 31-2003), pp 7–21

Pang Y, Zhang F, Qiu H et al (2007) Characteristics of microscopic pore structure and physical property parameter in acidic volcanic reservoir. Acta Petrolei Sinica 28(6):72–77 (in Chinese)

Shi LQ, Singh RN (2001) Study of mine water inrush from floor strata through faults. Mine Water Environ 20(3):140–147

Wang TT, Wang WL, Lin ML (2004) Harnessing the catastrophic inrush of water into new Yungchuen Tunnel in Taiwan. Tunn Undergr Space Technol 19:418

Zhang J, Peng S (2005) Water inrush and environmental impact of shallow seam mining. Environ Geol 48(8):1068–1076

Zhao Y, Li P, Tian S (2013) Prevention and treatment technologies of railway tunnel water and mud gushing in China. J Rock Mech Geotech Eng 5:468–477

105 Differential Settlement Control Technologies of the Long Submarine Tunnel Covered by Municipal Road

Cuiying Zhou and Zhen Liu

Abstract

Differential settlement control technologies for the long submarine tunnel covered by municipal road under complex geological condition are studied by depending upon a typical composite tunnel called the Western Corridor Connecting Project from Shenzhen to Hong Kong. Based on the analysis and generalization of the geological conditions of the research area, the computation sections and their corresponding parameters are selected, the differential settlement of all the selected sections in the tunnel area are calculated by the specification method and the nonlinear finite element method, the differential settlement isoline map of the tunnel box and the overlaying municipal road are obtained, the secondary consolidation settlements after construction are predicted. The optimization design and control schemes to the key control bids are proposed. And the results provide useful references for the design, construction and decision making of the project.

Keywords

Long submarine tunnel • Overlaying municipal road • Composite road structure • Differential settlement • Numerical simulation • Optimization design scheme • Control technologies

105.1 Introduction

Differential settlement of soft soil is one of the most difficult problems in geotechnical engineering. Differential settlement control for the long submarine tunnel covered by municipal road under complex geological condition is not only a complicated problem of differential settlement, but also a difficult problem needed to be solved both in engineering practice and theoretical research.

Currently, domestic and international researches on differential settlement mainly focus on the following aspects as (Wu Sheng-fa and Sun Zuo-yu 2005) forecast of the process of structure settlement, calculation of final settlement and specific measures to control differential settlement etc. Major methods to study and analyze consolidation settlement of foundation can be divided into the following three categories: empirical formula method (Li Guangxin 2004), model test method (Li Guangxin 2004), and numerical analysis method (Zhe Xueshen 1998). As a traditional method for calculation of consolidation settlement, empirical formula method includes: layer-wise summation, Skempton—Bjerrum three-dimensional stress effect, Huang Wenxi three-dimensional deformation and compression method, the Cambridge model, simulation curve method (Du and Zhang 2005; Lin and Cao 2006), Janbu tangent modulus method, Lambe stress path method etc. Empirical formula method is

C. Zhou (✉)
School of Engineering, Sun Yat-sen University, Guangzhou, 510275, China
e-mail: ueit@mail.sysu.edu.cn

C. Zhou · Z. Liu
Research Center for Geotechnical Engineering and Information Technology, Sun Yat-sen University, Guangzhou, 510275, China

Z. Liu
School of Marine Sciences, Sun Yat-sen University, Guangzhou, 510275, China

G. Lollino et al. (eds.), *Engineering Geology for Society and Territory – Volume 6*,
DOI: 10.1007/978-3-319-09060-3_105, © Springer International Publishing Switzerland 2015

aimed at simplifying engineering practice, with similar assumptions, and poor adaptability. It is usually used to solve simple structures. For complex structure, it has complicated calculation steps. Often as a result of geological and complexity of engineering factors involved, it is difficult to calculate the process of a more comprehensive reflection of the actual situation. There are often results of the error, and it is often difficult to find a reasonable solution. When the structure has a unique pattern, and loads as well as materials are very complex, model tests are needed to determine the mechanical behavior. Model test is divided into field test and lab test. Field test includes standard penetration test and pressuremeter test. Lab test includes seepage force model test and centrifugal model test (Han 2005, Zheng Yonglai et al. 2005, Zhan 2006), etc. However, model tests are often subjected to restrictions on site and equipment and only for small- scale tests. It is difficult to fully reflect the actual structure, also, the labor and material costs are expensive. Therefore, the calculation of a more comprehensive project to reflect the actual geological conditions has become the key to solve this problem. Numerical analysis (Xiong Chunbao et al. 2006; Hu et al. 2005; Du and Zhu 2005) represented by finite element method, is the product of research on modern soil mechanics. Since the 1970s, with the development of computer and application of finite element technology, complicated calculation of geotechnical problems have been compiled into finite element programs, and more accurate results have been obtained. Numerical simulation, as a "numerical experiments", can replace the expensive model test to some extent. What is more, if it is combined with the test results, more technical and economical benefits can be achieved. The use of numerical analysis can reach a comprehensive, more rigorous consideration on deformation properties of soil and boundary conditions.

Based on a typical composite tunnel called the Western Corridor Connecting Project from Shenzhen to Hong Kong, as well as in considering the unique compound road structures or various geotechnical structures which attach to the complexity of geological conditions, the causes of differential settlement have been analyzed. Furthermore, through calculation and analysis, this research has brought up the rule of differential settlement of the tunnel structure and its control measures, proposed the optimized proposals and specific control measures, and predicted post-construction settlement of the project, providing a significant reference for the project design, construction, and the owners in their decision-making. At the same time, this reference is also valuable to the study of differential settlement under complicated geological conditions and control measures for the differential settlement.

105.2 General Project Information

105.2.1 Brief Project Introduction

Shenzhen-Hong Kong western corridor connecting project is an important component of the Shenzhen-Hong Kong western corridor. This connecting project is a dedicated two-way six-lane passageway for the Shenzhen-Hong Kong transit vehicles, with designed driving speed of 80 km/h. It starts at YueLiangWan Road, passes through northern foot of NanShan Mountain, Dongbin Road and Houhai Bay, intersects with Industrial Road, Houhai Road, the planning Houhaibin Road and Keyuan Avenue, finally finishes at the export of Shenzhen-Hong Kong Port. The main line (ramps not included) is approximately 4.5 km (with underground structure 3.08 km and elevated structure 0.735 km). Blocks along the connecting project can be divided into three categories: Nanshan Mountain scenic spot, urban construction area and planning district. The general layout of the main line uses a combination scheme of whole-buried and sinking road structures.

105.2.2 General Information of Foundation Treatment

East from K2+837 (Contract section IV to contract section VII) was the section full of thick silt. Southeast from K3+820 was the region of large stone reclamation embankment, and northeast from K3+820 was the region of blasting toe-shooting reclamation. Due to the buried weak layer existed along the construction areas, a variety of foundation treatment options have been taken as followed: a major program of grouting reinforcement for contract section IV to V, particularly sleeve valve grouting reinforcement of mud and silt-based soil for the junction of the planning Metro line 2 and the connecting project. A diameter of 400 PHC (prestressed pipe pile) was applied to the thick mud layer in the tunnel area at reclamation section W-M. For the beach in the north of the reclamation area, foundation treatments such as stone pier replacement, riprap blasting toe-shooting method were applied. At the same time, dynamic compaction was used to the original seawall, as well as the stone pier replacement was taken to deal with the mud in the south. The excavation and support at section IV to W have taken a variety of treatment options: section IV to V mainly use vertical excavation, with secant piles applied to both sides of the pit. Step-slope was adopted to the north of section IV, when slope rate method and row piles were used as the support method. Section VII, that is, the ramps, took the method of step-slope in excavation, and the support method in this section was basically the same as section W. Among them, because of the step-slope used in the north of section VI, large slope in the north side led to a wide range of high-filling.

105.3 Analysis of Differential Settlement

105.3.1 Cause of Differential Settlement

According to the general information of the project, engineering geological conditions in section IV to VII, which are the key parts of the whole connecting project, are complex and changeable. Intricate and ever-changing foundation conditions as well as the wide range of high-filling caused by the large slope in the north side would lead to differential settlement of the cabinet. After the completion of the backfill in excavation, road fill and paving were carried out above the cabinet and within the width of the municipal road. Foundation treatment has been carried out in most of the cabinet coverage area, particularly in the reclamation area pre-stressed pipe piles applied, however, no foundation treatment has been applied to some parts of the municipal road which are beyond the cabinet area (Note: blasting toe-shooting or dynamic compacted stone pier was used only for the mud within the scope of the pit). As a result, when municipal road paving was completed, damages such as cracking or fluctuation would emerge as the result of large transverse differential settlement of the road. In summary, the causes of differential settlement can be attributed as: (1) Differential foundation condition caused differential settlement of the case structure. (2) Since different foundation treatments options were applied, longitudinal differential settlement of the case structure occurred as a result. (3) Negative friction, which was caused by single sided backfill load at the pit (8–11 m) and filling load over the cabinet (3 m), led to differential settlement of the case structure. (4) After the municipal road was constructed over the cabinet, transverse differential settlement occurred at the areas on either side beyond the scope of the cabinet. (5) Different foundation treatments brought about longitudinal differential settlement of the municipal road.

105.3.2 Section Selection

Considering many factors comprehensively, section selection should follow the principles as below: (1) In underlying stratum, which has great change, particularly weak strata (such as muddy loam) or relatively thick gravelly mild clay. (2) Both sides of conjoin in two different foundation treatment scheme in longitudinal direction (axis of tunnel). (3) In the excavation and backfilling area. (4) The area of different arrangement form of box. (5) The area of different distribution form of municipal road. (6) The section should be near to investigation drillings in order to keep aboriginality and authenticity of the parameters. If there is no investigation drillings nearby, or investigation drilling is so shallow that information of lower strata can't be obtained, the section should be selected in the principle of using investigation near it. (7) Select the section at a distance of 100 m in the other areas in order to ensure the data foundation of calculation analysis.

According to the principles above, based on *Working Drawings of Western Corridor Connecting Project from Shenzhen to Hong Kong* (bid IV-bid VIII) and *Geologic Survey Report of Western Corridor Connecting Project from Shenzhen to Hong Kong,* 26 sections are selected to calculate from bid IV to bid VII. There are four different types in overlying representative structure and distribution form as: (1) vertical excavation bid of the pit (shown in Fig. 105.1), (2) large-scale excavation bid of the pit (shown in Fig. 105.2), (3) single-box bid of ramps (shown in Fig. 105.3), (4) multi-box bid of ramps (shown in Fig. 105.4).

105.3.3 Calculation Parameters Selection

1. Determination of underlying soil layer parameters

This underlying soil layer in the site usually includes silt, gravel sand, clay, coarse (gravel) sand containing cohesive soil, coarse (gravel) sand containing organic matter, muddy loam, gravelly mild clay, loam containing gravel, gravelly clay, completely decomposed granite and so on. Analyzing the geological investigation data of connecting engineering comprehensively, the calculation of parameters is mainly obtained by *Geologic Survey Report of Western Corridor Connecting Project from Shenzhen to Hong Kong* (Hereinafter referred to *survey report*).

2. Backfill parameters selection

According to Road *Engineering Overall Arrangement Drawing of Western Corridor Connecting Project from Shenzhen to Hong Kong* and the design scheme provided by the owner, roadbed filling is designed to be sandy cohesive soil and gravel cohesive soil, which should be compacted by layered backfill. Referred to *Highway Roadbed Design Specification (JTGD30-2004),* the soil mechanics parameters are selected as calculation parameters of backfill and put into the model to calculate.

3. Parameters selection of dynamic compaction stone pier replacement area

Referred to *Composite Foundation Design* and *Construction Guide* edited by Gong Xiaonan, the enhanced body and the matrix are regarded as composite soil. Compressibility of composite soil is evaluated by composite compression modulus, also, referred to foundation treatment

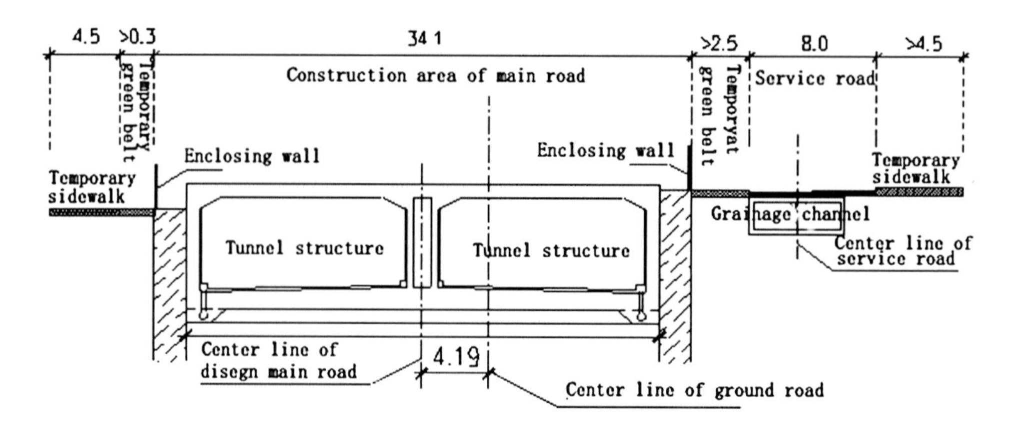

Fig. 105.1 Foundation vertical excavation section

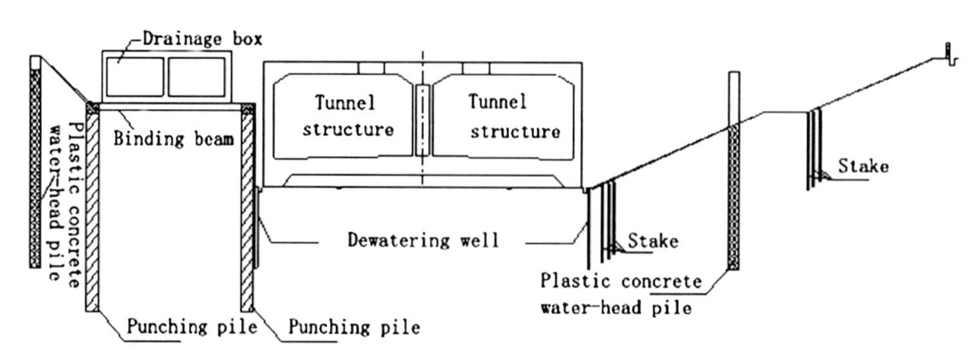

Fig. 105.2 Large-scale excavation bid to the north of reclamation area

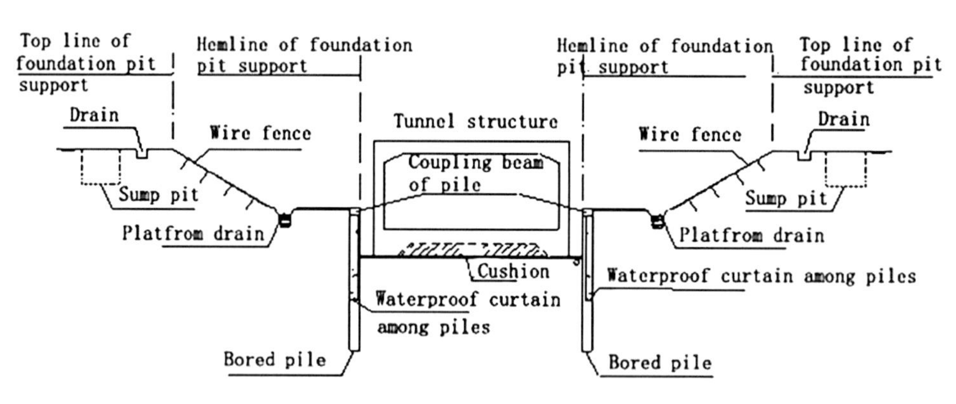

Fig. 105.3 Foundation pit section of single-box bid in the ramp

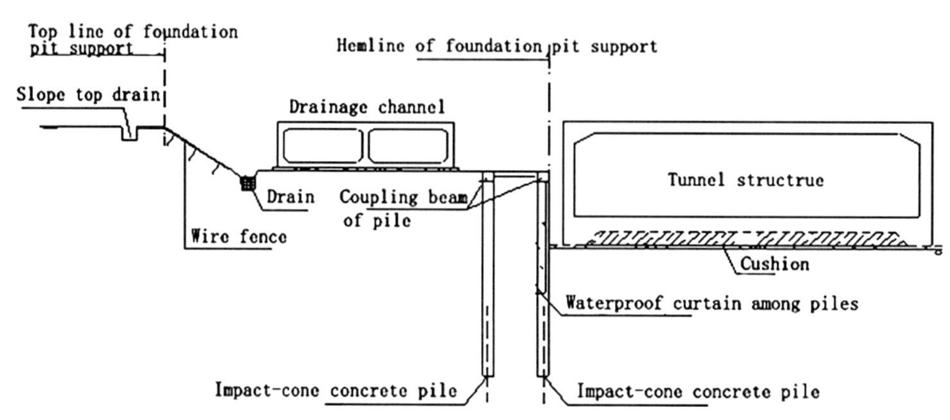

Fig. 105.4 Foundation pit section of multi-box bid in the ramp

requirements in *Road Foundation Treatment Construction Description in Reclamation Area of Western Corridor Connecting Project from Shenzhen to Hong Kong* (Hereinafter referred to *construction description*): deformation modulus of stone pier composite foundation should be more than 18 Mpa, the deformation modulus of Dynamic compaction stone pier replacement area is determined to be 19 Mpa.

4. Groundwater level determination

Referring to survey report, groundwater condition and the suggested values of design water lever to the east of K2+837 milepost, in order to fully reflect the impact of settlement on the road, the paper adopts the lowest anti-floating design water level. after pit backfill.

105.4 Results Analysis and Discussion

105.4.1 Analysis Methods and Calculation Results

Using the standard method to simplify the mathematical model and non-linear finite element model, combined with the section selected before and calculation parameters related, settlement values of 26 sections are calculated (shown in Figs. 105.5, 105.6, 105.7, 105.8)

105.4.2 Results Analysis and Discussion

105.4.2.1 The Differential Settlement of Box Structure Due to Differential Ground Conditions

By the compare of Figs. 105.5 and 105.6, the differential settlement exists in the structure apparently. Base on the geological data of section selected, foundation conditions is a principal factor, e.g. the differences of north and south thickness of gravelly mild clay layer under foundation of section ZXK2+750, ZXK3+280, ZXK3+365, ZXK3+730 are all 6 m exceeded. This obvious horizontal thickness maldistribution of the subjacent bed, in the additional stress (box structure load, municipal road embankment load, etc.), caused the transverse differential settlement of the box structure.

105.4.2.2 Box Structure Longitudinal Differential Settlement Caused by Different Foundation Treatment Plan

In Figs. 105.5 and 105.6, obvious differential settlement exists in the longitudinal of the box structure. It is due to the different foundation treatment plan. Sections of large settlement mainly distribute in the areas where the foundation treatment is natural foundation and gravelly mild clay layer is thick. E.g. where section ZXK3+550, ZXK3+640, ZXK3 +730 are all use the natural foundation treatment, and the thickness of gravelly mild clay layer are 16–22 m, In where 1AK0+100 is, although the prestressed pipe pile processing was taken, the thickness of gravelly mild clay layer under the pile. And small settlement section mainly located in ZXK82 +750, ZXK2+865, ZXK2+965, ZXK3+115 and ZXK3+215 of bid N, where its natural geological condition is good or it take the grouting strengthening foundation treatment.

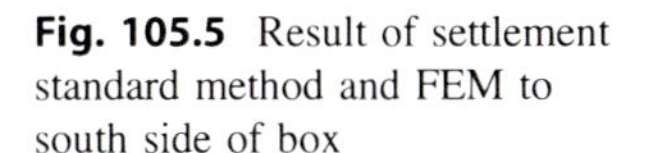
Fig. 105.5 Result of settlement standard method and FEM to south side of box

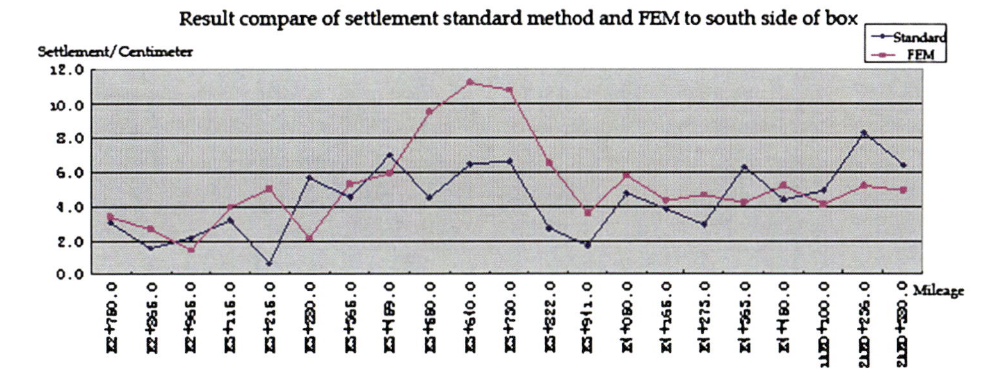

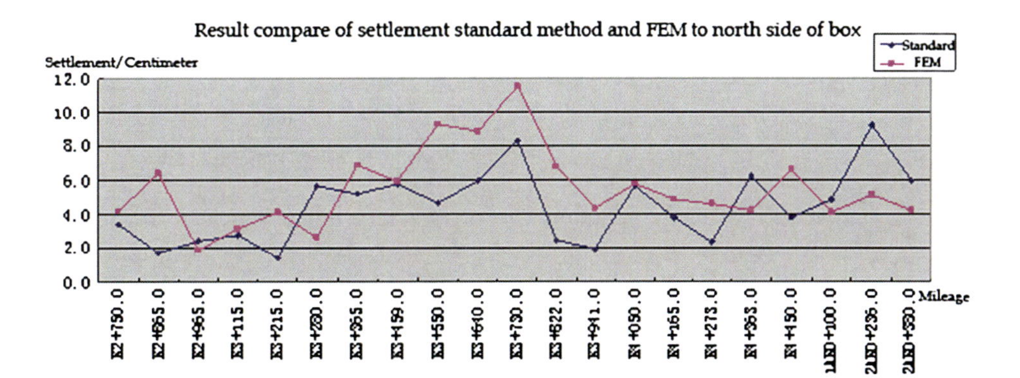

Fig. 105.6 Result of settlement standard method and FEM to north side of box

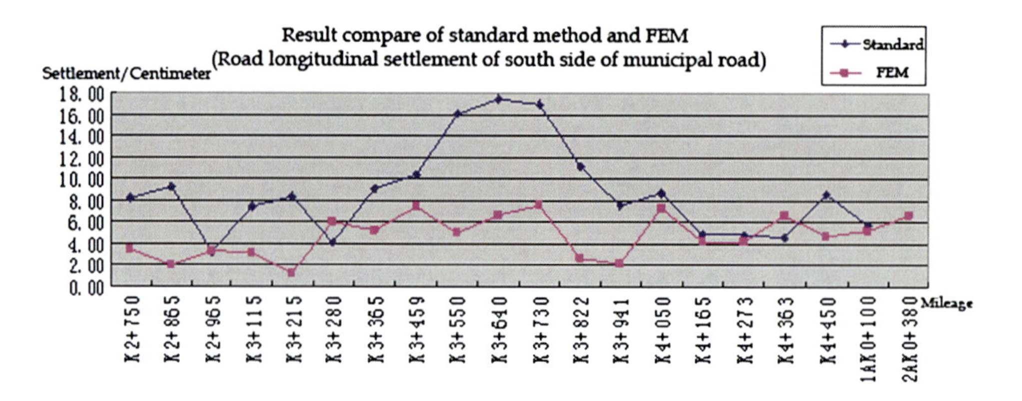

Fig. 105.7 (Road longitudinal settlement of south side of municipal road) Result compare of standard method and FEM

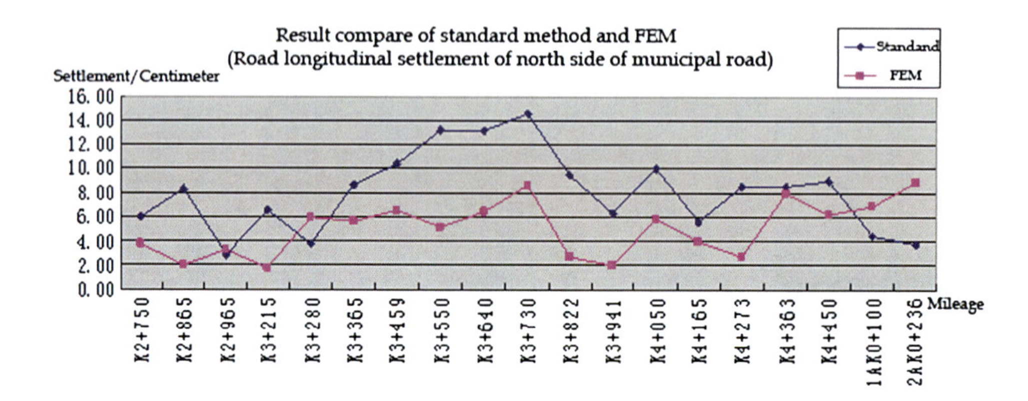

Fig. 105.8 (Road longitudinal settlement of north side of municipal road) Result compare of standard method and FEM

105.4.2.3 The Differential Settlement of the Box Structure is Caused by the Negative Friction, Which is Produced by the Pit Unilateral Backfill Soil Load (8–11 m) and Box Overlying Soil Load (3 m)

Based on the calculation results comparing analysis of the section in reclamation area from Figs. 105.5 and 105.6, lateral differential settlement is outstanding in the large slope. After the large slope construction completion, the north of the box will have 9–10 m high backfill soil, in addition, 3 m high municipal road embankment in the upper. The compression deformation of the north of box cuneiform new soil itself will produce large negative friction to the box, causing differential settlement, the north of box cuneiform new soil produce large additional stress in the bottom of the box. The differential distribution of the stress in both sides of the box causes the differential settlement.

105.4.2.4 The Municipal Road Across the Box, the Differential Settlement Beyond the Scope of the Both Sides of the Box

Form the comparing analysis of Figs. 105.7 and 105.8, the horizontal settlement difference of the municipal road is generally 1.1–2.2 cm, maximum 4.14 cm. But due to the difference occurs between the edge of the box and the municipal road boundary(part of municipal road beyond the scope of the box), plane distance is very small (most around 3.5 m), so that the difference is most around 3–6 ‰, some times arrives 10–14 ‰, maximum to 17 ‰. Besides related to the thickness of subjacent bed, the reason of lateral differential settlement is also related to the change of natural foundation condition.

105.4.2.5 The Municipal Road Longitudinal Differential Settlement Caused by Different Plans of Foundation Treatment

As Figs. 105.7 and 105.8 shown, the difference of longitudinal settlement of adjacent area from bid 4–7 is about 2–3 cm. The main reasons to longitudinal differential settlement are two: (1) Different foundation treatment of adjacent area, i.e. the juncture of natural foundation and grouting reinforcement, (2) Stratum are complicated and the thickness change greatly, i.e. the thickness of gravelly mild clay under ZXK3+730, ZXK3+640, 1AK0+100 is 16–22 m. The isoline map of the municipal road settlement is drawn by MapGis (Fig. 105.9), which clearly shows the differential settlement situation of the whole.

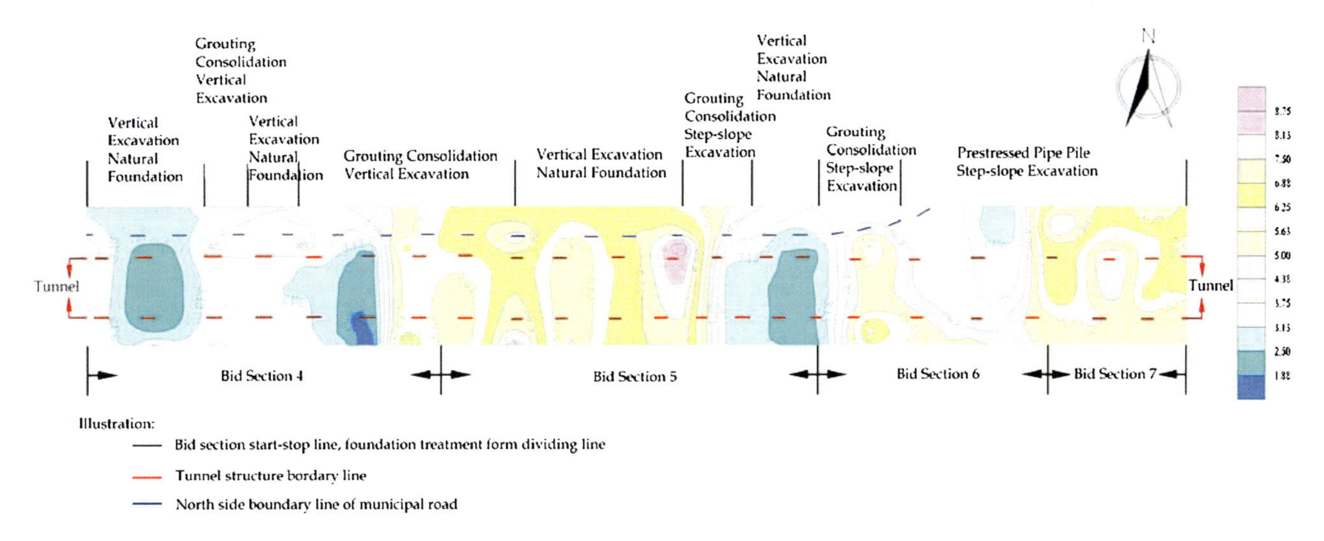

Fig. 105.9 Municipal road pavement longitudinal settlement isoline

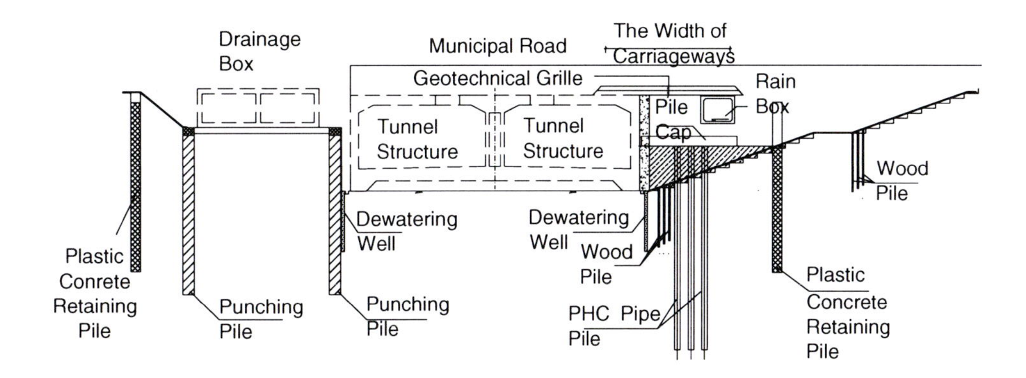

Fig. 105.10 Foundation pit of reclamation area follow-up treatment plan

105.5 Control Scheme

Base on the analysis, one main reason for differential settlement is the excavation and backfilling asymmetry (Fig. 105.2). Shanghai Municipal Design Institute bring up bearing the weight of high backfill soil by three rows PHC and flat bedplate in the slope of north reclamation area, in order to reduce the adverse effects of north highfill to tunnel structure stability and differential settlement. The design (shown in Fig. 105.10) laid down 20 cm thick plain concrete access board on the reinforced cushion, making road settlement even comparatively. This research was based on the *Standards in Guangdong Province—Building Foundation Design Specifications (DBJ15-31-2003)* and the *Technology Norms of Pile Foundation Construction (JGJ94-94),* puts forward optimum idea, which under the requirements in bearing, through expanding the pile spacing and shorten the pile length, achieve the effect of economic security (Fig. 105.11).

Through the comparing analysis we know, comparing with without special treatment of backfill scheme, after control measures, the absolute settlement of the box is reduced and the settlement difference of both sides of box

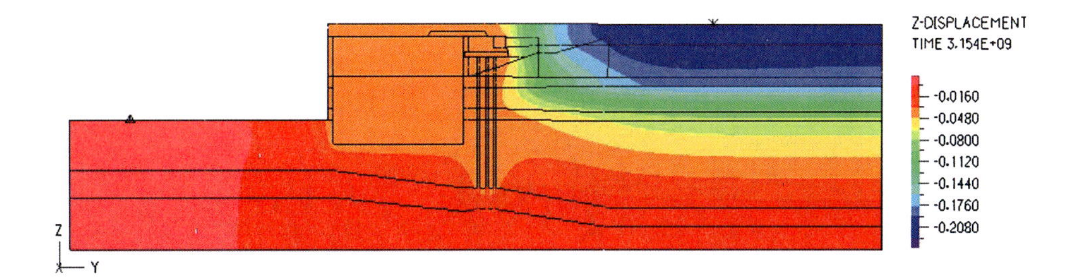

Fig. 105.11 Settling image of section ZXK4+165 after control measures

and the road decreased significantly. As the figure shows, pile and flat bedplate is used to bear the weight of high backfill in the northern slope. It plays an obvious role to improve the differential settlement of enclosure structure, making the settlement difference of box sides and road control effectively.

105.6 Conclusions

1. Through the comparative analysis, the overall trends of municipal road horizontal settlement by two methods are the same, but the settlement value of FEM are commonly less than standard method. And the analysis results of the two method both show that: the differential settlement of box are caused by uneven foundation conditions, difference of foundation treatment schemes, the asymmetry of foundation pit and excavation backfill, and the differential settlement of municipal road is caused by the different treatment schemes of road across the box and differential settlement of municipal road is mainly caused by the reasons of the road crossing over the cabinet and different proposals of foundation treatments. It should be pointed out that: the influences of the stress history, which the layers under the box went through, to the layers' physical and mechanical properties are not considered in the settlement calculation. If consider, the differential settlement will be larger.
2. Because the calculating parameters are selected according to the engineering geology survey report before construction, while the layer under the place had gone through complex stress historical process such as filling, excavation, structure and foundation backfill, the use of the parameters may cause certain effect to the results. The absolute calculation value may be too large while the differential settlement may be smaller. But according to the supplementary experimental results of later sampling of Shenzhen engineering institute (considered stress history comprehensively), the parameter values such as deformation modulus is far less than those of exploration stage. Considering comprehensively the statistical significance of the parameters of exploration stage, the parameters selection and calculation analysis are only based on the survey report. Some scenes in situ tests are suggested. Strengthening the field observation, obtaining reasonable calculating parameters and actual settlement situation to analyze and control.
3. In the future, the road will exist many other factors which can not be estimated very well, i.e. in the rain the filled soil moisture and subsidence may be caused, at the same time, unpredictable factors such as overloaded vehicles, impact load, vibration, etc. still can be existed. But these factors are not considered in this calculation, which makes the results tends to safety.

References

Du H, Zhang J (2005) The exponential curve fitting of the calculation of the subsoil settlement method and its application. Soil Eng Found 19(1):54–56

Du S, Zhu J (2005) Numerical analysis of interaction between underpass and soil across the railway. Undergr Space 1(2):242–246

Brand EW, Brenner RP (1981) Soft clay engineering. Elserier Scientific Publishing Company, Amsterdam

Fu H (2006) Optimizing back-analysis parameters for settlement prediction based on sensitivity analysis. J Changsha Univ Sci Technol (Nat Sci) 3(2):24–28

Han L (2005) Centrifuge model test and analysis on foundation settlement of metro river-crossing tunnel. Undergr Eng Tunn, 13–15

Hu Qi et al (2005) Three-dimensional numerical simulation analysis of settlements of buildings on soft soil. Rock Soil Mech (12):2015–2018

Li X (2005) Numerical analysis of the settlement of high fill embankment. Soil Eng Found 19(3):56–58

Lin Q, Cao X (2006) Discussion of soft soil foundation post-construction settlement prediction method. Subgrade Eng 2:78–80

Liu Y (2005) Application of improved genetic algorithm to predict the settlement of soft soil roadbed. Site Inv Sci Technol 1:29–31

Peng T et al (2005) Prediction of soft ground settlement based on BP neural network-grey system united model. Rock Soil Mech 26 (11):1810–1814

Shangai Municipal Engineering Design Institute (2005) Chongqing communications research and design institute etc. Construction Description of Road Foundation Treatments for the Reclamation in the Western Corridor Connecting Project from Shenzhen to Hong Kong

Xu F et al (2005) Research on the Final settlement prediction based on the first-hand observed data. West-China Explor Eng 1:1–3

Zhan X (2006) How to do the geotechnical engineering forecast and calculation. Guangdong Sci Technol 152:74–76

Zhang C et al (2005) Prediction of the Settlement of highway soft foundation by BP-NN model. Geol Sci Technol Inf (24):115–117

Zhang W, Li L (2005) The application of grey system theory in the rockfill subgrade settlement calculation. Jilin Sci Technol Commun 4:3–5

Zhou Y (2005) Embankment settlement deformation forecast based on the grey theory. West-China Explor Eng 1:52–53

106 Chontal HPP, Geological Features on Site Location and Dam Type

Cristina Accotto, Giuseppe Favata, Enrico Fornari, Nikolaos Kazilis, Marco Rolando, and Attilio Eusebio

Abstract

The Chontal Hydropower Plant is located along the Guayllabamba River, in the North-West of Ecuador, 100 km from Quito, in the districts of Pichincha and Imbabura. The geology of the dam area has strongly influenced the dam site location and the dam design: due to the geological characteristic of the area, not satisfactorily determined during a previous feasibility study, the layout of the plant has been entirely changed. During the review of the feasibility study, which foresaw two hydropower plants in cascade, new geological investigations have been carried out, acquiring more detailed information on the geology of the Guayllabamba River along the stretch of interest. In particular, a deep pervious old buried river channel has been detected and extensively investigated. Further to the new geological information, the inadequacy of the feasibility design, due to a deficiency of geological data, has been assessed, in particular with reference to the both reservoirs and dams site. Several design alternatives have been developed, changing both dams' location, type and number (1 or 2). A multi-criteria analysis and risk analysis have been carried out to identify, according to the Client, the design arrangement. Chontal Hydropower Plant is an example of the influence of geological features on the choice of dam site and dam design. The paper is focused on the dam design, and its relationship with the geology.

Keywords

Dam design • Reservoir leakage

106.1 Introduction

The Chontal Hydropower Plant is located along the Guayllabamba River, in the North-West of Ecuador, 100 km from Quito, in the districts of Pichincha and Imbabura. The plant incorporates a 142 m high RCC dam, a 870 m long headrace tunnel on the left bank of the river, a 14 m-diameter surge shaft, a 88 m high penstock shaft, a 200 m high pressure tunnel, an outdoor powerhouse equipped with 2 Francis turbines, and a tailrace channel. The plant has a total installed capacity of 194 MW, exploiting a hydraulic gross head of about 130 m.

The final arrangement of the Chontal HPP, and in particular the dam design, is completely different from the one foreseen in the feasibility design, because of strongly influence of the geology in the site area.

The paper is focused on the main geological features and its relationships with the design development, from the formulation of alternatives up to the final design.

106.2 The Feasibility Design Layout (1979–1980)

To understand the development of Chontal HPP, it is suitable an analysis of the layout selected during the feasibility design study, carried out in 1979–1980 by the former INECEL (Instituto Ecuatoriano de Electrificación).

C. Accotto · G. Favata (✉) · E. Fornari · N. Kazilis · M. Rolando · A. Eusebio
Geodata Engineering S.p.A., Turin, Italy
e-mail: gfa@geodata.it

G. Lollino et al. (eds.), *Engineering Geology for Society and Territory – Volume 6*,
DOI: 10.1007/978-3-319-09060-3_106, © Springer International Publishing Switzerland 2015

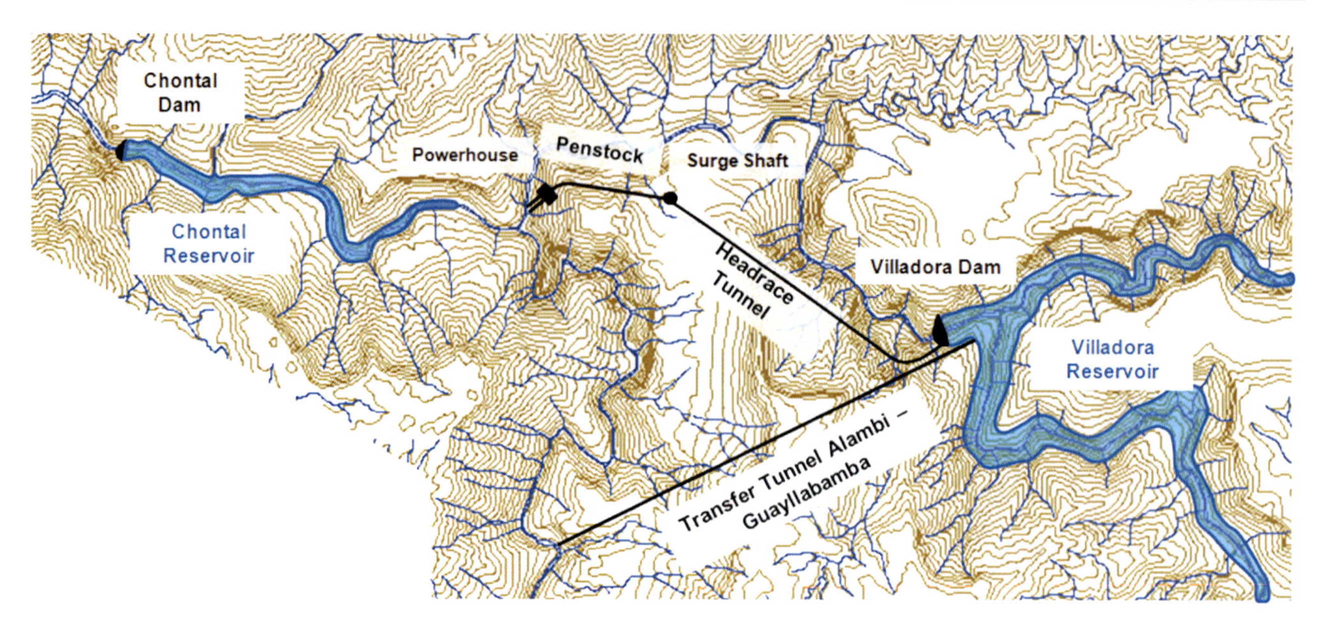

Fig. 106.1 Feasibility study layout

The feasibility design, carried out in 1979–1980, foresaw two plants, Villadora and Chontal, each one with a dam located on the Guayllabamba River. The upstream plant, Villadora, foresaw an arch dam 165 m high; the downstream one, Chontal, foresaw an arch gravity dam 68 m high with powerhouse at its toe. To increase the energy production, a hydraulic tunnel was foreseen to convey water from the Alambi River, a downstream tributary of Guayllabamba River, to Villadora reservoir. The general layout of the two plants is shown in Fig. 106.1.

In the dam area of Chontal only few site investigations, and in particular no boreholes, were carried out. However, in the feasibility study, it was foreseen the possible existence of an old buried channel (paleo-channel) along the right bank, identified by two seismic profiles. These two profiles show a bedrock with waves velocities of about 4,800 m/s. In correspondence of the buried channel, the waves velocities decrease up to 560 m/s. No more investigations were carried out during the feasibility design study, and therefore the conclusions of the feasibility study was the needed of more investigations, in particular boreholes and seismic profiles, to define the geometry of the old buried channel.

106.3 Geological Investigations

106.3.1 Investigation Plan

During the review of the feasibility design it has been recorded that the geophysical investigations carried out indicate the presence, along the right bank, of a depression filled with material through which seismic waves move at very slow velocities. The depth and extent of this depression, however, had not been clarified, since the geophysical surveys were not accompanied by verification through boreholes. The geological information therefore wasn't enough to develop a final design on the basis of the feasibility study, since characteristics, permeability and resistance parameters of the material in the depression were not available.

For the above reasons, it was suggested to carry out, and afterwards was carried out, a detailed geological survey in the project area, as well as field and laboratory investigations.

The following geological studies have been carried out:

- Geological survey at the dam site
- Geological survey of the old buried channel along the right bank
- Geological survey of the reservoir
- Detailed analysis of the joint sets at the dam site

Moreover, several geological investigations have been carried out to investigate the buried channel, such as geophysical profiles, boreholes and permeability tests.

106.3.2 Results of Geological Investigations

The detailed geological mapping and the field investigations allowed to identify the old buried channel, as well as the characteristics of its composing materials.

106.3.2.1 General Geological Features of the Area

The area considered for the development of hydroelectric projects is along Rivers Guayllabamba, Intag and Alambi. The river valleys are narrow, with high and steep slopes that would seem ideal for the construction of dams.

In the area of interest the rock formations are granodioritic rocks. In the outcrops that appear on the riverbed the rock appears sound and fresh, with relatively high resistance, close to the values of intact rock. Therefore it could be concluded that the rock mass in the studied area generally could be considered suitable for dam foundation.

106.3.2.2 The Old Buried Channel

The presence of the old buried channel has been identified through the detailed geological survey and field investigations, allowing the preparation of the map shown in Fig. 106.2.

The paleo-channel, as well as the actual riverbed, have been developed along zones of weakness of the rock mass. Occasionally, the old channel and the actual river bed coincide.

The paleo-channel materials are constituted by old alluvial deposits, lagoon deposits and laharitics deposits, with different values of permeability depending on the content of fine material. The overall permeability of the paleo-channel is difficult to estimate through conventional tests, due to the diversity of involved materials. Moreover, it is difficult to predict the long-term behavior of that materials under a high hydraulic gradient such the one that occurs in presence of high-head dams. The possible occurrence of internal erosion, with consequent increase of permeability and therefore serious risks related to plant functionality and stability, could not be discarded.

This critical situation has been considered and duly analyzed in all places where the paleo-channel is located within a possible reservoir area, since it could have direct or indirect communication with areas located downstream of the dam, with consequent water leakage from the reservoir. It should be considered that the leakage of water could have serious consequences on plant functionality, and therefore significant investments could be required to avoid it.

106.4 The Selection of Final Layout

106.4.1 The Study of Alternatives

On the basis of the geological features of the site and, in particular, taking into account the presence of the paleo-channel, it was necessary to select possible dam sites where the paleo-channel is coincident with the actual river bed and their axes are coincident or at least slightly sub-parallel, in order to avoid the risk of severe leakage from the reservoir. The study of alternatives, therefore, has been carried out following this principle.

Twelve alternatives (including the feasibility study layout) have been analyzed. Eleven of these alternatives foresaw two plants in cascade, as in the feasibility study layout, with different dam sites, while the last alternative foresaw only one high-head dam in the Chontal sector.

To facilitate the screening of the alternatives qualitative criteria were established, based on safety and risk analysis which allows to select or eliminate those alternatives that don't have adequate conditions, and to minimize future risks

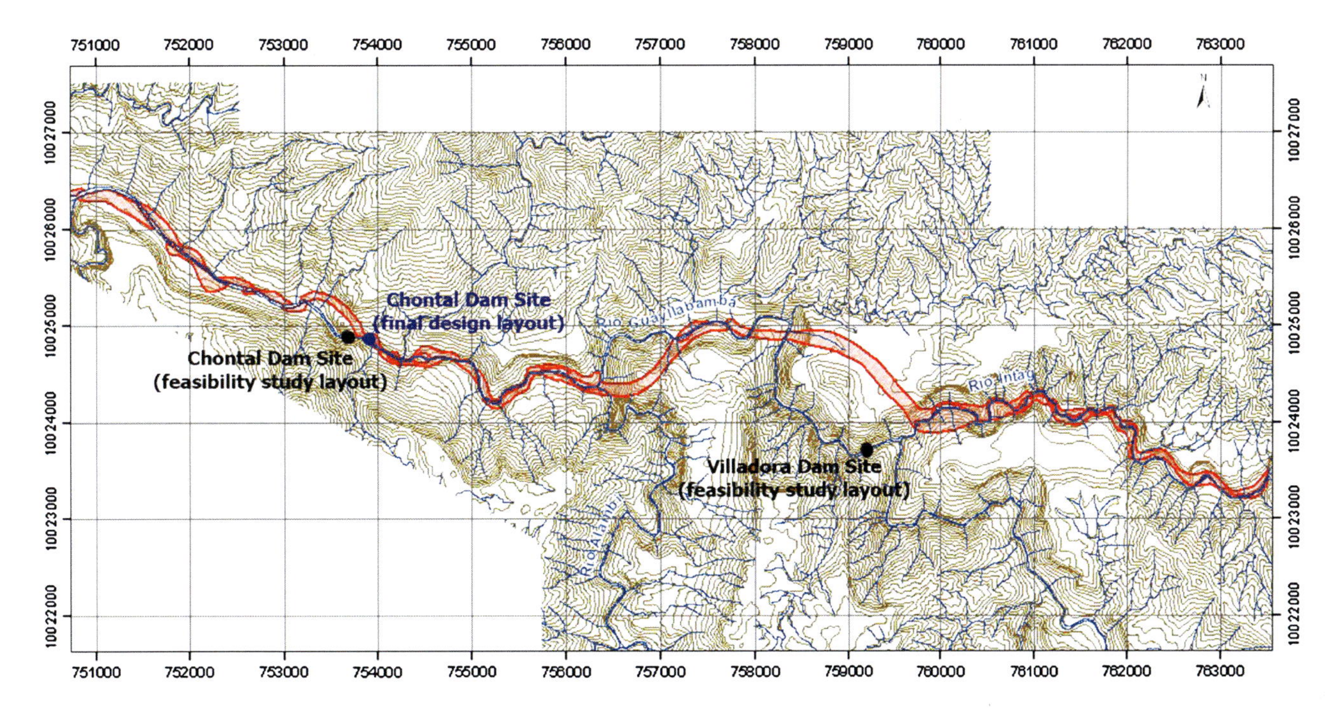

Fig. 106.2 Location of the old buried channel, after geological survey and field investigations

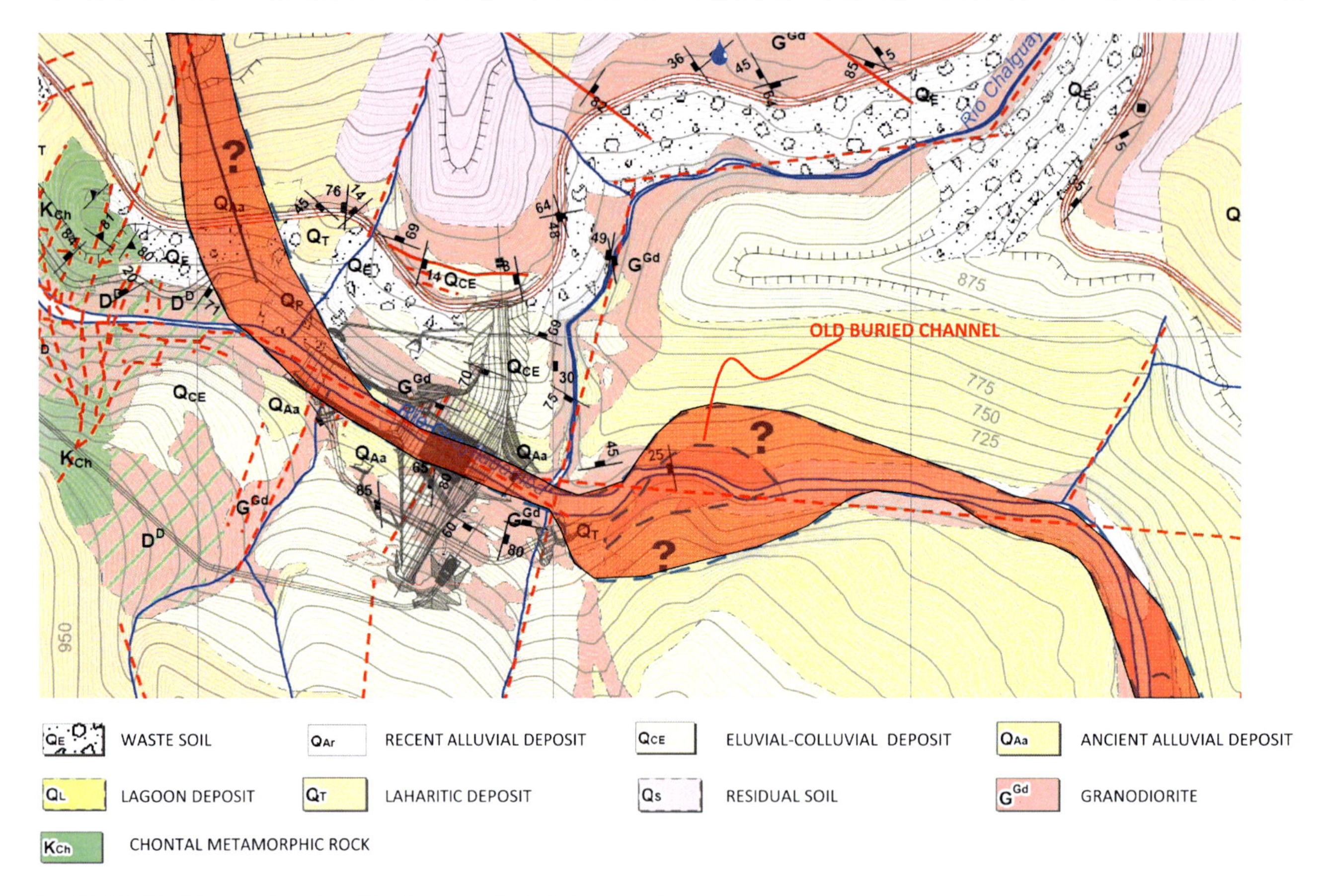

Fig. 106.3 Final layout, dam site with relevant geological characteristics

that will require additional repair works with subsequent related costs. For the screening of the alternatives six criteria have been used: (i) topographical and morphological; (ii) hydrological and sedimentological, (iii) geological and geotechnical, (iv) environmental, (v) economic, (vi) technical criteria related to civil works (e.g., dam stability).

As result of the preliminary screening, five viable alternatives have been selected. Afterward, a cost estimate for each viable alternative has been carried out, taking into account the prices for the civil works, as well as for electromechanical and hydromechanical equipment. Furthermore, the multicriteria analysis has been used to select the final layout. The identification of these criteria is organized in hierarchical levels, as follows:

- Feasibility criteria, focus on the issues relevant to general feasibility of the project
- Economical criteria, focus the attention on total cost and back discounted cost of the works
- Environmental criteria, focus the attention on environmental aspects during works execution as well as during plant operation

The technical feasibility considers geological criteria (seismic risk, rock mechanics, presence of faults, risk of landslides), complexity of the civil works, complex underground works, access availability, construction phases, expropriations, environmental authorizations.

The economic feasibility considers costs and benefits by energy production.

Finally, the environmental feasibility considers the impacts during construction, and impacts during plant operation.

For each alternative, the list of evaluation criteria was completed by assigning a rating to each criterion, ranging from “inadequate”, “unacceptable”, “acceptable” and “optimal” evaluation. The rating assignment was made according to expert judgment and quantitative criteria.

106.4.2 The New Layout: Feasibility Study and Final Design

106.4.2.1 The New Feasibility Study (2010)

At the end of the multi-criteria analysis, a new layout has been selected. The geological criterion allows to discard one of the five viable alternatives, for the risk of stability problems of the right bank near the dam site, related to the presence of the paleo-channel.

The selected alternative foresee a high-head dam (142 m) and an outdoor powerhouse, with a total installed capacity of

194 MW. The dam site is located about 120 m downstream of the Chalguayacu river confluence with the Guayllabamba river. At the dam site the paleo-channel is coincident with the actual river bed, as it is shown in Fig. 106.3. The selected dam site remove the risk of severe leakage from the reservoir, and the subsequent risk of internal erosion of the paleo-channel.

A new feasibility study has been realized for the selected layout. During this study, new geological investigations have been carried out. The geological investigations carried out on both banks of the dam site consist in geophysical seismic refraction (11 profiles on the right bank and 9 on the left bank, with a total length of 1,550.0 m), boreholes (7 on the right abutment, 8 on the left abutment and 3 along the river bed), with a total length of 963 m, geological surveys and geostructural surveys. The investigations confirm that the selected site has excellent characteristics for the dam foundation. The rock mass is a granodiorite, slightly fractured or sound, with permeability ranging from 10^{-5} to 10^{-7} cm/s, and local values of about 10^{-4} cm/s. The compressive strength of the rock is in the range 50.0–100.0 MPa, the elastic modulus is 64.1 GPa, and the equivalent friction angle, is ranging between 46° and 57°.

In some areas the rock is covered by alluvial and/or laharitic sediments, in particular along the right bank. In Figs. 106.4 and 106.5 there are shown respectively the geological profile along the dam axis, and a geological section in correspondence of the dam spillway.

The dam type has been selected considering geological morphological, hydrological and geotechnical aspects, as well as materials availability and economical aspects. The selected dam type is the RCC (roller compacted concrete) gravity dam.

106.4.2.2 The Final Design

The geological studies allowed to accurately determine the depth of alluvial materials, and therefore to determine the planimetric position of the dam and the excavations needed to reach the sound rock mass.

A curtain grouting is foreseen to reduce leakage through the dam foundation and to reduce the uplift pressures (in conjunctions with drain holes). A perimetrical plinth is foreseen to realize the grouting curtain, which will be

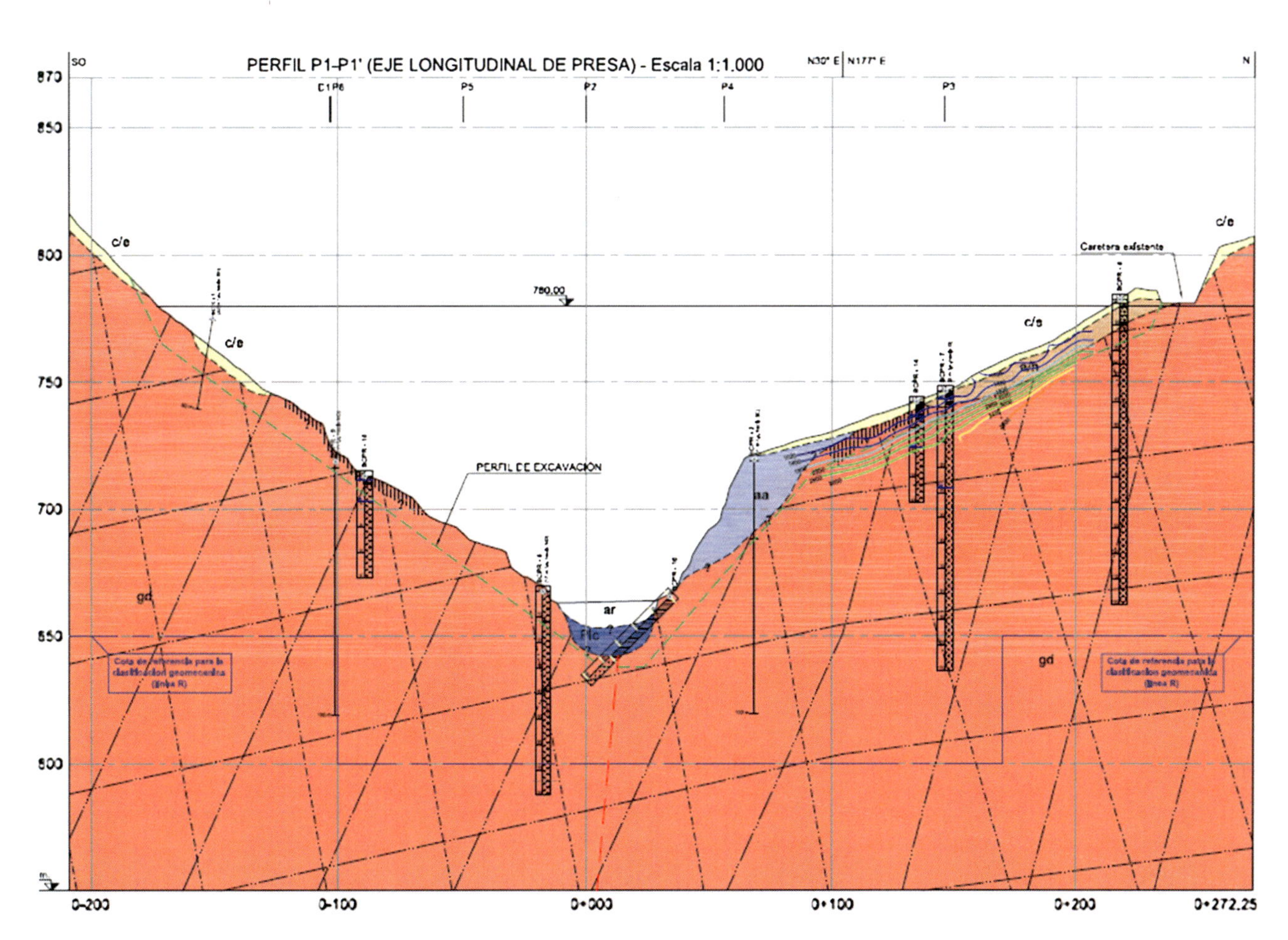

Fig. 106.4 Geological profile along the dam axis

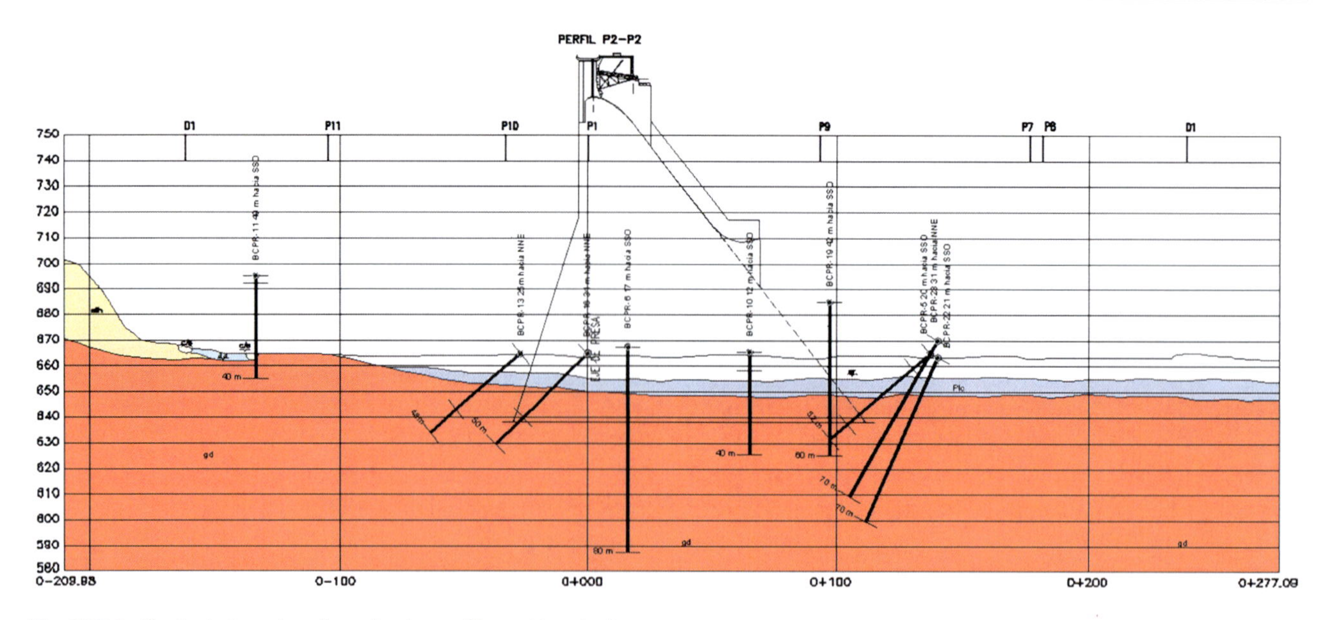

Fig. 106.5 Geological section along the dam spillway (river bed)

realized sequentially: the primary grout holes will be spaced 12 m, the secondary 6 m and the tertiary holes, if required, will be spaced every 3 m. According to the results of the Lugeon tests (51 tests on the left bank, 81 tests on the right bank and 29 tests along the river bed), the depth of the grouting curtain reaches values of 80 m in correspondence of the river bed.

106.5 Conclusions

The dam design is strongly influenced by the geology of the area, from the feasibility study to the final design. Adequate geological investigations and expert interpretation of the data are required in order to reach a good knowledge of the geological characteristics which could affect the design choices.

In the example of Chontal Hydropower Plant, reported in the present paper, the geological features of the project area have shown, during the review phase of the previous feasibility study, the inadequacy of the previous layout, and the need of more detailed studies. The acceptable dam sites have been controlled by the presence of an old buried channel, identified through a detailed geological survey. During the screening of the alternatives, the geological criterion, in conjunction with other technical, economical and environmental criteria, has guided the selection of the best alternative. Finally, during the new feasibility study and the final design of the selected alternative, the geological characteristics of the dam site determined some design choices, such as the exact planimetric position, the required foundation excavations, and the curtain grouting extension.

References

ASCE (1989) Civil engineering guidelines for planning and designing hydroelectric developments, vol 1

Barton N (2007) Rock quality, seismic velocity, attenuation and anisotropy. Taylor & Francis, UK

Bell FG (2007), Engineering geology. Elsevier, Amsterdam

Fell R, MacGregor P, Stapledon D, Bell G (2005) Geotechnical engineering of dams. Balkema Publishers, Boca Raton

Fetter CW (2001) Applied hydrogeology. Prentice Hall, US

Golzé AR (1977) Handbook of dam engineering. Van Nostrand Reinhold Co., New York

Hoek H, Palmieri A (1998), Geotechnical risks on large civil engineering projects, keynote address for theme I. International Association of Engineering Geologists Congress, Vancouver, Canada, 21–25 Sept 1998

Jansen RB (ed) (1988) Advanced dam engineering for design, construction and rehabilitation. Springer, Berlin

Londe P, Sabarly F (1967) Écoulements de percolation dans les massifs rocheux servant d'appui aux barrages. La Houille Blanche 1:37–46

Plata Bedmar A, Araguás Araguás L (2002) Detection and prevention of leaks from dams. Balkema Publishers, Boca Raton

USBR (1987) Design of small dams, 3rd edn

Verzani L, Soldo L, Uttini A, Vendramini M, Accotto C, Principi S, Ricci G, Eusebio A (2012) Lo studio geologico e geomeccanico applicato a progetti idroelettrici nelle Ande ecuadoriane. Geoingegneria Ambientale e Mineraria, XLIX 1:39–48

Volpe RL, Kelly WE (eds) (1985) Seepage and leakage from dams and impoundments. ASCE

Part X
Geological Model in Major Engineering Projects

Convener Mr. Erik Wunder—*Co-convener* Nicole Borchardt

Society is looking for a balance between quality of life and environmental protection. In this search, the wide spectrum of possibilities goes through discussions on renewable sources of energy and the use of underground space. The adequacy of infrastructure occurs through major engineering projects, where the geology applied to engineering plays a key role, from the conceptual design until its construction. In this context, the development of a realistic Geological Model is essential to describe the natural rock mass surroundings that should be worked to host the project in such a way that it gives subsidies to quality of design, feasibility of the project and for formatting contractual clauses regarding the assumption, contingency and sharing geological risk. The purpose of this session is to emphasize the role of the geologist in drafting the Geological Model and characterize the Geological Model as one of the main contributions of engineering geology for major engineering projects.

Evaluation of Geological Model in Construction Process of Sabzkuh Tunnel (Case Study in Iran)

107

Majid Taromi, Abbas Eftekhari and Jafar Khademi Hamidi

Abstract

Geological and geotechnical surveys, in general, should precede the tunnel excavation to ensure its safety. Also they should be continued during the tunnel excavation, because the geological condition can be changed unexpectedly while the tunnel is under construction. Tunnel face stability is one of the main issues in tunnel excavation in soft and alluvial grounds in order to keep the construction process stable and prevent localized face collapse. Choosing of an appropriate excavation method is influenced by geological and geotechnical studies and comprehensive and appropriate use of these data in tunnel designing. The Sabzkuh tunnel was excavated mainly in fault zones, alluvial, shale and limestones using the conventional and shielded tunnel boring machine (TBM) Methods. An advance length of 1.5 m without any support system caused many difficulties during the excavation of Sabzkuh tunnel. So, to overcome the complicated ground conditions and tunnel collapse, the excavation method were shifted from full face to sequential excavation in addition to decreasing the excavation step. This paper discusses some of the key geotechnical challenges faced in the tunnel design, including characterization of ground conditions, selection of appropriate design parameters, and evaluation of excavation and support installation sequence based on monitoring and analyzing ground behavior during construction.

Keywords

Geological model • Updating • Collapse • Sequential excavation • Monitoring

107.1 Introduction

The stability of underground structures is a key issue during design and construction. Depending on the geotechnical conditions and influencing factors, different failure modes can be expected. Also, depending on the potential failure modes, project specific requirements and boundary conditions, specific construction measures to ensure stability have to be chosen.

Submitting of a geological model, design parameters estimation, excavation and support system choosing with minimum risk, safety providing and cost reduction are major challenges for tunnel construction process in complicated ground conditions. Therefore, sufficient prior studies, interpretation, analysis, and process of data in an updatable model in order to planning for construction process are important in a tunnelling project. One of the most significant stages in tunnel construction is the displacement estimation of surrounding rock masses and support type regarding ground conditions and excavation method. The amount of tunnel wall displacement and required support is dependent upon the size of tunnel, available initial stresses, rock mass

M. Taromi (✉)
Department of Civil Engineering, Islamic Azad University, South Tehran Branch, Tehran, Iran
e-mail: majidtarami@ymail.com

A. Eftekhari
Department of Geology, Tarbiat Modares University, Tehran, Iran
e-mail: msc.eftekhari@yahoo.com

J. Khademi Hamidi
Department of Mining Engineering, Tarbiat Modares University, Tehran, Iran
e-mail: jafarkhademi@modares.ac.ir

G. Lollino et al. (eds.), *Engineering Geology for Society and Territory – Volume 6*,
DOI: 10.1007/978-3-319-09060-3_107,

properties, and tunnel excavation method Hence, the selected tunnel excavation method plays an important role in tunnel stability.

Sabzkuh–Choghakhor water conveyance tunnel with 10,617 m in length and 0.01 % in gradient is under construction in south-western Iran (Fig. 107.1). The tunnel is going to be excavated by a double shield tunnel boring machine (DS TBM). The information obtained from the site, including field reconnaissance, geotechnical and geophysical studies revealed that the tunnel from the beginning to chainage 0+390 km (T1) is located in the alluvium. So, taking into consideration the application range of DS TBM and its limitation for working in alluvial soils, this part of the tunnel is being excavated by conventional methods (Eftekhari et al. 2013a, b; Saeidi et al. 2012).

The results obtained from the numerical and sensitivity analysis on the soil geotechnical parameters demonstrated that a small part of tunnel face would provide a safe and stable excavation at the beginning of the tunnel drive. After full face excavation of 35 m and confronting unexpected ground conditions, a face collapse started and developed rapidly. This problem caused a failure in the tunnel portal area. In order to overcome this situation, Arch Support Technique (Eftekhari et al. 2013b) was employed. Various excavation methods such as open cut, DSM, freezing, jet grouting and sequential excavation method was used to excavate remaining part of tunnel.

Data obtained from previous studies usually are not sufficient for constructing process. In this situation, observation methods together with experience have an important role in tunnel construction. In Sabzkuh tunnel, after failure occurance in full face excavation, different applicable methods within similar geological conditions were chosen. Taking the strengths and weaknesses of each method as well as cost and time management into account, the sequential excavation method (SEM) due to its flexibility and compatibility with tunnel condition was selected.

A series of 3-D simulations using the finite different and finite element methods were performed in order to investigate the influence of the following aspects: (a) unsupported distance between the excavation face and the installation of support lining; (b) partial-face excavation.

Therefore, in this paper, the importance of geological studies in order to reduce the likelihood of unforeseen conditions, and monitor the design and construction with regard to geotechnical and geological interpretations are emphasized.

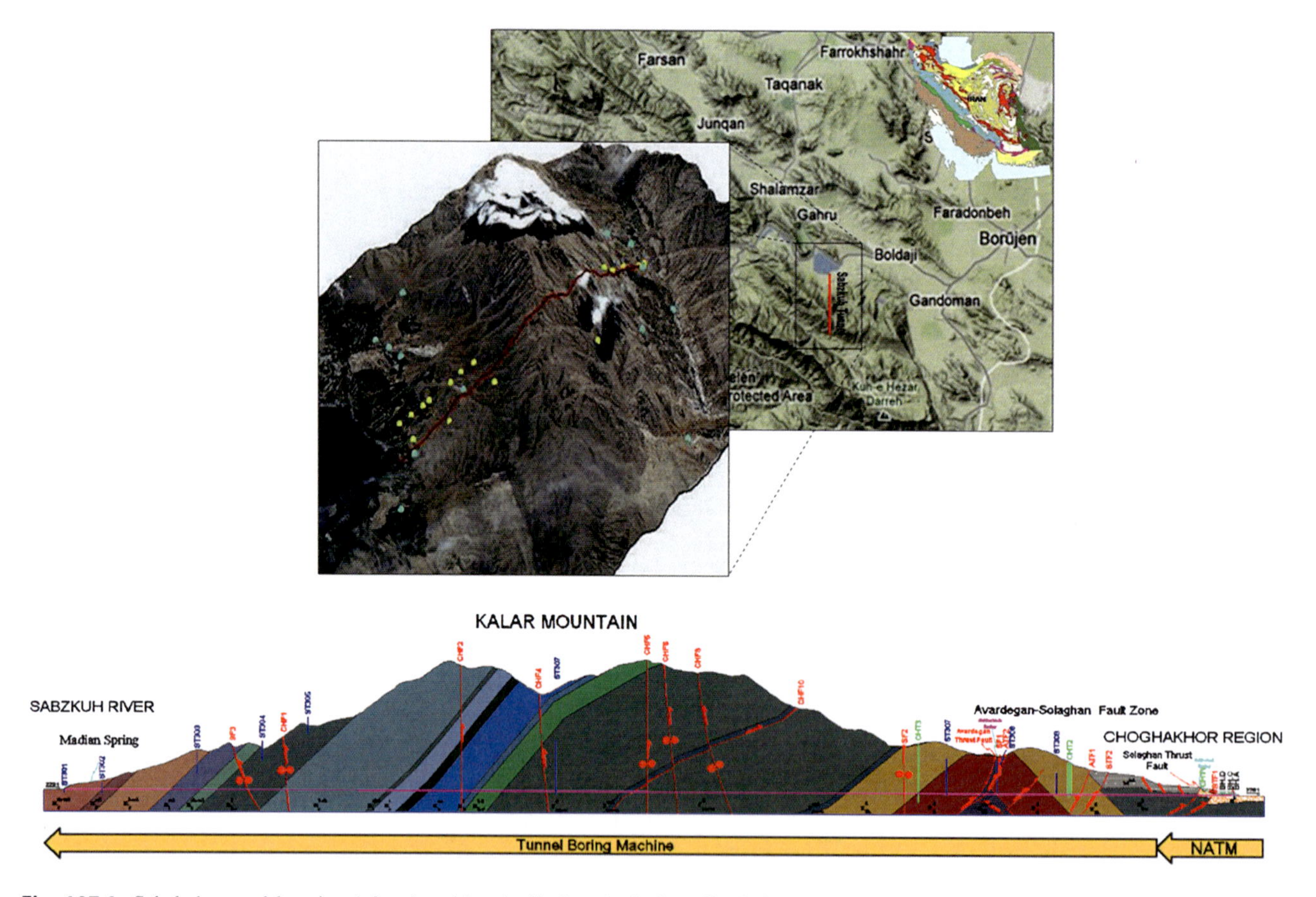

Fig. 107.1 Sabzkuh tunnel location (*above*) and longitudinal geological profile (*below*) (Taromi et al. 2013)

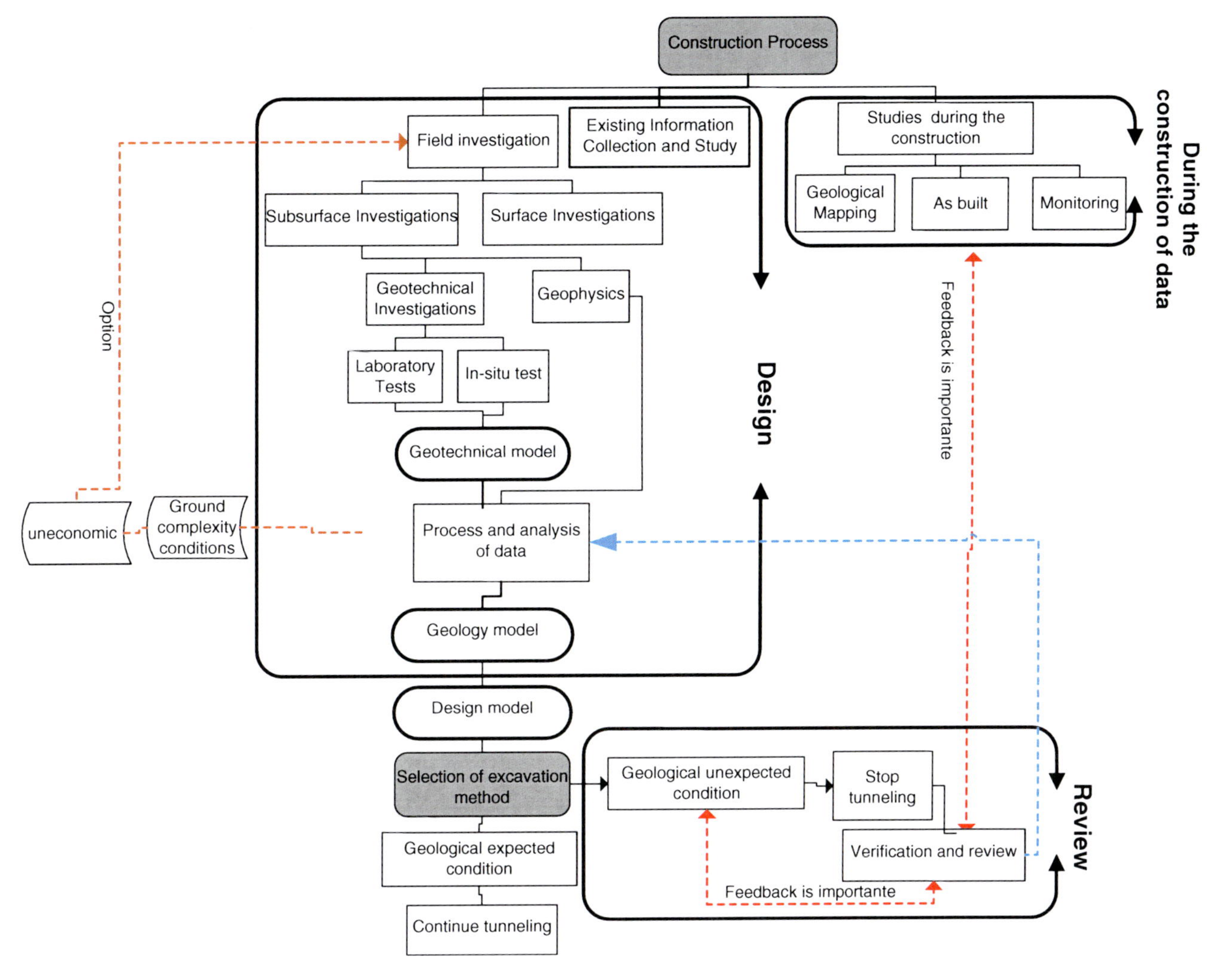

Fig. 107.2 Construction process algorithm (Taromi et al. 2013)

107.2 Construction Process

Due to the variation in the geotechnical conditions (the static system and the capacity of ground and supports), the design of an underground structure cannot be compared to a structural design of other buildings where the loads, the system, and the characteristics of the materials used are known.

In a tunnel constructing method, preparation of geological and geotechnical models and its design is very important (Fig. 107.2). This process contains analysing the data outcome from geological engineering, determination of ground design parameters, preparation of basic scheme for excavating and maintenance, evaluation of hazards, revising and correction of studies and risk management. These models may be changed or modified after reaching more accurate data during excavation. Respecting to complication of ground behaviour and uncertainty of available models, these models should be updatable and changeable in confront with unpredicted situation. Therefore, these models will be reliable if they can be simultaneously updated with new obtained data. Geological model proposing construction site with suitable engineering judgment has significant role in construction process.

107.2.1 Geological Model

Geological model is a basic prerequisite in tunnelling in order to improve safety and reduce costs. In tunnel excavation, field investigation cost is just about 3 % of total cost, which is considered unimportant by employer; however, it is very important due to probability, and uncertainty which exist in assumptions of modelling. Available experiences in United States shows that more than 55 % of contractor claims in tunnelling is about the ground condition that could reduce by more rigorous studies (Robinson et al. 2001). Regarding to experiences available in Iran, exploration cost is about 1–2 % of the total construction costs. Accordingly, appropriate data is not provided for designers.

In addition to site investigation studies at the feasibility and preliminary design phases, the geological studies carried out during construction have an important role in tunnelling process (Kolymbas 2005; Eftekhari et al. 2013a, b).

In Sabzkuh tunnel project, because of complicated ground conditions, geological reports does not cover all the ground parameters. So, geological models have been revised since the beginning of tunnel excavation. Face mapping and on-site modelling are followed at frequent intervals as the tunnel advances. This is done for the selection of the required support systems and optimum excavation pattern in different ground conditions.

107.2.2 Geotechnical Modelling

Increasing demand on geotechnical model in primary stages of a project in order to gather exploration operation information is felt more than ago. This model provides some initial information of ground properties and possibility of tunnel excavation for engineers and designers. These concepts can be updated and corrected in varied ground condition during excavation by additional studies.

Noticeable part in geotechnical studies is the correlation of these studies with geological ones, which is usually ignored in most projects. Subsequently, geotechnical studies are planned, performed and analyzed independently, regardless of the results of the geological studies. In other words, lack of engineering geology in correlation with geotechnical engineering can be felt clearly. In such this situation, complete expected studies about natural ground conditions are not done and lots of problem in interpolation of surface and subsurface data will be occurred. This process according to physical governing condition without regard to the nature of composition of soil in geological history is not only sufficient, but also causes great mistakes somehow. Hence, a geotechnical model can be expected to provide accurate results just when it is based on right knowledge of geological condition.

107.2.3 Design Model

In underground structure engineering there are two major aspects that must be addressed during the design phase. The first and most important is developing a realistic estimate of the expected ground conditions and their potential behaviours as a result of the excavation. The second is to design an economic and safe excavation and support system for the specified ground behaviours. The design process begins with the feasibility study and continues through the preliminary design, the detailed design, the tender design, and also the construction. The design is constantly updated during each stage, as more information is available. This requires the involvement of geological and geotechnical experts in all phases of a project.

During the design phases the inherent complexity and variability in many geological settings prohibits a complete picture of the ground structure and quality to be excavated. The geotechnical design is targeted to a continuous refinement of the models and decision criteria. Besides a high professional standard, a systematic and consistent, well documented evaluation and decision process is of paramount importance. Uncertainties in the ground model shall be considered in the design.

The variability of the geological architecture including the local ground structure, ground parameters, in-situ stress and ground water conditions requires employing a consistent and specific procedure during the design process. The key influences governing the geotechnical design are the ground conditions and behaviour.

Based on the ground characteristics and the expected ground behaviour, a feasible construction concept, consisting of excavation method, sequence of excavation, support and auxiliary methods, is chosen.

For stability analysis and reaching an appropriate tunnel model three analytical, observational and empirical methods are used. In recent decades, some numerical approaches such as FEM, BEM, DEM and FDM have been considerably spread. By using these methods, modelling of complicated geometry and different loading and employing of appropriate constitutive model is possible. Furthermore performance method can be modelled by this model.

A number of numerical methods have been developed in civil engineering practice. This methods are used extensively for analysis of underground excavation design problems.

For the analysis of tunnelling in soil, continuum analysis is generally accepted, where the domain can reasonably be assumed to be a homogeneous media. The continuum analysis includes Finite Element Method (FEM), Finite Difference Method (FDM), and Boundary Element Method (BEM).

Finite element or finite difference analysis has been used for a wide range of engineering practices for last several decades. Complex, multi-stage models can be easily created and quickly analyzed. The analyses provide complex material modelling options and a wide variety of support types can be modelled. Linear element, usually modelled as beam elements, can be applied in the modelling of shotcrete, concrete layers, and steel sets. Almost every project today requires numerical modelling to predict behaviour of structures and the ground.

107.3 Design and Construction of Tunnelling Work—Sabzkuh Tunnel

107.3.1 Loading

Unlike the rock tunnel in which the rock mass characteristics are used in estimated pressure imposed on support system (e. g. confinement-convergence analytical method, Ground Reaction Curve (GRC) and rock mass classification systems), in shallow tunnels and in particular in soil or soft ground tunnels, soil shear strength based analytical methods are used for estimation of support load. The support load estimated from various methods in Section T1 along the Sabzkuh tunnel is summarized in Table 107.1. The width and height of the tunnel are 4.9 and 5.4 m, respectively.

The vertical and horizontal loads on the support system estimated from various methods are given in Table 107.1. Terzaghi calculation model considers the soil cohesion and friction angle simultaneously, and consequently provides more accurate and more reasonable results compared to other methods. Meanwhile, for safety reasons, the lining horizontal and vertical pressures were selected 0.09 and 0.2 MPa, respectively.

107.3.2 Analysis of a Composite Liner with the 'Equivalent Section' Approach

Use of sprayed concrete or shotcrete as primary support is a standard practice in tunnel design and construction (see, for example, Hoek and Brown 1980; American Society of Civil Engineering 1984; Eisenstein et al. 1991; Franzen 1992). Steel arches can also be used with or without additional support or reinforcement to stabilize blocky or deformable ground. If the magnitude of loads transmitted by the ground to the support is large enough to preclude shotcrete alone or if squeezing or raveling behavior requires complete surface coverage, steel sets are commonly used in combination with shotcrete. This combination can be in the form of a complete composite annulus or may be a semi-circular or partial arch configuration (Table 107.2).

107.3.3 Stability Analyses

In Sabzkuh tunnel, designers used FEM as the stability analysis method in the design model with input data given in Table 107.3 (Itasca Consulting Group 1997). The obtained results are seen in Fig. 107.3. It can be revealed that the tunnel is stable with 1.5 m excavation step and 5.5 cm vertical displacement. Accordingly, full face excavation of tunnel section was chosen referring to the results of design model (Fig. 107.4).

After 35 m advance of the tunnel in full face excavation, a critical zone appeared and tunnel collapsed due to the redistribution of in-situ stresses around the excavation and formation of wide plastic zone as well as excessive axial displacements (Fig. 107.5).

The incident of tunnel collapse showed that the reliability of analytical methods in prediction of engineering behaviour in soil mechanic is largely dependent on the certainty of input data. Accurate measurement of in situ soil geotechnical parameters is difficult and sometimes impossible. In other hand, due to anisotropicity, the results of in situ soil tests cannot be used directly as design input data. In order to overcome these troubles, regarding to appropriate application of support system in collapses, excavation method should be optimized by considering available condition, outcome data and data which is gained during construction. Therefore, in a tunnel design model many uncertainties have geological and geotechnical source (Anagnostou and Kovari 1992).

107.4 Sequential Excavation Method

The Sequential Excavation Method (SEM), also commonly referred to as the New Austrian Tunnelling Method (NATM), is a concept that is based on the understanding of the behaviour of the ground as it reacts to the creation of an underground opening. This method needs minimum ground deflection and avoiding of softening and strength reduction. In this method, dividing tunnel section into smaller headings and their sequence excavation control the stress-strain condition and prevent it from collapse.

To be able to determine the encountered ground type, the geological documentation during construction has to be targeted to collect and record the relevant parameters specified in the design. Additional observations, like indications of overstressing, deformation and failure mechanisms, as well as results from probing ahead and the evaluation of the

Table 107.1 Estimation of support load in Sabzkuh tunnel, section T1

Load	Ref.	Equation	Value (MPa)
Vertical	Terzaghi	$\frac{B_1\left(\gamma-\frac{2c}{B_1}\right)}{k\tan\varphi}\left[1-e^{\frac{KD\tan\varphi}{B_1}}\right]$	0.136
	S.F.		0.20
Horizontal	Terzaghi (1)	$0.3\gamma(0.5m+h_p)$	0.046
	Terzaghi (1)	$\frac{B_1\left(\gamma-\frac{2c}{B_1}\right)}{k\tan\varphi}\left[1-e^{\frac{KD\tan\varphi}{B_1}}\right]$	0.059
	S.F.		0.09

Table 107.2 Input data for the analysis of a semi-circular liner comprised of shotcrete and steel sets according to the 'equivalent section' approach (Carranza-Torres and Diederichs 2000)

Geometry data	Shotcrete properties
$R = 2.60$ m (arch radius) $b = 0.75$ m (with of composite section) $s = 0.75$ m (spacing between steel set) $n = 1$ (number of steel sets along width)	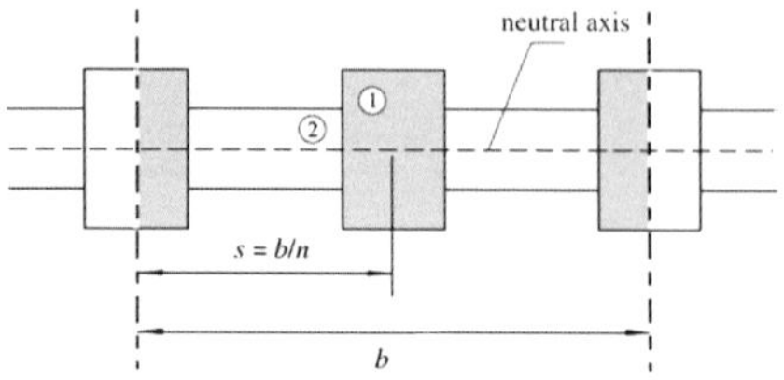
IPB140 Steel Set Properties	(1) Steel set, (2) Shotcrete
$t_s = 0.133$ m (height of the section) $A_s = 3.14 \times 10^{-3}$ m^2 (area of the section) $I_s = 1.030 \times 10^{-5}$ (moment of interia of the section) $E_s = 200000$ MPa (Young s Modulus) $\upsilon_s = 0.2$ (Poisson s ratio) $t_c = 0.2$ m $E_c = 22{,}900$ MPa (Young s Modulus) $\upsilon_c = 0.15$ (Poisson s ratio) $\sigma_c^c = 21$ MPa (tensile strength) $\sigma_t^s = -240$ MPa (tensile strength)	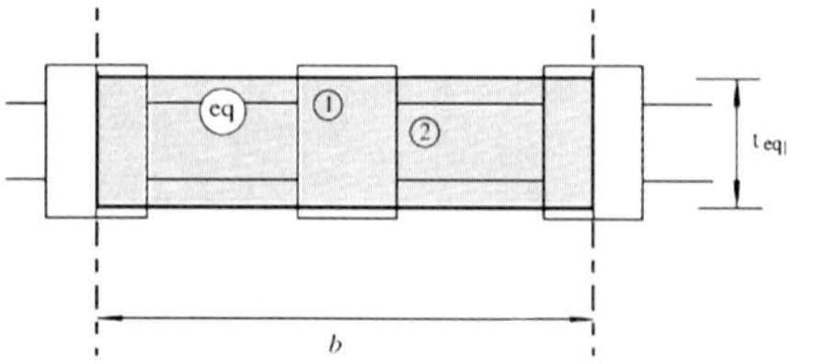
	Equivalent section
$D_{eq} = 3612.38$ MN $K_{eq} = 12.076$ MNm^2	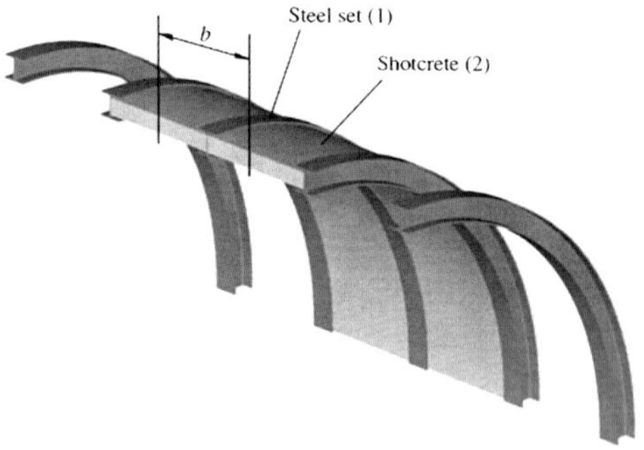 $t_{eq} = 20$ cm $E_{eq} = 24.05$ GPa

Table 107.3 The soil properties as input data in FEM

Parameter	Value	Unit
Type of material behaviour	Drained	-
Total unit weight (γ)	18.5	KN/m^3
Young's modulus (Es)	40	MPa
Poisson's ratio (ν)	0.25	-
Cohesion (c)	35	KN/m^2
Friction angle (ϕ)	25	degree

geotechnical monitoring are used to update the ground model and predict the conditions ahead of the face.

Based on the predicted ground conditions, the system behaviour in the section ahead has to be assessed under consideration of the influencing factors, and compared to the framework plan. Particular attention has to be paid on potential failure modes.

Hence, as a solution regarding to geological mapping and by using monitoring reports and revising the geological model, behaviour and classification of ground was investigated again in order to assess the excavation method (Terzaghi 1950; Heuer 1974). A prediction of the distribution of the excavation classes based on the results of the exploration and of stability analyses and on experience. The excavation sequence, round lengths and support measures for excavation classes B are given in Fig. 107.6 and Table 107.4.

FEM analyses in conjunction with empirical methods were used to evaluate potential ground behaviour upon tunnel excavation and to determine the required excavation sequence and support measures.

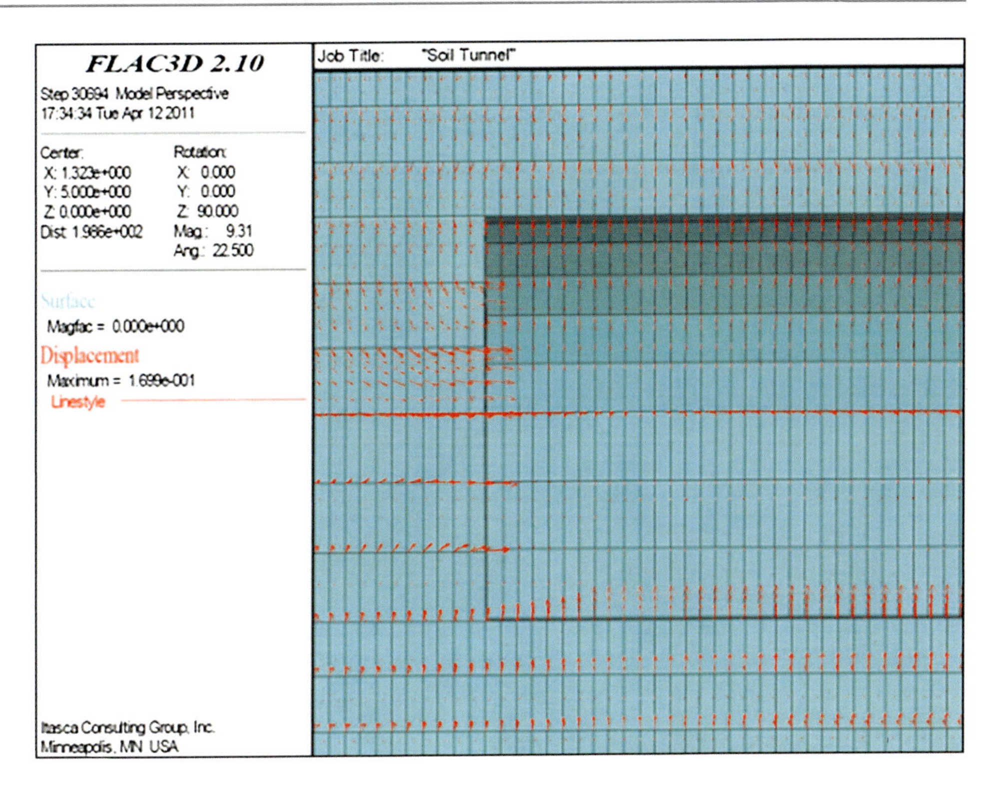

Fig. 107.3 Displacement vectors at the tunnel face

Excavation and Support Class (ESC)		Class A: Top, Bench Heading	Class A: Invert
Excavation Method		Mechanical Excavation (Backhoe)	
Unsupported Round Length (m)		1.5	15
Lining	Shotcrete (C25)	t=30cm	-
	Reinforcement	Steel set+WWF	
	Concrete (C21)	-	t=40cm
Tunnel Face Support		None	
Advance Support		None	
Trailing Distance		Continuous	

Fig. 107.4 Excavation and Support Class "A" for the tunnel

107.4.1 Stability Analysis for the Stages of Construction

Application of the NATM is based on empirical knowledge and local experience and may be adjusted according to observation. The use of numerical analyses with techniques such as the finite element method (FEM), could be of great value for this type of design. The use of numerical analyses to help tunnel design is becoming more popular both in industry and academic environments. In order to analyse the deformations and stability of the tunnel face excavated by NATM/SEM, the Plaxis 3D Tunnel software was used (Fig. 107.7) (Brinkgreve and Vermeer 2001).

The tunnelling construction process is simulated sequentially in several stages. Each construction stage may involve soil excavation and/or support lining construction.

– **Geometry**

In this study, the tunnel geometry and construction stages were simulated. The tunnel diameter and depth are 5 and 30 m, respectively. With regard to the symmetric geometry

Fig. 107.5 Collapse in initial part of Sabzkuh tunnel (T1) (Eftekhari et al. 2013b)

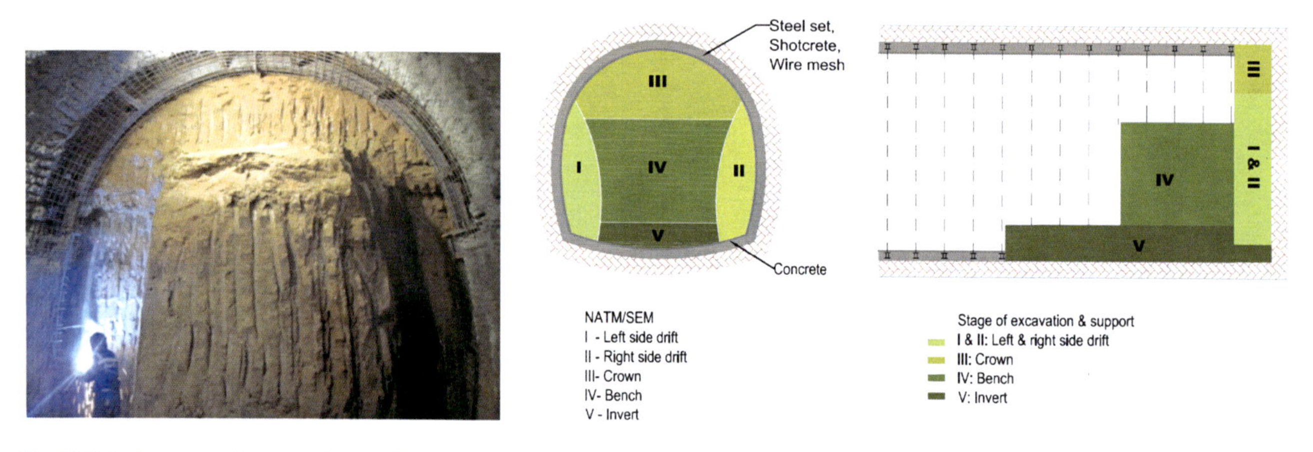

Fig. 107.6 Sequence of construction work

Table 107.4 Excavation and support class B for tunnel (after collapse)

Excavation and support class (ESC)		Class B				
		Side drift (I)	Side drift (II)	Crown (III)	Bench (IV)	Invert (V)
Excavation method		Mechanical Excavation (Backhoe)		Workman	Mechanical excavation (Backhoe)	
Unsupported round length (m)		0.75–0.85			3	7.5
Lining	Shotcrete (C25)	t = 25 cm				-
	Reinforcement	Steel Set+WWF				
	Concrete (C21)	-				t = 40 cm
Tunnel face support		A layer of flashcrete may be required				None
Advance support		Typically none; Locally pre-spilling or grouted pipe arch canopy (L = 4–6 m, D = 7.5 cm); alternatively ground improvement				
Trailing distance		Depending on the monitoring results, the encountered geotechnical conditions and result of stability analyses				

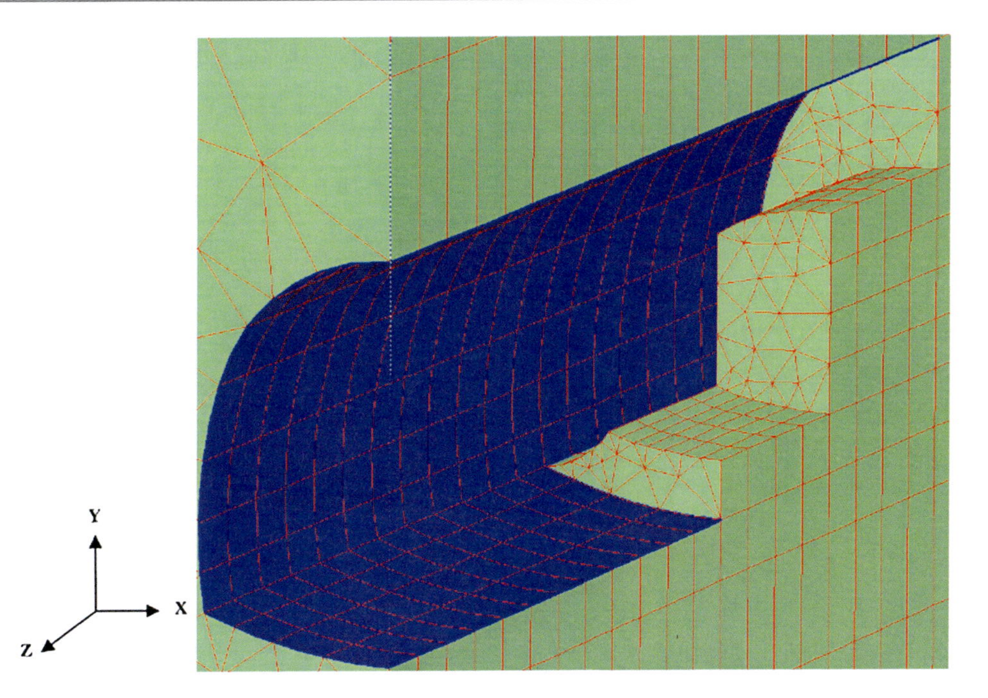

Fig. 107.7 Staged construction preview of SEM method

Table 107.5 Material properties of the soil

Parameter	Name	Sandy clay	Unit
Material model	Model	Mohr-Coulomb	-
Type of material behaviour	Type	Drained	-
Soil weight above phr. level	γ_{unsat}	15	KN/m^3
Soil weight below phr. level	γ_{sat}	18.5	KN/m^3
Young's modulus (constant)	E_{ref}	5×10^4	KN/m^2
Poisson's ratio	ν	0.3	-
Cohesion (constant)	c_{ref}	17	KN/m^2
Friction angle	ϕ	28	°
Dilatancy angle	Ψ	0	°

of the tunnel, on either side or one half of the tunnel can be modelled. In this study, left half of the tunnel was selected. The model has dimensions of 25 × 50 × 50 m in X, Y and Z directions, respectively.

– **Material properties**

In modelling material behaviour, the constitutive model of Mohr-Coulomb was used. The elastic-plastic Mohr-Coulomb model involves five input parameters, i.e. E and v for soil elasticity; φ and c for soil plasticity (Table 107.5). This model represents a 'first-order' approximation of soil or rock behavior. It is recommended to use this model for a first analysis of the problem considered.

– **Deformations**

The results of the analyses indicate that the tunnel is quite stable and initiated deformations are relatively low and can be controlled in an acceptable range (Fig. 107.8).

107.5 Monitoring

Ground deformation monitoring in tunnelling is a common means for selecting and controlling the excavation and support methods among those predicted in design, ensuring safety during tunnel excavation (including personnel safety

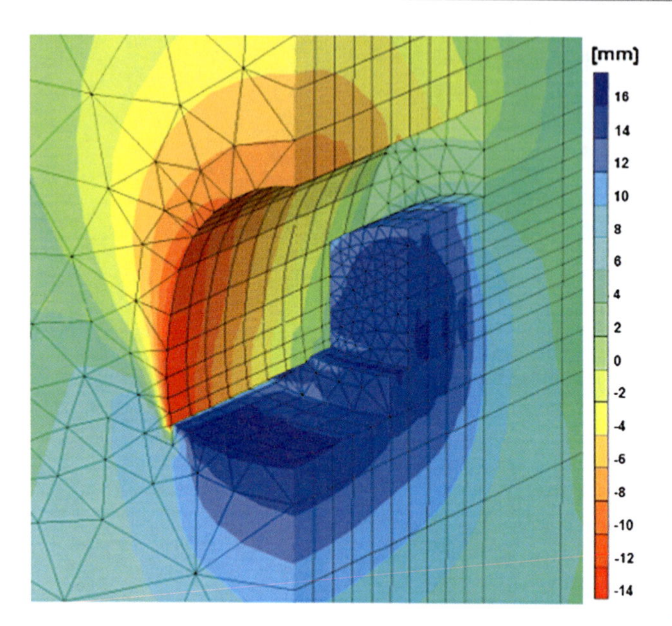

Fig. 107.8 Extreme displacements in tunnel face, Vertical displacements (Uy), Extreme Uy 13 mm (Crown)

inside the tunnel and safety of structures located at ground surface) and construction quality (Kavvadas 2005).

Ground deformation monitoring and its application in tunnel design and construction is illustrated using examples from the Jubilee Line Extension of London Underground (Burland et al. 2001; Mair 2001), from Lines 2 and 3 of the Athens Metro (Kavvadas 1997, 1999), from Resalat (Fakhimi et al. 2012) and Niayesh road tunnel project in Iran (Ghorbani et al. 2012), and from Dranaz tunnel, Sinop, Turkey (Aydin et al. 2004). An overview of the application of the observational method in tunnelling projects is given by Powderham (1994). Tsatsanifos et al. (1999) and Kontogianni et al. (2004) present interesting case studies of excessive deformations causing failure of mountain tunnels in Greece.

Ground deformation monitoring has different objectives in mountain and urban tunnels. In mountain tunnels, the main objective of deformation measurements during construction is to ensure that ground pressures on the temporary support system are adequately controlled, i.e., there exists an adequate margin of safety against roof collapse, bottom heave, failure of the excavation face, yielding of the support system, etc. Control of ground pressures ensures a safe and economical tunnel structure, well adapted to the inherent heterogeneity of ground conditions.

Typically, the majority of ground deformation takes place ahead and close to the tunnel face, from about one tunnel diameter ahead of the face up to about 1.5 diameters behind the face (e.g., Chern et al. 1998; Kavvadas 1999; Hoek 1999, 2001; Kontogianni and Stiros 2002). Thus, instruments placed on the tunnel wall (e.g., 3D optical reflector targets) or installed in the ground from the tunnel wall (e.g., rod extensometers) should be put in place as early as possible.

In Sabzkuh tunnel, a continuous geodetic monitoring began using two targets mounted on the tunnel wall and one target in tunnel crown at about 20 m intervals. The targets were mounted at least one meter behind the tunnel face. Recordings were followed in a regular and predetermined time intervals (Fig. 107.9).

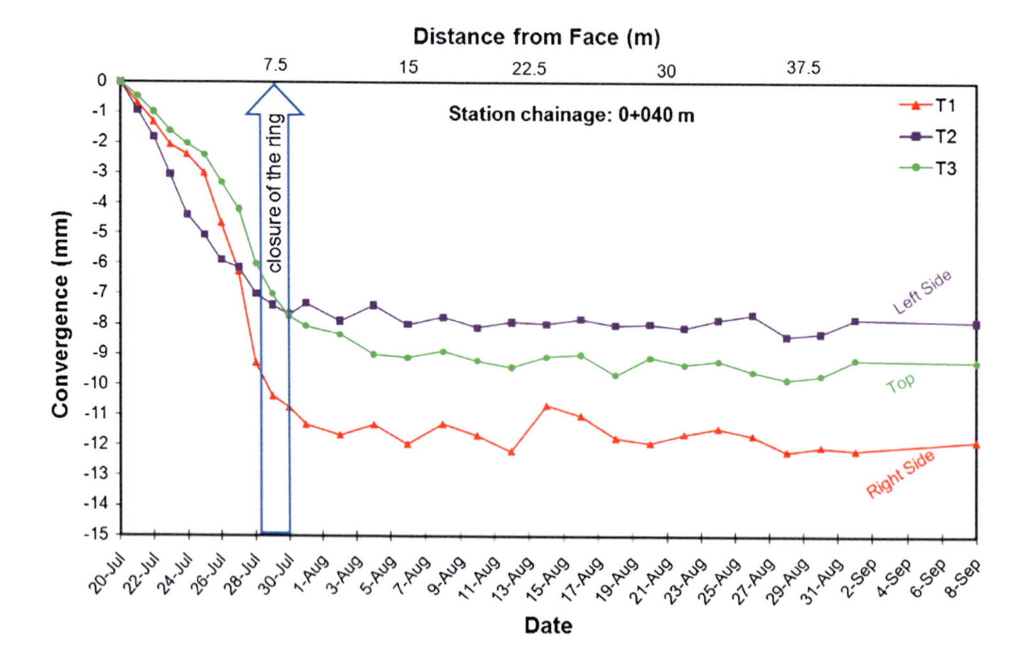

Fig. 107.9 Tunnel convergence at chainage 0+040

107.6 Conclusion and Summary

In construction process of Sabzkuh tunnel, a geotechnical model independently and without respect to geological model was planned, performed and analyzed. Lack of correlation between these two models and ground complicated situation and uncertainties provided some inappropriate information for designers and affected the analytical model as well as excavation method. After unworkability of full face excavation, a data server established in order to gather data of field studies and as built finding, regarding to previous experiences. This server helped to enhance prior construction method. The main specification of this server was its updatable in confront with new data.

Obtained results showed a complicated geological model, which needed frequent exploration, analysis and behaviour checking of surrounding soil moss in order to find the most economical and stable excavation model and support in face with different conditions. So, NATM/SEM method was selected due to its workability and flexibility in confront with tunnel site condition.

This case study demonstrates once again the importance of early detection or prediction of potentially problematic zones (via probe drilling and monitoring) in tunnelling, especially through mixed or difficult ground conditions characterized by alternating layers, faulting and localized zones of high water pressure. Because mechanical detection methods cannot be fully relied upon, availability of experienced personnel to predict and deal with such instability problems effectively and promptly is the best insurance for successful completion of tunnelling contracts.

References

American Society of Civil Engineering (1984) Guidelines for tunnel lining design. In: O'Rourke TD (ed) ASCE Technical Council on Research. Technical Committee on Tunnel Lining Design

Anagnostou G, Kovari K (1992) Ein beitrag zur static der ortsbrust beim hydroschidvortrieb.-in: Probleme bei maschinellen tunnivortrieben? Geratehersteller and anwender berichten. Beitrage zum symposium vom 22./23. Oktober 1992, TU Munchen

Aydin A, Ozbek A, Cobanoglu I (2004) Tunnelling in difficult ground: a case study from Dranaz tunnel, Sinop, Turkey. Eng Geol 74:293–301

Brinkgreve RBJ, Vermeer PA (2001) Manual of Plaxis 3D tunnel. A.A. Balkema, Rotterdam

Burland JB, Standing JR, Jardine FM (2001) " Building response to tunneling" case studies from construction of the Jubilee Line Extension. Thomas Telford Publishers, London

Carranza-Torres C, Diederichs M (2000) Mechanical analysis of circular liners with particular reference to composite supports. For example, liners consisting of shotcrete and steel sets. Tunn Undergr Space Technol 24(2009):506–532

Chern JC, Yu CW, Shiao FY (1998) Tunnelling in squeezing ground and support estimation. In: Proceedings of regional symposium on sedimentary rock engineering, Taipei, pp 192–220

Eftekhari A, Taromi M, Saeidi M (2013) Complexity of the ground conditions and non compliance with basic assumptions in the trench stability analysis: a Case Study in Iran. In: 7th SASTech 2013, Iran, Bandar-Abbas, 14–15 Mar 2013

Eftekhari A, Taromi M, Saeidi M (2013) Collapse zone passing through with IPE arc support technique (IAST): (a case study in Iran), 10th Iranian tunneling conference, "Underground space and the 3rd millennium development goals" Iran, Tehran

Eisenstein Z, Kuwajima FM, Heinz HK (1991) Behaviour of shotcrete tunnel linings. In: Proceedings of rapid excavation and tunnelling conference, Seattle, pp 47–57

Fakhimi A, Salehi D, Mojtabai N (2012) Numerical back analysis for estimation of soil parameters in the Resalat tunnel project. Tunn Undergr Space Technol 19:57–67

Franzen T (1992) Shotcrete for underground support: a state-of-the-art report with focus on steel-fibre reinforcement. Tunn Undergr Space Technol 7(4):383–391

Ghorbani M, Sharifzadeh M, Yasrobi S, Daiyan M (2012) Geotechnical, structural and geodetic measurements for conventional tunnelling hazards in urban areas—The case of Niayesh road tunnel project. Tunn Undergr Space Technol 31:1–8

Heuer RE (1974) Important ground parameters in soft ground tunneling. In: Proceeding of specialty conference on subsurface exploration for underground excavation and heavy construction, ASCE, New York

Hoek E, Brown ET (1980) Underground excavations in rock. The Institute of Mining and Metallurgy, London

Hoek E (1999) Putting numbers to geology—an engineer's viewpoint. Q J Eng Geol 32:1–19

Hoek E (2001) Big tunnels in bad rock. J Geotech Geoenviron Eng 127 (9):726–740

Itasca Consulting Group (1997) FLAC3D, fast Lagrangian analysis of Continua in 3 Dimensions. Version 2.1. Minneapolis

Kavvadas M (1997) Analysis and performance of the NATM excavation of an underground station for the Athens Metro. In: Proceedings of 4th international conference on case histories in geotechnical engineering, St. Louis, Missouri USA, March 1998 (paper No 6.11)

Kavvadas M (1999) Experiences from the construction of the Athens Metro project. Proceedings of 12th European conference of soil mechanics and geotechnical engineering, Amsterdam, June 1999. Invited Lect 3:1665–1676

Kavvadas M (2005) Monitoring ground deformation in tunnelling: current practice in transportation tunnels. Eng Geol 79(2005):93–113

Kontogianni V, Stiros S (2002) Shallow tunnel convergence: predictions and observations. Eng Geol 63(304):333–345

Kontogianni V, Tzortzis A, Stiros S (2004) Deformation and failure of the Tymfristos tunnel, Greece. J Geotechn Geoenviron Eng, ASCE 130(10):1004–1013

Kolymbas D (2005) Tunnelling and tunnel mechanics, a rational approach to tunnelling. Springer, Berlin. ISBN-13 978-3-540-25196-5

Mair RJ (2001) Research on tunneling-induced ground movements and their effects on buildings—lessons from the Jubilee Line Extension. In: Proceedings of international conference on response of buildings to excavation-induced ground movements, Imperial College, London, 17–18 July, CIRIA Special Publication, vol 199. CIRIA Publications, London, UK, pp 3–26. http://www.ciria.org.uk/index.html

Powderham AJ (1994) An overview of the observational method: development in cut and cover and bored tunnelling projects. Geotechnique 44(4):619–636

Robinson RA, Kucker MA, Gildner JP (2001) Levels of geotechnical input for design—build contracts. In: Proceedings of rapid excavation and tunnelling conference, Society for Mining, Metallurgy, and Exploration, Inc., Littleton, Colorado, pp 829–839

Saeidi M, Eftekhari A, Taromi M (2012) Evaluation of rock burst potential in Sabzkuh water conveyance tunnel, IRAN: a case study. In: 7th Asian rock mechanics symposium, ARMS 2012, Seoul, Korea, 15–19 Oct 2012

Taromi M, Eftekhari A, Saeidi M (2013) Adoption of an appropriate excavation method in construction process (case study: Sabzkuh tunnel, Iran). In: International conference on civil engineering, architecture and urban sustainable development, Tabriz, Iran

Terzaghi K (1950) Geologic aspects of soft ground tunneling. Chapter 11 in applied sedimentation. In: Task R, Parker D (eds) Wiley, New York

Tsatsanifos CP, Mantziaras PM, Georgiou D (1999) Squeezing rock response to NATM tunnelling. A case study. In: Proceedings of the international symposium on geotechnical aspects of underground construction in soft rock, Tokyo, Japan, pp 167–172

Geological Design for Complex Geological-Structural Contest: The Example of SS 125 "Nuova Orientale Sarda"

108

Serena Scarano, Roberto Laureti and Stefano Serangeli

Abstract

The following work is an example of road design in geologically complex environments, which requires a particular accuracy in the definition of the Geological Reference Model. In the studied area the rock masses involved in the plan are strongly deformed by tectonic activity that occurred over time. For this reason, the analysis of geological formations has been addressed, above all, to the study of geomechanical features, such as strength resistance and elastic properties of the rock mass. The integrated analysis of the data coming from geostructural and geomechanical surveys carried out on rocky outcrops in the area, and of the data obtained from site investigation and laboratory test on rock samples, has allowed to improve the characterization of rock masses and the definition of Rock Quality Indexes. The study provided the values of GSI for both formations, used to obtain the geotechnical parameters, adopted for the design of project interventions. The project included the study on the reuse of soil and rocks coming from the excavations, ahead of the ascertainment of their environmental characteristics by means of chemical analysis. Abnormalities in the chemistry of some samples are interpreted as due to the nature and evolution of the geological formations, and not to environmental pollution.

Keywords

Rock masses • Geological reference model • Geostructural relief • Reuse of soil • Chemical analysis

108.1 Introduction

A fundamental role in order to define the geotechnical characterization of soils and the geomechanical characterization of rock mass, is played by the identification of the geostructural and geomechanical features, and the Geological Reference Model, of the geological units outcropping in the area where the road plan is located. All this factors affects the resulting choice of design solutions.

An example is represented by the design of the new SS 125 "Nuova Orientale Sarda" Tronco Tertenia—San Priamo 1° Lotto—1° e 2° Stralcio.

The track of the whole parcel, developed in the south-east part of the island, 100 km far from Cagliari, along the Torrente Quirra Valley, is long more or less 13 km and is composed by 11 viaducts for a whole length of 750 m, and 3 tunnels (1 artificial and 2 natural).

108.2 Geological and Structural Framework

The road plan is located along the lower part of the right slope of the Rio Quirra Valley, in a complex geological contest, that outcrops along it. This valley follows the development of a transtensive tectonic element with regional importance, of which the path follows the development. The

S. Scarano (✉) · R. Laureti · S. Serangeli
Direzione Centrale Progettazione, Anas S.p.A, Rome, Italy
e-mail: se.scarano@stradeanas.it

G. Lollino et al. (eds.), *Engineering Geology for Society and Territory – Volume 6*,
DOI: 10.1007/978-3-319-09060-3_108, © Springer International Publishing Switzerland 2015

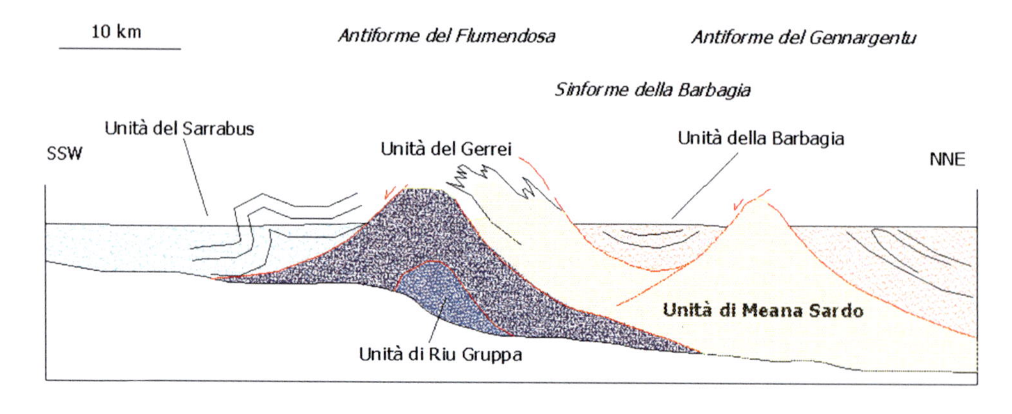

Fig. 108.1 Schematic geostructural diagram of the Paleozoic basement in the south-eastern Sardinia

valley is bordered by mountains composed by the Paleozoic basement units, represented by an alternations of metasandstones, metapelites, metavulcanites, metaepiclastites, metaconglomerates and greenish-gray shales, affected by polyphasic deformation, schistosity and by a medium-low degree of metamorphism. The terrains interested by the construction of the road belong to the Meana Sardo tectonic Unit, overlapped on Gerrei Unit; the Meana Sardo Unit is overlapped by the Genn'Argiolas unit (Fig. 108.1) (Calvino F. 1959).

108.2.1 Local Stratigraphic Succession

The units, that are directly involved by the project, have been identified in the literature, in the formational stratigraphic nomenclature, with the terms of S. Vito Sandstones and Monte Santa Vittoria Formation.

108.2.1.1 S. Vito Sandstones

This formation is placed at the base of the Meana Sardo Unit and is composed by an association of metarenites and quartz-mica metapelites, with shiny appearance, and, rarely, by metasandstones and thin levels of gray and greenish-gray metaquarzoarenites, with intercalations of gray or black metapelites and metasiltstones. In the middle part of the formation there are often metaconglomerates with elements of metasiltstones and metasandstones. In the section covered by the plan of the new SS 125, the outcrop formation is affected by schistosity, levels and lists of quartz, reflecting phenomena of *boudinage* or elongation in the stresses direction.

The basis of this lithostratigraphic unit does not outcrop; it rests with tectonic contact on the Gerrei Unit and is covered, in unconformity, by metaconglomerates of Muravera or, directly, from the Monte Santa Vittoria Formation.

The unit is referred to the Cambrian—middle/lower Ordovician (Calvino F. 1963, 1972).

108.2.1.2 Monte Santa Vittoria Formation

This formation is composed by two different lithofacies, represented by: metaepiclastites (cfr. Manixeddu and Monte Corte Cerbos Formations of Bosellini and Ogniben, 1968) to prevailing volcanic matrix, of various granulometry, with intercalations of metagraywackes, metasandstones and metaconglomerates with quartz pebbles, identified with the acronym MSVa; metagraywakes and metandesites (cfr. Serra Tonnai Formation of Bosellini and Ogniben, 1968), known as MSVb and represented by volcanic greenish metagraywakes, with intercalation of greenish-gray metavulcanites with composition from basaltic to andesitic. The formation is related to an effusive activity from intermediate to basic composition and deposition of graywakes resulting from the rearrangement of the volcanic deposits. The unit is referred to the Middle Ordovician (Pertusati et al. 2001a, 2001b).

These two formations have been involved in a series of plicative structures, which lead the San Vito sandstones to

Fig. 108.2 S. Vito sandstones. Small fold deformations in the micaceous metapelites

Fig. 108.3 Monte Santa Vittoria formation. Metagraywakes and compact metandesites

outcrop into the antiform cores, and the Monte Santa Vittoria Formation in the sinform cores.

In addition to these two formations, the tectonic unit includes the Muravera Metaconglomerates Formation, made up polygenic, heterometric, often coarse metaconglomerates, in quarzoarenitics matrix, not outcropping in the study area.

The bedrock just described is, at times, covered by ancient conoid deposits, including blocks and heterometric pebbles of metamorphic substrate, mixed with sandy-silty matrix reddened and well thickened; eluvio-colluvial blankets, consisting of angular blocks in silty matrix, without sorting; recent alluvional deposits, stabilized and thickened, and current alluvional deposits, formed by blocks and pebbles with poor matrix (Figs. 108.2 and 108.3) Carmignani et al. (2001a, 2001b).

108.3 Geostructural and Geomechanical Characterization

The road design in such rock masses has required a study of the geomechanic quality features Bieniawski (1974, 1976), that lead to the definition of their strength and of the elastic parameters of the rock masses (Clerici et al. 1986).

The site investigation data come from different campaigns, subsequently held in trust by ANAS, and include 71 boreholes, 10 geomechanical survey points, 38 geognostic trenches and 11 refraction seismic bases.

Therefore, the classification of rock masses has been made starting by data obtained during the geostructural and geomechanical survey of the rock outcrops (ISRM 1974, 1978, 1981), obtaining the GSI index (Hoek and Brown, 1997). Then a comparison between the results coming from survey and those obtained from site-investigation and laboratory tests (RQD, Point Load Test, uniaxial compression of rock samples) was made (Fig. 108.4).

This comparison led to the attribution of geotechnical parameters on the basis of which the project interventions have been designed.

Based on the obtained values of the index GSI, the two studied formations have been included in the "Geological Strength Index for Jointed Rocks" (Marinos and Hoek 2000) (Fig. 108.5) diagram. In this diagram the GSI values, obtained from survey data (oval empty areas) are drawn

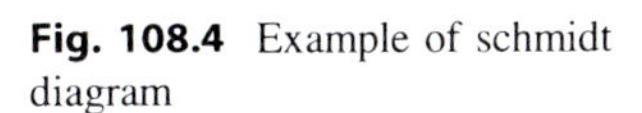

Fig. 108.4 Example of schmidt diagram

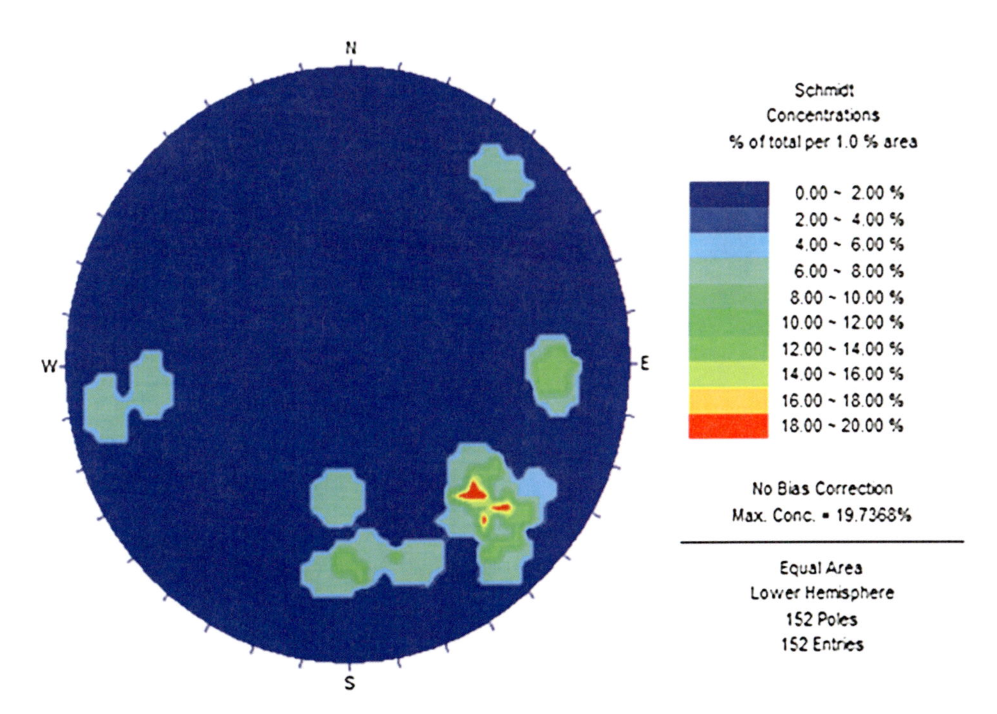

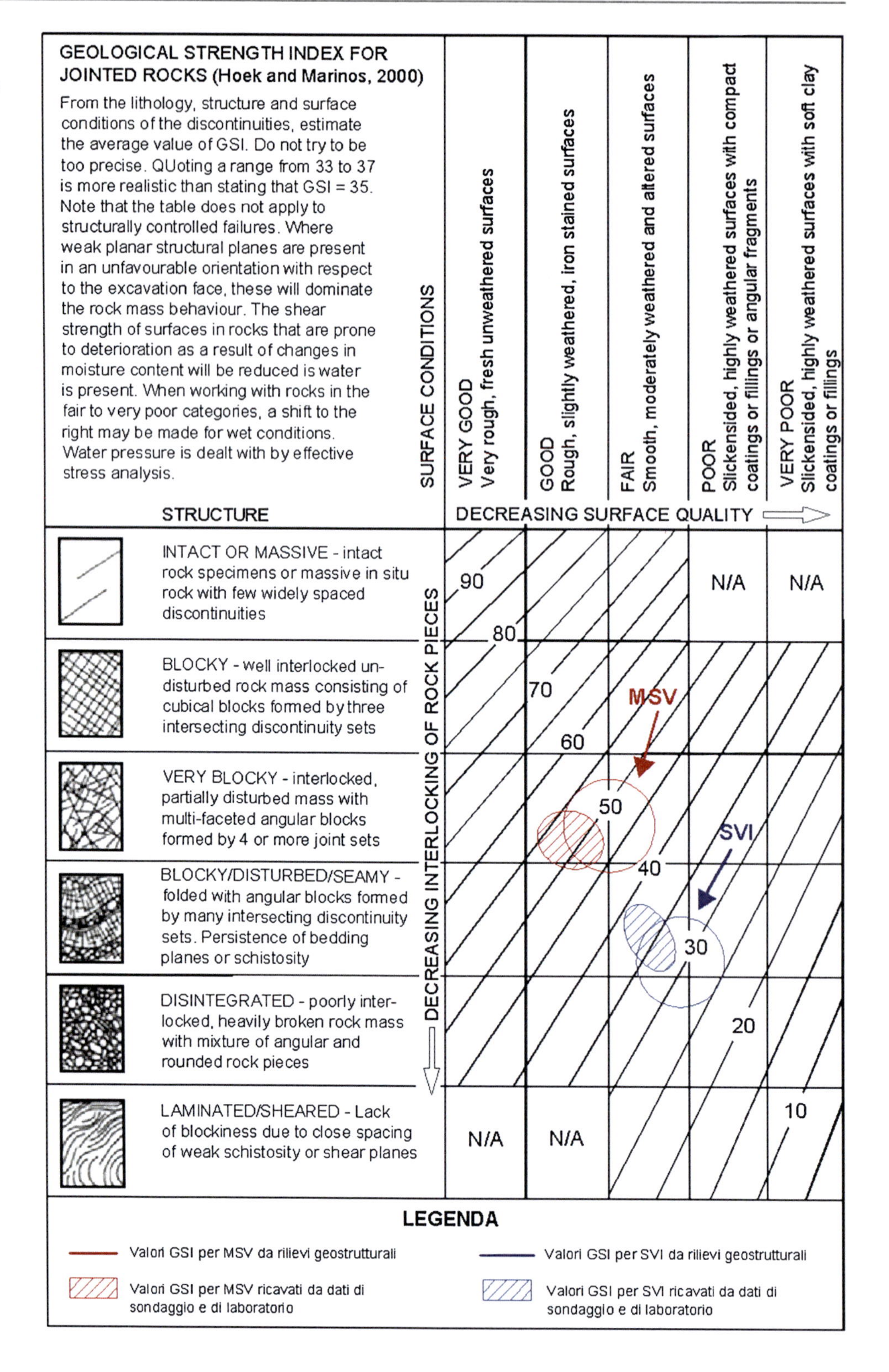

Fig. 108.5 Diagram for the estimate of the GSI index from geological observations (Marinos and Hoek 2001)

together with the ones obtained from site investigation and laboratory tests (oval hatched areas).

In general, it shows a good correspondence between the two sets of data, with an appreciable overlap between the areolas, which represents the index variability fields for the two units.

108.4 Environmental Study

The project, in addition to the precise definition of MGR, with a high degree of reliability, includes the study on the reuse of soil and rocks coming from the excavation, preceded by their environmental characterization works.

Through chemical analysis of environmental characters, in fact, anomalies in the soil samples chemical parameters coming from the layers of surface alteration of the substrate, and from alluvial deposits, found along the path, have been found.

In some samples, the values of arsenic, zinc, mercury and cobalt are higher than Contamination Threshold Concentration. This situation, in an area characterized by the presence of a mining site (Baccu Locci), at the foot of Mount Cardiga, can be refers, however, to the so-called "natural background". In fact, during the mining activity, metal sulfides (zinc and lead overall) and arsenopyrite were extracted, so it confirms the presence of those chemical elements inside the minerals founded into the rock mass.

The lithologies forming the bedrock, by which dismantling/alteration alluvial deposits and eluvio-colluvial accumulations originated, contain mineralization. Thus, there is a correlation between the mineralogical and petrographic composition of the sediments accumulated and the nature of the geological formations outcropping in the area of immediate concern, in which widespread circulation of hydrothermal fluids led, locally, to the formation of appreciable concentrations of metal-bearing metal sulphides.

108.5 Conclusions

The road design, especially in presence of infrastructure of considerable extent, presupposes a thorough knowledge of the geological and geotechnical characteristics of the soils outcropping in the area of roadway location and interacting with the planning civil works.

This is possible to obtain through direct studies of the area, through surface surveys, integrated with the data coming from geological site investigations planned on the basis of the design elements.

In the present case, in particular, the geological units belong to the ancient metamorphic substrate, whose behavior is difficult to achieve only following the laboratory characterization. Therefore we proceeded through the geostructural and geomechanical characterization of rock masses, by which we have obtained the geotechnical design parameters. In particular, the work involved the evaluation of geomechanical characters from both the examination of the direct geological analysis of the outcrops present, and the systematic analysis of data obtained by site investigations (boreholes), integrated with laboratory testing on rocks. The results, obtained through two different ways, indicate a significant convergence.

Another aspect that is often necessary to consider, is the environmental one, mainly if the design choices allow the recycling of materials resulting from excavations. The environmental characteristics of these soils are determined by chemical analysis regulated by national legislation.

References

Bieniawski ZT (1974) Estimating the strength of rock materials. J S Afr Inst Min Metall 74:312–320

Bieniaswski ZT (1976) The geomechanics classification in rock engineering applications. In: Proceedings of 4th international congress on rock mechanics, vol 2, pp 51–58

Calvino F (1959) Lineamenti strutturali del Sarrabus-Gerrei (Sardegna sud-orientale). Boll. Servizio Geologico d'Italia, Roma 81:489–556

Calvino F (1963) Carta Geologica d'Italia alla scala 1:100.000, "Foglio 227 Muravera". Servizio Geologico d'Italia, Roma

Calvino F (1972) Note illustrative della Carta Geologica d'Italia, "Foglio 227 Muravera", 1–58

Carmignani L, Conti P, Pertusati PC, Barca S, Cerbai N, Eltrudis A, Funedda A, Oggiano G, Patta ED (2001) Carta Geologica d'Italia alla scala 1:50.000, "Foglio 549 Muravera". Servizio Geologico d'Italia, Roma

Carmignani L, Conti P, Pertusati PC, Barca S, Cerbai N, Eltrudis A, Funedda A, Oggiano G, Patta ED, Ulzega A, Orrù P (2001) Note illustrative della Carta Geologica d'Italia alla scala 1:50.000, "Foglio 549 Muravera". Servizio Geologico d'Italia, Roma, 140 pp

Clerici A, Griffini L, Pozzi R (1986) Procedura per l'esecuzione di rilievi strutturali geomeccanici di dettaglio su ammassi rocciosi a comportamento rigido. Geologia Tecnica 3(88):21–31

Hoek E, Brown ET (1997) Practical estimates of rock mass strength. Int J Rock Mech Mining Sci Geomech Abstr 34:1165–1186

ISRM—International Society of Rock Mechanics—Commission on Standardization of Laboratory and Field Tests (1974) Suggested methods for determining shear strength. Committee on field tests, Doc. n.1

ISRM—International Society of Rock Mechanics—Commission on Standardization of Laboratory and Field Tests (1978) Suggested methods for the quantitative description of discontinuities.in rock masses. Inte J Rock Mech Mining Sci Geomech Abstr 15 (6):319–368

ISRM—International Society of Rock Mechanics—Commission on Standardization of Laboratory and Field Tests (1981) Rock characterization testing and monitoring. Suggested methods. In: Brown ET (ed). Pergamon Press, Oxford

Marinos P, Hoek E (2000) GSI: a geologically friendly tool for rock mass strength estimation. In: Proceedings of the GeoEng2000 at the international conference on geotechnical and geological engineering, Melbourne, Technomic publishers, Lancaster, pp 1422–1446

Marinos P, Hoek E (2001) Estimating the geotechnical properties of heterogeneous rock masses such as flysch. Bull Eng Geol Environ 60:82–92

Pertusati PC, Sarria E, Cherchi GP, Carmignani L, Barca S, Benedetti M, Chighine G, Cincotti F, Oggiano G, Ulzega A, Orrù P, Pintus C (2001)a Carta Geologica d'Italia alla scala 1:50.000, "Foglio 5419 Jerzu". Servizio Geologico d'Italia, Roma

Pertusati PC, Sarria E, Cherchi GP, Carmignani L, Barca S, Benedetti M, Chighine G, Cincotti F, Oggiano G, Ulzega A, Orrù P, Pintus C (2001)b Note illustrative della Carta Geologica d'Italia alla scala 1:50.000, "Foglio 5419 Jerzu". Servizio Geologico d'Italia, Roma, 168 pp

The "A12—Tor dè Cenci" Motorway: Geological Reference Model and Design Solutions in Presence of Soft Soils

109

Stefano Serangeli, Roberto Laureti, and Serena Scarano

Abstract

The link that connects the A12 "Roma-Civitavecchia" with the "Roma (Tor de 'Cenci)—Latina" motorway (today designed but not built) is a road infrastructure of considerable size and importance. It needs a very detailed Geological Reference Model, defined in each design phase through the analysis of numerous geological surveys and site investigation data. In the preliminary phase of the plan we found very useful to keep account of the considerable amount of data concerning the area surrounding the designed motorway. At a later stage we realized a very detailed geological survey and site investigation activity, especially referred to the scale of the main viaducts. This study has allowed us to define the sequence of the different lithofacies and their geometric relationships. The presence, in the subsoil, of highly deformable organic soils, with peat layers, characterized also by a low shear strength, has influenced the design solutions. In fact, throughout the development of the road axis, different interventions were used with the aim to reduce the settlements of the road body. So, the study of the plan was marked by the reduction of the applied load, keeping as low as possible the project level and the height of the embankments. At the same time it was necessary to design some embankments providing the use of very light materials (expanded clay, Polystyrene) and particular solutions for the foundations design of the main bridges and viaducts.

Keywords

Geological reference model • Geological survey • Organic soils • Design solution • Low resistance

109.1 Introduction

The design of the new connection between the A12 "Roma-Civitavecchia" and the "Roma (Tor de 'Cenci)—Latina" motorways (the last one still in project) was developed, on behalf of "ADL—Autostrade del Lazio SpA", by "Direzione Centrale Progettazione" of ANAS SpA (National Public Roads Company). It represents an example of a particularly detailed definition of the Geological Reference Model, as a prerequisite to the accomplished identification of geological problems and, finally, of consequent adoption of appropriate design measures.

This aspect is fundamental in order to build this kind of infrastructure, characterized by very important civil works (especially bridges and viaducts). The deepening of the MGR, achieved through the different stages of the project, has allowed us to define the sequence of the different lithofacies and their evolution, and their geometrical relationships. Starting from this modeling, the Geotechnical Reference Model, necessary for a correct evaluation of related design issue, was completely defined.

The road project extends for about 16 km, through the Roman countryside (Fiumicino Plain) and the hinterland. It is composed by 4 viaducts with considerable development, of which the longest exceeds 2.7 km and another one crosses the Tevere River, and one artificial tunnel.

S. Serangeli (✉) · R. Laureti · S. Scarano
Direzione Centrale Progettazione, Anas S.p.A., Rome, Italy
e-mail: s.serangeli@stradeanas.it

G. Lollino et al. (eds.), *Engineering Geology for Society and Territory – Volume 6*,
DOI: 10.1007/978-3-319-09060-3_109,

109.2 Studies and Geological Surveys

The project corridor and the surrounding areas were already covered by existing infrastructure (Fiumicino Airport and Motorway, "Rome Trade Fair District", Roma-Pisa Railway). For this reason, it was interested by several site-investigation activities; so that its subsoil appears, therefore, well-known from the geological and geotechnical point of view. Particularly, the bad features of the recent organic soils are just well-known.

During the preliminary design, the geological reference model was, therefore, essentially defined by using a massive amount of available data (65 boreholes; 27 static penetration tests (CPT and CPTU); 2 trenches and other 31 boreholes coming from preexisting investigations).

During the later stage of the design, it was realized a very detailed geological survey, together with a specific site-investigation campaign, especially referred to the scale of the main viaducts and bridges, in order to improve both the geological and geotechnical models.

This site investigation, carried out in 2012, consists of:

- 12 boreholes, including undisturbed sampling for laboratory testing;
- 11 static penetration tests with piezocone for interstitial pressure measurements (CPTU);
- 5 geophysical tests (Down Hole).

During the execution of borehole, a total of 75 dynamic penetration tests SPT and 38 undisturbed samples were carried out.

109.3 Local Stratigraphic Succession

The studies described above, together with bibliographic data, ISPRA EX APAT gave the opportunity to focus on the geological context of the project, in which various depositional and erosional stages in different genetic environments overlap each other, so that the sedimentary prevulcanic substrate consists first of marine units, then transitional ones and, finally, continental formations.(Faccenna et al. 1995; Ventriglia U 1971, 1990, 2002)

The oldest geological formation, as identified in the studied region, is represented by "*Monte delle Piche Formation*" (*MDP*), a marine-clayey deposit over which recent pyroclastic and alluvial sediments have been settled. These units are represented by "*Ponte Galeria Formation*": in particular, in the area, there are the "*Membro della Pisana*", characterized by three lithofacies (1. conglomeratic-sandy lithofacies, PGL3a; 2. clayey-sandy lithofacies, PGL3b; 3. sandy lithofacies, PGL3c). These elder units have been splitted by a system of small faults with Apenninic directions, which have lowered the substrate according to a system of steps, with other antiapenninic system and north-south direction (Funiciello R and Parotto 1978).

Along Tiber'plan, with an unconformity, there are more recent soils of alluvial and marsh and lacustrine environment ("*Sintema del fiume Tevere*"), with high organic content, divided into different lithofacies: SFTa, sandy gravel and gravelly coarse sand; SFTb, sands, silty sands and sandy silts; SFTc, organic cohesive deposits and peats (Conato 1980).

In correspondence of the hills, other unconformities separate the succession MDP/PGL from subsequent pyroclastic soils, of the Colli Albani volcanic apparatus ("*Tor dè Cenci Unit*", *TDC*, "*Pozzolane Rosse*", *RED*, "*Villa Senni Formation/Pozzolanelle*", *VSN*$_2$). Alternating with these terms there are continental and fluvial soils, called: "*Valle Giulia Formation*" (*VGU*), "*Fosso del Torrino Formation*" (*FTR*), "*Castelporziano Unit*" (*CLZ*) (De Rita D et al. 1988, 1989, 1995).

The upper fluvial-lacustrine deposits, with an high content of organic matter, have features of high deformability and low resistance (Molin et al. 1995; Servizio Geologico d'Italia 1967; Società Geologica Italiana 1990).

109.4 The Major Viaducts ("Tevere Viaduct" and "Interconnessione Viaduct")

The "Tevere Viaduct" is 1,424.86 m long; it is placed above the alluvial deposits belonging to Tiber River System, covered by recent alluvial soils. Here the substrate is characterized by a "steps" conformation, due to the presence of a series of faults, aligned NNE-SSW, that have displaced it, causing its deepening from a depth of about 25–30 m ad the edge of the main valley, near to the confluence of Fosso del Torrino, towards west, where it lies regularly at more than 65–70 m.

The viaduct is divided into two parts, a first composed by 13 spans variable from 30 to 150 m, while the second one is composed by 11 spans from 30 m (for the spans of the shore) to 40 m (for intermediate spans).

For the first part of the bridge, the foundations are direct compensated or indirect with driven piles, outside the embankment of the river; they are indirect, with diaphragms, inside the levees.

For the last part all foundations are indirect with large diameter bored piles (D = 1,500 mm); piles are circular for hydrodynamic problems, related to the presence of a River Tiber tributary (Fig. 109.1) (Calu).

The "Interconnessione Viaduct" stretches for about 2,250 m. The geological units under the viaduct are almost entirely represented by the alluvial deposits that extend in a uniform manner, for the whole extension of the viaduct. Here the substrate is regular and it's located below, at depths greater than 70 m.

This viaduct is composed by 62 spans on the northbound carriageway and 65 spans on the southbound carriageway, with variable ports from 26 to 126 m.

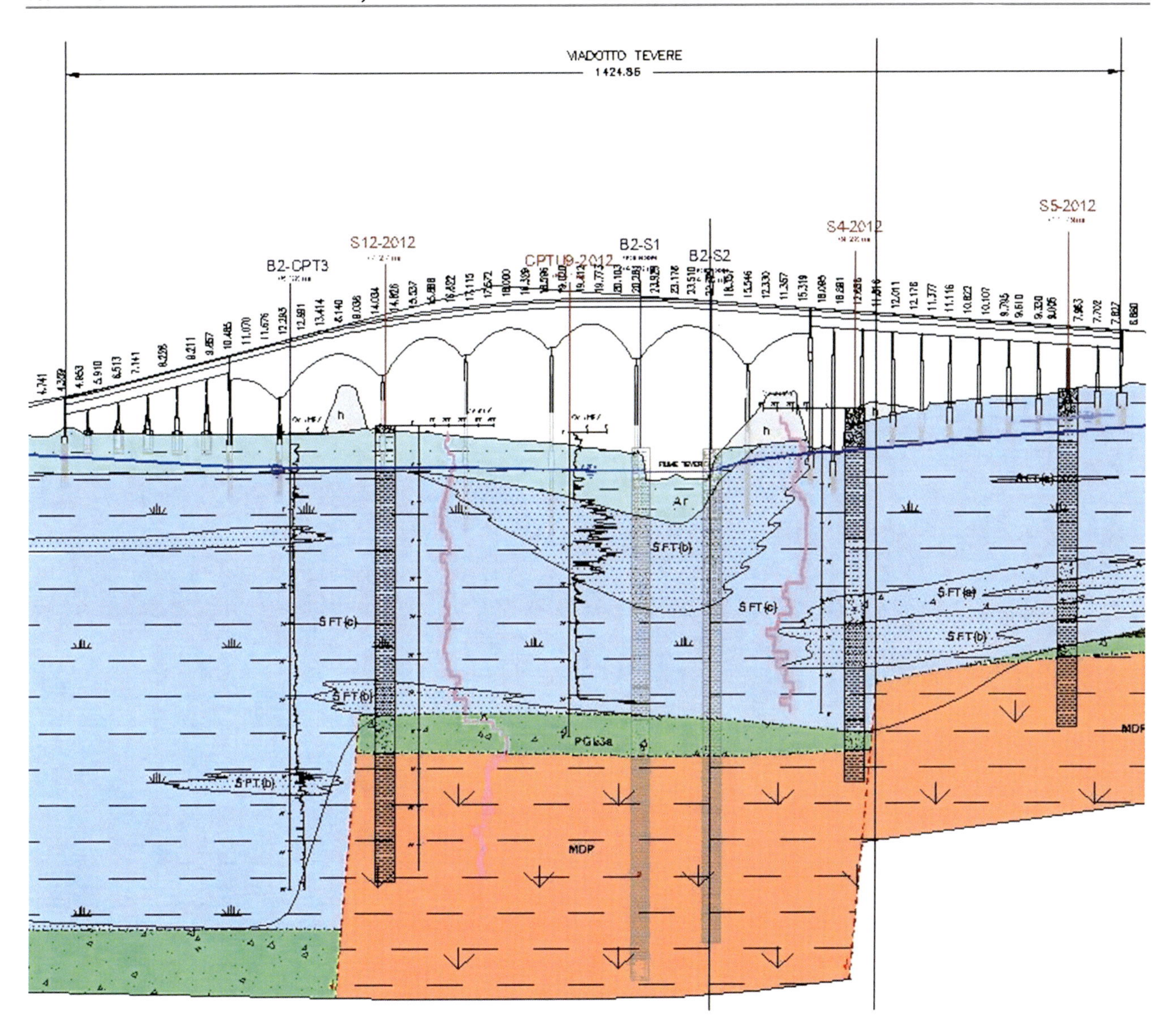

Fig. 109.1 Tevere Viaduct

In general, the foundations are direct, compensated, with protruding plinth on the terrain surface; in correspondence of spans more than 40/45 m wide, the foundations are not compensated, but deep, with piles of 70 m, for obvious load problems (Fig. 109.2).

109.5 Design Solutions

The high deformability and low resistance features of the upper deposits have influenced the design solutions developed along the axis road.

Therefore, different types of intervention, aimed at reducing the probable subsidence of the road body, also delayed in time, have been used. So, the study of geometry road body was marked by the reduction of overload transmitted, lowering the project level and the height of the embankments. At the same time some of the hightest embankment have been made of lightened material, in different ways for different traits: (1) insertion of metal pipes ARMCO type; (2) adoption of sintered expanded polystyrene (EPS) in preformed blocks; (3) use of expanded clay for the construction of the embankment. The application of a geogrid reinforcement has been envisaged to improve the resistance of the laying surface, below the remediation layer.

The foundational solutions, adopted for the principal viaducts, are also differentiated, related to the local context: compensated direct foundations, deep foundations of diaphragms, deep foundations of large diameter piles, deep foundations on beaten precast piles (Fig. 109.3).

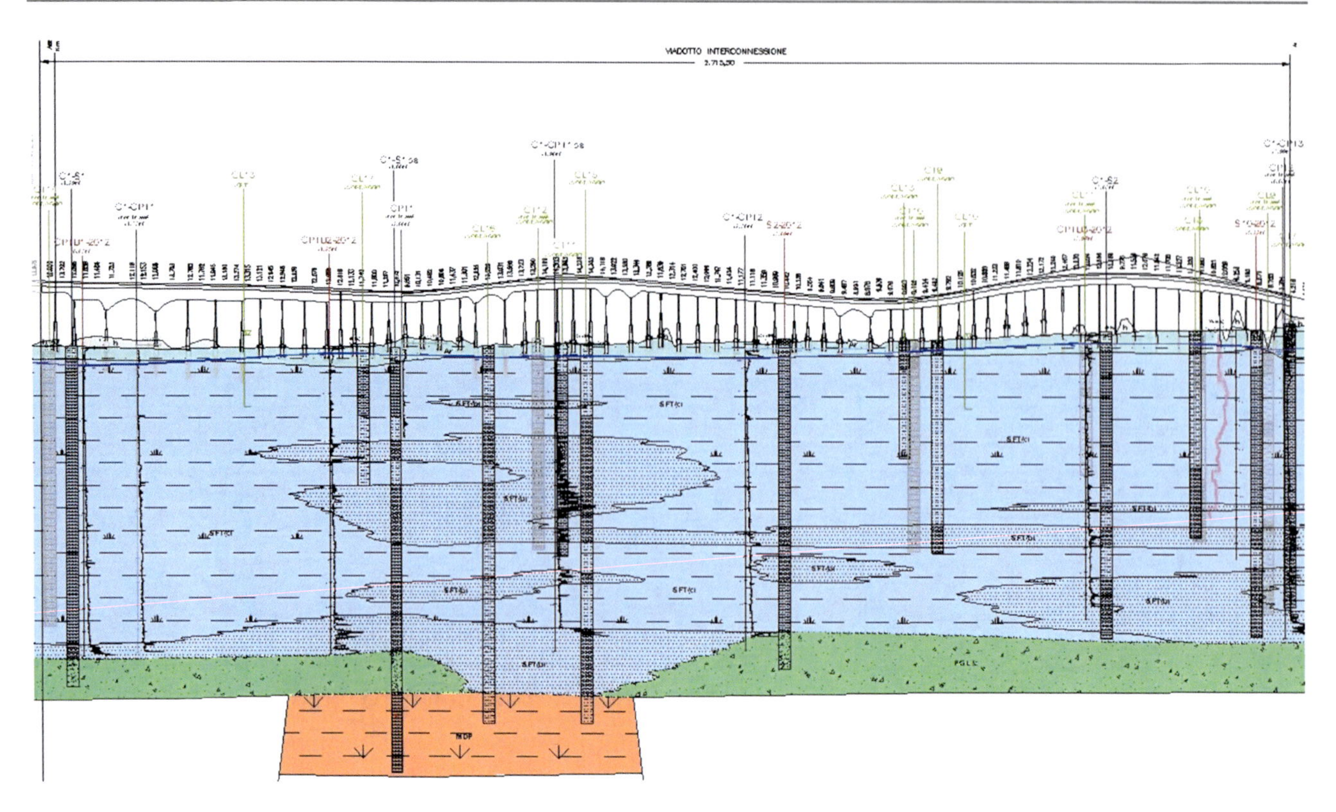

Fig. 109.2 Interconnessione Viaduct

Fig. 109.3 Example of section used

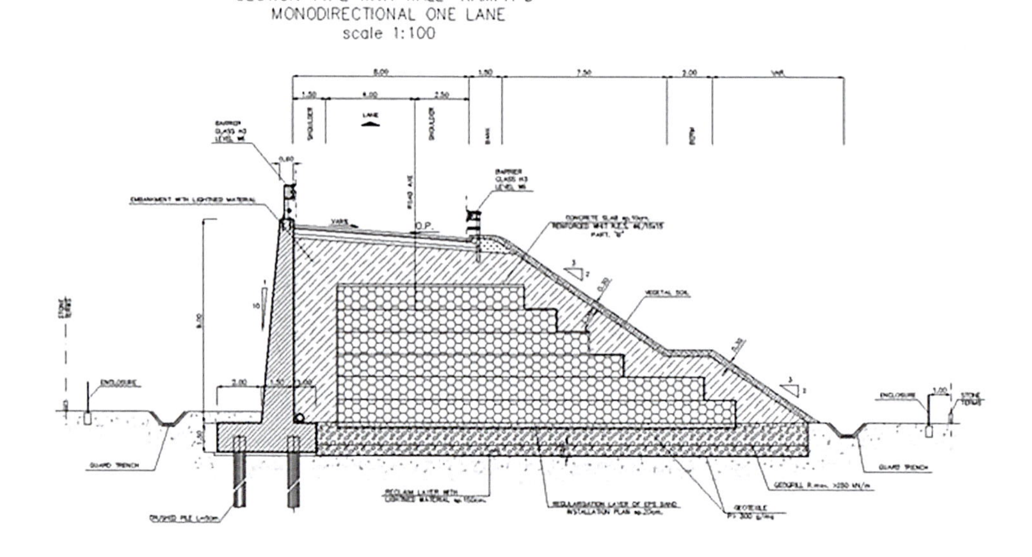

109.6 Conclusions

As is known, the evaluation of the characteristics of the soils on which a road infrastructure will be located has a fundamental importance. This is because, very often, there are geological complex situations and geotechnical soils with specific responses to stress.

The characterization of these soils derived from progressive cognitive insights and careful site investigation and testing on site and on laboratory on samples taken during the execution of different boreholes.

The study shows that, in the development of the various phases of the project, we have reached to a progressive deepening of the Reference Geological Model, based on the implementation of the knowledge framework. The modeling thus defined, according to the evolutionary context in terms of the gestructural and stratigraphic structure, has allowed to define in detail the particular aspects of the subsoil involved by the road project and its relationship with the marine units of the substrate. The subsoil, in fact, is characterized, mainly, by significant thicknesses of recent soils with poor geotechnical characteristics (low strength and high deformability). It has finally allowed to transfer to the geotechnical design all the necessary elements for the proper engineering design of interventions.

References

Conato V, Esu D, Malatesta A, Zarlenga F (1980) New data on the Pleistocene of Rome. Quaternaria 22:131–176

De Rita D, Faccenna C, Funiciello R, Rosa C (1995) The volcano of the Alban hills: stratigraphy and volcano-tectonics. In: Trigila R (ed) The volcano of the Alban hills

De Rita D, Funiciello R, Parotto M (1989) Carta geologica del complesso vulcanico dei Colli Albani ("Vulcano Laziale"). Progetto Finalizzato "Geodimamica"—Gruppo Naz. Vulcanologia, SELCA, Firenze, CNR Roma

De Rita D, Funiciello R, Rosa C (1988) Caratteristiche deposizionali della II colata piroclastica del Tuscolano-Artemisio (Complesso vulcanico dei Colli Albani, Roma), Bollettino GNV, IV, 278–297

Faccenna C, Funiciello R, Marra F (1995) Inquadramento geologico strutturale dell'area romana. Mem. Descr. Carta Geol. d'It., vol 50, La Geologia di Roma—Il centro storico, a cura di R. Funiciello

Funiciello R, Parotto M (1978) Il substrato sedimentario nell'area dei Colli Albani: considerazioni geodinamiche e paleogeografiche sul margine tirrenico dell'Appennino centrale. Geologica Romana, vol 17, pp. 233–287

ISPRA ex APAT—Università Roma Tre. Servizio Geologico d'Italia. Carta Geologica d'Italia in scala 1:50.000. Foglio 374 Roma

ISPRA ex APAT—Università Roma Tre. Servizio Geologico d'Italia. Carta Geologica d'Italia in scala 1:50.000. Foglio 387 Albano Laziale

Molin D, Castenetto S, Di Loreto E, Guidoboni E, Liberi L, Narcisi B, Paciello A, Riguzzi E, Rossi A, Tertulliani A, Traina G (1995) Sismicità di Roma. Mem. Descr. Carta Geol. d'It., vol 50, La Geologia di Roma—Il centro storico, a cura di R. Funiciello

Servizio Geologico d'Italia (1967) Carta Geologica d'Italia 1:100.000. Foglio 150—Roma

Società Geologica Italiana (1990) Guide Geologiche regionali, Volume V, Lazio. BE—MA Editrice

Ventriglia U (1971) La Geologia della città di Roma. Amministrazione Provinciale di Roma

Ventriglia U (1990) Idrogeologia della Provincia di Roma, Volume IV, Regione orientale. Amministrazione Provinciale di Roma

Ventriglia U (2002) Geologia del territorio del comune di Roma. Amministrazione Provinciale di Roma

The Geological Reference Model for the Feasibility Study of the Corredor Bioceanico Aconcagua Base Tunnel (Argentina-Chile Trans-Andean Railway)

110

Mattia Marini, Giuseppe Mancari, Antonio Damiano, Marta Alzate, and Michel Stra

Abstract

The Corredor Bioceanico Aconcagua base tunnel (CBA-BT) is a planned 53 km-long double pipe railway tunnel which will cross the Principal Cordillera of the Andean Thrust System under a maximum topographic cover of 2,000 m. The structural framework of the tunnel corridor is represented by a domino-style array of thrusts with ramp-flat geometry which detach along weak gypsum levels and stack diverse volcano-sedimentary terranes tectonically. As part of the CBA-BT feasibility study, the terranes crossed by the tunnel were mapped in detail and their component units were framed in a stratigraphic and palaeogeographic perspective overlooked by the earlier Geological Reference Model (GRM). Main results are: (i) the recognition of three different stratigraphic successions originally deposited in contiguous palaeo-domains of the Mesozoic Aconcagua Basin and (ii) the understanding of their mutual relationships; (iii) the description of thickness and lithofacies changes of the alluvial deposits of the Tordillo Formation; (iv) the re-interpretation of some gypsum levels, formerly referred to the Auquilco Formation, as part of the Mulichinco Formation (Mendoza Group). The implications of such advances on the GRM for a deep underground infrastructure are vast as they can help improve prediction of geological, structural and geomechanical data to excavation level, thereby contributing to implementation of future subsoil investigation, economic risk assessment and design optimization.

Keywords

Corredor Bioceanico Aconcagua • Base tunnel • Geological reference model • Andean Principal Cordillera • Volcano-sedimentary succession

110.1 Introduction

In feasibility-preliminary design of deep underground infrastructure the Geological Reference Model (GRM) is mostly intended to: (i) give a comprehensive picture of the infrastructure geology; (ii) share geological uncertainties with other professionals of the design team; (iii) contribute to decision making upon nature and location of integrative geognostic investigation aimed at tackling uncertainties. Although the reliability of the GRM of this early planning phases is generally low due to scarcity of direct geognostic investigations, the model consistency mainly resides in good understanding of surface geology and its successful extrapolation to excavation level.

Especially in sedimentary terranes with brittle tectonics, the unravelling of subsoil geology greatly benefit from detailed field mapping and correlation of terranes across thrusts with significant shortening. This is mainly because correlating consists in crudely restoring original spatial-temporal relationships between type-successions, thereby providing useful constraints for structural geology reconstructions and prediction of spatial heterogeneities in rock masses to be excavated.

M. Marini (✉) · G. Mancari · A. Damiano · M. Alzate · M. Stra
SEA Consulting s.r.l., Corso Bolzano, 14, 10121, Turin, Italy
e-mail: mattia.marini@uniroma1.it

G. Lollino et al. (eds.), *Engineering Geology for Society and Territory – Volume 6*,
DOI: 10.1007/978-3-319-09060-3_110, © Springer International Publishing Switzerland 2015

The case study of the Corredor Bioceanico Aconcagua base tunnel (CBA-BT hereafter; Argentina-Chile Trans-Andean railway) is an example of a deep-underground infrastructure crossing a thrust-belt which involves a complex and understudied volcano-sedimentary succession. This paper presents the results of geological-structural surface mapping conducted in the frame of the feasibility study of the CBA-BT along with implications for project implementation.

110.1.1 Project Description and Framework Geology

The CBA-BT is a 53 km-long double-pipe tunnel (presently under feasibility study) which will cross the Andean Thrust System along a East-West transect between the Santiago del Chile and Mendoza districts (Fig. 110.1).

The structural framework of the CBA-BT is related to the Andean orogenesis which started during the early Cretaceous in response to subduction of the oceanic crust of the Nazca Plate (eastern Pacific Ocean) underneath the South American Plate. Convergence of these plates caused tectonic inversion of pre-existing extensional faults related to rifting of the proto-Atlantic ocean which during the Mesozoic delimited a number of sedimentary basins, among which the Aconcagua-Neuquen Basin (Vincente 2006).

The study corridor of the CBA-BT is located in the central segment of the Andean Thrust System which from west to east can be subdivided into three main sectors, the thin-skinned Aconcagua Belt of the Principal Cordillera and the thick-skinned Frontal Cordillera and Precordillera (Ramos et al. 2004). The CBA-BT remains entirely within the Aconcagua belt, which consists of a domino structure of thrusts with ramp-flat geometry detaching the Mesozoic succession from its basement along weak levels of gypsum.

110.1.2 Methods and Materials

A geological-structural field survey was carried out during Spring 2009–Summer 2010 over a 4 km-wide corridor straddling the tunnel axis and the Rio del Las Cuevas Valley (Fig. 110.1). All the field data were collated in a geographic information system repository which allowed spatial queries and accurate drawing of geo-structural maps and cross-sections.

The study corridor was first divided into four homogeneous geological-structural domains (Fig. 110.1) according to surface geology and then characterized acquiring a number of geomechanical stations, thereby providing a comprehensive picture of the full range of rock mass types at excavation level.

The GRM was then improved thanks to the 3D visualization of the project geology achieved building a fence-diagram (Fig. 110.2) from longitudinal and serial transverse cross-sections.

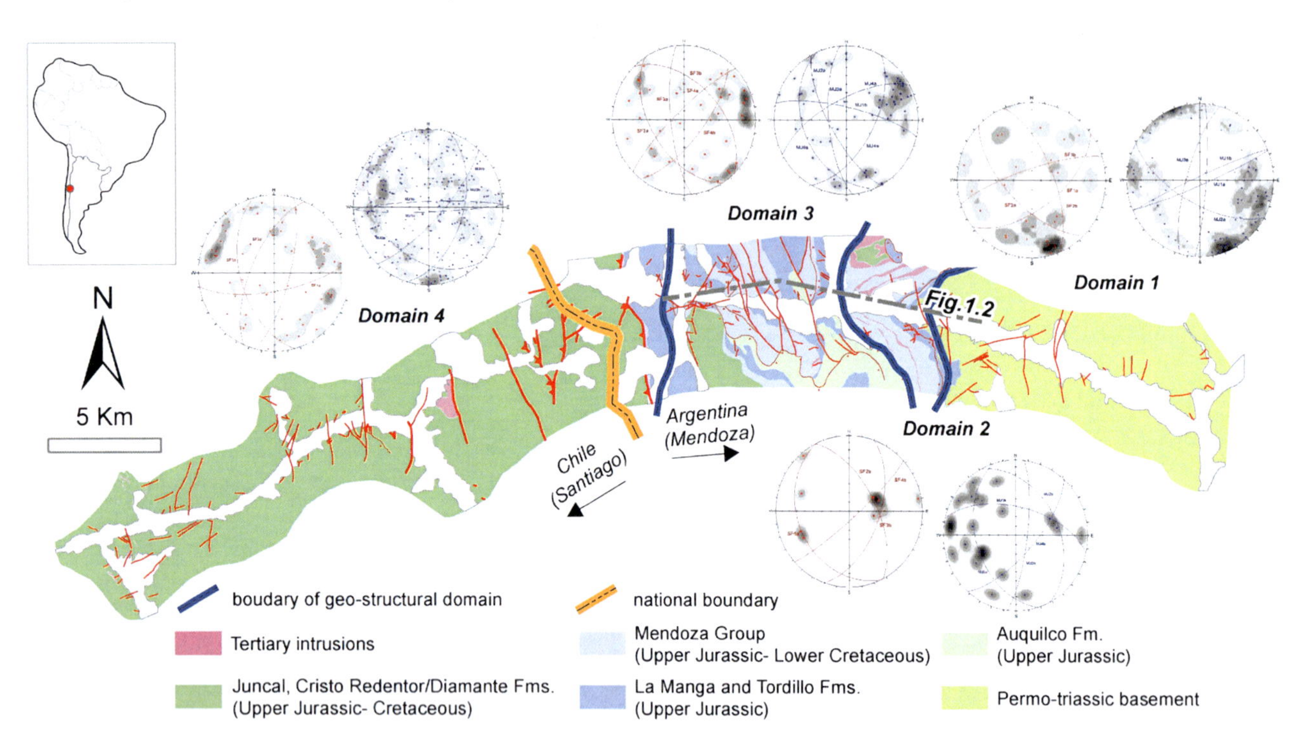

Fig. 110.1 Simplified geological map of the study area and stereographical projections of main faults (*left*) and master joints (*right*) for each geo-structural domain

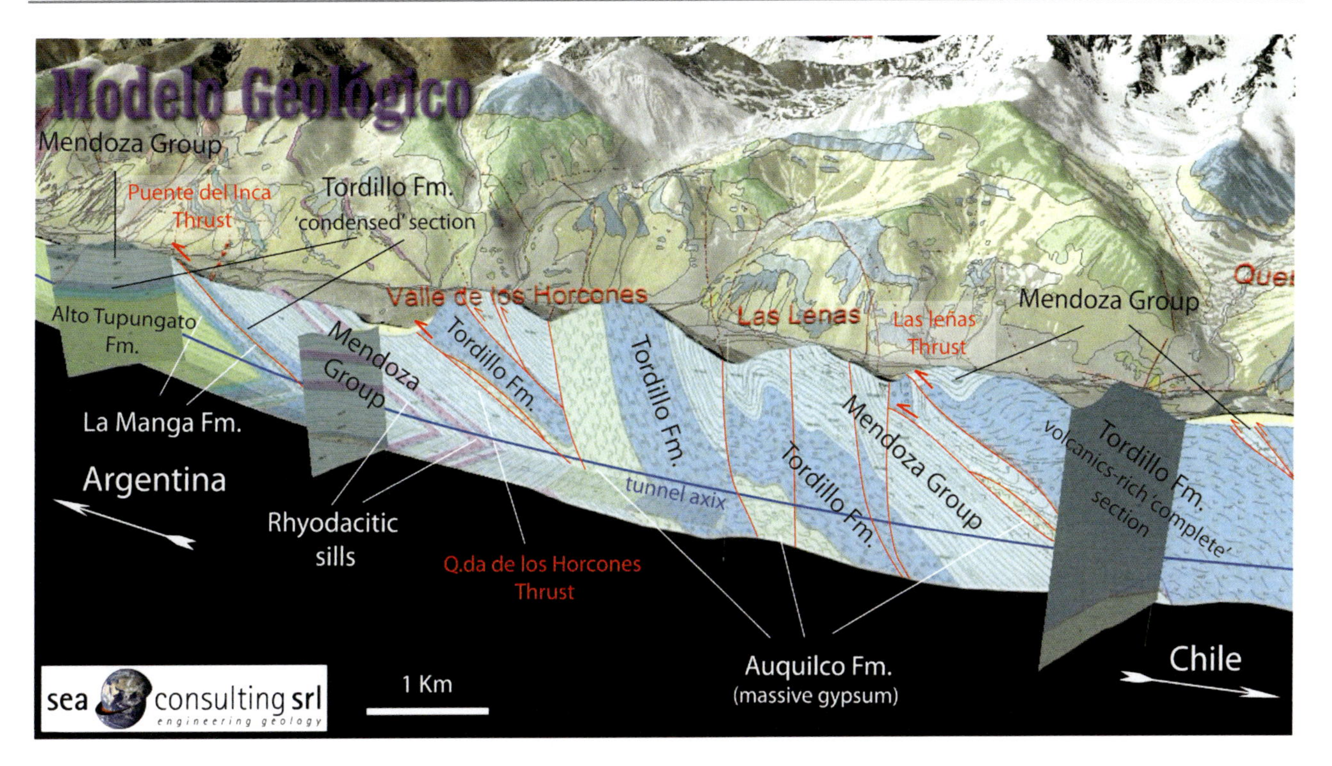

Fig. 110.2 Geo-structural fence diagram of the Argentinean stretch of CBA-BT. Note the thickness and lithofacies changes of the Tordillo Fm. across thrusts and the structural position of the Auquilco Fm

To assess the potential for water inflow in the tunnel, the geology was typified into hydrogeological complexes (i.e. fractured or porous media with homogeneous permeability characteristics) and elements (permeability fairways/boundaries, infiltration/recharge areas etc.) based on literature data from analogue scenarios (Dematteis et al. 2006; Perello et al. 2007). In addition, the reconstruction of deep flow paths was conditioned to the geochemistry of selected springs, including S_{34} and ^{18}O isotopic analysis on samples from the highly-mineralized waters of Puente del Inca and Agua Salada springs.

110.2 Results

Three different sedimentary successions deposited in as many palaeo-domains of the Aconcagua Basin were recognised in the study area making an advancement on the earlier GRM. These are, from east to west and from lower to upper in the thrust belt edifice: (i) a 'condensed', autochthonous succession (Puente del Inca Thrust sheet, domain 2 in Fig. 110.1) sitting on its Permo-Triassic basement (domain 2) and comprising, from bottom to top, marls and marly limestones (30 m; La Manga Fm., Oxfordian), alluvial conglomerates and sandstones (50–70 m; Tordillo Fm., Kimmeridgian) and a thick package (800 m; Mendoza Group, Tithonian-Neocomian) composed of shelfal limestones (Quintuco Fm.) alluvial conglomerates and marginal-marine limestones (Mulichinco Fm.) and thin-bedded marls, silts and sandstones (Agrio Fm); (ii) an allocthonous succession (Quebrada des Horcones and Las Leñas Thrust sheets, domain 3 in Fig. 110.1) detached along the massive gypsum of the Auquilco Fm. (Kimmeridgian) and comprising a thicker, volcanic clasts-rich Tordillo Fm. (300–500 m), which passes upward to restricted-marine organic-rich marls and shales intercalated in their uppermost part by gypsum and andesitic lavas; (iii) a clastic-volcanic succession (Quebrada Navarro and Las Quevas Thrust sheets, domain 4 in Fig. 110.1) composed of alluvial to marginal marine sandstones and volcanic clast-rich conglomerates with frequent lava intercalations (Tordillo, Quintuco and Mulicinco Fms.), shelfal limestones (Agrio Fm.), andesitic volcanics (Juncal Fm. Upper Jurassic-Cretaceous) and continental clastics (Cristo Redentor/Diamante Fm., Upper Cretaceous). In the Chilean side of the study corridor the Mendoza Group is not present as it passes laterally to the Juncal Fm.

By a structural standing point, the lowermost thrust in the structural edifice is the Puente del Inca Thrust (left-hand side of Fig. 110.2) which doubles the La Manga Fm.-Tordillo Fm. section and shows no evidence of gypsum at surface. Moving upward in the edifice, evidences of the Auquilco Fm. gypsum as detachment level are found in the Quebrada des Horcones and Las Leñas Thrusts (central part of

Fig. 110.2). Between these two thrusts, the geological structure is further complicated by NW-SE sinistral transpressive faults. Further up in the edifice, the tectonic shortening due to the Quebrada Navarro Thrust and Las Cuevas Thrust is relatively small and encompasses for the tectonic thickening of the Tordillo Fm. and its overthrusting by the Juncal Fm., respectively (right-hand side of Fig. 110.2). In the Chilean side of the project, due to the rheology of the Juncal Fm. the total shortening is even smaller and mainly accommodated by conjugated systems of NE-SW and NW-SE transpressive faults. At surface, the rock mass around these faults generally show evidences of hydrothermal alteration, which suggests they may potentially have a major role in underground water circulation. Also, these faults are locally intruded by trachi-dacitic and tonalitic lavas of Miocene age. Intrusive bodies are present on the Argentinian side as well in form of a granitoid dome at Los Penitentes and gently discordant rhyodacitic sills crosscutting the Puente del Inca Thrust sheet.

Lastly, geochemistry of water samples from Puente del Inca and La Salada springs confirm the model of Ramos (1993) which entails circulation of meteoric waters along relatively deep flow paths within gypsum and anidrithe masses associated to thrusts and resurgence along faults.

110.3 Discussion and Conclusions

Mapping of the lithofacies and thickness changes within the Mesozoic sedimentary succession of the Aconcagua Basin was undertaken as part of a detailed geological-structural field survey of the CBA-BT project corridor. The results are consistent with stratigraphic and palaeogeographic data from previous literature. In particular, two different sections within the Tordillo Fm. were recognized along the tunnel trace consisting of: (i) a easterly, 'condensed' section dominated by reddish conglomerates and sandstones (Puente del Inca Thrust sheet) and (ii) a westerly 'complete' section, previously attributed to the Juncal Fm., which thickens and is enriched in volcanic clasts toward the west. Moreover, the uppermost conglomerates of the Puente del Inca Thrust sheet were reinterpreted as part of the Mulichinco Fm. and correlated to the gypsum-bearing deposits of the Quebrada des Horcones Thrust sheet. These amendments to the earlier GRM allowed revaluating some tectonic features (e.g. inverse faults vs. thrusts) as well as the extrapolation of geology to excavation level (e.g. the occurrence of massive gypsum of the Auquilco Fm. and conglomeratic facies of the Tordillo Fm.) and deep flow paths.

The case study of the CBA-BT presented in this paper shows how a structural-geological survey can help improve prediction of the along-tunnel distribution of rock mass quality, deformation styles and water inflow. Geological-structural field studies can therefore provide a key contribution to the implementation of future subsoil investigation, economic risk assessment and design optimisation in deep underground infrastructure.

References

Aguirre-Urreta MB, Lo Forte GL (1996) Los depósitos tithoneocomianos. Geología de laregión del Aconcagua, provincias de San Juan y Mendoza, República Argentina. Dirección Nacional del Servicio Geológico, Anales 24:179–229, Buenos Aires

Dematteis A, Perello P, Torri R, Marini M, Venturini G (2006) Characterisation of permeability distribution for deep tunnel projects: examples from the Brenner and Lyon-Turin basis tunnels. In: Abstract volume, PANGEO AUSTRIA 2006, Innsbruck, Sept 2006

Perello P, Burger U, Torri R, Marini M (2007) Hydrogeological characterisation and water inflows forecasts for the Brenner Basistunnel. In: Abstract volume. Brenner Basistunnel und Zulaufstrecken, Innsbruck 1–2 Mar 2007

Ramos VA (1993) Geología y estructura de Puente del Inca y el control tectónico de sus aguas termales. Simposio sobre Puente del Inca. 12° Congreso Geológico Argentino and 2° Congreso de Exploración de Hidrocarburos (Mendoza), Actas, vol 5, pp 8–19

Ramos VA, Zapata T, Cristallini E, Introcaso A (2004) The Andean thrust system—latitudinal variations in structural styles and orogenic shortening. In: McClay KR (ed) Thrust tectonics and hydrocarbon systems, vol 82. AAPG Memoir, pp 30–50

Vincente JC (2006) Dynamic paleogeography of the Jurassic Andean basin: pattern of regression and general considerations on main features. Revista de la Asociación Geológica Argentina 61 (3):408–437

UHE Belo Monte: Geological and Geomechanical Model of Intake Foundation of Belo Monte Site

111

Jose Henrique Pereira and Nicole Borchardt

Abstract

The hydroelectric plant of Belo Monte, on the Xingu River, is located in northern part of Brazil, in the state of Pará and is characterized by a complex of dams, canals, dikes and reservoirs. The main set Intake /Power House will be deployed on site called Belo Monte. This paper presents the geological and geomechanical model developed for the foundation of the Intake in the Belo Monte Site. These studies have been developed within the consortium Intertechne-Engevix-PCE. For the development of the Project of Belo Monte was necessary to improve the geological and geomechanical model preliminarily designed in the previous steps of design to the structure of Intake in Belo Monte Site. The conception and development of this model was fundamental in the development of the stability studies for Intake, with reference to the presence of discontinuities that form the main systems of joints that cut the rock mass in that location. In developing this model were considered the data obtained from drill holes executed at various stages of design, geological and geotechnical mapping of the exposed foundations and field observations of excavation of provisional and definitive slopes. That information was the reference for the determination of the main systems of discontinuities that cut the rock mass in the region of the Intake structure. Subsequent analyzes have defined the main parameters that characterize such systems, particularly with regard to their spatial distribution geomechanical characteristics, persistence and spacing. It should be emphasized that the model presented in this work aims to define in general terms the behavior of the rock mass that serves as the foundation for Intake. However, the stability analysis of the structure was conducted block by block, as the foundations are exposed, and based on surface mappings and identifying the main features potentially conditioners of the stability of these blocks.

Keywords

Geological model • Geomechanical model • UHE Belo Monte • Intake

111.1 Geological Characterization of the Intake Foundation

According to the design criteria established, the Intake of Belo Monte Site must be founded on sound migmatite to moderately altered (A1/A2) belonging to stratigraphic unit Complexo Xingu. Thus, the occurrence of sedimentary rocks

J.H. Pereira (✉) · N. Borchardt (✉)
Intertechne Consultores S.A, Av. Iguaçu, 100, Curitiba, Brazil
e-mail: jhp.box@hotmail.com

N. Borchardt
e-mail: nb@intertechne.com.br

G. Lollino et al. (eds.), *Engineering Geology for Society and Territory – Volume 6*,
DOI: 10.1007/978-3-319-09060-3_111, © Springer International Publishing Switzerland 2015

(rhythmites) deposited over the crystalline basement and the migmatitic rocks with higher levels of alteration must necessarily be removed for the deployment of the structure.

111.1.1 Partitioning of the Rock Mass

The main systems of discontinuities occurring in the Intake region consists, mainly, of geological discontinuities with high dip associated with sub-horizontal discontinuities, which were characterized at various previous stages of project.

In these previous studies were identified three sets of joints potentially influential on the stability of the Intake, denoted by S1, S1′, and S6, for which were established, respectively, the following preferential attitudes: horizontal to 20°; EW/20–40°N and NW–EW/65–90°NE–SW.

On the final design with the exposure of bedrock of the crystalline basement, resulting from the excavations required for implementation of the project, preliminary and definitive mappings contemplating structural surveys and new boreholes were conducted to characterize and confirm the major systems of discontinuities and their preferential directions.

Based on these last data was prepared a stereogram of poles frequency for the discontinuities identified, and is presented in Fig. 111.1.

The analysis of the stereogram indicates that four major sets of joints occur in the region being called by S1, S2, S3, and S5. Addition to these main sets was identified other, less expressive, called by S1′. The spacial distribution of such sets, obtained from surveys carried out, can be summarized as shown in Table 111.1 below.

Table 111.1 Summary of the directions of the main sets of discontinuities

Sets of joints	Attitude
S1	Horizontal to 20°
S1′	NW/20°–45° NE
S2	NW/60°–90° NE or SW
S3	ENE/60°–90° NW or SE
S5	~NS/70°–90° E or W

111.1.2 Characterization of the Main Sets of Joints

111.1.2.1 Subhorizontal Set of Joints: S1

This set of joints is characterized by features of low angle (horizontal to 20°) with dips preferentially toward NE. The features mapped on surface generally consist of discontinuities with millimeter openings to sealed, sometimes with the presence of altered material in thicknesses typically ranging from millimeter to centimeter. The walls of these features usually show up sound to slightly altered (oxidized), especially in the lower elevations, as can be seen on the discontinuities exposed in the ramp of penstocks, excavated between units 1 and 8. The joints of this system, identified in surface mapping and borehole analysis, shows walls usually rough with roughness index (JRC) usually around 10.

At the foundation of the Intake "steps" are likely formed in the rock mass for these discontinuities, most often as a result of the constructive process of excavation to obtain the dimensions of the project.

Generally, the discontinuities associated with this system have persistence of the order of 10–15 m, average spacing of

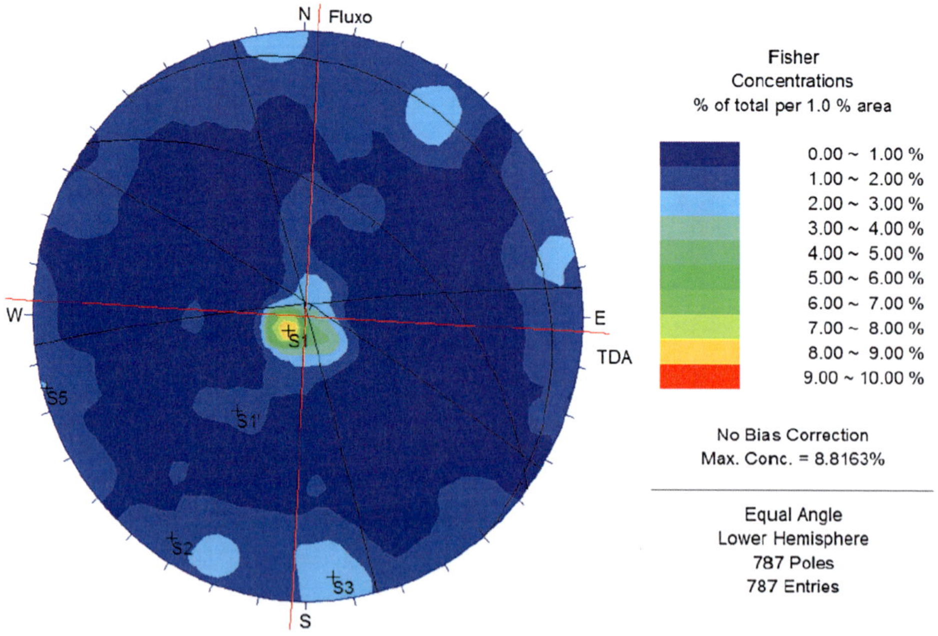

Fig. 111.1 Intake of Belo Monte site—overall stereogram of poles frequency

about 2 m, with local variations to lower numbers, especially near the top of excavated rock.

111.1.2.2 Vertical Set of Joints: S2 E S3

These sets have dips between 60° and 90° with ranging directions between NE–SW (S2) and NNW–SSE (S3). Are characterized by discontinuities generally planar to slightly undulating, with walls usually sound to slightly altered, rock to rock contact and roughness index (JRC) ranging between 8 and 12. The surveys performed in boreholes and surface mappings on the Intake foundation indicate that the features associated with this system have persistence between 8 and 10 m, with average spacing between the planes of discontinuities around 3.5 and 4.0 m.

The mappings performed in the Intake region have also identified some transcurrent faults with direction equivalent to set of joints S3 (ENE), but with persistence more expressive and high upstream dip (above 60°), present in the founding Intake and Power House region.

111.2 Geological and Geomechanical Model

The main sets of joints that influenced the model definition to Intake of Belo Monte Site are denoted by S1, S2 and S3, whose main features have been detailed in previous sections.

In Fig. 111.2, below, a isometric model of Intake excavation is presented, which are schematically represented the three main sets of joints that cut the rock mass. This guidance model indicates the spatial position of these sets and does not reproduce the persistence observed in the surveys carried out.

In this figure is observed that the discontinuities system S1 are arranged with main dips to northeast, which results in an attenuation of these angles in the flow direction. Whereas that the most frequent dips directions for this system vary between azimuths 40–60° (see Fig. 111.1), the apparent dips in the flow direction for this system range from 0 to 16°, taking as reference the true dips between 0 and 20°.

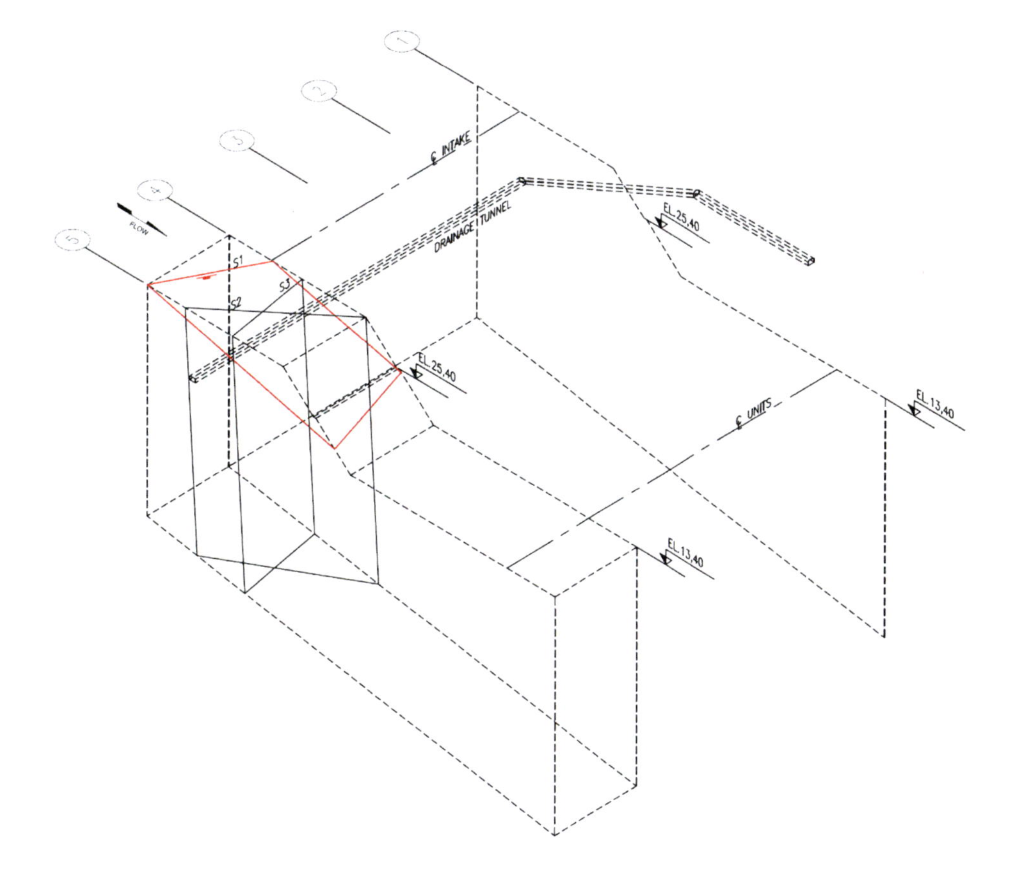

Fig. 111.2 Isometric model of intake excavation—spatial arrangement of the major sets of joints

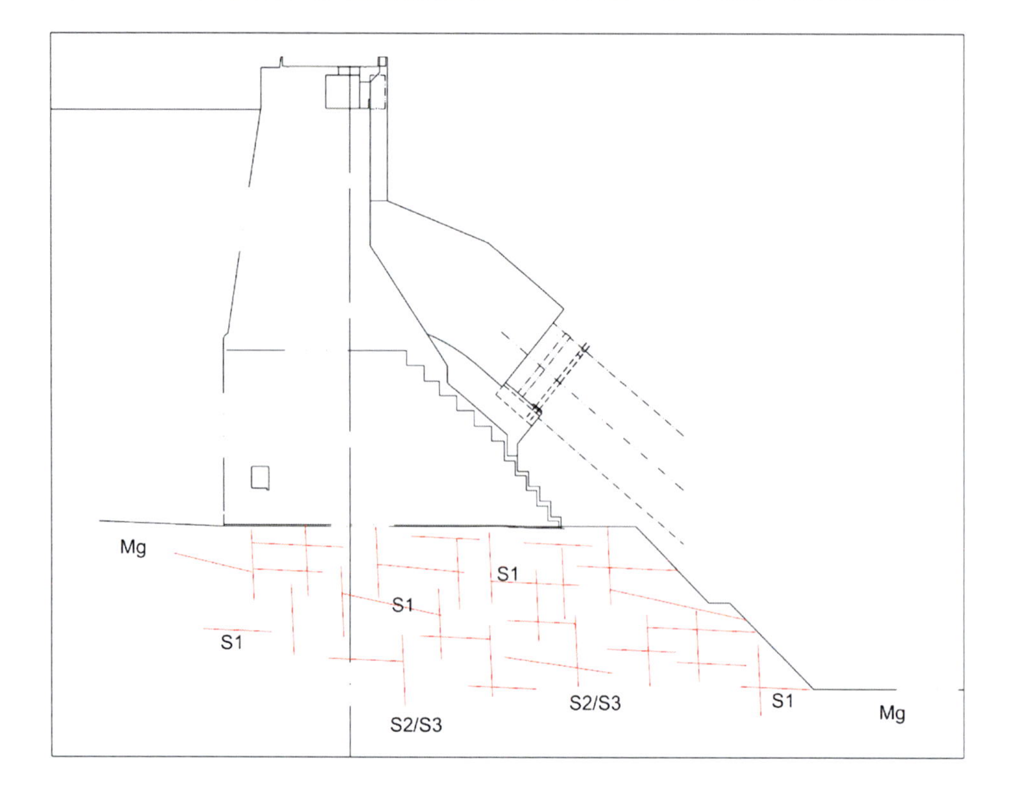

Fig. 111.3 Typical geological section of the founding of Intake showing the massive migmatitic cut by discontinuities systems S1 (subhorizontal) and subvertical sets S2/S3

Still based on surveys carried out, it was possible to establish the persistence and spacing of the main systems of discontinuities occurring in the Intake, as already detailed above.

The analysis of this information leads to the development of a geological model for the foundation of Intake where the rock mass is compartmented for vertical and subhorizontal discontinuities with persistence between 8 and 15 m and spacing ranging from about 2 m, for system S1, and about 4 m for the verticals (S2 and S3).

The spacing and the relative small persistence of discontinuities that cut the rock mass development a model as shown in Fig. 111.2, where the joints are interrupted and discontinuous along the projection of the structure of Intake, occurring that way significant "bridges" of sound rock interspersing the joint plans, giving to the rock mass a high strength envelope (Fig. 111.3).

Based on information currently available, from the mappings, and knowledge of the rock mass behavior, was definitely a strength envelope for the rock mass with reference to the failure criterion proposed by Hoek-Brown (2002). Based on that criteria, and using the computer program developed by Hoek (RocLab) was adopted the following parameters:

σ_{ci} = Uniaxial compressive strength of intact rock = 150 MPa (*)

GSI = Geological Strength Index = 70

mi = constant of the material (migmatito) = 29

D = disturbance factor in the face of the massive detonation process = 0, 7

Ei = modulus of deformation of the intact rock = 80 GPa (*)

(*)—average values obtained in tests on migmatite samples

Based on these parameters was obtained following strength envelope for the rock mass:

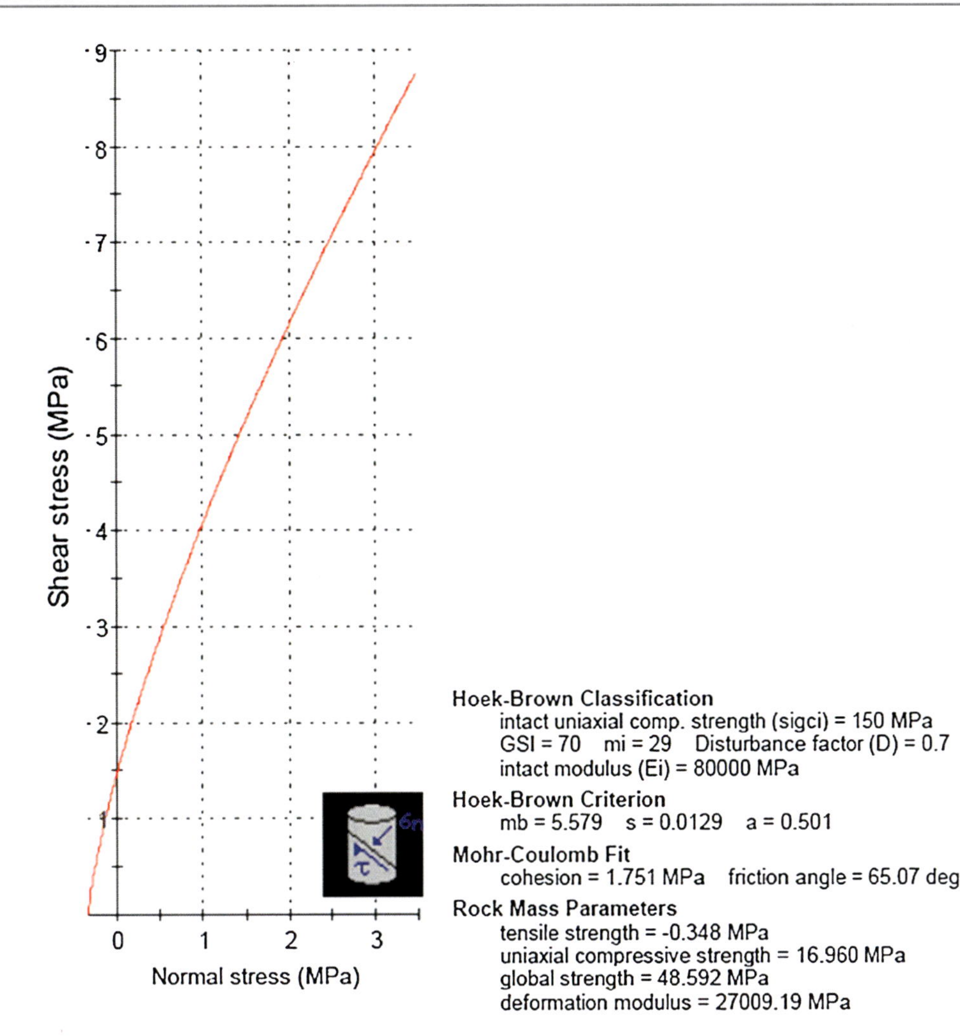

Based on this analysis the strength envelope for the rock mass is represented by the following equation:

$$\tau = 1,7\ \text{MPa} + \sigma_n.\text{tg}\ 65^\circ$$

For the evaluation of influence of the most persistent discontinuities identified by the mappings on Intake stability, should be considered specific parameters for the materials that constituting these features. In these analyzes should also consider the geological the Intake stability model and the results of evaluations of persistence and spacing defined for each system individually.

111.3 Conclusions

The conception of geomechanical model of the Intake of Belo Monte was of great value for characterization of the main systems of discontinuities present in the foundation of the intake and the definition of the main cinematic possibilities of rupture.

The subhorizontal systems were identified since the beginning of these studies as potentially destabilizing for the structure of intake and were object of special attention during the whole process of excavation and exposure of the foundation.

In order to ensure the safety factors adopt by the Project for the structure of the Intake, stability analyzes through the rock mass were conducted block by block, having as reference the discontinuities identified and characterized by geological mapping performed after exposure of the end surfaces of excavation.

The geomechanical characteristics of discontinuities that comprise those systems had variations with respect to the filler, opening, rock walls conditions and persistence, defining thereby changing parameters of cohesion and friction angle for each particular feature.

To estimate these parameters along the plane of the mapped features, was considered the formulation presented by Hoek (2007) for determining the instantaneous friction angle and cohesion, which were defined based on the estimates of the residual friction angle, joint coefficient roughness (JRC), joint compressive strength (JCS) and the normal stress applied in the plane of discontinuity.

Such analyzes have shown that the features of low angle, although potentially unstable, presented parameters and geomechanical characteristics appropriate to ensure the safety factors required by the project, even when considering their association with the vertical set of joints S2 and S3.

References

Barton N (1976) The shear strength of rock joints. Int J Rock Mech Min Sci Geomech. Abstr 19:255–279

Barton N, Choubey V (1977) The shear strength of rock joints in theory and practice. Rock Mech 1(2):1–54

Hoek E (1994) Strength of rock and rock masses. Int Soc Rock Mech News J 2(2):4–16

Hoek E, Carranza-Torres C, Corkum B (2007) Hoek-Brown failure criterion, 2002 edn. In: Proceedings of NARMS-TAC conference, Toronto, 1, 267–273

Hoek E (2007) In: Practical rock engineering. Shear strength of discontinuites, Chap. 4, 60–72

Geological Reference Model in the Design of the SS 182 "Trasversale Delle Serre": Ionian Calabria

112

Roberto Laureti, Serena Scarano, and Stefano Serangeli

Abstract

In road design, the Geological Reference Model plays an important role, especially in presence of complex geological conditions. In these areas, especially where is evident the role played by tectonics in the determination of geological-stratigraphic framework, the Geological Reference Model can be completed during various phases of the project, together with the improvement of the knowledge context. This case-history represents an example of application of this process. The first reconstruction of the model has been developed through the analysis of the available literature, especially referred to the geological, geomorphological and site-investigation data. After, a first survey phase has been completed. It represented the conceptual base to plan specific site-investigation, geophysical and laboratory survey, especially referred to the most geological and design critical situations. So that the Geological Reference Model has been gradually improved, highlighting particular tectonic structures, hardly detectable only by field geological and geomorphological survey. Finally, the design solutions, adopted for the civil works situated along the path, have been determined according to the identified geological conditions.

Keywords

Geological reference model • Geostructural reconstruction • Site-investigation • Serre chain

112.1 Introduction

As part of the upgrading of southern Italy road infrastructures, ANAS started the project to build the new S.S.182 "Trasversale delle Serre", located in the central-southern Calabria. The aim is to connect, from west to east, the two great coastal roads: the A3 Salerno—Reggio Calabria motorway, which runs along the tyrrenian coast, and the S.S. 106 "Jonica", along the Ionian sea.

The project road shown in this work is composed by two lots, 4 and 5, with a total length of 7.5 km, connecting the towns of Gagliato and Soverato.

R. Laureti (✉) · S. Scarano · S. Serangeli
Anas S.p.A., Design Management, Geological and Geotechnical Unit, Rome, Italy
e-mail: r.laureti@stradeanas.it

The land orography is, mainly, characterized by a series of mountain, which constitute the "Serre Chain". The geological and geomorphological characterization of the soils influences, considerably, the development of the civil works along the road and the design solutions used. These are represented by: 12 viaducts, 3 bridges, 3 artificial and 1 natural tunnels.

112.2 Reconstruction of the Geological Reference Model

During the design of the road the Geological Reference Model has been developed, and it has been continuously improved through several phases of deepening of the knowledge framework of the area, from the geological point of view (geological, geostructural, geomorphological, hydrogeological survey, and site-investigation). At the beginning of the road design activity, a first survey phase has

G. Lollino et al. (eds.), *Engineering Geology for Society and Territory – Volume 6*,
DOI: 10.1007/978-3-319-09060-3_112, © Springer International Publishing Switzerland 2015

been completed. It represented the conceptual base to plan specific site-investigation, geophysical and laboratory surveys, especially referred to the most geological and design critical situation

112.2.1 Site-Investigation Program

The site-investigation campaign consisted of 31 boreholes, 17 trenches, 13 Down-Hole tests and 2 Cross-Hole tests for seismic characterization of soils. On undisturbed and reworked samples, geotechnical laboratory and chemical tests, for the purpose of environmental characterization according to the new DM 161/12,[1] were carried out. In addition, in order to optimize the management of the soils coming from the excavation, with undeniable environmental advantages, the design has provided the reuse, in the construction of embankments, not only of the sandy-gravel formations, but also of the silty-clay soils, after lime-treatment. So that, in an experimental way, we conducted a study on the suitability of that soils, determining the performance characteristics of the mixtures studied.

112.2.2 Local Stratigraphic Sequence

The geological-structural model, as defined, is characterized by the presence of a Paleozoic granitic substrate, on which Plio—Pleistocenic continental and marine formations lie with unconformity.

The upper part of the road crosses a mountain ridge, elongated in east-west direction. It is composed by conglomeratic deposits, generated from the dismantling of the crystalline substrate and deeply affected by extensional tectonics, which has faulted the structure, causing a framework made by blocks lowered towards the Ionian sea. The conglomeratic deposits (Cg) consist of large and eterometric granitic and tonalitic blocks, with dimensions of 2–3 m, generally well rounded, embedded in a dark/brown sandy and sandy-coarse matrix of arkose nature. The conglomerates are moderately cemented and have interbedded by coarse sands.

Above, in stratigraphic continuity with them, there are the hold sands and sandstones (Ps) consisted of quartz—feldspathic/quartz-mica sands, from fine to coarse, with color varying from yellowish to brown, plane-parallel stratified, as well thickened to variously cemented.

At the top, the Plio-pleistocenic sequence ends with blue-gray hard clays, sometimes marly (Pa), which, mainly, outcrop in the lower part of the road, characterized by small hills, which often shows, above their slopes, landslides of various proportions (Fig. 112.1).

112.2.3 Particular Tectonic Situations

Situations of particular interest have emerged by the detailed geological analysis of the road design. Of course they affected, consequently, the design solutions. Along the fluvial incision of the Fosso Turriti, for instance, in correspondence of the confluence with the River Ancinale, the crystalline bedrock turns out stepped by a complex fault system, that created a series of buried structural blocks, elongated in WNW-ESE direction. The main buried horst consists of granite and it is covered by the pliocenic clay and sandy formations, laying under a thick alluvial deposit. In this stretch of the road design, there is a viaduct with a wide curve of about 120°, intersecting the tectonic structure along multiple directions.

The central piles of the viaduct are placed above the buried horst, made by the crystalline substrate (granite), whose roof, highlighted by the survey, is positioned at a depth of about 20 meters. Along an hypothetical section that cuts transversally the valley, it is evident that this block represents a tectonic "step", structurally lowered in comparison with the SW ridge, but raised compared to the NE sector, which forms the basis of the graben, where the river has set its course. Above the granite, the marine series is represented by the gray-blue clays (Pa) and, under that, by the sandy formation (Ps) with a maximum thickness of 6 m. Obviously the series appears here condensed and reduced, because of the control practiced by the morphology substrate on the sedimentation. The bedrock and the marine formations are covered by a thick alluvial deposit (above all sands and gravels) (Fig. 112.2).

112.3 Design Solutions

The critical situations derived from different local geological settings, highlighted by a detailed geological analysis, have suggested the adoption of different design solutions for the civil works along the road in project.

The most important viaducts, such as the "Turriti Viaduct" and the "Ancinale Viaduct", set to the tectonic structure analyzed before, have been provided with deep foundations in relation to the not exactly optimal geotechnical characteristics of the outcropping deposits.

For this purpose, the use of foundational wells, instead of large diameter piles, turns out to be more efficient also from the operational point of view. In fact, the strong heterogeneity of soils, especially of the alluvial and conglomeratic deposits, is a severe limit to executive technology of the "classics"

[1] Decreto del Ministero dell'Ambiente e della tutela del Territorio e del Mare 10 agosto 2012, n. 161 - Regolamento recante la disciplina dell'utilizzazione delle terre e rocce da scavo (G.U. n. 221 del 21 settembre 2012).

Fig. 112.1 Outcrop of the hold sands and sandstones

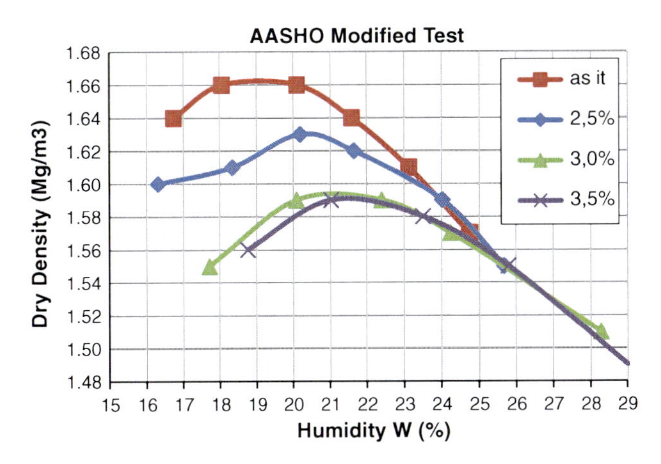

Fig. 112.3 Curve compaction of the AASHO mod. test performed on mixtures

structural foundational elements, without forgetting that the movement of machinery on the slopes requires the prior construction of site roads and non-trivial size temporary civil works. The geometric conditions of use of supporting structures, which are mostly set on strong transversal escarpment, require the use of flexible structures such as reinforced soil. Furthermore the achievement of the global stability conditions requires the realization of overseeing works on the foot, composed by "light" and resistant elements as the micropiles.

112.4 Lime Treatment of Clay Coming from Excavations

Clay materials, coming from the trenches excavations, will be reused in embankment, after their stabilization with lime, based on the studies specifically realized.

This study, in addition to determining the suitability to the treatment of the excavation soils, has compared the performance of different blends, comparing the variability fields of the most significant geotechnical parameters with the variation of lime percentage and the ripeness degree. The percentage of lime to be included was assessed from initial consumption of lime (ICL) +0.5 % up, forming 3 different mixtures with different lime content: 2.5–3.0–3.5 %. Each mixture was also tested for two different maturation states (after 1 day and after 14 days). In particular, from the AASHO mod. test performed on mixtures, we can observe a significant flattening of the compaction curve with the increase of the content in lime. It constitutes one of the main benefits for machinability of the treated material. The same test has shown how the intermediate mixture with 3.0 % of lime is those that provides the best performance benefit in relation to the amount of binder added, because the transition to the richer mixture does not provide an appreciable improvement of quality performance (Fig. 112.3).

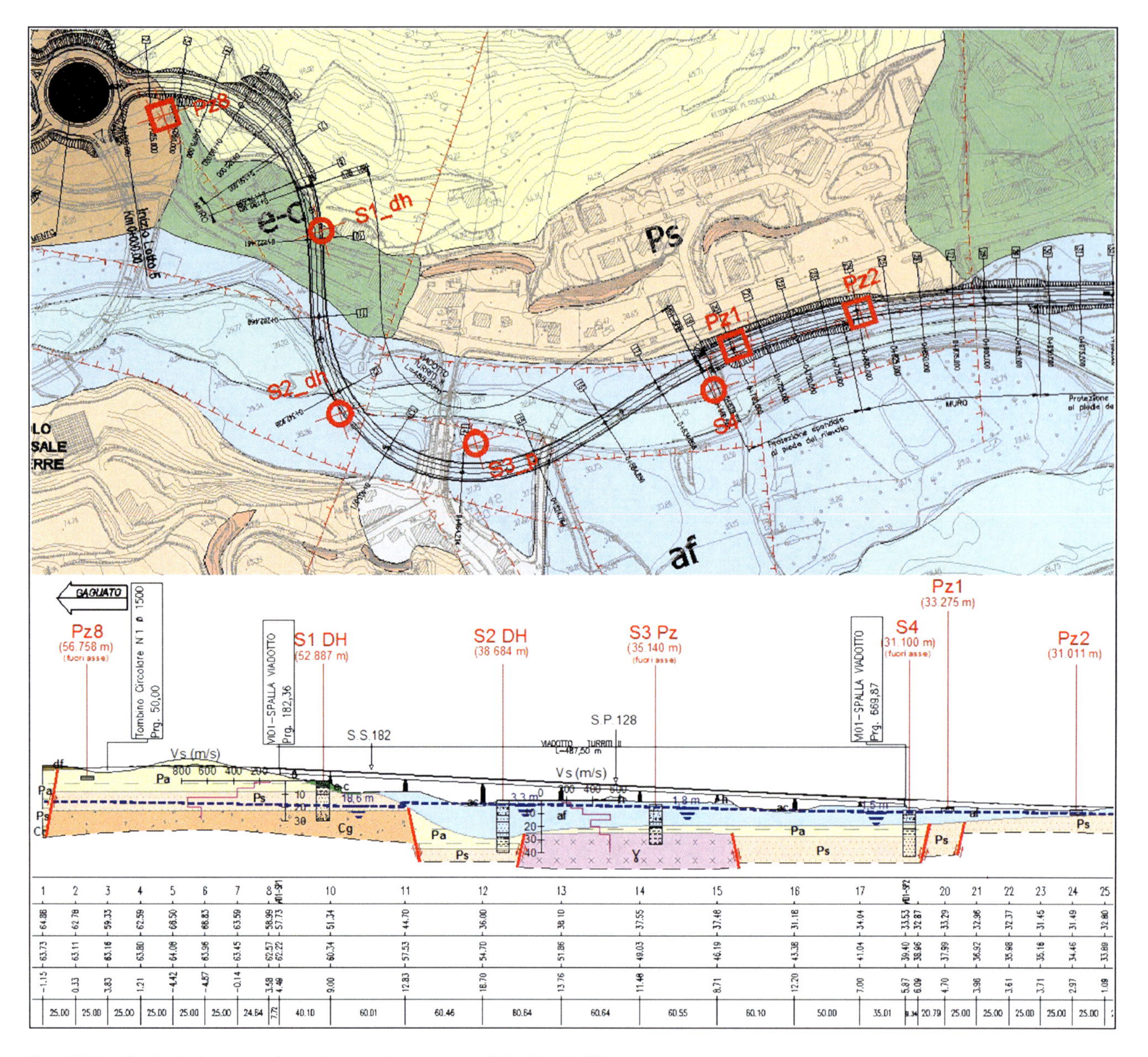

Fig. 112.2 Geological section along the tectonic structure of the Turriti Viaduct

112.5 Conclusions

Of course the design of a road infrastructure, of whatever extent, needs, from the early stages of study, a careful reconstruction of the reference geological model.

It allows, starting from the general geologic framework of the wide area, to have a detailed framework that permits to get the geotechnical characteristics of the soils involved by the road project and to make appropriate choices about the design elements to be introduced.

The definition of the geological reference model is obtained from the analysis of data provided by the geological surveys in the area or those suitably programmed. Particular situations, such as the geostructural setting through the Ancinale river, have been descripted by the deepening of the model, due to the site investigation data.

As described in this paper, sometimes, on the basis of specific design choices, specific studies are very important, such as the study of lime treatment, useful for the evaluation of reuse in embankment of clayey materials from excavations in the trenches.

References

AA.VV Capitolato speciale d'appalto parte II: Norme Tecniche—ANAS

Amodio Morelli L, Bonardi G, Colonna V, Dietrich D, Giunta G, Ippolito F, Liguori V, Lorenzoni S, Paglionico A, Perrone V, Piccaretta G, Russo M, Scandone P, Zanettin Lorenzoni E, Zuppetta A (1976) L'arco Calabro-Peloritano nell'orogene appenninico-maghrebide. Memorie della Società Geologica Italiana 17:1–60

APAT (2007) Rapporto sulle frane in Italia: il Progetto IFFI. Metodologia, risultati e rapporti regionali. Rapporti APAT, 78

Barbano M, Casentino M, Lombardo G (1980) Isoseismal maps of Calabria and Sicily earthquake. Work no. 341, CNR—Progetto Finalizzato Geodinamica, Catania

Cariboni Luigi (1996) L'impiego della calce nelle infrastrutture viarie —Le Strade, 9/1996

Cotecchia V, Guerricchio A, Melidoro G (1986) The geomorphic crisis triggered by the 1783 earthquake in Calabria (southern Italy). Atti del congresso "Engineering geology problems in seismic areas", Potenza, Napoli, 13–19 aprile 1986, 6, 245–304

Galadini F, Meletti C, Vittori E (2001) Stato delle conoscenze sulle faglie attive in Italia: elementi geologici di superficie. Risultati del progetto 5.1.2. "Inventario delle faglie attive e dei terremoti ad esse associabili", Gruppo Nazionale per la Difesa dai Terremoti (G.N.D.T.)

Ghisetti F, Vezzani L (1981) Contribution of structural analysis to understanding the geodynamic evolution of the Calabria arc (Southern Italy). J Struct Geol 3:371–381

Ghisetti F (1981) L'evoluzione strutturale del bacino plio-pleistocenico di Reggio Calabria nel quadro geodinamico dell'arco calabro. Bollettini della Società Geologica Italiana 100:433–466

Miyauchi T, Dai Pra G, Sylos Labini S (1994) Geochronology of pleistocene marine terraces and regional tectonics in the Tyrrhenian coast of south Calabria, Italy. Il Quaternario 7:17–34

Ogniben L (1973) Schema geologico della Calabria, in base ai dati odierni. Geol Romana 12:243–585

Scandone P, Giunta G, Liguori V (1974) The connection between the Apulia and Sahara continental margins in the Southern Apennines and in Sicily. Memorie della Società Geologica Italiana 13:317–323

Tortorici L, Monaco C, Tansi C, Cocina O (1995) Recent and active tectonics in the Calabrian arc (Southern Italy). Tectono-physics 243:37–55

Valensise G, Pantosti D (1999) The database of potential sources for earthquakes larger than magnitude 5.5 in Italy. EUG 10, volume degli abstract, pp 542–543

Part XI

Impacts of Environmental Hazards to Critical Infrastructures

Convener Dr. Paolo Frattini—*Co-conveners* Kyriazis Pitilakis, Carmelo Dimauro, Matteo M. Montini, Giovanni B. Crosta, Serena Lari

Territorial economic and social integration requires secure accessibility of people and goods through safe infrastructures, such as roads, railways and oil-gas pipelines. These are affected by a high number of hazards (earthquakes, floods, landslides, industrial explosion and transportation of dangerous goods) that can damage existing and planned infrastructures. Damage to infrastructures has also a clear transnational dimension, since the consequences of transportation disruptions in one region can propagate to other regions. However, this dimension is poorly accounted for in risk assessment and emergency response during disasters, due to the absence of a comprehensive and integrated resilience methodology and the lack of a joint strategy for risk management. The general objective of this session is to present contributions related to the assessment, prevention and efficient management of geological and geotechnical risks along the main existing and future critical infrastructures.

The Vulnerability Shadow Cast by Debris Flow Events

113

M.G. Winter

Abstract

Even in the absence of serious injuries and fatalities, rainfall-induced debris flow events can have significant socio-economic impacts when roads are affected. These include the cost of clean-up and repair, and of diversion. Costs also include those associated with the severance of access to/from remote communities for services and markets for goods; employment, health and educational opportunities; and social activities. The extent of such impacts depends upon the vulnerability shadow cast, which can be extensive and its geographical extent may be determined by the road network rather than the small footprint of the event itself. Indeed, such small events may (at most) directly affect a few tens of metres of road but cast a vulnerability shadow amounting to thousands of square kilometres.

Keywords

Landslides • Debris flow • Vulnerability • Area • Socio-economic impacts

113.1 Introduction

The risks associated with landslide hazards affect many parts of the world and many different cultures. The elements at risk may include infrastructure, public service buildings, commercial property and residential property as well as the occupants and users of such facilities. The type of element at risk and the vulnerability of those elements determine what might be described as a reasonable and proportionate response to a given risk profile.

In Scotland, rainfall-induced debris flow events often affect the strategic road network. Even in the absence of serious injuries and fatalities, such events have significant socio-economic impacts. These include the severance of access to and from relatively remote communities for services and markets for goods; employment, health and educational opportunities; and social activities in addition to clean-up, repair and delay costs. The types of economic impacts were summarized by Winter and Bromhead (2012) who also pointed out that the vulnerability shadow cast can be extensive and that its geographical extent can be determined by the transport network rather than the relatively small footprint of the event itself. The vulnerability shadow is presented here as a means of articulating the geographically dispersed impacts of landslide events.

M.G. Winter (✉)
Transport Research Laboratory (TRL), 13/109 Swanston Road, Edinburgh EH10 7DS, UK
e-mail: mwinter@trl.co.uk

113.2 Debris Flow in Scotland

Rainfall-induced debris flows affect the Scottish strategic road network on a regular basis (Winter et al. 2005, 2006, 2009). Figure 113.1 shows the 2007 event at the A83 route. The photograph is taken from the opposite side of the valley and even in this small area evidence of numerous past events can be clearly seen.

The mass movement comprised two discrete but related parts. First, the flow above the road commenced with a relatively small slide (or slides) into an existing drainage channel. This then triggered the movement of a large amount of marginally stable material in and around the stream channel of which an estimated 400 tonnes was deposited at road level. Second, this material blocked the open drain

G. Lollino et al. (eds.), *Engineering Geology for Society and Territory – Volume 6*,
DOI: 10.1007/978-3-319-09060-3_113,

Fig. 113.1 Debris flow of 03:00 h on Sunday 28 October 2007 at the A83 rest and be thankful. The head scar is at 370 m, A83 at 240 m and the old road at 180 m. Both recent (*lighter area*) and older (*darker*) erosion is visible below the main road and to the left of the recent scar

which carries water along the road to a series of culverts beneath. While the material from above the road had limited impact upon the slopes below the road, water diverted from the drain was channelled across and over the edge of the road causing some significant undercutting of the slope below and associated deposition further down the hill as can be seen in Fig. 113.1.

The culvert at this location was around 400 mm diameter and thus marginal for water flows let alone for debris. It is also clear that the culvert does not follow a straight path, which would reduce its capacity and increase the potential for blocking. Additionally, water flowed from the culvert at an angle to the hillside of considerably less than ninety degrees and caused the older erosion visible in Fig. 113.1.

The A83 site has long been very active and debris flow events and associated closures in 2007, 2008, 2009, 2011 and 2012 had an adverse effect on the travelling public. The area has become the focus of extensive landslide management and mitigation activity (Anon 2013; Winter and Corby 2012).

Major injuries have not resulted from rainfall-induced debris flow events that have affected the Scottish strategic road network for some years. The impacts of such events are socio-economic, particularly the severance of access to/from remote communities. Substantial disruption is experienced by local and tourist traffic, and goods vehicles as a result of such events. The maximum (July/August) daily traffic levels at the A83 Rest and be Thankful are 5 to 6,000 vehicles (AADT). The minimum traffic levels in January and February are roughly half the maxima.

The major contribution that tourism and related seasonal industries make to Scotland's economy means that the impacts of summer events are particularly serious, but the impacts of winter events should not be underestimated.

113.3 Economic Impacts

The economic impacts of landslide events as they relate to roads or other forms of transport infrastructure were summarized by Winter and Bromhead (2012) in three categories as follows.

Direct economic impacts include the direct costs of clean-up and repair/replacement of lost/damaged infrastructure in the broadest sense and the costs of search and rescue. These are relatively straightforward to estimate for any given event.

Direct consequential economic impacts relate to disruption to infrastructure and are really about loss of utility. For example, the costs of closing a road (or implementing single-lane working with traffic lights) for a given period with a given diversion, are relatively simple to estimate using well-established models. The costs of fatal/non-fatal injuries may also be included here and may be taken (on a societal basis) directly from published figures. While these are set out for the costs of road traffic accidents, or indeed rail accidents, there seems to be no particular reason why they should be radically different to those related to a landslide as both are likely to include the recovery of casualties from vehicles. Indeed, for events in which large numbers of casualties may be expected to occur data relating to railway accidents may be more appropriate.

Indirect consequential economic impacts relate to the loss of opportunity and potential loss of confidence. The economies of remote rural areas can be particularly dependent upon transport and if a route is closed for a long period then confidence in local business can be affected. Manufacturing and agriculture are a concern as access to markets is constrained, the costs of access are increased, and business profits and viability are affected. There may also be impacts on tourist (and other service economy) businesses. It is important to understand how the reluctance of visitors to travel to and within landslide areas is affected after an event that has received publicity and/or caused casualties and how a period of inaccessibility (reduced or complete) affects short- and long-term travel patterns to an area for tourist services. Such costs are a fundamental element of the overall economic impact on society and are thus important to governments as they should affect the case for the assignation of budgets to landslide risk mitigation and remediation activities. However, these are also the most difficult costs to determine as they are generally widely dispersed both geographically and socially.

113.4 Vulnerability Shadow

The vulnerability shadow (Winter and Bromhead 2012) is closely linked to economic impacts and determines their extent and magnitude. The shadow cast can be extensive and its geographical extent can be determined by the transport network rather than the relatively small footprint of the event itself.

In the case of the A83 the event itself was of the order of around 400 m^3 with a footprint that closed a few tens of metres of the road. The vulnerability shadow has been evaluated using knowledge of the local transport networks and the socio-economic activity associated with the network that has been built up over a period of 25 years. This includes an

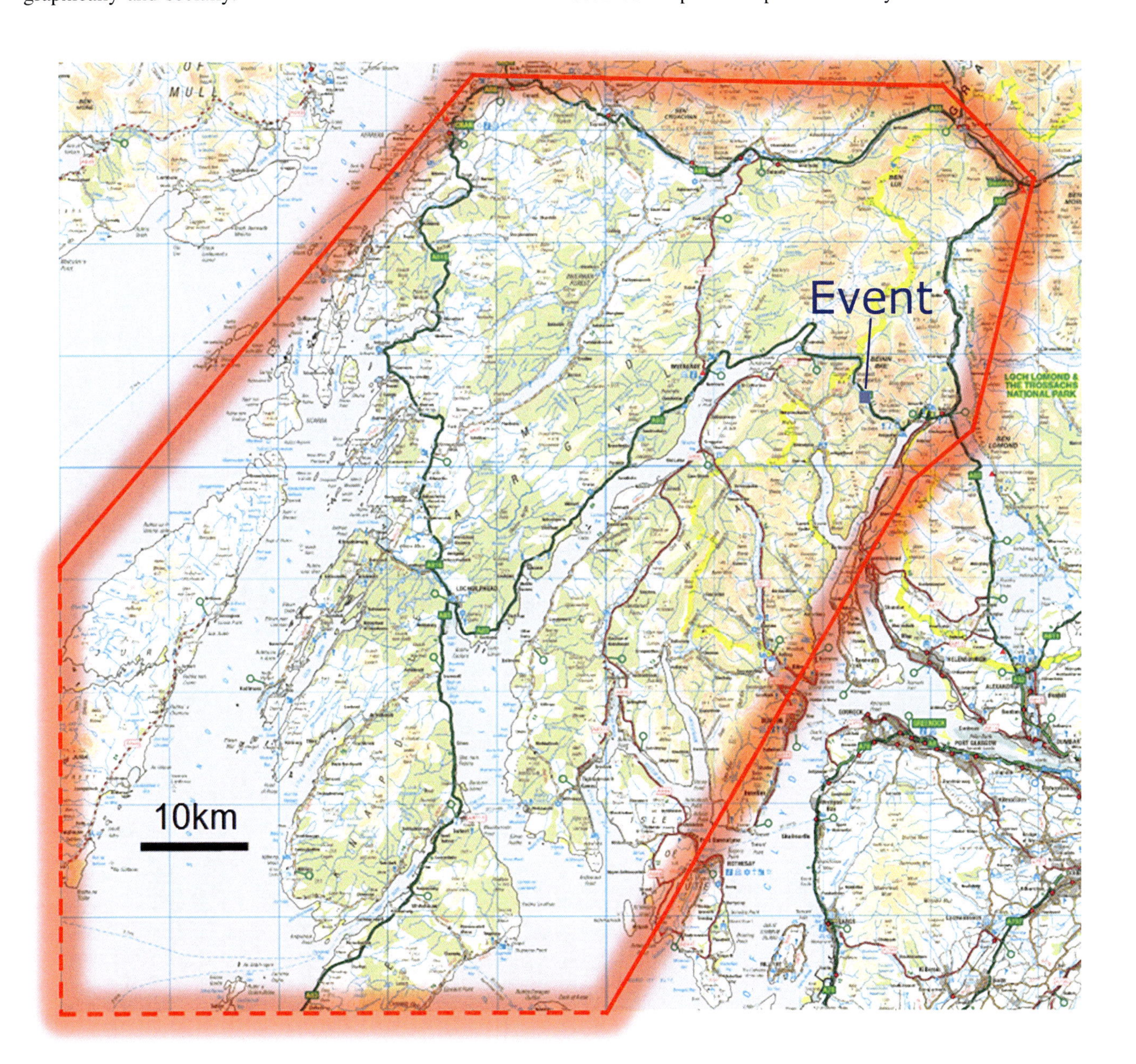

Fig. 113.2 A relatively small debris flow event (*blue square*) closed the A83 at the rest and be thankful in October 2007; the vulnerability shadow cast (bounded in red) was extensive (Winter et al. 2013; Winter 2013)

holistic evaluation of major nodes, origins and destinations and includes both experience and knowledge gleaned from formal surveys (e.g. Winter et al. 2013). The vulnerability shadow was thus estimated (Fig. 113.2) to be of the order of 2,800 km^2 (allowing 20 % for areas of sea). The area has a population density of approximately 13/km^2 (www.argyll-bute.gov.uk). Tentatively, the event may have had an economic impact upon approximately 36,400 people in Argyll and Bute, plus any transient (e.g. tourist) population.

Hong Kong SAR has an average population density of around 6,500/km^2 (www.gov.hk) which dictates a much greater transport network density. Thus, and for the sake of comparison, in order to have an economic impact on the same number of people the vulnerability shadow cast need only be approximately 5.6 km^2 (2 km by 2.8 km). It is not suggested that the economic impacts would be similar for events with vulnerability shadows of these diverse sizes in Argyll and Bute and Hong Kong. However, it is clear that the low density/dispersed network in Argyll & Bute dictates a large vulnerability shadow while the much more dense/less dispersed network in Hong Kong means that vulnerability shadows will be small, with the exception of events that affect critical infrastructure corridors, as more alternatives routes will exist and will be more proximal to the event.

A landslide on the B1 route in the Blue Mountains of Jamaica effectively severed the local coffee production industry from the most direct route to the international market for this high value product. As such a single landslide event placed severe constraints on the economy of the Blue Mountains. Again, while the footprint of the actual event was relatively small, the vulnerability shadow was projected over a much greater area creating tangible economic and social losses.

113.5 Conclusions

Linear infrastructure such as services, communications and transport networks presents significant risk factors including the near-continuous temporal occupation of the hazard zone, high vulnerability to damage, adverse orientation (as often the design demands limited gradients). The construction of the infrastructure itself may also increase landslide susceptibility (e.g. cuts, fills, interruption of groundwater flow, concentration of water).

The vulnerability shadow for socio-economic factors may be projected over a much greater area (orders of magnitude) than the event footprint. It is largely the nature, extent and density of the transport network that determines the extent of the vulnerability shadow cast. This, in turn, determines the magnitude of the consequential economic impacts. In a sparsely populated area such as western Scotland the transport network is dispersed and the vulnerability shadow is likely to be large, while in an area with a dense network, such as Hong Kong, a similarly sized event is likely to produce a vulnerability shadow orders of magnitude smaller.

The vulnerability shadow is presented here as a means of articulating the geographically dispersed impacts of landslide events.

References

Anon (2013) A83 Trunk road route study: part A—A83 rest and be thankful. Final report. Report prepared by Jacobs for Transport Scotland, 212 p. http://www.transportscotland.gov.uk/road/maintenance/landslides. Accessed Mar 2013

Winter MG (2013) A classification scheme for landslide management and mitigation. Int J Landslide and Environ 1(1):123–124

Winter MG, Bromhead EN (2012) Landslide risk—some issues that determine societal acceptance. Nat Hazards 62:169–187

Winter MG, Corby A (2012) A83 rest and be thankful: ecological and related landslide mitigation options. Published Project Report PPR 636. Transport Research Laboratory, Wokingham

Winter MG, Macgregor F, Shackman L (eds) (2005) Scottish road network landslides study, 119p. The Scottish Executive, Edinburgh

Winter MG, Heald A, Parsons J, Shackman L, Macgregor F (2006) Scottish debris flow events of August 2004. Q J Eng Geol Hydrogeol 39(1):73–78

Winter MG, Macgregor F, Shackman L (eds) (2009) Scottish road network landslides study: implementation. Transport Scotland, Edinburgh 278 p

Winter MG, Kinnear N, Shearer B, Lloyd L, Helman S (2013) A technical and perceptual evaluation of wig-wag signs at the A83 rest and be thankful. Published Project Report PPR 664. Transport Research Laboratory, Wokingham

Active Faults at Critical Infrastructure Sites: Definition, Hazard Assessment and Mitigation Measures

114

Alexander Strom

Abstract

A definition of active faults crossing critical lifelines that uses both time interval during which movements have occurred and the threshold displacement/rate value that require special technical measures to ensure construction safety, is proposed. Accuracy of active faults delineation and criteria used for their identification depends on type of hazard—while active faults considered as causative tectonic structures can be derived based on indirect evidence, faults passing directly across construction site or lifeline route must be proved by direct evidence of past offsets. Ability of different types of engineering structures to sustain fault displacements should be considered both for site investigations and mitigation measures planning. The proposed approach requires better interconnection of earth scientists and design engineers al all stages of project implementation.

Keywords

Active fault • Surface rupture • Displacement • Lifeline • Trunk pipeline

114.1 Introduction

One of the strict requirements when selecting engineering structures' placements or routes is to avoid sites that can be crossed by active faults. While buildings, dams, or factories can be shifted to safer site if such fault would be identified, routes of critically important lifelines such as trunk pipelines, roads, railroads often can not bypass active faults and special measures should be anticipated to ensure structures' safety and/or operating capacity. Effectiveness of such measures was proved at the Trans-Alaska oil pipeline crossed by the active Denali fault. Special studies provided reliable input data allowing proper design of fault crossing that sustained ~6-m lateral offset during large (M 7.9) 2002 Denali earthquake (Haeussler et al. 2004).

From the engineering point of view, an active fault is a boundary between more or less rigid blocks moving against each other. Motion could be either continuous with constant or variable rate, or episodic, associated with earthquakes and have dual effects on engineering constructions and lifelines. First, active faults are the main type of causative tectonic structures responsible for large earthquakes which strong motion affects vast areas and must be considered for hazard assessment even if they pass distant from the construction site. On the other hand, they can affect engineering utilities directly. In both cases definition of what should be considered as an "active fault" is critical, since delineation of such feature at or close to the construction site or lifeline route implies expensive protection measures. Inadequate attribution of faults to the "active" category could complicate construction significantly or even make it economically inefficient. Though methods of active faults' identification and study are well developed, nevertheless, in actual practice their definitions are still inconsistent. It results in different approaches to their identification, quantification, and hazard assessment, which, in turn, are interconnected with engineering protection measures.

A. Strom (✉)
Geodynamics Research Centre, Branch of JSC "Hydroproject Institute", Moscow, Russia
e-mail: strom.alexandr@yandex.ru

G. Lollino et al. (eds.), *Engineering Geology for Society and Territory – Volume 6*,
DOI: 10.1007/978-3-319-09060-3_114, © Springer International Publishing Switzerland 2015

114.2 Definition

A generally applied definition for active fault in "active tectonics" literature is: "*A fault that has been ruptured at least once during given period of time must be considered as active*" (Allen 1975; Trifonov 1985; Hancock et al. 1991; Nikonov 1995). The duration of this "reference period of time" ranges, most often, from ~10,000 years up to ~1,000,000 years. In some "extreme" definitions time interval decreases up to decades (period of instrumental geodetic observations) (Kuzmin 2004) or increases up to 2,000,000 years (Lunina et al. 2012). Considering typical recurrence intervals between large earthquakes accompanied by surface faulting that range from several hundreds years to several tens thousands years, the most reasonable time interval for hazard assessment of lifelines is from 50,000 to 100,000 years. Besides formal probabilistic reasoning this interval is best supported by commonly used dating methods (^{14}C, OSL).

However, from engineering point of view such definition is incomplete, since it does not determine the threshold displacement value. Faulting recurrence estimates are necessary for risk assessment mainly, which is the economic issue, while engineering itself requires single-event or cumulative offset value and fault kinematics that predetermine design of the protection measures. Besides, discrimination of the potentially dangerous active faults at the structures' foundation depends at a large extent on the type of the designed structure. While some rigid (e.g. concrete) structures could be sensitive even to minimal displacement, flexible steel structures such as pipeline could accommodate much larger offset. Moreover, as for pipelines, pipe thermal or hydrodynamic deformations considered for the design could be large enough (SNiP 1996; American Lifelines Alliance 2005) to provide additional constraints of the threshold value. With due regard to the above considerations we proposed following definition of active fault for the new version of Russian Guidelines on seismic design of hydraulic facilities: "*A fault must be considered as active one if there is evidence of permanent or periodical displacements in Late Pleistocene—Holocene (during last 100,000 years), with such single-event offset value or displacement rate that pose a threat to the construction and require special engineering or placement measures to ensure its safety*". Such approach requires close co-operation between earth scientists studying hazardous natural phenomena and design engineers. Designers should provide basic characteristics of the proposed construction and its ability to resist deformations, which threshold value must be considered by earth scientists.

The following reasoning relates to trunk pipelines mainly. Here the main goal is to prevent pipeline leakage and environmental pollution, rather then maintain its working capacity. For such infrastructure elements as roads and railroads it seems that it is much more efficient to monitor them regularly and to repair in case of rupturing event rather than to protect from unacceptable deformations.

114.3 Hazard Assessment

If a hazard (in the probabilistic meaning of this term) of rupturing and corresponding risks are high enough, the following parameters must be provided to design a fault crossing: possible rupture location and its accuracy; expected single-event displacement (for seismic ruptures) or displacement rate (for creeping faults); fault kinematics. Additional input data includes width of secondary deformations zone, style and values of these deformations.

While for causative faults kilometres-scale accuracy of their location can be acceptable, active faults that can affect structures directly must be localised much more precisely (Besstrashnov and Strom 2011). Accuracy predetermines usage of direct and indirect evidence of activity. Delineation of causative faults is often based on a set of indirect evidence of fault activity, but only direct evidence allows tracing of active faults that cross a construction site or critical lifeline route. Only direct evidence of past surface rupture allows estimating of future offset value and direction with an accuracy acceptable for engineering purpose. Indirect evidence of fault activity, indicating high permeability for fluids, does not provide information on the mechanical mobility of fault sides. Hazard predetermined by permeability might require different protection measures than those against mechanical displacements provided by active faults.

Probabilistic seismic hazard assessment (PSHA) that is widely used in engineering practice implies that shorter the considered time interval is, lower strong motion parameters (intensity, acceleration, velocity) should be. It follows from the fact that strong motion at a particular construction site can be caused by seismic waves from multiple earthquake sources, both local and distant. Recently, the probabilistic approach was applied for fault displacement analysis (PFDHA) (Youngs et al. 2003). However, its applicability for particular fault crossing seems to be problematic. First, unlike PSHA, it deals with a unique feature—a particular active fault (or active fault segment) for which it is difficult if not impossible to collect statistically representative data. Second, according to pipeline construction codes (Honegger and Nyman 2004), seismic design is performed for two levels of seismic loading—Strength-Level Earthquakes (SLE) with recurrence of 200 year and Ductility-Level Earthquakes (DLE) that can occur once in 1000 years. But a similar approach for surface faulting assessment is not applicable, since usually surface faulting occurs once in more than 1000 years (such offset could be equalised to the DLE), while no rupture at all along the same fault occur more often. If a rupturing event would occur during pipeline lifetime, displacement would be most likely equal to a characteristic value (Schwartz and Coppersmith 1984).

114.4 Mitigation Measures

Safety measures for critical infrastructures, which cross active faults, strongly depend on the lifeline type. For pipelines aim of these measures is to exclude leakage which can be provided by special engineering efforts. Crossing of the Trans-Alaska ground-surface oil pipeline with the Denali that sustained fault rupturing during the 2002 Denali fault includes special compensator placed on rails with Teflon coating to facilitate its slipping against the ground in case of earthquake. This construction proved its efficiency during the 2002 Denali earthquake (Haeussler et al. 2004).

Significant efforts were undertaken to ensure safety of the Sakhalin-1 and 2 Projects in Russia. Their additional complexity was conditioned by the strict requirement to have buried pipelines that reduce the degree of pipe freedom at a large extent in comparison with ground-surface pipelines. Nevertheless, special "dog leg" tracing of pipeline close to its crossings with active faults, use of trapezoid-shape trenches with drainage aimed to avoid freezing of trench fill during winter period, and other technical measures allowed construction of fault crossings that could accommodate offsets half as much as the design displacement values—$1.5 \times D_{des}$ (Mattiozzi and Strom 2008; Strom et al. 2009).

As mentioned above, railways and highways safety can be achieved, mainly, by regular monitoring and obligatory run-time inspections after large earthquakes, before traffic reopening. Any engineering solutions aimed to retain operational integrity of railroads and highways crossed by active faults after large earthquakes seem to be too complex and economically inefficient.

114.5 Conclusions

The proposed definition of active faults that cross critical lifelines routes includes assessment of the displacement value and/or rate of permanent fault movements (tectonic creep)—parameters really used as an input data for engineering design of fault crossings. The traditionally determined parameter—i.e. recurrence of displacements and/or mean rate calculated for the entire earthquake cycle can be used for risk assessment mainly.

Ability of different types of engineering structures to sustain fault displacements should be considered both for site investigations and mitigation measures planning to avoid exceptional selection of sites that, in fact, do not require special technical measures to ensure structures' safety. Such approach need better interconnection of earth scientists and design engineers at all stages of project implementation. Special analysis of cases studies is needed to investigate if similar approach could be applied for other geological processes associated with mechanical displacements, such as block slides, in particular (Hungr et al. 2013).

References

Allen CR (1975) Geological criteria for evaluating seismicity. Bull Geol Soc Am 86:1041–1057

American Lifelines Alliance (2005) Guidelines for assessing the performance of oil and natural gas pipeline systems in natural hazard and human threat events

Besstrashnov VM, Strom AL (2011) Active faults crossing trunk pipeline routes: some important steps to avoid disaster. Nat Hazards Earth Syst Sci 11:1433–1436

Hancock PL, Yeats RS, Sanderson DJ (eds.) (1991) Characteristics of active faults. J Struct Geol 13:1–240

Haeussler PJ, Schwartz DP, Dawson TE, Stenner HD, Lienkaemper JJ, Sherrod B, Cinti FR, Montone P, Craw PA, Crone AJ, Personius SF (2004) Surface rupture and slip distribution of the Denali and Totschunda faults in the 3 November 2002 M 7.9 Earthquake, Alaska, BSSA, 94, 1 No. 6B, pp 23–52 (2004)

Honegger DG, Nyman DJ (2004) Guidelines for the seismic design and assessment of natural gas and liquid hydrocarbon pipelines

Hungr O, Leroueil S, Picarelli L (2013) The Varnes classification of landslide types, an update. Landslides. doi:10.1007/s10346-013-0436-y

Kuzmin YO (2004) Modern geodynamics of fault zones. Phys Earth 10:95–111 (in Russian)

Lunina OV, Gladkov AS, Gladkov AA (2012) Systematization of active faults for seismic hazard assessment. Pac Geol 31:49–60 (in Russian)

Mattiozzi P, Strom A (2008) Crossing active faults on the Sakhalin II onshore pipeline route: pipeline design and risk analysis. In: Santini A, Moracittie N (eds) 2008 seismic engineering conference commemorating 1908 Messina and Reggio Calabria earthquake, pp 1004–1013

Nikonov AA (1995) Active faults, definition and problems of their identification. Geoecology 4:16–27 (in Russian)

Schwartz DP, Coppersmith KJ (1984) Fault behavior and characteristic earthquakes—examples from the Wasatch and San Andreas fault zones. J Geophys Res 89:5681–5698

SNiP 2.05.06.85* (1996) Russian State construction code "trunk pipelines" (in Russian)

Strom A, Ivaschenko A, Kozhurin A (2009) Assessment of the design displacement values at seismic fault crossings and of their excess probability. In: Huang R, Rengers N, Li Z, Tang C (eds) Geological engineering problems in major construction projects. Proceedings of 7th Asian regional conference of IAEG, 9–11 Sept 2009, Chengdu, China

Trifonov VG (1985) Peculiarities of active faults evolution. Geotectonics 2:16–26 (in Russian)

Youngs RR, Arabasz WJ, Anderson RE, Ramelli AR, Ake JP, Slemmons DB, McCalpin JP, Doser DI, Fridrich CJ, Swan FH III, Rogers A, Yount JC, Anderson LW, Smith KD, Bruhn RL, Knuepfer PLK, Smith RB, dePolo CM, O'Leary DW, Coppersmith KJ, Pezzopane SK, Schwartz DP, Whitney JW, Olig SS, Toro GR (2003) A methodology for probabilistic fault displacement hazard analysis (PFDHA). Earthq Spectra 19:191–219

Foundation Damage Responses of Concrete Infrastructure in Railway Tunnels Under Fatigue Load

115

L.C. Huang and G. Li

Abstract

The concrete foundation in railway tunnel settles gradually following long-term exposure to repeated loads by trains on the track. We established two foundation numerical models of a railway tunnel and their structure responses under the loads by trains and verified the methodology by field testing. With these models, we examined the relationship between the train velocity and the vertical dynamic coefficient, demonstrated that the damages in the concrete foundation resulted from fatigue damage accumulation due to repeated loads. In concrete foundations with rebar, the attenuation rate of stress was higher than those without rebar, thus having better structure responses. In conclusion, we advise to increase the reinforcement ratio of the concrete foundation of tunnel railways to improve its structure responses.

Keywords

Concrete foundation • Railway tunnel • Structure response • Simulation model

115.1 Introduction

The concrete foundation of a railway tunnel settles gradually following long-term exposure to repeated loads by running trains on the track, which is more obvious in the west of China with the soft soil foundation. So there are many engineering problems in the foundation of railway tunnel, but it is pity that some engineers do not pay much attention in the design and construction, and the design codes are often times outdated. Because of the difficulty to simulate the foundation structure under railway live load, the calculation results can not reflect the real stress state. Jin et al. (2005) used the Newmark method to solute the dynamic equilibrium equation, and then calculated the vibration load history. Jones et al. (2003) simulated the vibration propagation on tunnels. Moghimi and Tonagh (2008) performed a parametric study to identify the effects of road surface roughness on various parameters, such as vehicle speed, aspect ratio of steel girders, stiffness of neoprene, type of vehicle, vehicle lane eccentricity and initial bounce of the vehicle, on dynamic load allowance using non-linear dynamic simulation. Jiang et al. (2010) analyzed the seismic response of underground utility tunnels through shaking table testing and finite element method (FEM) analysis. It showed that the numerical results matched the experimental measurements very well. A theoretical study of the stability of a two-mass oscillator moving along a beam on a visco-elastic half-space was analyzed by Metrikine et al. (2005), who, using Laplace and Fourier integral transforms and expressions for the dynamic stiffness of the beam derived the point of contact with the oscillator. The propagation of vibration generated by a harmonic or a constant load moving along a layered beam which was resting on the layered half-space was investigated theoretically by Sheng et al. (1999).

L.C. Huang · G. Li (✉)
School of Engineering, Sun Yat-sen University, Guangzhou, China
e-mail: badboy955@163.com

L.C. Huang
e-mail: hlinch@mail.sysu.edu.cn

G. Lollino et al. (eds.), *Engineering Geology for Society and Territory – Volume 6*,
DOI: 10.1007/978-3-319-09060-3_115, © Springer International Publishing Switzerland 2015

The aim of our study is to find out the reason of tunnel foundation cracks, to descript the damaging effects of the loads by trains on the concrete foundation of railway tunnels, and to explore a better modality of building concrete foundation in railway tunnels.

115.2 Computational Model

A railway tunnel in the west of China is chosen as the research model in this paper, with the three-dimensional simulation analysis, shown in Fig. 115.1.

In the computation, the surrounding rock is classified as IV by the design standard of railway tunnel in China. Table 115.1 shows the corresponding parameters for these materials.

In order to find out the reason of tunnel foundation cracks by trains live loads on the concrete foundation of railway tunnels, two computational models have been implemented here:

Model I—the concrete (C20) foundation is the plain concrete structure, and the depth is 30 cm;

Model II—the concrete (C20) foundation is the reinforced concrete structure, and the depth is 20 cm.

115.3 Field Testing Model

In order to verify the actual stress responses with the numerical simulation results, a field testing of the foundation base was implemented in this tunnel.

In both models, five inbuilt pressure cells were embedded at the top of the concrete foundation that is exactly located under the sleeper, as shown in Fig. 115.2. Seven accelerometers embedded at the top of the concrete foundation, which are located under the sleepers and exactly located in between them; the specific dimensions are shown in Fig. 115.2.

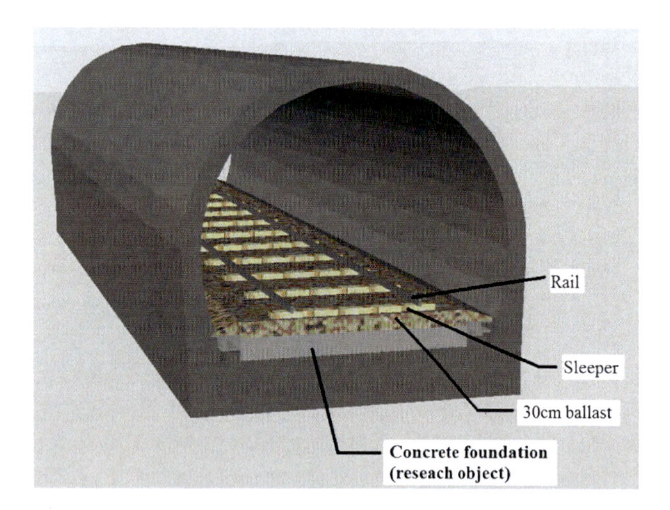

Fig. 115.1 Sketch map of the cross section of the tunnel structure under the rails (*unit* cm)

Being identical with the simulation conditions, the load used in the field test was Shaoshan III locomotive. It was stopped only at specific locations in order to measure the static stresses. In the dynamic tests, the data pertaining to the load of the running train, the Shaoshan III locomotive, were recorded as the effective measured results.

As the speeds of the train in the field testing varied from 15 to 28 km/h, we took an average speed of 22 km/h in the calculation. The average stress ($\overline{\sigma}$), mean-variance (δ_σ), and statistical maximum values (σ_{maximum}) are shown in Tables 115.2 and 115.3.

These results showed that the maximum dynamic pressure stress in the top of foundation base was only 0.293 MPa under the live load of the train, which is also considerably less than the compression strength of the concrete material of the tunnel concrete foundation. In other words, the concrete foundation base of the tunnel cannot be damaged under the vertical pressure stress of the running train.

Tables 115.2 and 115.3 suggests that the average dynamic stresses are approximately 1.2–1.4 times of the static stresses in the foundation base, which implies that the live load of the train clearly affects the foundation base.

115.4 Comparison Results

The structure responses and the vertical dynamic coefficient at an average speed of 22 km/h are compared between the measuring results from field testing and the theoretical results from numerical simulation (Table 115.4). The distribution of the stresses at the top of the foundation base is shown in Fig. 115.3.

It shows that the numerical simulation results are in good agreement with the results of the field testing. In the entire concrete foundation base, the maximum stress is located exactly under the two rails and the minimum stress at the middle of the sleeper. It can be concluded that the damage in the concrete is likely to occur not from attaining an elastic limit under the static loads but more likely from the damage accumulation in fatigue due to repeated loadings. At this point, we conclude that the disrepair and damage in the concrete foundation are usually detected in the longitudinal direction, immediately below the two rails, and thus, the concrete foundation surface should be reinforced in the transverse direction. At the same time, in order to enhance the tensile capability and to decrease the formation of cracks in the concrete foundation, it is reasonable to adopt the compact reinforce bar.

Table 115.1 Material parameters

No.	Material	E (GPa)	γ (kN/m^3)	Φ (°)	μ	C (kPa)
1	Surrounding rock	6	21	50	0.30	25
2	Lining (C20 concrete)	26	23	N	0.20	N
3	Bedding (C20)	26	23	N	0.20	N
4	Ballast	0.15	24	30	0.27	5
5	Sleeper (C40)	36.0	28	N	0.15	N
6	Rail	210.0	77	N	0.28	N

Note E is the Young's modulus, γ is the bulk density, φ is the inter friction angle, μ is the Poisson's ratio, and C is the cohesion force

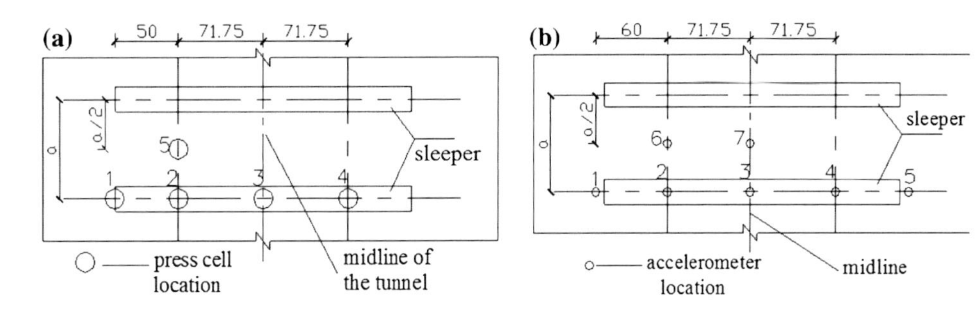

Fig. 115.2 Arrangement plan of the testing locations at the upper of the concrete foundation (*unit* cm)

Table 115.2 Statistic values of the dynamic stresses in Model I (MPa)

Stress item	1	2	3	4	5
$\overline{\sigma}$	0.087	0.232	0.052	0.161	0.123
δ_{σ}	0.017	0.047	0.010	0.022	0.010
σ_{maximum}	0.112	0.293	0.072	0.198	0.137

Table 115.3 Statistic values of the dynamic stresses in Model II (MPa)

Stress item	1	2	3	4	5
$\overline{\sigma}$	0.097	0.200	0.038	0.236	0.107
δ_{σ}	0.006	0.030	0.006	0.029	0.014
σ_{maximum}	0.112	0.293	0.048	0.284	0.131

Table 115.4 Comparison between testing values and the computational values

Locations			1		2		3	
			Stress (MPa)	Dynamic coefficient	Stress (MPa)	Dynamic coefficient	Stress (MPa)	Dynamic coefficient
Model I	Testing	Static	0.063	1.381	0.171	1.889	0.040	1.300
		Dynamic	0.087		0.232		0.052	
	Computation	Static	0.070	1.086	0.075	1.080	0.060	1.083
		Dynamic	0.076		0.081		0.065	
Model II	Testing	Static	0.075	1.293	0.168	1.190	0.030	1.267
		Dynamic	0.097		0.200		0.038	
	Computation	Static	0.070	1.086	0.074	1.081	0.059	1.085
		Dynamic	0.076		0.080		0.064	

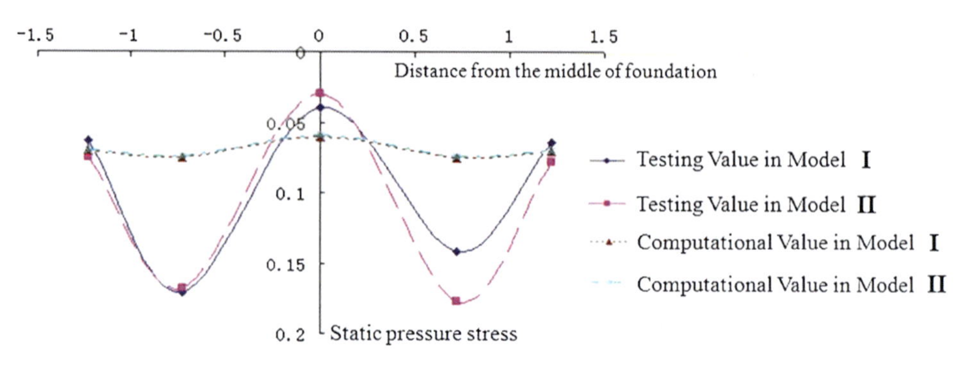

Fig. 115.3 Stresses distribution on the top of the of the concrete foundation

115.5 Main Conclusions

(1) The thicker the foundation base, the greater the stresses are. However, it is not advisable to improve the tunnel foundation base by simply increasing the thickness of the concrete foundation; it is more advisable to improve the structure responses of the tunnel foundation base by adopting the compact reinforced concrete structure.
(2) Lateral tensile stress is the main factor that causes damages to the foundation base.
(3) The damages in the concrete foundation occur more likely from fatigue damage accumulation due to repeated loadings rather than from attaining an elastic limit under static loads.

Acknowledgments This work has been supported by the National Natural Science Foundation of China (No. 51108472 & 51309261), the Natural Science Foundation of Guangdong Province, China (No. S2011040005172 & S2012010010446), and the Specialized Research Fund for the Doctoral Program of Higher Education of China (No. 20110171120012), these supports are gratefully acknowledged.

References

Fiala P, Degrande G, Augusztinovicz F (2007) Numerical modeling of ground-borne noise and vibration in buildings due to surface rail traffic. J Sound Vib 301(3):718–738
Huang LC, Xu ZS, Zhou CY (2009) Modeling and monitoring in a soft argillaceous shale tunnel. Acta Geotechnics 4(4):273–282
Jones CJC, Thompson DJ, Petyt M (2003) A model for ground vibration from railway tunnels. Transport 153(2):121–129
Jiang LZ, Chen J, Li J (2010) Seismic response of underground utility tunnels: shaking table testing and FEM analysis. Earthquake Eng Eng Vibr 9(4):555–567
Jin LX, Zhang JS, Nie ZH (2005) Dynamic inverse analysis for vibration-load history of high-speed railway. Chin J Traffic Transport Eng 5:27–32
Moghimi H, Tonagh HR (2008) Impact factors for a composite steel bridge using non-linear dynamic simulation. Int J Impact Eng 35:1228–1243
Madshus C, Kaynia AM (2000) High-speed railway lines on soft ground: dynamic behavior at critical train speed. J Sound Vibr 231 (3–5):689–701
Metrikine AV, Verichew SN, Blaauwendraad J (2005) Stability of a two-mass oscillator moving on a beam supported by a visco-elastic half-space. Int J Solids Struct 42(3–4):1187–1207
Sheng X, Jones CJC, Petyr M (1999) Ground vibration generated by a load moving along a railway track. J Sound Vibr 228(1):129–156

Assessment and Risk Management for Integrated Water Services

116

Loretta Gnavi, Glenda Taddia, and Stefano Lo Russo

Abstract

Drinking water supply and waste water networks and treatment systems provide essential services to the community. These infrastructures are vulnerable and subject to a wide range of risks; damage and operation interruptions that require immediate restoration and management of the emergency, may occur. The vulnerability of water supply systems is nowadays the focus of attention of everyone responsible for their security and good performance. Direct effects caused by disasters involve the physical damage to the infrastructure while indirect damage is linked to the additional expenses that the water companies need to incur in order to respond to the emergency, as well as the loss of revenue due to the interruption of their services. The analysis and risk assessment for the Integrated Water Services relating to natural hazards and/or human origin threats is based upon the equation that correlates the risk to the probability of occurrence of an adverse event, the vulnerability and the consequences. The paper presents an overview of the general framework and the risk tools useful for the risk management for Integrated Water Services.

Keywords

Water services • Risk assessment • Risk management • Vulnerability

116.1 Introduction

Drinking water supply and waste water networks and treatment systems are vulnerable and subject to a wide range of risks. In the third edition of the Guidelines for Drinking-water Quality, the World Health Organisation (WHO 2008) has emphasized that a comprehensive risk management approach represents the best way to ensure safe drinking water supply, considering the entire supply system, from source to tap. As part of risk management, WHO recommends preparation of Water Safety Plans (WSPs), including system assessment, operational monitoring and management plans.

The knowledge of the different sources of risk and the interrelationships between them have not yet been systematized. The final objective of this study will be to define some guidelines applicable to the national context and useful for the development of an integrated methodology that allows the evaluation, quantification and management of natural and human origin hazards for the Integrated Water Services. Models in this sense have been developed for national contexts in Australia (NHMRC/NRMMC 2004), New Zealand (Ministry of Health 2005a, b), Sweden (SNFA 2007), Denmark (DWWA 2006) and Norway (NFSA 2006).

L. Gnavi (✉) · G. Taddia (✉) · S.L. Russo (✉)
Department of Environment, Land, and Infrastructure Engineering (DIATI), Politecnico di Torino, Corso Duca degli Abruzzi 24, 10129 Turin, Italy
e-mail: loretta.gnavi@polito.it

G. Taddia
e-mail: glenda.taddia@polito.it

S.L. Russo
e-mail: stefano.lorusso@polito.it

G. Lollino et al. (eds.), *Engineering Geology for Society and Territory – Volume 6*,
DOI: 10.1007/978-3-319-09060-3_116,

116.2 Risk Assessment for the Integrated Water Services

The analysis and risk assessment for the Integrated Water Services relating to natural hazards and/or human origin threats is based upon the equation that correlates the risk to the probability of occurrence of an adverse event, the vulnerability and the consequences. This analysis must include:

- a procedure to identify all possible hazardous events (floods, landslides, earthquakes, terrorist attacks, intrusions and sabotage to critical components, etc.). Among the natural hazards, earthquakes, floods and droughts are the three most significant hazards that can cause water utilities damage;
- a methodology for the analysis of response mode and the vulnerability of the system and its components (Fig. 116.1), in case the adverse event occur;
- a procedure for the evaluation of the consequences in case the occurrence of an adverse event should compromise, for example, the continuity of drinking water supply. The consequences should be evaluated from the environmental, social, economic and legal point of view.
- a procedure for the assessment and risk management, which, once identified and quantified, place to confirm whether this is tolerable and if the control and prevention measures are adequate, and to identify appropriate strategies for mitigating risk.

The vulnerability assessment is essentially a four-step procedure (AWWA 2001):

1. identify and describe the separate components of the water supply total system;
2. estimate the potential effects of probable disaster hazards on each component of the system;
3. establish performance goals and acceptable levels of service for the system;
4. if the system fails to operate at desired levels under potential disaster conditions, identify key or critical system components responsible for the condition.

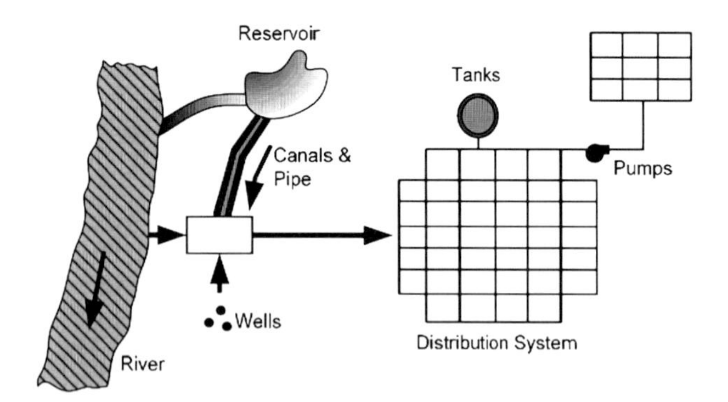

Fig. 116.1 Elements and vulnerable points in a general water supply system (*source* Haestad et al. 2003)

The risk assessment can be conducted by splitting the system into sub-systems in turn subdivided into components that, in the case in which they reach a given limit state that produces breakage or malfunction, the crisis of the entire system may occur. The division of the system into components should take into account, as much as possible, the interactions between the different components and the subsystems.

Nowadays the methods of risk analysis (Table 116.1) can be more or less detailed, qualitative and/or quantitative, based on the use of standard indicators and implemented with inductive and deductive logical processes.

The identification of hazards can be based on experiences from the past, checklists and structured methods such as Failure Modes and Effects analysis (FMEA).

The qualitative methods generally requires less input data; the most common qualitative risk assessment is risk ranking. Risk estimation with risk matrices is a useful and efficient tool and it is easy to understand and present data.

The aim of quantitative methods is to provide an estimation of the risk level in absolute terms. A wide range of quantitative methods exists.

Among the methods that can be both qualitative and quantitative, the Event tree analysis (ETA) and the Fault tree analysis (FTA) should be mentioned. A fault tree is a logic diagram modelling failure events that may occur in a system and the interactions between these events. The top event may represents a failure in delivering the required quantity and/or quality of drinking water.

The availability of a system that, through the mapping of the global and local fragility of the networks system, might provide indications on the most likely "break" points, represents a useful tool for the planner that can ensure effectiveness and efficiency in the processes of network implementation and management.

116.3 Example on Risk Assessment for Water Utilities

In order to provide some examples on risk assessment practice the integrated water production and management services of Società Metropolitana Acque Torino (SMAT) S.p.A. can be taken into account. SMAT S.p.A. manages some of the largest, most advanced drinking water supply and waste water networks and treatment systems in Europe. It produces every year more than 260 millions cubic metres of quality drinking water. About 70 % of the supplied water comes from wells which are fed by one or more from 40 to over 100 m deep groundwater tables and 14 % is taken from Pian della Mussa, Sangano and other spring sources. The water taken from the Po river totals 16 % of the water introduced into the network. Through an approximately

Table 116.1 Overview of possible risk analysis methods for use in water safety plans (*source* Rosèn et al. 2007)

Risk analysis method	Stage in the risk analysis process	Qualitative/ quantitative	Part of water supply system	Water quality/ water quantity	Data requirements	Need of training
Hazard and operability analysis techniques (HAZID)	Hazard identification	Qualitative	All	Both	Low	Novice
Hazard and operability analysis (HAZOP)	Hazard identification	Qualitative	Treatment distribution	Both	Medium	Specialist
Preliminary hazard analysis (PHA)/ risk and vulnerability analysis (RVA)	Hazard identification risk estimation	Qualitative	All	Both	Medium	Novice
Failure modes, effects, and criticality analysis (FMECA)	Hazard identification Risk estimation	Qualitative	Treatment Distribution	Both	High	Specialist
Fault tree analysis	Risk estimation (causes)	Qualitative/ quantitative	All	Both	High	Expert
Reliability block diagram	Risk estimation (causes)	Qualitative/ quantitative	All	Both	High	Expert
Event tree analysis	Risk estimation (consequences)	Qualitative/ quantitative	All	Both	High	Specialist
Human reliability assessment (HRA)	Risk estimation (causes)	Qualitative/ quantitative	Treatment Distribution	Both	High	Expert
Physical models	Risk estimation (consequences)	Quantitative	All	Both	High	Expert
Quantitative microbiological risk assessment (QMRA)	Risk estimation (consequences)	Quantitative	All	High	Quality	Expert
Barriers and bow-tie diagram	Risk estimation	Qualitative/ quantitative	All	Both	Low	Specialist

12,000 km long network, SMAT S.p.A. distributes drinking water to about 2.2 million people, and collects urban waste water through 8,000 km of sewers. 412 small-, medium- and large-sized sanitation systems constantly run and sanitise about 350 million cubic metres of waste water every year.

Currently the SMAT crisis plan for water supply plants provides an assessment based on the Failure Modes and Effects analysis. For each event (flood, earthquake, snow or ice, wind, contamination, electricity supply failure, vandalism, terrorism, electrical fault, mechanical/hydraulic failure, human error, etc.) the following parameters are evaluated: (1) the probability of the event (from 1 remote to 10 very high); (2) the areas of the facility involved; (3) the potential failure modes; (4) the potential failure effects; (5) the severity of the failure effects (from 1 barely perceptible to 10 extremely severe; (6) control measures and containment activities; (7) evaluation of the effectiveness of control measures and containment activities (from 1 high to 10 unlikely).

On the basis of the previous evaluations a Priority Risk Index (PRI) is defined and should be compared with the severity of the failure effects. If the effect of the failure is more serious, then the intervention threshold is lower, in order to contain the effect itself. For example, situations with high severity (grade 9–10) require the implementation of measures already with PRI > 40, while situations with low gravity (1) require interventions with PRI > 150.

A further step forward could concern the analysis through quantitative methods; the main advantage of a quantitatively estimated risk is that it facilitates comparison with other risks and acceptable levels of risk in absolute terms.

A logical approach could be to first perform a qualitative risk assessment, covering the entire system, and later to apply a quantitative method for a more detailed assessment.

To fulfil this objective, it may be necessary to carry out risk analysis at various levels: (1) Integrated analysis of the total water supply system, (2) Specific analysis of the raw water source, the treatment system, the distribution network and/or the plumbing system, (3) Specific analysis of technical systems/operational activities and (4) Analysis of subfunctions within the system.

Fault tree analysis can be identified as a suitable method to quantitatively estimate risk; it makes possible to model failures as chains of events and thus consider interactions between components and events.

To analyse the entire drinking water system, it can be divided into three sub-systems: raw water, treatment and distribution. The main reason for this division is to make it possible to calculate which part of the system contributes most to the total risk. Another reason is that it shows more easily how different parts of the system may compensate for failure in other parts.

The overall failure event is the supply failure and may arise due to: (1) quantity failure, i.e. no water is delivered to the consumer; or (2) quality failure, i.e. water is delivered but is unfit for human consumption.

The fault tree analysis cannot include health effects of drinking water not fulfilling the quality standards. To also include health effects (assessing the risk of human infection) the fault tree method may be combined with a Quantitative Microbial Risk Assessment (QMRA).

The fault tree analysis furthermore provides a detailed system description that can be used to identify options to reduce the risk.

116.4 Conclusions

To facilitate risk management of water utilities, including preparation of Water Safety Plans, suitable methods and tools for analysing systems and comparing risk-reduction measures are necessary. Since water utilities are very diverse and exhibit different types of risks, one single method cannot be used in all cases; both qualitative and quantitative methods can provide useful information.

An effective risk analysis requires basic knowledge about possible risks, characteristics of potential hazards, and comprehensive understanding of the associated cause-effect relationships within the water system.

The generic framework described aims at providing support and structure for risk management in preparation of Water Safety Plans. The analysis of some examples on risk assessment for water utilities will form the basis for the choices and decisions on control measures to increase the security of the Integrated Water Services. The continuation of the study aims to provide useful results in order to draw the attention of the managers of Integrated Water Services on the priorities and possible alternatives to be considered in the organization and in the choice of maintenance or updating of the system which they are responsible.

References

AWWA (2001) emergency planning for water utility management. American Water Works Association, Manual of water supply practices-M19, 4th edn

DWWA (2006) Guidelines for safe drinking water quality (in Danish), Danish Water and Wastewater Association, Skanderborg

Haestad M, Walski TM, Chase DV, Savic DA, Grayman W, Backwith S, Koelle E (2003) Advanced water distribution modeling and management. Haestad Press, Waterbury

Ministry of Health (2005a) Drinking-water standards for New Zealand 2005, New Zealand Ministry of Health, Wellington

Ministry of Health (2005b) A framework on how to prepare and develop public health risk management plans for drinking-water supplies, New Zealand Ministry of Health, Wellington

NFSA (2006) Improved safety and emergency preparedness in water supply: Guidance (in Norwegian), Norwegian Food Safety Authority, Oslo

NHRMC/NRMCC (2004) National water quality management strategy: Australian drinking water guidelines. National Health and Medical Research Council and Nature Resource Management Ministerial Council, Australian Government

Rosèn L, Hokstad P, Lindhe A, Sklet S, Røstum J. (2007) Generic framework and methods for integrated risk management in water safety plans, TECHNEAU report. Deliverable no. D 4.1.3, D 4.2.1, D 4.2.2, D 4.2.3

SNFA (2007) Risk and vulnerability analysis for drinking water supply (in Swedish), Swedish National Food Administration, Uppsala

WHO (2008) Guidelines for drinking-water quality [Electronic Resource]: Incorporating first and second addenda, vol 1, Recommendations, 3rd edn. World Health Organization, Geneva

Multi-risk Assessment of Cuneo Province Road Network

117

Murgese Davide, Giraudo Giorgio, Testa Daniela, Airoldi Giulia, Cagna Roberto, Bugnano Mauro, and Castagna Sara

Abstract

Natural risk assessment for urban areas and infrastructures is important for the definition of management and prevention plans against consequences of natural events. In this paper we present the results of a multi-risk assessment for the Cuneo Province road network. The study defined specific risk levels with regard to landslides, floods, torrential floods, debris-flows, snow-avalanches, earthquakes and forest fires. Consequences for infrastructures were assessed by quantifying exposed elements value and vulnerability. All acquired data are then combined in order to produce specific hazard and risk maps. Specific risk levels were then processed to produce a multi-risk map for the Cuneo province road network. Landslide runout was numerically simulated in a GIS environment, for a comparison with hazard assessment results obtained following the methodology here proposed. Multi-risk assessment represents a valuable tool for enhancing scheduling activities related to the implementation of mitigation structure/measures and for supporting the coordination of risk management procedures at a cross border level.

Keywords

Multi-risk • Natural hazard • GIS • Roads • Vulnerability

117.1 Introduction

Road networks represent a key factor in the emergency phase, for they are the most effective way for reaching and providing assistance to affected areas. Also, they are exposed elements and it is important to provide a reliable and updated risk assessment for these infrastructures. Road networks often extend over areas with variable morphological conditions and, for this reason, different risk scenarios have to be considered. In order to perform a reliable risk evaluation it is important to collect all data that allow the accurate assessment of all risk components: hazard, vulnerability and value of exposed elements. These parameters are often expressed following different approaches (e.g. qualitative, quantitative) and by using different measurement systems. Risk assessment requires specific procedures for the homogenization of processed data. Hazard and risk maps are a powerful tool for providing risk information to the population and all authorities involved in risk management and civil protection activities. Reliable risk assessment results allow the audit of risk management procedures, the implementation/correction of existing approaches/structures/organizations, the revision of maintenance programs and the improvement in funds allocation for strategic interventions. Natural risk assessment is a complex activity and the

M. Davide (✉) · T. Daniela · A. Giulia
SEA Consulting srl, 4e-arth (for environment, for ecology, for education for earth our planet), Turin, Italy
e-mail: murgese@4e-arth.eu

G. Giorgio
Cuneo Province, Public Works Sector Civil Protection Services, Cuneo, Italy

C. Roberto · B. Mauro · C. Sara
Polithema Società di Ingegneria srl, Turin, Italy

G. Lollino et al. (eds.), *Engineering Geology for Society and Territory – Volume 6*,
DOI: 10.1007/978-3-319-09060-3_117,

adopted methods have to be clearly described to allow a correct implementation of risk management procedures. This is particularly relevant when risk management is performed at a cross-border level by authorities from different regions and countries, as in the area of Cuneo and Imperia Provinces (Italy) and the Provence-Alpes-Côte d'Azur region (France). This paper illustrates the results of natural risk assessment for the Cuneo Province road network, based on the methodology defined by the technicians of Cuneo and Imperia Provinces (Giraudo et al. 2012) the project was part of the Strategic Program **Risknat** a cross-border project in the framework of the program Alcotra France-Italy (2007–2013), developed between 2009 and 2012, focusing on natural hazards in mountain areas.

117.2 Methods

The multi-risk assessment of Cuneo Province road network was referred to the following processes: landslides, floods, debris-flows, torrential floods, snow avalanches, forest fires and earthquakes. Risk assessment was performed according to the methodology defined by Cuneo and Imperia Provinces (Giraudo et al. 2012). Considered infrastructures extend over areas with different morphological conditions (flood-plains, hills, mountain areas), with a total length of 862 km. Each road was divided into segments; each segment was defined by nodes corresponding to road intersections or other critical points (e.g. towns crossed by the examined road). Risk level was calculated for each segment. The study was divided into three main phases:

Phase (1)—hazard elements (features to be stored as shape files) were acquired from (i) existing databases (e.g. IFFI, SIFRAP, PAI, SIVA, Forest territorial Plans) owned by public authorities, such as Piedmont Region, Piedmont Environmental Protection Agency (ARPA Piemonte), Cuneo Province and Po Basin Authority, and (ii) public reports on past events (e.g. ARPA and or Piedmont Region event reports). Acquired data were used to produce inventory maps for each natural hazard. Road network data were provided by Cuneo Province and ANAS (Italian National Road Agency). Road segments were identified along each considered infrastructure, based on the presence of intersections with other roads and the distribution of urban centres and public services. The study area (1,277 km^2 wide) was defined according to the distribution and dynamics of hazard phenomena and the distribution of exposed elements.

Phase (2)—collected information were validated by means of aerial photographs interpretation and field survey. Data related to hazard-mitigation structures and other structures along the roads (bridges, embankments, tunnels) were also collected in the field and stored into a specifically designed database.

Phases (3)—**(3a)** data on hazard phenomena were stored as shape files in a GIS geo-database, each file corresponding to a specific hazard type. Each feature of the shape files was then characterised in terms of hazard level. Hazard levels are expressed as numbers ranging from 1 (low hazard level) to 4 (very high hazard level). Parameters used for hazard calculation were those derived from features geometry and other hazard characteristics, according to the method proposed by Giraudo et al. (2012). **(3b)** Each road segment was characterised in terms of specific vulnerability and exposed value, both expressed as numbers ranging from 0 to 1 (Giraudo et al. 2012). Specific vulnerability was calculated as the combination of the efficacy level assigned to hazard-mitigation structures and damage proneness in case of e specific natural event. The value of exposed elements was determined according to the following parameters: road network hierarchy, existence of alternative paths to the considered segment and segments representing the unique access to relevant territorial elements stored into the database of Civil Protection Department of Cuneo Province (e.g. strategic factories, small urban centres, etc.). **(3c)** Specific and multi-risk levels were calculated for each segment, on the basis of the relationships with hazard features, which were established by means of specific algorithms in the GIS environment. According to the distribution of risk numeric results (natural breaks of series values defined in a GIS environment), four risk classes were defined (low, moderate, high, and very high). Specific risk and multi-risk maps were produced for the analysed road network of Cuneo Province. Also, numerical modelling of runout for 20 critical landslides (shallow, rotational and complex) was performed (Fig. 117.1). This activity provided a detailed risk scenario in case of propagation of mobilised materials towards the examined roads. Modelling was made by using the software RASH (Pirulli 2005). The model calculates the runout based on 10 × 10 m grid size DTM according to Voellmy (1955) equation.

117.3 Results and Discussion

Hazard assessment results were organized in form of specific hazard maps, where each process was represented according to its hazard level. Studied phenomena are generally characterised by "high" to "very high" hazard levels, exception made for earthquakes, whose hazard levels range from "moderate" to "high". Forest fires are usually characterised by "low" hazard level (Table 117.1).

Landslide hazard: shallow landslides and rock falls are characterised by the highest number of feature with "very high" hazard level. This result is mainly related to the expected velocity of the phenomena. Hazard levels recorded for the other types of landslide (Table 117.1) generally range

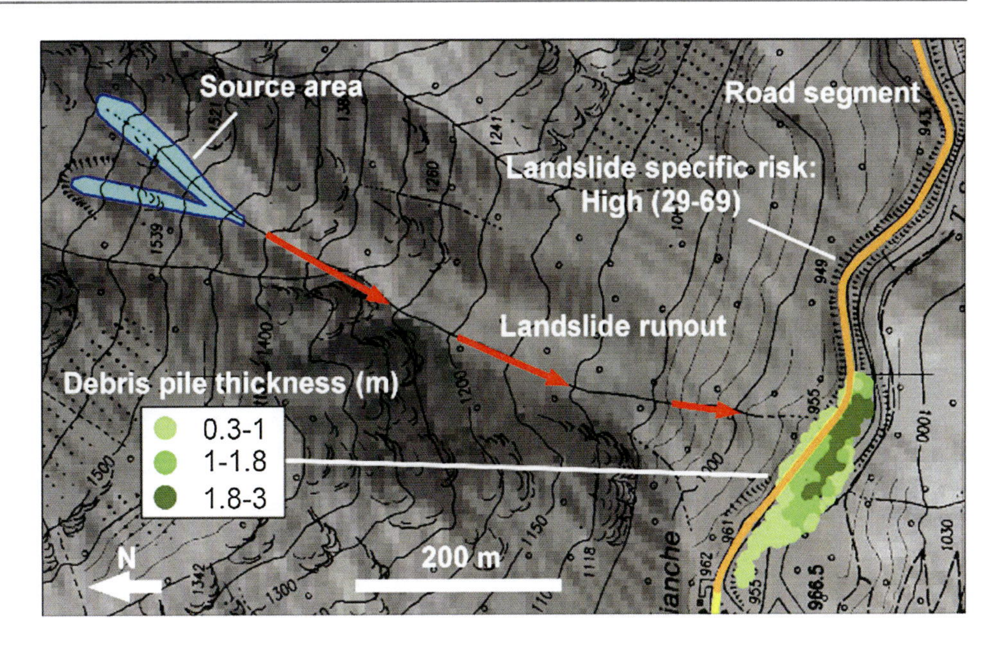

Fig. 117.1 Example of landslide runout modelling results (landslide type: shallow landslide; location: Viandio Municipality, Stura Valley). Landslide specific risk for the road segment is also indicated

Table 117.1 Hazard levels calculated for each phenomenon

Natural Hazards		Low (%)	Moderate (%)	High (%)	Very high (%)
Landslide	AC/CR	0.04	0.07	0.48	5.59
	AS	0.15	0.07	1.75	29.39
	CL	7.27	1.38	4.14	12.46
	CM	0.56	1.04	2.42	0.60
	DG	0.71	0.86	0.00	0.00
	SR	0.34	9.44	9.14	0.30
	ST	0.04	1.53	8.17	1.75
	X	0.30	0.00	0.00	0.00
Floods		31.00	1.00	23.00	45.00
Torrential floods		25.00	10.00	31.00	34.00
Debris flows		7.00	11.00	32.00	50.00
Snow avalanches		0.40	9.80	29.80	60.00
Forest fire		90.86	9.09	0.04	–
Earthquakes		19.00	50.00	31.00	–

Landslide codes *AC/CR*—rock fall (*AC* punctual r.f.; *CR* areal r.f.); *AS* shallow landslide; *CL* creeping; *CM* complex movement; *SR* rotational slide; *ST* translational slide; *DG* deep seated gravitational slope deformation; *X* sinkhole. Percentages are referred to the number of features considered for the representation of each studied process

from moderate to high. *Flood hazard*: sectors with "very high" hazard levels are those characterised by processes with high energy levels (Fluvial zones A and high hazard level areas according to the Po Basin Hydrogeological management Plan—PAI) and average recurrence intervals between 20 and 50 years. *Torrential flood hazard*: "very high" or "high" hazard levels are principally related to: (1) more than 5 critical events in 100 years; (2) large volume of sediment that can be mobilised in case of intense precipitations. *Debris-flow hazard* is referred to alluvial-fan sectors. In this case the assessment takes into account the hazard level of torrential floods combined with the channel conditions along the alluvial fan. Debris-flow hazard ranges from "high" to "very high" and is associated to: (1) critical flow conditions for water and/or solid load, mainly due the presence of obstacles on the path along the alluvial fan; (2) streams characterised by "very high" hazard level. *Snow avalanches* with "very high" or "high" hazard levels are those characterised by an extension greater than 50,000 m^2, with short average recurrence interval (less than 1 year). Value of exposed elements was calculated for each one of the 361 considered road segments (Table 117.2a). Nearly 90 % of

Table 117.2 **a** Road segments falling in the different exposed value classes. **b** Number of road segments potentially affected by studied natural hazards and average vulnerability level to each considered natural process

A) Number of road segments	Exposed value (E)			
	Low	Moderate	High	Very high
361	**0**	**42**	**176**	**143**

B) Natural hazards	Number of hazard features affecting road segments	Average specific vulnerability (V)
Landslides	677	0.77
Floods	616	0.32
Torrential floods	140	0.34
Debris flows	149	0.77
Snow avalanches	119	0.96
Forest fires	6,631	0.60
Earthquakes	361	0.50

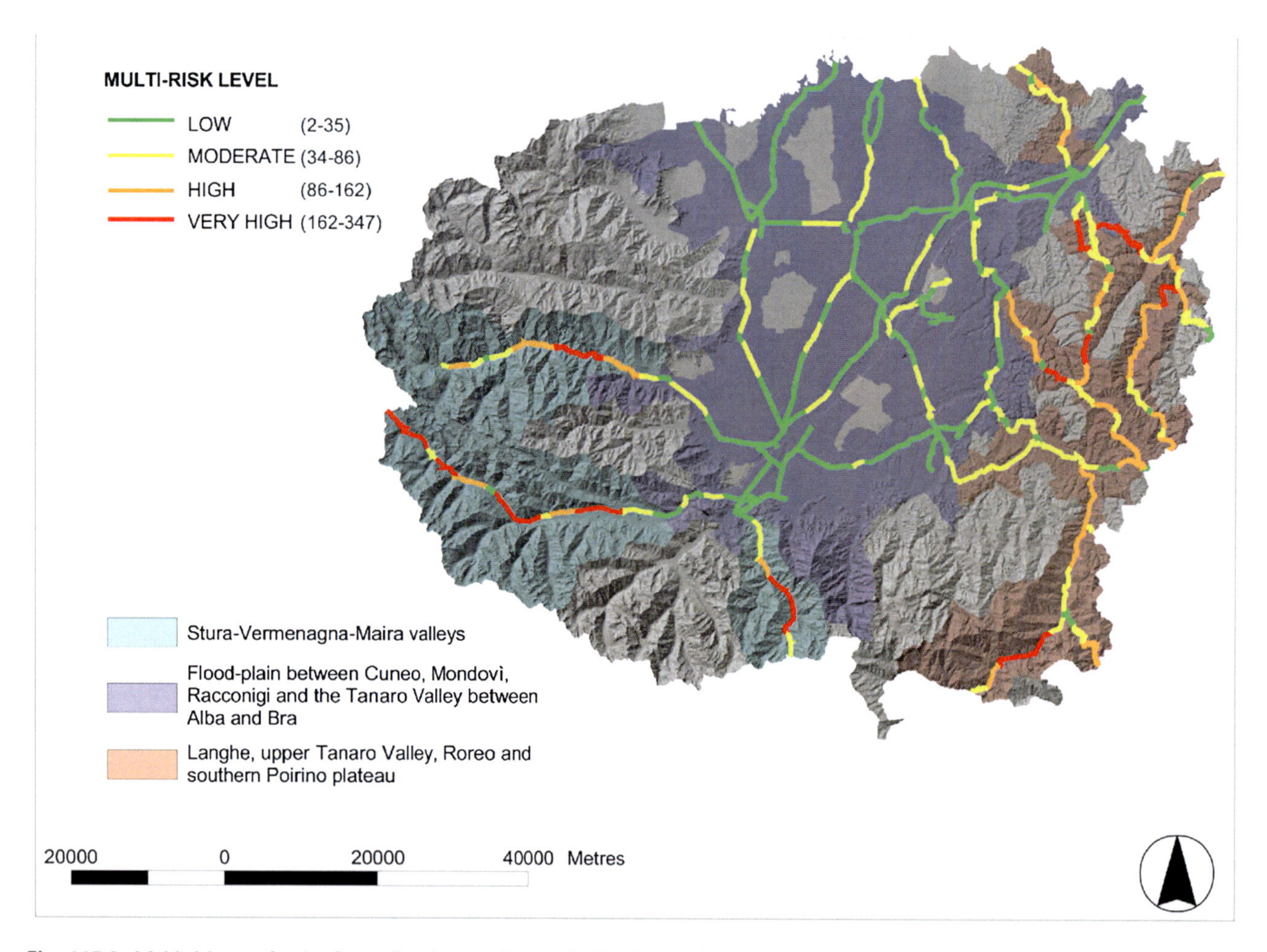

Fig. 117.2 Multi-risk map for the Cuneo Province road network. Numbers in brackets indicate the limits of the four multi-risk levels

road segments has a value belonging to the classes "high" or "very high". Segments characterised by "very high" exposed value commonly belong to the primary network system, for which no alternative connections are present. "High" exposed values mainly relate to secondary road-network segments. Vulnerability was assessed for road segments potentially affected by the inventoried processes: a specific vulnerability level was calculated for each studied natural phenomena (Table 117.2b). Highest values are related to the occurrence of landslides, debris-flows and snow avalanches, providing useful support for scheduling the implementation of mitigation measures.

The combination of hazard levels, exposed values and vulnerability resulted in specific risk and multi-risk maps (Fig. 117.2). Based on multi-risk levels, three homogenous sectors were identified:

(a) Stura-Vermenagna-Maira valleys: risk-level ranges from high to very-high; this sector is characterised by the potential occurrence of all considered natural phenomena.
(b) Flood-plain between Cuneo, Mondovì, Racconigi and the Tanaro Valley between Alba and Bra: risk level ranges from low to moderate; it is related to the potential occurrence of floods, forest fires and earthquakes;
(c) Langhe, upper Tanaro Valley, Roero and southern Poirino plateau: roads in this sector are characterised by risk levels mainly ranging from moderate to high, because of the potential hazard related to landslides, torrential and alluvial-fan activity, floods, earthquakes and bushfires.

Twenty landslides were selected for runout modelling to assess the reliability of landslide hazard assessment method adopted in this study. Numerical simulation reproduced the expected conditions in case of landslide activation. Obtained hazard scenarios following the two approaches indicated that the procedure proposed in this project provides a reliable hazard classification (Fig. 117.1).

117.4 Conclusions

This study provided the assessment of specific natural hazard and natural risk for the Cuneo Province road network. Specific risk were combined in order to produce a multi-risk map to provide the technician of Cuneo Province a reference document during the risk management process. The quantitative assessment of all risk elements (hazard, value and vulnerability of exposed elements) allows the enhancement of scheduling procedures for the implementation of mitigation measures to be adopted.

References

Giraudo G, Faletto C, Grosso GF, Galli M, Spano M, Isoardi P, Pozzani R (2012) Metodologie di analisi del multi-rischio. http://www.provincia.cuneo.it/protezione-civile/progetti/progetto-risknat

Pirulli M (2005) Numerical modelling of landslide runout, a continuum mechanics approach, PhD Thesis Geotechnical engineering, Politecnico di Torino, Italy

Voellmy A (1955) Über die Zerstörungskraft von Lawinen. Schweiz. Bauzeitung 73:159–165, 212–217, 246–249, 280–285

118 Superficial Hollows and Rockhead Anomalies in the London Basin, UK: Origins, Distribution and Risk Implications for Subsurface Infrastructure and Water Resources

Philip E.F. Collins, Vanessa J. Banks, Katherine R. Royse, and Stephanie H. Bricker

Abstract

Recent findings in London show that the subsurface is much more complex than expected, with a number of apparently anomalous features that present a direct hazard to infrastructure development and a risk to ground water management. Of these features, one of the least understood are the large superficial hollows which occur in the rockhead—in much of the London Basin, this is the top of the London Clay Formation—and which are infilled by a range of Quaternary deposits, principally alluvial sands and gravels deposited by the River Thames and its tributaries. The hollows range in size and shape. Several are a few hundred metres across and can be up to 40–50 m deep, though determining their exact form is problematic. The soil and sediment infill of the hollows differs substantially from the surrounding ground in terms of strength and drainage, as well as some differences in chemistry. This presents a real hazard to infrastructure as there is a potential for vertical and horizontal movement, flooding, as well as increasing the risk of contamination of the deeper aquifer. In the paper, the locations and characteristics of known hollows and deformed strata are reviewed and evidence for how they formed is reassessed, systematically considering different hypotheses (scour, ground ice, karst subsidence, seismo-tectonic). From this we consider the implications for continued development of subsurface infrastructure development, and for water resources.

Keywords

Quaternary • Deformation • Risk • Infrastructure • Hydrogeology

118.1 Introduction

Increasingly, infrastructure development in large cities is exploiting the subsurface for key resources and space. Unexpected ground conditions present key risks to projects during construction and operation e.g. due to subsidence and groundwater contamination.

In the London area of south east England, a major source of risk is unexpectedly deep sequences of permeable, often unconsolidated Quaternary sediments and soils (sand, gravel, clay, silt, fractured/puttied chalk and peat). Typically, these occur beneath the floodplain and low terraces of the River Thames and some of its tributaries. Interest in these infilled hollows has increased because of major infrastructure projects e.g. CrossRail and the Olympic Park. There is also new interest in the hydrogeology of the London Basin as groundwater flow is both a resource and area of risk. In addition, evaluation of existing and new borehole and exposure data suggests a much greater degree of structural complexity in the geology of the London Basin. Several of the known hollows occur at historic water sources and near to newly-identified geological faults.

P.E.F. Collins (✉)
Brunel University, Kingston Lane, Uxbridge, London, UB8 3PH, UK
e-mail: philip.collins@brunel.ac.uk

V.J. Banks · K.R. Royse · S.H. Bricker
British Geological Survey, Keyworth, Nottingham, NG12 5GG, UK

G. Lollino et al. (eds.), *Engineering Geology for Society and Territory – Volume 6*,
DOI: 10.1007/978-3-319-09060-3_118,

While the general location of many of the hollows is known, their full geometry is uncertain. A large number of borehole records exist for the London Basin, but these are geographically clustered, often shallow, and often of low quality. As a result, it is probable that several, perhaps many undiscovered hollows remain. There is a clear need to be able to predict the probability of encountering these hollows, or the deformed ground associated with them. It is also important to determine the cause of the hollows, both in terms of helping to predict their occurrence, but also to assess if they are potentially active features.

118.2 Locating Superficial Hollows in the London Basin

Numerous hollows have been encountered in the London Basin over the past 150 years. These were examined in the 1970s (Berry 1979; Hutchinson 1980), and this catalogue has been updated and digitized by the British Geological Survey, linked to the development of the Joint London Basin Model (BGS 2013 http://www.bgs.ac.uk/londonBasinForum/JLBM.html) (Fig. 118.1). Work is ongoing, linked to the activities of the London Basin Forum.

The digitization of the dataset has enabled the distribution of the hollows to be considered in their geological and geographical contexts (Banks et al. in prep). GIS analysis has identified key geographical contexts that, in combination, are associated with known hollows: proximity to present-day river channels, artesian ground water conditions, thickness of London Clay, proximity to known geological structures. Based on this analysis, a provisional risk evaluation has been produced. This is a major step forward, but still needs to be refined, not least because knowledge of the location of the known hollows is influenced by a spatially skewed borehole data set. Despite this limitation, this GIS analysis has helped define some of the major contexts which will have controlled the formation of the hollows. These include availability of water, confining layers, geological structures and geomorphological position.

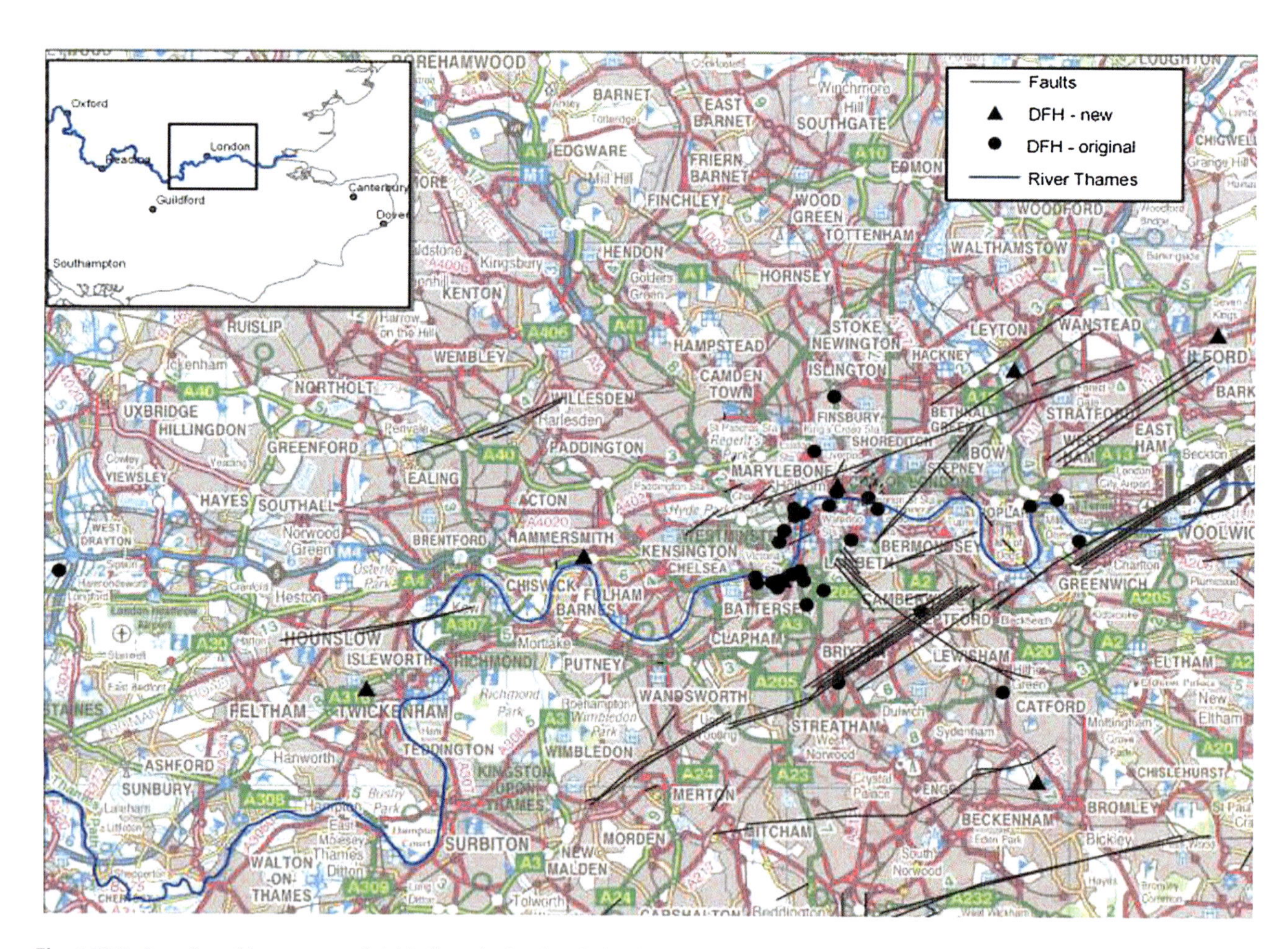

Fig. 118.1 Location of known superficial hollows in London. Point data British Geological Survey ©NERC. Contains Ordnance Survey data ©Crown Copyright and database rights 2014

118.3 Characteristics of Superficial Hollows in the London Basin

The full characteristics of the superficial hollows and the surrounding strata are generally unknown. Those that have a reasonable number of high quality borehole records, including some encountered during the CrossRail project, appear closed with no obvious inlet/outlet. Several are a few hundred metres across and can be up to 40–50 m deep. Most appear to have elongate forms.

The majority in central London are inaccessible for re-examination as they are located beneath buildings, roads and other infrastructure. Similar features, however, occur elsewhere in the London Basin, most notably in the valley of the River Kennet, a major tributary of the Thames. In this area, several hollows and their surroundings have been studied, both in quarry exposures and using boreholes. These include sites at Woolhampton, Brimpton and Ashford Hill (Fig. 118.2).

A detailed reconstruction of the London Clay rockhead at Woolhampton reveals a deep closed hollow with steep margins (Collins et al. 1996). Infilling strata showed evidence of post-depositional tilting towards the centre of the hollow that occurred between ~15 and ~11 ka BP. At nearby Brimpton, a similar hollow was infilled between ~100 and ~74 ka BP (Bryant et al. 1983; Worsley and Collins 1995), though no evidence of tilting was reported. No data are available for the strata below the rockhead at Woolhampton or Brimpton.

A feature at Ashford Hill has been described (Hawkins 1953; Hill 1985), based on boreholes, many of which extend below the base of the hollow. These permit a provisional 3 dimensional reconstruction of the subsurface stratigraphy (Fig. 4). The hollow is closed, but extends along the valley floor. Sediments in the hollow, indicate local lacustrine, marsh and fluvial conditions as the hollow formed, followed by subsidence of the hollow's centre and mass movements from the over-steepened margins. Deeper boreholes indicate that a mass of brecciated and puttied Cretaceous Chalk, has penetrated upwards through up to 60 m of Tertiary strata (Fig. 118.3).

118.4 Possible Origins

Several hypotheses have been proposed to explain the presence of superficial hollows and associated features in the London Basin. These vary from simple to complex, and most rely on assumptions of former conditions and limited data. Based on the features from the Kennet valley, hypotheses for the origin of the hollows can be assessed.

a. 'Pingo' (or related ground-ice form). Regional palaeoclimatic reconstructions suggest that permafrost is likely to have existed at various times, though the local evidence for it is ambiguous. The available data show no evidence of the ramparts that surround many relict and active pingos. Work on active pingos (Mackay 1998) suggest a ± planar base associated with the maximum depth reached by the massive segregated ground ice—this would not be likely to leave a deep hollow on melting. The infilling sediments suggest ongoing subsidence after hollow formation—at Ashford Hill, this may be continuing to present.
b. River scour. Some hollows may be due to locally deep erosion in the past. The Kennet hollows do not occur at confluence points, where scour is potentially at a maximum. Several of the hollows are also very deep, penetrating beyond the likely maximum depth of scour.
c. Localised consolidation settlement. The depth of the hollows is too great and the underlying strata are already over-consolidated.
d. Karst subsidence. Dissolution-prone puttied Chalk is present beneath the hollow at Ashford Hill. Boreholes penetrating solid Chalk nearby show the presence of cavities at depth. Both may have contributed as micro-faults, tilting and breccias in the hollow infill suggest that both slow and rapid collapse occurred.

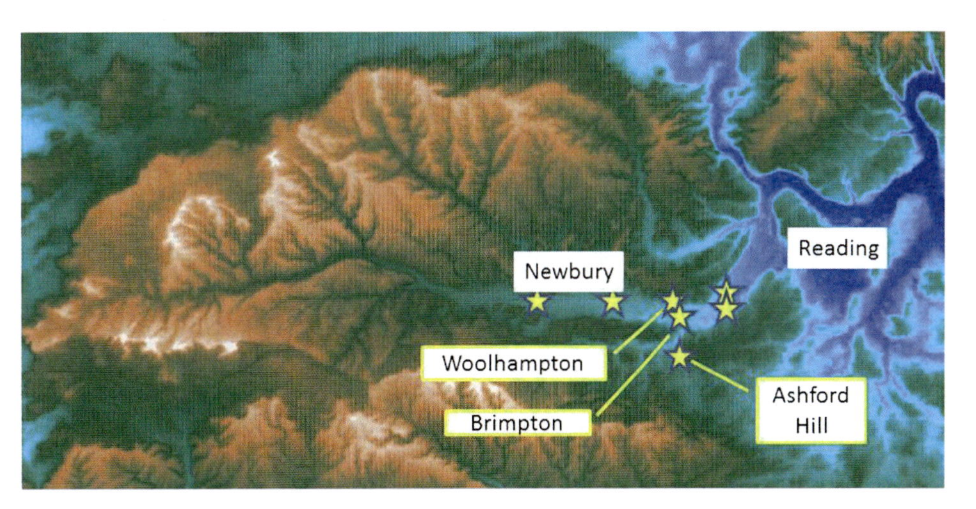

Fig. 118.2 Digital elevation model of the Kennet valley region, western London Basin (*dark blue* low elevation; *white* high elevation) showing location of known superficial hollows

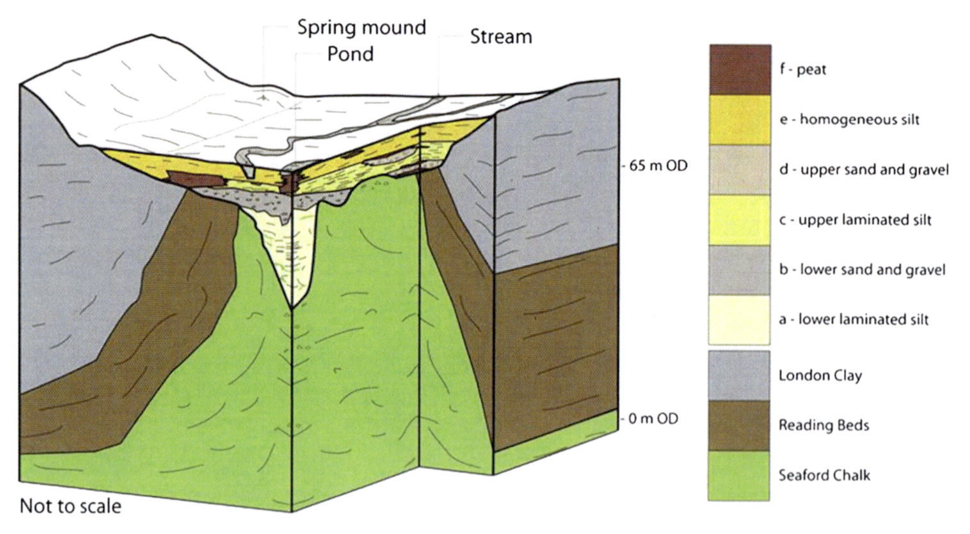

Fig. 118.3 Conceptual ground model for Ashford Hill, showing the superficial hollow and associated deformed strata

The origin of the Ashford Hill Chalk intrusion (similar features are known in central London), is uncertain. Freeze-thaw associated with the development and decay of permafrost may have been involved, though permafrost depth in former cold stages is uncertain. There is also a lack of experimental studies of how freeze-thaw affects heavily loaded Chalk.

However the Chalk was broken up, it was then mobilized. Loading was almost certainly involved as movement was towards the valley centre—this may have been through creep as the horizontal gradients are shallow. Upwards movement may have been driven by high groundwater pressures (Hutchinson 1980). A confining permafrost layer might have enhanced this, though cold stage regional groundwater tables may have been lower than present. The Ashford Hill valley appears structurally controlled and upwards movement of the Chalk may have exploited a pre-existing fault or joint. Additional loading to drive this might have come through a seismic event, though this is speculative. Flow structures were found in Chalk samples collected in the 1930s and 1980s, but these might have been due to sampling.

118.5 Conclusions

The superficial hollows and associated deformations remain problematic. The 'pingo' hypothesis, as a single causal mechanism at least, seems unsupported by the available evidence. Hollow formation can be explained by a simple hypothesis involving dissolution-driven subsidence, though uncertainty remains over the extent of areas affected by this, and whether it remains a significant hazard. A more complete understanding of the risk will only come when the nature and extent of Chalk intrusion is better understood.

References

Berry FG (1979) Late quaternary scour-hollows and related features in central London. Q J Eng Geol 12:9–29

BGS (2013) Joint London basin model. British Geological Survey. http://www.bgs.ac.uk/londonBasinForum/JLBM.html. Accessed 21 Oct 2013

Bryant ID, Holyoak DT, Moseley KA (1983) Late Pleistocene deposits at Brimpton, England. Proc Geol Assoc 94:321–343

Collins et al (1996)

Hawkins HL (1953) A pinnacle of chalk penetrating the Eocene on the floor of a buried river-channel at Ashford Hill, near Newbury, Berkshire. Q J Geol Soc Lond 108:233–260

Hill DT (1985) Quaternary geology of large-scale superficial features at Ashford Hill, Hampshire, England. Unpubl. PhD thesis, University of Reading

Hutchinson JN (1980) Possible late quaternary pingo remnants in central London. Nature 284:253–255

Mackay JR (1998) Pingo growth and collapse, Tuktoyaktuk Peninsula Area, Western Arctic Coast, Canada: a long-term field study. Geographie Physique et Quaternaire 52:271–323

Worsley P, Collins PEF (1995) The geomorphological context of the Brimpton late pleistocene succession (south central England). Proc Geol Assoc 106:39–45

Analysis of the Interaction Between Buried Pipelines and Slope Instability Phenomena

119

Lisa Borgatti, Alessandro Marzani, Cecilia Spreafico Margherita, Gilberto Bonaga, Luca Vittuari, and Francesco Ubertini

Abstract

Economic and social integration across Europe requires secure lifelines, such as roads, railways and pipelines. Existing and planned lifelines may come across a large number of different natural and anthropic hazards. For instance, past catastrophic events have dramatically shown that steel pipelines may be highly vulnerable to permanent ground deformation due to earthquakes and landslides. Therefore, their behaviour when exposed to processes that can generate large displacement and strain (co-seismic deformation and faulting, liquefaction, earth and rock slides and flows, rock falls) needs to be assessed. In fact, and in particular if toxic and/or flammable materials are transported, structural damage with eventual leakage might result in a severe risks for both human life and the environment, with associated relevant economic costs. To such purpose, in this work, a methodology for the assessment, prevention and efficient management of geological risks, mainly landslides, in steel buried pipelines will be presented. The proposed procedure aims at reducing the risk of environmental disasters and the subsequent huge financial and environmental losses.

Keywords

Landslide • Steel buried pipelines • Permanent ground deformation

119.1 Introduction

Industrialized countries have widespread and intricate networks of infrastructures that support and ensure the effectiveness of many crucial services for our society. Among the different lifelines, important pipeline networks for transportation of oil, gas and water are generally buried below ground surface for technical, aesthetic, safety, economic and environmental respect reasons. As pipeline systems are spread over large areas, they may come across a variety of possible natural hazards. In mountainous areas, for example, past catastrophic events have shown that pipelines are highly vulnerable to permanent ground deformation due to landslides. Therefore, the behaviour of these infrastructures when exposed to slope movements has to be assessed, in particular if toxic and/or flammable materials are transported. In fact, in case of structural damage, with eventual leakage, the risk for human life, for the environment and for the economy could be severe.

The assessment of these problems, the modelling and the mitigation measures have been tackled with in different studies (among others: Liu et al. 2010; Magura and Brodniansky 2012; Rajeev and Kodikara 2011; Zheng et al. 2012). In this context, the main purpose of this study is to determine the stress state induced by slope movements on buried steel pipelines, in order to develop tools and strategies for the monitoring and the prediction of residual life of the pipeline, together with new design solutions for their protection.

L. Borgatti (✉) · A. Marzani · C. Spreafico Margherita · G. Bonaga · L. Vittuari · F. Ubertini
Department of Civil, Chemical, Environmental and Materials Engineering DICAM, Alma Mater Studiorum, Università di Bologna, Viale del Risorgimento 2, 40136, Bologna, Italy
e-mail: lisa.borgatti@unibo.it

G. Lollino et al. (eds.), *Engineering Geology for Society and Territory – Volume 6*,
DOI: 10.1007/978-3-319-09060-3_119, © Springer International Publishing Switzerland 2015

119.2 Methodology

The method proposed by the Indian Institute of Technology Kanpur (2007) "Guidelines for seismic design of buried pipelines", was adopted to compute the maximum deformations in pipes due to ground deformation. Such methodology was selected among the available ones being very flexible with respect to possible different scenarios (longitudinal and transversal pipe crossing, liquefaction, faults, seismic action), as well as because simple and straightforward to be applied. The main idea is that the limit deformations in tension and compression caused by ground movements, namely Permanent Ground Deformation (PGD), should be lower than the allowable ones. Such limit deformations can be computed by proper formulae for two limit cases, i.e. the pipeline crossing longitudinally and transversally the PGD zone, and as a weighted sum of the two limit cases for pipelines crossing the PDG and an arbitrary angle. To such scope (i) the geometrical and mechanical data of pipeline, (ii) the geotechnical parameters of the soil, (iii) and the extension and magnitude of the PGD zone, must be known.

At this purpose, a georeferenced database was created (in UTM-ED50), elaborating data of different layers, within a GIS environment. In particular, the considered layers are: the database of the elements of the pipelines network (location, geometrical, physical and operational data) provided by the company operating the gas distribution service; the landslides inventory provided by Regione Emilia-Romagna geological survey; the geotechnical parameters of the soils inferred from lab data on typical landslide material in this geological context (clayey-silty soil: cohesion c = 10,000 Pa; specific weight γ = 20,000 N/m^3; friction angle ϕ = 18°).

At first the landslides were classified on the basis of the state of activity (dormant landslides and active landslides) and on the type of movement. In the investigated area, the landslides are about 800; 520 of them classified as dormant, while the others 280 defined as active (or suspended); they are generally characterized by slow movements (mm/year—cm/year). The majority of these movements are rockslide or earth slides and flows, which are also the most common types of landslides in the northern Apennines. In this geomorphological context, pipeline damage caused by landslides is mainly due to the intermittent reactivation of existing landslides generally due to intense and/or prolonged rainfall.

Next, superimposing the data of the different layers, the pipelines involved in instability phenomena were identified. In particular, 57 out of 103 pipeline elements of the so-called fourth type (i.e., with pressure in the interval 1.5–5 bar) were found in zones of prominent slope instability. For such pipes, the direction and magnitude of the ground movement with respect to the pipe direction was analysed, to distinguish among the two cases of longitudinal and transversal crossing, and so to allow a proper selection of the formula provided in the guidelines.

119.2.1 PGD Estimation

The PGD was estimated by means of two different approaches. The first one estimated the PGD occurred since the pose of the pipeline on the basis of geomorphological evidence and expert knowledge by assigning to each type of landslide a range of velocity (with minimum, average and maximum values, e.g. for very slow landslides, 0.005, 0.010 and 0.015 m/year respectively), determined according to the classification of Cruden and Varnes (1996), and considered to act at the depth of the pipe. The area and the length of the landslide bodies were obtained from the landslide inventory.

Available data from in situ geotechnical monitoring systems (e.g., inclinometers) have been used for the estimation of slope movements on the ground surface and at depth in the vicinity of pipelines. Inclinometer data were analysed to calculate the actual displacements. In general, the data appeared to be scarce, the instruments often placed far away from the pipeline and the monitored period not appropriate.

The second method to define the PGD used velocity data from satellite interferometry; the analysis procedures described in PST-A Guidelines (MATTM 2009) were followed. Ascending and descending ENVISAT images were collected between 2003 and 2008 in the frame of PST-A (Extraordinary Plan of Environmental Remote Sensing) project; they were processed by the companies T.R.E. S.r.l. and e-GEOS with the PS InSAR techniques. These datasets were resampled on a square grid mesh in order to make them comparable: the average velocity value of the Permanent Scatterers (PS) located inside each grid cell was assigned to the centroid of the cell. The velocity values recorded along the ascending and descending orbits were geometrically combined to obtain the velocities along the vertical and E-W horizontal direction; it must be assumed that the N-S horizontal velocity of deformation is negligible (one of the main limitations of the PS technique is, in fact, the difficulties in recording horizontal movements along the N-S direction). These velocities were combined obtaining an averaged value to be used in the formulas.

119.3 The Case Study

The municipality of Santa Sofia (Forlì-Cesena, Emilia-Romagna, Italy) was chosen as case study. One-sixth of the area of this municipality is affected by landslides of various types and state of activity and it can be considered as representative of the geological and geomorphological setting of the northern Italian Apennines. The considered steel pipe network for gas distribution has an overall length of 42 km and about a quarter of the pipelines of the network are exposed, in different ways, to landslides. The pipelines are of the so-called continuous type, i.e. the various segments are welded together, in steel, and are classified, according to the operative pressure, in pipes of the fourth type (i.e., pressure 1.5–5 bar) and of the sixth type (i.e., pressure 0.004–0.5 bar). In this work, pipelines of the fourth type, characterized by a larger diameter, as they generally imply a greater danger in case of catastrophic events, were considered. The most common causes of failure for these pipes are related to the high tensile stresses in correspondence of the welded joints (yielding and ruptures) as well as to the relevant axial compression on the walls of the pipeline (local instabilities). An examples is presented, describing a fourth type pipeline crossing transversely a dormant landside (Fig. 119.1). Velocity data available from 2003 to 2008 from four PS positioned inside the landslide area and rather close to the pipelines were analysed. Results show a very good correspondence with the estimated range of expected velocities from Cruden and Varnes (1996). In fact, the PS results provided an average velocity of about 5 mm/year along E-W direction, toward east, and about 2 mm/year in the vertical direction, whereas Cruden and Varnes assume 6 mm/year for slow landslides. An average velocity of 5.84 mm/year was assumed for the computation of the PGD.

For the transversal crossing the maximum axial strain in the pipe due to the PGD was calculated as the minimum resulting from two pipe-soil models: the first refers to a large width of PGD zone and pipeline is assumed to be flexible; whereas the second refers to narrow width of PGD zone and pipeline is assumed to be stiff. In both models the initial stresses in the pipelines due to internal pressure and temperature change, were added to the value of maximum axial strain due to PGD and then compared to the allowable strains for continuous pipelines according to the guidelines proposed by ASCE (1984), ALA (2005) and JSCE (2000).

119.4 Discussion

In Fig. 119.2 the results for the considered steel pipeline are shown. In particular, the PGD at year 1998 was considered null and only strains due to operative pressure and temperature were considered. From 1998 to 2080 the maximum pipe strain was computed according to the "Guidelines for seismic design of buried pipelines" (2007) by adding to the initial value the strain due to the PGD. This last term was computed by assuming an average velocity of 5.84 mm/year, as the one computed from ENVISAT data in the period 2003–2008. By doing so, the estimated residual life of the pipelines can be assessed by comparing the maximum strain with the allowable strain that in this case is related to a

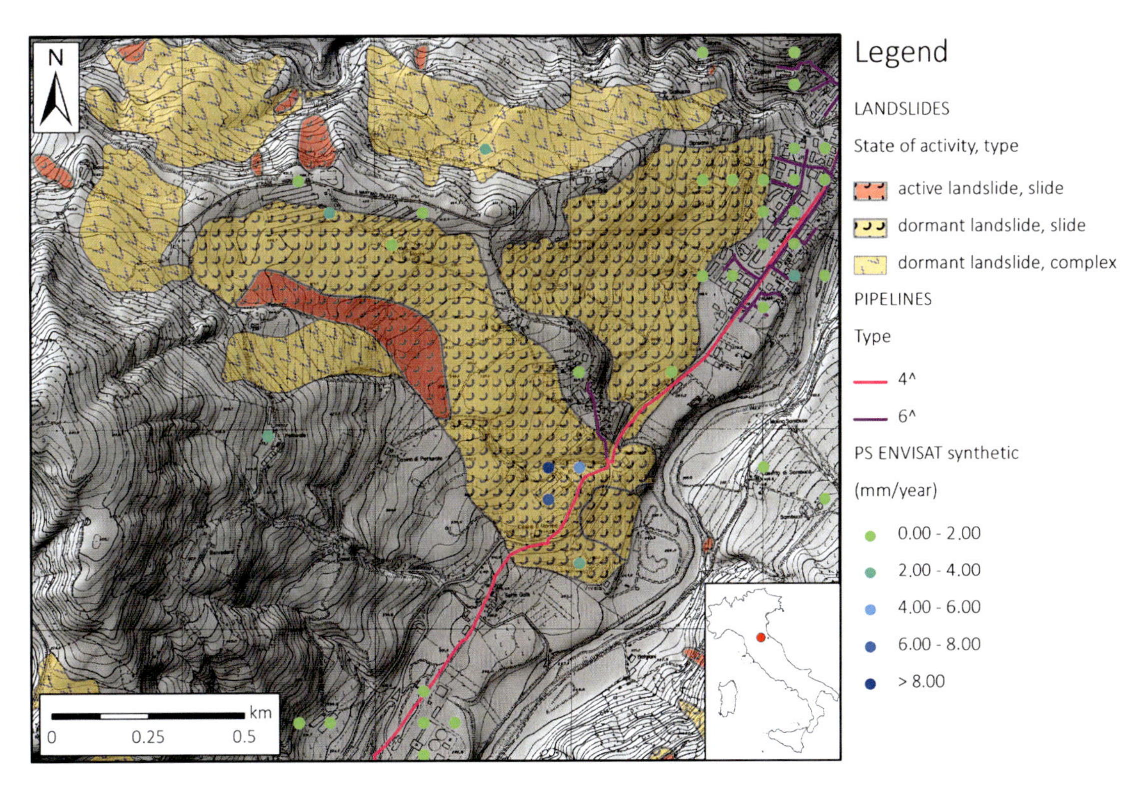

Fig. 119.1 Geographical location and sketch of landslide 35101, in Santa Sofia municipality (Forlì Province, northern Apennines, Italy) and pipelines affected by slope instability processes

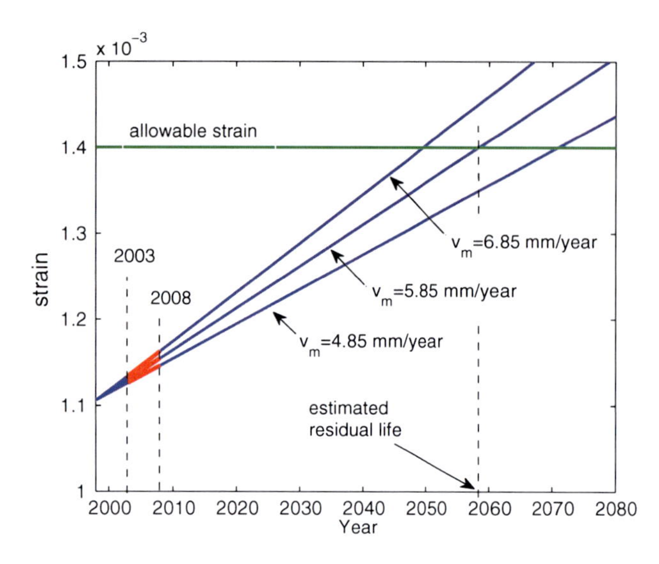

Fig. 119.2 Maximum pipe strain due to temperature and pressure (year of pose 1998) along with strain due to PGD (velocities extrapolated from ENVISAT data measured in the period 2003–2008)

compressive strain that can yield pipe wrinkling. Of course, this estimation strongly depends on the assumption of an active landslide body displaying steady-state displacements during the considered period. Further calculations could consider actual velocity and acceleration (i.e., rapid dynamic actions), as well as depth of movements, based on in situ geotechnical instruments. Moreover, differential velocities could be accounted for inside the landslide body.

119.5 Conclusions

The results show that, on the basis of technical data on the pipeline and of the available data on slope instability phenomena, it is possible to calculate the tensions that have acted and that are acting on the pipeline, due to PGD. This computation can be performed easily and eventually updated in near-real-time. In addition, the projection of the PGD in time allows for an estimation of the pipelines residual life. The outlined procedure is a suitable tool in a proper strategy for assessing and managing the risk under both ordinary and emergency conditions in mountainous areas prone to slope instability phenomena. Moreover, the procedure, by assigning to the various sections and components of the network a value of relative vulnerability, allows targeted strategies of network maintenance to be programmed.

References

ALA (2005) Seismic guidelines for water pipelines. A report by public-private partnership between Federal Emergency Management Agency (FEMA) and National Institute of Building Sciences (NIBS), American Lifelines Alliance (ALA)

ASCE (1984) Guidelines for the seismic design of oil and gas pipeline systems. Committee on gas and liquid fuel lifelines, American Society of Civil Engineers (ASCE), US, Nyman DJ (Committee chairman)

Cruden DM, Varnes DJ (1996) In: Turner AK, Schuster RL (eds) Landslides: investigation and mitigation. Transportation Research Board Special Report 247. National Academy Press, WA, p 36

Indian Institute of Technology Kanpur (2007) IITK-GSDMA Guidelines for seismic design of buried pipelines (Indian Institute of Technology Kanpur, Gujarat State Disaster Management Authority National Information Center of Earthquake Engineering—NICEE Publication)

JSCE (2000) Recommended practices for earthquake resistant design of gas pipelines. Earthquake resistant codes in Japan. Japan Society of Civil Engineering (JSCE), Japan Gas Association

Liu PF, Zheng JY, Zhang BJ, Shi P (2010) Failure analysis of natural gas buried X65 steel pipeline under deflection load using finite element method. Mater Des 31(3):1384

Magura M, Brodniansky J (2012) Experimental Research of Buried Pipelines. Proc Eng 40:50

MATTM (2009) Direzione Generale per la Difesa del Suolo del Ministero dell'Ambiente e della Tutela del Territorio e del Mare. Linee guida per l'analisi di dati interferometrici satellitari in aree soggette a dissesti idrogeologici. http://www.pcn.minambiente.it/GN/leggi/LINEE%20GUIDA%20PER%20ANALISI%20DI%20DATI.pdf. Accessed 5 May 2014

Rajeev P, Kodikara J (2011) analysis of an experimental pipe buried in swelling soil. Comput Geotech 38(7):897

Zheng JY, Zhang BJ, Liu PF, Wu LL (2012) Failure analysis and safety evaluation of buried pipeline due to deflection of landslide process. Eng Fail Anal 25:156

The Importance of Rockfall and Landslide Risks on Swiss National Roads

120

Philippe Arnold and Luuk Dorren

Abstract

To obtain standardized information on the type, frequency, intensity and location of natural hazards that threaten national roads, the federal roads office FEDRO initiate a Swiss-wide project in 2008. This paper presents the methodology used in this project and presents a summary of the monetarised risks of the evaluated road sections. The natural hazards that need to be assessed are snow avalanches, rock- and icefall, flooding, debris flows, landslides (permanent, spontaneous and slope type debris flows) and collapse dolines. Risk hot spots mainly occur due to road closure related to rockfall or bank erosion. Damage to infrastructure represents generally only up to 20 % of the total calculated risk; person risks (casualties) add up to 8 % of the total risk. Rockfall is responsible for 35 % of the total calculated risk, rock avalanches for 8 %, permanent landslides for 5 %, spontaneous landslides for 3 % and slope-type debris flows for 1 %.

Keywords

Risk analysis • Infrastructure • Landslides • Rockfall • Switzerland

120.1 Introduction

Natural hazards such as avalanches, rockfall, landslides and flooding persistently threaten Alpine regions (e.g., BUWAL 1999; Rudolf-Miklau et al. 2006; Bezzola and Hegg 2007). The rockfall event of June 2006 on the Gotthard highway, the road destructing flooding and landsliding events in August 2005 or the numerous snow avalanches in the winter of 1999 show that road infrastructure, its users and its availability are vulnerable to the impact of natural hazards. Since January 2008, the federal roads office (FEDRO) is responsible for the Swiss national road network (highways and the main alpine passes). Before then, the national roads were managed by Cantonal road services. As a result, Swiss-wide, standardized information on the type, frequency, intensity and location of natural hazards that threaten national roads, as well as the costs of required protective measures, was not available. The FEDRO therefore decided to initiate a swiss wide project, called “national hazards on national roads—NHNR” with the technical support of the Federal Office for the Environment (FOEN), aiming at quantifying and mapping all risks due to natural hazards threatening Swiss national road network (total length = 1892 km) comparable to the work of Roberds (2005). At the moment of writing, almost all alpine national roads have been analysed, which allows to analyse the importance of rockfall and landslide related risks on national roads. This paper will present the methodology used in the project and presents a summary of the monetarised risks of the evaluated road sections.

P. Arnold (✉)
Federal roads office FEDRO, Bern, Switzerland
e-mail: philippe.arnold@astra.admin.ch

L. Dorren (✉)
Bern University of Applied Sciences, Zollikofen, Switzerland
e-mail: luuk.dorren@bfh.ch

L. Dorren
Federal office for the environment FOEN, Bern, Switzerland

G. Lollino et al. (eds.), *Engineering Geology for Society and Territory – Volume 6*,
DOI: 10.1007/978-3-319-09060-3_120, © Springer International Publishing Switzerland 2015

120.1.1 Project Organisation

The FEDRO is in charge of the project coordination; technical support and expertise on hazard and risk analysis is provided by the division of hazard prevention of the Federal Office for the Environment (FOEN). The field and modeling studies needed for the hazard and risk analysis are being done by consortiums of collaborating geotechnical firms. They work on sections of the Swiss national road network with a length of 30–70 km. In general, each consortium consists of an interdisciplinary project leader with experience in natural hazards, an avalanche expert, one or two geological experts, a hydraulic engineering/flooding expert and a risk analysis expert. Approximately one and a half year is available to complete the natural hazard assessment in a given highway section and subsequently, the risk analysis is finalised within the following 3–4 months.

The developed project methodology, which will be described in the subsequent section, allows for a very detailed risk analysis. The key challenge is to ensure a similar level of detail in the hazard assessments of the different geotechnical bureaus. Our approach therefore obliges a discussion on the hazard scenarios for the different hazard sources with the experts in the terrain in the first half year of each subproject per highway section. The results of the hazard assessment proved that this is essential and effective. Also, a thorough proof of the hazard assessment results and the underlying, transparently presented, calculations and assumptions has shown to be of key importance. In total, 45 subprojects for different highway sections are defined; 20 of these are finished.

120.2 Project Methodology

The detailed project methodology is published in FEDRO (2012). This methodology describes in detail the following 4 main parts: (1) hazard assessment, (2) risk analysis, (3) risk evaluation and (4) planning of protective measures. As such the methodology defines the natural hazards to be studied, the study perimeter, the standards to be used, the risk equations, and parameter values to be used, as well as the products to be delivered. The methodology does not prescribe models for simulating the different natural hazards to be assessed; it only prescribes the required products in detail, as well as a maximum transparency and traceability of the methods, models and assumptions used.

The natural hazards that need to be assessed are:

- snow avalanches
- rock- and icefall, rock avalanches
- flooding and debris flows
- landslides (permanent, spontaneous and slope type debris flows)
- collapse dolines

To aim for a Swiss-wide homogeneous and comparable dataset, 4 event-size scenarios (return period 0–10 yrs, 10–30 yrs, 30–100 yrs, 100–300 yrs and intensity classes low, medium and high) should be defined for each the potential hazard source area. The so called damage potential perimeter that is to be taken into account is the area covered by the highway with a 10 m buffer, as well as surrounding facilities (e.g., parking places, technical tunnel installations, ...). The risk analysis is carried out on one or two lines that represent the road axes and on surrounding facilities.

To standardise the risk analysis, we developed an internet based risk calculation tool called RoadRisk (http://www.roadrisk.admin.ch/). This tool is WebGIS-based and intersects the intensity maps of all studied natural hazards for the defined return periods with the damage potential. It calculates the total risk (cf. Fig. 120.1) based on the following "damage" types:

- Direct impact (Rdirect),
- Collision with deposits on the road or with cars that are impacted by natural hazards (Rcollision),
- Damage to infrastructure (Rdamage),
- Precautionary road closure (Rpreclos),
- Road closure after an event (Rpostclos).

Casualties are also expressed in costs and is based on a value of 5 million CHF per human life. Therefore, variables required are the maximum speed defined at the highway section, the average number of cars passing daily, the probability of having a traffic jam, the daily costs for road closure (varying between 150,000 and 4,000,000 CHF/day), the lethality of the people in a car being hit or colliding with deposits, etc. To calculate the daily costs due to road closure, we use a model developed by IVT-ETH (Erath 2011). This model allows quantifying the costs caused by the closure of a road section in the national road network taking into account:

- time loss due to detours for passenger cars and lorries
- increase of costs due to an increase in the number of accidents
- cancelled professional and leisure trips

Both a summer and winter scenario for road closure have been calculated for all section in the national road network. The winter scenario accounts for the closure of alpine passes due to snow in wintertime. The final step in the Swiss-wide national hazards on national roads project is the risk evaluation. For that purpose, three evaluation criteria have been defined to identify risk hot-spots.

Criterion (1) is the individual probability of death. For this FEDRO defined a threshold of 1 $*$ 10–5 per year. The value applies to road users who regularly drive on a national road section. If this threshold is exceeded due to natural hazards, all possible protective measures for the problematic hazard source have to be evaluated. The limit value of the

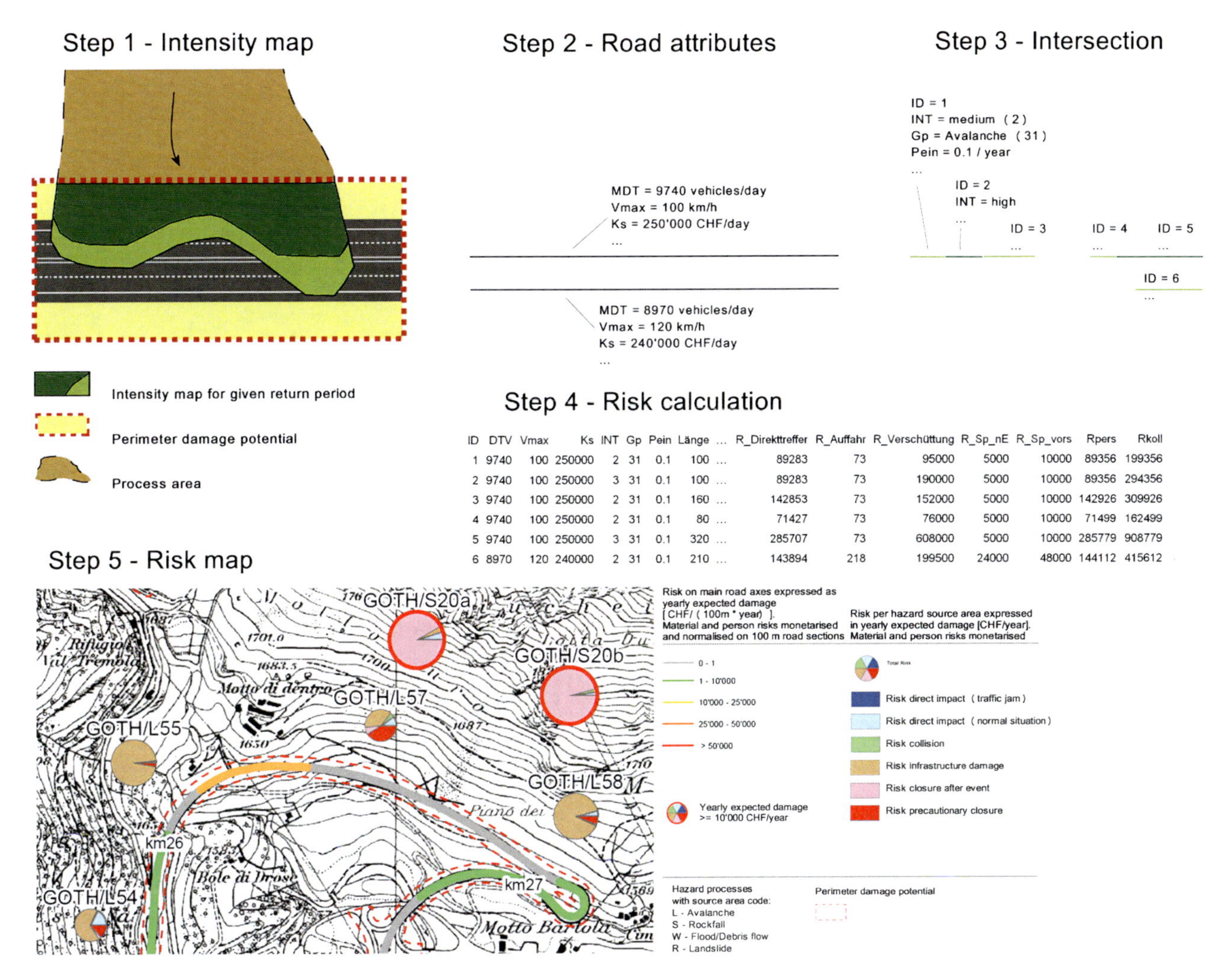

ID	DTV	Vmax	Ks	INT	Gp	Pein	Länge	...	R_Direkttreffer	R_Auffahr	R_Verschüttung	R_Sp_nE	R_Sp_vors	Rpers	Rkoll
1	9740	100	250000	2	31	0.1	100	...	89283	73	95000	5000	10000	89356	199356
2	9740	100	250000	3	31	0.1	100	...	89283	73	190000	5000	10000	89356	294356
3	9740	100	250000	2	31	0.1	160	...	142853	73	152000	5000	10000	142926	309926
4	9740	100	250000	2	31	0.1	80	...	71427	73	76000	5000	10000	71499	162499
5	9740	100	250000	3	31	0.1	320	...	285707	73	608000	5000	10000	285779	908779
6	8970	120	240000	2	31	0.1	210	...	143894	218	199500	24000	48000	144112	415612

Fig. 120.1 Explanation of the major steps in the risk calculation tool. *1* Import of the intensity map for a given process and return period; *2* Importation of the road axis lines and accompanying attributes; *3* Intersection of the road axes and the intensity map; *4* Creation of a joined attribute table and risk calculation per damage type; *5* Display of the risk values in a map

probability of death of 1 ∗ 10–5 per year is derived from the average probability of fatalities of all 15-year-olds in Switzerland. Here, a risk of death 10–100 times lower is assumed to be acceptable by society for deaths caused by natural hazards. The higher value (1 ∗ 10–5) is selected based on PLANAT (2009). The other two evaluation criteria are (2) total risk per road section >10,000 CHF/100 m ∗ year and (3) total risk per hazard source or per secondary facility >10,000 CHF/year.

No protection goals with absolute limit values are defined for the collective or person risks. At locations where one of these three criteria is exceeded, possible risk reduction measures will be studied. These measures will subsequently be implemented only if those measures prove to be cost-effective. This means that the yearly risk reduction should be equal or larger than the yearly expected damage.

120.3 Results

The results of the 20 completed subprojects (approx. 30–70 km highway sections) show that potential damages per 100 m highway section due to gravitational natural hazards can add up to several 100,000 CHF/year, summing up to several millions CHF/year for entire road sections. Figure 120.2 shows the risk hot spots on the completed highway sections.

In most cases risk hot spots occur due to road closure in areas that are strongly affected by rockfall or bank erosion. Details (as shown in Fig. 120.3) show that direct damage to infrastructure represents generally only up to 20 % of the total potential damages, but can reach up to 40 % for some hazard sources. Person risks (victims) are mostly to be expected due to falling rocks, rock avalanches, debris flows

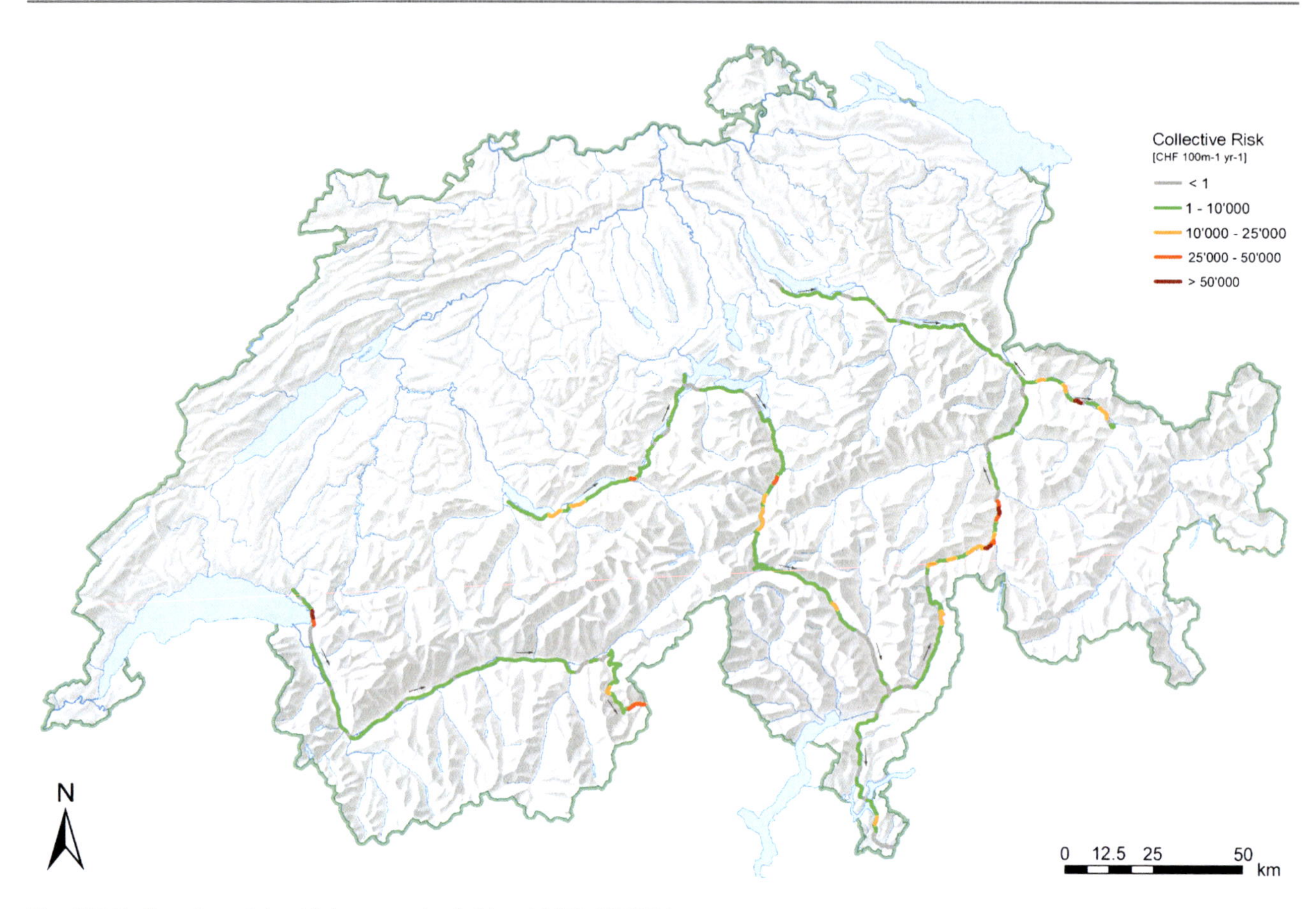

Fig. 120.2 Overview of the risk hot spots (total risk > 10,000 CHF/100m * year) on the Swiss national road sections that have been analysed between 2008 and 2014

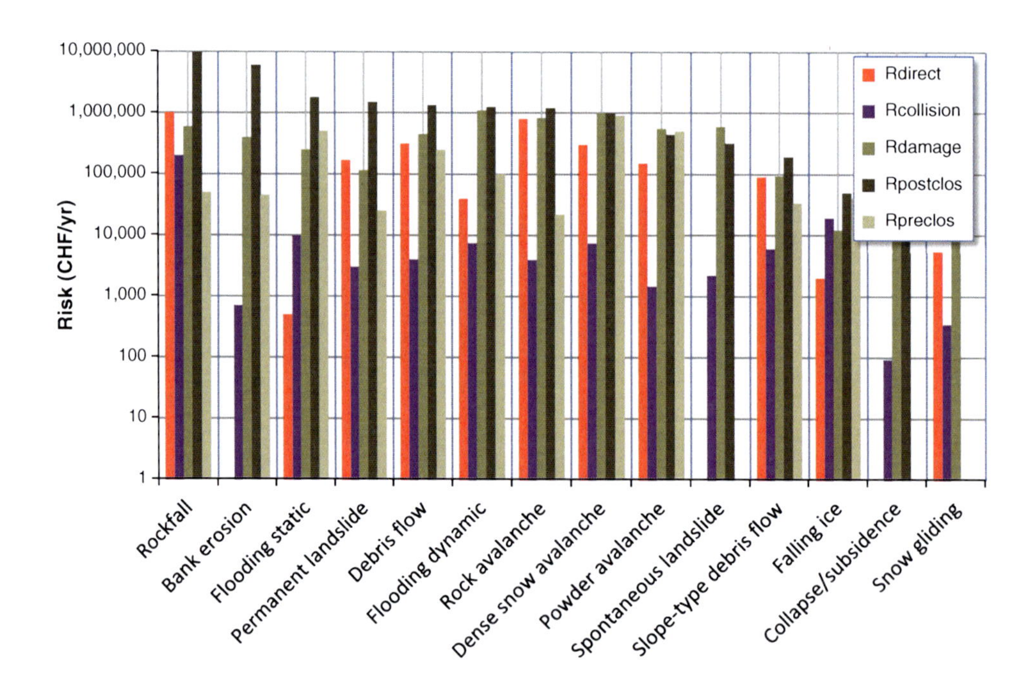

Fig. 120.3 Risk details per hazard and damage type for all analysed road sections

and dense snow avalanches, which add up to 8 % of the total risk. Road closure is responsible for 75 % of the total calculated risk on the national roads that have been assessed so far. Regarding geological mass movements, the results show that rockfall is responsible for 35 % of the total calculated risk, rock avalanches for 8 %, permanent landslides for 5 %, spontaneous landslides for 3 % and slope-type debris flows for 1 %. Finally subsidence processes due to doline collapse account for 0.1 %.

References

Bezzola G-R, Hegg C (2007) Ereignisanalyse Hochwasser 2005. Teil 1 – Prozesse, Schäden und erste Einordnung. Bundesamt für Umwelt BAFU, Bern und der Eidgenössischen Forschungsanstalt für Wald, Schnee und Landschaft WSL, Birmensdorf

BUWAL (1999) Leben mit dem Lawinenrisiko - Die Lehren aus dem Lawinenwinter 1999. Bundeamt für Umwelt, Wald und Landschaft (BUWAL), Bern

Erath A (2011) Vulnerability of road transport infrastructure. Ph.D. Thesis, ETH Zürich, Zürich

FEDRO (2012) Natural hazards on national roads: risk concept. Methodology for risk-based assessment, prevention and response to gravitative natural hazards on national roads. ASTRA Documentation 89001. Federal Roads Office, Bern. http://www.astra.admin.ch/dienstleistungen/00129/00183/01156/index.html?lang=en

PLANAT (2009) Risk concept for natural hazards, strategy natural hazards Switzerland (in German, www.planat.ch)

Roberds W (2005) Estimating temporal and spatial variability and vulnerability. In: Hungr O, Fell R, Couture R, Eberhardt E (eds) Landslide risk management. Taylor and Francis Group, London, pp 129–158

Rudolf-Miklau F, Ellmer A, Gruber H, Hübl J, Kleemayr K, Lang E, Markart G, Scheuringer E, Schmid F, Schnetzer I, Weber C, Wöhrer-Alge M (2006) Hochwasser 2005 - Ereignisdokumentation. Bundesministerium für Land- und Forstwirtschaft, Umwelt und Wasserwirtschaft, Sektion Forst, Abteilung Wildbach- und Lawinenverbauung, Wien

Part XII

Innovative Methods in Characterization and Monitoring of Geotechnical Structures

Convener Prof. Anna Maria Ferrero—*Co-conveners* Marco Barla, Andrea Segalini, Gessica Umili

Technological and scientific developments have contributed to improve geological-geotechnical characterization and monitoring of geotechnical structures. This is due to the use of miniaturized sensors, new materials, data acquisition, transmission and analysis by web-oriented platforms. Although the measured quantities remain the same (displacements, rotations, strains, stresses, water pressures, temperatures, etc.) monitoring and data acquisition systems are dramatically changed. The quality and quantity of the data made available "real-time" contribute to the assessment of the actual behaviour. According to Eurocode7 the observational method in which the design is reviewed during construction is a consolidated practice in geotechnical engineering. This is strictly connected to measurement to ascertain that the actual behaviour is within the acceptable limits. Since the nature of the acquired data is fundamental, the innovations will significantly improve the design approach.

TLS Based Determination of the Orientation of Discontinuities in Karstic Rock Masses

121

Th. Mutschler, D. Groeger, and E. Richter

Abstract

One of the major tasks in quantitative description of discontinuities in rock masses is the determination of their spatial orientation. In recent years, Terrestrial Laser Scanning (TLS) allows a three-dimensional point mapping of surface points with high accuracy. By thinning and meshing the mapped points with a specific software tool, the natural surface can be modeled by planar elements and analyzed in standard CAD software. If the surface represents discontinuity planes and if it is not heavily distorted by natural processes such as erosion, a simple algorithm can be used that allows to identify discontinuities and to determine their spatial orientation. The paper describes the application of this method in a karst cave. TLS measurements were carried out under very harsh conditions. The results of the procedure are presented and their validity is proven on a chosen surface area by comparing the results to in-situ measurements with a stratum compass.

Keywords

Discontinuity systems • Dip and strike • TLS • Karst cave

121.1 Introduction

The behavior of rock mass is dominated by its discontinuities; hence the information about their spatial orientation is an important issue (ISRM 1978). Usually, the spatial orientations are measured with a stratum compass. If several sets of discontinuities exist and the area of interest is large, such measurements are highly time-consuming. Several authors have applied different remote sensing methods, such as stereophotography and digital image processing (Gaich et al. 1999) and later the processing of point clouds generated by Light Detection And Ranging (LiDAR, Kemeny 2005) or TLS, to determine the spatial orientation of discontinuities. Most of the applications of remote sensing methods were related to above-ground problems. Subsequently, the usage of remote sensing methods in a karst cave in Indonesia is presented.

121.2 Location, TLS Hardware and Scanning Technique

The cave "Gua Seropan" is located in the south of Middle Java, Indonesia. The lime rocks in this region are intensively karstified and hold huge caves systems. Ongoing solution and erosion processes mainly caused by a cave river have created canyon-like karst structures with varying morphology. The area of highest interest is located around a waterfall, where the three most dominating sets of discontinuities are clearly visible (Fig. 121.1).

In this part of the cave the canyon-like cross section changes to a hall-like shape. The walls of the cave consist of joint faces and the roof is formed by stratification faces.

Th. Mutschler (✉)
Institute of Applied Geosciences, Karlsruhe Institute of Technology, Karlsruhe, Germany
e-mail: thomas.mutschler@kit.edu

D. Groeger
Institute of Geotechnical Engineering and Mine Surveying, Clausthal University of Technology, Clausthal, Germany

E. Richter
COS Systemhaus, Ettlingen, Germany

G. Lollino et al. (eds.), *Engineering Geology for Society and Territory – Volume 6*,
DOI: 10.1007/978-3-319-09060-3_121, © Springer International Publishing Switzerland 2015

Fig. 121.1 Dominating sets of discontinuities (*left*); eroded discontinuity faces (*right*)

However, not all parts of the faces around the waterfall are suitable for the analysis presented in this paper. At the lower sections of the walls the solution and erosion processes resulted in wash-outs of the joint faces (Fig. 121.1) and the detection as flat planes is not possible anymore.

Within the scope of the IWRM-project—Indonesia (http://www.iwrm-indonesien.de), the surface of the karst cave was mapped by TLS in 2009 (Schmitt 2010). The TLS system deployed in Gua Seropan was a Leica HDS 6000, which works with the phase-shift-method. In contrary to the time-of-flight-method (TOF), the phase-shift method works much faster. Consequently it enables higher resolution on the object surface in the same time. Moreover, it provides higher short-range accuracy. The distance accuracy of a single measurement is 5 mm at 25 m at a surface reflecting of only 18 % of the incident light. Its main disadvantage of a shorter adjustment distance was of no significance. The high resolution scans with lateral point distances of <10 mm covered even small details and discontinuities. Compared with photogrammetric systems the presence of heavy spin-drift and poor light conditions were no problem for TLS.

The high, narrow and stretched geometry of the cave demanded closely neighbored scanner positions to catch all details. The lateral orientation of the scans was determined with the help of targets (spheres and black and white targets) (Fig. 121.2). Whenever possible, the targets were placed on survey points with known coordinates. Despite a maximum range of 79 m, the mean distance between two scanner positions was chosen to be about 15 m. Thus, the cave was mapped from 19 different scanner positions.

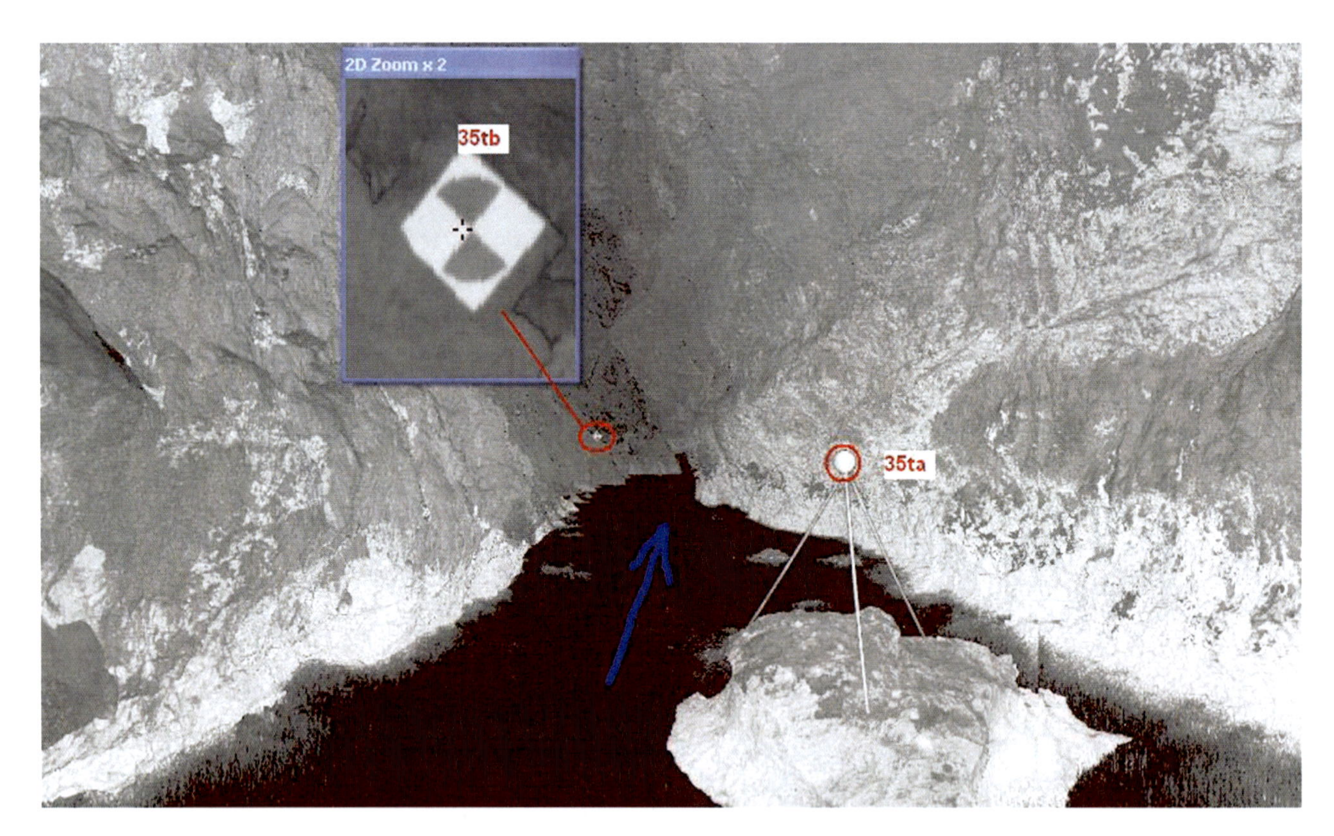

Fig. 121.2 Intensity image of a point cloud with a *flat black* and *white* target as well as a *sphere* on a *tripod*

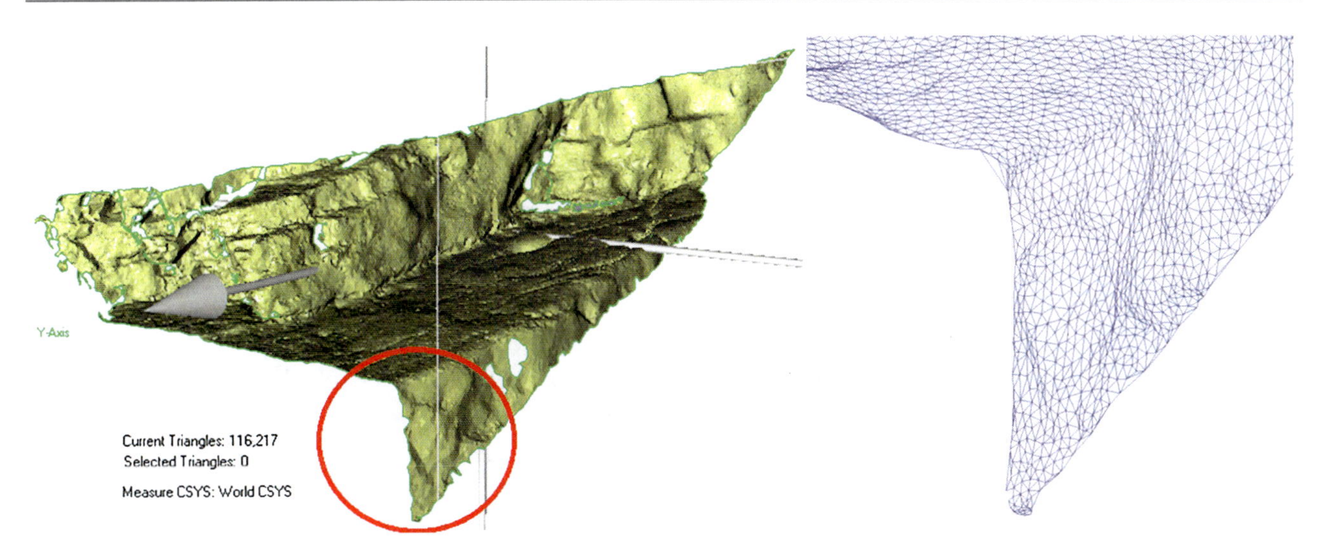

Fig. 121.3 Section of the meshed point cloud, 8 m long (*left*); view of the area inside the *red circle*, *triangle* with 2 cm edge length (*right*)

121.3 Processing of Point Clouds and CAD Models

Pre-processing of TLS measurements was done in several steps, beginning with the removal of erroneous and surplus measurements in each point cloud, followed by the positioning of the point clouds relatively to each other. The point clouds were oriented by both matching of overlapping parts and artificial targets (Fig. 121.3) that are part of the point cloud. After the relative orientation and the merging of the point clouds, the absolute orientation of all scans inside a superior coordinate system was calculated.

The modeling of the natural cave surface was done by calculating a mesh of triangles or polygons. For further data processing, the point clouds were converted to a CAD data format. The focus, hereby, lay in the thinning of the data sets without changing or losing the orientation of the natural surface of the cave.

There are different kinds of algorithms to reduce the number of the triangles of the mesh. The simplest approach, the uniform sampling, is re-calculating the point cloud or mesh, so that the points are equidistant to each other. This does not necessarily preserve all discontinuities and results in a huge amount of data. Yet, if smooth surfaces are given this approach is better suited. Here the requirement of smooth surfaces was only granted for some parts of the cave walls and the roof. Therefore, the so called curvature-based filter was used. It allows the preservation of the discontinuities by reducing the number of points for meshing the surfaces with triangular elements. As a result, the generated meshes possessed a low density inside flat areas and became higher near and across edges.

For the whole cave 15 CAD section models were generated of which 4 were used for the most interesting area around the waterfall. Figure 121.4 shows one model which is divided in three parts: the roof (blue), the sidewalls (silver) and the area which was not investigated because of coverage with rock fall material (black). In addition to the analysis of a whole model, specific areas, which are mainly dominated by discontinuity faces, were chosen to validate the results of the whole model.

The coordinates of each triangular element were extracted from the CAD models to a spreadsheet. By applying a simple algorithm, the dip angle and the dip direction were calculated. The fact, that the curvature-based filter led to different areas area sizes of the triangles, was handled by a weighting factor that incorporated the area of a single triangular element to the over-all area of the model. If the area of the discontinuity faces dominates the over-all area of the cave surface, an analysis of the spatial orientation of the discontinuities can be carried out.

121.4 Results

In order to validate the procedure, the algorithm was applied to a surface area with dominant discontinuity faces and the results were compared to field measurements with a stratum compass. Figure 121.5 shows the results of both the field measurements and the analysis of the CAD the Lambert azimuthal equal-area projection.

It can be seen, that a good match between the field measurements (263°/72°) and the analysis of the CAD model (250°/78°) is achieved. Besides the spatial orientation of the joint face, the spatial orientation of the bedding can be seen from the analysis of the CAD model.

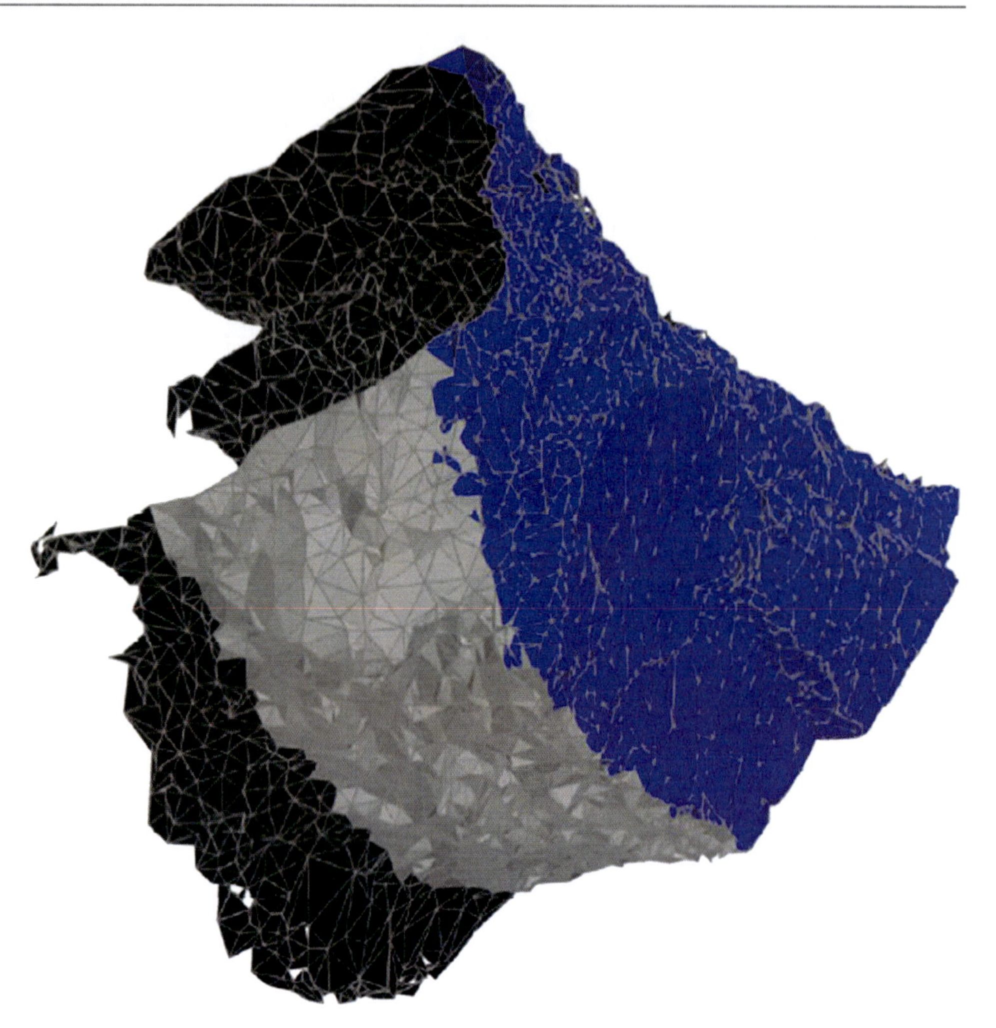

Fig. 121.4 CAD model around the water fall

By splitting the 4 models into roof and wall, the joint sets and the bedding set could be considered separately (Fig. 121.6). The spatial orientation of the bedding has a dip direction of 20° and a dip angle of 10° approximately. In addition, the right figure indicates two joint sets. One has a dip angle of nearly 90° and a dip direction of around 250° and the other dips with 60° in a direction of 130°.

From the analysis it can be seen that if the discontinuity faces dominate the surface of the cave, a further revision of the models by excluding wash-outs or other phenomena changing the natural discontinuity faces must not be considered. As a result of the genesis of a karst cave this requirement is usually satisfied, so that the analysis introduced is supposed to hold in general.

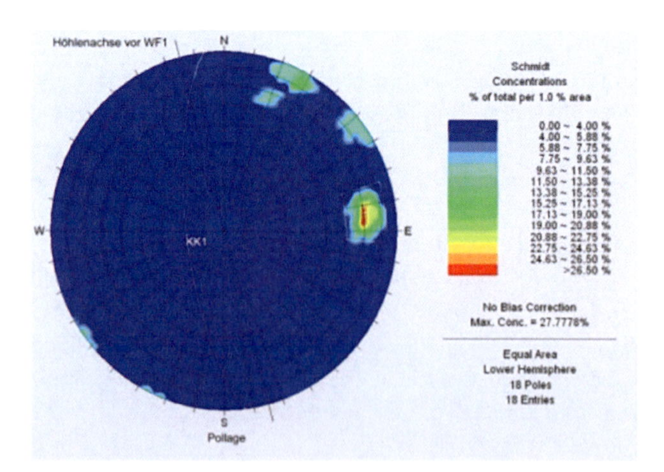

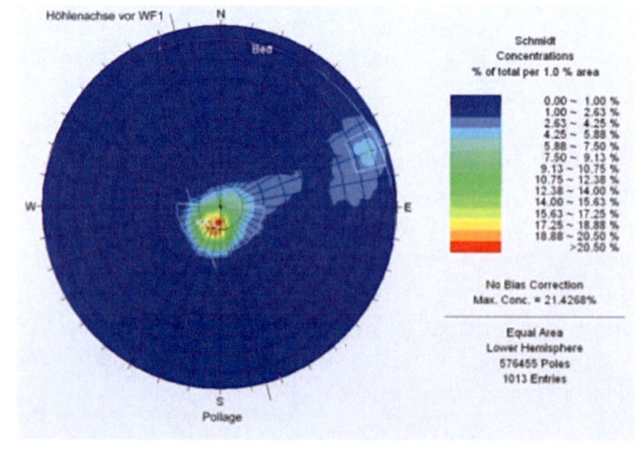

Fig. 121.5 Field measurement (*left*), CAD model (*right*)

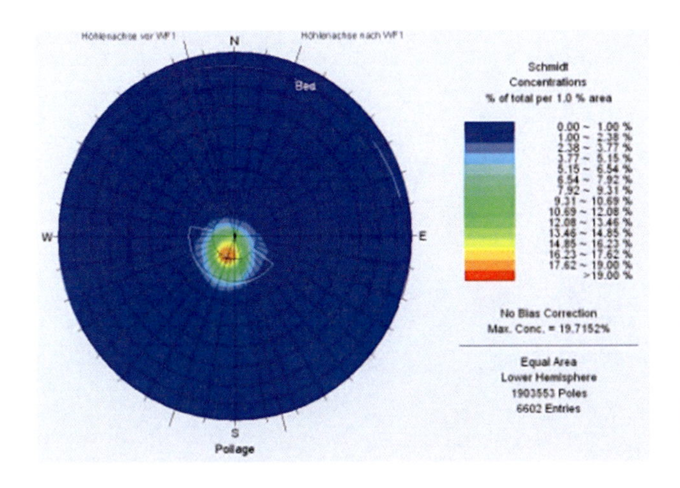

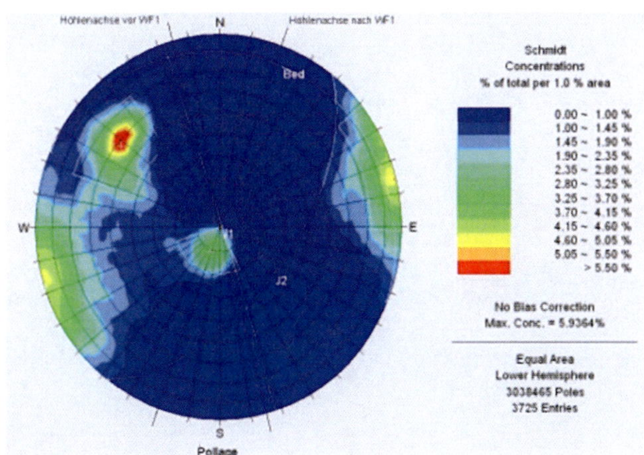

Fig. 121.6 Cave roof (*left*), cave walls (*right*)

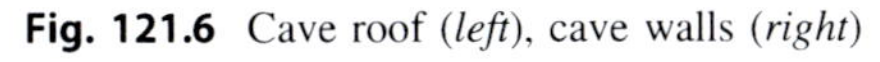

121.5 Conclusions

The following conclusions can be drawn:

- TLS is suited to provide the data for the determination of the spatial orientation of discontinuities even under harsh conditions in a karst cave.
- Areas which are almost inaccessible can be fully covered.
- Curvature-based filtering is able to reduce the data amount significantly without losing detailedness.
- The validation on a specific area shows good agreement.
- Due to the area-weighted evaluations, an exclusion of eroded and altered areas can be neglected if the surface is dominated by discontinuity faces.

Acknowledgements The IWRM-project is funded by the German Federal Ministry of Education and Research (BMBF).

References

Gaich A, Fasching A, Gruber M (1999) Stereoscopic imaging and geological evaluation for geotechnical modelling at the tunnel site. Felsbau 17 Nr 1:15–21

ISRM Commission on Testing Methods (1978) Suggested methods for the quantitative description of discontinuities in rock masses. Int J Rock Mech Min Sci 15:319–368

Kemeny J, Turner K, Norton B (2005) LIDAR for rock mass characterization: hardware, software, accuracy and best-practices. In: Tonon F, Kottenstette J (eds) Workshop laser and photogrammetric methods for rock face characterization, ARMA

Schmitt G (2010) Geodetic contributions to IWRM-projects in middle Java. Indonesia J Appl Geodesy 4:167–175, de Gruyter 2010

122 Characterization of the Dagorda Claystone in Leiria, Portugal, Based on Laboratory Tests

A. Veiga and M. Quinta-Ferreira

Abstract

The Hetangian Dagorda claystone Formation occupies the core of a diapir outcropping in part of the city of Leiria, exhibiting a complex geological structure. The geological and geotechnical characterization is presented based in field observation and laboratory tests allowing to conclude that the Dagorda clay soils exhibit an unfavourable behaviour for urban occupation, due to the presence of expansive clay minerals. When partially saturated these soils can lead to cracking of walls and floors, or even to endanger the stability of buildings, if they have not been strengthened to resist to expansive soil. The presence of soluble minerals, mainly gypsum and seldom halite, can allow the formation of voids and eventually the deformation or collapse of the ground. Suitable safety procedures, in order to prevent hazards should be used.

Keywords

Dagorda formation • Claystone • Laboratory tests • Leiria diapir • Land occupation

122.1 Introduction

The Dagorda claystone Formation occupies the core of a diapir, outcropping partially in the city of Leiria, exhibiting a complex geological structure (Fig. 122.1). This Formation is dated from the Hettangian and is constituted by clay, silt, occasionally with gypsum or even halite, and intercalations of limestone and marls. The presence of gypsum and halite, both soluble minerals, must be taken in account during the engineering works. Saline water can also be aggressive to concrete and iron components of the structures.

Laboratory tests performed on more than 30 samples collected in outcrops and shallow excavations, corresponding to decompressed soils, allowed to determine their geotechnical parameters and to characterize the materials variability. Identification tests (grain size analysis, Atterberg limits, specific gravity, methylene blue value), characterization tests (Proctor compaction and CBR) together with mechanical tests (direct shear tests and oedometric tests) were done. In order to identify the types of clay present in these materials, X-ray diffraction tests were performed.

122.2 Lithological Characterization

According to Teixeira et al. (1968) the nucleus of the diapir is constituted by red or grey marls and clays with gypsum, and by dolomite and marly limestone. Based on the field observations in outcrops, it was possible to identify red and grey clays, silty or marly clays and claystone, and thin layers of marly limestone. This lithological diversity was influenced by the diapiric tectonics that affected the area in the Mesozoic, generating quite heterogeneous terrains.

A. Veiga (✉)
Geosciences Center, Polytechnic Institute of Leiria, Campus 2, Morro do Lena—Alto Vieiro, 2411-901, Leiria, Portugal
e-mail: anabela.veiga@ipleiria.pt

M. Quinta-Ferreira
Department of Ciências da Terra, Geosciences Center, University of Coimbra, Largo Marquês de Pombal, 3000-272, Coimbra, Portugal
e-mail: mqf@dct.uc.pt

G. Lollino et al. (eds.), *Engineering Geology for Society and Territory – Volume 6*,
DOI: 10.1007/978-3-319-09060-3_122, © Springer International Publishing Switzerland 2015

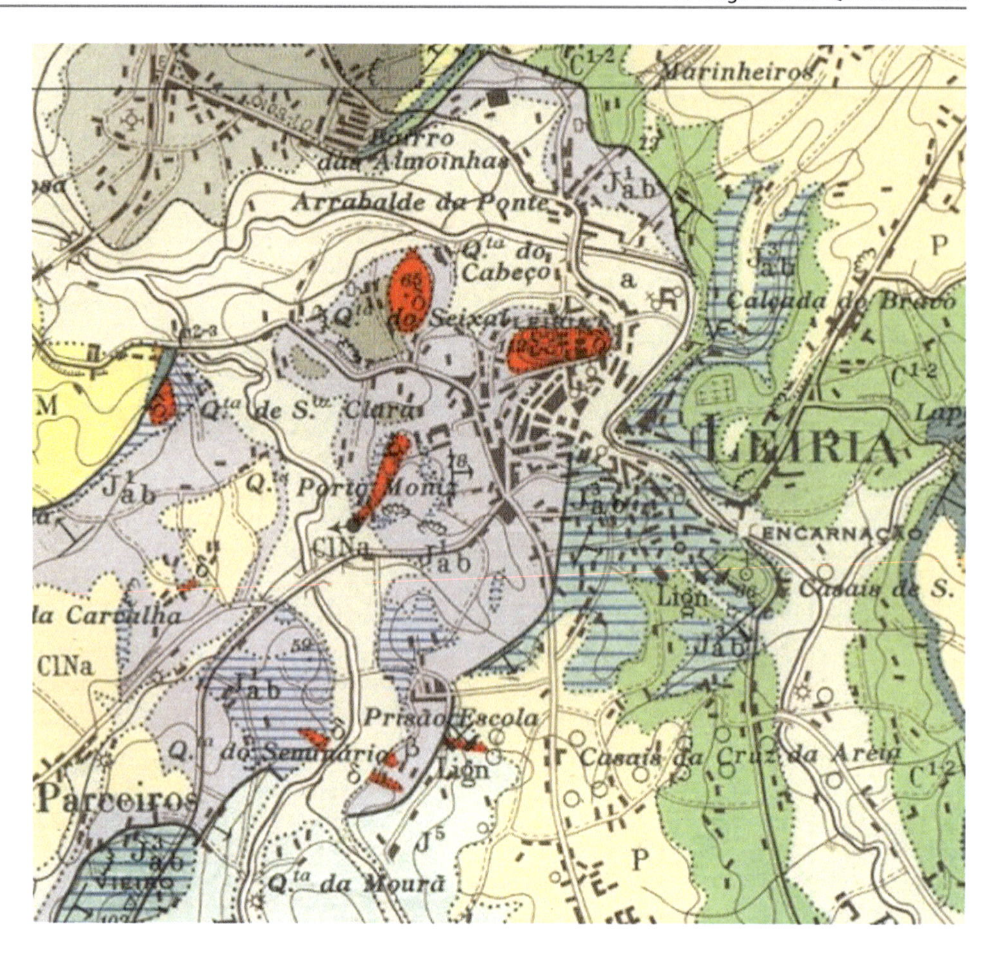

Fig. 122.1 Geological map of Leiria (in Teixeira et al. 1968)

122.3 Laboratorial Characterization

122.3.1 Tests for Classification, Identification and Description

The laboratory tests were done using samples collected in outcrops or in small excavations corresponding to decompressed soils. In order to classify the soil samples, a few laboratory tests were done: grain-size distribution using dry sieving and sedimentation, consistency limits, specific gravity and methylene blue test.

Figure 122.2 presents the grading curves of the samples, showing that fine grained soils are predominant. Table 122.1 presents a summary of the samples characteristics.

The average value of G_s is characteristic of fine soils. The I_p values correspond to soils of low to medium plasticity and the Ic values classify these soils from soft to stiff,

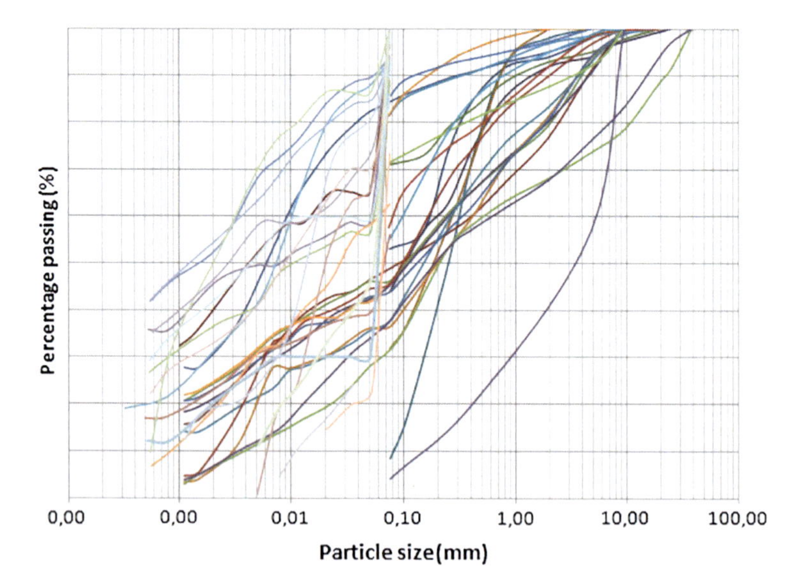

Fig. 122.2 Grain size distribution *curves* (Veiga 2012)

Table 122.1 Summary of the samples characteristics (Veiga 2012)

	Max	Min	Av	Med	Mo	S	n
sWater content (%)		3.0	11.3	10.9	14.5	6.64	24
G_s	2.95	2.46	2.68	2.65	2.6	0.12	28
Liquid limit (W_L) (%)	49	17	32	30	30	7.68	42
I_P	30	4	13.7	14	16	5.5	41
Consistency index (I_c)	4.24	0.28	1.9	1.9	2.36	0.87	38
Ac	0.87	0.21	0.5	0.5	–	0.17	23
MBV (g/100 g)	5.6	0.70	2.6	2.7	2.8	1.12	32

Ac—Activity, G_s—specific gravity, I_p—plasticity index, MBV—methylene blue value, Max—maximum, Min—minimum, Av—average, Med—median, Mo—mode, S—standard deviation, n—number of samples

predominantly hard. A_c values correspond to inactive normal clay. The average of MBV corresponds to water sensitive soils.

122.3.2 X-Ray Diffraction Tests

To identify the types of clay present in these materials, X-ray diffraction tests were performed in 6 samples. Two techniques were used: the proposed by Biscay (1965) and the proposed by Thorez (1976). Three of the samples are grey, showing predominantly illite, but also kaolinite and chlorite. These clays are characterized by low to very low ion exchange capacity, and therefore no expansibility, corresponding to soils presenting a stable behaviour. The two red samples have a high content of smectite (54 and 29 %), and a significant content of interstratified clays and illite. The smectite has a high adsorption capacity for water molecules, leading to higher volumetric variation. Although the number of samples is small, the results showed a relationship between colour and their clay composition, with predominance of illite in the grey clay, and the predominance of smectite in the red clay.

Amado et al. (2003) studied the chemical and mineralogical composition of these soils concluding that they have a quite variable chemical composition, ranging from highly carbonated with low Fe content soils, to silicate-aluminium with significant Fe content. The Mg content presents a large variation related with the presence of dolomite limestone and clay minerals. Clay minerals (smectite, montmorillonite, illite and kaolinite), calcite, dolomite, quartz and gypsum minerals were also identified, using X-ray diffractometry.

122.3.3 Geotechnical Classification

The laboratory tests, allowed to do the geotechnical classification of the samples. The obtained soil classification is presented in Fig. 122.3.

According to the unified soil classification, these soils are mostly classified as low plasticity clay (CL—62 %), while according to the AASHTO soil classification, they belong mainly to group A-6. Concerning the behaviour, the predominance is for materials that are impermeable to semi-permeable when compacted, presenting fair shear strength and fair compressibility when compressed and saturated, and reasonable to good workability as building material. Most of the clay materials present a very bad behaviour in layers under pavements.

122.3.4 Compressibility Testing

Oedometer tests were conducted on intact and remoulded samples. The tests were performed with induced stresses from 25 to 800 kN/m^2. After determining the parameters defining the stress-strain relationships, it was possible to obtain the preconsolidation stress (σ_p'), overconsolidation

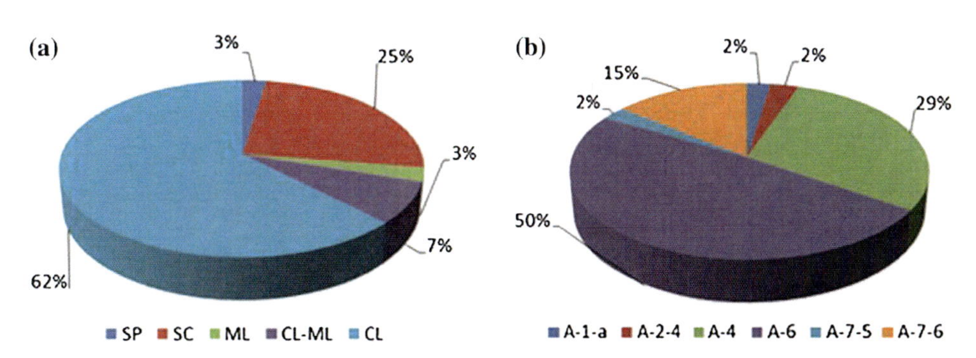

Fig. 122.3 Unified soil classification (**a**) and AASHTO soil classification (**b**)

Table 122.2 Statistical results from oedometer tests (Veiga 2012)

	Max	Min	Av	Med	s	n
σ'_p (kpa)	250.0	22.0	60.0	103.2	83.8	13
OCR	97.0	1.4	24.2	36.0	32.9	13
C_c	0.278	0.029	0.171	0.166	0.077	16
c_v (cm^2/min)	4.6	0.1	1.9	1.8	0.009	16

ratio (OCR), compression index (C_c) and coefficient of consolidation (c_v) (Table 122.2). The oedometer tests revealed overconsolidated soils, medium compressibility index and average vertical c_v of 1.9 cm^2/min. These tests were used to estimate permeability, revealing soils with very low permeability (between 8.2×10^{-10} and 7.5×10^{-9} cm/s).

122.3.5 Strength and Compaction

Direct shear tests were performed on remoulded samples. The results showed a wide scatter of the peak shear strength, with 65 % of the results between 29° and 36°, with a median of 30.4°.

The uniaxial compressive strength tests were done during the construction of the Leiria municipal stadium. The samples were obtained during drilling at depths between 14.5 and 19.5 m. The uniaxial compressive strength values range between 0.7 and 21.4 MPa (Tecnasol 2001), corresponding from stiff soils to moderate strength rocks. The median value is 2.72 MPa corresponding to very soft rocks, being unattractive as foundation materials.

From nine Proctor compaction tests it was possible to obtain γ_{dmax} and $w_{ópt}$ values. The upper $w_{ópt}$ is very high (19 %) conferring high deformability to these materials when used as embankment. The CBR index and expandability were obtained from eight CBR tests at 95 % compaction grade. The very low values of the CBR index (between 0 and 7) and the fine character of these materials give them a poor ability for use as backfill material, and the low strength makes them unattractive as foundation material.

122.4 Conclusions

The results allow to conclude that the Dagorda clay soils exhibit an unfavourable behaviour for urban occupation due to the presence of expansive clay minerals. When partially saturated these soils can lead to cracking of walls and floors, or even to endanger the stability of buildings, if they have not been dimensioned to resist to expansive soil. The presence of soluble minerals, mainly gypsum and seldom halite, can allow the development of voids and eventually the deformation or even the collapse of the ground. These cavities are difficult to identify, requiring special prospection techniques.

Despite the bad geotechnical behaviour of the Dagorda claystone, the city of Leiria is expanding to this area because of its high economic value. The geological and geotechnical characterization allowed to foreseen the unfavourable geotechnical behaviour upon occupation, requiring suitable safety procedures, in order to prevent engineering hazards.

Acknowledgments This work was funded by the Portuguese Government through FCT –Fundação para a Ciência e a Tecnologia under the project PEst-OE/CTE/UI0073/2011 of the Geosciences Center. We would like to thank the Polytechnic Institute of Leiria the conditions for carrying out the laboratory work.

References

Amado C, Veiga A, Quinta-Ferreira M, Velho J (2003) Análise de problemas de Geologia de Engenharia em zonas diapíricas: os casos de Belo Horizonte, Santa Clara e Morro do Lena (Leiria). [M. P. Ferreia]. A geologia de engenharia e os recursos geologicos. s.l.: Imprensa da U C, Coimbra, 2003 (in Portuguese)

Biscaye PE (1965) Mineralogy and sedimentation of recent deep-sea clay in the Atlantic Ocean in the adjacent seas and oceans. Geol Soc Am Bull 76:803–832

Tecnasol FGE (2001). Estádio Dr. Magalhães Pessoa. Projecto de execução. Relatório geológico-geotécnico do maciço de fundação. Câmara Municipal de Leiria. Setembro 2001 (in Portuguese)

Teixeira C, Zbyszewski G, Torre de Assunção C, Manuppella G (1968) Carta Geológica de Portugal, na escala 1/50 000. Notícia explicativa da folha 23-C (Leiria). (S. G. Portugal, Ed.) Lisboa (in Portuguese)

Thorez J (1976) Practical identification of clay minerals. In: Lelotte G (ed) vol 1. Belgium, p. 90

Veiga AQV (2012) Caracterização geotécnica dos terrenos do vale tifónico Parceiros-Leiria. Coimbra. www: http://hdl.handle.net/10316/19906. PhD thesis (in Portuguese)

Acoustic Monitoring of Underground Instabilities in an Old Limestone Quarry

123

Cristina Occhiena, Charles-Edouard Nadim, Arianna Astolfi, Giuseppina Emma Puglisi, Louena Shtrepi, Christian Bouffier, Marina Pirulli, Julien de Rosny, Pascal Bigarré, and Claudio Scavia

Abstract

The presence of abandoned and unstable underground cavities can give rise to a potential risk of surface collapse, particularly alarming when the presence of buildings is attested on the surface. The INERIS (Institut National de l'Environnement Industriel et des Risques, France) has carried out many studies, in recent years, investigating the effectiveness of several methodologies, devoted to monitor the evolution of the damaging mechanisms, reducing the risks of surface instability. Among the investigated methodologies, the acoustic monitoring has appeared as a suitable technique to detect damaging processes. During 2012 the Brasserie quarry, an old limestone mine located in the South-East of Paris (France) was instrumented with microphones to detect the acoustic waves generated by the collapse of blocks from the roof and the walls of the cavity. A series of tests were carried out with the aim of determining the propagation characteristics of sound and the attenuation of the acoustic waves inside the complex geometry of the quarry. Preliminary data processing concerned the classification of the recorded signals: the analysis of the signal-to-noise ratio (SNR) evidenced that the quarry can be subdivided into three main areas on the base of different attenuation attitude. The presence of critical areas and paths influencing the wave propagation also emerged. The research is still in progress. Specific analyses will be carried out to deepen the aspects related to the source localisation, simulating the acoustic behaviour of the quarry to obtain maps of sound attenuation and reverberation times.

Keywords

Acoustic monitoring • Quarry instabilities • Rock block collapse • Sound attenuation

123.1 Introduction

Many anthropogenic (mines, quarries, storage facilities, etc.) or natural (as karsts) underground cavities are responsible for surface instabilities caused by different mechanisms. These mechanisms depend on the cavity and can be global like the subsidence or the collapse of the overburden, or more local like the sinkholes (Didier 2008).

In France, especially in the northern regions, the closure and abandonment of many mines, largely exploited in the XIX and XX centuries, has evidenced the necessity of the management of these phenomena: monitoring the evolution of the triggering mechanisms plays a major role in the risk

C. Occhiena (✉) · M. Pirulli · C. Scavia
Department of Structural, Geotechnical and Building Engineering, Politecnico di Torino, Torino, Italy
e-mail: cristina.occhiena@polito.it

C.-E. Nadim · C. Bouffier · P. Bigarré
Institut National de lEnvironnement Industriel et des Risques (INERIS), Verneuil-en-Halatte, France

A. Astolfi · G.E. Puglisi · L. Shtrepi
Department of Energy, Politecnico di Torino, Torino, Italy

J. de Rosny
Laboratorie Ondes et Acoustique, Institut Langevin, Grenoble, Italy

G. Lollino et al. (eds.), *Engineering Geology for Society and Territory – Volume 6*,
DOI: 10.1007/978-3-319-09060-3_123, © Springer International Publishing Switzerland 2015

reduction, which is especially important when the presence of stakes is attested on the surface (Nadim 2009).

Local phenomena as sinkholes cannot be easily monitored neither by the classical geotechnical or geophysical methods. Therefore, the INERIS (Institut National de l'Environnement Industriel et des Risques, France) carried out several studies in recent years to identify the correct methodologies aiming at detecting any damaging process that could lead to consequences at the surface.

Recent studies evidenced that, among other possible methodologies, the acoustic monitoring can be adopted as a useful technique to detect and record rock falls, thus to monitor the evolution of sinkholes and/or localised collapses inside quiet cavities.

The acoustic technique has been applied to some case studies in order to define a suitable methodology to collect and analyse data, with the objective of locating the most active areas and identifying the critical periods characterised by an intense activity.

The Brasserie quarry was instrumented with 5 microphones on January 2012.

Later, on November 2012, a campaign of measurements was planned to better understand the propagation of sound inside of the quarry. The main objective of the experiments was twofold. On the one hand it was the determination of the attenuation of the acoustic waves due to the action of the existing surfaces of the quarry, like pillars and boundaries. On the other hand it was the investigation on the source localisation process.

To this aims, the data collected during the measurement campaign were classified on the base of the signal-to-noise ratio (SNR) and preliminarily analysed for the acoustic characterisation in this phase of study.

In the present paper, the preliminary results and the description of the upcoming works and expectations will be outlined after the description of the tests carried out in the quarry.

123.2 Site Description

The site under study is the Brasserie quarry, an old limestone mine located in the South-East of Paris (France) (Fig. 123.1). The quarry has been mainly exploited using the room and pillar technique, over approximately 4.5 Ha. The depth of the quarry does not exceed 25 m and several starts of sinkholes and extended roof falls have been observed. The quarry is open to visitors and administrative buildings lie above the quarry surface, so the quarries inspection office (Inspection Générale des Carrières de Paris) decided to set up a monitoring system within the quarry to reduce the risk for both surface and underground instabilities.

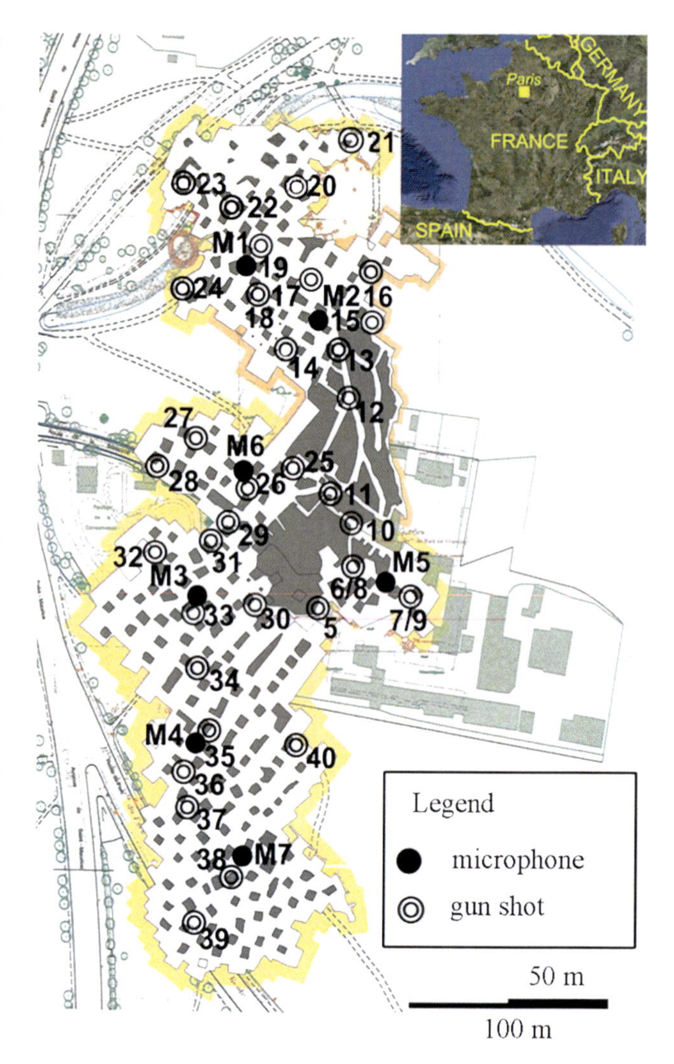

Fig. 123.1 The microphone array and the gun shot spatial distribution inside the quarry

On January 2012 the site was instrumented by the INERIS through the installation of an acoustic monitoring network which consisted of 5 microphones and 2 geophones, spread all over the quarry.

On November 2012 a campaign of measurements was carried out in order to provide a dataset for the determination of the acoustic energy attenuation inside the quarry. For that matter, the geophones were temporarily removed and 3 more microphones were installed, and one of them could be easily moved. Three types of acoustic sources were emitted, referring to the emission of artificial sounds simulating the block falls and the firing of 40 gunshots in different positions in the quarry (Fig. 123.1).

The final purpose of the experiment was the characterisation of the source event: to this aim the different types of sources used in the measurements campaign helped in localising, dimensioning and evaluating the attenuation effect.

In order to investigate the wave propagation and the attenuation attitude inside the quarry, the impulse responses generated by the gun shots were first analysed and classified.

123.3 Analysis of the Gun Shot Recordings

The recorded signals were processed with a Matlab®-based code for the computation of the SNR, defined as the ratio between a signal level and the corresponding background noise level. Theoretically, the background noise should be detected by recordings of some seconds for each microphone of the array. These recordings should be used to assess the ratio between the event sound level and the background noise in each sector of the quarry. Since this was not possible, a method to elude the lack of the background noise recordings at each microphone was used: a procedure based on the short-term average/long-term average (STA/LTA) method was implemented. The difference stands in computing the ratio of two average amplitudes between a short-term window and a long-term window on the same recording, instead of calculating the ratio between two tracks of equal time duration.

The main function of the code to the SNR computation is to perform a loop that scans the signal isolating a portion of the track within a floating window. The selected portion is sent to a secondary function, which subdivides the signal portion into 2 sections and computes the SNR as:

$$SNR = 20\log_{10}\left(\frac{A_{s2}}{A_{s1}}\right) \quad (123.1)$$

where A_{s2} and A_{s1} are the root mean square amplitudes of the two sections of the signal, respectively.

A SNR value is computed at each step while the floating window scrolls the signal. The trend of the SNR exhibits a maximum in correspondence to the first time arrival of sound; in fact, before the sound arrival there should be only noise while after the arrival time the components due to the recording of the acoustic wave should be prevailing.

The SNR is a good indicator of the acoustic attenuation and of the quality of the audio recordings. The attenuation is a complex phenomenon that generally increases with the distance from the source, i.e. loss of the intensity of the sound energy. Moreover, it is sensitive to other environmental physical properties such as absorption and diffusion. SNR was chosen in alternative to the signal sound pressure level only since it is a relative measure, and therefore enables the comparison of the acoustic condition between different measuring positions that may be affected by various background noise levels.

In Fig. 123.2 is shown the SNR versus the source-microphone distance, first considering all the microphone positions (Fig. 123.2a) and then for each microphone positions (Fig. 123.2b–h). From this evaluation it is possible to

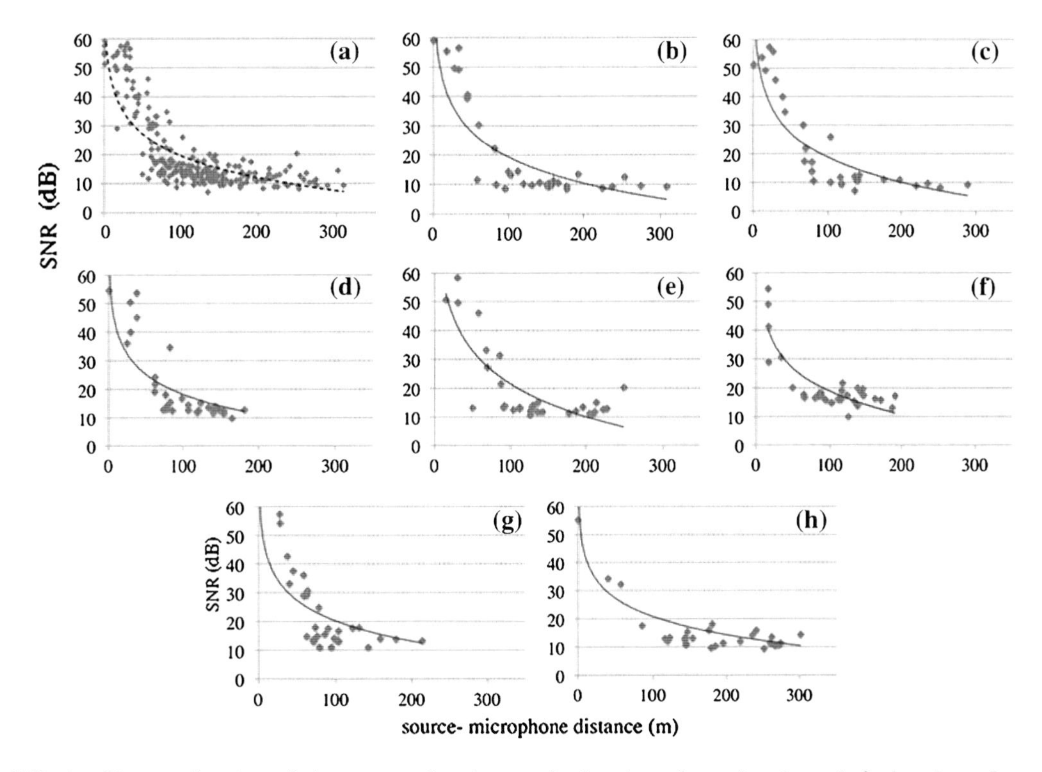

Fig. 123.2 SNR, in dB, as a function of the source-microphone distance for: **a** all the microphones; **b** microphone 1; **c** microphone 2; **d** microphone 3; **e** microphone 4; **f** microphone 5; **g** microphone 6; **h** microphone 7

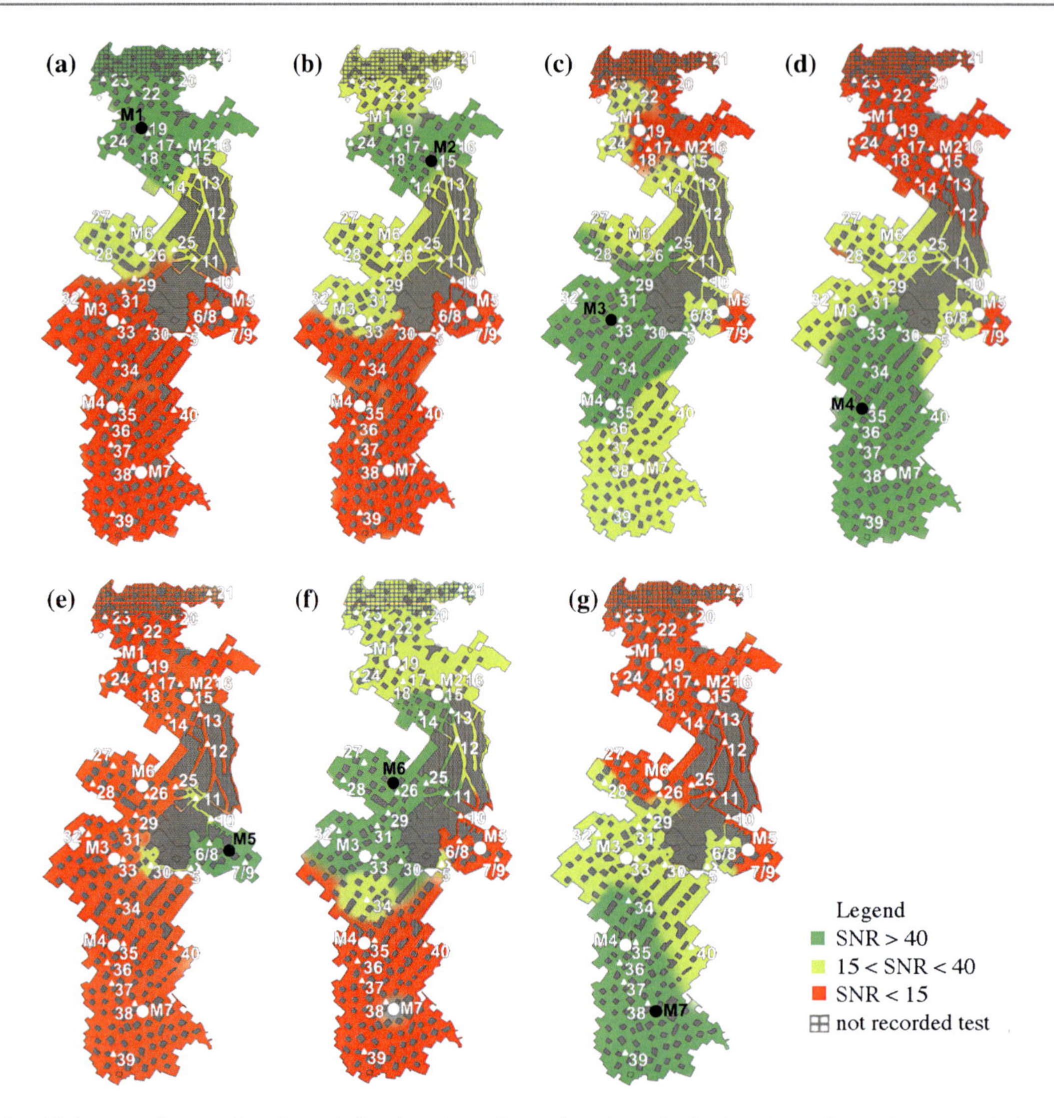

Fig. 123.3 Sensitivity map for: **a** microphone 1; **b** microphone 2; **c** microphone 3; **d** microphone 4; **e** microphone 5; **f** microphone 6; **g** microphone 7

consider that the source signal is fairly repeatable. The graphs of Fig. 123.2 are an attempt to obtain a relationship between the measured acoustic pressure level and the source-microphone distance. A more deep slope in SNR decreasing up to about 100 m is shown for all the microphones, and afterward a constant behaviour with an increase in the distance from the source.

The analyses described above have led to the creation of a ‘sensitivity’ map for each microphone position: the different areas of the Brasserie quarry have been coloured according to the SNR computation (Fig. 123.3). The green areas identify the zones where signals with SNRs higher than 40 dB were registered; the yellow areas identify the origin zones of signals with SNR between 15 dB and 40 dB, and the red areas the zones where the signals produced from the gunshots could not be recorded by the microphone under analysis (SNR < 15 dB).

A detailed analysis of the colour distributions evidenced that there are critical corridors (dotted line in Fig. 123.4a) and a zone characterised by a complex geometry (orange area in Fig. 123.4a), which are responsible for the low communicability between the northern and the southern portions of the quarry, and for the total isolation of the sector surrounding microphone 5 (blue area in Fig. 123.4a). In fact, all the sources fired in the area close to microphone 5 (blue area in Fig. 123.4b) were recorded with a high SNR and a good signal quality only by microphone 5.

On the base of the above mentioned observations, the quarry volume could be divided into four main areas (areas 1, 2 and 3 and 4 in Fig. 123.4b), characterised by different acoustic behaviour within each area. Moreover the propagation between the identified areas is difficult and affected by high attenuations.

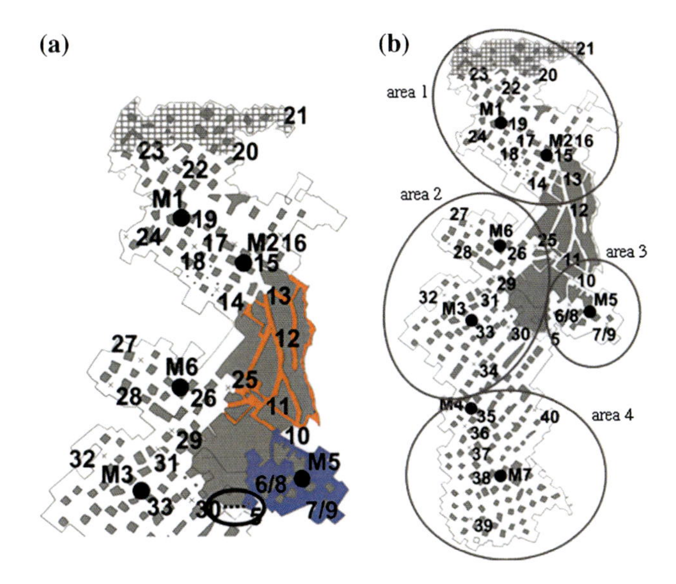

Fig. 123.4 **a** Identification of the critical corridor and areas; **b** identification of the four main areas characterized by specific attenuation attitudes

123.4 Conclusions and Further Developments

The present paper describes the preliminary results obtained from the processing of the data collected by an acoustic monitoring system during a measurement campaign inside an old limestone quarry.

The simple observation of the quarry map evidenced the presence of four isolated areas characterised by different geometries and volumes. This observation was confirmed and deepened by the analysis of the SNR computed for the signals recorded during a set of tests with gunshots. The classification of the tracks on the base of the SNR has led to the drawing of 'sensitivity' maps indicating each microphone's response to the acoustic stimuli of gunshots (i.e. for each sensor the quarry volume was subdivided in areas where the gunshots resulted to be with high, medium or very low SNR). The maps allowed to conduct a primary evaluation of the acoustic attenuation for the firings and could be a useful support to the assessment of the array implantation and to plan possible modifications to the sensor distribution. Moreover it emerged that critical corridors and a complex geometry area delimit four sectors, isolating the northern, the middle, the southern and the eastern portions of the quarry.

Further developments of the research are considered: specific analyses will be carried out in cooperation with the Department of Energy of the Politecnico di Torino through the use of the commercial acoustic software ODEON®. This will allow to deepen the aspects related to the source localisation (i.e. natural block fall), simulating the acoustic behaviour of closed spaces to obtain maps of sound attenuation and reverberation. To this aim the gunshot recordings will be used for the evaluation of the impulse response and the computation of the reverberation time. The data collected will be employed to calibrate the prediction model of the quarry.

The use of a code developed by the Institute Langevin of the CNR of Paris (France) for the simulation of the acoustic wave propagation is also planned. A comparison between the obtained results could allow a more accurate determination of the sound propagation characteristics in a site with complex geometries and heavy uncertainties.

References

Didier C (2008) The French experience of post mining management, Post Mining 2008, Nancy, 6–8 Feb 2008

Nadim CE (2009) Surveillance en grand de cavités souterraines complexes par méthode acoustique. Analyse préliminaire de données de référence: rapport d'étape, Internal Report, DRS-09-99345-10707A, INERIS, Paris, France

Investigative Procedures for Assessing Subsidence and Earth Fissure Risk for Dams and Levees

124

Kenneth C. Fergason, Michael L. Rucker, Bibhuti B. Panda, and Michael D. Greenslade

Abstract

Investigative procedures for assessing land subsidence and earth fissure risk for dams and levees have been developed for projects in arid regions of the USA. The overall assessment includes evaluation of the overall subsidence experienced in the vicinity of a subsidence-impacted structure, detailed investigation to search for earth fissures, prediction/modeling of future subsidence and related earth fissure development, delineation of risk, and recommendations for moving forward, such as engineered solutions and monitoring. Utilizing the findings of subsidence investigation, additional investigative methods for earth fissure search include photogeologic (lineament) analysis, assessment of the capability of near-surface soils to develop an earth fissure, assessment of the degree of ground disturbance, detailed site inspection, seismic refraction profiling for concealed earth fissures, and excavation of trenches. Such an investigation must include investigative techniques capable of detecting earth fissures that lack significant surficial expression. Satellite-based interferometry by repeat pass synthetic aperture radar (InSAR) provides unique information about active land subsidence over large areas. The subsidence or deformation image known as an interferogram can also, with proper interpretation, reveal some preliminary subsurface information about alluvial basin geometry, lithology and hydrology where active land subsidence is interpreted. Effective subsidence risk assessment and mitigation requires understanding and quantifying historic subsidence, and estimating potential future subsidence that could impact the dams and levee infrastructure. Basin alluvium and bedrock interface geometry, and basin alluvium lithology changes and interfaces, profoundly influence patterns and degree of subsidence. Characterization includes collection and synthesis of historic survey and well data, surface geophysical methods for basin and bedrock characterization, and when available, InSAR to document recent or current subsidence patterns. Utilizing a synthesis of this information, subsidence modeling matching

K.C. Fergason (✉) · M.L. Rucker · B.B. Panda
AMEC Environment and Infrastructure, Inc., 4600 East Washington Street, Suite 600, Phoenix, AZ 85034, USA
e-mail: ken.fergason@amec.com

M.L. Rucker
e-mail: michael.rucker@amec.com

B.B. Panda
e-mail: bibhuti.panda@amec.com

M.D. Greenslade
Flood Control District of Maricopa County, 2801 West Durango Street, Phoenix, AZ 85009, USA
e-mail: mdg@mail.maricopa.gov

G. Lollino et al. (eds.), *Engineering Geology for Society and Territory – Volume 6*,
DOI: 10.1007/978-3-319-09060-3_124, © Springer International Publishing Switzerland 2015

documented historic subsidence and estimating potential future subsidence can be developed to assess potential impacts on dam and levee infrastructure. Utilizing the results of site characterization and subsidence modeling, a finite-element stress-strain model can be developed to estimate past and future ground strain. Estimated tensional strain values can be utilized to predict where earth fissures are likely to initiate with future subsidence and reduce the risk of failure.

Keywords
Subsidence • Earth fissure • Dam • Levee • Guidelines

124.1 Introduction

Depletion of groundwater resources in many deep alluvial basin aquifers in the Western USA is causing ground subsidence. Ground subsidence can severely and adversely impact infrastructure by changing the ground elevation, ground slope (grade) and through the development of ground cracks known as earth fissures, develop into large gullies. Earth fissures have the potential to undermine the foundations of dams, levees, and other pertinent structures and cause system failure.

Earth fissures that have been exposed to flowing water will most likely have observable surficial expressions such as ground cracking, piping holes, vegetative and tonal lineaments, and similar features. However, uneroded earth fissures often do not have surficial expression.

Earth fissures are presumed to have caused the failure of Picacho Dam in South-Central Arizona in 1955. In addition, earth fissures have been identified in close proximity to or underlying the foundations of McMicken Dam and Powerline Flood Retarding Structure (FRS) in Central Arizona (AMEC 2003, 2008). Both structures are earthfill embankments with maximum height of 25–35 feet that are founded on alluvial fill. A segment of McMicken Dam was rehabilitated in 2006 and construction of an interim dam safety measure at Powerline FRS was completed in 2013.

124.2 Investigative/Procedural Methods

Existing literature providing recommendations for and/or descriptions of investigative techniques and procedures for land subsidence and earth fissure investigations for dams and levees is sparse. The Arizona Land Subsidence Group, through the Arizona Geological Survey published a document titled Suggested Guidelines for Investigating Land-Subsidence and Earth Fissure Hazards in Arizona (Arizona Land Subsidence Interest Group 2011). A similar publication from the Utah Geological Survey is currently in publication limbo as stake-holders work out issues associated with the publication. The Flood Control District of Maricopa County (District) contracted AMEC Environment & Infrastructure, Inc. (AMEC) to develop procedural guidelines for investigating land subsidence and earth fissuring with input from an independent consultant (GeoSouthwest LLC) and the regulatory agency, Arizona Department of Water Resources (ADWR) Dam Safety (AMEC 2011). The investigative/procedural methods summarized in this paper are adapted from these procedural documents. More detailed publications that discuss these methods include Fergason et al. (2013), Rucker et al. (2013), and Panda et al. (2013).

Subsequent to performance of an evaluation of overall subsidence experienced in the vicinity of a subsidence-impacted structure, a detailed investigation to search for

• Repeat-Pass Synthetic Aperture Radar Interferometry (InSAR)	• Deep refraction microtremor (ReMi) • Shallow ReMi
• Photogeological lineament analysis	• Fissure detection by seismic refraction signal trace analysis
• Geological reconnaissance of photolineaments and terrestrial search for earth fissures	• Subsurface characterization by seismic refraction • Trench investigation
• Deep resistivity soundings	• Future subsidence prediction • Stress–strain modeling

earth fissures must be performed. Such an investigation must include investigative techniques capable of detecting earth fissures that do not have significant surficial expression. Procedural documents have been developed for the following geohazard investigation and analysis techniques to perform such an investigation (AMEC 2011):

Each procedural document includes detailed descriptions of each geohazard investigative/assessment method inclusive of supporting documentation; qualifies the level of uncertainty associated with each method; allows for the transfer of methods to future similar projects; and establishes procedures with which all interested agencies and stakeholders will concur.

These procedures are modified and updated from those developed by AMEC with input from GeoSouthwest, LLC and the ADWR Dam Safety Division for use by the District (AMEC 2011). These guidelines have not been developed as a prescriptive recipe for planning. The approach, its associated protocols and recommended methods are presented with the intent of providing a generic basis for independently developing site-specific strategies. A project-specific plan should provide sufficient flexibility to allow interim adjustments to the analysis as the process evolves.

124.3 Overview of InSAR

InSAR data can detect relative terrain elevation changes to a possible 0.2–0.4 inch (5–10 mm) resolution under favorable radiometric conditions, where the ground surface is divided into individual 'pixels' with a size of about 100 feet (30 m) across. InSAR can be utilized for general subsidence geometry, detailed deformation data, input data for modeling and other applications, and monitoring. Additionally, interpreted subsidence patterns aid in determining the subsurface geology in areas where active subsidence is observed.

124.4 Subsidence Estimation

A primary subsidence mechanism is increasing effective stress due to groundwater level decline within saturated compressible basin alluvium. Ultimate subsidence magnitude at a given location is a function of change in effective stress, compressible alluvium thickness and material modulus. Modulus is typically a function of depth and effective stress. Subsidence rates are assumed to largely be a function of rate of groundwater level decline, alluvium permeability (or hydraulic conductivity) and distance from groundwater level stress points (i.e. pumping wells). Basin alluvium and bedrock interface geometry, and changes and interfaces in basin alluvium lithology, profoundly influence patterns and magnitude of subsidence. Characterization includes collection and synthesis of historic survey and well data, surface geophysical methods for basin and bedrock characterization, and when available, InSAR to document recent or current subsidence patterns. Utilizing a synthesis of this information, subsidence modeling matching documented historic subsidence and estimating potential future subsidence can be developed to assess potential impacts on dam and levee infrastructure.

As in other geotechnical applications addressing ground settlement, several approaches may be applied to the problem of estimating subsidence due to groundwater withdrawal. Empirical approaches such as presented by Bell (1981) provide a 'rule-of-thumb' based on a ratio of subsidence for a given amount of groundwater decline; for the Las Vegas area, this ratio is reported to be about 20:1 (water level decline to subsidence) for fine-grained alluvium and about 40:1–60:1 for coarse-grained alluvium. Subsidence estimation based on traditional soil mechanics principles and soil parameters suffer from the general inability to obtain relevant, and relatively undisturbed, alluvium material samples for laboratory consolidation or triaxial testing. Such undisturbed sampling would have to be successful at depths of several hundred feet to in excess of 1,000 feet. Even if such sampling were to be successful, a relatively very few data points would have to suffice to characterize huge volumes of heterogeneous basin alluvium.

The approach to subsidence estimation summarized in this paper is, at a given location, to estimate modulus through the column of compressible alluvium, apply the increase in effective stress through that column from a given groundwater level decline, and calculate the vertical displacement (subsidence) from the resulting strain. Initial and final groundwater levels, or beginning and ending water levels for a distinct time period, are needed to calculate change in effective stress. Compressible alluvium thickness or depth is needed to which to apply change in effective stress. A compressible alluvium modulus profile, with modulus increasing with depth, must be estimated. Alluvium modulus is also a function of the alluvium lithology; surface geophysics provides means to assess alluvium thickness and lithology when useful well data is not available. Finally, to estimate time-dependent lateral propagation of subsidence from pumping centers, alluvium mass permeabilities or hydraulic conductivities must be estimated and distance to pumping centers assessed.

124.5 Stress-Strain Model

The relationship between pore fluid pressure changes and aquifer system compression is based on the principle of effective stress proposed by Terzaghi (1925), where effective stress is the difference between the total stress and the pore fluid pressure. The total stress represents the geostatic load. Under this principle, when the total stress remains constant, a change in pore fluid pressure causes an equivalent change in effective stress within the aquifer system. This results in a small change in volume in an aquifer system that is governed by the compressibility of the aquifer system skeleton. Conceptually, the change in pore pressure with time is related to the change in porosity. Subsidence is estimated by multiplying the change in porosity by the vertical thickness of the medium.

The amount of compaction and fissure location are closely related to the thickness and skeletal compressibility of fine-grained sediments within the aquifer system. In a region having thick compressible aquitards (fine sediments) of low vertical conductivity, the pore fluid pressure does not simultaneously equilibrate with the head in the surrounding aquifer. Instead, a pore fluid pressure gradient develops across the aquitards, driving the slow drainage of water from the aquitards into the aquifer. Because the compaction of an aquitard results from head change in the aquitard itself, the subsidence of the land surface lags behind the head decline measured in wells tapping more permeable sediments within the aquifer. This time delay leads to continuing subsidence despite static or recovering hydraulic heads observed in the field (Hoffmann et al. 2003). Modeling subsidence and deformation of the alluvium in response to changes in groundwater levels in the aquifer system requires addressing displacements and pore water pressure changes simultaneously in a coupled manner. A fully coupled analysis requires that both the stress-deformation and seepage dissipation equations be solved simultaneously. This coupling is typically achieved by using finite element-based computer programs that link seepage modeling with stress–strain deformation modeling.

References

AMEC Earth and Environmental, Inc. (AMEC) (2003) Earth Fissure Investigation Report, McMicken Dam, Maricopa County, Arizona. Prepared for the Flood Control District of Maricopa County, Contract FCD 2000C006—Work Assignment Nos. 4 and 5, April 11

AMEC Earth and Environmental, Inc. (AMEC) (2008) Supplemental Earth Fissure/Ground Subsidence Investigation Report, Powerline Flood Retarding Structure, Pinal County, Arizona. Prepared for the Flood Control District of Maricopa County. Contract FCD 2006C020, Work Assignment No. 2. AMEC Job No. 7-117-001082. June 4

AMEC Earth and Environmental, Inc. (AMEC) (2011) Procedural Documents for Land Subsidence and Earth Fissure Appraisals. Prepared for the Flood Control District of Maricopa County, Contract FCD2008C016—Work Assignment No. 13. May

Arizona Land Subsidence Interest Group (2011) Suggested Guidelines for Investigating Land-Subsidence and Earth Fissure Hazards in Arizona. Arizona Geological Survey Contributed Report CR-11-D. August

Bell JW (1981) Bulletin 95, Subsidence in the Las Vegas Valley, Nevada Bureau of Mines and Geology, Mackay School of Mines, University of Nevada, Reno

Fergason KC, Rucker ML, Greenslade MD (2013) Investigative procedures for assessing earth fissure risk for dams and levees, United States Society on Dams, 33rd annual meeting and conference, Phoenix, AZ, 11–15 Feb 2013

Hoffmann J, Galloway DL, Zebker HA (2003) Inverse modeling of interbed storage parameters using land subsidence observations, Antelope Valley California. Water Resour Res 39(2):1–13

Panda BB, Rucker ML, Fergason KC (2013) InSAR as a subsidence characterization tool for flood control dam studies, United States Society on Dams, 33rd annual meeting and conference, Phoenix, AZ, 11–15 Feb 2013

Rucker ML, Fergason KC, Greenslade MD, Hansen LA (2013) Characterization of subsidence impacting flood control dams and levees, United States Society on Dams, 33rd annual meeting and conference, Phoenix, AZ, 11–15 Feb 2013

Terzaghi K (1925) Principles of soil mechanics, IV, settlement and consolidation of clay. Eng News Rec 95(3):874–878

125 An Integrated Approach for Monitoring Slow Deformations Preceding Dynamic Failure in Rock Slopes: A Preliminary Study

Chiara Colombero, Cesare Comina, Anna Maria Ferrero, Giuseppe Mandrone, Gessica Umili, and Sergio Vinciguerra

Abstract

Rock slope monitoring is a major aim in territorial risk assessment and mitigation. The high velocity that usually characterizes the failure phase of rock instabilities makes the traditional instruments based on slope deformation measurements not applicable for early warning systems. On the other hand the use of acoustic emission records has been often a good tool in underground mining for slope monitoring. In this paper the design and installation of a monitoring system based on acoustic emission aimed at interpret and forecast a large rock instability phenomenon is reported together with some preliminary geophysical and geomechanical studies performed.

Keywords

Rock slope instability • Slope monitoring • Acoustic emission • Cross hole seismic tomography

125.1 Introduction

Rock slope instabilities are usually preceded by slow small entity deformations that can precede a dynamic fast failure. Small deformations, which can be recorded with standard devices (e.g. extensometer, etc.) and new remote sensing technologies (e.g. Lidar, GBInSAR, etc.), can be very significant to forecast instability development. Prior to the ultimate fracture, the rock also releases energy and it determines the generation of microtremors. The record and monitoring of acoustic emission can be therefore an alternative strategy for forecasting dynamic ruptures and is widely applied in mine monitoring (Kwiatek and Ben-Zion 2013).

For these reasons we propose to monitor and detect small signals of impending failures and mitigate natural hazards, by:

1. quantification of critical damage thresholds triggering dynamic failure, throughout the 'in situ' identification of characteristic slow deformation signals and accelerating patterns before impending 'large scale' failure events;
2. setting up of early warning models for forecasting the time of rupture with application to natural hazards;
3. transferring knowledge between multi-scale signs of slow deformation before dynamic failure from the laboratory to field.

To do this, the installation of a series of devices based on acoustic emission/microseismic approaches is planned in a test site where standard monitoring systems have been installed several years ago and, consequently, a set of data is already available. The installation of the monitoring network will be accompanied by a detailed geophysical and geomechanical characterization of the test site, in order to establish the best nodes position; to define the seismic velocity field of the rock mass, which is a fundamental parameter for the following monitoring step; to define the internal characteristics of

C. Colombero · C. Comina (✉) · A.M. Ferrero · G. Mandrone · S. Vinciguerra
Department of Earth Sciences, University of Turin, Turin, Italy
e-mail: cesare.comina@unito.it

A.M. Ferrero
e-mail: margheritaferrero@yahoo.it

G. Umili
Department of Civil, Environmental and Territory Engineering, University of Parma, Parma, Italy

G. Lollino et al. (eds.), *Engineering Geology for Society and Territory – Volume 6*,
DOI: 10.1007/978-3-319-09060-3_125, © Springer International Publishing Switzerland 2015

the monitored landslide and to image the fracturing state and the relative variation in seismic velocities between altered and intact rock to be related with geological observations (both sounding results and overall fracturing state).

Moreover rock physical and mechanical characterization will be carried out in laboratory on the main lithologies, throughout measurements of basic parameters such as elastic constants, fracture strength, density, porosity and elastic wave velocities. This will allow a direct comparison with the same parameters registered in the field.

In this paper the preliminary results of both geophysical tests and geological characterization are presented together with plans for the installation of the acoustic emission/ microseismic network.

125.2 Geological Framework

The site that has been chosen for the study is the rock slope of the Madonna del Sasso, that is affected by a rock instability phenomenon, highlighted by neat and long lasting episodes of slow deformation recorded by standard measurement devices such as inclinometers, topographic measurements and fissurometers.

The cliff of Madonna del Sasso is located along the western shore of Orta Lake and takes its name from the eighteenth-century sanctuary located at about 650 m a.s.l. In this area a granitic rock mass, called Granito di Alzo, outcrops; this is part of Lower Permian granitoid masses and, in the past, it was subjected to mining activity. It is late-Hercynian magmatic intrusion, not metamorphosed and generally little deformed (Boriani et al. 1992; Giobbi Origoni et al. 1988) that occurs along the contact between the "Serie dei Laghi" and the Ivrea-Verbano Zone. These granites, commonly known as "Graniti dei laghi", constitute a large batholith elongated in NE–SW direction that intrudes both the "Scisti dei laghi" and the Stron Ceneri zone. This batholith includes five plutons among which, that of study area, the Alzo-Roccapietra Pluton, of granite-granodiorite composition, outcropping between the lower Sesia Valley and the Orta Lake.

A preliminary geomechanical characterization (based on previous data and brief surveys) leads to define the rock mass on which the Sanctuary is built as intact or massive, with widely spaced discontinuities (GSI > 70) characterized by good surface quality. The rock mass characterization has been carried out by means of a conventional survey (Lancellotta et al. 1991) that has identified (Fig. 125.1) four main joint sets (K1 (110/75), K2 (0/80), K3 (150/15) and K4 (50/75)). Particularly along the K4 discontinuity there is a clear evidence of movement: a decimetric step is visible on the yard and on the small walls in front of the sanctuary.

125.3 Geophysical Site Investigation

A cross-hole seismic tomography has been performed between the two available inclinometric boreholes S1 and S2 in the area in front of the Sanctuary (Fig. 125.1). In cross-hole seismic tomography, seismic sources are located both in well and on the surface and are shot into receivers located in a nearby well or on the same surface. The travel times of the first arrivals are then used to produce a tomographic velocity cross-section of the subsurface between the two wells (Bregman et al. 1989; Calnan and Schuster 1989; Lines and LaFehr 1989; McMechan et al. 1987). Crosshole tomography is expected to provide better resolution than surface based seismic methods, since most of the energy does not travel through the highly attenuating near surface and the travel distances are shorter. In addition, the resolution of cross-hole tomography is not depth-limited since most of the energy travels between the wells and a trans-illumination of the imaged medium is achieved. This is even more important in a fractured medium, as the one object of the study, where waves can have complicated travel paths, not easily interpreted by the surface alone.

To perform the tests, a Borehole Impacter Source by Geotomographie GmbH has been used as in-hole source in the S2 borehole in three different locations till a depth of about 6 m (after this depth an obstruction of the hole casing didn't allow further penetration) while an hammer, impinging both vertically and horizontally, has been used as surface source in three different locations along the line connecting the two holes. A prototype borehole string equipped with 8 three component geophones (10 Hz) stiffly connected by a bar, that allows to control geophone orientation, has been progressively lowered (with a 2 geophone superposition each subsequent positioning) in the S1 sounding at different depths till the maximum available depth of 27 m. On the surface 4 three component geophones (2 Hz) have been moreover used. First break picking has been performed on the acquired seismic traces to allow for both P and S wave velocities imaging. Data have been inverted to obtain a tomographic image by the use of GeoTom® software.

The P wave velocity seismic image depicted by cross-hole tomography is reported in Fig. 125.2 together with the acquisition scheme, traced rays after the inversion and ray coverage. The seismic image correlates well both with the evidence of the soundings logs and with the expected fracture state. An high velocity of the intact granite formation is revealed and two main fractures are also evidenced showing a velocity reduction and a major localization of seismic rays. These fractures compare well with the K4 system which, between the two boreholes, shows also a surface manifestation (Fig. 125.1).

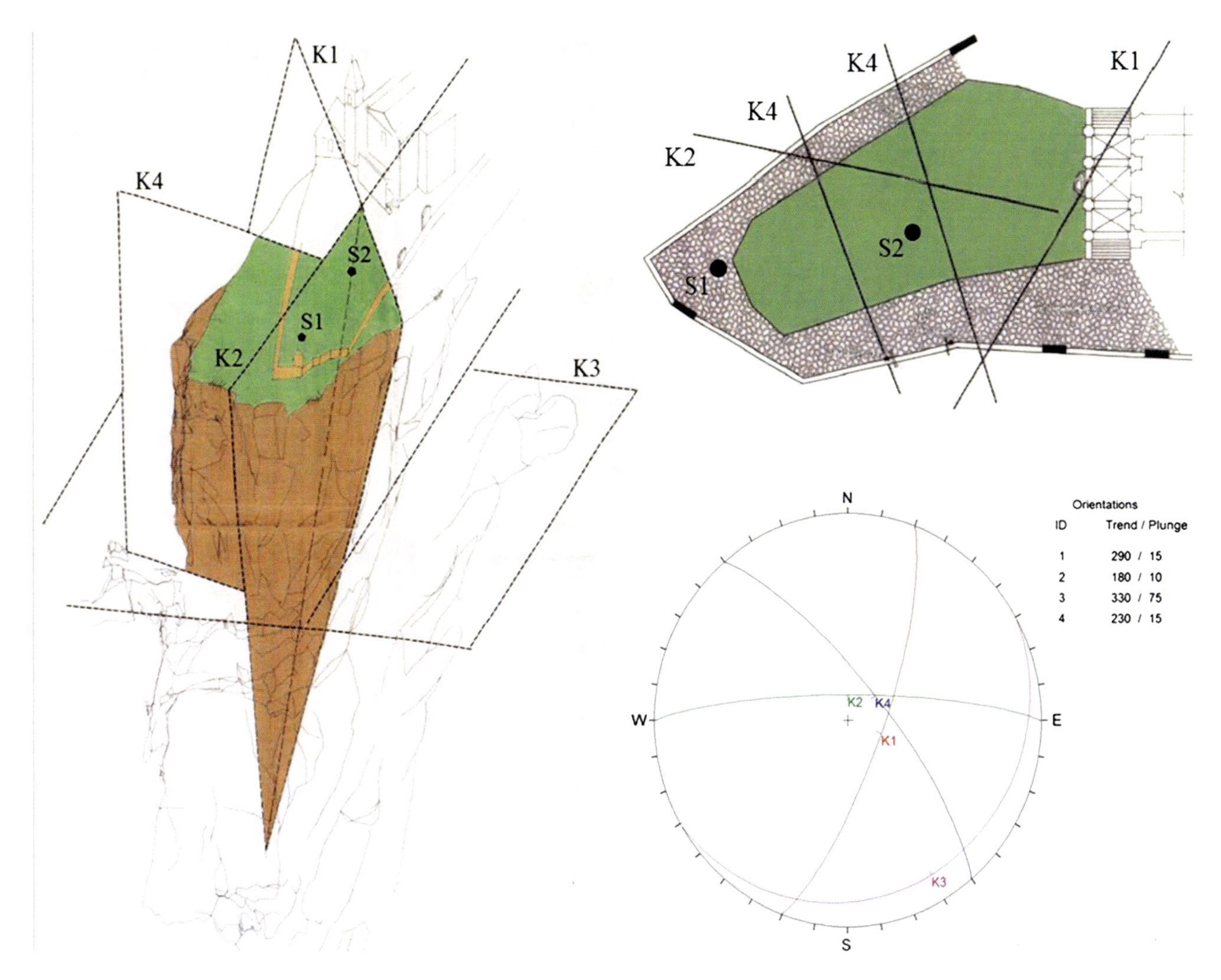

Fig. 125.1 Slope schematic structure: **a** assonometric and **b** plan views together with **c** surveyed joint sets after Lancellotta et al. 1991

125.4 Evaluation of the Slope Stability

On the basis of the geological and geophysical characterization, a preliminary kinematic analysis of the slope has been performed, by means of a software implemented by the authors, to better understand the possible instability phenomena that are occurring in the rock mass. Input parameters were the granite density (2.6 g/cm^3), the rock mass height (about 200 m) and the orientation of the main joints sets (see Figs. 125.1 and 125.2). This analysis has outlined 2 different kinds of sliding phenomena and a toppling phenomenon that could be compatible with rock mass structure: planar sliding along K4, wedge sliding along the intersection between K1 and K2 and toppling on K4. In order to identify the stability conditions of the slope some parametrical analysis have been carried out with the limit equilibrium approach. Both discontinuity persistence and the friction angle have been varied between realistic values determined by in situ observations and bibliographic data. Different water contents in the discontinuity have also been considered. In Fig. 125.3 the slope factor of safety for planar sliding for different discontinuity persistence is shown.

For all the considered failure mechanisms the factor of safety can go below one even if the presence of a small amount of rock bridges can guarantee a global stability. Water can play a major role in all the examined cases, however for toppling it can be the triggering effect even for a discontinuity partial saturation (around the 50 %). Rock bridges failure is consequently a very strategic phenomenon that needs to be surveyed.

125.5 Concluding Remarks and Future Works

The paper shows an ongoing work concerning the study of the possible application of innovative monitoring systems based on acoustic emissions to rock slope instability. The characterization studies based on geomechanical and

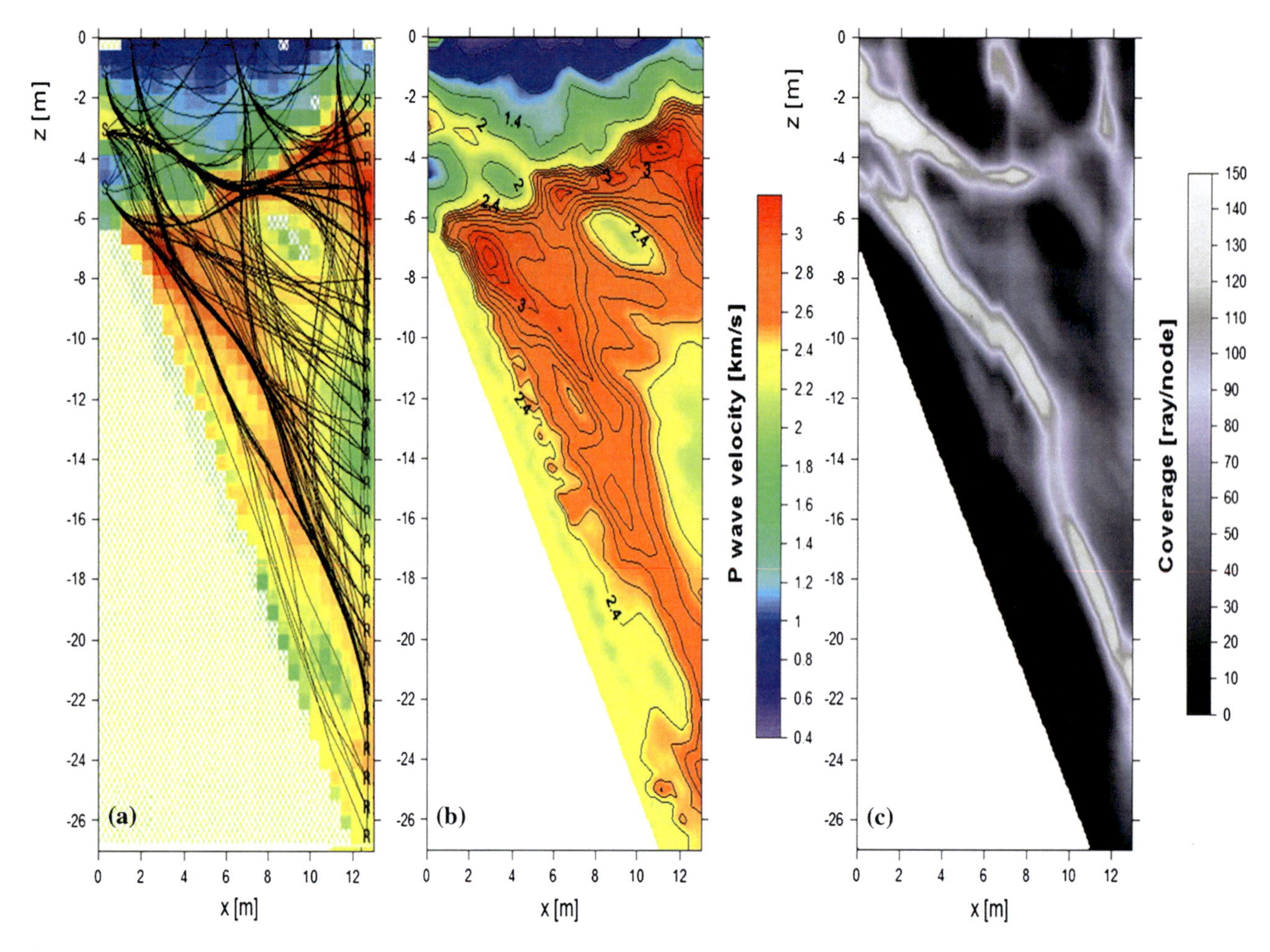

Fig. 125.2 Result of the cross-hole seismic tomography: **a** acquisition scheme (S are seismic sources and R receivers) and traced rays after the inversion; **b** P wave seismic imaging and **c** ray coverage

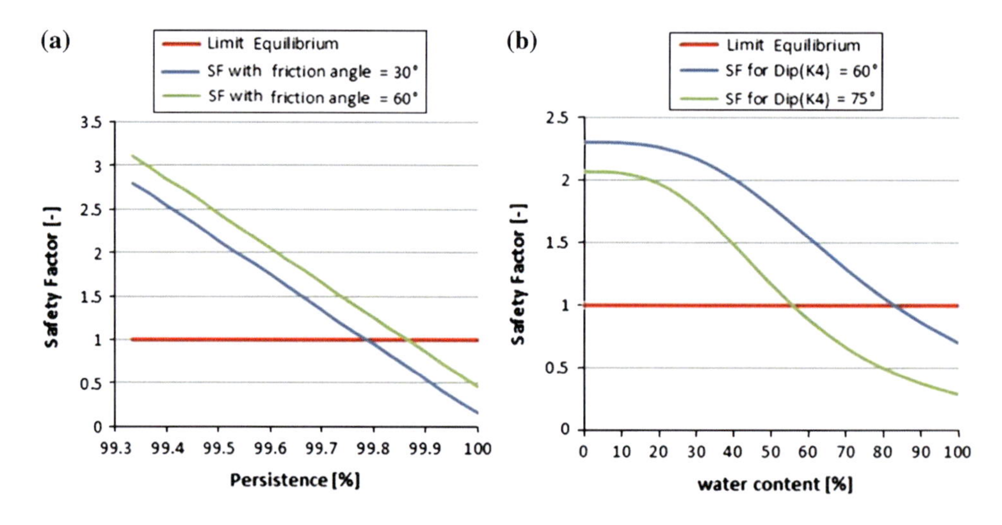

Fig. 125.3 Factor of safety versus persistence for **a** planar sliding and factor of safety versus water content for **b** toppling

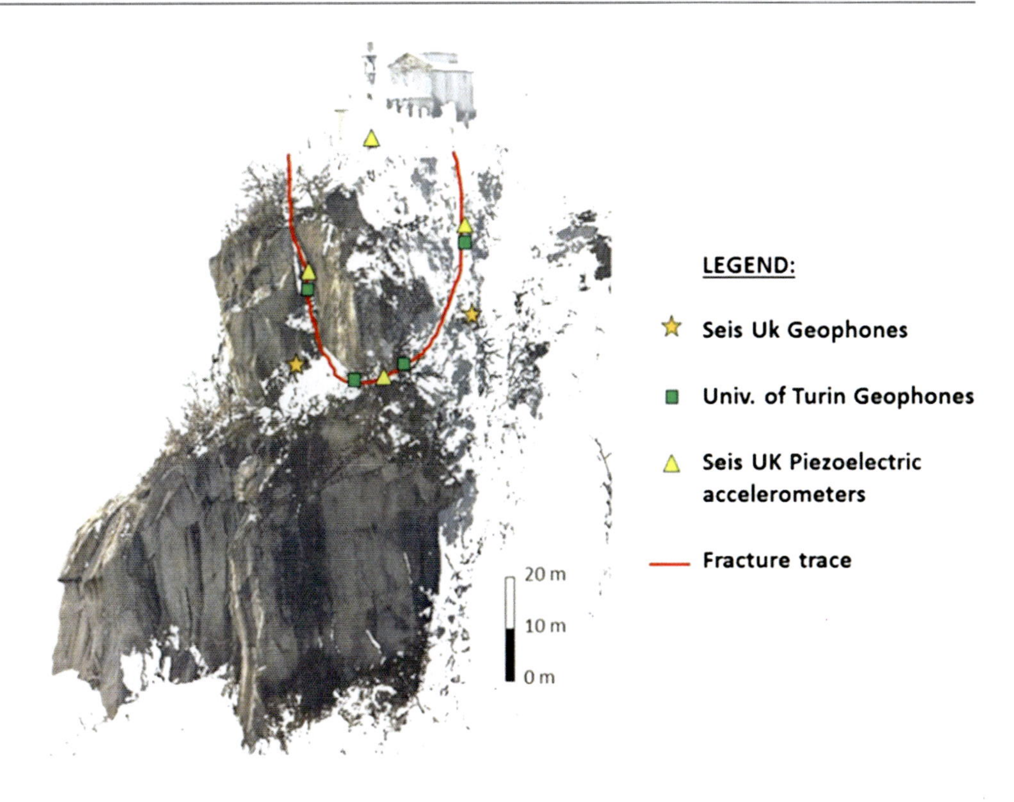

Fig. 125.4 Instrument location in relation to the fracture

geophysical tests allowed to carry on simple parametrical analysis to preliminary better understand the instability mechanism and to design the monitoring systems that will be placed and verified in the next future. Stability analysis showed that the stability of the slope is due to rock bridges. Their failure progress can results in a global slope failure. Consequently we propose to install a monitoring system to record the energy realised by the rock bridges dynamic ruptures. The devices layout is reported in Fig. 125.4. We aim to identify the characteristic signs of impending failure, by deploying an array of instruments designed to monitor subtle changes of the mechanical properties of the medium and installed as close as possible to the source region. A "site specific" micro-seismic monitoring system to detect micro-seismic events which reflect the subtle changes of the mechanical properties of the medium, made of 4 triaxial piezoelectric accelerometers operating at frequencies up to 23 kHz with a conventional monitoring for seismic detection (4.5 Hz seismometers) and ground deformation (strainmeters), provided by the University of Turin and SEIS-UK is proposed. The high-frequency equipment will allow us to develop a network capable of recording events with Mw < 0.5 and frequencies between 4.5 and 20 kHz. Sensors will be installed within short boreholes (2–4 m) adjacent to the probable slip surface, in order to maximize the coupling and improve the transmissivity. Waveforms will be stored for detailed analyses.

References

Boriani A, Caironi V, Giobbi Origoni E, Vannucci R (1992) The Permian intrusive rocks of Serie dei Laghi (Western Southern Alps). Acta Vulcanol 2:73–86

Bregman ND, Bailey RC, Chapman CH (1989) Crosshole seismic tomography. Geophysics 54:200–215

Calnan C, Schuster GT (1989) Reflection and transmission cross-well tomography: Presented at the 59th annual international meeting, society exploration geophysicists, expanded abstracts, pp 908–911

Giobbi Origoni E, Bocchio E, Boriani A, Carmine R, De Capitani L (1988) Late-Hercynian mafic and intermediate intrusives of Serie dei Laghi (N-Italy). Rend Soc It Mineral Petrol 43:395–410

Kwiatek G, Ben-Zion Y (2013) Assessment of P and S wave energy radiated from very small shear-tensile seismic events in a deep South African mine. J Geoph Res: Solid Earth 118:3630–3641. doi:10.1002/Jgrb.50274

Lancellotta R, Gigli P, Pepe C (1991) Relazione tecnica riguardante la caratterizzazione geologico-strutturale dell'ammasso roccioso e le condizioni di stabilità della rupe. Private comunication

Lines LR, LaFehr ED (1989) Tomographic modeling of a cross-borehole data set. Geophysics 54:1249–1257

McMechan GA, Harris JM, Andesron LM (1987) Cross-hole tomography for strongly variable media with applications to scale model data. Bull Seismol Soc Am 77:1945–1960

Combining Finite-Discrete Numerical Modelling and Radar Interferometry for Rock Landslide Early Warning Systems

126

Francesco Antolini and Marco Barla

Abstract

A new methodology for rock landslides Early Warning Systems is presented in this paper. The methodology is based on the integration between monitoring, thanks to the Ground-Based Interferometric Synthetic Aperture Radar technique, and advanced numerical modelling, with the combined Finite-Discrete Element Method. The integration procedure converges to a final decisional algorithm that represents the continuous and real time verification protocol for the monitored landslide.

Keywords

Numerical modelling • Radar interferometry • Landslide • Early warning

126.1 Introduction

The assessment of the evolution scenarios of rock landslides, as well as the associated hazard and risk, is, in general, a complex task. When it is not possible to reduce the hazard with cost-effective slope stabilization works or decrease the vulnerability of the exposed elements, real time monitoring associated to well defined Early Warning Systems (EWS) can be used for the mitigation of the risks (Di Biagio and Kjekstad 2007). In these situations the monitoring systems should be able to measure continuously over time physical quantities that can be used to predict the landslide short-term behaviour. EWS are then adopted to allow elements exposed to risk (e.g. the population) to evacuate the hazardous areas as a consequence of an alarm. For large rock landslides, the physical quantities that have proved to be most interesting for EWS are displacements and velocities.

The Ground-Based Interferometric Synthetic Aperture Radar (GBInSAR) and the combined Finite-Discrete Element Method (FDEM) were respectively selected to be part of an EWS for landslides. This EWS is based on the integration between the real-time monitoring of landslides surface displacements and velocities and on the realistic numerical prediction of their behaviour. An extensive description of the two techniques can be found respectively in Atzeni et al. (2014); Barla and Antolini (2012), Barla and Beer (2012), Barla et al. (2010, 2012), Luzi (2010), Mahabadi et al. (2012).

126.2 The Proposed Integration Methodology

The integration methodology is constituted of four main components, here called "modules", as shown in Fig. 126.1:

- radar monitoring module;
- conventional monitoring module;
- characterization and modelling module;
- verification module.

The first three modules are the source of input data for the verification module (or the decisional algorithm) of the process. The algorithm allows for the continuous assessment of the alert levels and determines the respective actions to be undertaken in order to keep an adequate level of safety for the elements exposed to risk. The integration methodology proposed can be hence considered as an EWS. The surface displacements, velocities and accelerations have been selected as the main quantities to be taken into account and measured. Modules 1, 2 and 3 (radar monitoring,

F. Antolini (✉) · M. Barla
Politecnico di Torino, Corso Duca degli Abruzzi 24, 10129 Turin, Italy
e-mail: francesco.antolini@polito.it

G. Lollino et al. (eds.), *Engineering Geology for Society and Territory – Volume 6*,
DOI: 10.1007/978-3-319-09060-3_126,

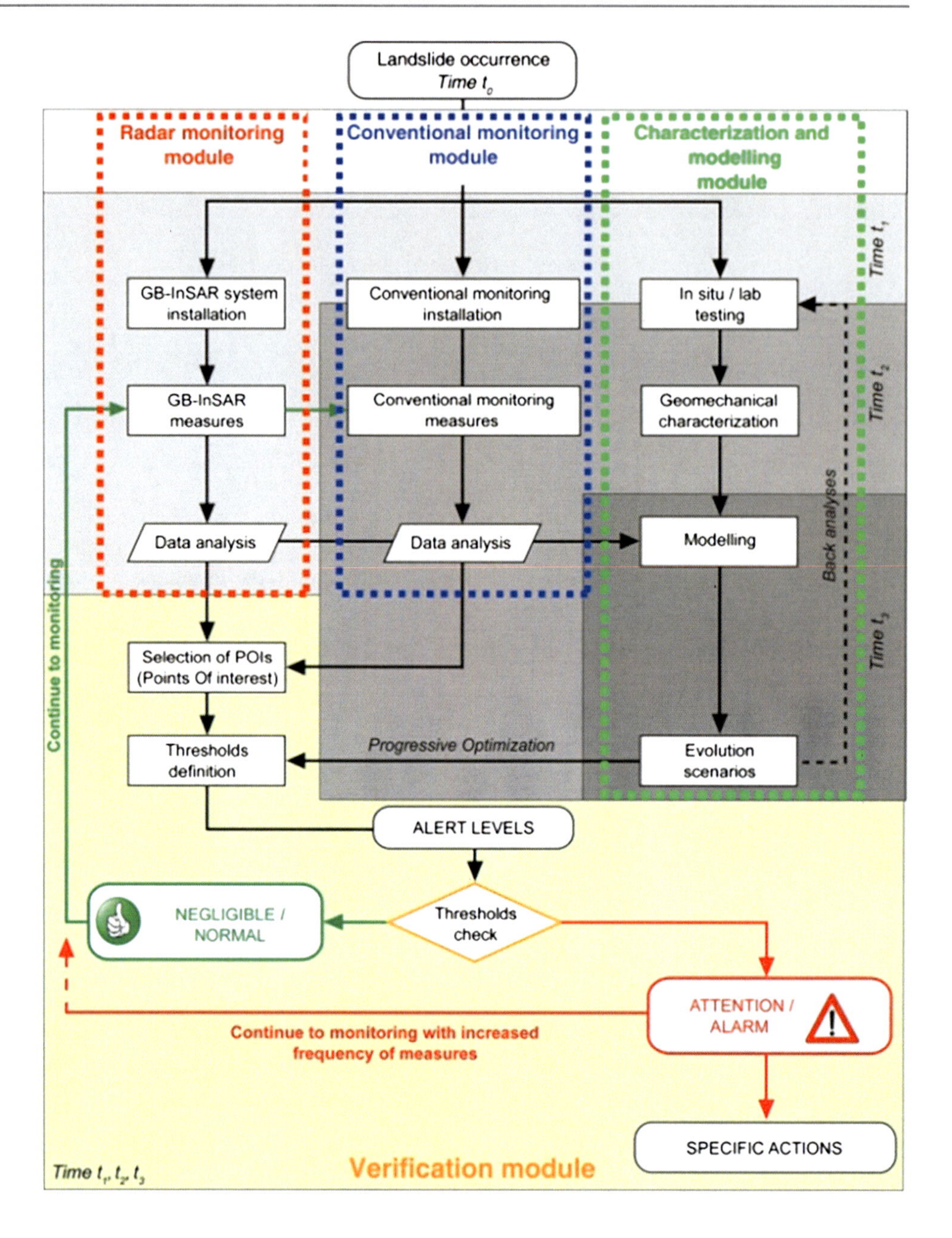

Fig. 126.1 Flow chart of the proposed integration methodology

conventional monitoring and characterization and numerical modelling) act in parallel with some of their phases related each other by transversal connections. The modules are also characterized by specific temporal constraints which indicate the time needed to perform all the listed activities.

The temporal constraints are intended as follows:

- t_0: is the initial reference time; for a first time failure, this is the time of landslide occurrence while for a dormant or potential landslide this represents the time of reactivation;
- t_1: from the time immediately next to t_0 to the following 3 days;
- t_2: from 3 to 20 days following t_0;
- t_3: more than 20 days from t_0.

126.2.1 Radar and Conventional Monitoring Modules

The first step to adopt in the proposed methodology, in emergency and just after a landslide occurrence or reactivation, is the installation of a GBInSAR system. The capability of the radar to measure very quickly (near real time) and continuously, over large areas, displacements and velocities with a sub-millimetric accuracy and in almost all weather conditions without the need to install targets on the slope, makes this tool unique among the slope monitoring systems. These features allow for obtaining displacement and velocity maps of the monitored scenario just few hours after the system installation (t_1). Other monitoring systems,

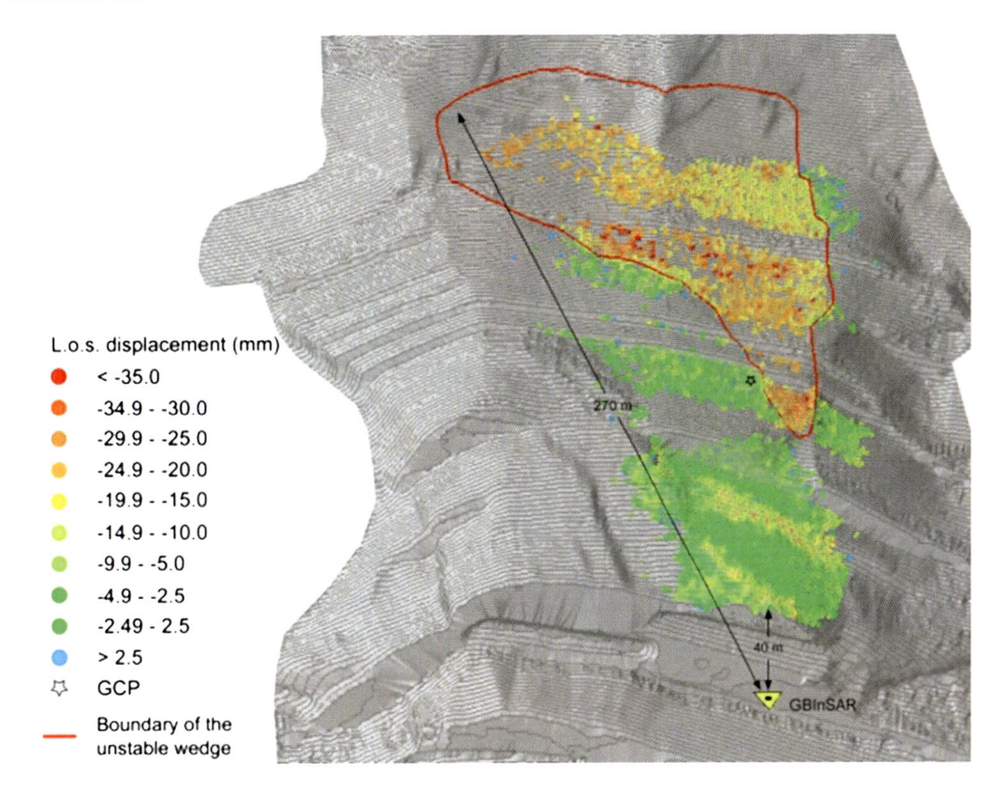

Fig. 126.2 GBInSAR displacement map projected over a digital elevation model

Fig. 126.3 Results of a FDEM numerical analysis

which for the sake of simplicity have been here indicated as conventional, even remote-sensing based, very often require days or weeks to became fully operative. Consequently they have been considered in a different temporal frame (t_2).

After the installation of a GBInSAR system it is possible to obtain a nearly continuous displacement and velocity field of the observed area. Figure 126.2 shows an example of displacement map obtained with the GBInSAR projected over a Digital Elevation Model of the monitored area.

Another important aspect that should be emphasized concerns the data acquisition frequency (sampling rate) of a radar system. The possibility to obtain a displacement map updated every 10 min or less, fully satisfies the requirements for a real time monitoring system and, especially in emergency conditions, represents an important added value for EWSs.

126.2.2 Characterization and Modelling Module

This module includes in situ investigations, field surveys (geological, geomorphological, structural, geophysical), laboratory tests for the geomechanical characterization of intact rock and discontinuities and the numerical modelling of the instability phenomenon. Here, a combined finite-discrete element method (FDEM) is used to model triggering and run out of a landslide. This method allows one to investigate brittle failures of slopes from initiation through transportation and deposition as shown in the example of Fig. 126.3. Moreover the back analysis of already occurred failures, along with a continuous calibration process over monitoring data, as they become available, allows for the results of the numerical models to be used in scenario-based analysis.

All the listed activities are typically time demanding and should be considered in a reference time spanning from few days to few weeks (t_2–t_3) after the landslide occurrence.

126.2.3 Verification Module

The last section of the proposed algorithm is the verification module which is the core of the integration methodology. It represents the set of operations needed to determine continuously and in near real time the alert levels related to a particular instability phenomenon, using the data provided from at least one of the modules previously described. The decisional module does not present any particular time constraint as it can be reached in a time which is variable from t_1 (few hours after the landslide occurrence) to t_3 (few weeks) depending on the particular combination of the modules chosen.

The decisional module is composed of three different phases:

- selection of Regions of Interests (ROIs);
- definition of thresholds and alert levels;
- verification.

The ROIs are portions of a landslide characterized by an homogenous kinematic behaviour (i.e. type of motion, direction, trend of displacement, degree of activity). The adoption of ROIs allows to take into account the heterogeneities in the geomechanic behaviour of complex and compound landslides.

When the decisional module is reached directly through the radar module (t_1), the thresholds can be only determined on the basis of the direct observation of the time series of the radar displacements maps, due to the lack of other information. As a consequence, the thresholds thus defined will be markedly conservative. From the results of the FDEM modelling, by means of a progressive optimization procedure, the initial conservative thresholds will be eventually modified to be used for the medium and long term monitoring of the landslide.

A typical set of 3 alert levels and 2 thresholds (attention and alarm) are adopted. Each alert level is then associated to a state of activity of the landslide (normal or seasonal activity, increased activity, possible collapse) and to a list of actions to be engaged for the mitigation of the risk. Alert levels are triggered by exceeding the relative threshold for one, or a combination, of the monitored parameters. In the proposed scheme the responses associated to each alert level imply also a feedback loop to the frequency of measurements in the radar and conventional monitoring modules. Finally, once the alert levels and the respective thresholds have been selected, the verification phase of the algorithm simply concerns the continuous and real time monitoring of the selected parameters over the ROIs and the comparison with the pre-defined thresholds values.

126.3 Conclusions

In this paper a novel contribution to the set-up of a cost-effective EWS for rock landslides was illustrated. This was obtained through the combination of innovative remote sensing techniques (i.e. GBInSAR) with advanced numerical modelling (i.e. FDEM).

Acknowledgments The work described in this paper is partially funded by the National Research Project PRIN 2009 "Integration of monitoring and numerical modelling techniques for early warning of large rockslides".

References

Atzeni C, Pieraccini M, Barla M, Antolini F (2014) Early warning monitoring of natural and engineered slopes with Ground-Based Synthetic Aperture Radar. Rock Mechanics and Rock Engineering, doi:10.1007/s00603-014-0554-4

Barla M, Antolini F (2012) Integrazione tra monitoraggio e modellazione delle grandi frane in roccia nell'ottica dell'allertamento rapido. In: Barla G, Barla M, Ferrero A, Rotonda T (eds) MIR 2012 - Nuovi metodi di indagine e modellazione degli ammassi rocciosi. CELID, Torino

Barla M, Beer G (2012) Editorial: special issue on advances in modelling rock engineering problems. Int J Geomech 12(6):617

Barla G, Antolini F, Barla M, Mensi E, Piovano G (2010) Monitoring of the Beauregard landslide (Aosta Valley, Italy) using advanced and conventional techniques. Eng Geol 116:218–235

Barla M, Piovano G, Grasselli G (2012) Rock slide simulation with the combined finite discrete element method. Int J Geomech 12(6):711–721

Di Biagio E, Kjekstad O (2007) Early warning, instrumentation and monitoring landslides. In: 2nd regional training course, RECLAIM II, 29 Jan–3 Feb 2007

Luzi G (2010) Ground based SAR interferometry: a novel tool for geoscience. Geosci Rem Sens New Achievements, p 26

Mahabadi O, Lisjak A, Munjiza A, Grasselli G (2012) Y-Geo: a new combined finite-discrete element numerical code for geomechanical applications. Int J Geomech 12(6):676–688

A Tool for Semi-automatic Geostructural Survey Based on DTM

127

Sabrina Bonetto, Anna Facello, Anna Maria Ferrero, and Gessica Umili

Abstract

Tectonic movement along faults is often reflected by characteristic geomorphological features such as linear valleys, ridgelines and slope-breaks, steep slopes of uniform aspect, regional anisotropy and tilt of terrain. In the last years, the remote sensing data has been used as a source of information for the detection of tectonic structures. In this paper, we present a new approach for semi-automatic extraction and characterization of geological lineaments. The overall positive aspects of this semi-automatic process were found to be the rapidity of preliminary assessment, the possibility to identify the most interesting portions to be investigated and to analyze zones that are not directly accessible. This method has been applied to a geologically well-known area (the Monferrato geological domain) in order to validate the results of the software processing with literature data. Results obtained are discussed and preliminary remarks are put forward.

Keywords

Geological lineaments • DTM • Semi-automatic extraction • CurvaTool

127.1 Introduction: State of the Art

In geology the satellite remotely sensed data has been used as source of information for the detection of tectonic structures such as faults, large-scale fractures, and fracture zones (Wladis 1999; Morelli and Piana 2006; Hashim et al. 2013). Geological lineaments are parameters that can be used in assisting mineral prospecting, hydrogeology studies, tectonic studies for the delineation of major structural units, analysis of structural deformation patterns and identification of geological boundaries.

Generally, in literature, the extraction of geological lineaments can be grouped into three main approaches: (i) manual extraction (Jordan and Schott 2005), (ii) semi-automatic extraction (Lim et al. 2001; Jordan et al. 2005), and (iii) automatic extraction (Masoud and Koike 2011; Saadi et al. 2011). Manual and semi-automatic approaches are greatly influenced by the experience of the analyst, while automatic extraction depends on the algorithms efficiency and on the information content in the image (Hashim et al. 2013).

Normally, lineaments can be detected due to their geomorphological features, such as morphotectonic elements, drainage network offsets and stream segment alignments, and/or spectral criterion, such as tonal change, pattern and textures, using (stereo-) aerial photographs and other remotely sensed imagery.

S. Bonetto · A.M. Ferrero
Department of Earth Sciences, University of Turin, Valperga Caluso 35, 10125 Turin, Italy

A. Facello (✉)
CNR-IRPI, Strada delle Cacce, 73, 10135 Turin, Italy
e-mail: facelloanna@gmail.com

G. Umili
Department of Civil, Environmental and Territory Engineering, University of Parma, Parco Area delle Scienze 181/A, 43124 Parma, Italy

G. Lollino et al. (eds.), *Engineering Geology for Society and Territory – Volume 6*,
DOI: 10.1007/978-3-319-09060-3_127, © Springer International Publishing Switzerland 2015

Regarding the Digital Terrain Model (DTM), this has been used as shaded relief model either alone or in combination with remotely sensed images on a regional scale. Moreover, three-dimensional view with image drape and digital cross sections have been used for morphotectonic investigations (Jordan et al. 2005).

In this paper, authors propose the use of an innovative method for the extraction of geological features using a semi-automatic approach. The method will be discussed and presented in the following sections. The overall positive aspects of this semi-automatic process were found to be the rapidity of preliminary assessment, the possibility to identify the most interesting portions to be investigated and to analyze zones that are not directly accessible.

127.2 Software

The method is based on the assumption that a geological lineament can be geometrically identified as a convex or concave edge of the surface of a DTM, particularly in presence of a structural control of the geomorphological evolution of the analyzed areas.

The code CurvaTool (Umili et al. 2013) was originally developed to automatically detect edges on Digital Surface Model (DSM) of natural rock mass outcrops, assuming that they represent the discontinuity traces. In this work the code CurvaTool has been applied to DTM of large portion of territory in order to automatically detect edges which represent potential geological lineaments. As natural outcrops, also the earth surface can have an infinite variety of shapes whit different dimensions, but a common characteristic is generally their non-planar surface. In fact, the surface has often edges that can be both asperities or depressions.

The code quantifies the non-planarity by means of principal curvature values associated to the DSM/DTM points; the user is asked for two thresholds on principal curvature: the first one to detect significant convex edges and the second one to detect significant concave edges. After the identification, edges are segmented in order to obtain the segments that better interpolate the obtained polyline.

Post-processing operations are required in order to filter and to classify segments representing items of interest among all the reconstructed edges: therefore specific algorithms, called Filter in the following, have been created to perform these operations. The user is asked for the minimum edge length and the orientations of the expected lineaments clusters (expressed by an angle respect to the North and the relative standard deviation). Filter code deletes edges shorter than the fixed length and classifies them attributing each edge to the correspondent input cluster. Non-classified edges are recorded as "others".

127.3 The Test Area: The Monferrato Geological Domain

The Monferrato has been selected as the area-test to verify the software application in the lineaments identification on a large scale.

The Monferrato is located in correspondence of the Alps-Apennines Junction Zone; it is an highly deformed geological domain, but is also rife with literature data which are very helpful for a suitable validation of the results in software applications. Most of the information about faults trending and their distribution result either from field evidences and small-scale kinematic observations or they have been verified by the geomorphological and spectral analysis with remote sensed imagery (Morelli and Piana 2006). In Monferrato, because of a rich vegetation and human activity, the substratum is poorly exposed and a small number of outcroppings is present. Therefore, direct evidences of the structural lineaments are not easily detectable on the field and, sometimes, their presence is just supposed for stratigrafical reasons.

In the Monferrato, geological succession is divided in a lower part of strongly deformed Apennine calcareous flysch (late Cretaceous to middle Eocene age) and an upper terrigenous succession (middle Eocene—Pliocene) resting unconformably on the previous one (Clari et al. 1995). In particular, the sedimentary sequence is mainly composed of marls, arenites, siltstone (with locally interbedded sandstone), evaporates, mudrocks and sandstones. The stratigraphic succession is characterized by lateral thickness variations and by the occurrence of local unconformities; it is poorly folded, but highly tilted and deformed in reason of a continuous uplift (also recent) which caused a structural control during geomorphological evolution. Steep slopes, well-organized drainage network, fractures and faults are the evident consequences of that interaction.

Tectonic boundaries divide the Monferrato geological domain in tectonostratigraphic units which are characterized by distinctive sedimentary evolution, stratigrafical sequence, geometries and amount of deformation. Despite the structural complexity and the presence of many deformation zones, four main systems faults have been recognised in Monferrato (Piana 2000; DeLa Pierre et al. 2003). They are oriented NW–SE, NE–SW, E–W and N–S respectively.

127.4 Data Discussion

A preliminary approach to the software application consisted in the lineaments identification in a geologically well-known area in order to validate the results of the software processing with literature data. Since the structural setting is

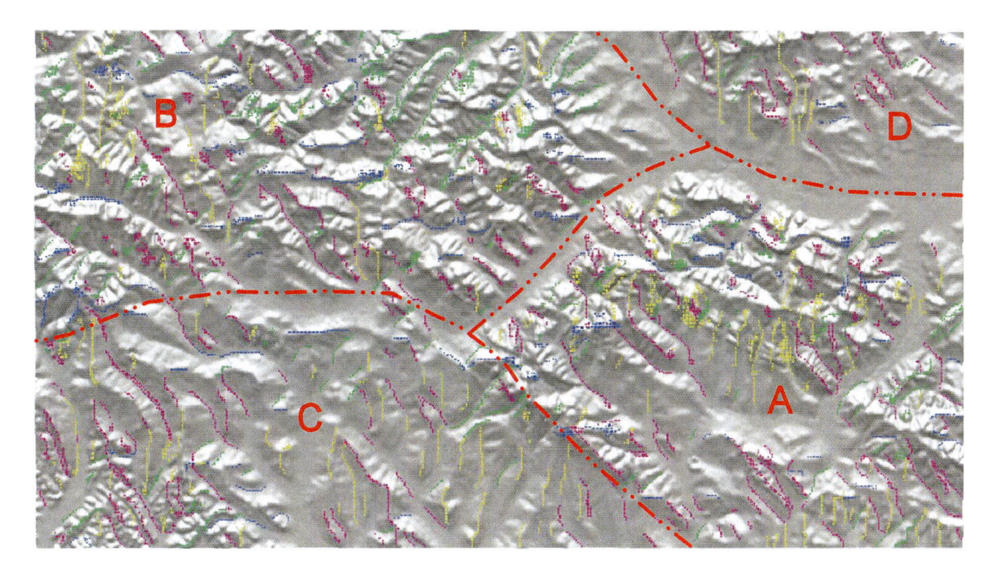

Fig. 127.1 Image of Monferrato DTM with lineaments extracted by CurvaTool and processed with Filter. Four sets are visible: L1 (fuchsia), L2 (*green*), L3 (*blue*) and L4 (*yellow*)

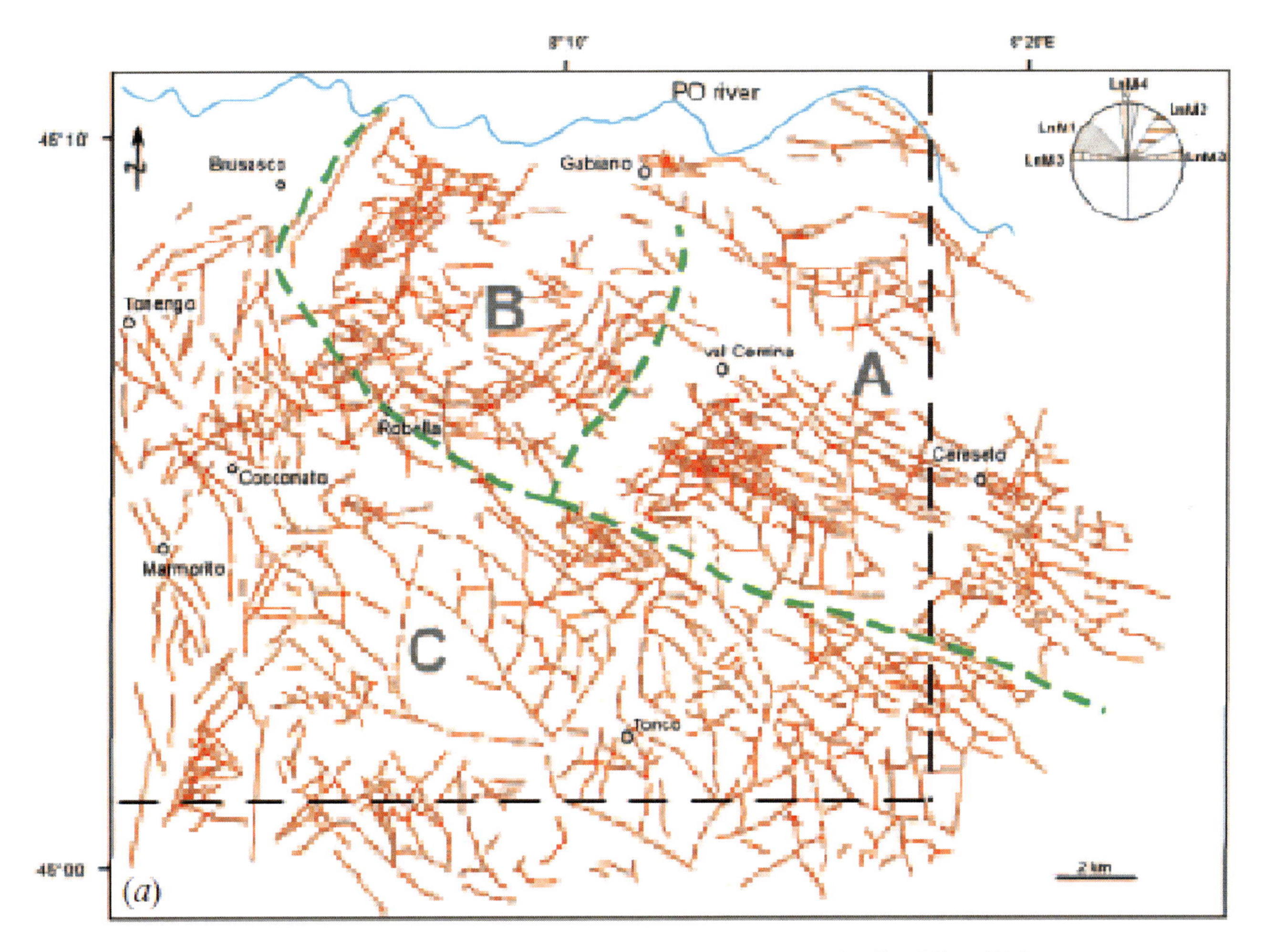

Fig. 127.2 Image of Monferrato DTM with lineaments extracted by Spot HVR data (*source* Morelli and Piana 2006)

well-known, a DTM of the Monferrato domain was processed by CurvaTool software, furnishing a preferential range of directions as a geometrical filter to simplify the lineaments identification. A large number of lineaments were found and different filters were tested to obtain a final simplified image.

Geological structures manually extracted through photo-interpretation, remote sensing and analysis of geological maps were compared with the semi-automatic outputs of CurvaTool and Filter. According to literature data (Piana 2000), the CurvaTool processing of the DTM of the Monferrato area identified several lineaments oriented coherently with the main four lineament systems.

The image representing the results (Fig. 127.1) displays a greatest lineaments length and frequency on NW–SE direction (L1), whereas short and intermediate lengths generally occur on the E–W (L3) and N–S (L4) average direction. The NE–SW lineaments (L2) are usually short, the longest ones are concentrated in the north-western part of the map.

In particular, the NW–SE striking system is uniformly distributed and shows a regular geometry and direction, generally with long or intermediate-length lineaments. The N–S striking system is present in the whole area with a different distribution and length and it is particularly frequent in the southern area of the map.

The distribution and length of the lineaments were compared with the geological structures collected at different scales, resulting by field work (DeLa Pierre et al. 2003) and Spot and SAR image analysis (Morelli and Piana 2006).

As for Morelli and Piana (2006), according to geometry and spatial distribution of the lineaments, different sectors should be recognized (Fig. 127.2).

From the centre to the north-western part of the map (sector B), all lineament systems were recognized. The NW–SE and the NE–SW striking systems are predominant; in particular, the sector B is the only one within the NE–SW striking system is particularly distributed and shows the longest lineaments. The L1 consists of long and intermediate-length lineaments; the L3 is well represented, particularly with intermediate length. The N–S lineaments are less frequent and their geometrical features are not so clear in that sector.

The south–eastern part of the map (sector A) is characterised by long and evident L1 lineaments and L3 lineaments of intermediate length (particularly distributed in the north and north–western part of the sector). The L2 system is poorly represented, whereas the L4 system has a wider distribution than in sector B, particularly in the centre of the sector.

The south western part of the image (sector C) is the poorest deformed area. It is characterised by the lowest frequency of the lineaments: L1 system is still present with a few intermediate-length lineaments, L2 and L3 systems are poorly represented and are concentrated principally on the northern and north–eastern border of the sector respectively. Unlike the other sectors, the L4 system is clearly represented and regularly distributed in the whole sector.

The north–eastern area (sector D) is too small to make statistical analysis, even if L1 and L4 systems seem to be predominant.

127.5 Conclusion

Through this work, the authors the authors propose and prove the use of an innovative method for the extraction of geological features using a semi-automatic approach. This method has applied in the identification of lineaments features in an area geologically known (Monferrato domain) in order to validate the results obtained with the data of literature.

By comparing, it is possible to note a good correspondence between literature data and the preliminary CurvaTool results, as regard to geometry and distribution of the main lineaments systems.

According to the final remarks, the first results of software application are good. Software should be improved, at any rate the correspondence between lineaments and geological structures are mainly not direct and a post-processing subjective interpretation by the user is anyway necessary.

References

Clari P, DeLa Pierre F, Novaretti A, Timpanelli M (1995) Late Oligocene-Miocene sedimentary evolution of the critical Alps/Apennines junction: the Monferrato area, Northwestern Italy. Terra Nova 7:144–152

DeLa Pierre F, Fioraso G, Piana F, Boano P (2003) Foglio 157 'Trino' della Carta Geologica d'Italia alla scala 1: 50,000. APAT, Dipartimento Difesa del Suolo, Roma

Hashim M, Ahmad S, Johari MAM, Pour AB (2013) Automatic lineament extraction in a heavily vegetated region using Landsat Enhanced Thematic Mapper (ETM+) imagery. Adv Space Res 51 (5):874–890

Jordan G, Schott B (2005) Application of wavelet analysis to the study of spatial pattern of morphotectonic lineaments in digital terrain models. A case study. Remote Sens Environ 94(1):31–38

Jordan G, Meijninger BML, Van Hinsbergen DJJ, Meulenkamp JE, Van Dijk PM (2005) Extraction of morphotectonic features from DEMs: development and applications for study areas in Hungary and NW Greece. Int J Appl Earth Obs Geoinf 7:163–182

Lim CS, Ibrahim K, Tjia HD (2001) Radiometric and geometric information content of TiungSat-1 MSEIS data. In: TiungSAT-1: from inception to inauguration, pp 169–184

Masoud A, Koike K (2011) Auto-detection and integration of tectonically significant lineaments from SRTM DEM and remotely-sensed geophysical data. ISPRS J Photogr Remote Sens 66:818–832

Morelli M, Piana F (2006) Comparison between remote sensed lineaments and geological structures in intensively cultivated hills

(Monferrato and Langhe domains, NW Italy). Int J Remote Sens 27 (20):4471–4493

Piana F (2000) Structural setting of western Monferrato (Alps-Apennines junction zone, NW Italy). Tectonics 19:943–960

Saadi NM, Abdel Zaher M, El-Baz F, Watanabe K (2011) Integrated remote sensing data utilization for investigating structural and tectonic history of the Ghadames Basin, Libya. Int J Appl Earth Obs Geoinf 13:778–791

Umili G, Ferrero AM, Einstein HH (2013) A new method for automatic discontinuity traces sampling on rock mass 3D model. Comput Geosci 51:182–192

Wladis D (1999) Automatic lineament detection using digital elevation models with second derivative filters. In Photogrammetric Engineering & Remote Sensing 65(4):453–458

New Perspectives in Long Range Laser Scanner Survey Focus on Structural Data Treatment to Define Rockfall Susceptibility

128

Andrea Filipello, Leandro Bornaz, and Giuseppe Mandrone

Abstract

Laser scanning techniques are nowadays more and more used in engineering geology in order to describe slope instability assessment. In this study it is described an application of a new Terrestrial Laser Scanner that offers an extremely long measurement range and the procedure applied for geological data treatment. Point cloud, 3D models, "solid images" and DEM generated from Terrestrial Laser Scanner survey were managed with Ad Hoc, a software platform with specific tools for geological and geomechanic analysis.

Keywords

Laser scanner • Rockfall • Geomechanical classification • GIS tools

128.1 Introduction

The use of Terrestrial Laser scanners (TLS) are very useful where the starting point to evaluate stability properties is a morphological investigation. Laser scanners are nowadays successfully used also to study stability problems by measuring geomechanic parameters, such as distance measurements, angles, DIP and DIP direction of discontinuities, volumes. The introducing of new very long range laser scanner open a series of new very interesting perspectives.

In order to obtain a correct geometrical information, laser scanner technique requires special management during acquisition and data treatment. This is more important if very long range acquisition is carried out due to geodethic problems (Biasion et al. 2005; Bornaz and Dequal 2003; Fricout et al. 2007).

A. Filipello · G. Mandrone (✉)
Department of Earth Sciences, University of Torino, Via Valperga Caluso 35, 10125 Turin, Italy
e-mail: giuseppe.mandrone@unito.it

A. Filipello · G. Mandrone
AG3 S.r.l., Via Valperga Caluso 35, 10125 Turin, Italy

L. Bornaz
Ad Hoc 3D Solutions S.r.l., Fraz. La Roche 8, 11020 Gressan, AO, Italy

The study site is located in the lower Aosta Valley (NW Italy), near the village of Hone. The planimetric extension of the study area is approximately 63 ha, the slope has a vertical height of about 700 m, from the bottom valley (345 m a.s.l.), up to the ridge, located at about 1,060 m a.s.l.

128.2 Laser Scanner Data Acquisition

Acquisition should be correctly planned and executed in order to obtain a valuable final result and a real time data check is suggested. This operation depends on the characteristics of the laser scanner used too: in our case it is the new RIEGL VZ4000. This instrument is very innovative and offers a lot of new possibilities and a very long range measurement performance. It can acquire points at more than 4,000 m with high accuracy and a very small laser beam. Despite this, the data can be very noisy and they must be used and treated very carefully due to "distortions" introduced by the very long range measurements, the echo digitization and the online waveform processing techniques on which the laser is based.

When an object has a complex shape or when a single scan cannot record the whole object, a series of scans must be done. These series has to be correctly planned to avoid hidden areas. To plan the survey we have to pay attention not only to the shape of the object, but also to the instrument

G. Lollino et al. (eds.), *Engineering Geology for Society and Territory – Volume 6*,
DOI: 10.1007/978-3-319-09060-3_128, © Springer International Publishing Switzerland 2015

characteristics (such as vertical and horizontal field of view, maximum range,...). In this test site we had the problem that the morphology of the study area is very steep and instrument, unlike other Riegl instruments, cannot be tilted and has a limited field of view along the vertical axes.

To measure at 4–6 km with a laser rangefinder it is necessary to digitally model and consider a lot of atmospheric parameters. If the very long range acquisition (4–6 km) is carried out from a airplane (ALS), these parameters are quite easy to be considered because the laser beam through all different atmospheric layers perpendicular each to the other. In TLS instead, the laser ray can have (depending on the laser range measurement, the laser emission angle, and the earth curvature) a very high and changeable incident angle with the atmospheric layers. So the laser ray can be strayed in a different way along its path.

In our case this aspect has greatly complicated the acquisitions. We had to carry out acquisitions only from the valley floor due to the impervious slopes on each side of the valley, not achievable by walking. The relationship between the height of the rock wall and the planar distance between laser position and the upper part of the cliff was not enough to acquire the whole site using the VZ4000 vertical FOV. So, we had to carried out 4 scans using the VZ4000 and a set of acquisition using the VZ400 (used in the lower part of the cliff). Moreover, a GPS and topographic surveys have to be carried out because the acquired data must be geo-referenced. In our case, 5 GPS GCP was acquired.

128.3 Data Processing

RIEGL's V-Line technology is based on echo digitization and online waveform processing. The VZ4000 in addition is equipped with multiple-time-around capability (MTA). This combination allows users to benefit from the high pulse rate also from very long range and thus to achieve high measurement densities on the object (Fig. 128.1). The elaboration (ambiguity resolution) of multiple-time-around capability (MTA) is accomplished using the algorithms and software developed by RIEGL in addition to classical data treatment. The result of this operation is a point cloud with more information in the remote part. The operation known as "preliminary data treatment" concerns data filtering, vegetation and artifacts removal, point cloud registration, multiple scan triangulation operations.

The noise reduction is one of the preliminary fundamental operations because laser scanner data always have noise that is lower than the tolerance of the used instruments. In order to obtain a "noise free" model of the object, it is necessary to use specific algorithms able to reduce or even eliminate the acquisition errors that can be found in the point clouds. Simply visualizing the data acquired with the VZ4000 after the MTA processing, it is possible to see a lot of noise in the 3D point cloud (Fig. 128.2). Sometimes it is also necessary to remove scattered points that do not belong to the object. In the particular case of rocky faces with irregularities that could produce rock falls, it is common practice to reinforce the stability of the walls using protection barriers. In environmental surveys the vegetation has to be considered too. Usually vegetation and scattered points removal operation is performed manually by the operator. In our case the data was elaborated using specific automatic algorithms developed by Ad Hoc 3D Solutions (Fig. 128.3). The used algorithms have been suitably modified to properly consider the new VZ4000 data. The result of these procedures is a complex, "noise free" point cloud (without any outliers, gross or systematic errors, vegetation and artifacts).

Usually, more than one scan is needed to completely describe a rock wall. In these cases each scan has its own reference system: the reconstruction of the 3D model of the surveyed object requires the registration of the scans in a single reference system. As Dip and Dip directions have to be measured and integration with regional DSM, the reference system has to be properly choose (X along East, Y along North and Z along the vertical direction).

If the 3D acquired data must be geo-referenced in a cartographic reference system some problems due to cartographic projection deformations have to be considered to correctly estimate the alignment and the real displacement.

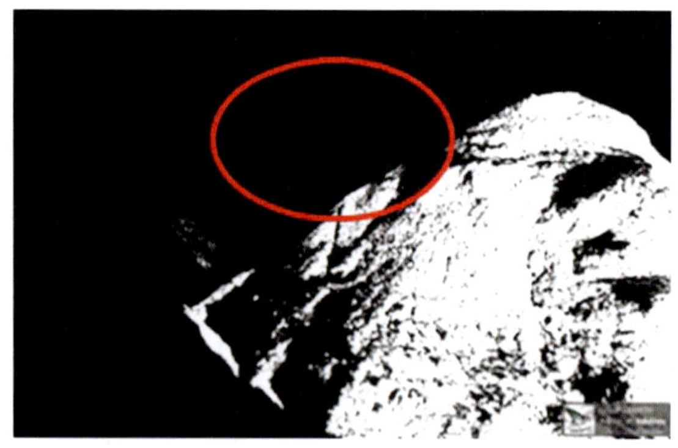

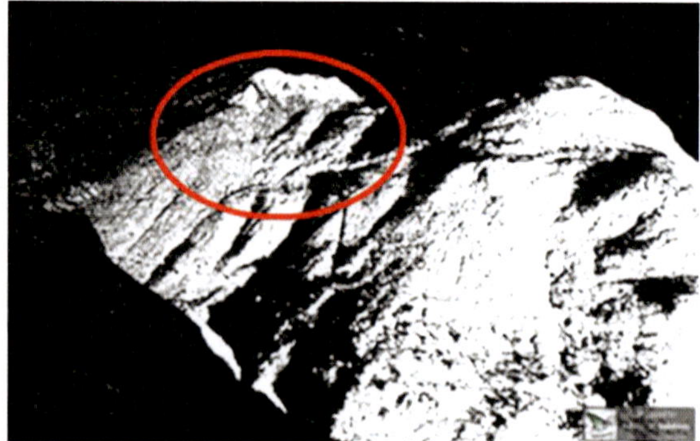

Fig. 128.1 Original data (*left*). Original data with MTA pre-processing (*right*)

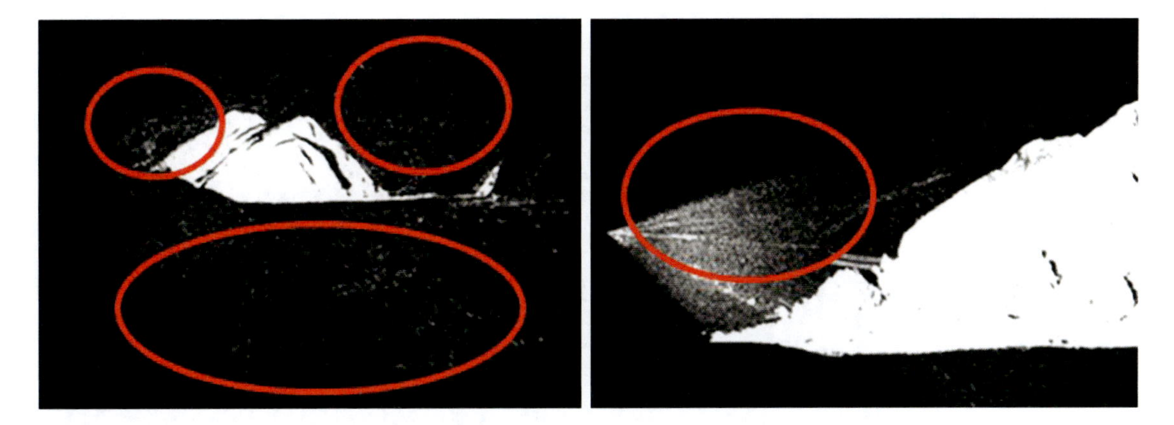

Fig. 128.2 Original VZ4000 data. Noise in the acquisition

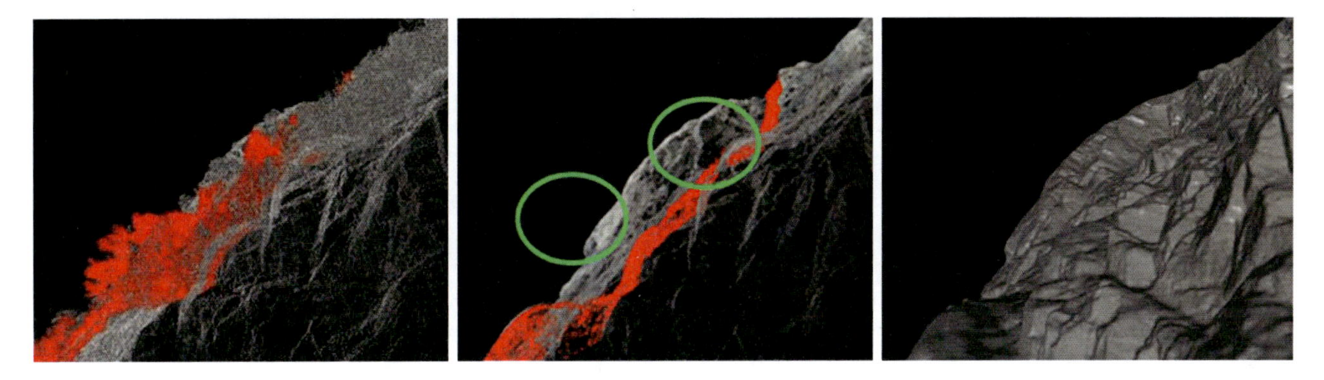

Fig. 128.3 Vegetation filter: original 3D point cloud (*left*), filtered point cloud (*center*) and 3D surface (*right*)

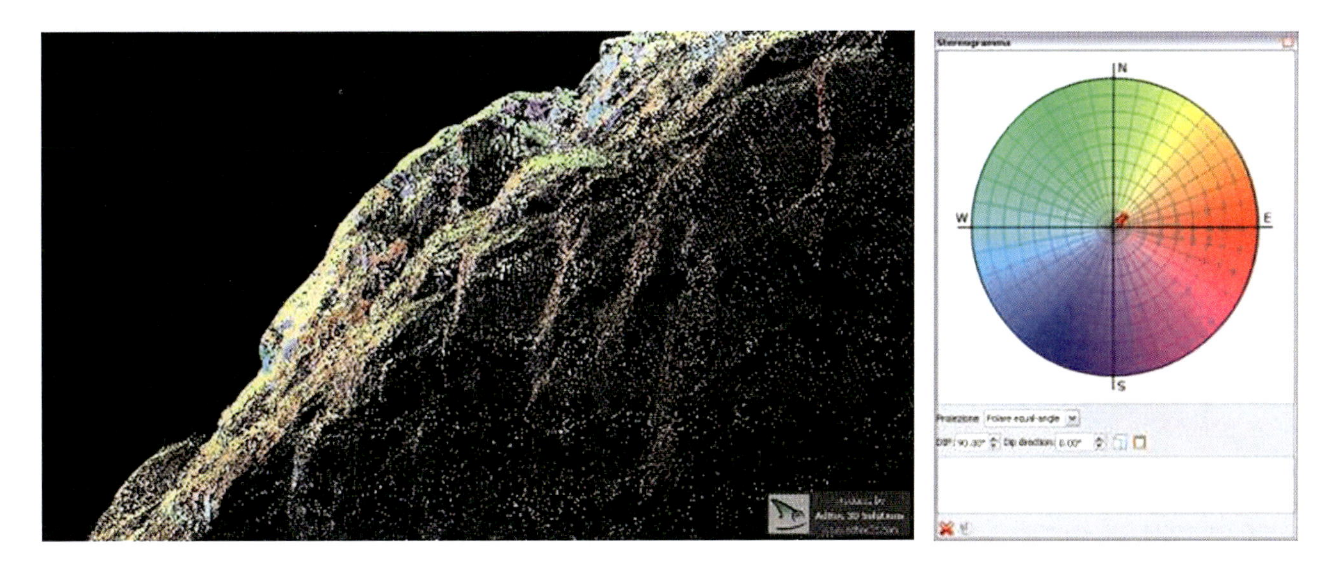

Fig. 128.4 Final noise free point cloud. DIP DIPDIR colored

The problems related to reference systems is generally underestimated or completely overlooked. Certainly it can be neglected if you are acquiring and georeferencing a very small site (considering the topographic field and the earth curvature).

If we try to align some scans in a cartographic system the result (errors on GCPs) can be poor due to the cartographic deformations, the earth curvature and the geoid undulation. In particular if the rock wall is in mountain as the deformations groove with the height. But rules are essential when

Fig. 128.5 Final noise free surface of the rock wall. 19.7M points—39.4M facets

you want to integrate data acquired with different methodologies and techniques. In order to integrate TLS, aerial laser scanner, GPS, photogrammetry, geophysics data a.s.o, we have to consider the different reference systems. The troubles come when you try to merge the data coming from the two instruments (e.g. TLS and ALS), that are not suitable. This operation is generally carried out using ICP, thus minimizing alignment errors but forgetting completely the deformations induced by the reference systems and thus committing a serious error from a topographical point of view. Ad Hoc 3D Solutions developed a proprietary and patented technique able to solve the merging problem between TLS and cartography. Our technique uses a rigorous geodetic approach, that is the basis for monitoring of rock walls or glaciers.

The same technique allow us to acquire TLS data in a local reference system and extract every information from the point cloud without deformations, convert all in the 3D model and extract features in the cartographic reference system by rigorous transformations. The system allow to collect, merge and correctly manage laser scanner data coming from different sources, also in different epochs (very useful for monitoring). This approach was used to correctly georeference the VZ4000 3D data with a very high global accuracy (~ 5 cm). This accuracy is checked on site on a big set of control points. The georeferencing and alignment algorithm used as homologous and georeferencing points, to obtain the final point cloud, both GCP that natural points having particular geometric characteristics and consider all the geodetics rules in a correct way. The final result is a noise free, correctly aligned and georeferenced point cloud, available in all reference systems, fully integrated with the regional UTMED50 DSM.

128.4 Conclusion

Detailed studies for rock fall hazard are necessarily based on knowledge of the morphological and mechanical characteristics of the cliff. When the morphology of the cliff makes difficult a direct access the use of climbing techniques for access to the wall is recommended, but not always possible for economic or safety reasons. The use of terrestrial laser scanner (TLS) in addition to the traditional techniques of investigation is becoming more widespread. The result of a terrestrial laser scanning is a points cloud that reproduces the morphology of the investigated target.

In this study a complete 3D survey of the area is carried out, using a methodology that integrates Terrestrial Laser Scanner, GPS and photogrammetry. In particular, a new long range laser scanner has been tested (Riegl VZ4000) in steep and large area along the right flank of the lower Aosta Valley. The 3D VZ-Line laser scanner offers very long range, up to 4000 m. In fact for this case, was developed a specific procedures and methodologies for the treatment and correct management of a terrestrial laser scanner survey of a site larger than 2 km. This kind of instruments open new perspectives on environment applications, but it is very important to balance real capabilities and real limits.

The data treatment carried out by Ad Hoc 3D Solutions allow to obtain not only a simple point cloud, but a 3D true colored 3d model (using an independent HDR photogrammetric survey, Fig. 128.4) where each point have a series of additional information (e.g. dip, dip dir,...). Finally, using a specific surface reconstruction algorithm, the surface of the rock wall have been generated (Fig. 128.5). A set of solid images were generated too in order to allow experts to extract information not only from the 3D model but using a more simple interface to use the 3D.

Only at this point it is possible to start measuring geo-mechanic information focused to slope stability analysis.

References

Biasion A, Bornaz L, Rinaudo F (2005) Laser scanning applications on disaster management. First international symposium on geo-information for disaster management (Gi4DM), Jan 2005

Bornaz L, Dequal S (2003) A new concept: the solid image. Int Arch Photogram Remote Sens Spat Inform Sci 34:78–82

Fricout B, Villemin T, Bornaz L (2007) Remote analysis of cliff outcrops using laser DDSM and digital images. Geophys Res Abs 9 (08194):200

Structural Data Treatment to Define Rockfall Susceptibility Using Long Range Laser Scanner

129

Andrea Filipello, Giuseppe Mandrone, and Leandro Bornaz

Abstract

Laser scanning techniques are nowadays more and more used in engineering geology in order to assess slope stability. In this study it is described an application of a new Terrestrial Laser Scanner that offers an extremely long measurement range and the procedure applied in geologic data treatment. Point cloud, 3D models, "solid images" and DEM generated from Terrestrial Laser Scanner survey were managed with Ad Hoc, a software platform with specific tools for geologic and geomechanic analysis. The assessment of rockfall prone area has been developed according to two approaches: deterministic and empirical. In the first case the detection of potentially unstable blocks, typically done with field surveys in case of outcrops of limited extension, were performed directly on the point cloud and on solid images. In the second case GIS tools were used allowing to apply the geomechanic classification SMR Romana for each pixel. Results obtained from both approaches is compared as a further data validation procedure.

Keywords

Laser scanner • Rockfall • Geomechanical classification • GIS tools

129.1 Introduction

Detailed studies for rock fall hazard are necessarily based on knowledge of morphologic and mechanic characteristics of rock masses. When morphology of slopes make difficult field surveys, the use of climbing techniques for access to the wall is needed but not always possible for economic or safety reasons. The use of Terrestrial Laser Scanner in addition to the traditional techniques of investigation is becoming more widespread. The result of a terrestrial laser scanning is a point clouds that reproduces the morphology of the investigated target.

In this study a complete 3D survey of the area was carried out, using a methodology that integrates Terrestrial Laser Scanner, GPS and photogrammetry (Fig. 129.1). In particular, a new long range laser scanner has been tested (Riegl VZ4000) in a large (about 2 km^2) and steep area along the right flank of the lower Aosta Valley. The 3D VZ-Line laser scanner offers very long range, up to 4,000 m. Specific procedures and methodologies for the treatment and correct management of the terrestrial laser scanner survey of the test site were developed. These kind of instruments open new perspectives in environmental applications but it is very important to balance real capabilities and technical limits.

In addition, a set of solid images were generated in order to allow experts to extract information not only from the 3D model but also using a more simple interface.

Both automatic data extraction procedure from point cloud and interpretation on solid images have been applied. The assessment of rock fall prone area has been developed according to deterministic and empirical approaches. In the

A. Filipello · G. Mandrone (✉)
Department of Earth Science, University of Torino, Via Valperga Caluso 35, 10125 Turin, Italy
e-mail: giuseppe.mandrone@unito.it

A. Filipello · G. Mandrone
AG3 S.r.l., Via Valperga Caluso 35, 10125 Turin, Italy

L. Bornaz
Ad Hoc 3D Solutions S.r.l., Fraz. La Roche 8, 11020 Gressan, AO, Italy

G. Lollino et al. (eds.), *Engineering Geology for Society and Territory – Volume 6*,
DOI: 10.1007/978-3-319-09060-3_129, © Springer International Publishing Switzerland 2015

Fig. 129.1 Laser scanner survey: the study area is about 1,500 m in elevation and about 2,000 m in length

first case, the detection of potentially unstable blocks, typically done with direct surveys in case of outcrops of limited area, were evaluated also on the point cloud and on the solid images. In the second case, GIS tools that allow to apply the geomechanic classification SMR Romana for each pixel were used.

129.2 Study Area

The study site is located in the lower Aosta Valley (NW Italy), within the municipality of Hone. The planimetric extension of the studied area is approximately 63 ha (much less than the survived one), the slope has a vertical height of about 700 m, from the bottom valley (345 m a.s.l.), up to the ridge, located at about 1,060 m a.s.l.

Lithologies identified are mainly lithoids: in most of the study area soil covers are thin or non-existent. Tree and shrub vegetations are mainly located on alluvial fans or on terraces. The study area consists mainly of Gneiss Minuti of the Sesia Lanzo Zone and it is frequently affected by rock falls. During a recent collapse in the 20 Feb 2012, the municipal Waste Recycling Centre—located at the base of the slope—was involved in rock falling and it was closed as a precautionary measure.

129.3 Geomechanical Data Collection

The geomechanic and morphostructur characterization of discontinuities has been carried out by three different approaches (Fig. 129.2):

(a) standard geomechanical field survey of the rock masses;
(b) automatic extraction of dips and dip directions from laser point clouds;
(c) manual measurement of the discontinuities on "solid images" by selection of an area, computation of the mean plane that best fit the selected points; from the plane equation, dip and dip direction can be extracted.

The field surveys were done following ISRM "Suggested Methods" (ISRM 1978). The rock mass characterizations fallow the Geological Strength Index (GSI) index (Hoek and Marinos 2007), that is based on the description of two factors, rock structure and block surface conditions. Physical access to higher parts of the slope was difficult and only 5 scan lines were measured, mainly located at the base of cliff. Discontinuities characteristics were investigated through the description of surface morphology, roughness, persistence, block size and rock fracturing degree.

An objective of traditional geomechanic surveys was also to demonstrate the capabilities of laser scanner in geometric measure. In fact, an automated extraction procedure was applied to obtain dip and dip direction of every point belong the cloud of laser scanning. The extraction of surfaces derives from a segmentation of the point cloud based on criteria of proximity and points similarity. A normal vector for every point is calculated and the corresponding plane equation is derived. The dip/dip direction data sample was very large (over 250.000 measures) and can not be used directly in a stereo net analysis program. For this reason the test site has been spitted into four slices, according to a vertical geometric pattern. The choice of dividing the area of study in vertical zones depends on the orientation of schistosity (next to verticality) that force to distinguish vertical morpho-structural domains.

It is not always easy to recognize structural features in a textured Dense Digital Surface Model (DDSM), especially if the study area is large, while the interpretation of morphostructural elements is easier when performed on high-resolution digital photos. The solid image approach (Bornaz and Dequal 2003) simplifies this task, it keeps the image in its original geometry and resolution, and reprojects point clouds obtained from laser scanning systems on the image itself (Fig. 129.3). In this way it's possible to select a discontinuity directly on the photo and obtain the

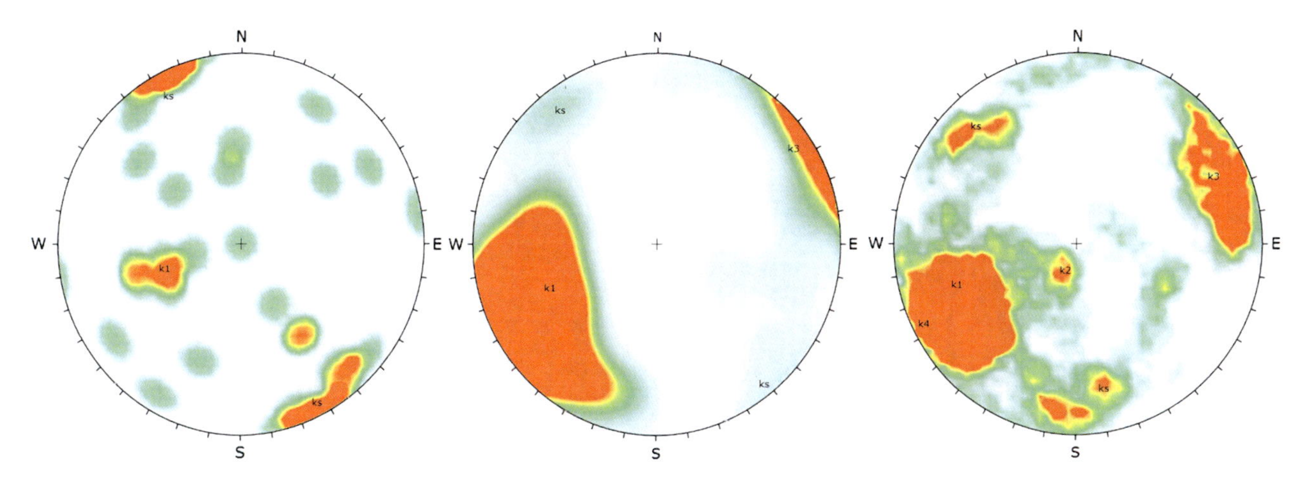

Fig. 129.2 Different approaches in geomechanic surveys: from *left* to *right*, example of the 3 approaches are shown

Fig. 129.3 Example of measurement on solid image of discontinuities surfaces

corresponding orientation that is derived computing the best fitting surface of the selected points. The high quality of solid images has allowed to interpret rock mass quality according to the GSI index also for impervious areas.

129.4 Rockfall Susceptibility

Every code for rock fall simulation requires a starting locations: it can be a single and well definite block or an area with diffuse sources. The identification of potential rock fall source areas requires susceptibility mapping. In this study, the susceptibility evaluation was obtained according to two different approaches: deterministic and empirical (Fig. 129.4).

Point cloud, 3D models and solid images have been used to identify rock fall sources. For each source area a monograph card was filled containing data on position (coordinates x, y, z), volume, shape, an indication on more probable collapse mechanism and on a kinematic analysis based on discontinuity orientation that isolate the block. Overall 39 local rock fall sources have been identified in the area. About half of the blocks have an expected volume less than 25 cubic meters, but situations with potential volumes that can exceed hundreds of cubic meters were observed. The information about the block volume, very important for rock fall simulation, were computed as the difference between DDSM and the plans obtained from the projection in depth of discontinuities.

To identify areas with an high rock fall susceptibility and with an diffuse instability, the r.SMR GIS tool (Filipello et al. 2010) has been applied. The calculation procedures of r.SMR is implemented as raster modules of a Open-Source GIS. The r.SMR module is based on geomechanic classification and uses the continuous equations introduced by (Tomás Jover et al. 2007). The input data are: DEM, dip and

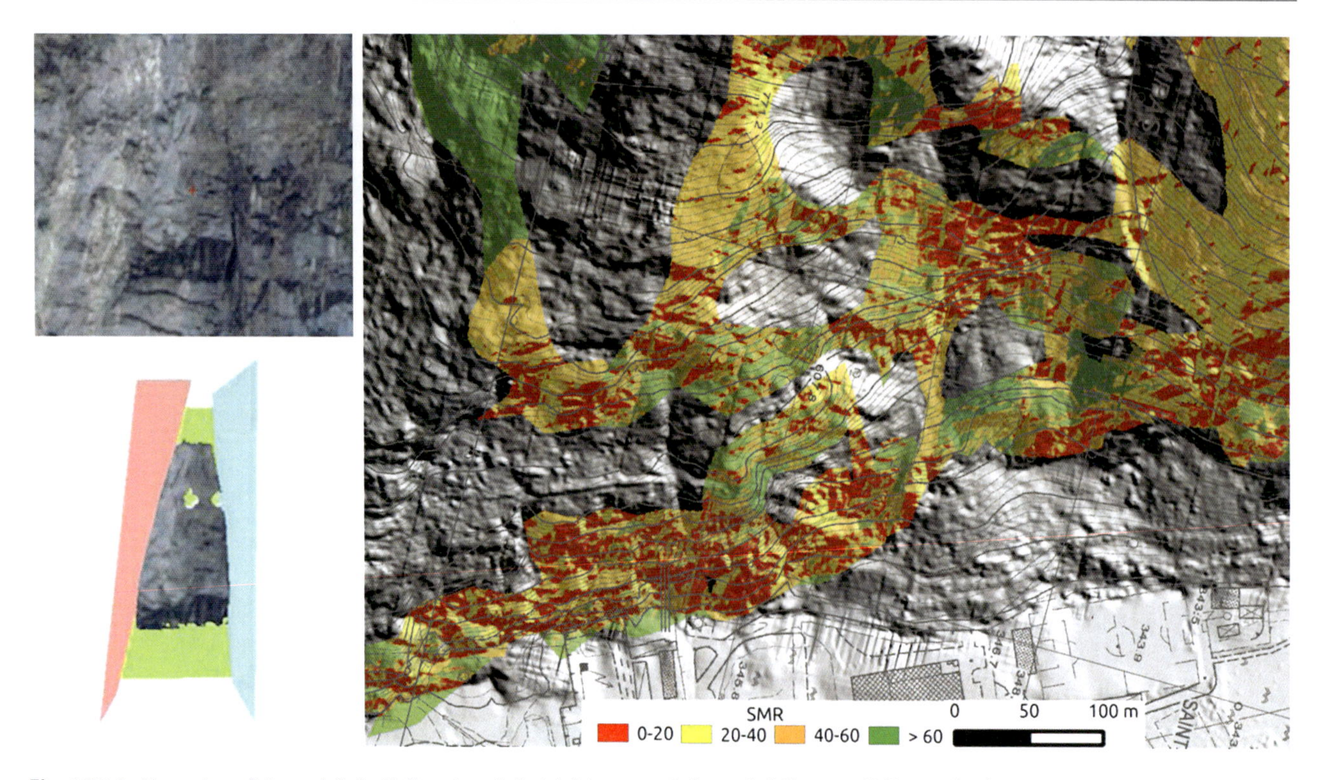

Fig. 129.4 Examples of deterministic (*left*) and statistic (*right*) approach for rock fall susceptibility evaluation

dip direction of discontinuity, the F4 index (Romana 1993) and the RMRb index (Bieniawski 1989).

The result is a raster map showing the distribution of the Slope Mass Rating (Romana 1993) for the analyzed system of discontinuity. For each discontinuity code returns two raster maps, one for planar sliding and the other for toppling.

129.5 Discussion and Conclusion

The test site is a challenging rock face in the lower Aosta valley periodically affected by collapses. The direct access to most of the cliff is virtually impossible and, therefore, we had to resort to alternative techniques, and in particular the laser scanner surveys, in order to measure some parameters essential for the assessment of the propensity to collapse.

In this article has been verified the reliability and consistency of measurements made directly on site and those made using remote sensing. This allowed to develop, with greater reliability over the input data, two different approaches for mitigation of landslide risk: the first deterministic (identifying the individual unstable blocks), the other statistic (testing the intersection between systems of discontinuity and rock wall). Both agree in identifying most critical areas and allow a operation plan aimed minimizing risk of rock falling on valley floor.

References

Bieniawski ZT (1989) Engineering rock mass classifications: a complete manual for engineers and geologists in mining, civil and petroleum engineering, 1st edn. Wiley, New York, p 251. ISBN 978-0-471-60172-2

Filipello A, Giuliani A, Mandrone G (2010) Rock slopes failure susceptibility analysis: from remote sensing measurements to geographic information system raster modules. Am J Environ Sci 6(6):489–494

Hoek E, Marinos P (2007) A brief history of the development of the Hoek-Brown failure criterion. Soils Rocks 2:1–8

ISRM (1978) Suggested methods for the quantitative description of discontinuities in rock masses. Int J Rock Mech Min Sci Geomech Abstr 15:319–368. doi:10.1016/0148-9062(78)91472-9

Neteler M, Bowman MH, Landa M, Metz M (2012) GRASS GIS: a multi-purpose open source GIS. Environ Model Softw 31:124–130

Romana MR (1993) A geomechanical classification for slopes: slope mass rating, vol III, Cap. 23. In: Hudson JA (ed) Comprehensive Rock Engineering. Imperial College-Pergamon Press, p 45. ISBN:0-08-042066-4

Tomás Jover R, Marchal JD, Serón Gáñez JB (2007) Modification of slope mass rating (SMR) by continuous functions. Int J Rock Mech Min Sci 44:1062–1069. doi:10.1016/j.ijrmms.2007.02.004

Artificial Neural Networks in Evaluating Piezometric Levels at the Foundation of Itaipu Dam

130

Bruno Medeiros, Lázaro Valentin Zuquette, and Josiele Patias

Abstract

Itaipu Dam is an engineering work of high importance. Located at the border between Brazil and Paraguay in the Paraná River and with approximated geographical coordinates 25°24′29″S, 54°35′21″W, it feeds these two countries with electrical energy and has to be constantly monitored in order to maintain its levels of quality and security. Over two thousand instruments, including more than 650 piezometers, have been installed for the monitoring of the dam and they provide continuous data about several characteristics of its foundation and structure. The evaluation of piezometric levels in dams is important for it reflects the values of the uplift pressure that acts on the structure of the dam. The utilization of new methods in such an analysis can provide agility to decisions-taking by the security team of the dam. Depending on the method applied, a better comprehension of the phenomenon in time and space may be achieved. This study employs Artificial Neural Networks (ANN) to simulate the behavior of the piezometers installed in a geological discontinuity in the foundation of Itaipu Dam. It considers different types of entry data in a Multilayer Neural Network and determines the best ANN architecture that is closest to the real situation. Some parameters have a higher weight in the variation of the piezometric levels, whereas some others do not affect it considerably. A geological geotechnical model of the foundation rock fractures would be helpful to improve the entry data and achieve better results.

Keywords

Piezometer • Artificial neural network • Itaipu dam

130.1 Introduction

The study of the acting forces on dams has high importance in the evaluation of their security. New methods for that analysis are required in order to provide agility in decisions-taking by the security teams who are in charge of maintaining such structures.

Itaipu Binacional is the world's second largest dam, and first in generation of energy. It has a crucial paper in the economy of Brazil and Paraguay and even for its surrounding territories. By the time of its construction it was severely instrumented as to ensure great levels of quality and its stability.

B. Medeiros (✉) · L.V. Zuquette
São Carlos School of Engineering, University of São Paulo, São Paulo, Brazil
e-mail: brunomedeiros.eng@gmail.com

L.V. Zuquette
e-mail: lazarus1@sc.usp.br

J. Patias
Itaipu Binacional, Foz do Iguaçu, Brazil
e-mail: jpatias@itaipu.gov.br

G. Lollino et al. (eds.), *Engineering Geology for Society and Territory – Volume 6*,
DOI: 10.1007/978-3-319-09060-3_130, © Springer International Publishing Switzerland 2015

In order to attend international codes and local laws, it was established a program of control of the instrumentation of Itaipu. Most instruments are read every month, while piezometers and flow meters are read every 15 days (Itaipu Binacional 1994). Such a work has created a database of about 30 years of continuous readings, which has to be handled and analyzed.

The use of Artificial Neural Networks (ANN) allows the insertion of such data into computational programs together with other characteristics of the dam, and provides new ways of obtaining and analyzing data.

In this context, this study proposes the utilization of ANN to evaluate the behavior of 48 piezometers situated at a discontinuity in Itaipu foundation.

130.2 Study Area

Itaipu is located at the border between Brazil and Paraguay with approximated coordinates 25°24'29"S, 54°35'21"W. It is compound of several types of structures as can be seen in Fig. 130.1.

This study focuses on the structure 3, which is above a discontinuity called Joint D that is intensely instrumented. 48 piezometers were installed on Joint D and measure its hydraulic pressure.

130.3 Data Preparation

For the use of ANN, it is necessary to provide the network data that represent the behavior wanted to be reproduced. Some characteristics of the Joint D were not directly available for this use.

By the time of Itaipu construction its foundation was vastly investigated and the Joint D was mapped. Using its contour map and the installation elevation of the piezometers, a digital elevation model (DEM) was generated in ArcGIS® software with *Topo to Raster* tool. From this model, were extracted values of elevation of the joint topography, slope and orientation for the exact location of each of the 48 piezometers. This data were used together with the piezometric readings as entry data of the ANNs.

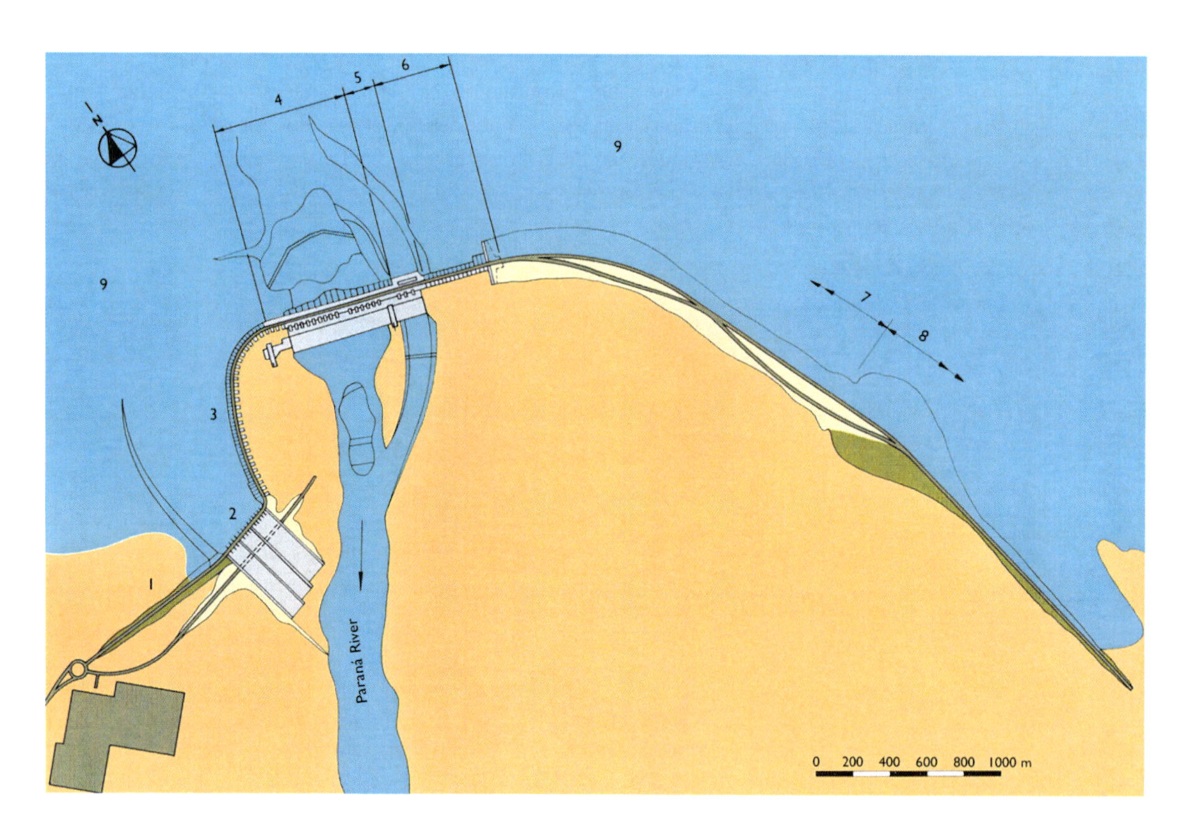

Fig. 130.1 Types of dams in Itaipu hydroelectrical complex. *1*—*right* earth dam; *2*—spillway; *3*—*left* side dam (buttresses); *4*—main dam (buttresses); *5*—diversion structure (concrete); *6*—*left* binding dam (buttresses); *7*—rock fill dam; *8*—*left* earth dam; *9*—reservoir. Modified from: Itaipu Binacional (1994)

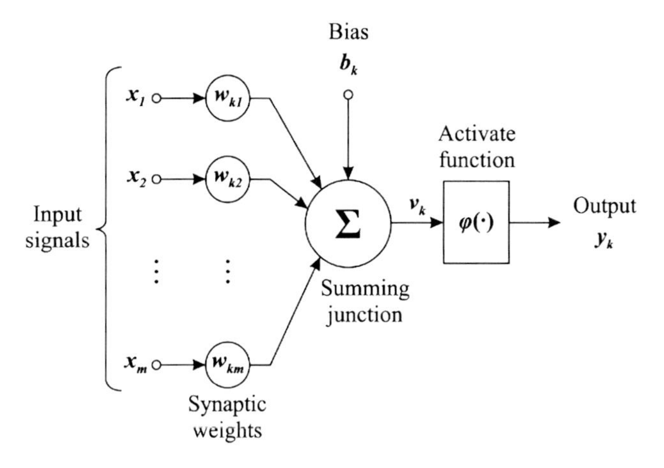

Fig. 130.2 Parts of an artificial neuron. Font: Haykin (1999)

130.4 Artificial Neural Networks

The idea of creating an artificial system that would represent the concept of functioning of the human brain has been studied for many years, since late 19th and early 20th centuries (Hagan et al. 1996). Modern thought on this matter were developed by McCulloch and Pitts (1943), and since then it has been widely spread through the sciences, being used in engineering, medicine, stock market, and others.

The artificial neuron works like the human neuron receiving signals, weighting them and compiling into one or more answers. The Fig. 130.2 shows a concept of how it is organized.

130.4.1 Piezometric Readings Preparation

In order to organize all data available from the 30 years of readings of the instruments, electronic worksheets were elaborated. In this study, the 48 piezometers that monitor the Joint D were selected. The worksheets showed the identification of each instrument, its location in UTM coordinates, the installation elevation of the piezometer, the date of the reading, and the measured value of piezometric elevation.

Not all readings were executed on the same days for all 48 instruments. A work was performed to correct this on the worksheets and normalize the amount of data for each piezometer. It was only maintained the readings that were executed for all instruments on the same days. This resulted in an amount of 733 readings for each of the piezometers of Joint D, totalizing 35,184 values.

Each of the series of readings of the piezometers was put into graphics. Analyzing the behavior presented it was seen that some piezometers behaved similarly. From this observation it was possible to establish groups that could generate

Table 130.1 Groups of piezometric same behavior

Group	Code	Group	Code
Orientation north	ON	Elevation D	ED
Orientation east	OE	Elevation E1	EE1
Orientation south	OS	Elevation E2	EE2
Orientation west	OW	Slope 1	S1
Elevation A1	EA1	Slope 2	S2
Elevation A2	EA2	Slope 3	S3
Elevation B	EB	Slope 4	S4
Elevation C	EC	–	–

different neural networks, and ease their process of learning. These groups received the nomenclature as seen on Table 130.1.

Some groups were seen to represent a most coherent and homogeneous behavior between the piezometers, as can be seen on Fig. 130.3. These groups were chosen to be used in ANN learning process. They are: EA1, EA2, EB, EC, ED, EE1, EE2, OE, OS, S1, S2 and S3.

130.4.2 Choice of the Best ANN Architecture

Some tests were performed in order to check the best architecture of ANN that would well represent the behavior expected. These input data were used: installation elevation of the piezometer, orientation, declivity, topographic elevation of the DEM, and the last two readings of each piezometer. The worksheets of each group was randomized and separated into two parts: 60 % for training and 40 % for validation.

On MatLab® software, the ANNs were programed to use 20 neurons in the input layer, and 1,000 epochs of training. For each of the groups analyzed several training sessions were performed. It was seen that it was necessary to increase the number of epochs, so it was adjusted for 2,000, and the sessions were run again. After the end of a session, the algorithm was adjusted to use 5 more neurons in the input layer. It was made until the number of neurons was 40.

130.5 Conclusions

The results were satisfactory for it showed an average percentage error of 0.132 % between the expected values and the obtained with the simulations. For each group the average, minimum and maximum percentage error can be seen on Table 130.2.

The Fig. 130.4 shows a comparison between the values expected for the group EB, and those obtained by the ANN simulation, and below the percentage error.

Fig. 130.3 Behavior of piezometers of group EB

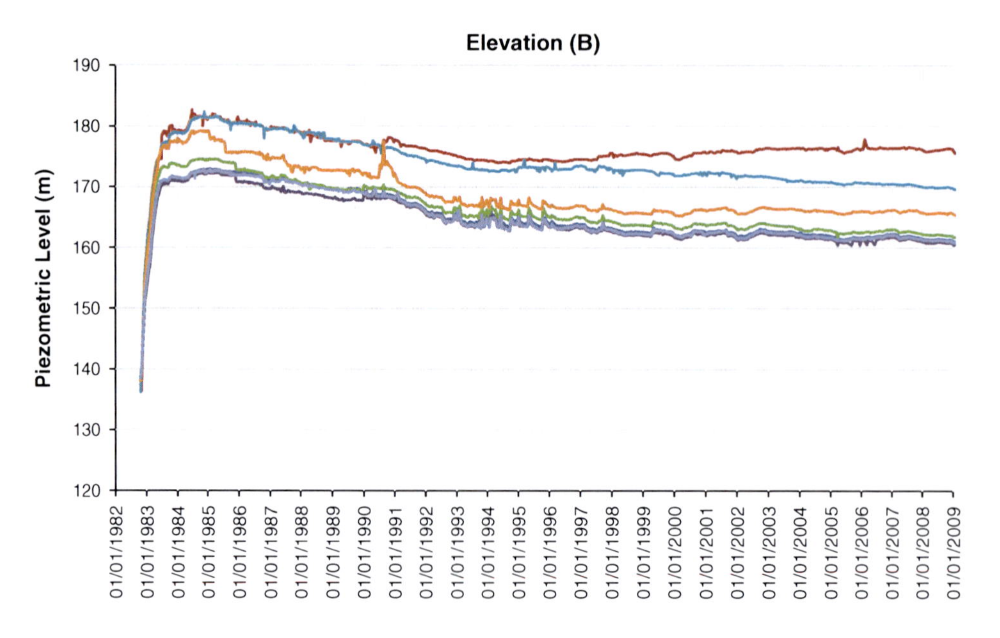

Table 130.2 Percentage error of the ANN simulation for each group

Group	Average percentage error	Minimum percentage error	Maximum percentage error
D1	0.174	0.000	12.561
D2	0.221	0.000	16.974
D3	0.139	0.001	13.672
EA1	0.143	0.000	6.971
EA2	0.057	0.000	1.285
EB	0.167	0.000	10.820
EC	0.159	0.001	3.658
ED	0.143	0.000	13.987
EE1	0.083	0.000	0.818
EE2	0.049	0.000	7.460
OE	0.191	0.000	13.944
OS	0.053	0.000	5.266
Average	**0.132**	**0.000**	**8.951**

Fig. 130.4 Values expected and obtained, and percentage error for the group EB

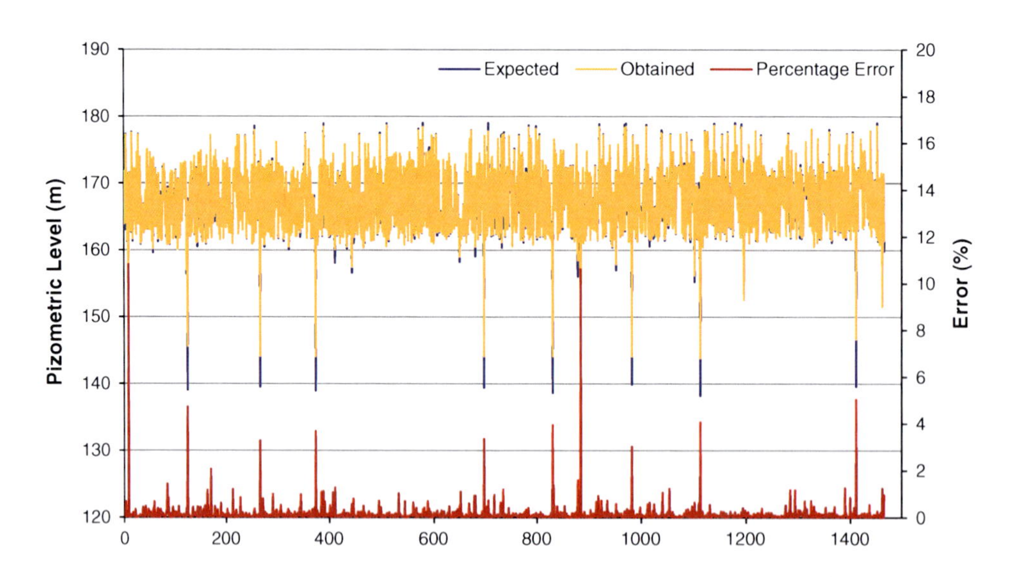

It is seen that Artificial Neural Networks was a useful tool for the comprehension of piezometric behavior in the foundation of Itaipu Dam. With the correct input data, it can be consistent in presenting a well-adapted model that corresponds reality. With the results of this study it is possible to apply these created ANNs for spots that are not instrumented and possibly receive good response of how would its behavior be in case there was an instrument installed there.

References

Hagan MT, Demuth HB, Beale M (1996) Neural network design. PWS Publishing Company, Boston

Haykin S (1999) Neural networks: a comprehensive foundation, 2nd edn. Pearson Prentice Hall, New Jersey

Itaipu Binacional (1994) Itaipu hydroelectric project: engineering features. Curitiba

McCulloch WS, Pitts WH (1943) A logical calculus of the ideas immanent in nervous activity. Bull Math Biophys 5:115–133

Use of an Advanced SAR Monitoring Technique to Monitor Old Embankment Dams 131

Giovanni Nico, Andrea Di Pasquale, Marco Corsetti, Giuseppe Di Nunzio, Alfredo Pitullo, and Piernicola Lollino

Abstract

The work mainly discusses the use of the Ground-Based Synthetic Aperture Radar (GBSAR) interferometry technique to observe and control the behavior of earthfill or rockfill embankments for dam impoundments. This non-invasive technique provides overall displacements patterns measured with a sub-millimeter accuracy. The need of reliable monitoring of old embankment dams is rapidly increasing since a large number of these structures are still equipped with old monitoring devices, usually installed some decades ago, which can give only information on localized areas of the embankment. A case study regarding the monitoring of an earthfill dam embankment in Southern Italy by means of GBSAR interferometry is presented.

Keywords

Synthetic aperture radar (SAR) • Ground-Based SAR (GBSAR) • SAR interferometry • Old embankment dams

131.1 Introduction

The present work mainly discusses the use of the Ground-Based Synthetic Aperture Radar (GBSAR) interferometry technique, as integrated with the available in situ traditional monitoring data as well as Finite Element analysis, to observe and control the behavior of earthfill or rockfill embankments for dam impoundments. The need of reliable monitoring of old embankment dams is nowadays rapidly increasing since a large number of these structures are still equipped with old monitoring devices, usually installed some decades ago, which are generally capable to provide only information on local areas of the embankment.

A Synthetic Aperture Radar (SAR) is an active microwave sensor used to produce 2D microwave images of the observed scene (Massonnet and Feigl 1998). The main advantage of microwave images is their capability to observe a scene without the need of solar illumination and in any weather condition. In the last decade, Ground-Based SAR (GBSAR) systems have gained an increasing interest in different applications such as monitoring of dams, landslides, buildings, bridges, or to extract information on terrain morphology (Guccione et al. 2013; Di Pasquale et al. 2013). The SAR interferometry (InSAR) technique relies on the processing of two SAR images of the same scene obtained by almost the same position. The phase difference φ between the corresponding pixels of two coherent complex-valued SAR images provides displacements measurements of structures and natural scenes with a sub-millimetre precision. The GBSAR system is installed at a distance from the observed object ranging from less than one hundred metres

G. Nico (✉) · P. Lollino
Consiglio Nazionale delle Ricerche, Istituto di Ricerca per la Protezione Idrogeologica, Via Amendola 122/I, 70126 Bari, Italy
e-mail: g.nico@ba.iac.cnr.it

A. Di Pasquale
DIAN srl, Z.I. La Martella, III Trav. G.B. Pirelli sn, 75100 Matera, Italy

M. Corsetti
D.I.C.E.A, Università la Sapienza, Via Eudossiana 18, 00184 Bari, Italy

G. Di Nunzio · A. Pitullo
Consorzio di Bonifica della Capitanata, Corso Roma 2, 71121 Foggia, Italy

G. Lollino et al. (eds.), *Engineering Geology for Society and Territory – Volume 6*,
DOI: 10.1007/978-3-319-09060-3_131,

up to four kilometres. The interferometric processing of two coherent SAR images results in a map of displacements occurred between the acquisitions of the two SAR images. Radar measurements give the projection along the radar line-of-sight of the 3D displacement vector. The main advantage of GBSAR interferometry with respect to traditional techniques is its capability to provide 2D information on the overall displacement field rather than measurements of displacements in only a few points.

131.2 Test-Site

The experiment described in this paper has been carried out at the Occhito's earth dam managed by the Consorzio di Bonifica di Capitanata (CBC). This 432 m-long and 60 m-high earth dam is one of the biggest ones in Europe. At present this dam is monitored with an array of piezometers, inclinometers and sensors of pressure. Furthermore, the position of a set of marks along the crest and the downstream surface is measured by topographic techniques (see Fig. 131.1).

131.3 GBSAR Results

In this section we present the results obtained by GBSAR technique. Two 2-h measurement campaign have carried out from two different positions. The two installation positions, A and B, are reported in Fig. 131.1. The two measurement positions are located at the basis of the dam, on opposite sides, in order to reconstruct the true displacement vector. In fact, interferometric SAR measurements can only provide the line-of-sight (LOS) component of the displacement. The aim the campaign has been to set a "zero" measurement with respect to which to compare radar measurements to be acquired in future campaigns.

Figure 131.2 shows a picture taken from the installation position of GBSAR system denoted as A in Fig. 131.1. From this position the radar observed more than a half of the dam surface and the concrete structural elements represented on the right hand side of the picture. A 2-h measurement campaign has been done. The amplitude of one of the SAR images acquired on this dam is represented at the top of Fig. 131.4. Pixels with a higher value of dB correspond to patches on the dam surface with a higher scattering capability of radar signal. The comparison with the optical picture in Fig. 131.3 helps to easily recognize the different portions of the dam. The map of surface displacements measured by the radar during the 2-h campaign is displayed at the bottom of Fig. 131.4. As mentioned before, radar measurements can provide only the line-of-sight (LOS) component of the displacement vector, which that shown in Fig. 131.4. It is worth noting that this due this property of radar measurements, the GBSAR system should be properly installed in order to avoid a line-of-sight perpendicular to the expected displacement vectors. Figure 131.5 reports the

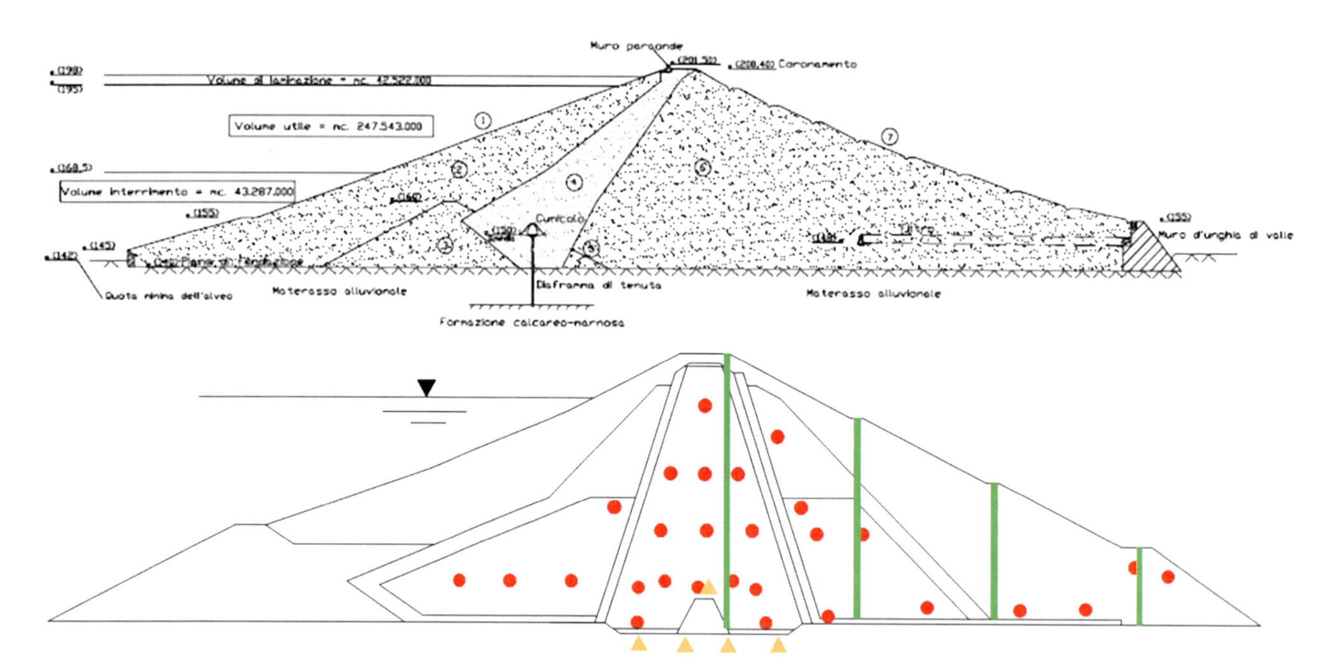

Fig. 131.1 Occhito's earth dam: (*top*) Geometry; (*bottom*) localization of sensors: piezometric cells (*circles*), pressure cells (*triangles*), inclinometers/assestimeters (*vertical columns*)

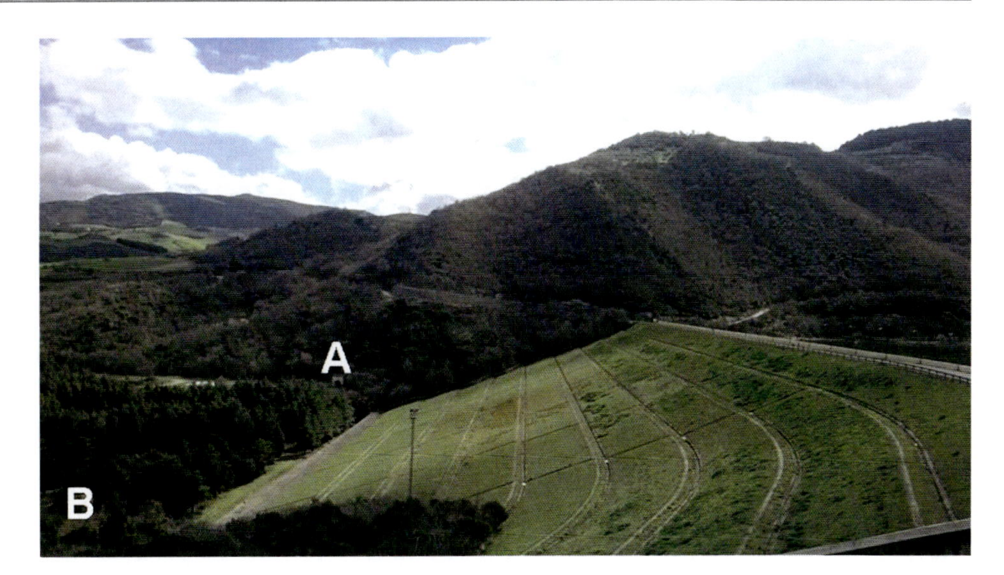

Fig. 131.2 The two installation positions of the GBSAR sensor

Fig. 131.3 Picture of the Occhito dam as seen by the installation position A of the GBSAR sensor

frequency distribution of displacement measurements. The mean and standard deviation values are, respectively, 0 and 0.3 mm. Due to the short time interval of monitoring this statistical figures provide the precision and accuracy of the interferometric GBSAR measurement. Further assessments of the GBSAR capability to provide unbiased sub-millimetre precision measurements of surface displacements can be found in (Di Pasquale et al. 2013).

Figure 131.6 shows a picture taken from the installation position of GBSAR system denoted as B in Fig. 131.1. From this position the radar observed the remaining part of the dam not observed form position A. The amplitude of one of the SAR images acquired on this dam is represented at the top of Fig. 131.7. The map of surface displacements measured by the radar during the 2-h campaign is displayed at the bottom of Fig. 131.7.

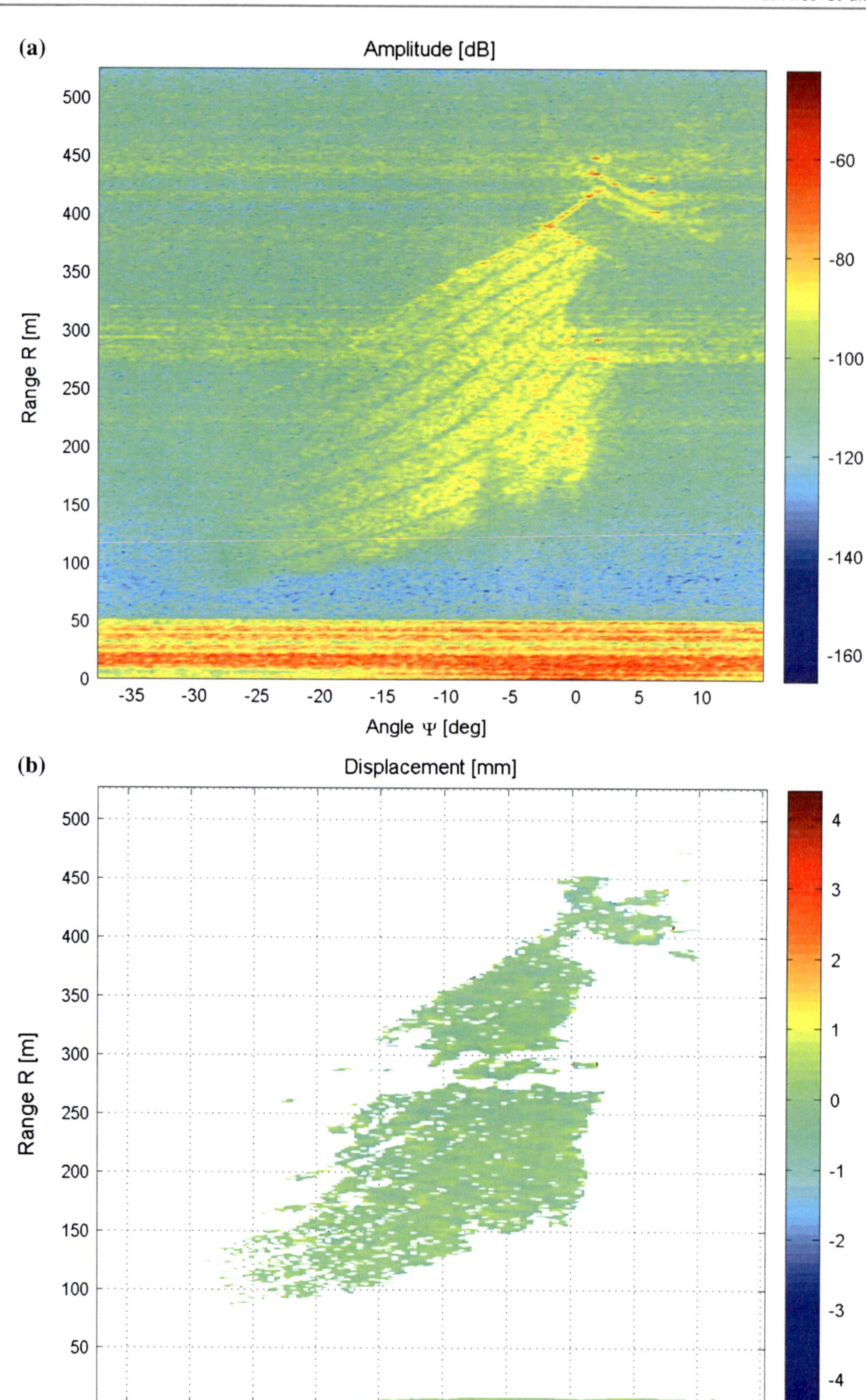

Fig. 131.4 (*Top*) amplitude of the radar image; (**b**) LOS displacement map in radar coordinates as measured from position A

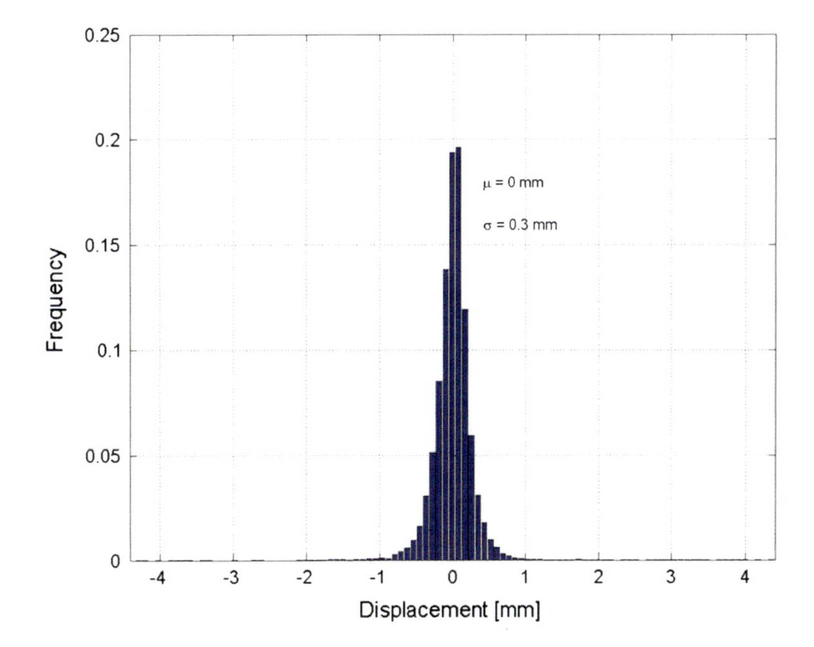

Fig. 131.5 Frequency distribution of dam surface displacement values

Fig. 131.6 Picture of the Occhito dam as seen by the installation position B of the GBSAR sensor

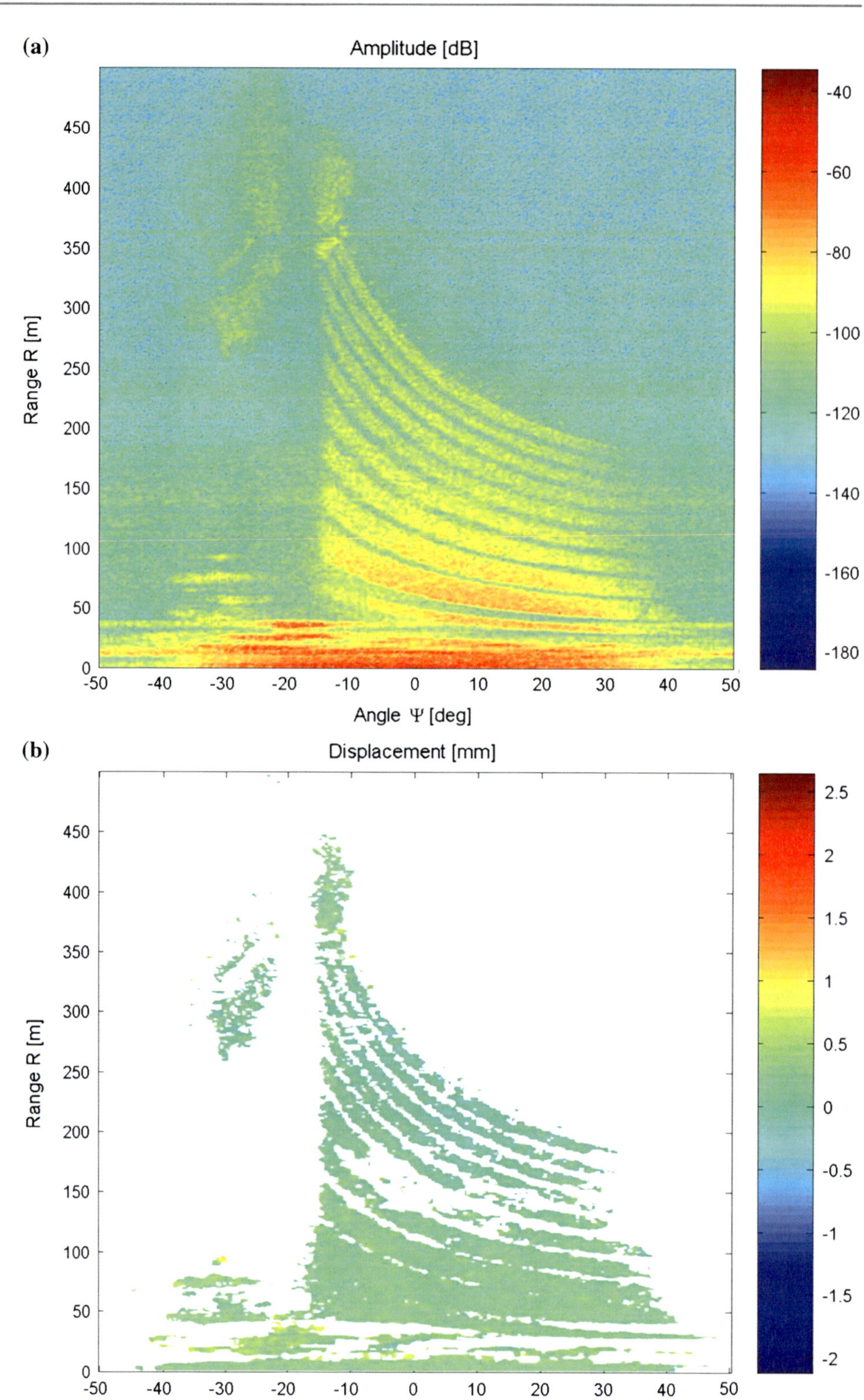

Fig. 131.7 (*Top*) amplitude of the radar image; (**b**) LOS displacement map in radar coordinates as measured from position B

References

Massonnet D, Feigl KL (1998) Radar interferometry and its applications to changes in the Earth's surface. Rev Geophys 36:441–500

Guccione P, Zonno M, Mascolo L, Nico G (2013) Focusing algorithms analysis for ground-based SAR images. In: Proceedings of the international geoscience and remote sensing symposium, Melbourne, Australia, 21–26 July

Di Pasquale A, Corsetti M, Guccione P, Lugli A, Nicoletti M, Nico G, Zonno M (2013) Ground-based radar interferometry as a supporting tool I natural and man-made disasters. In: Proceedings of the 33rd annual EARSeL symposium, Matera, Italy, 3–6 June

Modelling and Optimization of the Biological Treatment in the Conception of Water-Treatment Plants Whith Activated Sludge

132

Moncef Chabi and Yahia Hammar

Abstract

The pond of aeration, or airy biological reactor, is the major element of a water-treatment plant with activated sludge. His functioning packages the quality of the treatment, the network mud and the energy expenditure. The good knowledge of the hydrodynamics of these reactors is an essential stake to improve their conception and thus to optimize their functioning. Our study is interested in the use of the code of calculation FLUENT for the simulation of the flows in ponds of aeration the case of the water-treatment plant of Souk-Ahras, where the functions of aeration and admixture are separated. This software uses the method of the volumes finished to solve the equations of Navier-Stockes in turbulent regime. The final goal lives in the definition of technical capacities to improve the functioning of the work. Finally, we highlight the influence of the aeration on the speeds of circulation and the phenomena of ascending convection of the water (spiral spring—flows), responsible for a decrease of the transfer of oxygen in the pond. The impact of a horizontal speed on certain types of spiralflows is studied. Big spiral spring - flows disappear totally from a speed of 0,3 m.s^{-1}. These simulations are led for various geometries of ponds

Keywords

Channel of oxidation • Flow • Hydrodynamics • Transfer of oxygen • Agitation

132.1 Introduction

The secondary treatment is essentially a biological oxidation of the dissolved materials. The agents of this oxidation are microorganisms, in particular aerobic bacteria, susceptible to feed on present organic matters in waste water. The secondary processing plants thus appear as ponds of culture where we put in touch a bacterial population and the effluent to be treated in the presence of oxygen.

M. Chabi (✉) · Y. Hammar
Badji Mokhtar University, Annaba Box 12, 23000 Annaba, Algeria
e-mail: moncef.chabi@univ-annaba.org

Y. Hammar
e-mail: yahia.hammar@univ-annaba.org

These ponds called also ponds of oxidation operate a free bacterial biomass associated in plops. These flakes of muds include heterotrophic microorganisms and autotrophes nitrifiants when the residence time of the mud is self-important so that their multiplication produces an active biomass in the treatment.

This reproduction of microorganisms intervenes in favorable conditions, when their growth is important is that bacterium beginning dividing. The exo-polymers which they secrete allow them to gather together in plops settlings.

The chosen conditions of operation are the ones which favor the settling of these plops. In this case, the biomass can be separated by a second settling where the extracted mud is recirculate towards the pond of aerobic treatment.

The aeration can be assured on surface by turbines or at the bottom by processes of banister of distribution of air bubbles. The efficiency on transfer can be improved by the increase of the height of water.

G. Lollino et al. (eds.), *Engineering Geology for Society and Territory – Volume 6*,
DOI: 10.1007/978-3-319-09060-3_132, © Springer International Publishing Switzerland 2015

The process with activated sludges is at present the most valid process of biological purge of residual waters. His big advantages are:

- Security as regards the degree of purge of handled waters, because the most important factors of influence, for example contribution of residual water, and bacterial mass, are controllable,
- A bigger efficiency than with the bacterial beds, than this process is much less dependent on the temperature,
- A shorter phase of starting up (less than 2 weeks) with regard to compared with the bacterial beds (4 in 6 weeks),
- total Absence of smells and flies.

This process allows to handle residual waters from 100,000 to 200,000 living equivalents, and it calls on all the same to certain notions important for the exploitation…

132.2 Display of the Zone of Study

132.2.1 Presentation of the Wilaya de Souk Ahras

The wilaya of Souk-Ahras localized in the east of the country covers a total surface of 4 359,65 km^2. The map below represents the demarcations of the wilaya as follows:

- the wilaya of Tarf in the North-East,
- the wilaya of Guelma in the North-west,
- the wilaya of Tebessa in the South
- the wilaya of Oum el Bouaghi in the South-west
- and the border Algéro-Tunisienne in the East (Fig. 132.1).

The wilaya of Souk-Ahras is situated in the extremity is of the zone tellienne, crossed from east to west by the oued Medjerda which is a cross-border stream, its paying pond covers a surface of 7,870 km^2 only 1,411 km^2 are distributed on the Algerian territory.

The valley of the wilaya of Souk-Ahras and known by a very heterogeneous geology, represented by sedimentary trainings the oldest age of which is Sorted out it and, constituted generally by limestones, wills, marls, gravel and alluviums, it forms a pivotal zone between the Tell Atlas in the North and Saharan Atlas in the South.

132.2.2 Presentation of the Treatment Plant for Souk-Ahras

The water-treatment plant of the city of Souk-Ahras is intended to handle domestic waste water before their discharge in the oued of Medjerda of among which:

- The first stage allows to handle the pollution resulting from a population of 150,000 equivalents living.
- The second stage will carry its capacity in 225,000.

The station is designed to work in a low(weak) mass responsibility, thus according to the process of a prolonged aeration. By this process, we obtain a good efficiency of elimination of the DBO, the stabilization "mineralization" of secondary muds will be made within the pond of aeration (Fig. 132.2).

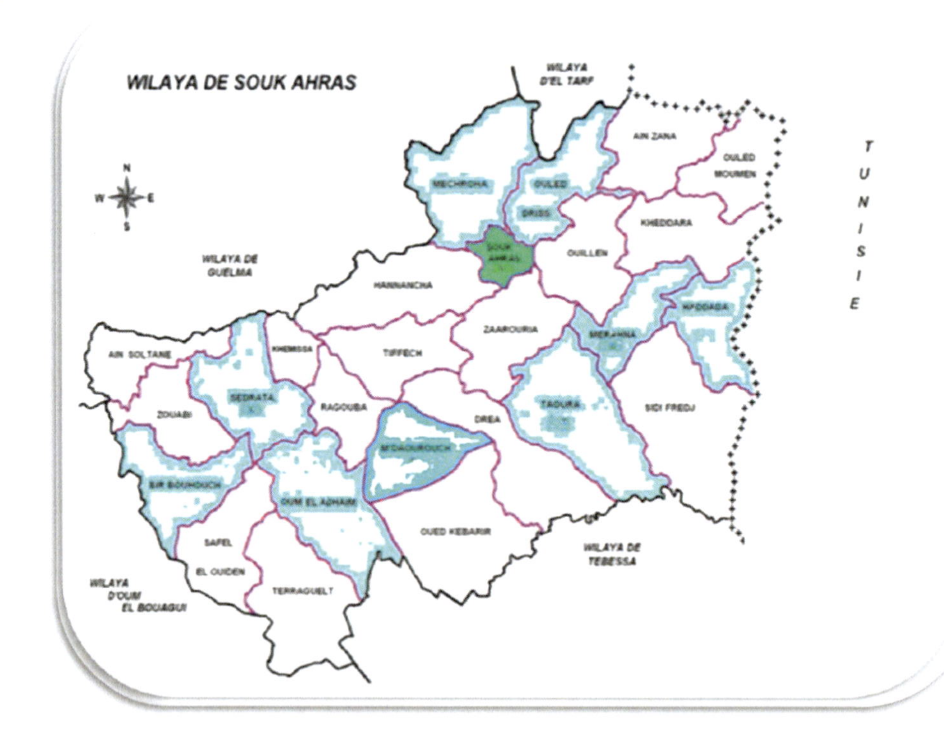

Fig. 132.1 Map of demarcation of the wilaya of Souk-Ahras

Fig. 132.2 View google—earth of the treatment plant for Souk-Ahras

Table 132.1 Characteristics of the trial of the installation

Hydraulic residence time	21 h
Residence time of muds	20 days
Concentration in mud	5 g/l
Nominal flow	30 000 m^3/d

The characteristics of the treatment plant are defined in the Table 132.1:

132.2.3 Biological Reactor

There are two different and identical ponds of aeration having the following characteristics:

- Length: 66 m
- Width: 44 m
- Height of water: 4.5 m
- unitarian Volume: 13,068 m^3
- total Volume: 26 136 m^3

Waters arriving in these ponds are brewed with the mixed liqueur and the muds of recirculation from the secondary decanter.

This pond is equipped with 12 ventilators of surface. Their characteristics are the following ones: of chap AVL 3,750 (supplier TMI), power 75 kW, tension 380/660 V 50 Hz (Fig. 132.3).

132.3 Tool of Simulation Gambit and Flow

132.3.1 The Tool Gambit

GAMBIT is a software which allows to create meshings which can be used in particular under FLUENT.

The mailleur is a preprocessing in the simulation software. He allows to generate a meshing structured or not structured in Cartesian, polar, cylindrical address and phone coordinates or axisymétriques. He can realize complex meshings in two or three dimensions with stitches of chap rectangle or triangle.

To build a meshing, it is good to follow the following approach:

- define the geometry.
- realize the meshing.
- define the parameters of computation zones.

132.3.1.1 Define the Geometry

The geometry includes the physical positions of characteristic points define the zone which we have to mesh: spatial address and phone coordinates of four summits of a square; of the starting point (Fig. 132.4)

132.3.1.2 Realize the Meshing

For structured meshings, we can make the connection between the geometry and the meshing, also, before realizing the meshing, it is necessary to specify in writing on one hand the geometry adopted to define the zone to be meshed, on the other hand the numbers of stitches corresponding to the characteristic points of the geometry. A knot can then be spotted by these physical address and phone coordinates x and y.

Besides, he can be advantageous to define intermediate points which are not essential to the definition of the geometry but which allow to bound the zones in which stitches will be adapted to refine the meshing near walls for example either to marry at best the shape of the obstacles, the walls or others (Fig. 132.5).

132.3.1.3 Define the Parameters of Computation Zones

When the geometry was created, when the conditions in the limits were defined, it is necessary to export the meshing. msh (mesh = meshing in English) so that Flow is capable of reading it and of using it (Fig. 132.6).

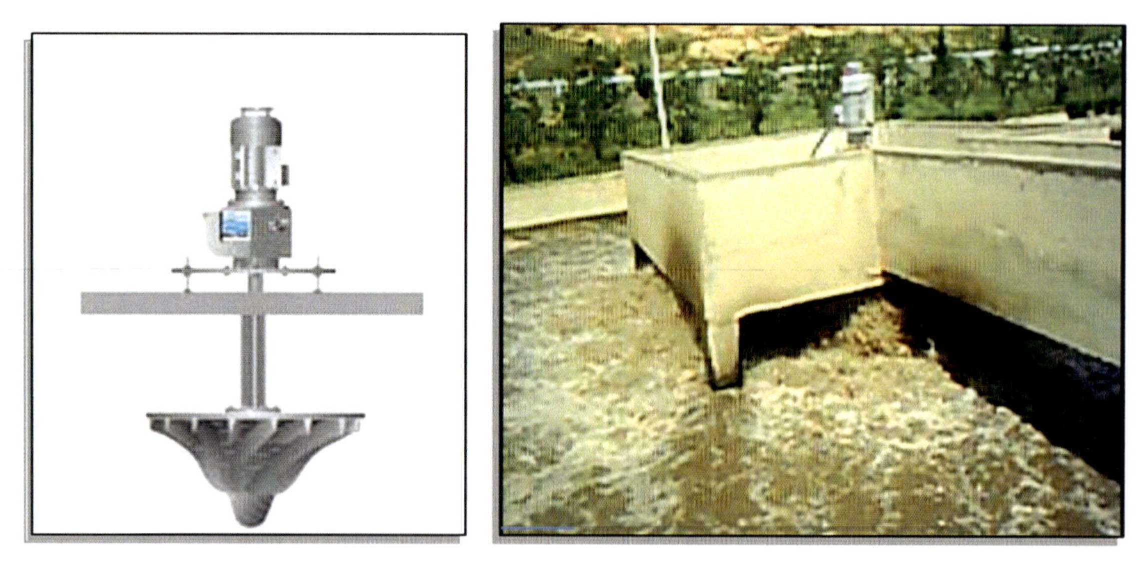

Fig. 132.3 The system of aeration of surface

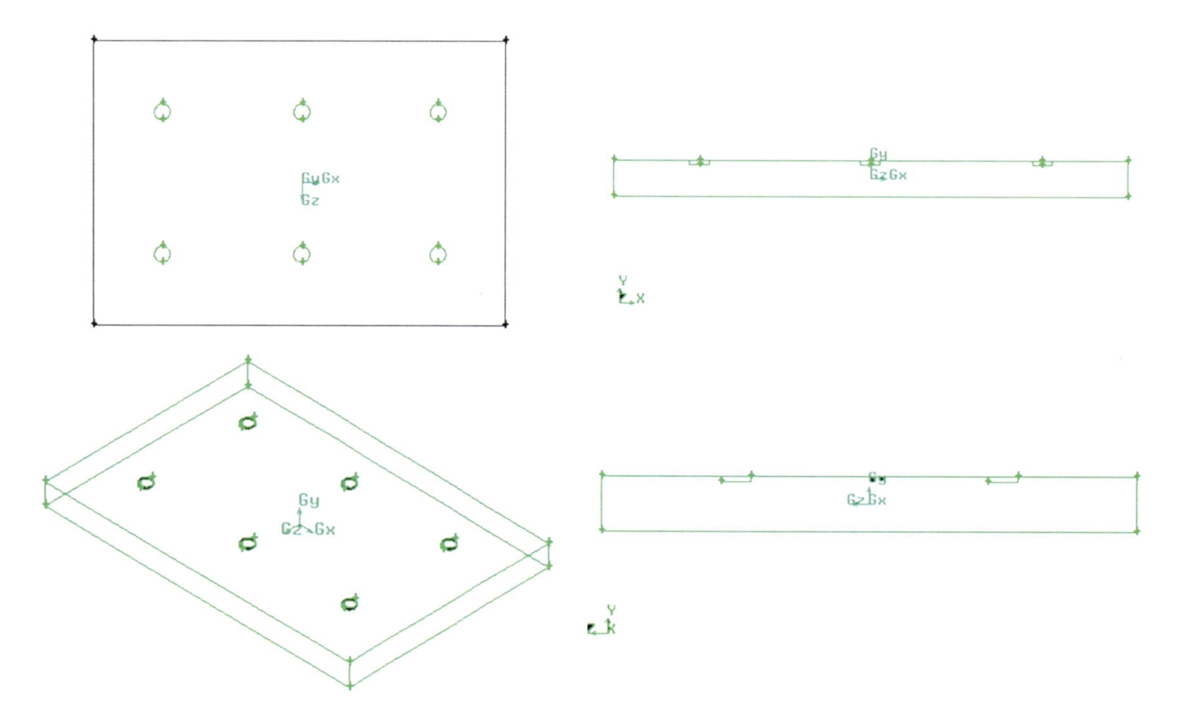

Fig. 132.4 The construction of the pond of aeration of the treatment plant for Souk Ahras

We can then close Gambit by protecting the session (if we wish to open it) and throw Fluent.

132.3.2 The Tool Fluent

Fluent is a commercial code CFD very used in the industry. He allows to solve the fluid sellings (fields of speeds, temperature), it is the reason why one it chosen to realize our simulation.

The simulation in Digital Fluid mechanics (Computational Fluid Dynamics: CFD), is used for the modelling, the visualization and the analysis of the fluid sellings and thermal transfers. She allows the users to optimize the performances of the new concepts, while reducing the cycle of marketing, the associated risks and the costs.

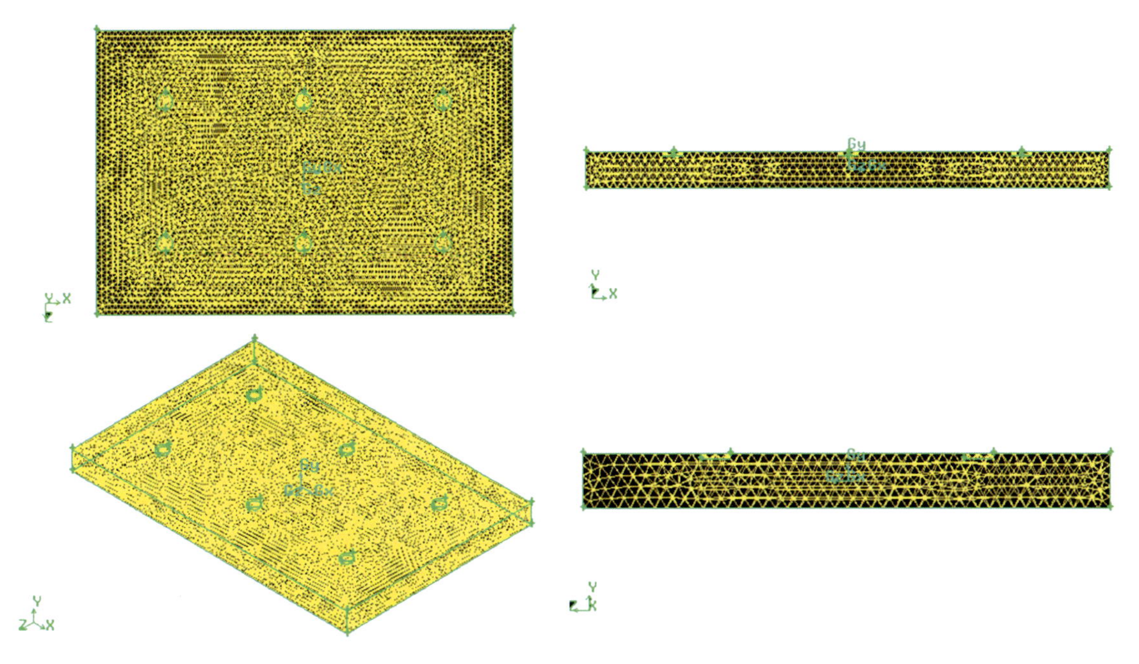

Fig. 132.5 The meshing of the pond of aeration of the treatment plant for Souk Ahras

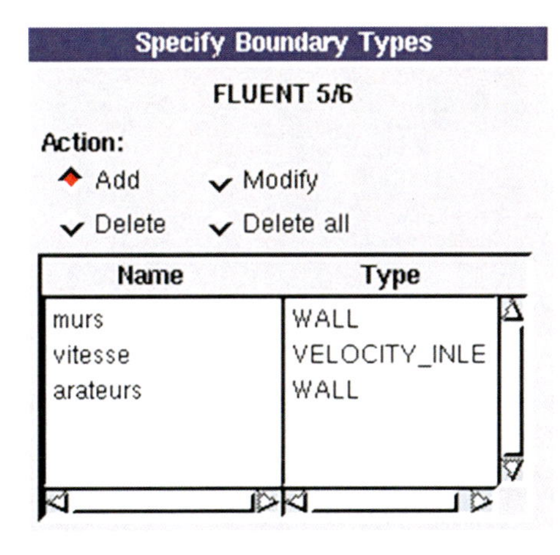

Specify Boundary Types
FLUENT 5/6
Action:
Add
Modify
Delete
Delete all

Name	Type
murs	WALL
vitesse	VELOCITY_INLE
arateurs	WALL

Fig. 132.6 The conditions on the verge of our treatment plant

For more than 10 years, Flow, world leader in digital Simulation of the Fluid software sellings and departments. The solution of the most adapted simulation strives to offer, whether it is for the whole industry, research or engineering education of tomorrow. This solution spends by services of technical audience, studies with way, and by personalized formations.

The siege of the group Fluent is situated to Lebanon, New Hampshire, the USA, and account of the subsidiaries in Belgium, England, France, Germany, India, Italy, Japan and Sweden. Its software are also sold worldwide thanks to joint-ventures, partnerships and distributors in Korea, Australia, Brazil, China, Taiwan, Czech Republic, the Middle East and in most of the European countries.

132.3.2.1 Choice of the Model of Turbulence

Because the geometry is in 3D, and because the geometry is a pond of aeration with 6 ventilators of surface of chap slog slow, compared with the rest of the domain of calculate the choice of 3D double precision seem the most suited to our simulation (Fig. 132.7).

Fluent proposes various modellings of the turbulent selling. Among which the not viscous, laminar, turbulent sellings etc. (Table 132.2).

The characteristics of materials are given in the following Table 132.3:

These properties are then loaded in signs materials of the software FLUENT.

132.4 Results

The method of the finished volumes is an approach very often used for the dynamic calculations in thermal because of its ease of implementation.

In the approach of the finished volumes, the value calculated in the center of the volume represents the approximation of the variable considered on all the volume. In the case of the use of a not structured meshing (triangulation of Delaunay), the "central" points of triangles and squares used to represent the approximation of the field of temperature do not belong inevitably to the orthogonal of the borders, but this does not prevent from having a correct evaluation of flows and temperatures

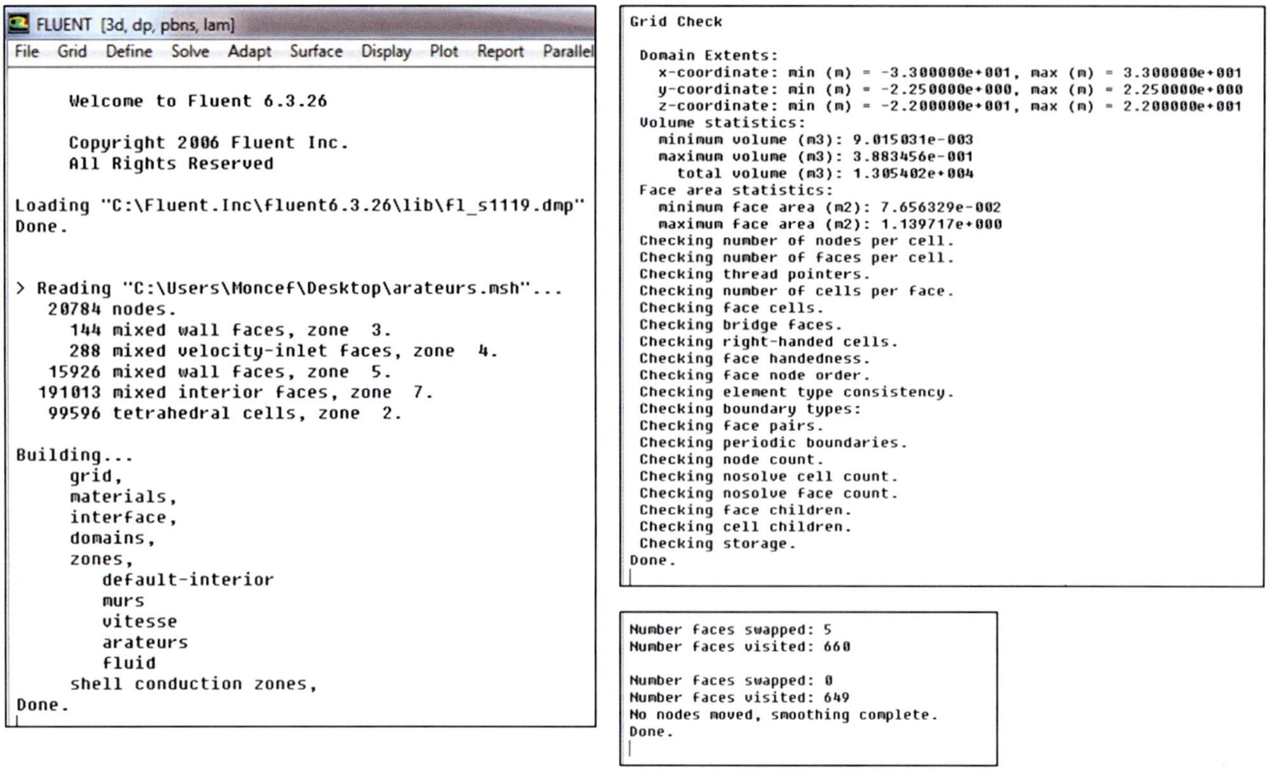

```
FLUENT [3d, dp, pbns, lam]
File Grid Define Solve Adapt Surface Display Plot Report Parallel

     Welcome to Fluent 6.3.26

     Copyright 2006 Fluent Inc.
     All Rights Reserved

Loading "C:\Fluent.Inc\fluent6.3.26\lib\fl_s1119.dmp"
Done.

> Reading "C:\Users\Moncef\Desktop\arateurs.msh"...
   20784 nodes.
     144 mixed wall faces, zone  3.
     288 mixed velocity-inlet faces, zone  4.
   15926 mixed wall faces, zone  5.
  191013 mixed interior faces, zone  7.
   99596 tetrahedral cells, zone  2.

Building...
     grid,
     materials,
     interface,
     domains,
     zones,
        default-interior
        murs
        vitesse
        arateurs
        fluid
     shell conduction zones,
Done.
```

```
Grid Check

 Domain Extents:
   x-coordinate: min (m) = -3.300000e+001, max (m) = 3.300000e+001
   y-coordinate: min (m) = -2.250000e+000, max (m) = 2.250000e+000
   z-coordinate: min (m) = -2.200000e+001, max (m) = 2.200000e+001
 Volume statistics:
   minimum volume (m3): 9.015031e-003
   maximum volume (m3): 3.883456e-001
     total volume (m3): 1.305402e+004
 Face area statistics:
   minimum face area (m2): 7.656329e-002
   maximum face area (m2): 1.139717e+000
 Checking number of nodes per cell.
 Checking number of faces per cell.
 Checking thread pointers.
 Checking number of cells per face.
 Checking face cells.
 Checking bridge faces.
 Checking right-handed cells.
 Checking face handedness.
 Checking face node order.
 Checking element type consistency.
 Checking boundary types:
 Checking face pairs.
 Checking periodic boundaries.
 Checking node count.
 Checking nosolve cell count.
 Checking nosolve face count.
 Checking face children.
 Checking cell children.
 Checking storage.
Done.
```

```
Number faces swapped: 5
Number faces visited: 660

Number faces swapped: 0
Number faces visited: 649
No nodes moved, smoothing complete.
Done.
```

Fig. 132.7 The data entry of meshings on Fluent

Table 132.2 Advantages and inconveniences of the various models of turbulences

Models	Advantages	Inconveniences
Spalart-Allumaras	Economic (1 equ). Well for the sellings averagely complex.	Is not widely tested.
STD k-ε	Strong, economic and relatively precise.	Results mediocre for complex sellings (strong gradient of pressure, rotation and swirl).
RNG k-ε	Well for sellings averagely complex (impact of jet, separation of sellings, secondary sellings)	Limited by the hypothesis of isotropic turbulent viscosity.
Realizable k-ε	Offer the same advantages as the RNG. Recommended in the case of turbomachines.	Limited by the hypothesis of isotropic turbulent viscosity.
Reynolds Stress Model (RSM)	The most complete model physically (transport and the anisotropie of the turbulence are kept account there)	Requires more weather CPU. The equations of momentum and turbulence are closely linked.
SST et standard k-ω	Model the most recommended for the problems connected to turbomachines, better than Realizable k-ε	Require a bigger resolution of the meshing on the borders (no laws in walls).

Table 132.3 Properties physical appearance of materials used in the simulation

Materials	Thermal conductivity (W/m.k)	Density (kg/m$^{3-1}$)	Mass thermal capacity (j/kg.k)	Viscosity (kg/m.s)
Water	0.6	998.2	4,182	0.001003

Residues are calculated from the corrections in variables; pressure, speed, temperature of the problem between the present iteration and the previous iteration.

In most of the cases, the convergence criterion by default in FLUENT, residual is self-important. The solution converges when residues affect 10^{-3}. However, in certain cases it is necessary to urge the calculations to 10^{-4} to see 10^{-6}. Thus There is no universal ruler.

To make sure of the convergence of the calculations, we use two visual criteria. The first one consists in observing the curves of residues defined by the equations, drawn by Flow, according to the iterations. When residues are weak (lower than 10^{-3} at least) and when curves become flat as illustrated below, we can consider that the solution is affected (Figs. 132.8, 132.9, and 132.10).

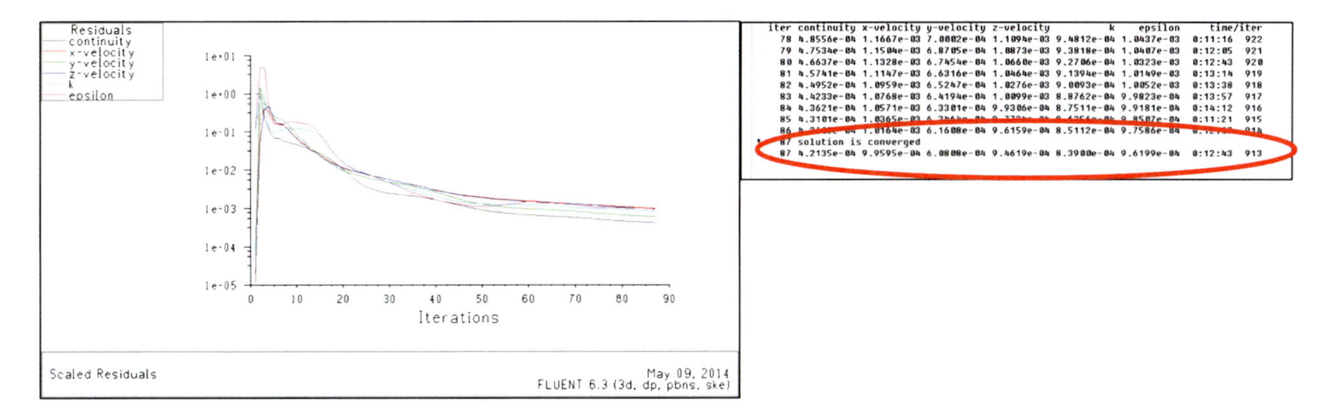

Fig. 132.8 Results of the convergence of the system of the treatment plant

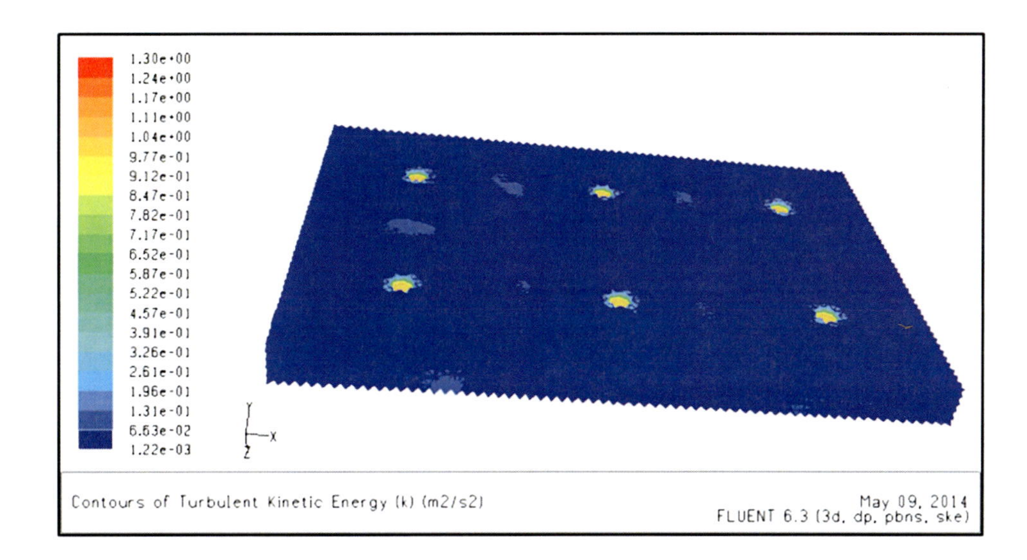

Fig. 132.9 The distribution of the turbulence in the treatment plant

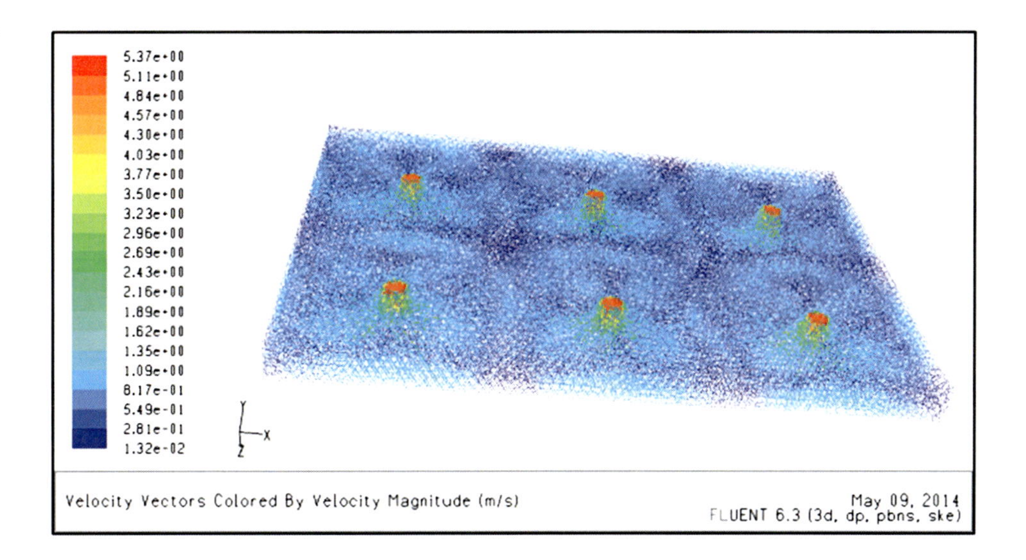

Fig. 132.10 The distribution of the speed in the treatment plant

132.5 Conclusion

The distribution of it We used for it the code of calculation Flow. At first, it was necessary to validate this digital model by means of experimental measures made on the biological reactor of the water-treatment plant of Souk-Ahras. We compared the measures of momenta experimentally on the site with those feigned by means of Flow. It was shown that the model of the biological reactor could be validated. The sellings in a biological reactor filled with clear water flow reproduced correctly thus.

References

Abusam A, Keesman KJ, Meinema K, Van Straten G (2001) Oxygen transfer rate estimation in oxidation ditches from clean water measurements. Water Res 35(8):2058–2064

Boumansour BE, Ounaïes F, Roche N, Vasel JL (1999) Tracer gas method characterization of oxygen transfer in dense membrane reactors. In: 3ème Congrès International de recherche-L'eau et sa réutilisation, INSA Toulouse, p 77–82. 9–10 nov 1999

Chatelier P (1991) Simulation de l'hydrodynamique des chenaux d'oxydation par l'utilisation des équations de Navier-Stokes associées au modèle k-ε : évaluation de la vitesse de circulation, Thèse de doctorat, INSA Toulouse, p 220

Duchène Ph, et Héduit A (1990) Aération et brassage en stations d'épuration : efficacité des divers systèmes, Inf Tech du Cemagref 78(2):7

Duchène Ph, et Héduit A (2009) Expérience et interprétation de mesures de performances d'aération en boues activées. 23e Symposium sur les eaux usées. Réseau environnement, Laval (Québec), p 6. 28–29 Nov

Part XIII

Large Projects Impact Assessment, Mitigation and Compensation

Convener Prof. Giuseppe Spilotro—*Co-convener* Canora Filomena

Infrastructure development, both in fast-developing and in more developed countries, is supported by innovative technologies, which makes every large project feasible. We think of bridges, dams and reservoirs, high-speed railways, pipelines across land and seas, energy farms, drilling and tunnelling activities, etc. The impact produced by such types of works should be predictable on the basis of the experience gained on similar projects, but it can also be surprisingly different, either due to inherent complexities in the concerned environment, as to the true extension of the impact, or, even for a new social awareness, supported by measuring capabilities, previously unavailable.

The assessment of the impacts is the first necessary step: it needs not only a reasonable evaluation of interaction between project and environment elementary components, but also of their reciprocal interaction. The assessment should be directed to the mitigation, which is in the space and in the time and should represent the relevant part of the project, while the compensation is outside of the space and the time. The question is: Can they be sufficient to maintain the overall balance as sustainable?

Mining with Filling for Mitigating Overburden Failure and Water Inrush Due to Coalmining

133

Wanghua Sui, Gailing Zhang, Zhaoyang Wu, and Dingyang Zhang

Abstract

This paper investigates the heights of overburden failure with cut-and-fill mining method using empirical formula, scaled model tests and numerical simulations. An equivalent cutting height was proposed for calculating overburden failure with formulas and from measurement data. Scale model tests and numerical simulations show that the heights of caving zone and water flow fractured zone due to mining with filling decreased obviously compared to those due to caving mining. The results also indicate that mining with filling method has protective function for overlaid unconsolidated clay and sand layers.

Keywords

Mining with filling • Water and sand inrush • Mining under water body • Water flow fractured zone

133.1 Introduction

Mining under water bodies will suffer from groundwater and sand inrush if fractures reach the aquifers and form flowing pathways. To reduce the fracture heights is a useful method to avoid water inrush from overlying aquifers. Filling mining can realize this and create safety for mining under water-bearing strata.

The paste filling technology was first used in 1979 in the Grund lead-zinc ore mine. It was applied to the Walsum Colliery (Mez and Schauenburg 1998) to mitigate surface subsidence in 1991. The first Chinese paste filling system was constructed in the Jinchuan mining area in 1990s. In 2006, filling method was first used in the Taiping Coalmine, Shandong province, China. By now, researchers have studied the properties of paste materials, filling methods, overburden failure and subsidence due to filling mining (Zhou et al. 2004). This paper presents the prediction, assessment, and practice using the paste backfilling method to mitigate overburden failure for preventing water and sand inrush through a case study in the Taiping Coalmine under unconsolidated aquifers.

133.2 Hydrogeological and Engineering Geological Conditions

The strata in the Taiping Coalmine include the Ordovician, the Carboniferous, the Permian, the Jurassic and Quaternary Systems. The sand aquifers at the bottom of the Quaternary System threaten coal mining safety. The specific capacity of the bottom aquifers ranges from 0.023 to 0.11 L/s·m. A clay layer with a thickness of 2–5 m distributes above Panel S02. The plastic index of this layer ranges from 12 to 24. During the first slice mining in Panel S02, the water inflow rate was less than 6 m^3/h. It indicates that mining the first slice weakly affects the integrity of this clay layer. It still plays an important role in preventing water inflow from the unconsolidated aquifers during backfilling mining the second slice.

W. Sui (✉) · G. Zhang · Z. Wu · D. Zhang
China University of Mining and Technology, 1 Daxuelu Rd, 221008 Xuzhou, China
e-mail: suiwanghua@cumt.edu.cn

G. Lollino et al. (eds.), *Engineering Geology for Society and Territory – Volume 6*,
DOI: 10.1007/978-3-319-09060-3_133, © Springer International Publishing Switzerland 2015

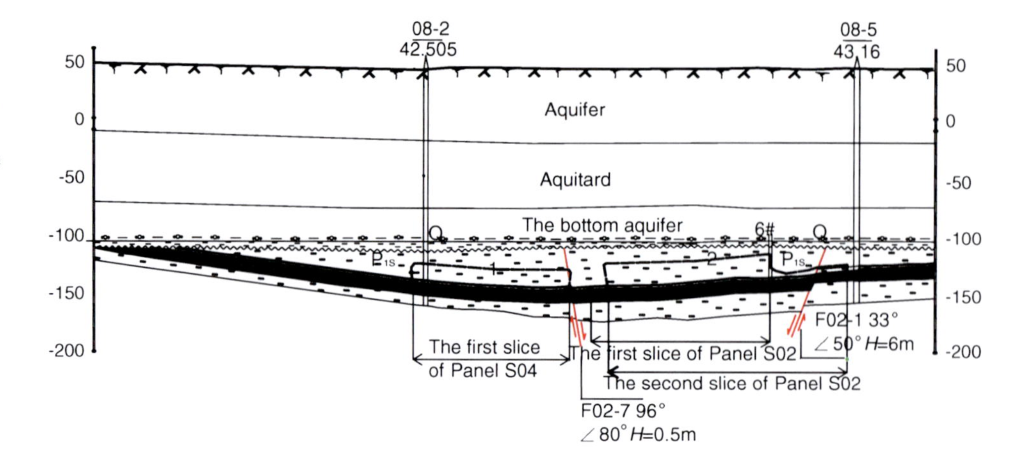

Fig. 133.1 Cross section of Panel S02 in the Taiping Coalmine. *1*—Water flow fractured zone due to mining the first slices of Panel S04; *2*—Water flow fractured zone due to mining the first and second slices of Panel S02

Table 133.1 Mechanical property of overburden and Seam 3

Types	*RQD* (%)	σ_c (MPa)	σ_s (MPa)	*E* (GPa)	λ
Fine sandstone	61	21.6–25.0	1.46–1.54	2.06–2.72	0.20–0.24
Mudstone	57	5.7	0.45	–	–
Seam 3	–	8.45	0.51	1.31	0.29

σ_c is the uniaxial compression strength, σ_s is the tensile strength, *E* is the modulus of deformation, λ is the Poisson's ratio

Table 133.2 Main influencing factors on safety mining of Panel S2

H_b (m)	H_o (m)	H_a (m)	*K* (m/d)	*q* (L/(s·m))
23.5–29.2	−116.5–−108.2	1.4–5.0	0.234–0.47	0.023–0.11

H_b = thickness of overburden; H_o = elevation of base rock surface
H_a = thickness of Quaternary bottom clay layer
K = hydraulic conductivity of the bottom aquifer; *q* = specific capacity

Figure 133.1 shows a cross section of Panel S02, which depicts the relationship between overburden and excavation. The Quaternary System is about 160 m. Table 133.1 lists the engineering geological types and characteristics of the bedrock above Panel S02 with a thickness ranging from 23.5 to 29.2 m. Table 133.2 lists the influencing factors on mining safety in Panel S02.

133.3 Heights of Overburden Failure

133.3.1 Empirical Method

The heights of the caving and the water flow fractured zones and the surface subsidence for the first slice with caving method have been studied through experiments and in situ detections. To calculate the heights of backfilling method, we proposed a concept of equivalent mining thickness, which equals to a uniform thickness of thin coal seam. This equivalent mining thickness can be estimated by $m_e = m - m_f$, where m_e is the equivalent cutting height, *m* is cutting height and m_f is thickness of filling body.

According to in situ measurements data during backfilling mining in the Taiping Coalmine, the convergence of roof and floor before backfilling is general 329 mm, non-backfilling account is 190 mm and the compression of backfilling body is estimated to be 63 mm for a cutting height of 2.2 m, therefore, the equivalent mining thickness is 0.58 m. Table 133.3 lists the heights of overburden failure for mining the first and second slices separately, which are calculated from empirical formulae proposed by Code (Coal Industry Bureau of PRC 2000).

133.3.2 Scale Model Test

The scaled model is designed according to the prototype of Panel S02 of the Taiping Coalmine. The length scale is selected as $b = 1/150$, the time constant $a = \sqrt{b} = 0.082$, the unit weight ratio $C_\gamma = 1.5/2.5 = 0.6$, and strength ratio $C_R = 1/250$. The model mining process, deformation and cracks were recorded by a high-speed digital camera and a fracture gauge.

Table 133.3 Results of heights of overburden failure by different methods

Slices and method	h (m)	Scaled model test		Empirical method		Numerical simulation	
		H_c (m)	H_f (m)	H_c(m)	H_f (m)	H_c (m)	H_f (m)
$1^{\#}$, caving	2	6.82	18.21	6.6	18.88	7.6	19.2
$2^{\#}$, filling	2	7.24	21.46	6.6	19.85	8.4	22.2
$2^{\#}$, caving	2	–	–	–	–	10.6	25.8

$1^{\#}$ = the first slice; $2^{\#}$ = the second slice; h = cutting thickness; H_c = caving zone; H_f = water flow fractured zone

Fig. 133.2 After mining the second slice. *1*—caving zone; *2*—water flow fractured zone

Figure 133.2 shows the development of overburden failure and deformation due to mining the first slice of Seam 3 with caving mining and the second slice using the back-filling mining.

After the first slice mined out, the caving zone and fractured zone formed a shape of "the saddle", the height of caving zone is 6.8 m, and the water flow fractured zone is up to 18.2 m.

In the mining process of the second slice, the overburden in the expanded region without mining the first slice was deformed weakly, only produced a few micro fractures. After finishing mining, the heights of the caving and the water flow fractured zones grow to 7.24 m and 21.46 m, respectively. Compared to the height of overburden failure after mining the first slice, it increases much smaller when mining the second slice using backfilling method. In the expanded region the height of the water flow fractured zone is 3.82 m and did not form the caving zone (Fig. 133.3).

Table 133.3 lists the results of overburden failure using different methods. The numerical simulation was conducted using FLAC (Wu 2013). If the second slice is excavated by caving method, its caving zone and water flow fractured zone can reach 10.6 m and 25.8 m, higher than those induced by backfilling method. It was inferred that the paste backfill mining can control the growth of fractures, and it is significant for mining under thick unconsolidated aquifers and thin bedrock.

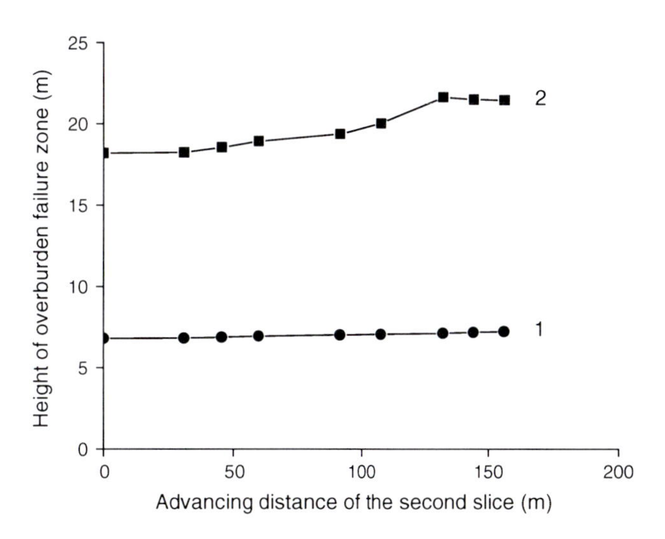

Fig. 133.3 The height of overburden failure after mining the second slice by backfilling method. 1—caving zone; 2—water flow fractured zone

133.4 Conclusions

This paper presents a concept of equivalent mining thickness for calculating the overburden failures of mining by paste backfilling. The overburden failure was studied using scaled and numerical simulations. The results show that the caving and water flow fractured zones due to the mining of the second slice with paste backfilling increased relatively small. And there was only the water flow fractured zone formed in expanded area without mining the first slice. Comparing the overburden failure results in different mining methods, we found that the paste backfilling was useful in inhibiting obvious damage of the overburden failure. The heights of the water flow fractured zone with theoretical calculation were in good agreement with that of scale test and numerical simulations. The paste backfilling used in coalmining under thin bedrock and unconsolidated aquifers provides an effective method to mitigate the overburden failure, and prevent panels from groundwater and sand inrush hazards.

Acknowledgments The authors would like to thank the financial support from the National Natural Science Foundation of China—Shenhua Group Jointly Funded Project under Grant No. 51174286.

References

Coal Industry Bureau of the PRC (2000) Regulations for setting coal pillar and mining under buildings, water bodies, railways and main roadways. China Coal Industry Publishing House, Beijing (in Chinese)

Mez W, Schauenburg W (1998) Backfilling of caved-in goafs with pastes for disposal of residues. Australasian institute of mining and metallurgy publication series no. 1, In: Proceedings of the 1998 6th international symposium on mining with backfill, Minefill, p 245–248

Wu Zhaoyang (2013) Overburden failure and surface subsidence due to mining with paste backfilling under thick unconsolidated layers and thin bedrock. China University of Mining and Technology, Xuzhou

Zhou Huaqiang, Chaojiong Hou, Sun Xikui et al (2004) Solid waste paste filling for none-village-relocation coal mining. J China Univ Min Technol 33(2):154–158

Empirical Cutting Tool Wear Prognosis for Hydroshield TBM in Soft Ground

134

Florian Köppl, Kurosch Thuro, and Markus Thewes

Abstract

The wear of cutting tools is a major issue in tunneling with Hydroshield TBM in soft ground, because it is a common reason for unplanned downtime. The replacement of cutting tools requires access to the excavation chamber, which is only possible with hyperbaric works or at fixed positions with prearranged grout blocks. In order to improve predictability of the maintenance stops, an empirical prognosis model for the required maintenance stops for changing of cutting tools and the amount of tools was developed. Based on the new Soil Abrasivity Index (SAI), the model helps to estimate distances between maintenance stops and the required amount of cutting tools to be changed. The authors validated the prognosis model based on the original reference projects. The validation lead to an adaption of the prognosis process and individual correction factors in the model itself. The article therefore describes the updated prognosis model.

Keywords

Hydroshield TBM • Soft ground • Excavation • Tool wear • Prognosis

134.1 Introduction

The replacement of cutting tools on Hydroshield TBMs is done with hyperbaric works in the excavation chamber or at fixed positions with grout blocks, in order to maintain face stability during the interventions. The position of the maintenance stops therefore and the obtainable length of the intervals between them are determined by the wear of the cutting tools. At the same time boundary conditions in the projects are often critical for hyperbaric works, particularly under buildings, at low overburden or very high water pressure. Köppl and Thuro (2013) developed an empirical prognosis model, to estimate the distances L_I in m between the maintenance stops I and the number of cutting tools $n_{c,I}$ to be replaced. The model enables foresighted planning, in order to avoid critical areas as described above. Validation of the prognosis model with reference to the original projects used in the data analysis allowed to introduce improvements to the prognosis process and to specify correction factors. The article summarizes the update model.

F. Köppl (✉)
Herrenknecht AG, Schlehenweg 2, 77963 Schwanau, Allmansweier, Germany
e-mail: koeppl.florian@herrenknecht.de

K. Thuro
Technische Universität München, Arcisstr. 21, 80333 Munich, Germany

M. Thewes
Ruhr-Universität Bochum, Universitätsstr. 150, 44780 Bochum, Germany

134.2 Empirical Wear Prognosis Model

134.2.1 Basic Data

The basic data for the prognosis model contain a complete geotechnical data set, as e.g. detailed by Scholz and Wendl (2010), the TBM design, e.g. the cutter head layout and expected values for TBM advance parameters.

G. Lollino et al. (eds.), *Engineering Geology for Society and Territory – Volume 6*,
DOI: 10.1007/978-3-319-09060-3_134, © Springer International Publishing Switzerland 2015

For the prognosis process, homogeneous geotechnical sections are formed along the tunnel axis, using the criteria recommended by Köppl and Thuro (2013) and assigned with a whole-numbered index z beginning at the start of the tunnel:

- Constant share of different soil types in the excavation face in % (±10 %)
- Constant thickness of the cover above the tunnel axis in m (±5 m)
- Constant water table above the tunnel axis in m (±5 m).

134.2.2 Estimation of the Cutting Distance S_c

The expected value for the cutting distance $s_{c,e(z)}$ in km of each individual tool on the cutter head within a geotechnical section is estimated using the correlation with the Soil Abrasivity Index $SAI_{(z)}$ which is estimated for each section by Eq. 134.1:

$$SAI_{(z)} = \left(\frac{EQC_{(z)}}{100}\right) \cdot \tau_{c(z)} \cdot D_{60(z)} \quad (134.1)$$

In Eq. 134.1 $EQC_{(z)}$ [%] is the Equivalent Quartz Content, $D_{60(z)}$ [mm] the grain size where 60 % of all grains are smaller than the given value and $\tau_{c(z)}$ [kN/m^2] the shear strength of the soil, estimated by the Mohr-Coulomb criterion, using the shear parameters of the soil and the vertical primary stress at the tunnel axis $\sigma_{n(z)}$ [kN/m^2]. In Eq. 134.2 h_i [m] is the thickness of the soil layers above the tunnel axis and γ_i [kN/m^3] unit weight, considering the actual dry or saturated conditions:

$$\sigma_{n(z)} = \sum_i h_i \cdot \gamma_i \quad (134.2)$$

The basic value of the cutting distance $s_{c,b(z)}$ [km] is estimated with the $SAI_{(z)}$ as shown by Köppl and Thuro (2013) for disc cutters (Eq. 134.3) and scrapers (Eq. 134.4):

$$s_{c,b(z)} = 288 + \exp\left(-0.004 \cdot \left(SAI_{(z)} - 1.640\right)\right) \quad (134.3)$$

$$s_{c,b(z)} = 271 + \exp\left(-0.004 \cdot \left(SAI_{(z)} - 1.634\right)\right) \quad (134.4)$$

For disc cutters, $s_{c,b(z)}$ is corrected by the tip width b_{SR} [mm] of the cutter ring. With reference to the most common value for b_{SR} of 19 mm, the correction factor f_b is calculated by (Eq. 134.5):

$$f_b = \frac{b_{SR}}{19} \quad (134.5)$$

For scrapers, $s_{c,b(z)}$ is corrected by the actual penetration rate $p_{a(z)}$ [mm/rot] of the individual scrapers. Köppl and Thuro (2013) subdivide this step with two separate factors f_p and f_k, considering the expected penetration rate $p_{e(z)}$ [mm/rot] and the number of identical scrapers ksc per cutting track. Both factors indirectly cover the actual penetration rate $p_{a(z)}$ of the individual scrapers, since $p_{e(z)}$ is split between all identical scrapers on a cutting track (ksc) depending on their angular distance δ_a [°]. The verification of the model proved it to be more appropriate to directly consider $p_{a(z)}$ for correction in Eq. 134.7.

Assuming a symmetrical cutter head layout, the actual penetration rate $p_{a(z)}$ [mm/rot] of a scraper can be roughly estimated by Eq. 134.6:

$$p_{a(z)} = \frac{p_{e(z)}}{k_{sc}} \quad (134.6)$$

$$f_p = \frac{1}{1.6^{\log_{0.5}\left(\frac{16}{p_{a(z)}}\right)}} \quad (134.7)$$

The expected values for the cutting distance $s_{c,e(z)}$ for disc cutters and scrapers are calculated by multiplication of the basic value $s_{c,b(z)}$ with the respective factors f_b or f_p. The expected value for the cutting distance $s_{c,e(z)}$ is valid within a defined range of additional influencing factors on the wear of the cutting tools which were qualified, but not quantified in the data analysis. A description of these factors and the respective range is given by Köppl and Thuro (2013).

134.2.3 Estimation of the Maintenance Stops I

For the prognosis process the maintenance stops I are assigned with a whole-number index k in ascending order beginning at the start of the tunnel. The chainage $L_{I,act(k)}$ [m] of the stops $I_{(k)}$ for replacement of cutting tools over the tunnel axis depends on the advance distance $L_{I(k)}$, [m] achieved between the stops. The maximum value $L_{I,max(k)}$ [m] of $L_{I(k)}$ is defined by the wear limits of the tools.

Starting with unworn tools, the maximum achievable advance distance $L_{I,c(k)}$ [m] for each individual cutting tool is estimated by Eq. 134.8 using the expected value of the penetration rate $p_{e(z)}$ [mm/rot] and the track radius r_s [mm].

$$L_{I,c(k)} = \frac{s_{c,e(z)} \cdot p_{e(z)} \cdot 1.000}{2 \cdot \pi \cdot r_s} \quad (134.8)$$

Considering all cutting tools on the cutter head, the maximum advance distance $L_{I,max(k)}$ of the TBM is given by the minimum value out of all values for $L_{I,c(k)}$:

$$L_{I,max(k)} = \min_c\left(L_{I,c(k)}\right) \quad (134.9)$$

For the planning of the actual advance distance, $L_{I(k)}$ can be selected lower or equal to $L_{I,max(k)}$ without compromising the wear limits of the cutting tools:

$$L_{I(k)} \leq L_{I,max(k)} \tag{134.10}$$

Within the section $L_{I(k)}$ each tool performs the partial cutting distance $s_{c(k)}$ [km], which is calculated using the actual penetration rate $p_{e(z)}$ [mm/rot] over $L_{I(k)}$:

$$s_{c(k)} = \frac{L_{I(k)} \cdot 2 \cdot \pi \cdot r_s}{p_{e(z)} \cdot 1.000} \tag{134.11}$$

Considering the expected value for the cutting distance $s_{c,e(z)}$ [km], the partial utilization factor $e_{c(k)}$ of each cutting tool over $L_{I(k)}$ is calculated by:

$$e_{c(k)} = \frac{s_{c(k)}}{s_{c,e(z)}} \tag{134.12}$$

The advance sections $L_{I(k)}$ are strung together consecutively from the start of the tunnel. The actual chainage $L_{I,act(k)}$ [m] of each maintenance stop $I_{(k)}$ is calculated by accumulation of all sections $L_{I(k)}$ excavated until $I_{(k)}$:

$$L_{I,act(k)} = \sum_{1}^{k} L_{I(k)} \tag{134.13}$$

The cumulative utilization factor $e_{c,act(k)}$ of each tool at a stop $I_{(k)}$ is calculated by accumulation of the partial utilization factors $e_{c(k)}$ in the advance sections $L_{I(k)}$:

$$e_{c,act(k)} = \sum_{L_{a(k)}}^{L_{I,act(k)}} e_{c(k)} \tag{134.14}$$

Cutting tools may go through more than one advance section $L_{I(k)}$, so the lower constraint of the sum in Eq. 134.14 is set at $L_{a(k)}$ which equals the chainage $L_{I,act(k)}$ of the stop $I_{(k)}$, where the individual tool was assembled on the cutter head.

The process in Eqs. 134.8–134.14 supposes, that the maintenance stops $I_{(k)}$ and $I_{(k+1)}$ are located in the same geotechnical section, represented by the expected values for $s_{c,e(z)}$ and $p_{e(z)}$ in Eqs. 134.8 and 134.12. In case $L_{I(k+1)}$ crosses a boundary between geotechnical sections, the process in Eqs. 134.8–134.14 needs to be subdivided, considering different values for $s_{c,e(z)}$ and $p_{e(z)}$ in the legs of the different geotechnical sections.

134.2.4 Estimation of the Tool Changes n_c

Starting at a given maintenance stop $I_{(k)}$ the process in Eqs. 134.8–134.10 allows to plan the advance distance $L_{I(k+1)}$ to the next stop $I_{(k+1)}$. The chainage $L_{I,act(k+1)}$ of $I_{(k+1)}$ and the cumulative utilization factor $e_{c,act(k+1)}$ result from Eqs. 134.11–134.14, using $L_{I(k+1)}$.

The calculation of the maximum achievable advance distance $L_{I,c(k)}$ [m] in Eq. 134.8 starting from the stop $I_{(k)}$ assumes unworn tools, independent of the actual cumulative utilization factor $e_{c,act(k)}$. Consequently Eq. 134.14 may produce values for $e_{c,act(k+1)}$ greater than 1 for individual tools. For these tools the wear limit would be exceeded during the planned advance distance $L_{I(k+1)}$, leading to potential damage due to excessive wear. Therefore they have to be changed at the actual stop $I_{(k)}$, in order to enable the selected advance distance $L_{I(k+1)}$. Accordingly the general criterion for tool changes during a maintenance stop $I_{(k)}$ can be formulated as:

$$e_{c,act(k+1)} > 1 \tag{134.15}$$

Following the criterion in Eq. 134.15, the number of cutting tools $n_{c,I(k)}$ to be changed during each maintenance stop $I_{(k)}$ can be accumulated over all cutting tools c on the cutter head by:

$$n_{c,I(k)} = \sum_{c} e_{c,act(k+1)} > 1 \tag{134.16}$$

The prognosis process focuses on planning of the advance distances $L_{I(k+1)}$. The criterion for replacement of cutting tools in Eq. 134.15 is designed to enable the selected value for $L_{I(k+1)}$. Consequently the criterion may also effect preventive tool changes at values of $e_{c,act(k)}$ lower than 1. This relation reflects the higher impact of the stops $I_{(k)}$ on TBM advance compared to $n_{c,I(k)}$. It also implies that any change in the advance distances $L_{I(k+1)}$ effects on $n_{c,I(k)}$ and vice versa.

134.2.5 Adaption Algorithm

The prognosis model as detailed in Eqs. 134.1–134.16 does not consider the given conditions of the project regarding accessibility of the excavation chamber. In order to ensure realistic results it is mandatory to check these conditions at each planned chainage $L_{I,act(k)}$ of the maintenance stops $I_{(k)}$. In case the chainage $L_{I,act(k)}$ of a stop $I_{(k)}$ is located in a critical section, the model allows for adaption by:

- Variation of the basic data (e.g. layout of the tools or penetration rate p_e).
- Selection of different values for $L_{I(k)}$ in Eq. 134.10.

The variation of theses parameters effects the chainage of all maintenance stops $I_{(k)}$. It is therefore recommended to do the adaption in small iterative steps. The same process may also be used to analyze the propagation of variances of the basic data in the model by developing different prognosis scenarios.

134.3 Conclusion

The prognosis model provides a comprehensive method to estimate the required maintenance stops $I_{(k)}$ for Hydroshield TBM in soft ground. This method can be effectively used to reduce the impact of unplanned demand for TBM maintenance, especially in projects with complex geotechnical conditions and in densely populated areas. Thus the overall impact of this tunneling method on the project environment may be reduced to a minimum already in the planning stage.

The prognosis elaborated in the planning stage of a project requires follow up during the advance phase. Primarily to validate the assumptions in the prognosis, but also for comprehensive documentation of additional data for development of the prognosis model, as for example demonstrated by Wendl et al. (2010), Düllmann et al. (2013) and Hollmann et al. (2013).

References

Düllmann J, Hollmann F, Thewes M, Alber M (2013) Analysis of soil-mechanic-interactions (Part 1): processing of TBM-machine-data and extraction of excavation-specific data. In: Proceedings of the EURO:TUN 2013. Aedificatio Publishers, Freiburg, pp 621–634

Hollmann F, Düllmann J, Thewes M, Alber M (2013) Analysis of soil-mechanic-interactions (Part 2): influences on the excavation-specific data of TBM-machine data. In: Proceedings of the EURO: TUN 2013. Aedificatio Publishers, Freiburg, pp 635–648

Köppl F, Thuro K (2013) Cutting tool wear prognosis and management of wear related risks for Hydroshield TBM in soft ground. In: Proceedings of the 18th ICSMGE. Presses du Ponts, Paris, pp 1739–1742

Scholz M, Wendl K (2010) Geological aspects of slurry shield drives. In: Proceedings of the 11th IAEG congress. CRC Press, London, pp 3507–3513

Wendl K, Scholz M, Thuro K (2010) A new approach to engineering geological documentation of slurry shield drives. In: Proceedings of the 11th IAEG congress. CRC Press, London, pp 3827–3834

Risk and Mitigation of the Large Landslide of Brindisi di Montagna

135

Giuseppe Spilotro, Filomena Canora, Roberta Pellicani, and Francesco Vitelli

Abstract

The *Brindisi di Montagna Scalo* landslide is located in Basilicata region (Italy), about 18 km south-east of Potenza. It consists on an active earth-flow with a longitudinal extension of about 700 m. The accumulation zone extends till the Basento's riverbed (on the left side), which is partially obstructed. This area is periodically fed with debris or mud moving along the flow channel, due to both the reactivations of the earthflow along the landslide channel caused by wintry rainfalls and the retrogression of the crown towards upstream and deconstruction of collapsed soil in different parts of the landslide body. The risk induced by this earth flow derives from the presence, on the right bank of Basento river, of a railway line and, a little further away, of the state road 407 "Basentana". The aim of this work is the analysis of the possible design solutions in order to mitigate the risk. The assessment of risk mitigation measures for the *Brindisi di Montagna Scalo* landslide was differentiated for the three main geomorphological zones of the landslide: alimentation area, landslide channel and accumulation zone.

Keywords

Landslide • Risk • Mitigation

135.1 Introduction

The *Brindisi di Montagna Scalo* landslide consists on an active earth-flow with can be classified as a "rotational slide-flow" (Cotecchia et al. 1986; Cruden and Varnes 1986; Bentivenga et al. 2006). Basically, it is a flow, deriving from the toe of an existing terrace of rotational landslide affecting Red Flysch formation, and it is located within a pre-existing watershed in Varicolored Clays formation.

G. Spilotro · R. Pellicani (✉)
Department of European and Mediterranean Cultures, University of Basilicata, Matera, Italy
e-mail: pelliro@libero.it

F. Canora
School of Engineering, University of Basilicata, Potenza, Italy

F. Vitelli
Freelance Geologist, Potenza, Italy

The instability of the area is related to the morphology of the slope, which is steep where Red Flysch outcrops and becomes less steep where Varicolored Clays outcrop.

The risk induced by the *Brindisi di Montagna* earth flow derives from the presence, on the right bank of Basento river, of a railway line and, a little further away, of the state road 407 "Basentana" (Fig. 135.1). The erosion of the dam, created by the landslide toe into the riverbed, may cause the outflow of significant amount of water and debris downstream. Along the slope, the landslide has already damaged a provincial road.

In the past years, gabions and retaining walls on piles were realized, on the left bank of the river, in order to contain the landslide toe. Subsequently, part of the retaining wall was damaged due to a reactivation of the earth flow, causing the mobilization of debris towards the riverbed and the consequent restriction of the outflow section of Basento (Fig. 135.2). Because of this, a reservoir has been created immediately upstream of the landslide dam. At the section of

G. Lollino et al. (eds.), *Engineering Geology for Society and Territory – Volume 6*,
DOI: 10.1007/978-3-319-09060-3_135,

Fig. 135.1 Accumulation area of the landslide occluding Basento riverbed

Fig. 135.2 Basento River occluded by landslide debris and parts of existing retaining wall

Basento blocked by landslide debris, the river flow has a higher velocity and, therefore, its erosive capacity increases.

In order to reduce the risk induced by the landslide to the infrastructures, located near the toe, the evolutions of landslide toe in relation to the groundwater income from the rear crown area, to the rainfalls and to the river floods have been analyzed.

Some floods, in February and March of the last year, produced an increase of the river level up to 1.5 m above the crown of the gravity wall retaining the railway embankment, removing the soil on the back of the wall and triggering instability processes at the toe of the same embankment.

135.2 Measures for Risk Mitigation

The aim of this work is the analysis of the possible design solutions in order to mitigate the risk. In general, the mitigation measures aim, on the one hand, to stabilizing the landslide phenomenon, in order to reduce the retrogression of the crown and the contribution of debris to the accumulation zone; on the other, to the protection of the infrastructures located near the landslide toe, on the right side of Basento river, by realizing engineering structures able to dispose safely the flood flows, in case of severe obstruction of Basento's riverbed.

The assessment of risk mitigation measures for the *Brindisi di Montagna Scalo* landslide was differentiated for the three main geomorphological zones of the landslide: alimentation area, landslide channel and accumulation zone.

The alimentation area develops in the "Red Flysch" formation and has a lobed-shape like a fan, with different minor scarps within the landslide body. The main scarp is located at 670 m a.s.l. and has a total width of about 500 m. In this area little landslide ponds and accumulations of saturated soil are present. The alimentation area is also characterized by the presence of groundwater which feeds the landslide basin. For this reason, in this area the main mitigation measure consists in avoiding the groundwater and surficial water feeding of the crown area by means of a drainage system, that intercepts the groundwater and is constituted by draining panels positioned transversely to the landslide crown and connected to a deep well with automatic pumping system. It is also needed to intercept the superficial waters with a hydraulic enclosure, realized in the stable area (upstream of the main scarp) through channels. Finally, in order to remodel and regularize the edge of the crown it is possible to use soil nailing, micropiles, gabions, etc. The main problem connected to this type of drainage system is the durability at long time within landslide body constituted by fine graded terrain.

In correspondence of the landslide cannel, the mitigation techniques are generally influenced by the thickness of the landslide debris. As in this case it is about 6–7 m, the suggested mitigation technique can consist of rows of large diameter piling, with piles not connected in head, and draining trenches for the drainage of surface and deep waters.

The stabilization of the accumulation zone of the landslide aims to make the riverbed of Basento stable, even in conditions of overflowing of the river. In this area, the protection of the railway embankment is a priority, as it is in complete erosion due to the periodic flooding of the river.

The proposal mitigation measures have been the following (Fig. 135.3):

- Removal of the landslide debris, which block the riverbed of Basento, from the accumulation zone.
- Realization of a diaphram, on the left side of the river, in order to reshape the stretch of the riverbed, upstream of the section blocked by the landslide toe.
- Realization of a diaphram, on the right side of the river and at a distance of about 10 m from the existing gravity wall, composed by piles of a diameter of 600 m and a length of 14 m, with a wall at the pile top of a height of about 3 m and longitudinal extension of 67 m.
- Filling with selected draining material between the existing wall and the diaphram.

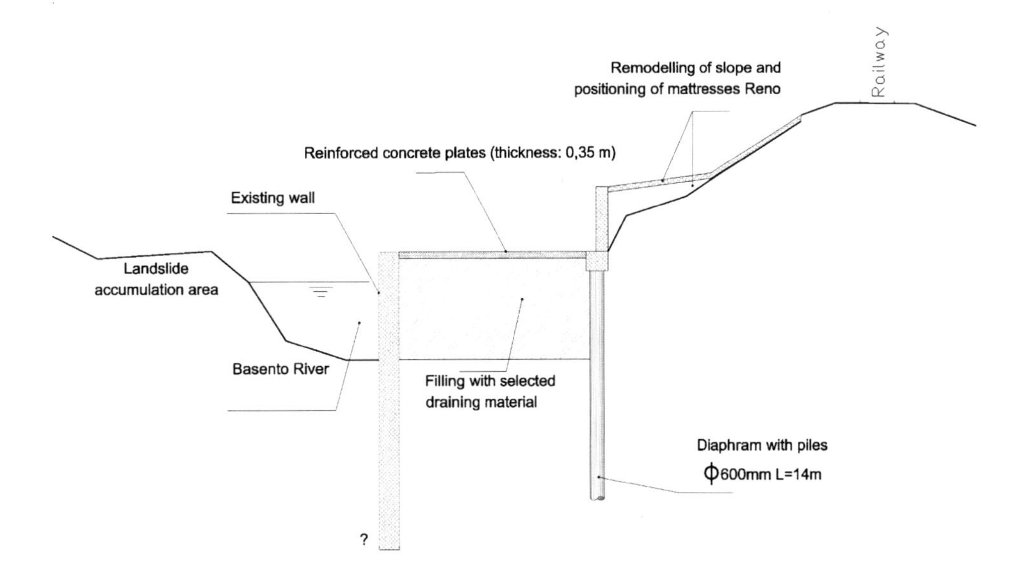

Fig. 135.3 Cross section of the area near the landslide toe with the structural measures for mitigating risk

- Positioning of reinforced concrete plates, of a thickness of about 35 cm, connecting the top of the existing gravity wall and the top of the piles of the diaphram.
- Positioning of Reno mattresses from the top of the new wall to the railway line, for the erosion control of the embankment.

135.3 Conclusions

As risk controlling is the process of measuring or assessing risk and then developing strategies to manage the risk, in addition to the design of mitigation measures, the following monitoring program has been recommended:

- Monitoring of the hydraulic flow through the Basento's section blocked by the landslide.
- Monitoring of the landslide superficial displacements at the alimentation and accumulation zones by means of GPS localizators, specifically designed for this purpose.
- Monitoring of the landslide deep displacements by means of TDR measurements.
- Monitoring of groundwater by means of piezometers.

References

Bentivenga M, Grimaldi S, Palladino G (2006) Caratteri geomorfologici della instabilità del versante sinistro del fiume Basento interessato dalla grande frana di Brindisi di Montagna Scalo (Potenza, Basilicata). Giornale di Geologia Applicata 4:123–130

Cotecchia V, Del Prete M, Federico A, Fanelli GB, Pellegrino A, Picarelli L (1986) Studio di una colata in formazioni strutturalmente complesse presso Brindisi di Montagna Scalo (PZ). Associazione Geotecnica Italiana-Atti del XVI convegno Nazionale di Geotecnica, pp 14–16 maggio, Bologna

Cruden DM, Varnes DJ (1986) Landslide types and processes. In: Landslides: investigation and mitigation, Turner AR and Schuster RL (ed) Sp. Rep. 247, Transportation Research Board, National Research Council, National Academy Press, Washington D.C., pp 36–72

Environmental Impact of a Motorway Tunnel Project on an Important Karst Aquifer in Southern Latium Region: The Case of Mazzoccolo Spring (Formia, Italy)

136

Giuseppe Sappa, Flavia Ferranti, and Sibel Ergul

Abstract

Due to the heavy traffic problems in Formia downtown, a motorway project was designed to reduce traffic congestion in the fastly growing city. The proposed motorway passes over one regionally an important karst aquifer feeding Mazzoccolo spring, in the southeast part of Latium Region. This paper deals with the analysis of the most important environmental impacts on this vulnerable karst aquifer and, as a consequence, on groundwater coming out form the Mazzoccolo Spring, which feeds one of the most important drinking water supply network in South Latium Region. A multisystem approach has been applied for vulnerability analysis using SINTACS method. The climatic, topographic, geomorphological and hydrogeological data and field investigations of previous works has been employed. On the other hand, this approach was also evaluated by geochemical and isotope tracers techniques of groundwater samples for the identification of environmental impacts. According to the proposed project, the Motorway tunnel has a significant environmental impact, on the vulnerability of the karst aquifer and hence, on the water supply networks. Thus, based on these analyses the highway investment strategy adapted to Variant of the SS 7 (Appian Way) project should ensure the protection of groundwater resources designing a new variant route avoiding the construction of Mola Mountain tunnel.

Keywords

Karst aquifer • Vulnerability • Protection • Infrastructure project

136.1 Introduction

The preservation and protection of groundwater is a topic of increasing technical relevance, due to the widespread of infrastructure designs for human settlements and technological development, which makes today executable any kind of civil work, regardless of the environmental impacts. On the other hand, water demand is increasing all over the world, and also in the south part of Latium Region, Central Italy, groundwater is the most important resource of drinking and agricultural purposes (Casa et al. 2008; Sappa et al. 2012). In this region, Mazzoccolo is the most important karst spring that feeds drinking water supply networks of Formia and other communities with a rate of 900 l/s (Ialongo 1983). For more than a decade, due to heavy traffic problems on the motorway SS 7—Appia, Formia (LT), a new design of a tunnel, passing thorough Mola Mountains, has been discussed. Besides, due to increasing urbanization, the downtown of Formia city has already moved to close to the catchment area of Mazzoccolo spring. Currently, because of the heavy traffic problems in SS 7 (Appia), close to the center of Formia town, a variant motorway project was proposed which has to be built far from the urban area. The performed designed works for the proposed variant will have several negative impacts on the groundwater systems both quantitatively and qualitatively. In the proposed project solutions, the possible interactions between motorway tunnel

G. Sappa (✉) · F. Ferranti · S. Ergul
DICEA, Department of Civil, Building and Environmental Engineering—Sapienza, University of Rome, Via Eudossiana 18, 00186, Rome, Italy
e-mail: giuseppe.sappa@uniroma1.it

G. Lollino et al. (eds.), *Engineering Geology for Society and Territory – Volume 6*,
DOI: 10.1007/978-3-319-09060-3_136, © Springer International Publishing Switzerland 2015

and the karst aquifer that feeds the Mazzocolo spring seem to be not considered in details. The designed tunnel, whose total length is 4 km has the potential to have an adverse effect on the karst aquifer. Here we try to suggest appropriate methodological approaches in the aim of highlighting the adverse environmental impacts on Mazzoccolo spring and some preliminary indications for designing a different, less impacting, layout of the motorway. Thus, the application of a vulnerability mapping as a multitracing approach (Civita 1998) was considered for the protection of this important groundwater system suggesting more detailed development of the design to reduce negative impacts and save costs. In the present work, the appropriate methodological approaches were applied for the protection and preservation of Mazzoccolo spring where groundwater emerges from the karst aquifer. Based on previous studies, for the identification of environmental impacts on the groundwater, data from different locations and field observations were employed. The new design solutions for proposed variant should ensure the reduction of potential negative impacts on the groundwater systems in the area, considering the interactions between the Mola Mountain tunnel and fragile karst aquifer.

136.2 Geological and Hydrogeological Setting

Mazzoccolo spring take place on the base of Pliocene conglomerates, which is tectonically in contact with the limestones. This spring is located at an altitude of 11.5 m above sea level, where the area is more fractured due to the intersection of numerous faults (Ialongo 1983). Mazzoccolo spring is caught at an elevation of 7.50 m by a drainage tunnel, whose layout is generally parallel to the slope (Fig. 136.1). The morphological setting of the study area is characterized by conglomeratic limestone on the hills (Mola Mountain), surrounded by clayey-arenaceous and alluvial deposits of debris (Di Nocera 1983). The abundance of groundwater is due to the permeability of the limestone (high fractured and deep karst), which stores a significant quantity of rainwater feeding perennial springs. The Pliocene conglomerates are strongly cemented, while the contacts with different formations show highly fractured structures. The karstified limestones stored the water and discharge Mazzoccolo spring and also some smaller pools occur to the west part. The underground reservoir of Mazzoccolo spring was formed by high permeable limestones for karstic fractures outcropping to the N–W and N–NW of the spring (Di Nocera 1983), while the bottom is composed of impermeable dolomites underlying the West part of the limestones.

136.3 Project Description

The preliminary design of the variant of the SS 7-Appia provides, bypassing the town of Formia, the excavation of a motorway tunnel in Mola Mountain with 4,578 m length as represented in Fig. 136.2. According to the proposed design, tunnel layout has a convex trend with respect to the surface and presents the highest elevation at about 119 m a.s.l. in a central position between the two entrances. The lowest planimetric distance between the tunnel axis and the Mazzoccolo group (springs-wells) is of 1 km, while the minimum height difference between the base of the gallery and the springs is of 80 m (Manfredini 1984).

136.4 Results and Conclusions

The present work deals with the environmental impact assessment of the designed motorway tunnel on karst aquifer, feeding Mazzoccolo spring, employing SINTACS method (Civita 1998). The hydrogeochemical monitoring activities were carried out, from May 2006 to January 2007, in the framework of this study. The stable isotopes of ^{2}H and ^{18}O in groundwater samples provided some important information about aquifer active recharge areas, aquifer morphology, spring discharge areas and the origin of rainfalls (Fig. 136.3). The elevation of the recharge area ranges between 600 and 800 m a.s.l. The geochemical analysis showed that all the sampled waters belong to the Ca-HCO_3 facies, reflecting the characteristics of the carbonate karst aquifer of western Aurunci Mts. It was also observed that one sample, taken in August, show very low mineralization with low electrical conductivity and a high pH value ($EC = 28\ \mu S/cm$ and $pH \sim 9$). For the evaluation of this phenomenon, the recharge elevation data was employed. Two different hypotheses have been developed and each based on a specific groundwater circuit. The pH value is related to scarce precipitations occurred in summer assuming that the source may return the "purging" of a previous meteoric recharge. However, considering the nature of karst basin, the low EC and the high value of pH in groundwater, taken in August, may be related to the presence of some sinkholes which are able to quickly convey the meteoric precipitation into the aquifer. In fact, between Ruazzo Mountain and the Mazzoccolo spring some dolines and ponors were found (Civita 1998).

The vulnerability analysis, carried out by SINTACS method (Fig. 136.4), shows that the karst aquifer feeding the Mazzoccolo spring has a high vulnerability, while presents very low capacity of groundwater protection.

Fig. 136.1 Drainage gallery of Mazzoccolo spring

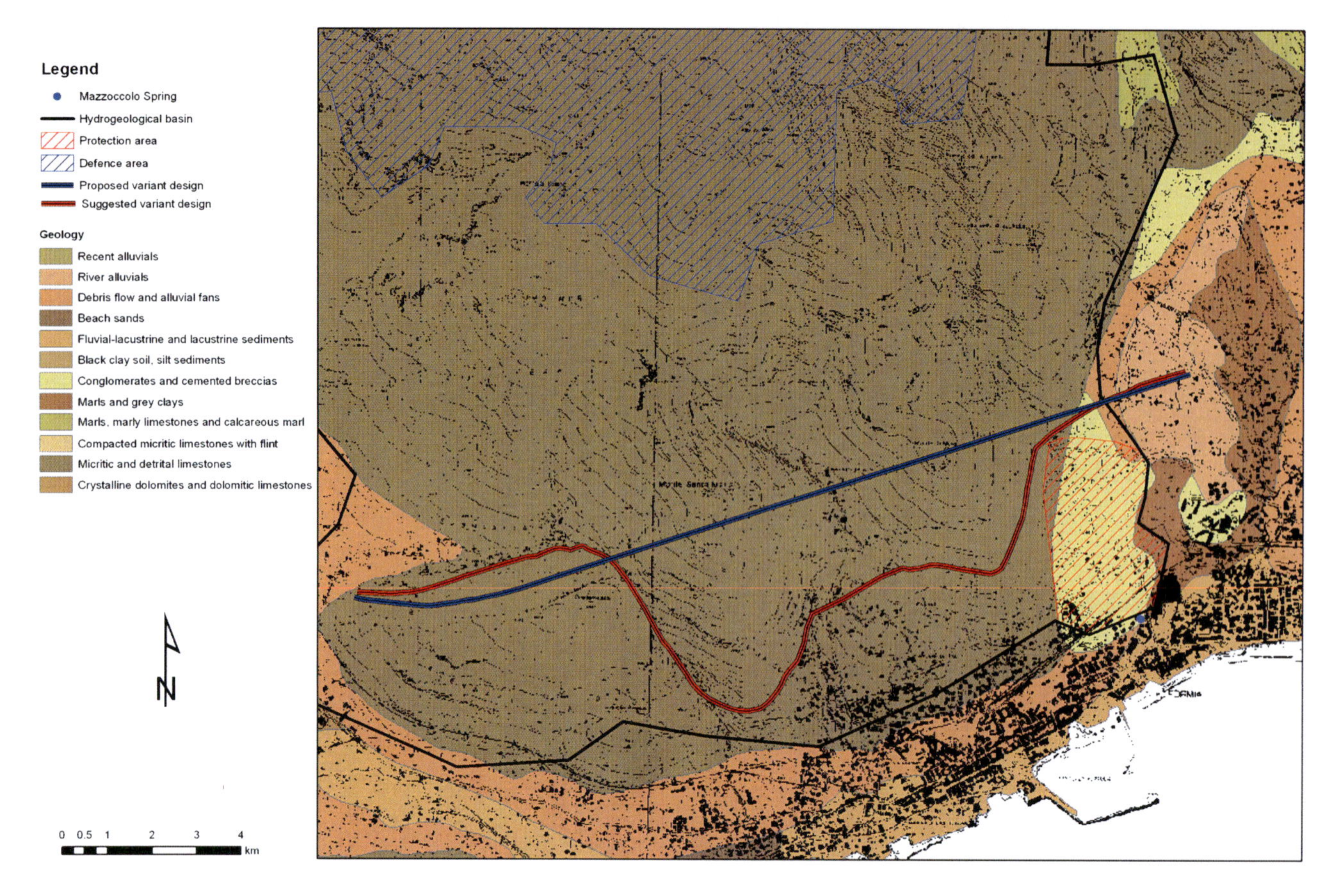

Fig. 136.2 Proposed variant design

According to the current knowledge, it is not possible to evaluate if the proposed design of Mola Mountain tunnel, located perpendicularly at the flow lines of the karst aquifer feeding the Mazzoccolo spring, will be held in the saturated or undersaturated zones of the aquifer. If the tunnel excavation passes through unsaturated zone of the aquifer, in any case it will meet tectonically fractured karstic zones draining the infiltrated water to the saturated zone. In the case of the saturated zone, the elevation difference between the base of the tunnel and Mazzoccolo spring may cause a simultaneous rise of the groundwater table of some dozen of meters during excavation (Civita 1998). These processes will have several negative impacts on the groundwater systems both quantitatively and qualitatively. Anyway previously studies, carried on in the late 70s (Civita 1998), have not clarify the hydrogeological system which was a useful tool for the evaluation of interactions between motorway tunnel excavation and vulnerable aquifer. Thus, in this study the geochemical and isotopic characterization of groundwater give us an important information about the vulnerability of this karst aquifer and hydrogeological regime of the area which contributes the environmental assessment strategies. On the base of these results, Mazzoccolo spring is partially fed by short residence time groundwater, with a short self-purification of rainwater effect due to the interception of karst cavities. At the same time, the long residence times lead us to consider the high vulnerability of Mazzoccolo spring without giving information about the purging of previous recharge. In conclusion, before to start the realization of the proposed project, it is necessary to perform some artificial tracer tests and a careful geomorphological investigation to identify the existing karst cavities with a detailed geostructural modeling. Consequently, we suggest a different motorway design, involving a superficial route,

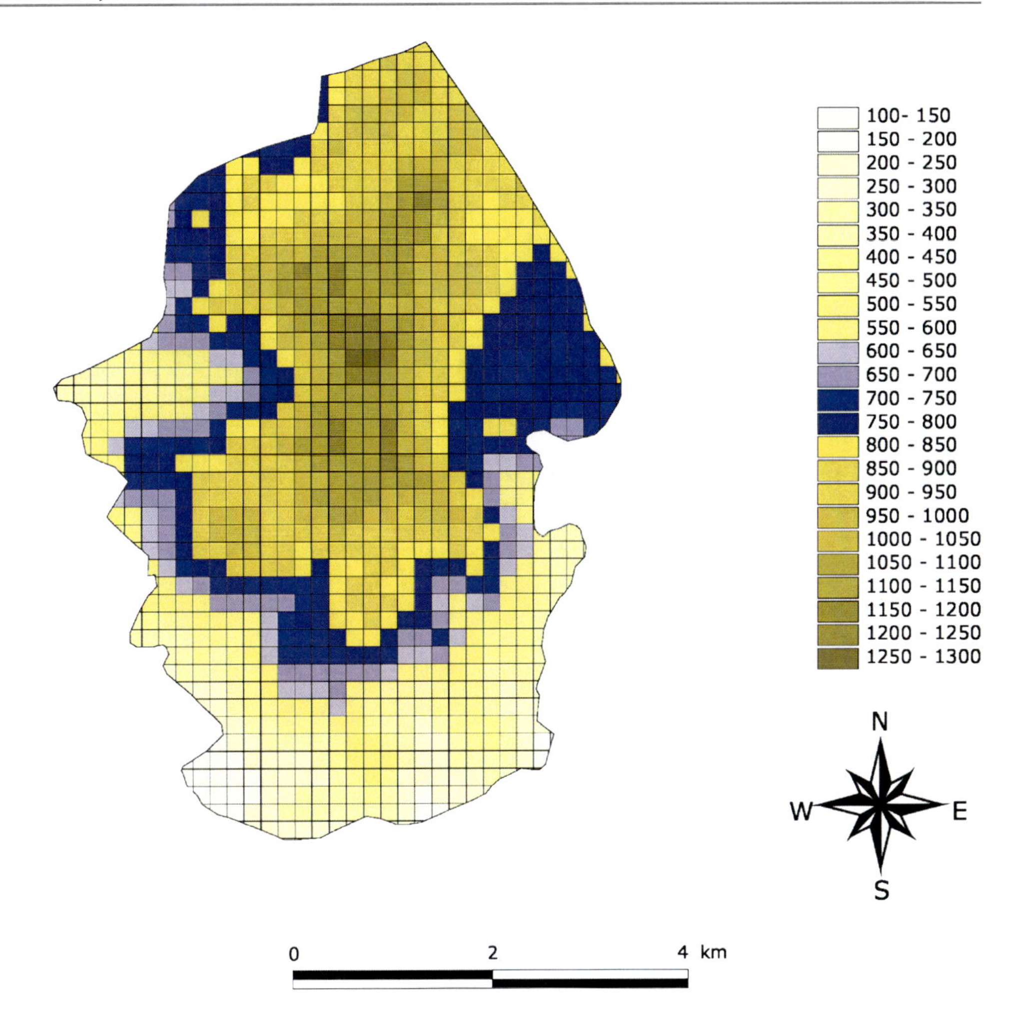

Fig. 136.3 Infiltration areas of Mazzoccolo spring

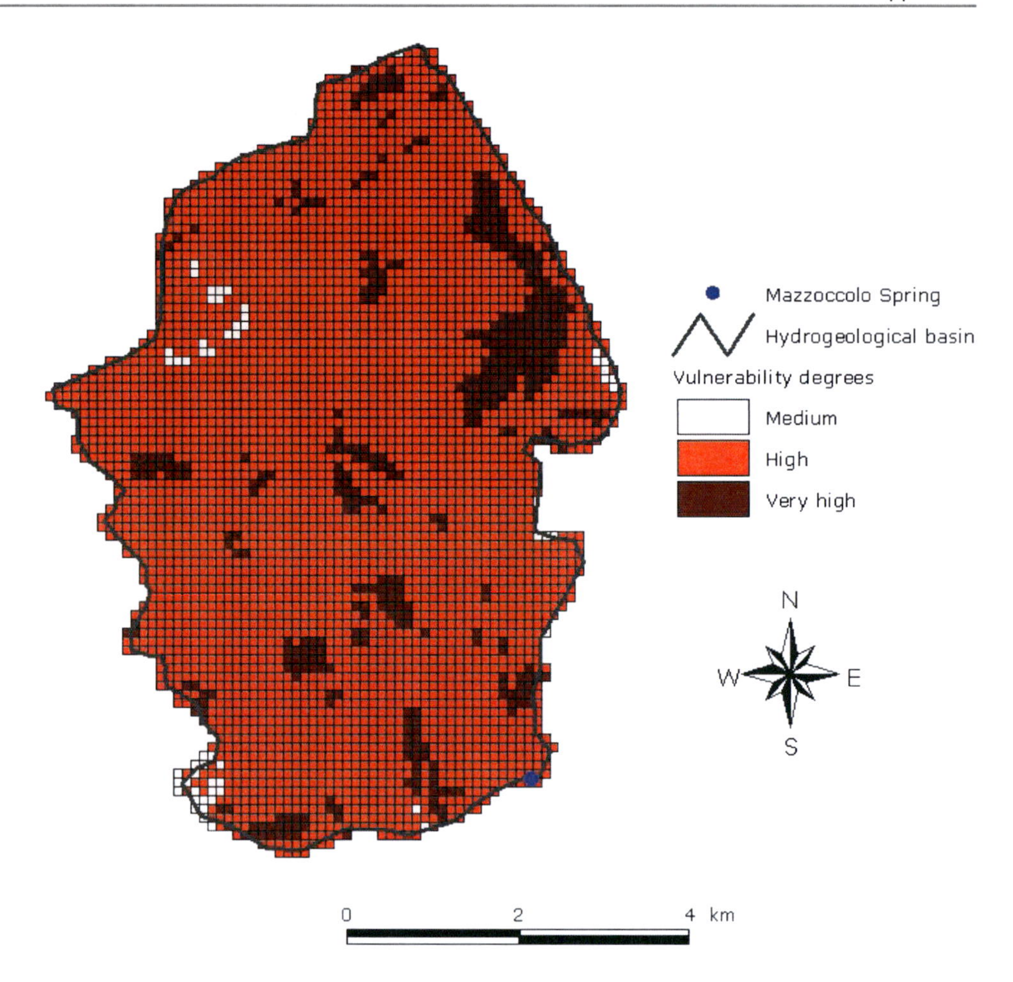

Fig. 136.4 Vulnerability map of the karst aquifer feeding the Mazzoccolo spring

even longer than the previous one, avoiding the construction of Mola Maountain tunnel and involving high self-purification capacity soils.

References

Casa R, Rossi M, Sappa G, Trotta A (2008) Assessing CropWater demand by remote sensing and GIS for the Pontina Plain, Central Italy, water resources Management, September 2008

Civita M. (1998) Parere tecnico sui possibili impatti della variante pedemontana alla SS 7 (Appia) sull'acquifero che alimenta le sorgenti Mazzoccolo in Formia (Latina). Comune di Formia (Provincia di Latina)

Di Nocera S. (1983) Studio idrogeologico Sorgente Mazzoccolo. Relazione geologica. Amministrazione comunale di Formia

Ialongo N (1983) Studio idrogeologico Sorgente Mazzoccolo. Relazione idrogeologica. Amministrazione comunale di Formia

Manfredini M. (1984) Studio sulle caratteristiche idrogeologiche e sui problemi di protezione della Sorgente Mazzoccolo. Comune di Formia

Sappa G, Barbieri M, Ergul S, Ferranti F (2012) Hydrogeological conceptual model of groundwater from Carbonate aquifers using environmental isotopes (^{18}O, ^{2}H) and chemical tracers: A case study in Southern Latium Region Central Italy. J Water Res Prot 4:695–716

Part XIV

Properties and Behaviour of Weak and Complex Rock Masses in Major Engineering Projects

Convener Dr. Vassilis Marinos—*Co-conveners* George Stoumpos, Petros Fortsakis

Numerical analysis and computational methods in geotechnical engineering are fields where great progress has been achieved. However, in the case of weak and complex rock masses, the results still involve uncertainties due to the difficulties in the reliable estimation of intact rock properties and the realistic quantification of rock mass properties and behaviour. The special features of these rock masses regarding both their structure and lithology impose a more specialized research. The weak rock masses could be cases with very low intact rock properties, highly tectonized or/and weathered rock masses, rock masses with members of low strength and/or inherent heterogeneity. This session may contain papers on weak and complex rock masses, regarding in situ and laboratory testing, characterization, geotechnical classification, design properties, behaviour, support measures and performance of the construction method adopted in the design approach according to the engineering project.

Rock Mass Quality Rating (RMQR) System and Its Application to the Estimation of Geomechanical Characteristics of Rock Masses

137

Ömer Aydan, Resat Ulusay, and N. Tokashiki

Abstract

In this study, a new rock mass quality system designated as Rock Mass Quality Rating (RMQR) is explained and its application to the estimation of geomechanical properties of rock masses is described. First, a brief outline of the input parameters of RMQR and their ratings are given. Then the unified formula proposed by the first author is used for estimating the rock mass properties as a function of intrinsic properties of intact rock material and they are compared with the results of the in situ tests carried out in Japan and those estimated from some other empirical relationships developed by some researchers.

Keywords

RMQR • Rock mass • Intact rock • Geomechanical properties • Empirical relation

137.1 Introduction

The qualitative description of rock masses by means of classification systems and subsequent correlation to establish engineering quantities has become one of the most challenging topics in rock engineering. However, many available rock classification systems have some repetitions such as RQD and discontinuity spacing resulting in essence doubles the influence of the spacing of discontinuities on the final rating. In addition, although the effect of water particularly on water-sensitive rocks plays an important role in decrease of their geo-mechanical properties, this effect is not adequately considered in the existing rock mass classification systems. By considering the scale effect for rock masses, laboratory testing on rock masses is not always easy and is very cumbersome. For this reason, in situ tests are generally preferred. But in situ tests are directly time consuming, expensive and quite cumbersome to conduct. Therefore, the recent tendency is to obtain rock mass properties from the utilization of properties of intact rock and rock classification indexes, which have some drawbacks. In this study, a new rock mass rating system designated as Rock Mass Quality Rating (RMQR) proposed recently by the authors (Aydan et al. 2013) is explained. Geomechanical properties of rock masses are estimated from the utilization of properties of intact rock and RMQR as one of applications of this system.

Ö. Aydan
Tokai University Shizuoka, Shizuoka, Japan
e-mail: aydan@scc.u-tokai.ac.jp

R. Ulusay (✉)
Department of Geological Engineering, Hacettepe University, Ankara, Turkey
e-mail: resat@hacettepe.edu.tr

N. Tokashiki
Department of Civil Engineering, Ryukyu University, Okinawa, Japan

137.2 Rock Mass Quality Rating (RMQR)

The parameters associated with discontinuities could be the discontinuity set number (DSN), discontinuity spacing (DS; spacing of dominant discontinuity set, mainly thoroughgoing discontinuity set) and discontinuity condition (DC). The weathering of rocks causes the weakening of bonds and decomposition of constituting minerals into clayey materials. The alteration process may act on rock mass in a positive or negative way. As the intact rock is one of the important

G. Lollino et al. (eds.), *Engineering Geology for Society and Territory – Volume 6*,
DOI: 10.1007/978-3-319-09060-3_137, © Springer International Publishing Switzerland 2015

elements influencing the mechanical response of rock masses, weathering and/or the negative action of hydrothermal alteration may be accounted as the degradation degree (DD) of intact rock. There are also cases, that some rocks may absorb groundwater electrically or chemically, resulting in the drastic reduction of material properties and/or swelling. In addition to seepage condition of groundwater (GWSC), the water absorption characteristics of rocks (GWAC) should also be taken into account.

RMQR has six basic parameters, which provides rating of each parameter, and ranges between 0 and 100 (Table 137.1). If detailed surveys on the conditions of discontinuities are carried out, a more detailed rating is necessary for characterization of rock discontinuities. For the evaluation of discontinuity condition from detailed surveys, Table 137.2 is recommended together with roughness concept of surface profiles by ISRM (2007). RMQR could be related to the well-known two rock mass rating systems, RMR (Bieniawski 1989) and Q-system (Barton et al. 1974), through some relations given in Fig. 137.1. It should be noted that the value of RMR is generally less than the value of RMQR. One reason may be such that the RMQR includes the effect of water absorption, which is not counted in RMR. The other reason may be the exclusion of intact rock strength in RMQR.

137.3 Relation Between Rock Mass Properties and RMQR

Aydan et al. (2013) provided relations for six different mechanical properties of rock mass using the proposed relation by Aydan and Kawamoto (2000) using RMQR as an independent parameter. It is given in the following form for any mechanical properties of rock mass in terms of those of the intact rock.

$$* R_{DC} = R_{DCA} + R_{DCI} + R_{DCR}$$

$$\alpha = \alpha_0 - (\alpha_0 - \alpha_{100}) \frac{RMQR}{RMQR + \beta(100 - RMQR)} \tag{137.1}$$

where α_0 and α_{100} are the values of the function at RMQR = 0 and RMQR = 100 of property α, β is a constant to be determined by using a minimization procedure for experimental values of given physical or mechanical properties. When a representative value of RMQR is determined for a given site, geomechanical properties of rock mass can be obtained using Eq. (137.1) together with the values of constants given in Table 137.3 and the values of intact rock for a desired property.

Table 137.1 Classification parameters and their ratings for rock mass quality rating (RMQR)[a]

Degradation degree (DD)	Fresh	Stained	Slight degradation	Moderate degradation	Heavy degradation	Decomposed
Rating (R_{DD})	15	12	9	6	3	1 − 0
Discontinuity set number (DSN)	None (solid or massive)	One set plus random	Two sets plus random	Three sets plus random	Four sets plus random	Crushed or shattered
Rating (R_{DSN})	20	16	12	8	4	1 − 0
Discontinuity spacing (DS) or RQD (%)	None or DS ≥ 24 m	24 > DS ≥ 6 m	6 > DS ≥ 1.2 m	1.2 > DS ≥ 0.3 m	0.3 > DS ≥ 0.07 m	0.07 m > DS
	100			100 > RQD ≥ 75	75 > RQD ≥ 35	35 > RQD
Rating (R_{DS})	20	16	12	8	4	1 − 0
Discontinuity condition (DC)	None	Healed or intermittent	Rough	Relatively smooth and tight	Slickensi-ded with thin infill or separation (t < 5 mm)	Thick fill or separation (t > 10 mm)
Rating (R_{DC})	30	26	22	15	7	1
Groundwater seepage condition (GWSC)	Dry	Damp	Wet	Dripping	Flowing	Gushing
Rating (R_{GWSC})	9	7	5	3	1	0
Groundwater absorption condition (GWAC)	Non-absorptive	Capillarity or electrically absorptive	Slightly absorptive	Moderately absorptive	Highly absorptive	Extremely absorptive
Rating (R_{GWAC})	6	5	4	3	2	1-0

[a] $RMQR = R_{DD} + R_{DSN} + R_{DS} + R_{DC} + R_{GWSC} + R_{GWAC}$

Table 137.2 Ratings for sub-parameters of discontinuity condition excluding "None" and "Healed or intermittent" classes

Aperture or separation		None or very tight, <0.1 mm	0.1–0.25 mm		0.25–0.5 mm		0.5–2.5 mm		2.5–10 mm		>10 mm
Rating (R_{DCA})*		6	5		4		3		2		1 – 0
Infilling		None	Surface staining only		Thin coating <1 mm		Thin filling 1 < t<10 mm		Thick filling 10 < t < 60 mm		Very thick filling or shear zones 60 mm < t
Rating (R_{DCI})*		6	5		4		3		2		1-0
Rough-ness	Descriptive	Very rough	Rough		Smooth undulat-ing		Smooth planar		Slicken-sided		Shear band/ zone
	ISRM (2007) profile No.	10	9	8	7	6	5	4	3	2	1 – 0
Rating (R_{DCR})*		10	9	8	7	6	5	4	3	2	1 – 0

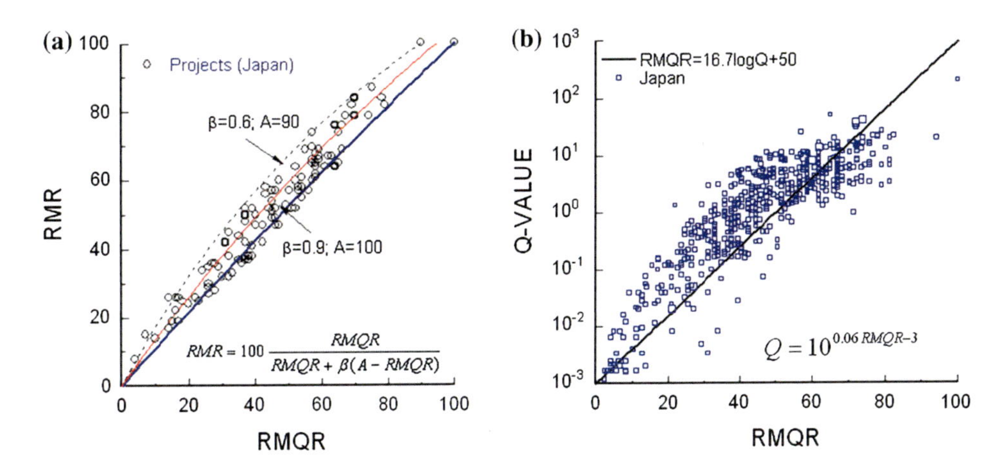

Fig. 137.1 The relations between **a** RMQR and RMR, and **b** RMQR and Q

Table 137.3 Values of α_0, α_{100} and $\beta$for various properties of rock mass

Property (α)	α_0	α_{100}	β
Deformation modulus	0.0	1.0	6
Poisson's ratio	2.5	1.0	0.3
Uniaxial compressive strength	0.0	1.0	6
Tensile strength	0.0	1.0	6
Cohesion	0.0	1.0	6
Friction angle	0.3	1.0	1.0

The empirical relations for normalized properties presented in the previous section are compared with the experimental results from in situ tests carried out at various large projects in Japan (Fig. 137.2). The experimental results on normalized elastic modulus of rock mass are closely represented by Eq. (137.1) together the values given in Table 137.3 and they are clustered around the curve with the value of coefficientβas 6.

Figure 137.3 compares the experimental results on various rock masses ranging from igneous rocks to sedimentary rocks with empirical relations for normalized uniaxial compressive strength (UCS) and tensile strength of rock masses by those of intact rock. The UCS of rock masses plotted in this figure are mostly obtained using rock shear test together with Mohr-Coulomb failure criterion. The experimental results generally confirm the empirical relation given in Eq. (137.1).

In literature, there is almost no in situ experimental procedure or experimental results for the tensile strength of rock mass to the knowledge of the authors. The authors (Aydan et al 2013) utilized back-analysed data on the stable and unstable (failed) cliffs using a theory based on the cantilever theory and fitted the inferred tensile strength of the rock mass normalized by that of intact rock using Eq. (137.1).

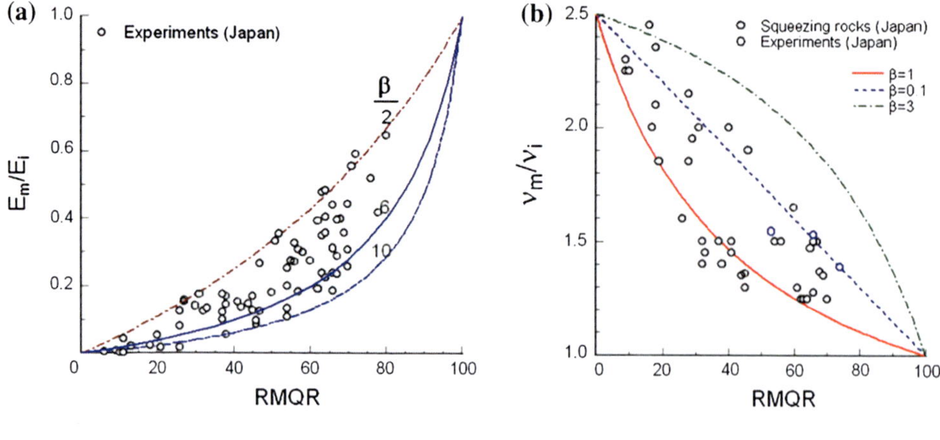

Fig. 137.2 Comparison of experimental data for **a** deformation modulus and **b** Poisson's ratio of rock mass with Eq. 137.1 with the parameters given in Table 137.3

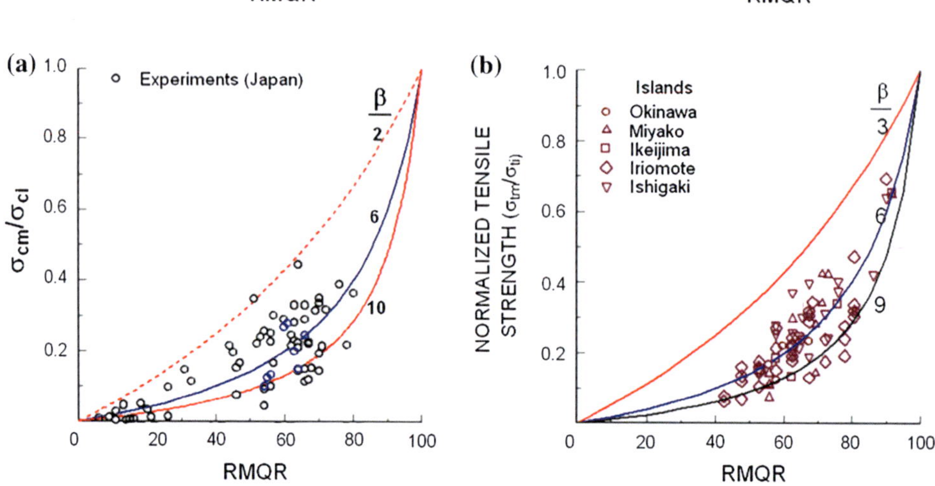

Fig. 137.3 Comparison of experimental data for **a** uniaxial compression and **b** tensile strengths of rock masses with Eq. (137.1) with the parameters in Table 137.3

The results are plotted in Fig. 137.2b by varying the value of empirical constant β between 5 and 7. It is found that the value of empirical constantβcould be designated as 6 in view of inferred tensile strength of rock mass.

137.4 Conclusions

In this study, a new rock mass rating system (RMQR) is described to assess the physical state of rock mass and used for the evaluation of engineering properties of rock masses utilising the unified empirical relation proposed by Aydan and Kawamoto (2000) and compared with actual measurements on rock masses in various sites of Japan. The comparison of the empirical unified formula together with the values of constants was found to be quite consistent with in situ experimental results for data compiled from Japan. The comparisons between experimental results and estimations indicate that the presented relations are quite promising. Therefore, the empirical relations used in this study should be quite useful tool for engineers involved in projects related rock masses. Furthermore the authors strongly suggest that the relations for normalized properties should be used for evaluating properties of rock mass using properties of intact rock and its rock mass rating.

References

Aydan Ö, Kawamoto T (2000) Assessing mechanical properties of rock masses by RMR rock classification method. In: Proceedings of GeoEng 2000 symposium, Sydney, Paper No. OA0926 (on CD)

Aydan Ö, Ulusay R, Tokashiki N (2013) A new rock mass quality rating system: rock mass quality rating (RMQR) and its application to the estimation of geomechanical characteristics of rock masses. Rock Mechanics and Rock Engineering (doi 10.1007/s00603-013-0462-z)

Barton N, Lien R, Lunde I (1974) Engineering classification of rock masses for the design of tunnel supports. Rock Mech 6(4):189–239

Bieniawski ZT (1989) Engineering rock mass classifications. Wiley, New York

ISRM (2007) The complete ISRM suggested methods for rock characterization, testing and monitoring: 1974–2006. Ulusay R, Hudson JA, (eds.), Suggested methods prepared by the ISRM commission on testing methods, compilation arranged by the ISRM Turkish National Group, Kozan Ofset, Ankara

Investigation and Treatment of Problematic Foundations for Storage Dams: Some Experience

138

Wynfrith Riemer and rer nat

Abstract

Certain geological conditions require specific efforts in exploration, sampling and testing to obtain representative material parameters. Whereas in many geological environments already routine classification systems offer a practical approach to initial site characterization, other geological environments cannot be treated in a standardized way. The paper selects two geological regimes to illustrate potential pitfalls affecting the comprehensive identification and characterization of a dam site: sites with important development of residual soils, sites in possibly karstic environment. Case histories exemplify concerns related to these environments.

Keywords

Dams • Residual soils • Karst • Rock mass classification

138.1 Introduction

Large dams count among the most impressive civil engineering structures but the success of their implementation and operation decisively depends on non-engineered elements: the rock or soil in the foundation of the dam and the materials forming the reservoir and its rims. Composition and properties of the respective geological materials are the product of a complex history. Clarifying the geological history allows to adjust the dam and its appurtenant structures to the foundation and to treat the foundation and basin so they will adequately respond to the stresses generated by the project. Proceeding from basic geological concepts as facies, lithology, tectonic history specific considerations and concerns can be formulated as guidance for the engineering geological, hydrogeological and geotechnical investigation and design of the project (cf. ICOLD 2009; Fookes et al. 2000). In the experience of the author, volcanic environments, residual soils and soluble, karstic formations required particularly close attention. Because of the limited space, the paper will briefly mention residual soils and will deal more in detail with karstic conditions.

138.2 Residual Soils

Only two aspects are mentioned in relation to residual soils:

(1) Mechanical and hydrogeological properties are largely controlled by texture and are not reliably assessed by soil classification and tests on disturbed samples (Fookes 1997; ICOLD 2005). Testing is preferably done in situ and, for this purpose, the excavation of pits and shafts is particularly useful for geological and geotechnical investigation of residual soils.

(2) Lateral variation in soil profile and thickness can be very significant. The case of the Amaluza dam (Ecuador) is mentioned. Up to the crest level of the dam the granodiorite provided an excellent foundation but silty-sandy residual soils, locally with large core stones, caused serious stability problems for the cut above the crest and upstream of the dam near the power intake. Exploration had concentrated on the area of the footprint of the dam and the immediately adjacent problems were not anticipated.

W. Riemer (✉)
Technische Universität Berlin, Berlin, Germany
e-mail: wynriemer@t-online.de

r. nat
Consultant, Trier, Germany

G. Lollino et al. (eds.), *Engineering Geology for Society and Territory – Volume 6*,
DOI: 10.1007/978-3-319-09060-3_138, © Springer International Publishing Switzerland 2015

138.3 Soluble Formations

138.3.1 General

Gorges formed in carbonate rocks offered morphologically attractive conditions for the construction of some of the highest existing dams: Vajont, Ingoury, Berke, Karun I and III, Bakhtyari, Cajon to name a few of them. The presence of karst was accepted and treated successfully.

Karst develops in soluble rocks like carbonates, sulfates and halites with engineering interest focusing on carbonates because of their wide distribution. The category also includes conglomerates with high carbonate content (e.g. the Tertiary conglomerates of Meteora and at the Kremasta dam site in Greece and along the Kopriicay in Turkey). Dolomite, in some regions acting as aquiclude (Milanovic 1981, 2000; Breznik 1998) displayed the most prominent karst development at the Ataturk site. Solution of dolomite tends to produce sandy, non- cohesive residue, easily eroded, potentially liquefiable and difficult to grout.

138.3.2 Specific Concerns

Notorious problems with these rocks are directly and indirectly associated with karst phenomena and cover a wide range:

- Loss of carbonate from the rock reduces shear strength and raises deformability (e.g. the marly limestones in the foundation of Khao Laem dam)
- Collapse of karstic voids under static and hydraulic loads. Prominent cases are the Keban dam in Turkey, the Lar dam in Iran. In Germany, the bottom of a tailings reservoir collapsed and all tailings, fortunately innocuous material, disappeared in the karstic voids of the Devonian limestone. The Perdikkas reservoir in Greece was abandoned when sinkholes formed in the valley floor and absorbed the run-off.
- Problems due to rapid dissolution. The effect becomes most critical where sulfates are present. At the Mosul dam in Irak the seepage water daily leached > 40 tons of gypsum from the foundation (Guzina et al. 1991). The headrace tunnel of Pueblo Viejo project in Guatemala failed following rapid dissolution of anhydrite.
- Erosion/suffusion of karst residues from dam foundations has in several cases rapidly increased seepage losses to unacceptable magnitudes. Remedial treatment became necessary at Mujib in Jordania and Aoulouz in Morocco (Mekboul et al. 1999). Progressive erosion of the karstic foundation eventually rendered the Henne dam in Germany useless and it was replaced by a new dam shifted upstream.
- Seepage losses from the reservoir can reach magnitudes affecting the viability of operation of the project. Before remedial treatment, the Keban reservoir lost up to 26 m^3/s and losses from the Lar reservoir (Iran) peak at about 16 m^3/s. Underseepage of about 12 m^3/s at the Ataturk dam was considered economically acceptable.
- Uplift resulting from underseepage and artesian conditions in karstic horizons endanger the stability of the foundation of the dam, of appurtenant structures and valley flanks downstream of the dam. Therefore, common practice provides extensive drainage systems at dam sites in karst. At Karun I in Iran, wells relieve the artesian head of a dolomite horizon at the toe of the dam and several galleries control the groundwater levels in the abutments downstream of the dam. At the Ataturk site, an internally drained tub assures the uplift stability at the powerhouse (Riemer and Andrey 1991) and when reservoir filling undesirably raised the pressure under the spillway chute, an additional drainage gallery was driven. At the Kremasta dam, drain galleries capture the seepage through karstic rock to keep the uplift in the right valley flank at a level which grants slope stability.

138.3.3 Investigations

The presence of karst conditions can be elusive and in many projects the importance of karst proved difficult to establish or was entirely overlooked. At the Perdikka site, the samples recovered from the boreholes had not given a distinction between clastic Neozoic sediments and karst residues of metamorphic limestone. At Ataturk, an exploratory adit eventually proved the karst development which a number of well performed core drillings had not clearly detected. At the Henne dam site, the karst hazard associated with the limestone layers alternating with calcareous keratophyr tuffs had not been recognized. Although the technology of geological exploration has improved with new developments in geophysics, in core sampling and borehole logging, the site investigations in formations with potentially karstic rocks always require additional efforts. Some aspects to be mentioned in this context are:

- Karst may have developed in past geological times in an environment significantly differing from the actual setting. Mantled or covered karst can be hidden from observation under more modern deposits. Paleokarst can exist in places where actual hydrogeological conditions would not imply solution effects.
- Karst groundwater gradients are flat (frequently on the order of %o) and levels can be very deep. At the Polifiton reservoir in Greece, piezometers had to be drilled to 800 m depth to reach the water table. At such depth,

accurate measurement of the water level becomes difficult and, with the flat gradients, readings can be misleading if deviation of the borehole is disregarded.

- Water level fluctuations in karst can be very rapid, propagating as waves in the permeable rock with low storage coefficient (cf. Yevyevich 1980). Thus, if readings are not taken at short intervals, hydrogeological transients are easily missed.
- Hydrogeological non-homogeneity of the karst aquifer can be misleading if insufficient data on the configuration of the water table are available.
- The permeability of karstic rocks is difficult to determine. Breznik (1998) recommends to apply high pressures in Lugeon tests to obtain more representative results. Even so, normal statistics of point permeability tests tend to be misleading. At Ataturk, some 1500 Lugeon tests in the initial series of grout holes gave a median of 7 and a mean of 14.5 Lugeon Units (Fig. 138.la). This would normally correlate to a permeability on the order of 10^{-6} m/s. Permeability estimated from pumping tests fell into the order of 10^{-4} m/s. Model simulations, calibrated to seepage observed after reservoir filling, confirmed the high range of permeability.
- Hydrogeological techniques, involving environmental and artificial tracers, usefully complement the more conventional explorations in karst environment.

138.3.4 Engineering Geological Assessment of Karst Conditions

Whereas in other types of rock masses the estimates of parameters related to mechanical strength and deformation essentially provide the basis for design, and these can in a first step be approximated by classification systems and empirical correlations, a substantially wider range of parameters has to be handled in soluble rocks. Hydrogeological considerations, chemical stability and resistance to erosion/suffusion may become decisive and have to be evaluated in conjunction with the mechanical properties of the rock.

Milanovic (1981) summarizes classification systems that have been proposed for karst formations. But these systems concentrate on morphological features and regional geological aspects, only indirectly related to engineering geology. As Milanovic points out, there are also restrictions to specific geological environments. The experience with many projects suggests that implications of the presence of soluble rocks for dams and reservoirs can be assessed considering mainly four parameters: (1) the mechanical strength of the rock mass, described e.g. by Ocm (Hoek 2005), (2) the size of voids created by solution which can range from mm to tens of m, (3) the large scale karst porosity, open or filled, typically on the order of a few percent, (4) the permeability, e.g. determined by Lugeon tests. Figure 138.2 illustrates the suggested approach. At project A, the rock mass has a significant proportion of large karstic voids, according to the permeability mainly open, and of a size where collapse is likely to occur. The caves will have to be stabilized by backfill. At project B, the hazard of collapse is marginal, the low permeability indicates filled voids, prone to complicate grouting. At projects C and D mechanical stability is adequate, medium permeability in conjunction with the low proportion of voids will facilitate grouting.

138.3.5 Design Considerations and Options for Treatment

- Separating karstic domain from Reservoir. A lateral valley with karstic limestone was separated from the Bigge reservoir (Heitfeld 1991).
- Blanketing. A bituminous facing seals a karst outcrop projecting into the Mornos reservoir (Heitfeld 1991) and

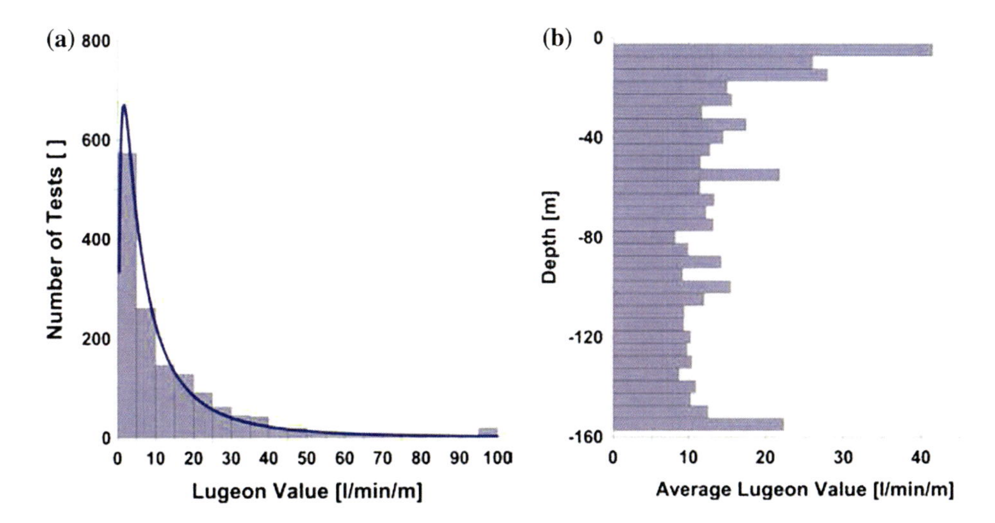

Fig. 138.1 Summary of Lugeon tests in primary holes of Ataturk grout curtain. **a** Histogram and fitted log-normal distribution. **b** Bar diagram Lugeon values against depth

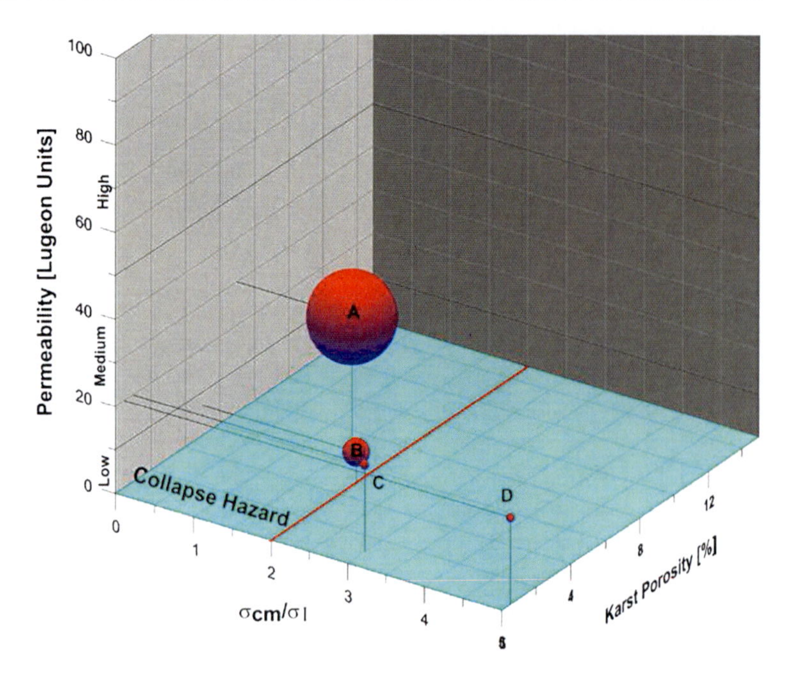

Fig. 138.2 Proposed scheme for classification of karst rocks for dam foundations. The size of the bubbles represent the dimensions of karst voids

a concrete blanket covers karstic limestone at the Planinsko Polje (Breznik 1998). In other cases blanketing with membranes or low permeability soils has been applied.

- Grout curtains and diaphragms have been constructed on many projects. But, whereas in other rocks the vertical and lateral extension of the curtain rarely exceed the height of the dam, some curtains in karst reach a multiple of the height of the dam into the underground and stretch laterally far into the abutments and reservoir rims. In the Dinaric karst, active solution is frequently found to 200 m below valley floor and paleo-karst may reach more than 1,000 m into the rock (Milanovic 1981, 2000; Breznik 1998). At the Ataturk site, the permeability does not decrease significantly with depth (see Fig. 138.lb) and the curtain was locally taken to 300 m below valley floor. Data collected for 30–40 dams on karst foundation show a typical depth of about 1.5 times the height of the dam, mean density of grout and test holes 0.6 m per square meter of the curtain and average grout absorption of 200 kg/m^2 (see Fig. 138.3).

Treatment of filled karst tends to be particularly demanding. At the Berke dam, compaction grouting at 60 bar pressure proved an economical alternative to construction of a diaphragm (Basar et al. 1999), whereas at many other dams diaphragms had to be constructed.

Even if initial treatment of karst foundations succeeded, the risk of deterioration remains (e.g. Aoulouz dam, Mekboul et al. 1999; Wolf Creek USSD 2013). Therefore, a comprehensive monitoring system and access for maintenance treatment should be provided (e.g. Riemer et al. 1995).

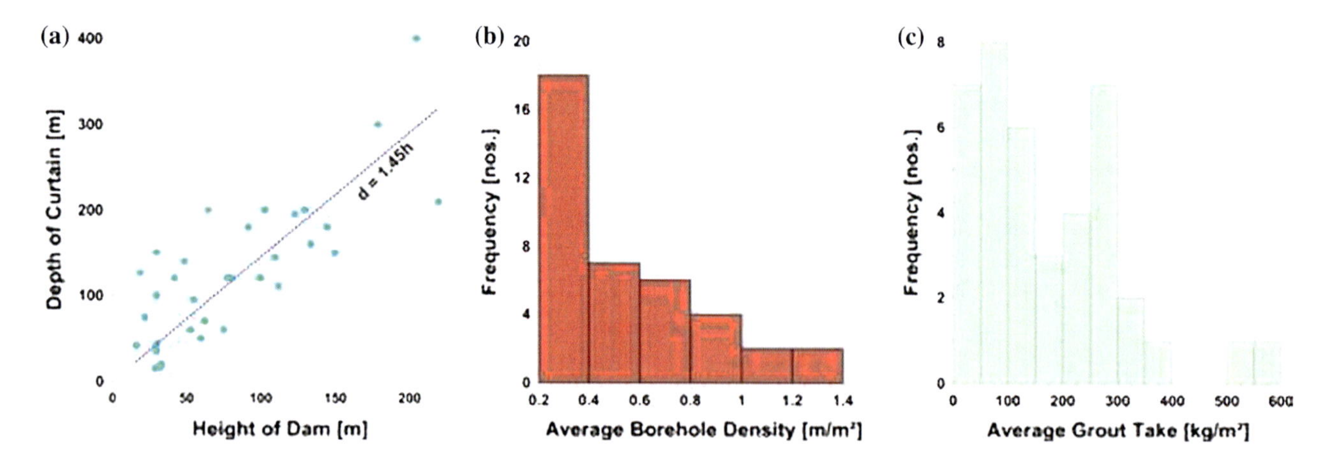

Fig. 138.3 Statistics of grout curtains in karst foundations of dams. **a** Depth of curtain versus height of dam. **b** Density of boreholes per square meter of curtain. **c** Grout absorption per square meter of curtain

138.3.6 Cost

Little information has been published on financial and economical aspects of karst treatment. But these aspects can attain decisive importance for the viability of a project. In this regard the Khao Laem project, Thailand, can be mentioned. The cost of treating dam foundation and reservoir rims was about three times higher than the construction cost of the 90 m high CFRD.

Acknowledgment The author gratefully acknowledges the discussions and collaboration with many colleagues who have assisted in the projects quoted in the text.

References

Basar M, Altug S, Vauloup L (1999) Berke arch dam and HEPP Turkey: foundation treatment of karstic limestone in suspended curtain. In: Turfan M (ed) ICOLD 67th Annual Meeting, Dam Foundations Problems and Solutions, pp 233–250

Breznik M (1998) Storage reservoirs and deep wells in karst regions. Balkema, 251 p

Fookes PG (1997) Tropical residual soils: a geological society engineering group working party revised report, 184 p

Fookes PG, Haynes FJ, Hutchinson JN (2000) Total geological history: a model approach to the anticipation, observation and understanding of site conditions. GeoEng 1:370–460

Guzina BJ, Saric M, Petrovic N (1991) Seepage and dissolution at foundations of a dam during the first impounding of a reservoir. In: Proceedings of 17th international congress of ICOLD, Q.66 R78

Heitfeld KH (1991) Talsperren. Lehrbuch der Hydrogeologie, Band 5. 468 p

Hoek E (2005) Uniaxial compressive strength versus Global strength in the Hoek-Brown criterion. RocScience web site

ICOLD (2005) Dam foundations, geologic considerations, investigation methods, treatment, monitoring. Bulletin 129

ICOLD (2009) Tropical residual soils as dam foundations and fill material. Bulletin 151

Mekboul M, Chraibi AF, Saidsallam M (1999) Aoulouz dam additional treatment in the grout curtain after reservoir filling. In: Turfan M (ed) ICOLD 67th annual meeting, Dam foundations problems and solutions, pp 117–132

Milanovic PT (1981) Karst hydrogeology. 434 p

Milanovic PT (2000) Geological engineering in karst. 347 p

Riemer W, Andrey JD (1991) Baugrundbehandlung am Ataturk Damm. Ber. 8. Nationale Tagung fur Ingenieurgeologie, Berlin, 1991, pp 167–174

Riemer W, Gavard M, Turfan M (1995) Ataturk dam—hydrogeological and hydrochemical monitoring of grout curtain in karstic rock. Verification of geotechnical grouting, ASCE geotechnical special publication no. 57, pp 116–126

USSD (2013) Dams of the United States. 205 p

Yevyevich V (1980) Investigation of karst hydrogeology, hydrology and water resources in southern Turkey. In: Proceedings of LRST international symposium on Karst hydrogeology, pp 55–117

Dissolution Influences on Gypsum Rock Under Short and Long-term Loading: Implications for Dams

139

Nihad B. Salih, Philip E.F. Collins, and Stephen Kershaw

Abstract

Dissolution of soluble substrates such as gypsum presents a major hazard to dams in many parts of the world. This research simulates hypothesised conditions beneath the Mosul Dam, northwest Iraq, where collapse of a karstic system associated with continuous fresh water supply from its reservoir is a recognised problem. Gypsum rocks from northern Iraq and similar rocks from Bantycock gypsum mine, UK, were analysed for short-term mechanical response following immersion (5–50 weeks) and long-term loading during immersion (maximum 50 weeks). New experimental devices were developed from a conventional oedometer. Cylinder samples provided a proxy for massive gypsum strata. Samples were permanently submerged at atmospheric water pressure, with groundwater recharge, flow and dissolution simulated by regular changes of water. Stress on each sample was progressively increased to a maximum of 2,688 kPa. Small increases in strain were recorded by the end of each test but no failures occurred within 60 days. However, notable failure due to atmospheric water pressure and axial stress occurred over long time periods. Visible physical changes included a decrease in sample mass and volume. Similar change was recorded in ultrasonic velocities. These indicate that gypsum collapse risk beneath dams requires prolonged exposure to dissolution. The modified device performed well and was robust, and demonstrates that such a modification can provide a simple low cost system for conducting laboratory creep tests on weak rocks.

Keywords

Gypsum rock • Dissolution • Short-term loading • Long-term loading • Dams

139.1 Introduction

Gypsum rock underly more than 20 % of the earth surface (Johnson 2005) with seven million km^2 underlain by highly soluble gypsum ($CaSO_4.2H_2O$) bearing rocks in a large number of countries (Cooper 2006). Its solubility in pure water is 2.531 g/l at 20 °C, around 140 times lower than halite and 4 times larger than $CaCO_3$ (Johnson 2006).

N.B. Salih (✉)
University of Sulaimani, Sulaimani, Iraq
e-mail: nihad_baban@yahoo.com

P.E.F. Collins · S. Kershaw
Brunel University, London, UK

Gypsum rock is common in Iraq, mainly in the Mid-Miocene Fatha (Lower Fars) Formation (Jassim and Goff 2006) and is associated with significant geotechnical problems. This is particularly the case in sites with high hydraulic gradients such as near dams (Salih 2013) e.g. the Mosul Dam, which is underlain by thick gypsum beds which are severely affected by dissolution and karstification, including general subsidence and sinkholes.

While cavity formation due to dissolution is known, mechanical change leading to creep and brittle failure has received little attention. This is significant as the formation of cavities will transfer load to surrounding, potentially weakened gypsum.

Gypsum resistance to compression is not high. Its average uniaxial compressive strength (UCS) is around 13.73 MPa, but it demonstrates considerable variability between 9.41 to

G. Lollino et al. (eds.), *Engineering Geology for Society and Territory – Volume 6*,
DOI: 10.1007/978-3-319-09060-3_139,

15.99 MPa (Kenneth 2005). While, others found their UCS is medium/moderately strength, varied from 24.1 to 40.8 and 18 to 36 MPa generally (Bell 1981; Karacan and Yilmaz 1997).

Clearly, exposure to water is a key control on gypsum dissolution (Jassim and Goff 2006). The chemical composition of this water is important and cause significant variability in rates. As a result, to control this, distilled water has commonly been used in experimental studies on dissolution. The degree of saturation also affects gypsum strength as measured in mechanical tests (Ali 1979; Elizzi 1976; Ergun and Yilmaz 2000; Gao et al. 2011; Sconnenfeld 1984), as may the soaking period (Gao et al. 2011; Sconnenfeld 1984).

139.2 Materials and Methods

The collection of gypsum in Iraq was limited by security concerns, though some samples could be collected from Bazyan, North of Iraq. Similar gypsum occurs in the highest strata of the Triassic Norian Mercia Mudstone Group (Cropwell Bishop Formation) at Bantycock mine (Worley and Reeves 2007). Both of sites feature thick gypsum rock layer, similar to those beneath the Mosul Dam.

NX cylinder samples (54 mm diameter and L/D = 2.5) were prepared following common practice (ASTM 2010; Bieniawski et al. 1978; Dreybrodt et al. 2002). A suitable loading rate of 0.025 MPa/s was determined following ASTM and ISRM standards (ASTM 2010; Bieniawski et al. 1978). A circumferential extensometer was used in the mid height of cylinders to calculate the radial strain (Fig. 139.2c). Ultrasonic observations were recorded for each sample in air dry state and after each of the immersion periods.

Unloaded gypsum samples were slowly saturated by vacuum, then soaked at atmospheric pressure at 5, 10, 15, 30 and 50 weeks. Three samples were used at each time interval for short-term loading tests. Continuous loading (2,688 kPa) using a modified oedometer was also applied to samples for 50 weeks. During soaking, the water was changed every 7 days and conductivity measurements taken. A full account of the methodology can be found in Salih (2013).

139.3 Results

139.3.1 Short-Term Results

See Figs. 139.1, 139.2, 139.3, 139.4 and Table 139.1, 139.2.

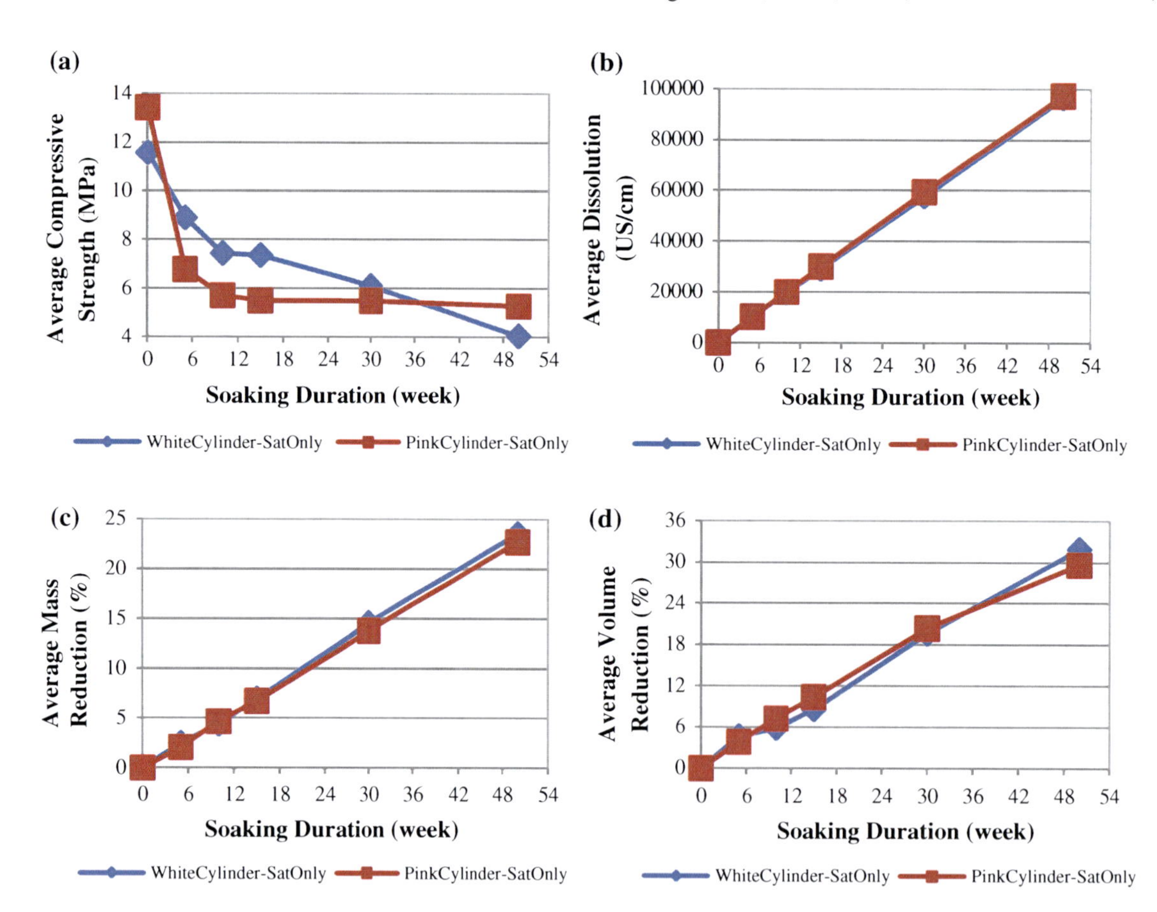

Fig. 139.1 Comparisons among short term loaded cylinders a is for compressive strength b is for dissolution c is for mass reduction and d is forvolume reduction. Each calculated values are the average of three tested cylinders

Fig. 139.2 **a** Gypsum outcrop, Bazyan/Iraq **b** Bantycock Mine/UK gypsum blocks **c** The extensometer in the mid-length of a cylinder

Fig. 139.3 The failed cylinders after short-term loading: **a** The white/Bantycock gypsum. **b** The pink/Bantycock gypsum. **c** The Iraqi gypsum

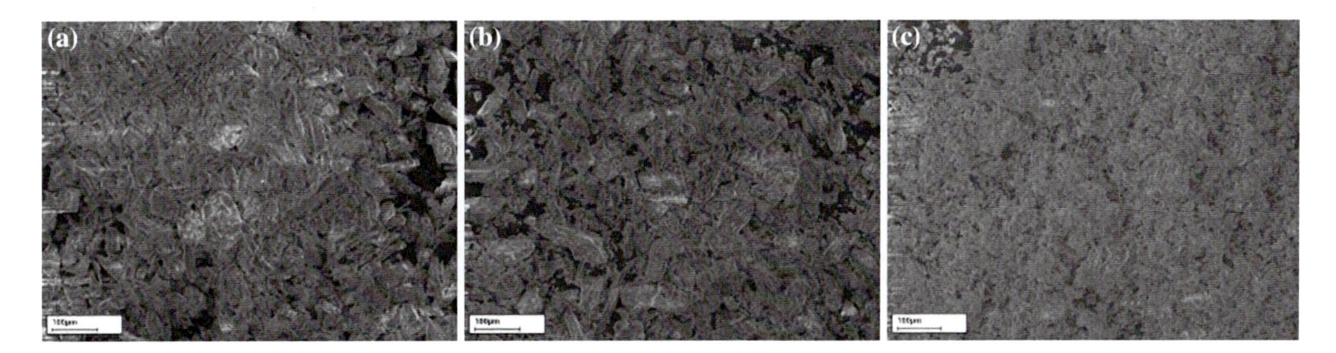

Fig. 139.4 Variations in gypsum crystal size, visualized using SEM: **a** The white/Bantycock gypsum. **b** The pink/Bantycock gypsum. **c** The Iraqi gypsum

139.3.2 Long-Term Results

See Fig. 139.5

139.4 Discussion

It is clear that saturation weakens gypsum rock, and this weakness progressively develops over time where there is a flow of water. This is relevant to the context of dams, and other large structures, where groundwater flow is enhanced, and where fresh water is constantly being introduced to the gypsum.

All the samples showed signs of dissolution, manifested as a change in shape over time. This change in shape does not, however, explain all the weakening that took place. There must also be a contribution from changes taking place within the samples.

Several characteristics of the gypsum influence the rate of dissolution weakening. Iraqi samples dissolved at a faster rate that the UK samples. This appears to be due to the size of the gypsum crystals, which were larger in the samples from Bantycock mine.

The larger dissolution values of Iraqi samples under atmospheric pressure than white/Bantycock samples in short-term datasets (Fig. 139.1) were correlated with the

Table 139.1 Description of the UK and Iraqi gypsum particles, the description based on Scanning Electronic Microscope (SEM) photos presents in Fig. 139.4

Gypsum sample type	Particle size (micron)	Particles gradation	Particles shape	Notes
White/Bantycock, UK (see Fig. 139.4a)	4–114	Poorly graded	Have three shapes randomly: longitudinal, semi-square and semi-circular/elliptical	Very sharp edges of particles, mostly medium size, some large and small sizes are found
Pink/Bantycock, UK (see Fig. 139.4b)	2–177	Not very well graded	Mostly semi-elliptical and longitudinal	Middle and large sizes are more than fine size
Iraqi (see Fig. 139.4c)	1–28	Looks well graded	Mostly semi-circular and semi-elliptical	Mostly fine particles, some concentration of fine particles with impurities/ other minerals together make dense look for some places

Table 139.2 Ultrasonic observation of white/Bantycock gypsum cylinders

Cylinder state	Transit time (Useos)	Velocity (m/s)	Path length (mm)	Elastic modulus (GPa)
Air-dry	27.77	18,870.67	108.67	159.7
Saturated under atmospheric pressure				
5 week	27.03	18,821.7	108	100.2
10 week	26.3	18,821.67	104.66	82.97
15 week	26.17	18,498.33	104.33	68.27
10 week	19.833	15,974.33	79.33	63.47
15 week	10.633	14,588.67	42	61.2

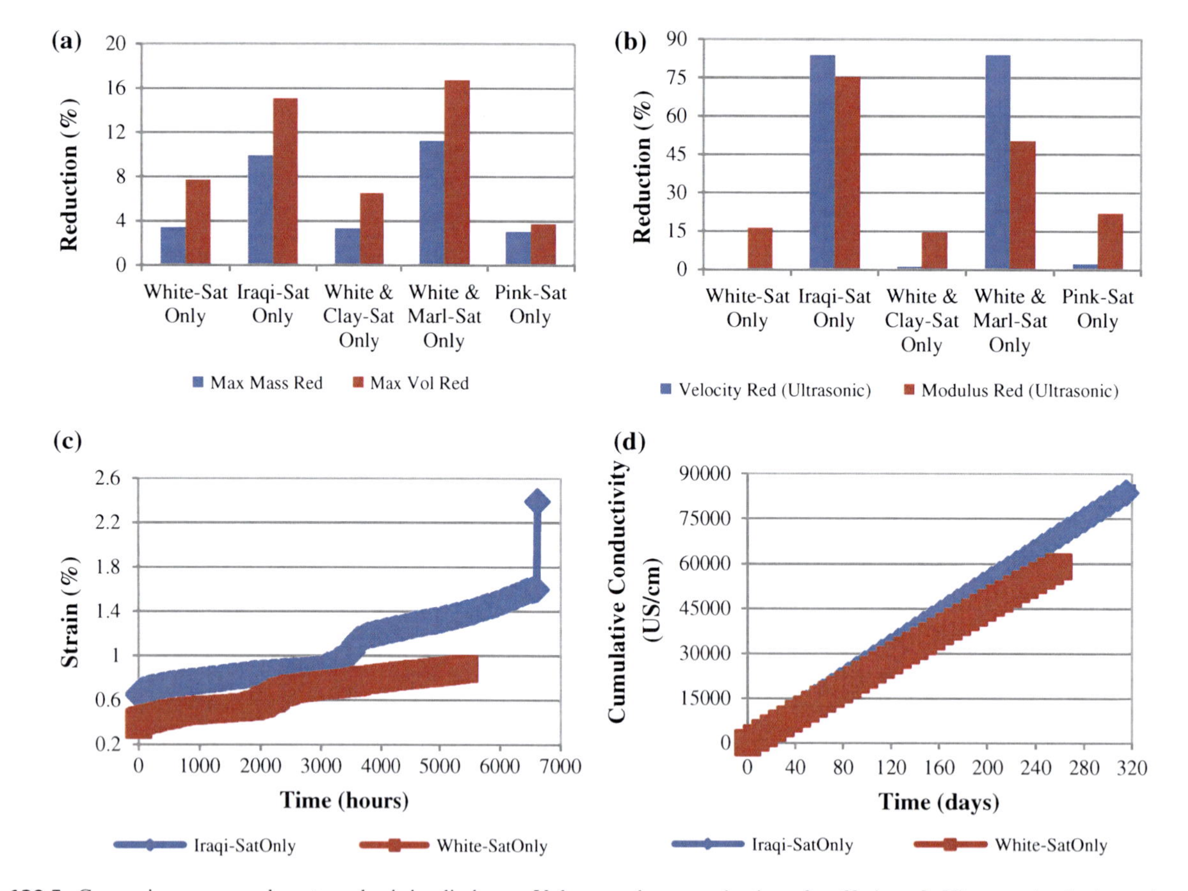

Fig. 139.5 Comparisons among long-term loaded cylinders. **a** Volume and mass reduction after 60 days. **b** Ultrasound velocity and modulus reduction after 60 days **c**, **d** are for 1 year loaded cylinders, Iraqi cylinder failed after 6,800 h as shown in part (**c**)

effects of gypsum crystal size and shape in characterizing the dissolution process i.e. larger crystals dissolve less quickly than smaller crystals. Slight differences in impurities within the gypsum, as reflected by the different colours of gypsum from Bantycock (white and pink) may also contribute to differences in response e.g. Fig. 139.1a where the strength of the pink gypsum declines much more quickly than the white gypsum. The shape of internal structures, determined by cracks and fissures, may also be a factor as these allow fresh water to penetrate the gypsum. In addition, many of the cracks contain impurities that have different mechanical and chemical properties to the gypsum. Samples that were soaked without loading showed more evidence of weakening through crack-enhanced dissolution than samples under constant load for 50 weeks. This may be because the high load caused the cracks to close through simple translational movement and through small amounts of creep. It should be noted that the water was under atmospheric pressure during these tests. Under pressure conditions similar to those found beneath large dams such as at Mosul, water may well be able to penetrate thin cracks.

The research has a number of implications for large structures than both load the ground and modify groundwater flow, such as large dams:

- The role of percolating fresh water in enhancing dissolution is confirmed. The significant head difference induced by a dam can be expected to increase the potential for groundwater flow and increase the likelihood of cavity formation. Similar effects could be induced by localized dewatering, or by wetter conditions caused by climate change.
- Importantly, the dissolution does not just generate macro-cavities. There is also alteration of apparently intact gypsum which increases its potential for failure.
- Sometimes subtle differences in gypsum, such as impurities and crack frequency can have a significant impact on the timing of failure.

139.5 Conclusion

The research has shown that the evolving behavior of weak soluble rocks such as gypsum can be investigated in the laboratory over extended periods. It highlights that the hazard presented by such rocks is not just simply one of dissolution induced cavities. It is also a progressive weakening of the rock mass itself.

Clearly, the findings suggest that the risk of failure in large dams is a result of a more complex set of processes than may have been expected. This has implications for mitigation solutions. If the problem is perceived as just being caused by cavity formation, then a solution might be to simply fill the cavities with grout on an *ad hoc* basis. This ignores the weakening of the 'solid' gypsum—and may make the problem worse. Localized injections of grout are likely to deflect and focus groundwater flow on the intact gypsum. This creates a risk of enhanced generation of new cavities, and of forcing more water into the gypsum and so reducing its overall strength.

Other than removing dams such as that at Mosul and elsewhere—an unattractive option given the need for water and power—the only solution is a systematic approach to creating a deep impermeable barrier that will significantly slow, or even stop, the movement of fresh water through the gypsum rock. Such a solution will inevitably be expensive, and challenging to achieve, but the cost must be weighed up against the risk of losing an essential piece of infrastructure.

References

Ali SA (1979) PhD thesis, University of Sheffield, UK

Bell FG (1981) Geotechnical properties of some evaporitic rocks. Bull Int Assoc Eng Geol 24:137–144

Bieniawski ZT, Franklin JA, Bernede MJ, Duffaut P, Rummel F, Horibe T, Broch E, Rodrigues E, Van Heerden WL, Vogler UW, Hansagi I, Szlavin J, Brady BT, Deere DU, Haweks I, Milovanovic D (1978) Suggested methods for determining the uniaxial compressive strength and deformability of rock materials. Int J Rock Mech Min Sci Geomech Abstr 16(2):135–140

Cooper AH (2006) Gypsum dissolution Geohazards at Ripon, North Yorkshire, UK. Field Trip Guide Ripon, IAEG

Dreybrodt W, Romanov D, Gabrovsek F (2002) Karstification below dam sites: a model of increasing leakage from reservoirs. Environ Geol 42:518–524

Elizzi MAS (1976) PhD thesis, University of Sheffield, UK

Ergun K, Yilmaz I (2000) Geotechnical evaluation of Miocene gypsum from Sivas-Turkey. Geotech Geol Eng 18:79–90

Gao H, Liang W, Yang X, Zhang G, Yue G, Zhang P (2011) Experimental study of mechanical property of gypsum rock soaked in hot saturated brine. Chin J Rock Mech Eng 30(5):935–943

Jassim SZ, Goff CJ (2006) Geology of Iraq. Prague and Moravian Museum, Brno, Dolin

Johnson KS (2005) Subsidence hazards due to evaporite dissolution in the United States. Environ Geol 48:395–409

Johnson KS (2006) Gypsum-Karst problems in constructing dams in the USA. Environ Geol 53:945–950

Karacan E, Yilmaz I (1997) Collapse dolines in Miocene gypsum: an example from SW Sivas (Turkey). Environ Geol 29(3/4):263–266

Kenneth SJ (2005) Subsidence hazards due to evaporite dissolution in the United States. Environ Geol 48:395–409

Salih NB (2013) PhD thesis, Brunel University/London, UK

Sonnenfeld P (1984) Brines and evaporates. Academic Press, London

ASTM Standards (2010) ASTM D7012-10 (Approved 15/01/2010)

Worley N, Reeves H (2007) Application of engineering geology to surface mine design, British gypsum, Newark Nottinghamshire. (Field Guide, 2007). http://nora.nerc.ac.uk/3225/1/Keyworth_Field_Trip.pdf. Accessed 21 Nov 2012

Underground Works in Weak and Complex Rock Mass and Urban Area

140

Serratrice Jean François

Abstract

This paper concerns the constructions of twin tunnels in a weak and complex rock mass and an urban area below the city of Toulon (France). The project is presented at fist time, then geological context and hydrogeological context. Both tunnels are built in thrusted grounds and under sea level. These grounds possess a high degree of heterogeneity and a high degree of fracturation at all scales inherited from their tectonic history. On a large part of the linear, the layers are arranged in reverse order of their depositional age. The second part of the paper addresses some geotechnical properties of the grounds. Then, some features about monitoring and surface settlements are presented. It seems that in some places, unusual responses of the ground appear at the surface, that reveals the strong influence of the local structure of the rockmass and, maybe, initial stress state. But the latter is unknown in all cases.

Keywords

Tunneling • Shallow tunnel • Urban area • Complex rockmass • Surface settlement

140.1 Introduction

Tunneling at shallow depth causes deformations in the rockmass and then deformations at the surface. These deformations are mainly represented by settlements, which are not uniform but are concentrated vertically above the tunnel axis in the form of a trough. The problem is particularly acute in urban area where settlements affect all components of the urban fabric such as buildings, structures, roads and networks. At depth, deformations can affect other constructions, especially when twin tunnels are built and the second interacts with the first, but more generally underground structures or deep foundations, which have not necessarily been designed to withstand excavation of a tunnel in their neighborhood. So tunneling in urban area requires control of ground movements to prevent excessive déformations on surrounding structures.

S.J. François (✉)
CETE Méditerranée, 13593 Aix en Provence, France
e-mail: jean-francois.serratrice@developpement-durable.gouv.fr

This paper concerns the constructions of twin tunnels in a weak and complex rockmass and an urban area below the city of Toulon (France) (Durand 1991; Gilbert et al. 2008). The project is presented at fist time, then geological context and hydrogeological context for this works built in thrusted grounds and under see level. The second part of the paper addresses example of geotechnical properties of the grounds. Then, some features about monitoring and surface settlements are présented. In some places, unusual responses of the ground appear at the surface, that seems to reveal the strong influence of the local structure of the rockmass and, maybe, initial stress state.

140.2 Project and Geological Context

The Toulon Underground Crossing project includes two road tunnels designed to establish continuity between the A 50 motorway in the west of Toulon (to Marseille) and the A 57 motorway in the east (to Nice). Each of the works extend over about 3,000 m with a central part drilled over

G. Lollino et al. (eds.), *Engineering Geology for Society and Territory – Volume 6*,
DOI: 10.1007/978-3-319-09060-3_140, © Springer International Publishing Switzerland 2015

1,800 m and then covered trenches at the ends. The works were carried out over several attacks from the ends and from intermediate shafts. The civil works of the north tunnel took place from the early 1990s until 2001, and between 2007 and 2011 fir the south tunnel. Both phases account for shutdown periods. The full-face excavation was carried out mechanically. The choice of type of pre-supports and supports to install was operated according to the nature and state of the ground encountered, underground deformations and surface deformations from monitoring and provisions of surface settlements.

Both tunnels are fit into old grounds very tectonised by origins of the Primary area and then Permian and Triassic grounds. The stratigraphic succession is as follows: Quaternary with colluvium, alluvium and fills; Keuper with clays, cargneules and clusters of gypsum; Muschelkalk with dolomitic limestones, marls and sandstones; Permian with sandstones and mudstones; Stephanian with sandstones and black shales (coal); Socle with strongly folded quartzo-phyllites. This stratigraphic sequence from Permian to Muschelkalk appears on the east side of the site. In contrast, grounds appear in reverse order of their age in the west and central part of the city of Toulon, which constitutes the singularity of the site (Rat and Serratrice 2004). At the regional level, this area is located at the northern edge of the thrust sheet of Cap Sicié. Thrusting in direction SW-NE dated Tertiary is responsible for the disruption of the stratigraphic sequence. The Triassic formations are overlain by Permian sandstones and mudstones and then topped themselves by Stephanian sandstones and coal shales in the form of shell structures. The ante-stephanian formations represented by quartzo-phyllites overlay this set under the Quaternary cover. It is likely that other tectonics movements have occurred later to achieve the mechanical initial state that prevails underground today. The site investigations showed three families of groundwater (Serratrice 2004).

Thus and for a large part, the diversity of the grounds encountered, their state and the complexity of their arrangement have prevented the geological and geotechnical knowledge of the site, which turned imperfect at every stage of the project, despite the geological and geotechnical site investigations during successive campaigns, the first of which began in 1971, and despite the investigations carried out with the progression of the works. Moreover, constraints imposed in urban areas are not in favor to site investigations.

140.3 Geotechnical Aspects

The data come primarily from core drilling and in situ testing. Shafts and exploratory tunnels were excavated too. Laboratory tests were carried out in large numbers, and pressuremeter tests, but also pump tests, plate tests, etc. Monitoring the digging of the north tunnel has been used to observe grounds under this scale, describe their state and fracturation, characterize water inflows, measure underground deformations, etc.

Before the beginning of work in south tunnel, new site investigation by drilling, analysis of test results and synthesis of all these new data with old data led to identify 21 families of grounds. All the state characteristics and mechanical properties of these families are dispersed due to the geological history of the rockmass which was destructured by thrusting.

The geotechnical site investigations during the successive campaigns by in situ testing favored the use of the pressuremeter. Numerous holes were drilled at various depths (up to 60 m). To the total, more than 3,500 tests were carried out to cover the different formations throughout the site. Figure 140.1 shows the histograms of pressuremeter modulus measurements E_M for Triassic marls, One layer of the 21 grounds families. In the first case of an arithmetic scale, the modulus distribution takes an exponential shape. In the second case of a logarithmic scale, the distribution takes a log-normal shape. The averages of E_M calculated in logarithmic scale are smaller than those calculated with an arithmetic scale and dispersion is less in the logarithmic scale.

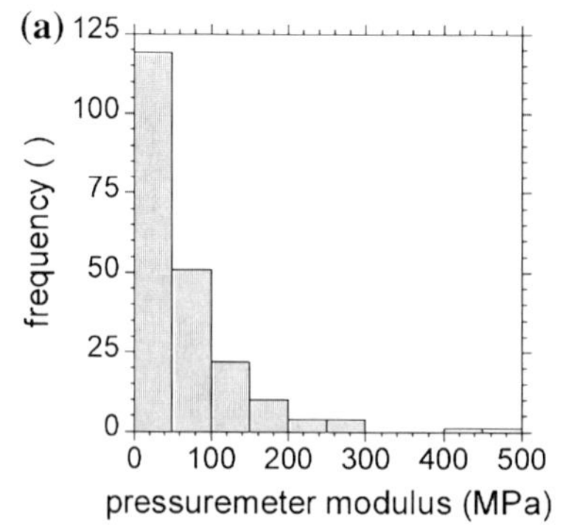

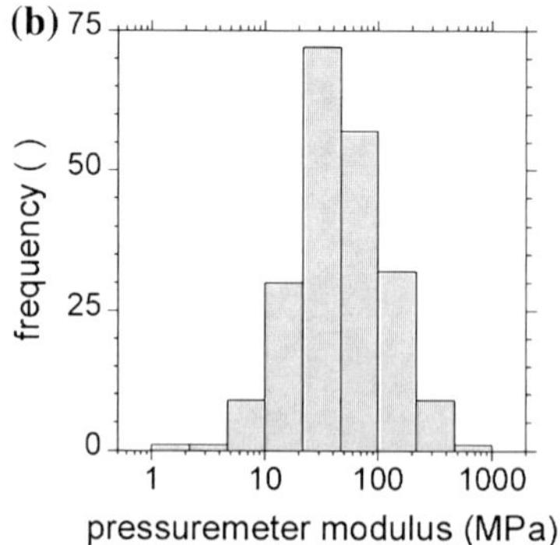

Fig. 140.1 Pressuremeter modulus in linear scale and logarithmic scale

140.4 Behaviour of the Rockmass

The Toulon Underground Crossing project is imbedded in a complex rockmass and below an urban area, which made the construction a very difficult works. Given the issues due to the presence of surrounding building at the surface, special provisions have been considered at the design stage and during construction works.

The methods have also evolved based on the lessons learned from the construction of the north tunnel. Expertise of the state and performances of all the constructions at the surface was carried out previously in the perimeter of influence of the tunnels (50 m at least on either side of each of the axes of the two tunnels). The survey program was constituted by an important process of underground deformations monitoring with measurement of convergences between the walls and extrusion at the tunnel faces. But it is especially at the surface that an important survey program has been developed, initially manual, and then automated using motorized theodolites (Serratrice and Dubois 2004; Caro-Vargas and Beth 2012). This monitoring is based on the topographic survey of targets sealed on the buildings or points carried by the roads. Measurements were collected in databases and then exploited. This monitoring was supplemented by periodic visits of buildings and networks. The results of measurements at the surface and underground associated with systematic survey by exploratory drilling during tunneling and then the analysis of these data made possible to drive the works and choose the profiles best suited to the grounds encountered and the rate of deformations observed. Innovative methods of exploitation of the data have been developed, which were used to examine carefully the evolution of longitudinal profiles of settlements or settlements troughs at the surface (Serratrice and Magnan 2002; Serratrice and Dubois 2004). Still based on clues warning often tenuous, these methods based on monitoring have proven effective in most cases to guide the project.

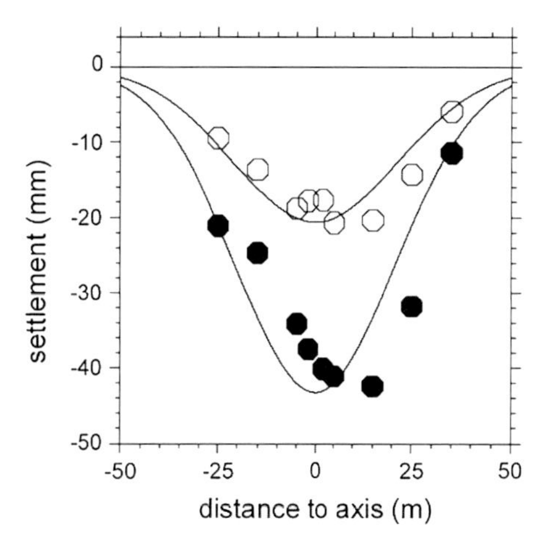

Fig. 140.3 Settlements profile

For each of the two tunnels, the final settlements were represented on a map of iso-values for the whole site. The map clearly shows pits and lumps of the final settlements trough along the tunnels axis. The lateral extension of the trough was greater than 50 m in some places. Sometimes the shape of the trough fits well with the structures encountered in the ground. For example, the map on the Fig. 140.2 shows the final settlements in the west part of the north tunnel. The tunnel is drawn at steps of 25 m. Iso-settlement curves are represented every 10 mm and every 2 mm. The grid spacing is 50 m. A particular trough appears which adopts an oblique direction relative to the tunnel axis. It happens that the trough coincides perfectly with a fault structure encountered by the tunnel at depth. The trough is set in the same direction that the fault F (N100).

An other example shows that settlements troughs are sometimes decentered in a transverse profile as illustrated on Fig. 140.3. The dots represent the measurements when the tunnel face crosses the sections (hollow dots) and then the final settlements when the tunnel face is far from the profile 3 months later (solid dots). The curves represent the

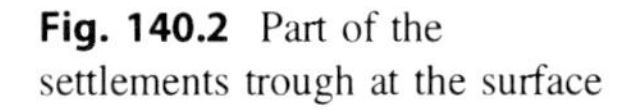

Fig. 140.2 Part of the settlements trough at the surface

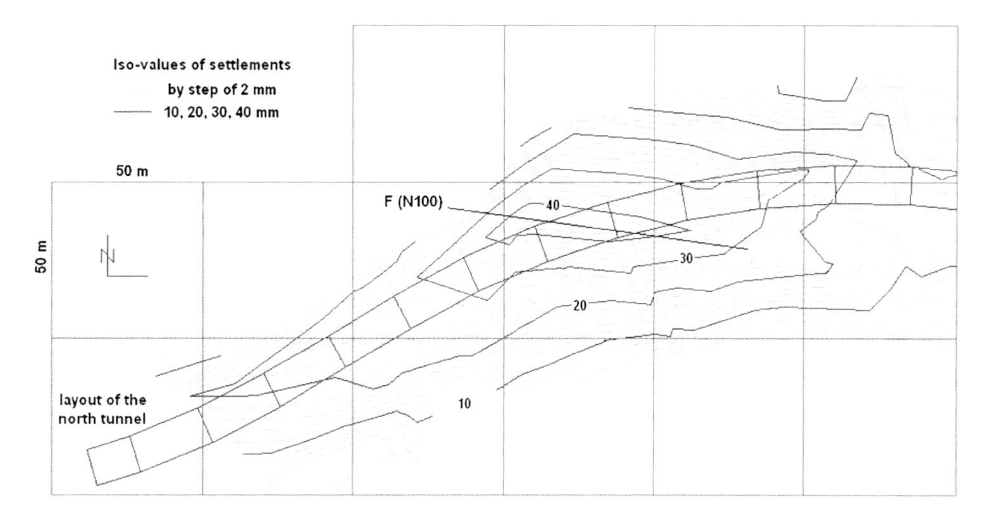

settlements profiles calculated for the centered Gaussian representations. In this example, the shift of the actual trough indicates a lateral heterogeneity of the rockmass. This heterogeneity is likely due to structural effects inherited from thrusting. But it is often difficult to establish a link between the nature, the structure, the mechanical properties of the ground observed during the excavation at the scale of the tunnels face and the final shape of the trough at the surface. It is even more difficult to detect these heterogeneities across a tunnel section during the site investigations and the design stage, at a time when it would be useful to provide reliable estimates of surface settlements.

140.5 Conclusion

The heterogeneity of the rockmass increases the difficulties for the construction of shallow tunnels in urban areas. To anticipate potential difficulties, the construction operations, which must remain based on a thorough knowledge of the ground behaviour and the techniques used, have only warning signs often tenuous. In this context, the deformation monitoring underground and at the surface plays an predominant role. On the surface, monitoring the dynamics of settlements in longitudinal profiles during the advance of the tunnel face, or better at three dimensions, gives a privileged means of observation. Overall, it seemed a fairly direct relationship between surface settlement and structure into the tectonized rockmass. But in some places along the works, it appeared a strong influence of the initial structural and mechanical states into the rockmass in the context of high tectonized media.

References

Caro-Vargas B, Beth M (2012) Observational method using real time surface settlement monitoring: The South Toulon Tunnel project. GeoCongress 2012, Oakland, California, pp 3109–3118

Durand JP (1991) Liaison A50/157, Traversée Souterraine de Toulon. Tunn Ouvrages Souterrains 106:177–179

Gilbert G, Serratrice JF, Thiebaut H, Marguet P, Le Bissonnais H, Vigouroux JP (2008) Deuxième tube de la traversée souterraine de Toulon. Congrès Internationnal AFTES, Monaco

Rat M, Serratrice JF (2004) Tunnel de Toulon. Contexte géologique, hydrogéologique et géotechnique. Rev Trav 806:25–30

Serratrice JF, Magnan JP (2002) Analyse des tassements de surface pendant le creusement du tunnel nors de la traversée souterraine de Toulon. Bull Lab 237:5–36 (P&C, Paris)

Serratrice JF (2004) Tunnel de Toulon. La piézométrie. Rev Trav 806:75

Serratrice JF, Dubois P (2004) Tunnel de Toulon. Interprétation des deformations en surface. Rev Trav 806:76–79

A Dam with Floating Foundations

141

Vinod Kumar Kasliwal

Abstract

In Engg Geology, a weak rock is considered as one whose deformation modulus is 1/2 to 1/5 of the enclosing surrounding rock. Such rocks are treated by replacing them with concrete plugs of suitable strength to depths as recommended by USBR Formula (Shasta and Friant Dams) and/or as by Japanese Team (Fumio Ishii et al.). These methods have their limitations since in each of them, it is assumed that the weak rock zone/fracture is bounded on each side/abutment by rocks that are strong having competent deformation modulus and that such rocks exist for a sufficient width so as to act as abutments of beam/bridge. But when the shearing, fracturing and faulting is so frequent and close-spaced that this condition is not available, then the problem is how to construct a gravity dam. One such 81 m high straight gravity dam across Jakham river in Chittorgarh distt, Rajasthan, India has been constructed by providing doubly-reinforced cantilever rafts for the foundations. The Jakham Dam is a 81 m high straight gravity dam across the river of same name Jakham, a tributary of Mahi river. The dam is located in a 150 m deep gorge with near-vertical escarpment slopes at the neck of a wide open valley, thereby making it an ideal topographical site. The rocks exposed are ferruginous quartzite belonging to pre-Aravallis (Archean). They are hard, brittle and have undergone extensive fracturing, folding, faulting and shearing to the extent that up to 65 % of the foundation rocks are affected by them so much so that while one abutment and the river bed shows faulting the other abutment shows presence of a cave. Further, while the fractures and joints are occupied by red ocherous clays, the bedding planes are frequently occupied by talcose bands. The paper highlights the geological conditions in the foundations by detailed (1:100) mapping so as to identify competent rock areas to which the load/stresses of the dam have been successfully transferred by means of cantilever rafts.

141.1 Introduction

Construction of the 81 m high and 235 m long straight gravity masonry dam in Chittorgarh (now Pratapgarh) district, Rajasthan, India (Fig. 141.1a) was commenced in the year 1970 and its 1.5 M cum reservoir was first filled in 1988.

Director (Retd) from Geological Survey of India. Now working as a Engineering Geology Consultant.

V.K. Kasliwal (✉)
Engineering Geology, A6, Hospital Road, Ashok Nagar, Jaipur, India
e-mail: vkkasliwal@hotmail.com

The dam site is located in the north-south band of a 150 m deep Z-shaped gorge. The foundations rocks of the dam are formed by pre-Aravalli (Archean) quartzites which are extensively folded, fractured and sheared. Due to difficulty in approaching the dam site and steepness of the walls of the gorge, very limited explorations were carried out. On the basis of these investigations, foundation grades for the various blocks were suggested and excavations commenced. However on reaching the suggested grades a great variance was observed. Even at this stage it was considered to abandon this site and look for an alternate, but after evaluating geological and engineering and socio-economic considerations it was decided to construct the dam at this site.

G. Lollino et al. (eds.), *Engineering Geology for Society and Territory – Volume 6*,
DOI: 10.1007/978-3-319-09060-3_141,

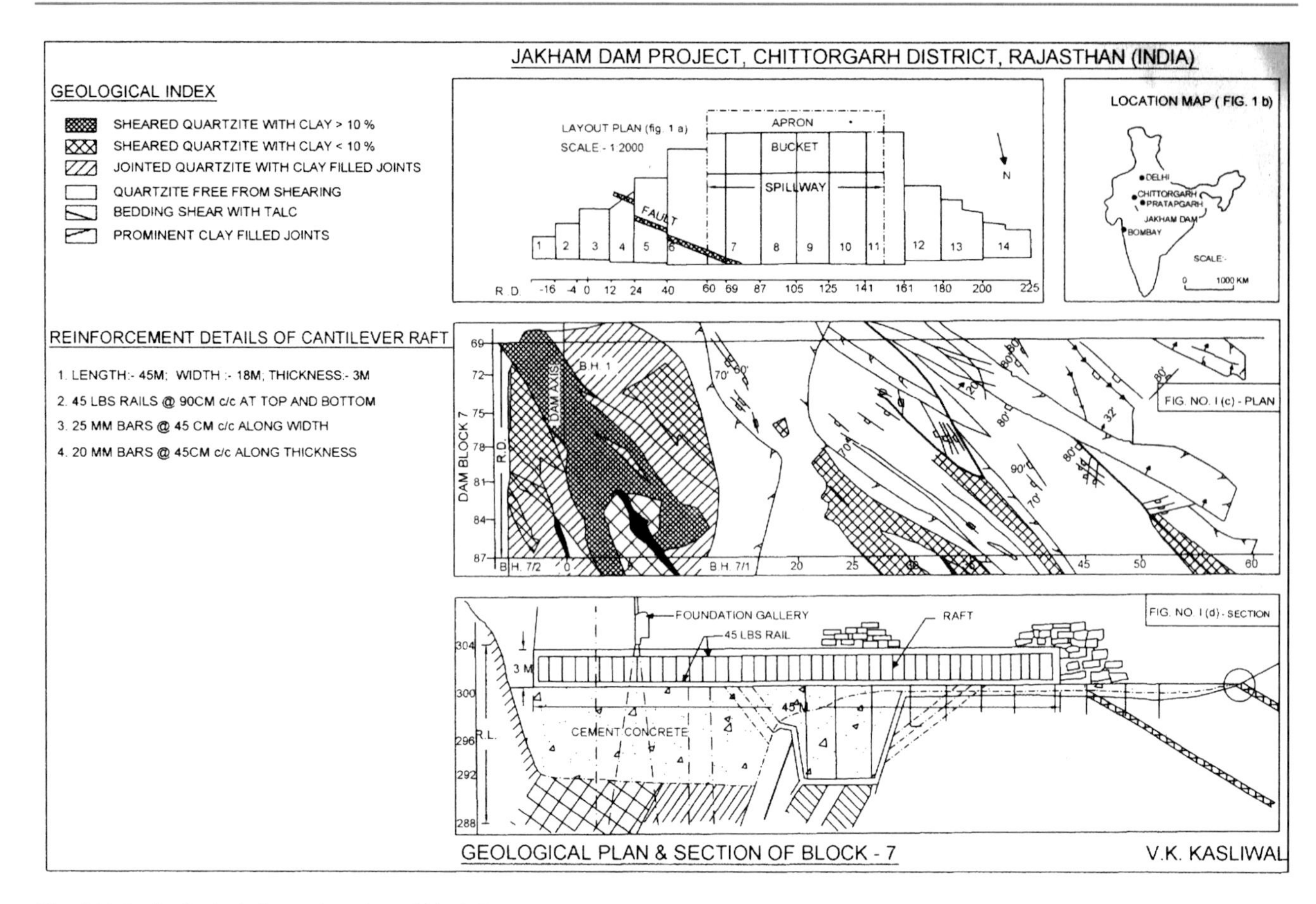

Fig. 141.1 Geological plan and section of block 7

To determine the competence of the various rock units, in situ testing by plate- bearing tests was carried out. These tests gave results which did not match the observed geological conditions (Kasliwal 1983). Hence a detailed geological appraisal of the foundation conditions distinguishing the fractured, faulted and sheared rock vis-a-vis sound rock capable of withstanding the stress conditions was undertaken for all the foundation blocks. This paper, as an example/ illustration, brings out by detailed mapping (1:100) of the geology of the 81 m high spillway block no 7, to arrive at foundation treatment.

141.2 Geology

So as to understand the complex geological set up at the dam site, background information thereon shall be helpful in understanding of the foundation problems.

141.2.1 Geology at Dam Site

At the dam site foundation rocks are pre-Aravalli (Archean-II; >2550 m.y) ferruginous quartzites frequently traversed by talcose bands. These generally occur along bedding. They strike askew to the dam axis in a NNW-SSE to N–S direction and dip at 30*–40* in a WSW to West. These quartzites are extensively folded, faulted and sheared. Non-overflow blocks 4–6 on the left abutment and parts of Overflow (spillway) blocks 6 and 7 in the river section are traversed by a 0.5–1.75 m thick fault zone (Fig. 141.1b). The fault zone consists of pulverized quartzite, red ocherous clay and talcose bands. Striking parallel to the fault zone there are a number of shear bands so much so that at times up to 65 % of foundation rocks are affected by them. One such similar 3.5 m thick shear zone has affected the toe-region of blocks 9–12 on the right abutment due to which a large cave with a natural water-spring has been formed Further, the foundation rocks are open jointed and have a Equivalent Permeability of 41.14–54.11 Lugeons,. It is in this complex geological background that the dam has been successfully constructed.

141.2.2 Geology of Block 7

Block 7, lies between R.D. (Reduced Distance) 69 and R.D. 87 m in the deepest part (R.L. 300 m).of river bed. Initially, only one NX borehole was drilled to investigate the rock conditions On the basis of this borehole it was anticipated that suitable foundation rocks shall be available at a depth of about 8.0 m (R.L. 292 m) but on actual excavations no substantial improvement in the rock conditions was observed. Four more shallow holes were put to find out the rock conditions at depth. These too showed no improvement even at 13 m depth (R.L. 287 m). Although the fault has not affected the foundations of this Block, yet at the Block-joint at R.D. 69 m the effect of the fault is manifested in the footwall of the fault. Here the rocks are broken up into highly sheared and broken bands/strips which are profusely filled with red ocherous clay (Fig. 141.1c). The thickness of this zone is about 7 m. Enclosing this zone were jointed quartzites with clay filled joints so much so that the entire heel region was occupied by them. They extended on the upstream and have an aggregate thickness of about 14 m (Kasliwal 1975). Hence it was not possible to provide a beam.

In situ plate bearing tests were carried out in the fault-affected abutment Block 5 and river bed Block 7 These showed that the values obtained for different rock conditions did not match the observed geology. The deformation modulus varied from 0.39 × 10 × 5 kg/cm × 2 for the fault zone material as compared to 0.28 × 10 × 5 kg/cm × 2 for the jointed rock. It was therefore decided to provide cantilever rafts, for which suitable rock conditions that could take the additional foundation stress needed to be identified.

Large scale (1:100) mapping undertaken to identify competent quartzites showed that between 10 m d/s (downstream) of dam axis to 40 m d/s, the quartzites are sound, free from shearing and contain few clay filled joints, whereas further on the downstream they are blocky jointed. Using this information, doubly acting cantilever raft was provided with the stresses of the heel region being transferred to the sound quartzites in the middle third area of the foundations.

141.3 Conclusions

The foundation rocks of the 81 m high Jakham Dam are quartzites, which contain talcose bands. These have undergone extensive shearing and faulting so much so that treatment for these zones could not be provided on the basis of the formula suggested by USBR or the Japanese Team. In situ tests carried out to determine the deformation modulus of the different rock members did not match the observed geology. Hence, a detailed mapping of the foundation rocks was carried out to delineate the extent of shearing vis-a-vis the sound rock. On the basis of this mapping, foundation area was identified which could be loaded. Hence doubly reinforced cantilever rafts were designed for load transfers using 45 lbs rails at 90 cm c/c with 25 mm distribution bars at 45 cm c/c covering the entire foundation area (Fig. 141.1d). This method has been adopted for all the Blocks 4–12. i.e. the Dam virtually floats on RCC raft foundation; A Dam with Floating Foundations. This solution to the complex and weak rock masses has been successful and the Dam has been operational for more than a decade now.

Reference

Kasliwal VK (1975) A note on the geology of block 7. Jakham Dam. Unpublished report of Geological Survey of India

Kasliwal VK (1983) A geotechnical appraisal of the in situ tests at Jakham Dam, Chittorgarh District, Rajasthan, India. International symposium on in situ testing, vol 2, pp 71–76, Paris

Classification of "Loosened Rock Mass" Based on Cases of Dam Construction

142

Takahiro Eguchi, Katsuhito Agui, and Yasuhito Sasaki

Abstract

In Japan, dams have been built in the mountainous area of many even now. However, sites that contain rock is not suitable as the basis of the dam is increasing recently. "Loosened rock mass" is one of the bedrock that is not suitable for the foundation of the dam. For this reason, if there is a "loosened rock mass" on the site, it is necessary to countermeasures. For example, excavation of loosened rock mass, modification of dam position, etc. In addition, the range and properties of looseness are different for each site by the terrain, the geology, the geological structure, the formation process of slope. Therefore, in the construction site of the dam, in order to understand exactly the properties and nature of looseness, it is necessary to careful survey. In this study, we tried to systematic classification of loosened rock mass in order to be able to compare the range and properties of the looseness of the site in dam construction in the future. When classifying systematically loosened rock mass, we focused on the following points as an element of the classification. "Slope movement pattern", "range regulation factor", "property regulating factor". We have set five slope movement pattern. And, combining "range regulation factor" and "property regulating factor", we have classified into 15 types and 9 pattern.

Keywords

Loosened rock • Classification • Slope movement pattern

142.1 Background and Purpose

In Japan, dams have been built in the mountainous area of many even now. However, sites that contain rock is not suitable as the basis of the dam is increasing recently. Because, the site that dam construction can be is limited.

"Loosened rock mass" is one of the bedrock that is not suitable for the foundation of the dam.

This research defines the state of rock mass called looseness as "situation in which deformation, volume increase, or density reduction etc. Caused by stress release, action of gravity, or action of weathering, etc., Which is called "looseness", causes, occurrence, opening, slippage etc. of cracks, and while the state of the rock mass is retained, overall it has become easily deformed and its non-elastic property becomes large" (Sasaki et al. 2005).

Loosened rock mass is high permeability and low strength as the basis of the dam.

Further, the stability of the slope is low when excavation.

In addition, the nature of the looseness is confirmed in multiple in the loosened rock slope in general.

For this reason, if there is a loosened rock mass on the site, it is necessary to countermeasures. For example, excavation of loosened rock mass, modification of dam position, etc.

In addition, the range and properties of looseness are different for each site by the terrain, the geology, the geological structure, the formation process of slope.

Therefore, in the construction site of the dam, in order to understand exactly the properties and nature of looseness, it is necessary to careful survey.

T. Eguchi (✉) · K. Agui · Y. Sasaki
Public Works Research Institute, 1–6 Minamihara, Tsukuba 305-8516, Japan
e-mail: eguchi@pwri.go.jp

G. Lollino et al. (eds.), *Engineering Geology for Society and Territory – Volume 6*,
DOI: 10.1007/978-3-319-09060-3_142, © Springer International Publishing Switzerland 2015

As a result, at dam construction sites where the rock mass has loosened, the cost of the survey and of countermeasure works tends to be far higher than at an ordinary dam construction site.

Thus, the investigation of loosened rock mass, idea to reduce the cost of investigation has been required.

If the range and properties of looseness are understood early on, more effective ways of the survey methods, the survey positions and the countermeasures are selected. As a result, the costs of surveys and countermeasures are reduced.

If there is a case that the range and properties of looseness are similar, it is possible to develop a survey plan by reference in the case.

However, such a comparison has not been done much. Because, many cases of loosened rock mass have not been systematically organized.

Therefore, in this study, we tried to systematic classification of l loosened rock mass in order to be able to compare the range and properties of the looseness of the site in dam construction in the future.

142.2 Concept and Result of Classification

When classifying systematically loosened rock mass, we focused on the following points as an element of the classification.

Processes of looseness: In order to estimate the properties and range of looseness in three dimensions, it is necessary to estimate the processes of looseness. Therefore, we focused it as an element of the classification.

But, the processes of looseness, are different in each site. Therefore, they can not be used as an element directly.

Therefore, We have focused as a classification element the type of affect slope movement (in this paper, this is called "slope movement pattern") in looseness of the slope.

Geological structure that characterizes the range of looseness (in this paper, this is called "range restriction factor"): Results of the analysis of the case, We have confirmed plurality of site that the range of the looseness is limited by geological structure (for example, fault).

If this structure is found, the range of looseness easily can be estimated by checking the distribution. Therefore, we focused it as an element of the classification.

Geological property or structure that characterizes the properties of the loosened rock (in this paper, this is called "property classification factor"): Results of the analysis of the case, We have confirmed plurality of sight that geology is related to the property of loose, for example, distribution areas of specific geology and range of crack opening is roughly consistent.

If this geological property or structure is found, the property or distribution areas of looseing easily can be estimated by checking the distribution. Therefore, we focused it as an element of the classification.

How to combine elements is as follows. First, We have classified the cases of looseness using the "slope movement pattern". Next, we have subdivided there using the "range restricting factor" and "property regulating factor".

Dam site number of the subject of the current study was 39.

In the 39 cases, "slope movement pattern" was confirmed five cases.

"Range restriction factors" was confirmed eight cases (example: fault, low angle cracs, etc.). "Property classification factor" was confirmed four cases (example: milonitization, directly under weathered granite, etc.).

Combining them, we have classified into 15 types and 9 pattern. Classification results are shown in Table 142.1.

142.3 The Contents of the Classification

Slope movement pattern are classified into the following five.

(a) Toppling
(b) Dip slope deformation
(c) Gravity deformation
(d) Crack opening of the shallow part of stress release
(e) Landslide

Here, the definition of (c) Gravitational deformation is as follows. There is the opening cracks or small deformation of rock mass in deep point of the slope, but there isn't toppling or slip. The definition of (d) Crack opening of the shallow part by stress release is as follows. There is opening cracks in shallow part of the slope, but there isn't a small deformation of the rock mass.

Figure 142.1 shows an exemplary four cases of slope movement pattern excluding (e).

(A) of slope movement pattern is toppling. And range restriction factor is low angle cracks. In this slope, chart falls down, and the influx of clay and the opening of the crack is happening. Just above the black slate, low angle cracks are well developed. There isn't looseing under the low angle cracks.

(B) of slope movement pattern is dip slope deformation. And range restriction factor is deterioration appears along the low angle layer, but does not extend overall. A part of pumice tuff is deteriorated. And sandy tuff at the top of the pumice is formed structure of the blocks.

(C) of slope movement pattern is gravity deformation. And the property classification factor is the link of the weak layer classifying degree of looseness. The river side of

Table 142.1 Result of classification of loosened rock

Forms of variation	Toppling			Dip slope deformation	Gravity deformation		Crack opening of shallow part by stress release		Landslide
Pattern name	T-1	T-2	T-3	D-1	G-1	G-2	O-1	O-2	L-1
Existence of range restricting factors	○	×	×	○	×	×	×	×	○
Example of range restricting factors	•Fault •Low angle crack			•Intersection of 2 faults •Fault and that with low-angle layer deterioration intersecting •Fault •That with low angle layer strength reduction •Deterioration appears along the low angle layer, but does not extend overall					•Slip surface
Existence of property classification factor	×	○	×	×	○	×	○	×	×
Example of property classification factor		•Myloniteization			•Link of weak layer classifying degree of looseness		•Directly under weathered granite •Along bedding plane		
Number of cases	3	1	6	6	4	8	2	5	4
Number of types	2	1	1	5	1	1	2	1	1

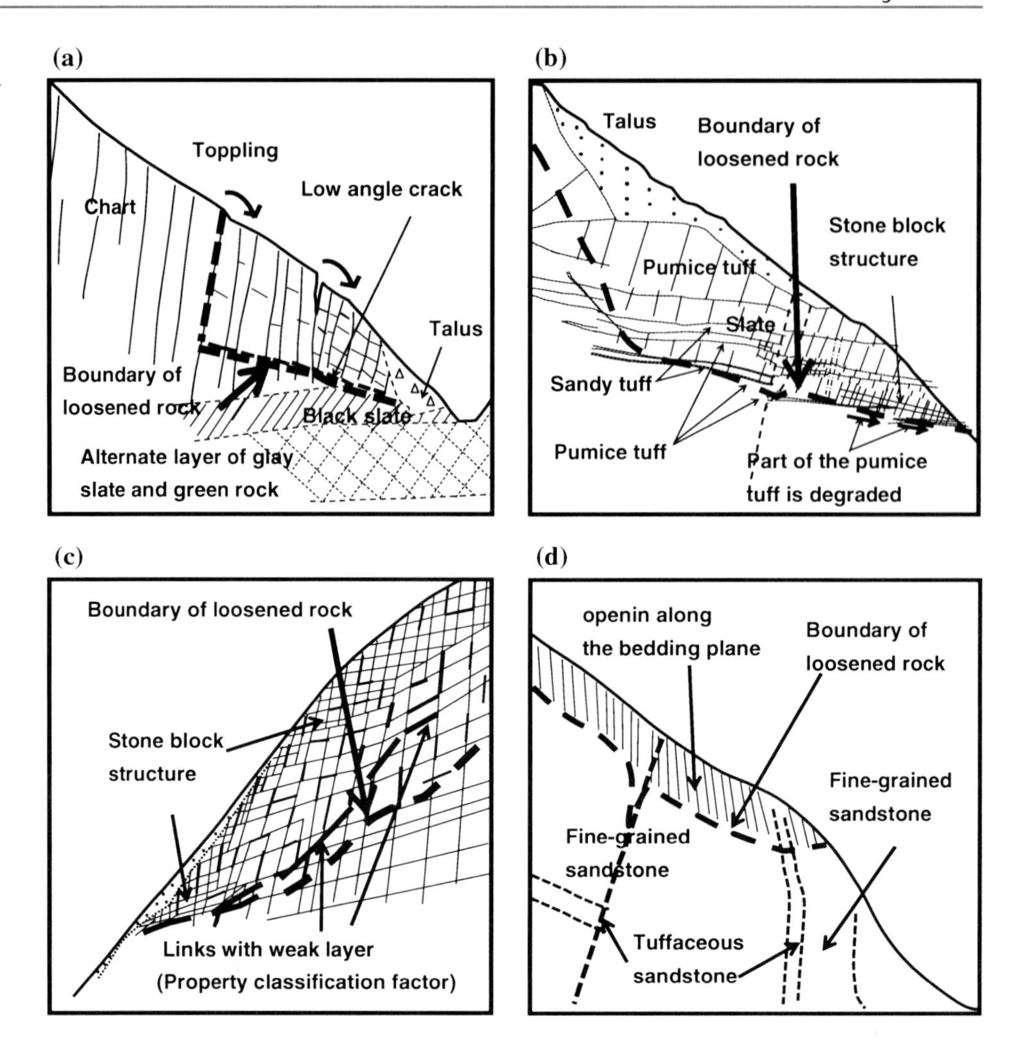

Fig. 142.1 Representative examples. **a** Example of toppling. **b** Example of dip slope deformation. **c** Example of gravity deformation. **d** Example of crack opening of shallow part by stress release

the weak layer, the phenomenon of opening cracks, the structure of the blocks, or the deformation of the rock mass occurs. The mountain side of the weak layer, the deformation degree of rock mass and the opening width of the crack is small.

(D) of slope movement pattern is gravity deformation. Range of looseness is limited to the surface layer portion. Crack is open to the bedding planes along the fine-grained sandstone.

142.4 Concept of Investigation by the Classification

It is summarized as follows the concept of loose rock survey based on the results of classification.

The toppling is subdivided into three types as shown below. In type 1, range restriction factor is present. In type 2, property classification factor is present. Both factors are not present in type 3. Therefore, the survey of the toppling, it is important to confirm the existence of "range restricting factor" and "property classification factor" firstly.

In all cases of dip slope deformation and landslide, range regulation factors were found. Therefore, in these slope movement patterns, it is advisable to look for a similar type by checking the contents of the range restriction factor.

In all cases of gravity deformation and crack opening of the shallow part of stress release, range regulation factor was not confirmed.

However, a few cases were found the property regulating factor. Therefore, in these slope movement patterns, it is advisable to look for a similar type by checking the contents of the property regulating factor.

142.5 Summary

In this study, we tried to systematic classification of loose rock in order to be able to compare the range and properties of the looseness of the site in dam construction in the future.

When classifying systematically loose rock, we focused on the following points as an element of the classification. "Slope movement pattern", "range regulation factor", "property regulating factor". We have set five slope movement pattern. And, combining "range regulation factor"and "property regulating factor", we have classified into 15 types and 9 pattern.

By using this classification, when you survey a rock looseness the future, we can select the cases that are similar to the classification case. And we can select the appropriate methods of countermeasures and survey position. As a result, it is expected that the cost related to survey and countermeasure for loose rock is reduced.

And, We want to continue this research and increase the classification pattern of loose rock.

References

Eguchi T et al (2012) Pattern classification of loose rock in dam construction case. In: Proceedings of annual meeting of JSEG, pp 85–86 (in Japanese)

Sasaki Y et al (2005) Actual situation and tentative classification of loose rock at dam site. Eng Dams 228:10 (in Japanese)

Sasaki Y (1999) Survey methods of loosened rock. Civ Eng J 41(2):8–9 (in Japanese)

Watari M (2005) Original birth of active landslide and "sagging" by gravitational deformation on mountain slope. J Jpn Landslide Soc Landslides 41(5):503–512 (in Japanese)

Numerical Analysis of a Crossover Cavern Excavated in a Complex Rock Mass as Part of the Hong Kong Express Rail Link Project

143

D.K. Koungelis and R. Lyall

Abstract

Kier Kaden OSSA Joint Venture is currently constructing Express Rail Link Contract No 824—Ngau Tam Mei to Tai Kong Po Tunnels for the MTR Corporation Limited. A critical review of the available ground investigation data and the regional geology was carried out to investigate the geological characteristics of the area surrounding Ngau Tam Mei shaft and the Crossover Cavern. The following paper outlines the methodology for designing the temporary support necessary for the Crossover Cavern to support the excavation during the construction period and until the permanent lining has been cast. The design process outlined below includes an assessment of the required support using the 'Q' method and empirical design charts developed by the Norwegian Geotechnical Institute, followed by a series of numerical analysis using UDEC to verify the suitability of the temporary support design.

Keywords

Udec • 'Q' method • Tunnelling • Crossover cavern

143.1 Introduction

Donaldson Associates Limited (DAL) was appointed by Kier Kaden OSSA Joint Venture (KKOJV) to design a section of tunnel to be constructed as part of the Guangzhou-Shenzhen-Hong Kong Express Rail Link (XRL). The contract (C824) involves design of the temporary and permanent linings for 2.6 km of running tunnel comprising twin bore single track SCL tunnels and a twin track Crossover Cavern (26.9 m wide by 16.6 m high by 115 m long). Permanent shafts (NTM and TKP) constructed at either end of the contract. DAL designed the temporary support for all tunnels and shafts and permanent lining for tunnels only. Preliminary ground investigation in the area of NTM shaft identified shearing features which meant that ground investigation data was limited due to difficult terrain resulted in a number of uncertainties about the ground conditions adjacent to the NTM shaft. DAL considered the overall rock quality in the area to be better than anticipated at tender stage and that shear zones recorded in the ground investigation could in fact be weathered hydrothermal zones which exhibit dramatically different characteristics to conventional shear zones. A number of locations for the Crossover Cavern were considered during a value engineering exercise and it was agreed to build the cavern 25 m south of the shaft, connected by a shortened length of tunnel to limit or negate any interaction between the Crossover Cavern and the NTM shaft from an engineering and technical point of view.

143.2 Geology and Ground Investigation

The geology at the location of the Crossover Cavern is Middle Jurassic Tai Mo Shan Formation of the Tsuen Wan Volcanic Group which comprises predominantly lapilli-lithic bearing coarse ash crystal tuff (GCO 1988). These are noted to be some of the oldest group of rocks within Hong Kong and were

D.K. Koungelis (✉)
Donaldson Associates Limited, Glasgow, Scotland
e-mail: d.koungelis@donaldsonassociates.com

R. Lyall
Donaldson Associates (Asia) Limited, Hong Kong, China
e-mail: r.lyall@donaldsonassociates.com

G. Lollino et al. (eds.), *Engineering Geology for Society and Territory – Volume 6*,
DOI: 10.1007/978-3-319-09060-3_143,

Fig. 143.1 Cores of Fe stained tuff adjacent to hydrothermal mineralized zones and zones of no recovery due to disturbance during drilling

formed prior to the most recent granitic intrusions which are dominant on Hong Kong Island and the Kowloon Peninsula; therefore the rock at NTM has been subjected to regional volcanic/hydrothermal activities. Additionally, the Geological memoir (GCO 1989) identifies some WSW-ENE striking bands of 'sandstone' (possibly volcaniclastic sedimentary units) along with a dyke of feldsparphyric rhyolite to the south of the proposed Cavern on the southern slope of Kai Keung Leng, the mountain which is being tunnelled through. The effect of hydrothermal activity in altering the geological conditions is highlighted by (Williams 1991). The majority of tunnels in Hong Kong have been built in granites and south of the Tai Mo Shan fault. Other than small water tunnels, no large diameter railway tunnels have been constructed within ground conditions similar to this site. Three phases of ground investigation were carried out at various points throughout the project design stage and prior to construction. These were: Preliminary Design Stage, Detailed Design Stage and Post Contract Award (AGI). The preliminary ground investigation identified zones of highly weathered tuff and no recovery which the borehole logs recorded as 'fault gouge'. Similar features were identified in the detailed design ground investigation which led to the interpretation of faulted or sheared ground adjacent to the NTM shaft. A number of quartz dykes and mineralisation associated with hydrothermal fluids such as pyrite and galena were also recorded in the boreholes. The DAL geotechnical team identified this mineralisation to be both 'fresh' and 'weathered' therefore used this as a basis for re-evaluating the geological conditions in the areas by considering the regional geology. The review concluded that shearing of the rock mass occurred in the past however mineral rich hydrothermal fluids had infiltrated the rock mass predominantly along the open fissures created by shearing. Over time, these fissures were filled by quartz and sulphides rich minerals such as pyrite and galena effectively 'healing' the rock mass and improving the overall rock quality. However, where water is still able to flow along joints or along the quartz veins, oxidation of the pyrite occurred causing the rock mass to become deeply iron stained and to form localised discrete weathering features which have been disturbed during drilling and recorded as "no recovery" (Fig. 143.1).

143.3 Analysis

A review of all available data was carried out to assess the rock mass characteristics and derive 'Q' (Barton et al. 1974) ranges to be used for the design of the temporary support for

Table 143.1 Temporary support requirements for the crossover cavern

Temporary support class	Mapped 'Q' range	Dowel spacing (m)	Dowel length (m)	Sprayed concrete thickness (mm)
CC-3	$1 \leq$ 'Q' < 2.5	1.70	8.00	125
CC-4	$0.1 \leq$ 'Q' < 1	1.30	8.00	250

Table 143.2 Summary of joint set orientations

Joint set number	Dip (°)	Dip direction (°)	Apparent dip (°)
1	61	148	339
2	03	148	355
3	67	355	329

the rock mass within the Crossover Cavern. Dip and dip direction information, was input into DIPS software to allow pole plots and contour plots to be produced. These plots have been used to assess the number and orientation of joints within the rock mass. Numerical modelling, using the discrete element code UDEC was carried out to assess the stability of the Crossover Cavern during the various excavation stages. Details of each of these design elements are discussed in the following sections. The initial stage in the determination of the temporary support for the Crossover Cavern was carried out using the empirical 'Q'-method. 'Q' value ranges (between $0.1 \leq$ 'Q' ≥ 6) for design were agreed with the main contractor.

Temporary support for all 'Q' ranges was determined for detailed design. This paper focuses only on the assessment of a combination of temporary support classes; i.e. CC-3 and CC-4 (Table 143.1), where a 5 m thick discrete zone (i.e. CC-4—Hydrothermal zone) exists above crown of the Crossover Cavern, within a rock mass of CC-3 (Fig. 143.2). Hence, numerical analysis predictions will be presented only for this case. For these analyses, discontinuity orientations (dip/dip direction) were collated and input into DIPS. Stereographical equal area, lower hemisphere projections were used to generate contour plots allowing the dominant joint sets and their relationships to be determined (Table 143.2).

Numerical modelling of the Crossover Cavern has been carried out using UDEC. The purpose of these calculations was to interrogate the structural forces imposed on the temporary support sprayed concrete and rock bolts, and resulting deformations, following each stage of construction. The aim of the numerical modelling was also to verify that the temporary support determined (Table 143.1) using empirical methods (Barton et al. 1974) is adequate to stabilise the open excavation until such a time as the permanent lining is installed. Predicted loads imposed on the bolts have been used to select dowel diameter and steel grade and that the bolt length extends beyond the anticipated yield zone. A deformable block model using Mohr Coulomb criterion has been adopted, assigning strength and deformation parameters to the blocks. Immediately surrounding the Crossover Cavern, a joint pattern was superimposed on top of the deformable block model to determine the effect of the jointing, which will create an anisotropy of movement and stress (Fig. 143.2). These discontinuities were assigned shear strength parameters (Table 143.4). The rock mass parameters are presented in Table 143.3. Design parameters for the analysis were based on intact rock properties. The rock mass was assumed to be dry during construction with a coefficient of earth pressure at rest Ko = 1.2. The overburden depth of 140 m has been modelled using an equivalent 'block load'. 50 % of stress relaxation was assumed prior to any temporary support installation. The tunnel profile and excavation sequence has been modelled, including the sequence of support installation (bolts and sprayed concrete) following excavation. To allow bolt support to be installed immediately following excavation of the individual headings, 4.0 m long Swellex friction dowels will also be installed on a staggered pattern between the 8.0 m long bolts. Each phase of excavation is allowed to run to equilibrium before the next dig and support phase is modelled.

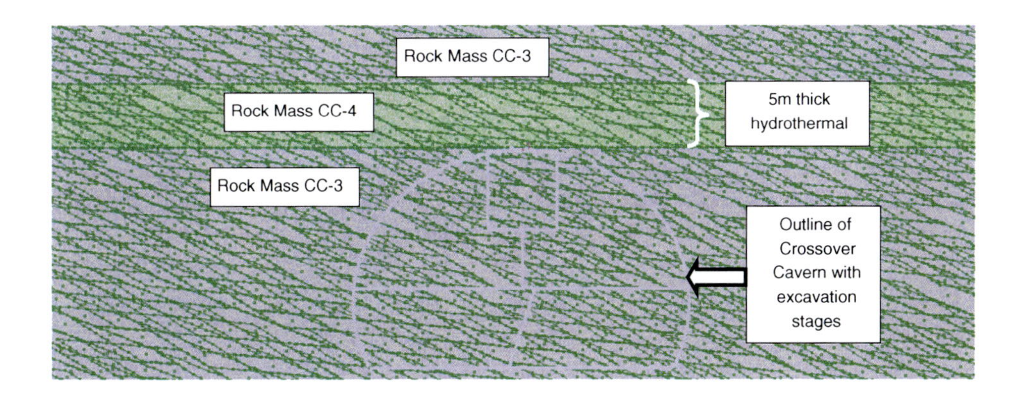

Fig. 143.2 UDEC block model

Table 143.3 Intact rock properties for UDEC model

Temp. Support class	Density (kg/m^3)	Bulk modulus (GPa)	Shear modulus (GPa)	Friction angle (°)	Cohesion (MPa)	Tension (MPa)	Dilation (°)
CC-3	2600	13.8	10.4	35	8.337	0	10
CC-4	2600	5.5	1.85	30	0.525	0	0

Table 143.4 Joint properties for UDEC model

Temp. support class	Normal stiffness (GPa/m)	Shear stiffness (GPa/m)	Friction angle (°)	Dilation (°)	Tension (kPa)	Cohesion (kPa)
CC-3	12	1.2	37	7	0	50
CC-4	7	1.0	30	0	0	0

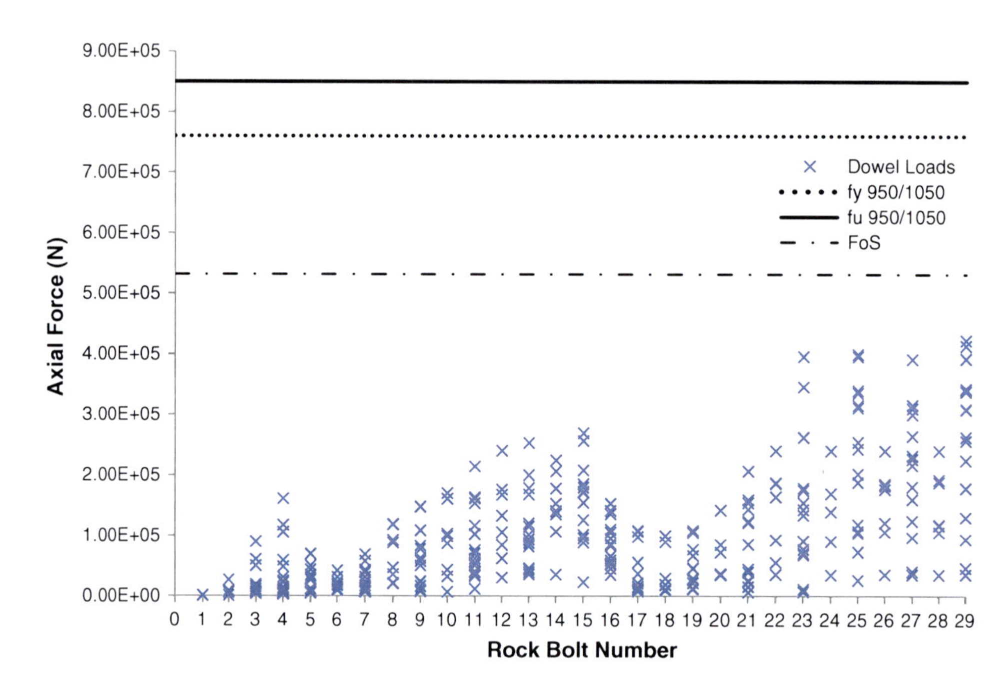

Fig. 143.3 Predicted rock dowel loads

143.4 UDEC Predictions

The final deformations and forces developed in the rock bolts following excavation and installation of temporary support from the UDEC analysis is presented in Figs. 143.3 and 143.4. The predicted maximum vertical displacements (occurred at the crown) were limited to <50 mm (Fig. 143.4). The factor of safety achieved (broken line on Fig. 143.3) on the loads developed in the high yield steel bars (solid line on Fig. 143.3 indicates the ultimate load of the rock bolts) was satisfactory (FoS $\geq$ 2.0 being achieved). Based on the UDEC output, the adoption of 26.5 and 32 mm diameter high yield grouted steel bar (Grade 950/1,050 N/mm^2) is considered to be appropriate for the Crossover Cavern for temporary support class CC-3 and CC-4 respectively. The analysis indicate that the recommended temporary support for the particular ground conditions, will be satisfactory and the excavation will remain stable throughout construction. Figure 143.5 shows the Cavern excavation finished with the required temporary support installed.

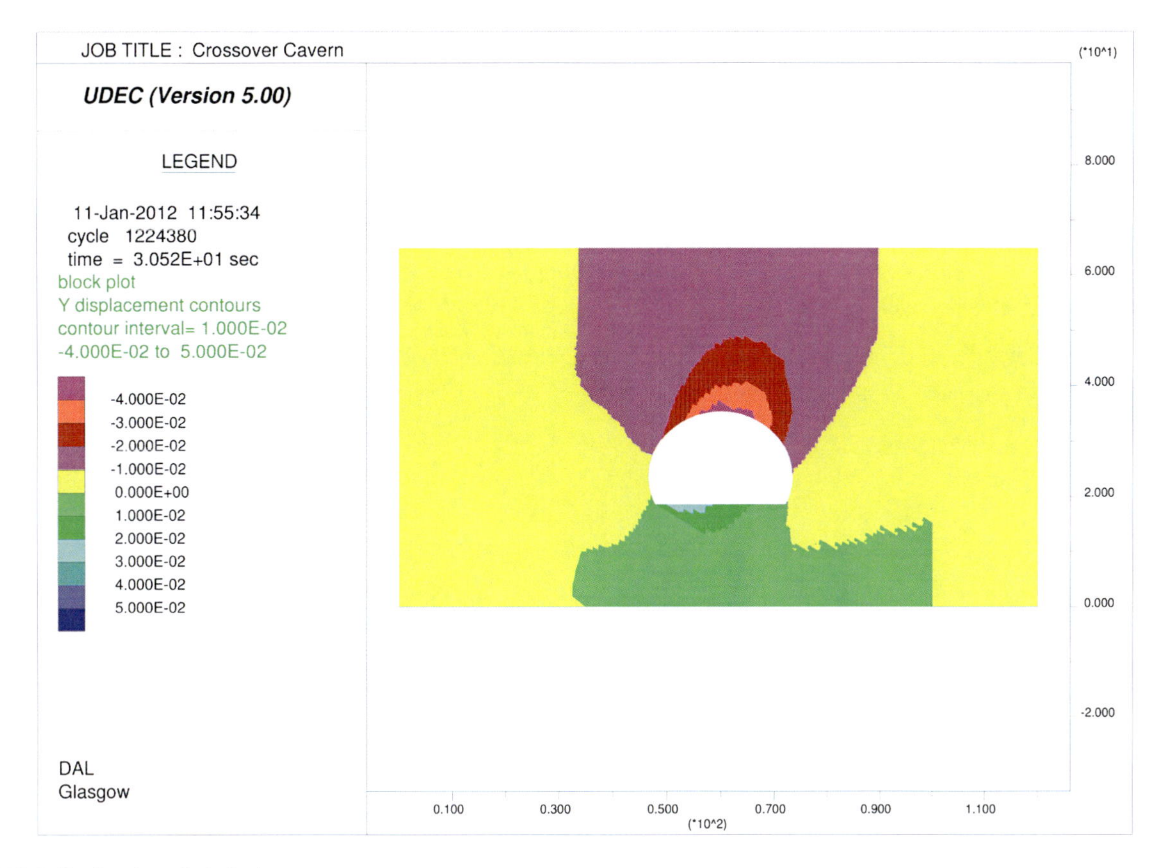

Fig. 143.4 Predicted vertical displacements

Fig. 143.5 Full face excavation of the Crossover Cavern

143.5 Discussion

A critical review of the available ground investigation data and the regional geology was carried out for C824 to investigate the geological characteristics of the Crossover Cavern. The ground conditions were found to be better than anticipated at tender stage due to hydrothermal minerailsation having a positive impact on the rock mass, confirming the 26.9 m span Crossover Cavern could be moved into an area previously thought to contain a number of shear zones.

Acknowledgements The authors would like to thank: KKOJV and the MTRCL for their permission to submit this paper and Donaldson Associates for their support in preparing this paper.

References

GCO (1988) Hong Kong geological survey sheet 6 (1:20,000). Solid and superficial geology for Yuen Long. Geotechnical Control Office, Hong Kong

GCO (1989) Hong Kong geological survey memoir no. 3. Geology of the Western New Territories. Geotechnical Control Office, Hong Kong

Williams T (1991) The story of Lin Ma Lead Mine, 1915–1962. Geol Soc Hong Kong Newslett 9(4):3–27

Barton NR, Lien R, Lunde J (1974) Engineering classification of rock masses for the design of tunnel support. Rock Mech 6(4):189–236

The Influence of Geological History on Preferred Particle Orientation and the Observed Anisotropy of Over Consolidated UK Mudrocks

144

Stephen Wilkinson and Clark Fenton

Abstract

One of the major aspects of mudrocks which influences their engineering behaviour is anisotropy, especially that of strength and stiffness. Anisotropy is caused by an underlying preferential alignment in particle orientations. Many factors contribute to the production of enhanced preferred particle orientations including: sedimentation process; particle shapes; bioturbation; burial depth; tectonism; weathering; and aging; in short the material's entire geological history. Initial structure is formed by sedimentation processes during deposition. These are then modified by post-depositional events. With burial, the particles of sediments tend to rearrange to allow for a decrease in void ratio which is combined with an expulsion of water from the soil. Much of the southern UK is underlain by mudrocks of Mesozoic and Cenozoic age displaying some degree of preferred particle orientation and hence anisotropy in their engineering behaviour. The degree of particle orientation is quantified by analysis using environmental scanning electron microscope imagery. Anisotropy in engineering behaviour has been quantified by a range of laboratory and field, static and dynamic, methods that allow the anisotropic elastic behaviour of the mudrocks to be investigated at very small strains. Generally good agreement is observed between four fully independent methods for evaluating the elastic G_{VH} stiffness mode. The results of both the image analysis and the laboratory testing build up a picture of microstructure anisotropy that results from the total geological history of each of the mudrocks investigated.

Keywords

Mudrocks • Microstructure • Anisotropy

144.1 Introduction

Microstructure is defined as the combination of the arrangement of particles which is known as "fabric" and the cementation between particles which is known as "bonding" (Gasparre and Coop 2008). Every element of the microstructure of mudrocks is formed as a result of events occurring during the mudrocks geological history. One of the most important aspects of soil fabric for engineers is anisotropy of properties such as strength and stiffness. Anisotropy is caused by the co-alignment of platy particles such as clay minerals. Although particle alignments can form during deposition (Barden and Sides 1971), alignments are often enhanced through an increase in the applied load i. e., burial. This process has been observed both experimentally in the laboratory (Vasseur et al. 1995), and through field observations (Ho et al. 1999). Other factors which influence the orientation of particles during burial are determined within the environment of deposition (Kim et al.

S. Wilkinson
Department of Civil Engineering, Xi'an Jiaotong-Liverpool University, Suzhou, China

C. Fenton (✉)
Department of Civil and Environmental Engineering, Imperial College London, London, UK
e-mail: c.fenton@imperial.ac.uk

G. Lollino et al. (eds.), *Engineering Geology for Society and Territory – Volume 6*,
DOI: 10.1007/978-3-319-09060-3_144, © Springer International Publishing Switzerland 2015

2001), the chemistry of pore fluids (Meade 1964). Thus many aspects of the geological history of a soil can influence its final degree of anisotropy.

144.2 Measurement of Preferred Particle Orientation

High quality, undisturbed block or wireline cored samples were obtained from several sites (Fig. 144.1) in the southern part of the UK to allow comparisons with previously analysed samples of London Clay (e.g. Gasparre et al. 2007). Each site was selected to minimise the influence of post-depositional tectonism (Wilkinson 2011). All of the samples were imaged using an environmental scanning electron microscope (E-SEM). Vertically and horizontally orientated broken surfaces were imaged separately (Fig. 144.2). Three different techniques were used to make measurements of preferred particle orientation. The results presented here (Table 144.1) are the ratio of the maximum and minimum eigenvalues (λ_a) of the orientation matrix:

$$OM = \begin{bmatrix} \frac{1}{n}\sum_{i=1}^{n} \cos\theta_i \cos\theta_i & \frac{1}{n}\sum_{i=1}^{n} \cos\theta_i \sin\theta_i \\ \frac{1}{n}\sum_{i=1}^{n} \cos\theta_i \sin\theta_i & \frac{1}{n}\sum_{i=1}^{n} \sin\theta_i \sin\theta_i \end{bmatrix} = \begin{bmatrix} a & b \\ c & d \end{bmatrix}$$

$$\lambda e = \frac{\lambda_{\max}}{\lambda_{\min}} \quad \text{where} \quad \lambda = \frac{a+d}{2} \pm \left(4bc + (a-d)^2\right)^{1/2}$$

where θ is the maximum ellipse angle for an ellipse which has the same moment of inertia as particles identified using, in this case, the upper 25 % of grey levels (shades from a light grey to white) within the electron microscope images taken of broken surfaces in a vertical orientation. As anisotropy of strength and stiffness is derived from the comparison of vertically and horizontally orientated samples, preferred particle orientation should theoretically also require the comparison of vertically and horizontally orientated broken surfaces. However, samples with high measured preferred particle orientations on vertically orientated surfaces in some cases show increased preferred particle orientation on horizontally orientated surfaces, often caused by steps in the surface between the horizontally orientated planes. Thus a comparison between preferred particle orientations from vertically and horizontally orientated samples has a tendency to reduce the measured level of preferred particle orientation in strongly aligned samples. Therefore the best measurements are obtained using vertically orientated samples alone.

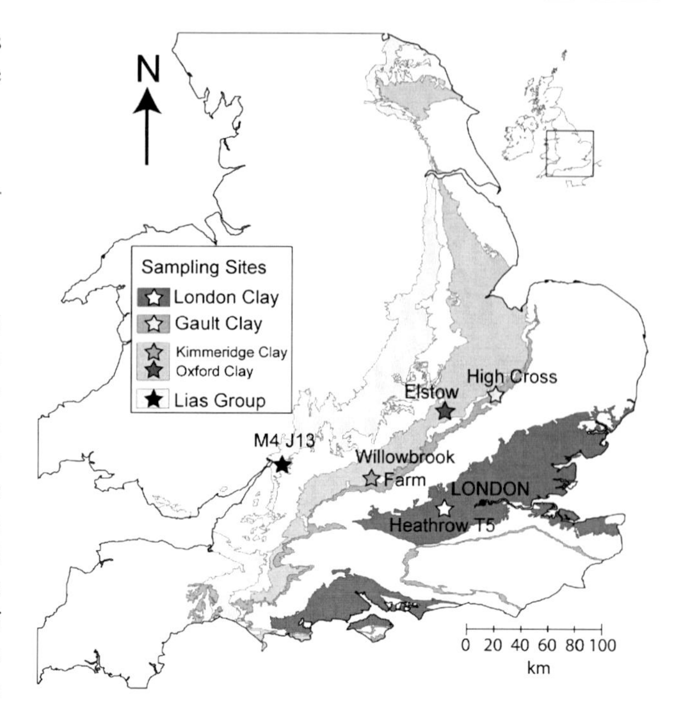

Fig. 144.1 Mudrock sampling locations

In addition to the method above, half circle addition and three-dimensional surface aspect analysis was also used to quantify anisotropy. All three methods produced similar results (Wilkinson 2011).

Comparison of the quantified measurements of preferred particle orientation and laboratory measurements of the anisotropy of elastic stiffness (Wilkinson et al. 2011), show a good correlation (Fig. 144.3).

144.3 Discussion

Since particle arrangements are a result of geological history, it is implied that geological observations may be used to predict/indicate the engineering behaviour of mudrocks. The preferred particle orientations in turn have a strong correlation with engineering behaviour. The depositional history of soils can be inferred through the field observations and classification of soils. Burial and uplift histories can be assessed by a combination of stratigraphic reasoning, measurement of the uplift of terraces and through apatite fission track analysis. By assuming constant geothermal gradients the maximum depths of burial can be assessed. High values of preferred particle alignment are obtained for the Gault, Oxford, and Lias Group Clays, spanning a broad range of burial depths. This shows that although burial depth is important for the development of individual mudrock

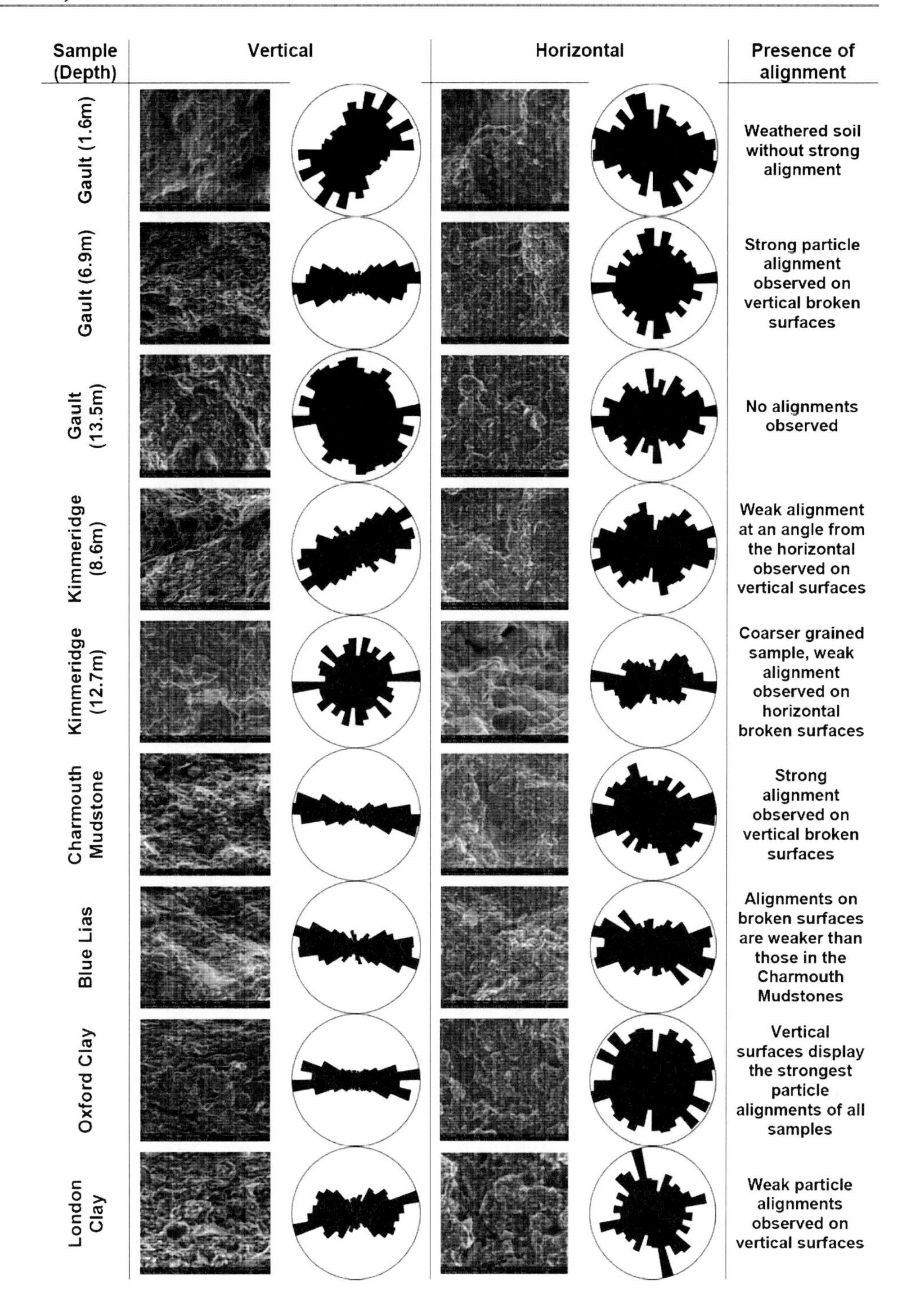

Fig. 144.2 Normalised rose diagrams of particle orientations for individual images for all sampled mudrocks. Note comparison of particle alignment between vertical and horizontal cut surfaces

Table 144.1 The average value of preferred particle orientation based on all images take from a single sample surface using magnifications between X600 and X1500

	Gault			Kimmeridge		Charmouth mudstone	Blue lias	Oxford clay	London clay
Depth	1.6	6.9	13.5	8.6	12.7				
λa	1.54	3.28	1.61	2.05	1.44	3.64	2.72	3.97	2.13
SD	0.19	0.38	0.41	0.43	0.19	0.65	0.61	0.78	0.58

Standard deviations (SD) from image measurement tend to be higher where preferred particle orientation is higher

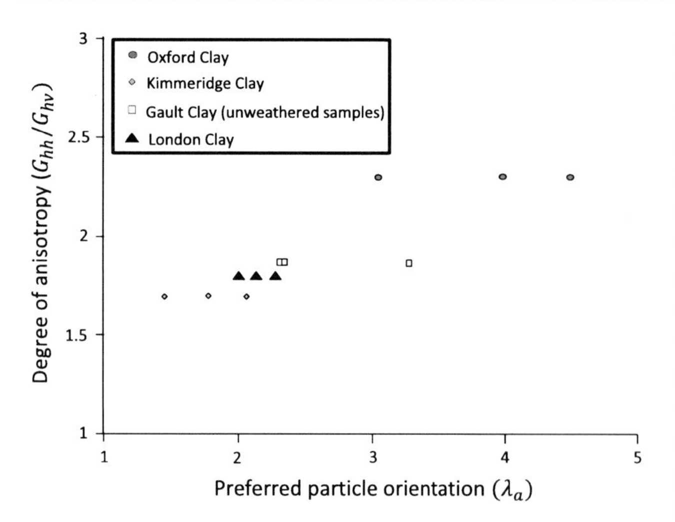

Fig. 144.3 Plot of the degree of anisotropy of stiffness and measurements of preferred particle orientation

structure, it cannot be used alone as an indicator of the preferred particle orientation. Arguably the depositional setting of each mudrock horizon is more important as it creates the initial conditions from which each mudrock evolves during its burial and uplift history. For example, the Kimmeridge Clay samples are from a shallow, coarse grained area of the depositional basin. These samples contain a large proportion of sand and silt grains. Some horizons have a relatively low measurement of preferred particle orientation, as the silt grains prevent the platy clay minerals from aligning (Barden and Sides 1971).

The final element for consideration in the geotechnical cycle is the weathering and erosion of soils. The evidence from the shallow samples of the Gault Clay is that weathering results in a reduction in preferred particle orientation. One out of five sample of the shallow Gault Clay samples displayed a much stronger preferred particle orientation. The enhanced weathering along joints, which as they are weaker zones are preferentially susceptible for forming the broken surfaces, completely replaces the orientations which are present in regions in the blocks between the joint surfaces. With continued weathering, depositional and burial structures may be completely removed.

144.4 Conclusions

Structural and strength anisotropy in mudrocks is determined by the geological history of the material. The full geological history of a sample can be defined by burial and uplift histories which can be well identified using apatite fission track analysis data. Although burial is important for the formation of preferential orientations in a single mudrock it cannot be used as a unique indicator to compare different mudrocks. Within an individual mudrock the burial history determines how structure evolves. The depositional variation within mudrocks is greater than that between any other sediment type and so most mudrock formations are not directly comparable. Within individual formations natural variations in the depositional environment can cause the structure and engineering properties of a mudrock to alter within the stratigraphic sequence. In deeply buried unweathered mudrocks, clay mineral and low silt content are good indicators of high levels of particle alignment. The most important factors which can reduce alignment in soils are the presence of silt/sand and weathering processes. The first is a result of the environment of deposition the second is a result of the present day environment.

References

Barden L, Sides G (1971) Sample disturbance in the investigation of clay structure. Géotechnique 21(3):211–222

Gasparre A, Coop MR (2008) Quantification of the effects of structure on the compression of a stiff clay. Can Geotech J 45:1324–1334

Gasparre A, Nishimura S, Coop MR, Jardine RJ (2007) The influence of structure on the behaviour of London clay. Géotechnique 57 (1):19–31

Ho N, Peacor DR, Pluijm BAVd (1999) Preferred orientation of phyllosilicates in Gulf Coast mudstones and relation to the smectite-illite transition. Clays Clay Miner 47(4):495–504

Kim JW, Lee YDE, Tieh TT (2001) Effects of laboratory consolidation on petrophysical properties of fine-grained marine sediments: electron microscope observations. Mar Georesour Geotechnol 18:347–360

Meade RH (1964) Removal of water and rearrangement of particles during the compaction of clayey sediments—review. US Geol Surv Prof Pap 479-B:23 p

Vasseur G, Djeran-Maigre I, Grunberger D, Rousset G, Tessier D, Velde B (1995) Evolution of structural and physical parameters of clays during experimental compaction. Mar Pet Geol 12(8):941–954

Wilkinson S (2011) The microstructure of UK mudrocks. PhD: unpublished PhD thesis, Imperial College London, 400 p

Wilkinson SA, Brosse CH, Fenton R, Kamal H, Jardine RJ, Coop MR (2011) An integrated geotechnical study of UK mudrocks. In: Anagnostopoulos et al (eds) Proceedings of the 15th European conference on soil mechanics and geotechnical engineering, part 1, pp 305–210

Mechanical Characterization of Weathered Schists

145

Thomas Le Cor, Damien Rangeard, Véronique Merrien-Soukatchoff, and Jérôme Simon

Abstract

Fractured and weathered rocks such as Brioverian schists are encountered in the West part of France. Their important variability, in terms of alteration and mechanical properties, is a problem regarding the geotechnical design of retaining structures and are foreseen when excavating the second subway line in the city of Rennes starting in 2014. Results of mechanical laboratory characterization carried out on schist samples extracted from various locations in Rennes are presented. Uniaxial compression test and direct shear test results show an important variability of the material and a great sensitivity to water.

Keywords

Shear test • Discontinuities • Weathered rock • Anisotropy

145.1 Introduction

Previous studies on the Brioverian schist (Jégouzo 1973; Le Corre 1978) have focused on the geological context of its formation whereas its mechanical behaviour is less acknowledged. The Brioverian massif is the occidental part of the European Variscan belt and is composed of sediments that formation age estimates between 750 and 520 Ma. These sediments were submitted to a greenschist type metamorphism. The numerous constructions in Brioverian formations in Rennes, including the excavation of the second subway line beginning in 2014, incites to improve the knowledge on the mechanical properties of this type of ground. The high variability in a matter of degree of fracturing, alteration, clay content (Le Cor et al. 2013) is an important problem when it comes to determining mechanical parameters used to design retaining structures. Considering the high density of fractures of the Brioverian schist, mechanical characterization requires to study the intact rock matrix and the discontinuities in order to have a global understanding of its behaviour.

To evaluate the mechanical strength of the rock matrix, uniaxial compression tests were carried out on plane-parallel samples extracted from different sites spread over Rennes (Fig. 145.1). The mechanical parameters of opened discontinuities, in the direction of the schistosity (cohesion and friction angle), corresponding to the weakest plane of the rock, were determined using direct shear tests on small samples (sheared surface of approximately 50 × 30 mm).

T. Le Cor (✉) · D. Rangeard
Civil Engineering and Mechanical Engineering Laboratory, INSA/IUT Rennes, European University of Brittany, 20 Avenue Des Buttes de Coësmes, CS 14315, 35043 Rennes Cedex, France
e-mail: thomas.lecor@groupe-dacquin.fr

D. Rangeard
e-mail: damien.rangeard@insa-rennes.fr

T. Le Cor · J. Simon
Design department, Groupe Dacquin, Parc d'activités des Portes de Bretagne Servon-sur-Vilaine, 35530 Brittany, France
e-mail: jerome.simon@groupe-dacquin.fr

V. Merrien-Soukatchoff
Equipe Génie civil & Géotechnique, Département Ingénierie de la Construction et Energétique, CNAM, 2, Rue Conté, Case Courrier 341, 75141 Paris Cedex 03, France
e-mail: veronique.merrien_soukatchoff@cnam.fr

V. Merrien-Soukatchoff
Laboratoire de Géodésie et de Géomatique (L2G), Le Mans, France

G. Lollino et al. (eds.), *Engineering Geology for Society and Territory – Volume 6*,
DOI: 10.1007/978-3-319-09060-3_145, © Springer International Publishing Switzerland 2015

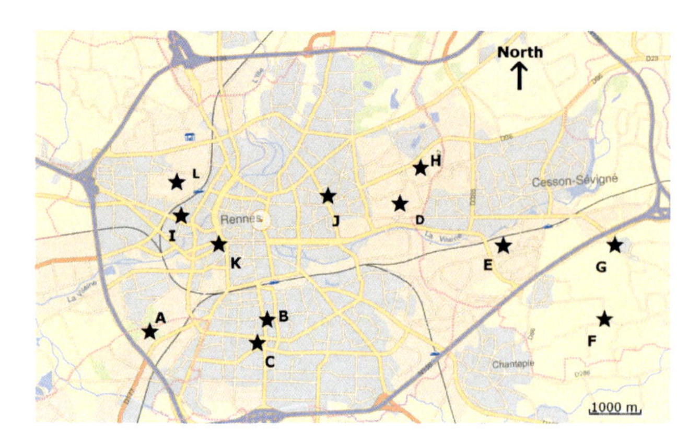

Fig. 145.1 Localization of the sampling sites (www.Géoportail.fr)

The results of these two types of tests are presented in the next sections of this paper.

145.2 Uniaxial Compression Tests

Uniaxial compression test is a simple method used to estimate the uniaxial compression strength (UCS) of a material (concrete, rock…) whether with isotropic or anisotropic mechanical properties (Nasseri 2003). Brioverian schists are foliated inducing a transverse isotropic behaviour. Due to the impossibility of drilling the rock without breaking it, the tested samples were plane-parallel and cut in an approximate size of 60 × 30 × 30 mm. Difficulties of sampling in weathered rocks is not unusual (Marques et al. 2010). Consequently, only specimens with a loading direction according to the schistosity have been tested to present. Samples with schistosity oriented perpendicularly to the loading could not be obtained because of losses during the cutting. In order to obtain a good planarity and parallelism of the loading faces of the samples, a plaster treatment was tested but gave the same results as the use of a ball joint between the loading plate and the loading axis which required much less preparation. Each site was tested using at least 6 samples stored in different conditions: 3 samples dried during a week at ambient temperature (water content lower than 1 %) and 3 samples immersed in water during a week (water content of the samples varied from 1 to 17 %). In all, 94 samples were tested: 48 "dry" samples and 46 "wet" samples. The results obtained are displayed on Figs. 145.2 and 145.3

An important influence of water condition is observed: the UCS decrease with the increase of the water content of the samples, when on the contrary, the UCS increase with the wet density. These variations are fitted to power laws which were observed by different authors on altered metamorphic rocks (Sousa et al. 2005; Marques et al. 2010; Vásárhelyi and Ván 2006). The variation of the water content after the same duration of immersion from a sampling site to an other can be explained mainly by the differences of clay content in the samples. Indeed, a clayey sample will have a greater sensibility to water than a sandstone type sample, particularly if the clays are of the swelling type.

As displayed on Fig. 145.3, a good correlation between water content and wet density was obtained, revealing that the samples with the lowest density are the ones reaching the highest water content and consequently the lowest UCS.

145.3 Direct Shear Test

Brioverian schists present, in situ, an important fracturing. These fractures constitute privileged plans of failure during excavation works. The characterization of these plans can be done using direct shear test on small samples of material presenting an opened joint. In order to perform these tests, we were able to use a direct shear test installation mainly used for soils (Fig. 145.4). Indeed, shearing opened joints require much less energy than necessary for a closed joint.

The experimental program consisted in testing samples from 11 different sites and in two different conditions: dry joint and joint immersed in water. Samples tested were of different sizes but the sheared surface was approximately about 60 × 30 mm. The preparation of the samples required

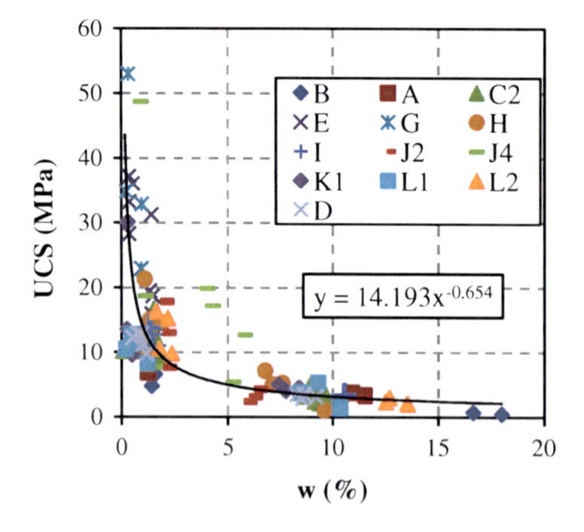

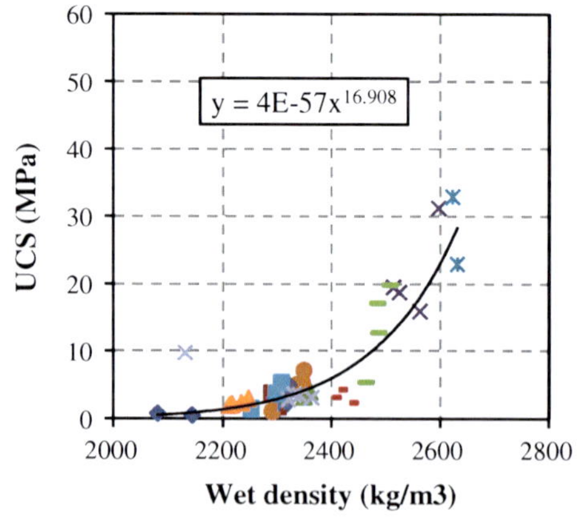

Fig. 145.2 Evolution of the UCS with the water content (*left*) and the density of the wet samples (*right*)

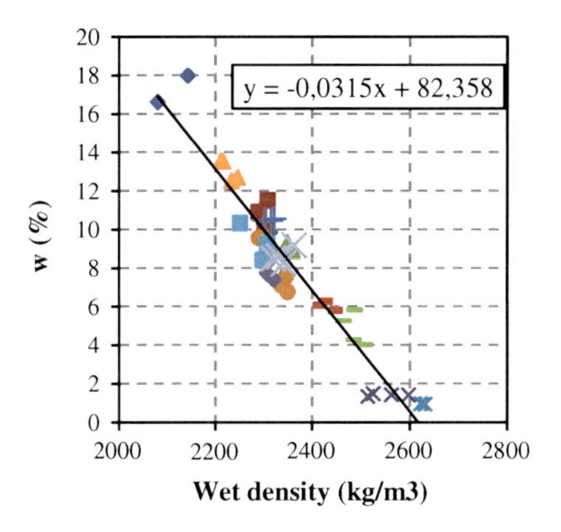

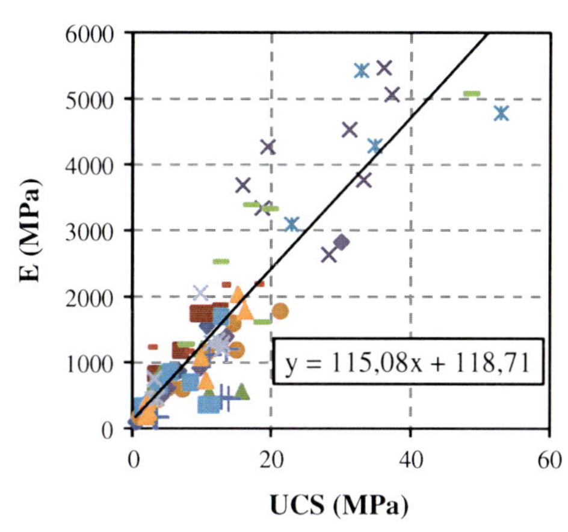

Fig. 145.3 Evolution of the water content with the wet density (*left*) and the tangent Young's modulus with the UCS (*right*)

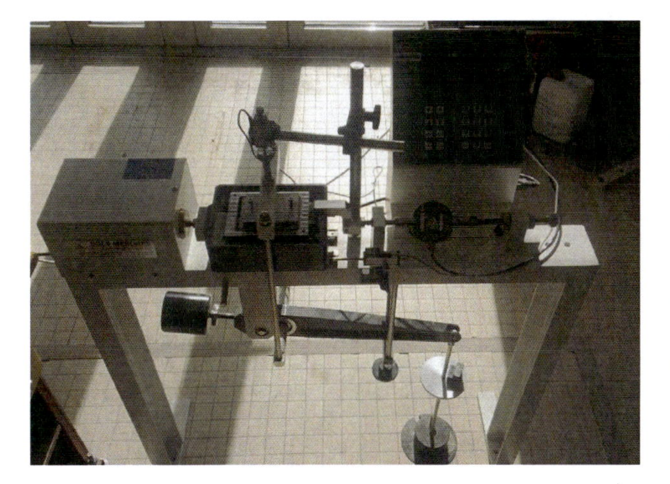

Fig. 145.4 Experimental shear test equipment

to seal both part of the samples in a mortar in order to maintain it during the test and also to facilitate the application of the vertical stress.

To compensate our limited stock of material, cyclic loadings were carried out on each sample (3–5 different normal loading were used) at a constant speed of 0.5 mm/min. Cyclic loading can lead to an important decrease of the mechanical characteristics of joints compared to joints sheared only once (Jafari et al. 2003). In our study an average decrease of 13 % on the cohesion and 7.5 % on the friction angle was observed when using cycling loading. For each site at least 3 samples were tested in dry conditions and for 6 sites 3 "wet" samples with the joint immersed in water 10 min prior to the test and before application of the normal stress. In the majority of the tests carried out, there was no distinct shear peak but a plateau, so that the couple (σ, τ) was determined by averaging this plateau. The results of these tests are displayed on Figs. 145.5 and the values of the Mohr Coulomb parameters for each tested site are summarized on Table 145.1.

As it can be observed on Fig. 145.5, despite the fact that the sites tested were all issued from the same geological formation, there is an important variability in the shear strength measured. Furthermore, the results presented on Fig. 145.3 indicates that the presence of water during shearing have an important influence on the behaviour of the joint, leading, in all the sites tested (except site A), to a decrease of the friction angle whereas two clear tendencies can be identified for the cohesion (an increase can be observed for the sites E, B and J whereas it's a decrease for the sites A, L, C2 and C3). However, if we consider the average cohesions and friction angles in both dry and wet

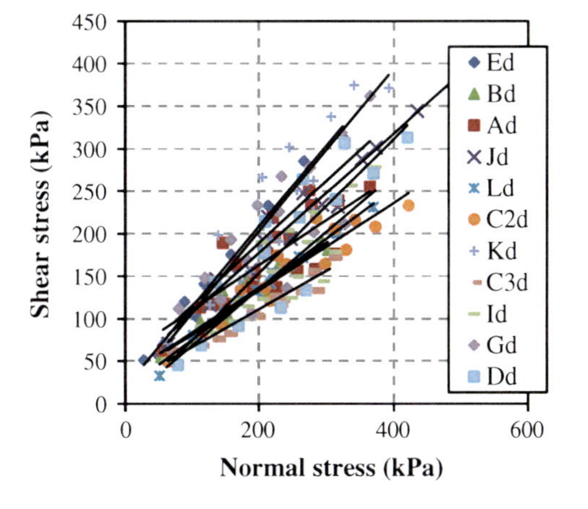

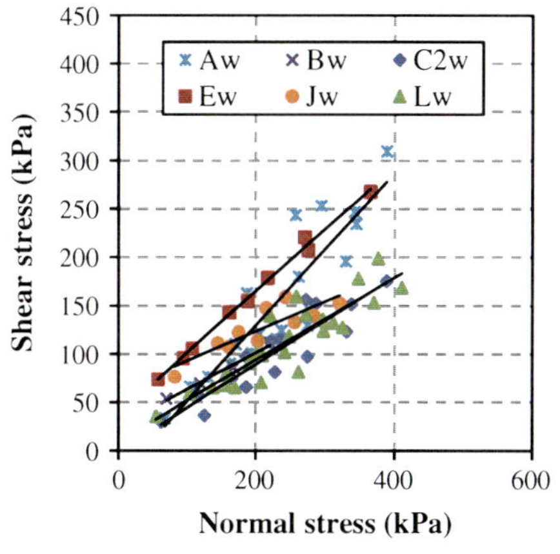

Fig. 145.5 Shear stress versus Normal stress in dry conditions and with joint immersed in water. "d" stands for dry and "w" for wet

Table 145.1 Mohr Coulomb parameters of the different sites tested

Site	Mohr Coulomb parameters				Dry-wet/dry	
	C′(kPa)		ϕ′(°)		ΔC′(%)	Δϕ(%)
	Dry state	Wet state	Dry state	Wet state		
E	18.7	42.5	41.1	34.8	−127.9	15.5
B	31.6	36.3	27.6	12.3	−15.0	55.6
A	57.3	−24.2	27.8	35.8	142.2	−28.7
J	36.5	62.2	35.2	17.1	−70.4	51.4
L	35.4	9.2	25.6	23.0	74.0	10.2
C2	29.9	−1.2	27.3	24.4	104.0	10.7
K	10.5	36.5	43.8	30.6	−247	30.2
C3	26.7	22.9	22.7	16.9	−59.2	26.2
I	0.9	/	33.9	/	/	/
G	32.5	/	39.0	/	/	/
D	−4.9	35.2	38.4	25.9	/	/
Average	**25**	**24.4**	**32.9**	**24.5**	**2.5**	**25.5**

conditions, a decrease of the two parameters is observed (2.5 % for the cohesion and 25.5 % for the friction angle) when tests occurred with the joint immersed. These observations have been made by other authors on clay rocks discontinuities (Pellet et al. 2013) but with a decrease of the mechanical parameters more important (about 50 %).

145.4 Conclusion

Brioverian schist, considering the low to medium mechanical strength obtained, can be classified as a HSSR type of ground (Hard Soil Soft Rock) if we refer to the classification proposed by Anon (1979a). The important impact of moisture on the mechanical strength of this material has been emphasized for each site tested.

UCS of the material tested varied in a large range from 5 to 50 MPa when the water content is lower than 5 %. However, for higher water content, the UCS ranges from 1 to 7 MPa. The results of the direct shear tests lead to the same results as the uniaxial compression tests: presence of water affected the shear strength of the opened joints tested. The decrease observed reached an average 3 % for the cohesion and 26 % for the friction angle.

To conclude, the sensitivity to water of this type of material should not be underestimated, especially when designing retaining structures such as planned for the second subway line of Rennes. Measuring water content is fast, affordable and gives good indications on the mechanical strength of the Brioverian schist.

References

Anon. (1979) Classification of rocks and soils for engineering geological mapping. Part 1—Rock and soil materials. Bull Int Assoc Geo 19:364–371

Jafari MK, Amini Hosseini K, Pellet F, Boulon M, Buzzi O (2003) Evaluation of shear strength of rock joints subjected to cyclic loading. Soil Dyn Earthq Eng 23(7):619–630

Jégouzo P (1973) Petrographical and structural study on lower Vilaine shales and granites. Phd thesis, Rennes University (in French)

Le Cor T, Rangeard D, Merrien-Soukatchoff V, Simon J (2013) Physical and mechanical characterization of weak schistose rocks. In: EUROCK 2013 symposium

Le Corre C (1978) Approche quantitative des processus synschisteux. L'exemple du segment Hercynien de Bretagne Centrale. Phd thesis, Rennes 1 University, Rennes, 382 pp (in French)

Marques EAG, Barroso E, Menezes Filho AP, Vargas EDA (2010) Weathering zones on metamorphic rocks from Rio de Janeiro-Physical, mineralogical and geomechanical characterization. Eng Geol 111:1–18

Nasseri M (2003) Anisotropic strength and deformational behavior of Himalayan schists. Int J Rock Mech Min Sci 40(1):3–23

Pellet FL, Keshavarz M, Boulon M (2013) Influence of humidity conditions on shear strength of clay rock discontinuities. Eng Geol 157:33–38

Sousa LMO, Del Rio LMS, Calleja L, Argandoña VGR, Rey AR (2005) Influence of micro-fractures and porosity on the physic-mechanical properties and weathering of ornamental granites. Eng Geol 77:154–168

Vásárhelyi B, Ván P (2006) Influence of water content on the strength of a rock. Eng Geol 84:70–74

Deformation of Soil and Rock Transition Belt Caused by the Mining Damage

146

Qinghong Dong, Fei Liu, and Qiang Zhang

Abstract

This study focuses on prevention and control of water issues when mining under the alluvium and thin bedrock in coalmines. At first a dual structure to describe the soil and rock transition belt in this situation is abstracted. Then a series of similar material simulations on considering the formation combination of hard/soft layers, fault and formation dip of certain degree are made to assess the affect caused by the dual structure to deformation and failure and each factor forms a specific deformation and failure characteristics. In addition, by obtaining cores from borehole and recording graphs of the wall of borehole, the deformation and failure characteristics and water blocking capability of the dual structure are verified. The research shows that the clay/rock structure has the coordination deformation characteristic, which is the deformation of the clay attaching to the ancient erosion surface of bedrock. If there appear no dramatic rock deformation and fracture, the clay layer will still has structural stability and continuity after mining.

Keywords

A dual structure • Clay • Thin bedrock • Coordination deformation • Continuity

146.1 Introduction

It has an important economic and social significance to mine the coal under water bearing layer and water body in China. The approach to keep mining safety is leaving coal or rock pillars above the coal seam in China like that in Britain, Japan, Australia and the former Soviet Union which have accumulated a lot of experiences for mining under water body (Liu 1982; Pertersion 1980; Zou 1998). The national regulation about underwater mining has been promulgated in 2000(State Bureau of Coal Industry 2000). Moreover, there are a lot of new understandings of the underwater mine have been gathered since 1950s (Liu 1998). When mining the coal seam under the water bearing layer of unconsolidated layers and shallow buried under the ancient erosion surface, the loose layer, bedrock composed of a dual structure which is a soft and hard contact or transition zone from up to down. This transition belt exhibits complex deformation and failure characteristics. Some works about the mechanical properties and deformation characteristics of deep clay has confirmed the continuity and impermeable stability of this bottom clay during mining (Sui et al. 2007; Huang 2009; Xu 2004; Ma et al. 2008; Li et al. 2000). The main work in the paper is to find the trends of the dual structural changing by model tests and in situ observations during or after coal mining.

146.2 A Dual Structure and Its Geological Conditions

In the eastern regions of China, the coal seams which lie under the unconsolidated layers of Quaternary or Neogene systems with water bearing layer in it, lake, river, and ocean water body and have a thin overlying rock pillars under extreme conditions are now being exploited because of the huge energy demand. As a result the upper boundary of mining area is rising accordingly.

Q. Dong (✉) · F. Liu · Q. Zhang
School of Resources and Geosciences, China University of Mining and Technology, Xuzhou, China
e-mail: dongqh@cumt.edu.cns

G. Lollino et al. (eds.), *Engineering Geology for Society and Territory – Volume 6*,
DOI: 10.1007/978-3-319-09060-3_146, © Springer International Publishing Switzerland 2015

As mentioned above, almost the thickness of the overlying rock is not greater than the height of the fractured zone. This kind of situation can conduct water flow easily. So the overlying bedrock is called thin bedrock for its relative thickness (Tu et al. 2004; Zhang 2004; Hu et al. 2006; Guo et al. 2006; Xuan 2008). In this situation, the soil-rock structure is a dual structure for water resistance. And its deformation and continuity are the most noteworthy features of mining safety. When the mining damage forming the caving zone, fractured zone and sinking/bending zone that extend to the bottom clay, the induced water passage would appear and cause water gushing. Actually there are many factors which have important influences on the deformation and failure of the dual structure.

Considering the most general case, because of the overlaying bedrock is composed by weathered sandstone, limestone, shale and etc., the rock mass is a combination of multi-engineering geological types of different weathering degree, different hardness or weakness. Therefore, the structure of overlying rock mass can be divided into 3 main kinds. They are the full of soft, full of hard and soft to hard from top to bottom. According to the inclination of strata, the situations of gently inclined (<35°) and moderately dipping (35°–55°) have been considered in the next experiments and analysis.

146.3 Affect of Formation Combination to the Deformation of Dual Structure

In order to observe the impact of stratum transformation and combination to the deformation of dual structure, a model without faults cutting from Yanzhou diggings, China is selected as the basic prototype. The basic parameters of the dual structure proportionately meet the actual site.

The observation is completed aided by a series model tests. Previous model tests about mining deformation and failure is mainly to reveal the changing process of caving zone, fractured zone and surface movement. But the discuss about the weak part of the dual structure is inadequate (Zhang et al. 2003). Obeying the regulations of model test, the uniaxial compressive strength of rock is taken as the main control parameter. Each similar model is equipped with the material equal to the 3 kinds of combination mentioned in part 2. The models all have a thickness of 30 cm, length of 150 cm, while the height is adjust according to the geometric scale and the depth of upper load. Figure 146.1 illustrates the overall trend of the relations between the formation combination and the deformation of the dual structure of gently inclined coal measures. The weathering zone and the clay layer covering on are inhibitions to the growth of the height of water flowing fractured zone. But each specific structure of the overlying strata gives a different deformation and failure phenomenon.

For the structure in Fig. 146.1a is a soft-hard combination. The hard roof forms a structure named masonry beams structure above the caving zone during mining. Then the soft part which is the weathered zone or mudstone attached to the masonry beam structure. This may play a coordinating role in the process of deformation. The clay at the bottom of Quaternary system will move toward the settlement direction by the driver of the overburden loads and is restricted by the edge of rock surface. The clay can maintain good continuity after large deformation.

If the overburden rock is soft-soft combination (Fig. 146.1b), it is difficult to form a masonry beams structure. As a result there appear a large range of overburden collapse but in general it converges within the weathered zone. Since the low strength and large deformation, the overlying clay is driven to keeping cover on the rock surface. But if the weathered zone is too weak or loose, the overlaying rock collapse may lead to ionospheric phenomena under the clay layer.

While the overburden rock is hard-hard combination (Fig. 146.1c), the masonry beams were formed during caving steadily. Above this is the upper rock mass under a structural stability situation. The deformation is too smaller for the overburden loads to drive the clay to move. This deformation almost has no affect on clay continuity.

These deformation characteristics indicate that the movement of the clay is limited by the bedrock surface deformation. The deformation of the dual structure determines its ability to maintain continuity and impermeable ability.

Fig. 146.1 Affect of formation combination to the deformation of dual structure. **a** Soft-hard combination. **b** Soft-soft combination. **c** Hard-hard combination

146.4 Affect of Faults to the Deformation of Dual Structure

In the coal mine with large faults density, mining equipment often makes its way by cutting the rock mass on the side of fault In order to avoid moving and re installation especially in the case of the fall <0.5 m or even more the fall <2 m in small faults density areas. Take a coal mine under unconsolidated layers in Yongcheng diggings, China for example, there are 4 faults lies in the area that the overlaying rock is no more than 30 m thick. All the faults are perpendicular to the direction of workface. The falls are 0.5–2 m and the dips are 45–75°. Similar material models are installed to simulate the situation of no fault, fault with dip angle 45° and 75°. The simulation reproduced the process of the deformation characteristics of the dual structure as shown in Fig. 146.2.

It can be seen from Fig. 146.2 that the fault affect to the dual structure cannot be ignore especially about the continuity of clay layer. The no fault condition model reveal a Normal mining subsidence basin in the bedrock surface. The clay layer attaching to the bedrock surface is still extending smooth and continuous during the subsidence or deformation, without obvious break (Fig. 146.2a). In the one fault mode, the bedrock surface subsidence basin affected by the fault. The subsidence around the fault zone is a little larger and the largest settlement appears in end of the fault (Fig. 146.2b, c).

It is obviously that the Clay layer attached to the bedrock surface appears an uneven settlement. The fault rupture may cause clay dislocation or destruction can be seen from the above results. The dual structure also has the synergistic deformation characteristics, but the continuity of clay layer is closely related to the fault activity. The aquifuge faces the risk of deterioration and even more being cut off or collapse. As a result in mining engineering, ground water in flow or underground quicksand will happen.

146.5 Affect of Inclination to the Deformation of Dual Structure

The dip of coal seam and inclination of the Erosion surface also have significant effect on the deformation and failure of the dual structure. It was reported in the paper that mainly discussed the deformation of overlaying rock relate to its combination of dip of coal seam and inclination of the Erosion surface (Dong et al. 2013). The works was done through a similar material model and a numerical model basing on the geological condition of a mine in Xuchang diggings, China. The basic parameters of the Dual structure proportionately meet the actual site as part 2.

The studies show that increasing of coal seam dip causes the overburden deformation and destruction offset towards the uphill. And the inclination of the erosion surface also has the same affect on overburden deformation and failure. But the affect induced by this offset on the covering layer of Neogene or Quaternary was almost rarely reported. The results of the similar material simulation are shown in Fig. 146.3. It can be seen that maximum height of overburden failure does not end within the bedrock surface, but there is a part of the top extending into the clay layer at the bottom of the Neogene systems. From the view of deformation, the main cause of the destruction occurred in the clay layer during the simulation should be part of mining

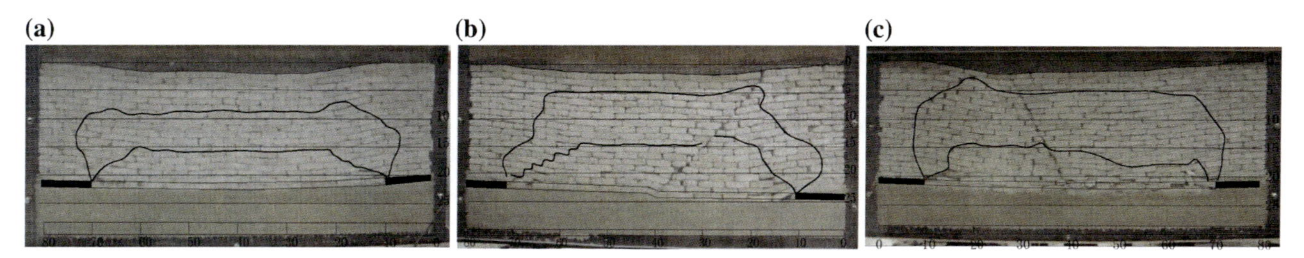

Fig. 146.2 Affect of faults to the deformation of dual structure. **a** No fault model. **b** One fault, dip = 45°, fall = 2 m. **c** One fault, dip = 75°, fall = 2 m

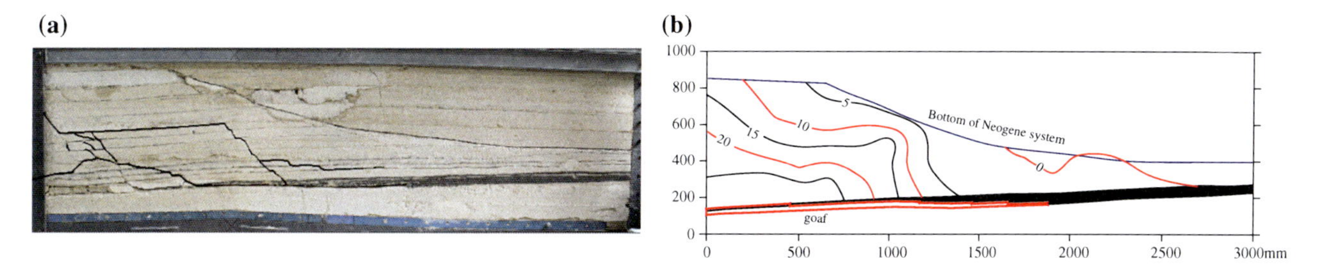

Fig. 146.3 Affect of inclination to the deformation of dual structure. **a** Deformation and failure. **b** Displacement offset of overlaying rock

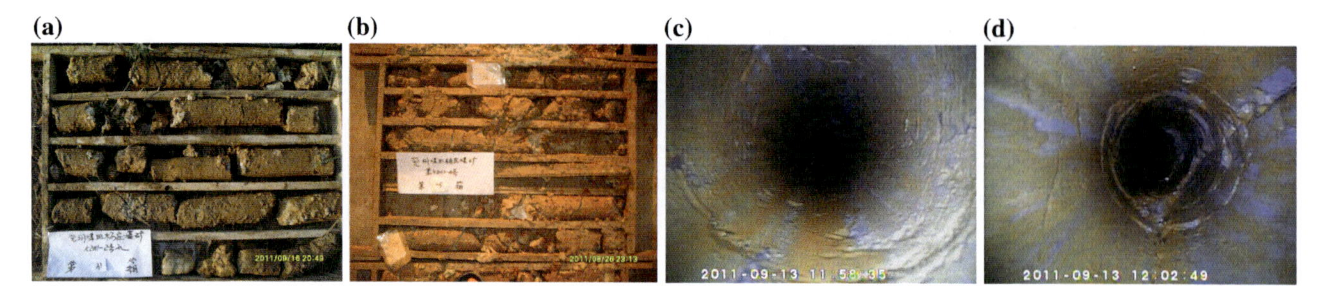

Fig. 146.4 Deformation and failure trends after mining. **a** Clay, **b** Sand, **c** overlaying rock (*top*) **d** overlaying rock (*middle*)

subsidence and horizontal movement. But the lack of confinement of the model is another reason that cannot be completely ruled out by now.

If the mining thickness is limited to a certain scale, the mining failure would do not cause dramatic damage and breakage to the clay at the bottom of Neogene system. And the impermeable role of the clay at the bottom of Neogene system can be controlled and protected.

146.6 Deformation and Failure Trends After Mining

In addition to the above regulations about the dual structure obtained by testing, An on-site observation was designed to evaluate the deformation and destruction of the dual structure after mining 14 years. The test site lies in a coal mine in Yanzhou diggings and the dual structure and the geological conditions are basically the same with the model shown in Fig. 146.1. This research area was recovered in 1997 and the first hierarchical is only 2.2 m in thickness at the bottom of the coal seam which is 9 m in thickness. In the on-site observation, all the soil samples from the drill hole have been photographed and all the wall of drill hole of overlaying rock was recorded by television imaging technique. The situation of the soil samples and overburden rock fissures are shown in Fig. 146.4.

From the photos and images of television, the clay at the bottom of Quaternary system is basically completely (Fig. 146.4a). But the sand layer looks like the loose sand without bonding (Fig. 146.4b). This may be due to the disturbances of drilling. This phenomenon is well proving the continuity of the clay layer after the first cutting for a long period of time. In other words, the fissures in the clay layer were bridged well because of the overburden load and the role of water-physical properties. According the borehole imaging, the fractured zone or disturbance area of low levels in the upper part of the bedrock the fractures are still can be seen but in the closed state (Fig. 146.4c). In the lower part of the bedrock, larger and steeply dipping fractures occurred. The walls of the Fissure are basically contacting with each other and partially opening as a small gap (Fig. 146.4d). The top of the hydraulic conductivity fractured zone after the first mining slice did not go through the weathered zone in the bed rock. The television imaging of four drill holes can explain the long time mining subsidence and consolidation of clay lead a series of changes to the dual structure. The fissures in the overlaying rock have been closing and the settlement is becoming to stabilize gradually. The clay layer keep covering on and attaching to the bedrock and its continuity did not change significantly after mining. It is suggested that, if the clay layer is thick enough, it was stable and impermeable. The resisting of this clay layer still has significance to further repeated exploration.

146.7 Conclusions

1. The dual structure is a typical geological model to analysis the deformation and failure of mining under unconsolidated layers and thin bedrock. The top edge of the cracking range is obviously reduced because of the thin bedrock and overlying clay. The synergistic deformation of the clay and overlaying rock is of great significance to analysis the continuity and impermeable ability of the dual structure.
2. The combination of overlaying rock is a main factor that affect on the deformation and failure of the dual structure. Hard-hard combination, soft-hard combination may lead little change to the continuity of the clay layer while soft-soft combination may lead destruction to the clay layer. Because of the plastic of the clay and the weak rock mass, the dual structure always appears a process of plastic deformation and this will cause the uniform, continuous deformation of the clay.
3. The fault and inclination of strata are main reasons to the offset of settlement, abnormal deformation and destruction in local areas.
4. It was found that the rock fissures have the tendency of compaction but having not been filled after a long time since the first cutting. The fissures in the clay layer were bridged well because of the overburden load and the role of water-physical properties.

Acknowledgments The authors want to acknowledge the financial support of the National Natural Science Foundation of China under Grant No.41272343, "A Project Funded by the Priority Academic Program Development of Jiangsu Higher Education Institutions" and "the Fundamental Research Funds for the Central Universities". We also would like to express our acknowledgements to the reviewers for their constructive comments.

References

Dong QH, Chen K, Du ZJ, Sui WH (2013) A case study of overburden failure and deep quicksand prevention. In: The international symposium & 9th Asian regional conference of IAEG

Guo WJ, Chen SJ, Li FZ (2006) Study on strip mining size under thick alluvium and thin bed rock. J China Coal Soc 31(6):747–751

Huang QX (2009) Simulation of clay aquifuge stability of water conservation mining in shallow-buried coal seam. Chinese J Rock Mech Eng 28(5):988–992

Hu BN, Zhao YX, Zhang HX (2006) Surface movement parameters and practice effect of strip mining under thick alluvium and thin bedrock. Coal Min Technol 11(1):56–58

Liu TQ (1982) Experiences about mining under the sea in Australia. Mine Survey 3:48–51

Liu TQ (1998) The optimization design theory and technology about the pillar for the outcrop coal seam. Coal Industry Publishing House, Beijing

Li WP, Ye JG, Zhang L, Duan ZH, Zhai LJ (2000) Study on the engineering geological conditions of protected water resources during coal mining action in Yushenfu mine area in the North Shaanxi Province. J China Coal Soc 25(5):449–454

Ma JR, Qin Y, Zhou GQ (2008) Research on triaxial shear properties of clay under high pressures. J China Univ Min Technol 37(2):176–179

Bureau National Coal Industry (2000) Regulations of resorting coal mining and pillars leaving under buildings, water bodies, railroads and main mines and entries. Coal Industry Press, Beijing

Pertersion HF (1980) Northeast submarine of British. World Coal Technol 3:40–43

Sui WH, Cai GT, Dong QH (2007) Experimental research on critical percolation gradient of quicksand across overburden fissures due to coal mining near unconsolidated soil. Chinese J Rock Mech Eng 26 (10):2084–2091

Tu M, Gui HR, Li MH, Li W (2004) Testing study on mining of waterproof coal pillars in thick loose bed and thick coal seam under ultrathin overlying strata. Chin J Rock Mech Eng 23(20):3494–3497

Xu YC (2004) Mechanic characteristics of deep saturated clay. J China Coal Soc 29(1):26–30

Xuan YQ (2008) Research on movement and evolution law of breaking of overlying strata in shallow coal seam with a thin bedrock. Rock and Soil Mech 29(2):512–516

Zou YS (1998) The practice and understanding of safety mining under water body in 20 years. Coal Sci Technol 1:53–55

Zhang GL, Cai R, Dong QH, Chen DJ, Zhao QJ, Liu ZX (2003) Engineering geologic mechanical model experimental study on the overburden failure induced by coal mining. Coal Geol China 15 (5):34–36

Zhang SK (2004) The law of mining pressure in shallow seam. Ground Press Strata Control 21(4):32–34

Geomechanical Assessment on a Metasedimentary Rock Cut Slope (Trofa, NW Portugal): Geotechnical Stability Analysis

147

M.J. Afonso, R.S. Silva, P. Moreira, J. Teixeira, H. Almeida, J.F. Trigo, and H.I. Chaminé

Abstract

This work emphasises the importance of a detailed geomechanical study to a better geotechnical understanding of a weak rock mass. In addition, Geographic Information Systems (GIS) tools for spatial analysis and rock database overlay were applied. The stability of a metasedimentary rock cut slope in Trofa region (NW Portugal, Iberian Peninsula) was studied. The main geomorphological, geological, geotechnical and geomechanical constraints were characterised and integrated along a highly deformed phyllitic rock mass. The scanline sampling technique of discontinuities has been applied to the study of basic geotechnical description of free rock mass surfaces. In order to classify the quality of the phyllitic rock mass, basic Rock Mass Rating (RMR), Slope Mass Rating (SMR) and the Geological Strength Index (GSI) were applied. The structural and geotechnical solution for the stabilisation of the cut slope was outlined. This methodology turned out to be valuable to understand the behaviour of slope stability in weak rock mass and it can be useful to the accurate estimation of future unstable rock slope.

Keywords

Cut slope • Weak rock mass • Geomechanical classification systems • Slope stability • NW Portugal

M.J. Afonso (✉) · R.S. Silva · P. Moreira · J. Teixeira · H. Almeida · J.F. Trigo · H.I. Chaminé
Laboratory of Cartography and Applied Geology (LABCARGA), DEG, DEC, School of Engineering (ISEP), Polytechnic of Porto, Porto, Portugal
e-mail: mja@isep.ipp.pt

R.S. Silva
e-mail: rui.m.santossilva@gmail.com

P. Moreira
e-mail: pfsm.82@gmail.com

J. Teixeira
e-mail: joaat@isep.ipp.pt

H. Almeida
e-mail: henriquemm.almeida@gmail.com

J.F. Trigo
e-mail: jct@isep.ipp.pt

H.I. Chaminé
e-mail: hic@isep.ipp.pt

M.J. Afonso · J. Teixeira · H.I. Chaminé
Centre GeoBioTec, University of Aveiro, Aveiro, Portugal

G. Lollino et al. (eds.), *Engineering Geology for Society and Territory – Volume 6*,
DOI: 10.1007/978-3-319-09060-3_147, © Springer International Publishing Switzerland 2015

147.1 Introduction

The design of slopes in heterogeneous weak rock masses presents a major challenge to the engineering design. The complex structure of these materials, resulting from their tectonostratigraphic features, means that they cannot easily be classified in terms of widely used rock mass classification systems (Hoek et al. 1998; Marinos and Hoek 2001). Rock mass outcrops need a correct in situ geotechnical evaluation on slopes for engineering purposes. Basic geotechnical description of rock masses requires the characterisation of its geological and geomechanical features, particularly discontinuities behaviour and intact rock properties (e.g., Priest 1993; Zenóbio and Zuquette 2009; ISRM 2007; Chaminé et al. 2010). It has been recognised that discontinuities have a major influence on the mechanical properties of a rock-mass, namely on slopes. This perception has major consequences for the assessment of the engineering behaviour of a rock mass. This work presents the results of a combined geotechnical and geomechanical studies of Trofa phyllitic cut slope (NW Portugal), which was excavated in order to construct a road.

147.2 The Study Area

147.2.1 Regional Background: Trofa Region

The northwest portion of the Iberian Massif is crossed by major deep crustal faults. Trofa region is located in the vicinity of regional fault/shear zones. The regional tectonic framework comprises a Palaeozoic metasedimentary highly fractured and folded basement rock (Fig. 147.1). It defines some main tectonic lineaments orientation, trending N-S and NW-SE. The geological units that outcrop in the study area are darkish phyllites and metagraywackes interbedded with metasiltites (Teixeira et al. 1965).

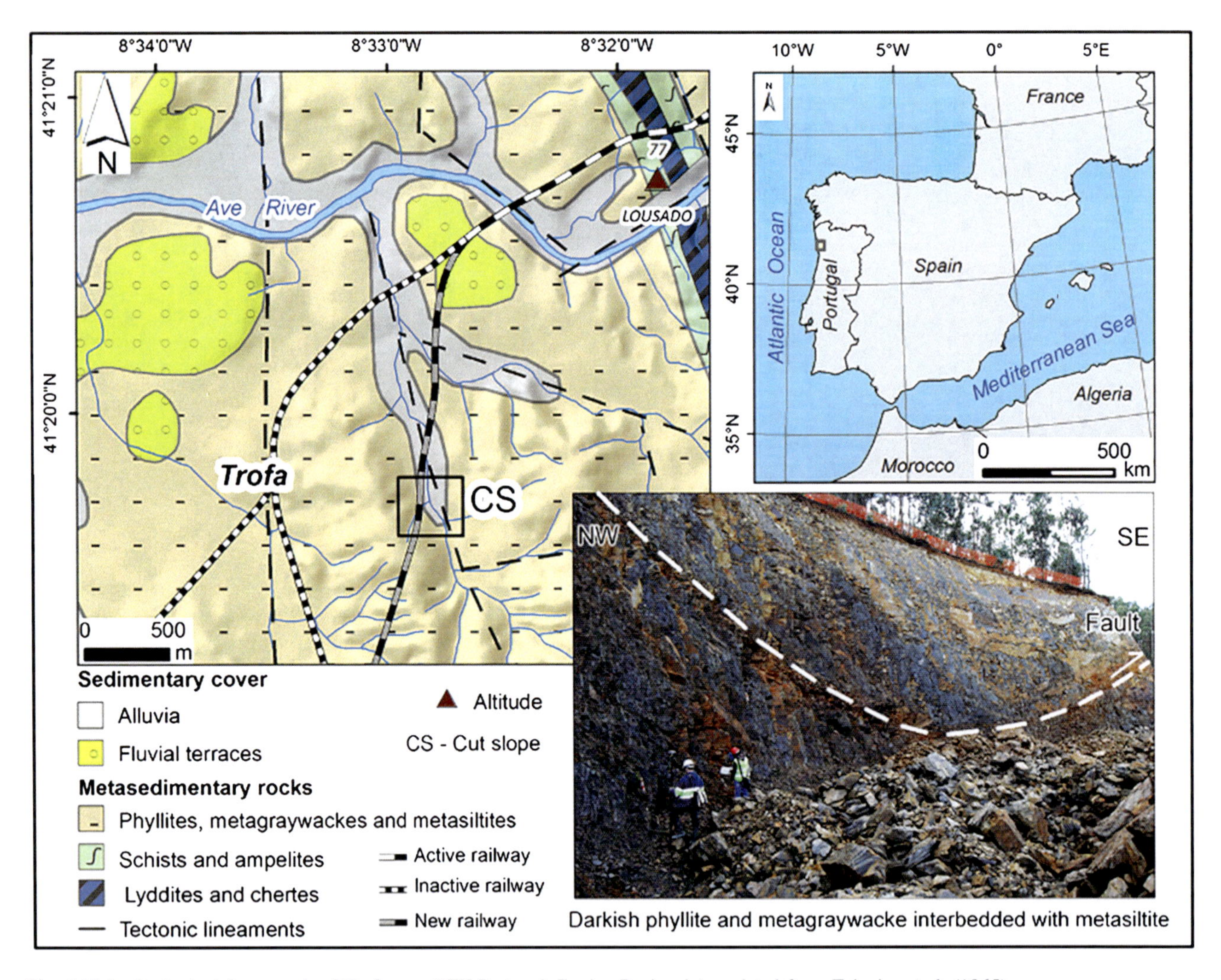

Fig. 147.1 Geological framework of Trofa area (NW Portugal, Iberian Peninsula), updated from Teixeira et al. (1965)

147.2.2 Methodology

This study involved, beforehand, a topographic, morpho-structural and geotectonical characterisation of the study area and, in a subsequent stage, a geotechnical and geomechanical assessment. The main features were compiled and integrated along the rock cut slope. A high-resolution GPS (Trimble GeoExplorer) was used for the fieldwork surveys. The scanline sampling technique was applied to the study of basic geotechnical description of free rock mass surfaces (ISRM 2007). To establish the main discontinuity sets, the structural geology data collected at the slope site were analysed with Dips 6.0 software (Rocscience). Weathering grade of rock material was used following the proposals by GSE (1995) and ISRM (2007). In addition, uniaxial compressive strength was assessed by Schmidt Rebound Hardness after the recommendations by ISRM (2007). In order to classify the quality of the tectonised phyllitic rock mass, the geomechanical classification systems and indexes, such as Rock Mass Rating (RMR; Bieniawski 1989) and Slope Mass Rating (SMR; Romana 1993) and Geological Strength Index (GSI; Hoek et al. 1998, 2013) were applied. Furthermore, SMR, Markland Test (Wyllie and Mah 2004), Slide 5.0 and Swedge 5.0 software (Rocscience), and the European Standard Eurocode 7 were used to the stability analysis and to identify potential slope failure mechanisms. Geomechanical modelling and the design of the reinforcement elements were developed with Phase^2 6.0 software (Rocscience). A structural and geotechnical solution for the stabilisation of the cut slope was presented.

147.2.3 Results and Discussion

The slope has a curvilinear outline (Fig. 147.2), varying from WNW-ESE to NNW-SSE orientation, and a dip with a steep angle (70–85°). Trofa cut slope is constituted mainly by micaceous-clayish phyllites and metagraywackes, with fine-grained quartz veins, which outcrop fresh to slightly weathered and highly tectonised.

A detailed geological and geomechanical characterisation was performed. Concerning the basic geotechnical description of rock mass, 368 discontinuity (mainly joints) measurements were taken and 71 geomechanical stations were carried out for

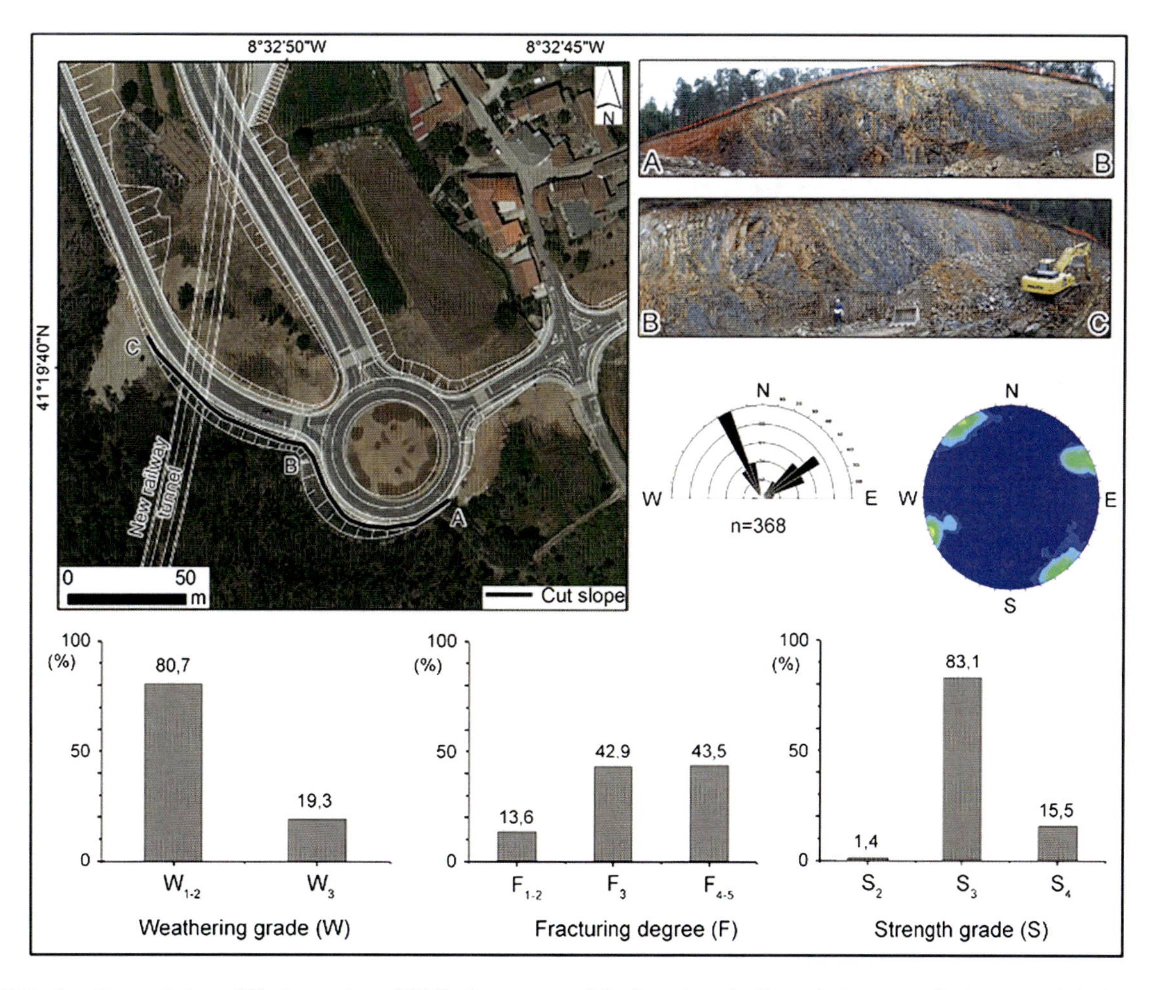

Fig. 147.2 Location and view of Trofa cut slope (ABC), (*upper part* of the figure); main discontinuity sets and other geotechnical parameters (*lower part* of the figure)

Table 147.1 Summary of the basic geotechnical parameters from the studied rock slope (Trofa)

Basic geotechnical description of rock mass parameters (after ISRM 2007)										
Slope (length/ height, m)	Weathering grade (W)	Discontinuity type	Main discontinuity sets	Aperture (mm)	Fracture intercept / Spacing (F)	Persistence (m)	Roughness (R)	Filling	Seepage	Uniaxial Compressive Strength (σc)
165.3 / 15	Fresh to slightly weathered (W_{1-2})	Predominantely joints, and faults	N155°E; 80° NE/SW N60°E; 75°NW/ SE (n = 368)	Closed (< 0.5)	Very close to moderate spacing (F_{4-5} to F_3) average value = 37 cm	Very low (< 1) to low (1–3)	Undulating, rough (R_{1-2}) to smooth (R_3)	None	Damp	Moderate to low (S_3 to S_4) 25–35 MPa

Table 147.2 Synthesis of the rock mass classifications and the geomechanical parameters

Slope features		Rock mass classification and index			Geomechanical parameters adopted		
Geological unit	Failure type	RMR_{basic} Bieniawski (1989)	SMR Romana (1993)	GSI Marinos and Hoek (2001); Marinos et al. (2005)	Cohesion, c (kPa) Bieniawski (1989); Design value (Eurocode 7)	Friction angle, ϕ (°) Bieniawski (1989); Design value (Eurocode 7)	Intact rock parameter (m_i) Marinos et al. (2005)
Phyllites, metagraywackes with metasiltites	Mainly wedge, also toppling and planar failures	25–35 IV Poor rock	25–35 IVb to IVa Poor rock	20–30 Poor to fair rock	40	22.4	9

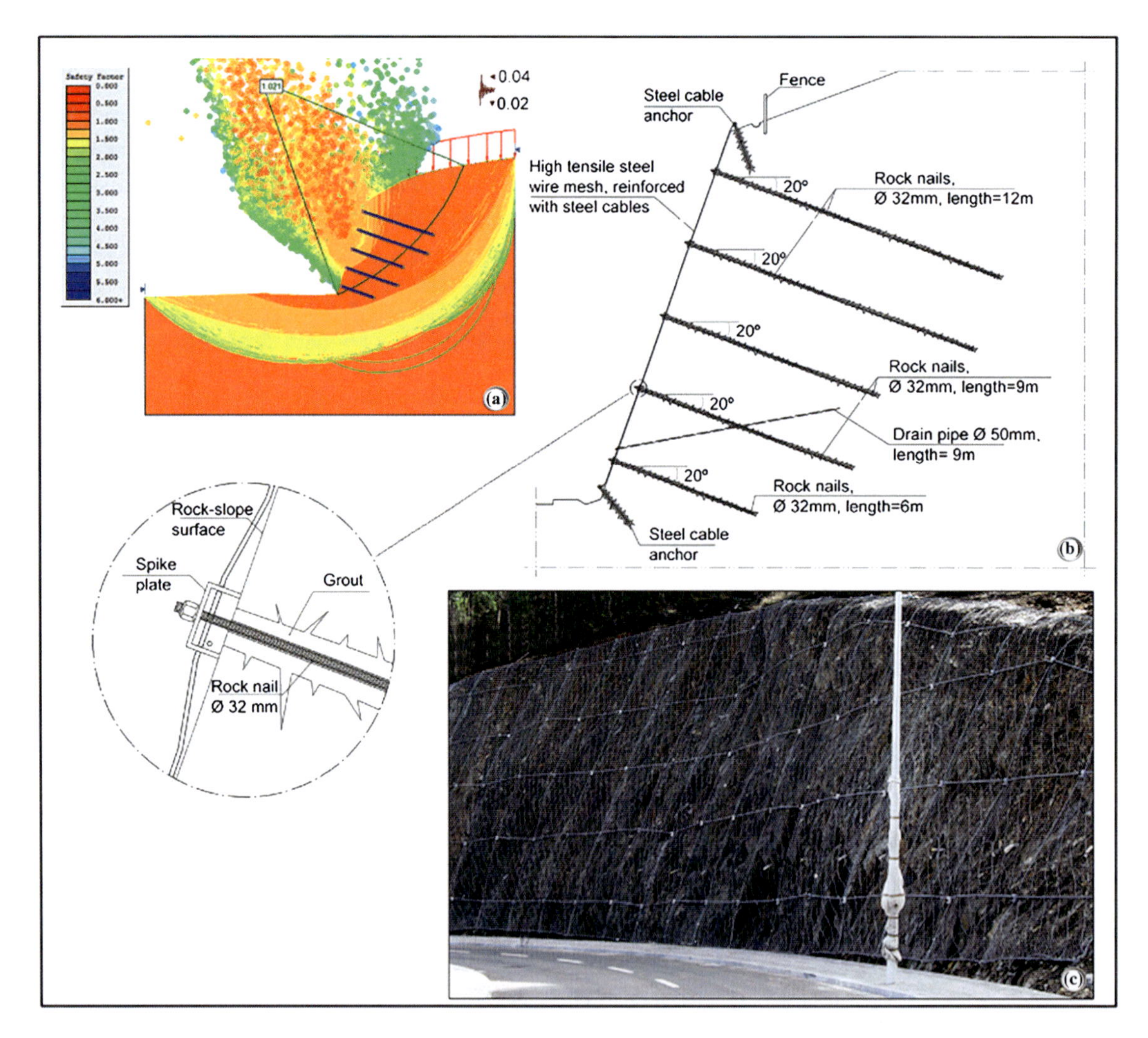

Fig. 147.3 Stabilisation solution design adopted for the studied rock slope. **a** example of slope stability analysis; **b** schematic cross-section with the main reinforcement features; **c** photo of the final stabilisation

the Schmidt Rebound Hardness technique. Table 147.1 summarises the most important geological and geotechnical parameters compiled and analysed in situ for the rock slope. Table 147.2 synthesises the adopted geomechanical parameters and rock mass classifications. The rock mass has dominantly a poor quality, seldom a fair quality. The GSI index appears, in general, to be more optimistic. The slope was classified as unstable and the potential failure types are mainly wedge slides, yet toppling and planar failures may occur.

A stabilisation program has been accomplished after the geological, geotechnical and geomechanical studies. A solution composed of high tensile wire mesh, reinforced with steel cables connected with rock nails and anchors, on the slope borders, was implemented. A level of drain pipes was installed near the base of the slope (Fig. 147.3).

147.3 Concluding Remarks

Geomechanical classification systems can be a way of evaluating the dynamic of rock cut slopes, based on the most important characteristic and structural parameters (Pantelidis 2010). Besides, these geomechanical indexes and classification systems are one of the techniques currently in use for slope stability analyses. The main advantage of using a rock mass classification scheme is that it is a simple and effective way of representing rock mass quality. The slope stability of rocks is an important issue in rock engineering. This holds for both the design and construction stages. Currently, a number of methods are being used for the assessment of slope stability (e.g., Goodman 1989; Romana 1993; Pettifer and Fookes 1994; Irigaray et al. 2003; Pantelidis 2010; Chaminé et al. 2010). Rock slope instabilities area major hazard for human activities often causing economic losses, property damages and maintenance costs, as well as for injuries or fatalities (Fernández et al. 2003; Pantelidis 2010). The combined geotechnical and geomechanical studies of Trofa metasedimentary rock cut slope based on several geomechanical classification systems and indexes proved to be a tool of great utility in the geotechnical slope stability assessment and also to the reinforcement design. This interdisciplinary methodology was vital to understand the geo-hazards in the weak rock mass of Trofa region and must be important to the accurate prediction of future slope instabilities.

Acknowledgements This work is under the framework of the LAB-CARGA-IPP-ISEP|PADInv'2007/08 and Centre GeoBioTec|UA (PEst-C/CTE/UI4035/2014). We thank to the colleague J. Correia (Tecnep) for all the support during the project. We acknowledge the anonymous reviewer for the constructive comments.

References

Bieniawski ZT (1989) Engineering rock mass classifications: a complete manual for engineers and geologists in mining, civil, and petroleum engineering. Interscience, Wiley, NY, p 272

Chaminé HI, Afonso MJ, Silva RS, Moreira PF, Teixeira J, Trigo JF, Monteiro R, Fernandes P, Pizarro S (2010) Geotechnical factors affecting rock slope stability in Gaia riverside (NW Portugal). In: Williams A et al (eds) Proceedings 11th congress. Auckland, CRC Press, IAEG, pp 2729–2736

Fernández T, Irigaray C, El Hamdouni R, Chacón J (2003) Methodology for landslide susceptibility mapping by means of a GIS: application to the contraviesa area (Granada, Spain). Nat Hazards 30:297–308

Goodman RE (1989) Introduction to rock mechanics, 2nd edn. Wiley, NY 576 p

GSE [Geological Society Engineering Group Working Party Report] (1995) The description and classification of weathered rocks for engineering purposes. Q J Eng Geol Hydrogeol 28(3):207–242

Hoek E, Marinos P, Benissi M (1998) Applicability of the geological strength index (GSI) classification for very weak and sheared rock masses: the case of the Athens Schist formation. Bull Eng Geol Environ 57:151–160

Hoek E, Carter TG Diederichs MS (2013) Quantification of the geological strength index chart. In: Proceedings geomechanics symposium 47th US rock mechanics. San Francisco, CA, ARMA 13–672, pp 1–8

Irigaray C, Fernández T, Chacón J (2003) Preliminary rock-slope-susceptibility assessment using GIS and the SMR classification. Nat Hazards 30:309–324

ISRM [International Society for Rock Mechanics] (2007) The complete ISRM suggested methods for characterization, testing and monitoring. In: Ulusay R, Hudson JA (eds) Suggested methods prepared by the commission on testing methods. ISRM, Ankara, Turkey, pp 1974–2006

Marinos P, Hoek E (2001) Estimating the geotechnical properties of heterogeneous rock masses such as flysch. Bull Eng Geol Environ 60:85–92

Marinos V, Marinos P, Hoek E (2005) The geological strength index: applications and limitations. Bulletin of Engineering Geology and the Environment, 64(1):55–65

Pantelidis L (2010) Rock slope stability assessment through rock mass classification systems. Int J Rock Mech Min Sci 46(2):315–325

Pettifer GS, Fookes PG (1994) A revision of the graphical method for assessing the excavatability of rock. Q J Eng Geol Hydrogeol 27 (2):145–164

Priest SD (1993) Discontinuity analysis for rock engineering. Chapman and Hall, London 473 p

Romana M (1993) A geomechanical classification for slopes: slope mass rating. In: Hudson J (ed) Comprehensive rock engineering. Pergamon, Oxford, 3: pp 1–45

Teixeira C, Medeiros AC, Assunção CT (1965) Notícia explicativa da Carta Geológica de Portugal, escala 1/50000, Folha 9□A (Póvoa de Varzim). Serviços Geológicos de Portugal, Lisboa

Wyllie DC, Mah CW (2004) Rock slope engineering: civil and mining, 4th edn. Spon Press, London 431 p

Zenóbio A, Zuquette L (2009) Geotechnical mapping of rock masses in natural slopes using geomechanical classifications. In: Culshaw MG et al (eds), Engineering geology for tomorrow's cities (IAEG'2006). Geological society, London, engineering geology special publications, 22 [on CD-ROM insert, Paper 173]

Geomechanical Characterization of a Weak Sedimentary Rock Mass in a Large Embankment Dam Design

148

Gian Luca Morelli and Ezio Baldovin

Abstract

The geomechanical characterization of a weak and tectonically undisturbed sedimentary rock mass ("molasse"), carried out for the design of a large embankment dam in the Kurdistan Region (Iraq), is presented. In view of the relevant dimensions of the dam and of its main ancillary works, the rock mass characterization has represented one of the basic focus of the geological and geotechnical investigations. However, various uncertainties have been considered to potentially affect the final estimate of the engineering properties of rock masses in the context of the present study. In particular, besides the natural variability of weak rocks characteristics and the well-known difficulties in their realistic estimation, uncertainties may also arise from the potential incompleteness of the basic data available, mainly due to time and budget limitations. In order to quantify such uncertainties and adequately incorporate their effects into the design process, a probabilistic approach based on Monte Carlo method has been adopted to determine the probability distribution functions describing the rock mass strength parameters. A practical example of application of the followed probabilistic approach to the design of large excavation rock slopes is briefly illustrated.

Keywords

Weak rocks • Geomechanical characterization • Probabilistic approach • Slope stability

148.1 Introduction

The Mandawa Dam is presently under construction for irrigation, hydropower production and drinking water supply, in the Kurdistan Region (Northern Iraq) along the main course of the Greater Zab River, a major left tributary of the Tigris River. The dam is a 65 m high embankment with gravel and sand shells and a central silty core. It extends straight for about 1,000 m from the two extreme abutments and includes, on the left side, a concrete structure composed of three blocks housing the main hydraulic structures for the temporary diversion, the bottom outlets and the inlets of the penstocks of the hydropower plant (Fig. 148.1). A very large spillway channel is located on the right side, separated from the dam. The subtended reservoir has a maximum storage capacity of about 520 million m^3.

The foundations of the left concrete part of the dam body and of the spillway channel have required the design of large excavations on both river banks, resulting in cut multi-bench rock slopes even more than 80 m high.

148.2 Site Geology and Geotechnical Investigations

The rock mass characterization was based on data gathered from an investigation geotechnical campaign carried out during the 2012, that has included a total of 1,700 m

G.L. Morelli (✉) · E. Baldovin
Geotecna Progetti S.r.l, Via Roncaglia, 14, 20146 Milan, Italy
e-mail: gianluca.morelli@geotecna.it; gl.morelli.geo@gmail.com
URL: http://www.geotecna.it

E. Baldovin
e-mail: ezio.baldovin@geotecna.it

G. Lollino et al. (eds.), *Engineering Geology for Society and Territory – Volume 6*,
DOI: 10.1007/978-3-319-09060-3_148, © Springer International Publishing Switzerland 2015

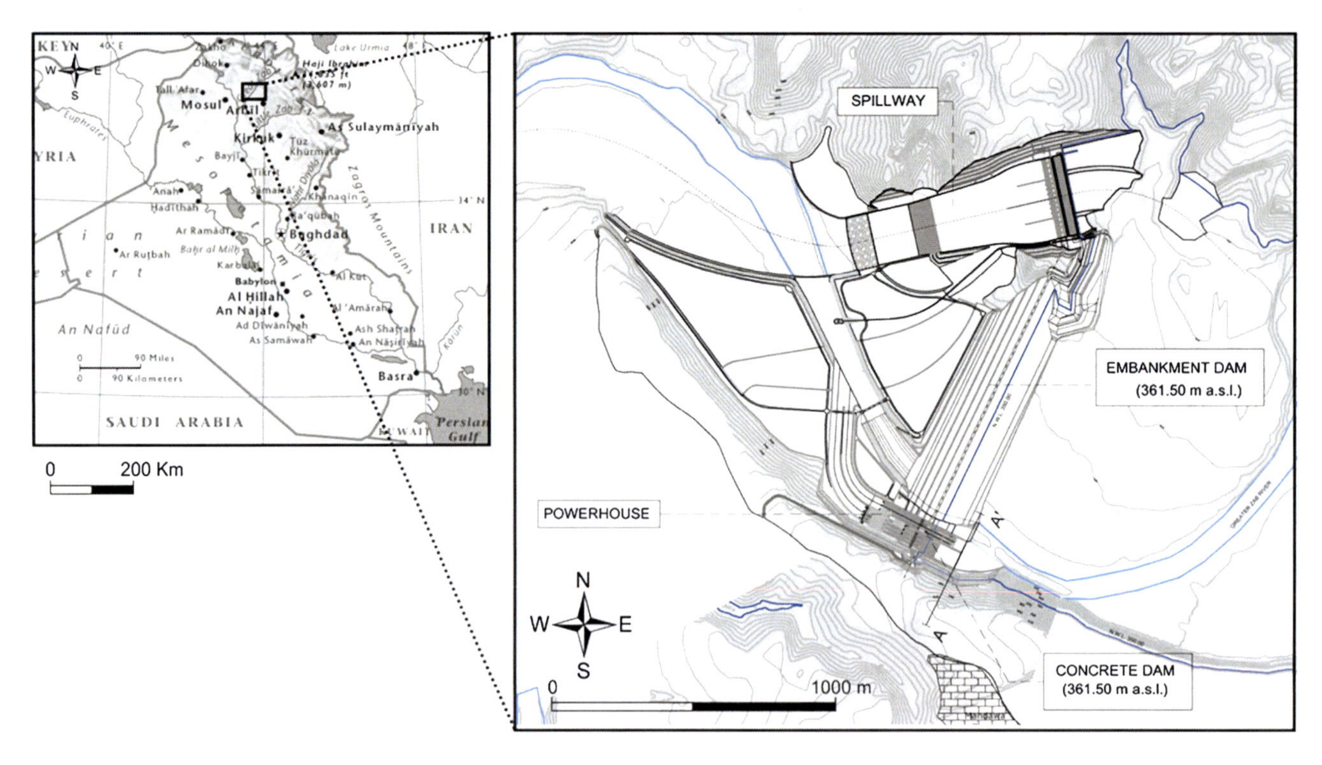

Fig. 148.1 Dam location map and general layout of works

borehole drillings, 60 geotechnical laboratory tests on selected rock cores (mainly uniaxial compression tests on specimens at natural water content and physical property determinations) and field geological mapping. In the dam area, the local bedrock consists of the Upper Miocene sedimentary Mukdadiyah Formation, composed of a clastic sequence of pluri-metric alternations of reddish siltstone and grayish silty sandstone members. It was formed in a mostly fluvial depositional environment in front of the Zagros Mountains belt and appears not significantly interested by orogenic deformations. Both sandstone and siltstone members are stratified and show grain size and induration degree rather variable among layers. Three discontinuity sets intersecting at right angles, one of which coincident with the sub-horizontal bedding and the others consisting of sub-vertical tensile joints striking sub-parallel and sub-orthogonal to the river course, were observed in the bedrock. Spacing and persistence of vertical joints result well correlated with the thickness of layers and usually varies in the range 0.5–3 m. Discontinuities in sandstone are mostly undulated, rough and free of infilling materials; those in siltstone (mainly the bedding surfaces) are frequently planar and moderately rough, sometimes with skinny silty coatings on the surfaces. Weathering effects are limited to the few superficial meters of the bedrock and disappear rapidly with depth.

148.3 Rock Mass Characterization

The rock mass has been characterized with reference to an equivalent-continuum model and the GSI classification system, combined with the strength properties of the intact rock, has been used to derive the Hoek-Brown strength parameters (Hoek et al. 2002). The uniaxial compressive strength (UCS) of the rock matrix has been evaluated through compressive tests on core specimens at varying natural water contents. The UCS values resulted sensitive to saturation degree (Sr, %) and correlated to the dry density (ρ_{dry}, Mg/m^3) of specimens. The average saturation degree of tested samples was about 30 %. Relationships linking UCS and the main index properties of intact rocks have been established by multiple linear regression analyses of the available laboratory test results, in the form

$$UCS_{sandstone}(MPa) = -66.61 - 0.107Sr(\%) + 41.282\rho_{dry}(Mg/m^3)\left[\mathrm{R}^2 = 0.69\right] \quad (148.1)$$

$$UCS_{siltstone}(MPa) = -45.27 - 0.116Sr(\%) + 28.876\rho_{dry}(Mg/m^3)\left[\mathrm{R}^2 = 0.64\right] \quad (148.2)$$

For studied rocks, established relationships allow to predict the UCS for specified values of dry density and degree of saturation expected in situ (e.g. saturated under reservoir water level and naturally wet above) and imply, for tested samples, an average reduction of the UCS between dry (Sr = 0 %) and fully saturated (Sr = 100 %) in the order of about 50 % for sandstone and 60 % for siltstone, rather in line with previous literature findings (e.g. Palmström 1995; Romana and Vásárhely 2007). Proposed relationships have, evidently, not general validity and should then be applied with extreme care for other rock types and in different geological contexts. The GSI index of the rock mass at the depths of the planned excavations has been assessed on the base of data collected from core logging and field mapping. The method reported by Hoek et al. (1995), based on traditional Q-System descriptor codes (Barton et al. 1974), has been used to quantitatively estimate the GSI through the relationship

$$GSI = 9\ln(RQD/Jn \times Jr/Ja) + 44 \qquad (148.3)$$

where RQD = Rock Quality Designation (%); Jn = rating for number of joint sets; Jr = rating for joint roughness and Ja = rating for joint alteration and filling. The GSI obtained for the unweathered bedrock with the described quantitative approach resulted in general accordance with the indicative range suggested by Hoek et al. (2005) for "blocky" molasses.

148.4 Probabilistic Approach in Rock Mass Characterization

To quantify the possible variability of rock mass properties and to incorporate its effects into the design process, a probabilistic approach based on the classical Monte Carlo (M-C) method has been used to derive the probability density functions (PDFs) of the Hoek-Brown (H-B) strength parameters. At this scope, probability distributions have been assessed for the input parameters used to calculate the UCS of intact rocks, i.e. ρ_{dry}, and the GSI of rock masses, i.e. the RQD and the Jr and Ja factors, by considering available field and laboratory data. On the base of such input distributions, the PDFs for UCS and GSI have been calculated by best fitting the results of M-C simulations conducted using Eqs. 148.1, 148.2 and 148.3 (Table 148.1 and Fig. 148.2). The input parameters considered for simulations have been implicitly assumed to be random and independent variables.

Table 148.1 Monte Carlo simulations inputs and outputs

	Parameters and probability distributions input into Monte Carlo simulations		Output parameters and PDFs from M-C simulations	
Intact rock	ρ_{dry}	LogNormal (Trunc: min and max values of lab tests)	UCS	Beta
	Sr	Constant = 30 % (naturally wet) and 100 % (saturated)		
Rock mass	RQD	Weibull (Trunc: min = 0 to max = 100)	GSI	Normal
	Jn	Constant = 9 (n. 3 joint sets)		
	Jr, Ja	Normal		

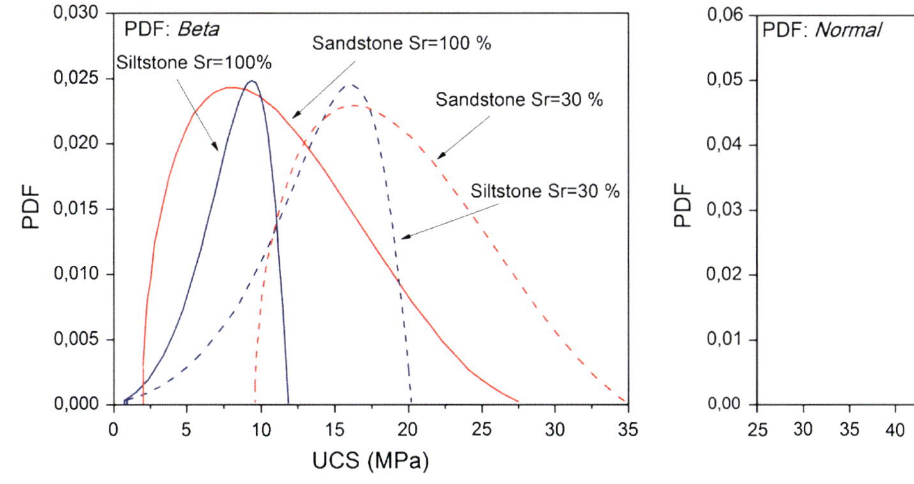

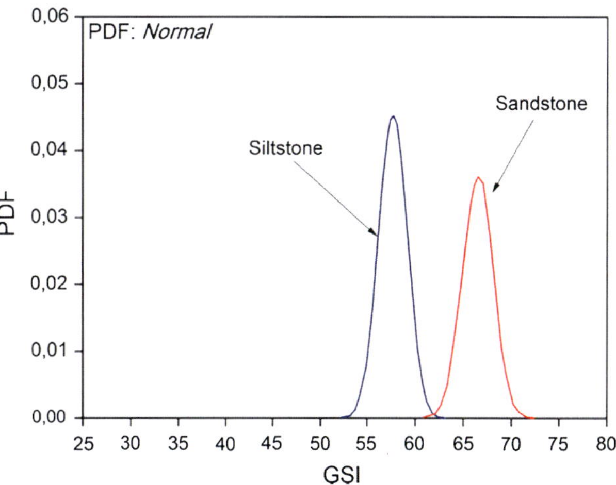

Fig. 148.2 Plots of the PDFs obtained from Monte Carlo simulations for UCS (*left*) and GSI (*right*)

The mean values of the H-B constant m_i have been derived from literature, arbitrarily assuming normal distributions with coefficient of variation (COV) of 10 %. By using the probability distributions previously defined for GSI and m_i, the PDFs of the H-B strength parameters of the rock mass, s, m_b and a, have been determined using the Monte Carlo method. The disturbance factor D has been set to zero in calculations.

148.5 Example of Application in Rock Slope Design

An example of application of the described probabilistic characterization approach in the Mandawa Dam project mainly relates to global stability analyses performed for the rock slopes to be excavated in the reservoir nearby the left side of the dam, where the HPP headrace inlets and the bottom outlets intakes have been planned (analysis section A–A' in Fig. 148.1). Since the structural setting of the rock mass does not control the stability of such slopes, global analyses were performed assuming Bishop's circular failure surfaces and using the probabilistic analysis option available in the program SLIDE (Rocscience Inc.). The possible presence of tension cracks, coincident with the joints striking nearly parallel to the river course, has also been considered. The input probability functions of main geotechnical parameters and tension crack depth derived from Monte Carlo simulations and used in the analyses, are summarized in Table 148.2.

The typical slip surface considered in the analyses and the probability of failure obtained by varying the overall slope angle α, are shown in Fig. 148.3.

Analyses were performed considering the most severe condition of rapid drawdown of the reservoir, from the maximum to the minimum level without the drainage of the slope, coupled with the maximum seismic action (defined from a peak design earthquake acceleration of 0.2 g). The overall slope angle of 50° finally chosen for the design, resulting from a multi-bench shaped profile of interposed 3 m benches with vertical drops of 15 m and scarps inclined 3v:2 h (Fig. 148.3-left), allows to advantageously reduce to zero the risk of global failure of the analyzed slope (Fig. 148.3-right), ensuring, in the long term, the full operability of the dam outlets and the HPP production.

Table 148.2 Main input parameters used for the slope stability analyses

Property		PDF	Mean	Std. Dev.	Min	Max
Siltstone	UCS(sat)[Mpa]	Beta	15.79(7.89)	2.91(2.54)	0.63(0.5)	20.26(12.14)
	H-B "s"	LogNorm	0.0096	0.0019	0.0044	0.0202
	H-B "a"	LogNorm	0.503	0.0001	0.502	0.505
	H-B "m_b"	LogNorm	0.8939	0.1047	0.508	1.312
Sandstone	UCS(sat)[Mpa]	Beta	19.70(12.21)	5.77(5.77)	9.65(2.16)	35.04(27.54)
	H-B "s"	LogNorm	0.021	0.01	0.01	0.05
	H-B "a"	LogNorm	0.502	0.0001	0.501	0.503
	H-B "m_b"	LogNorm	5.149	0.624	3.166	7.762
T. Crack	Depth [m]	Uniform	20		10	30

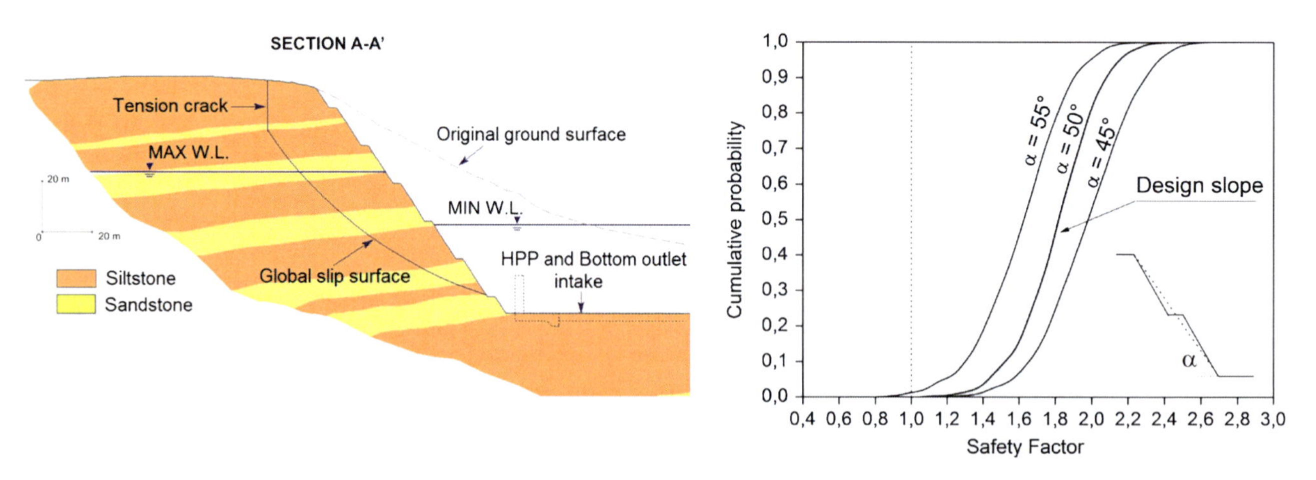

Fig. 148.3 *Left* Slope geology and typical slip surface analyzed; *Right* probability of failure obtained by varying the overall slope angles

148.6 Conclusions

A geomechanical rock mass characterization approach based on probabilistic analyses has been performed for a large embankment dam design, enabling to take into account and to quantify the possible variability of the engineering properties of the bedrock. A practical example showing how this approach has been applied to the design of cut rocky slopes has been presented. According to the wide literature available today on this subject, the adoption of a probabilistic design method proved to be suitable for managing uncertainties in rock properties assessment and can be particularly effective when, as in the presented case, the difficulties usually encountered in characterizing a weak rock mass have to be faced with time and budget limitations.

References

Barton NR, Lien R, Lunde J (1974) Engineering classification of rock masses for the design of tunnel support. Rock Mech 6(4):189–239
Hoek E, Kaiser PK, Bawden WF (1995) Support of underground excavations in hard rock. Rotterdam, Netherlands In: Balkema (eds)
Hoek E, Carranza-Torres CT, Corkum B (2002) Hoek-Brown Failure Criterion—2002 Edition. In: Proceedings of 5th North American rock mechanics symposium
Hoek E, Marinos PG, Marinos VP (2005) Characterisation and engineering properties of tectonically undisturbed but lithologically varied sedimentary rock masses. Int J Rock Mech Min Sci 42:277–285
Palmström A (1995) RMi—a rock mass characterization system for rock engineering purposes. PhD thesis, University of Oslo, Norway, p 400
Romana M, Vásárhely B (2007) A discussion on the decrease of unconfined compressive strength between saturated and dry rock samples. In: 11th Congress of the international society for rock mechanics—Ribeiro e Sousa, Olalla and Grossmann (eds), pp 139–142

Experimental Study of Anisotropically Mechanical Features of Phyllite and Its Engineering Effect

149

Meng Lubo and Li Tianbin

Abstract

Based on conventional triaxial compression tests for phyllite with various orientation angles, the relationships of mechanical features, confining pressure and anisotropy are discussed. The results indicate that the peak strength, residual strength, elastic modulus and Poisson's ratio increase with increasing confining pressure. The effect of confining pressure on the elastic modulus is relatively smaller than the peak strength and Poisson's ratio. The peak strength and Poisson's ratio decrease with orientation angle increases. The effect of anisotropy on the peak strength is larger than Poisson's ratio. The split and shear mechanisms are observed, their failures are brittle and ductile, respectively. The three modes of surrounding rock failure in deep tunnel are derived from the test results, which are split and buckling, split and shear, shear sliding.

Keywords

Phyllite • Anisotropy • Mechanism • Engineering effect

149.1 Introduction

Deep underground engineering's environments typically have one important feature: high geostress level. Research on soft rock mechanical property under high confining pressure conditions are very important, which can help understand the surrounding rock stability for deep underground engineering.

Anisotropy is an important property of rock masses, and is one of the focus and hotspot of research for many experts and scholars. Lots of anisotropy behaviors of transversely isotropic rock have been investigated by tests and theoretical analysis (Jaeger 1960; Ramamurthy et al. 1993), the mechanical properties of layered rock mass have been discussed by uniaxial compression tests (Li 2008; Liu et al. 2012), the elastic parameters and strength for transversely isotropic rocks have been investigated by triaxial compression tests (Liu et al. 2013). However, the mechanical features and failure modes of phyllite for deep tunnel is relatively less. So a phyllite conventional triaxial compression tests with various orientation angles is studied. The results of this research will provide a reference for phyllite surrounding rock stability analysis for deep underground engineering under high geostress conditions.

149.2 Test Scheme

Phyllite samples, which are collected from a drill hole of the ChengLan railway tunnel under high geostress conditions, are cylindrical specimens with a diameter of 45 mm and a length of 100 mm. The testing is performed with the professional standard of People's Republic of China (SL264-2001).

Defining the planes of weakness making an angle β with the axis of the major principal stress, the angle β is designated as the "orientation angle". The tests are conducted at

M. Lubo (✉) · L. Tianbin
State Key Laboratory of Geohazard Prevention and Geoenvironment Protection, Chengdu University of Technology, Chengdu, 610059 China
e-mail: menglubo@163.com

G. Lollino et al. (eds.), *Engineering Geology for Society and Territory – Volume 6*,
DOI: 10.1007/978-3-319-09060-3_149, © Springer International Publishing Switzerland 2015

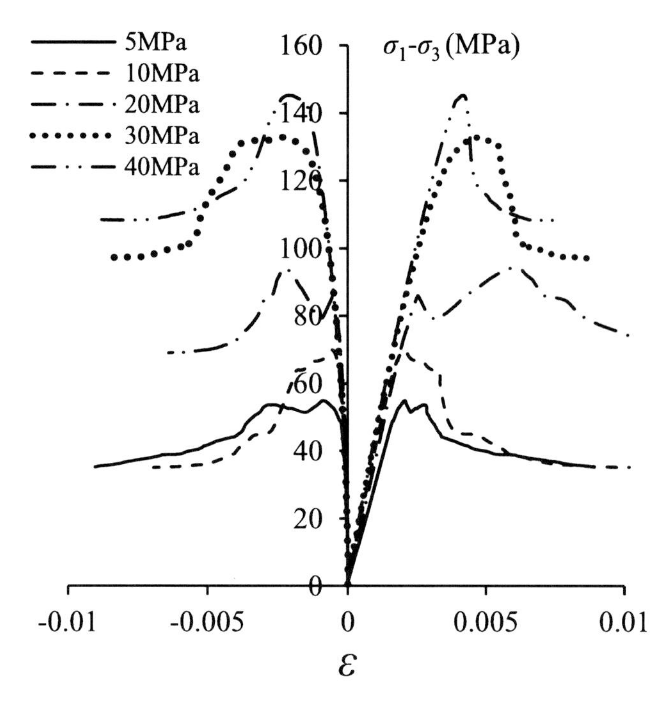

Fig. 149.1 Stress-strain curves of specimens (β = 15°)

specified orientation angles of 0, 15 and 30°. The MTS815, produced in the U.S.A., is used for testing. The confining pressure ($\sigma_2 = \sigma_3$) is controlled at 5, 10, 20, 30, and 40 MPa.

The testing sequence is presented as follows:

1. The specimen is installed according to the requirements of test machine.
2. The load control is adopted; the confining pressures are loaded with predetermined values at the rate of 5 MPa/min, the confining pressure is maintained at the predetermined pressure.
3. The axial deformation control is then adopted, and the rate is 0.1 mm/min, until the specimen is destroyed.

149.3 Results and Discussion

149.3.1 Mechanical Features and Confining Pressure

The stress-strain curves for typical samples are shown in Fig. 149.1. The rock deformation of samples can be divided into three phases: the elastic deformation, yield and failure phases. The peak strength increases with increasing confining pressure from 5 to 40 MPa. Contrast the peak strength at confining pressure of 5 MPa, the peak strength increased by 26.8 % (σ_3 = 10 MPa), 69.9 % (σ_3 = 20 MPa), 121.2 % (σ_3 = 30 MPa), 162.7 % (σ_3 = 40 MPa), respectively.

The elastic modulus (E_{50}) and Poisson's ratio (μ_{50}) can be determined with the following equations (Gao et al. 2005):

$$B = \varepsilon_3/\varepsilon_1 \tag{149.1}$$

$$\mu_{50} = (B\sigma_1 - \sigma_3)/((2B-1)\sigma_3 - \sigma_1) \tag{149.2}$$

$$E_{50} = (\sigma_1 - 2\mu_{50}\sigma_3)/\varepsilon_1 \tag{149.3}$$

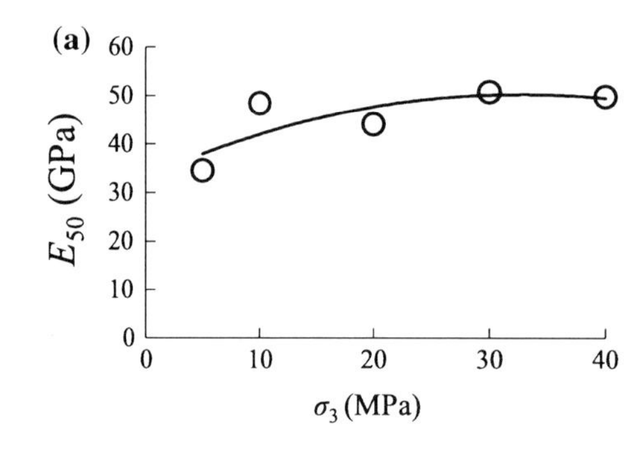

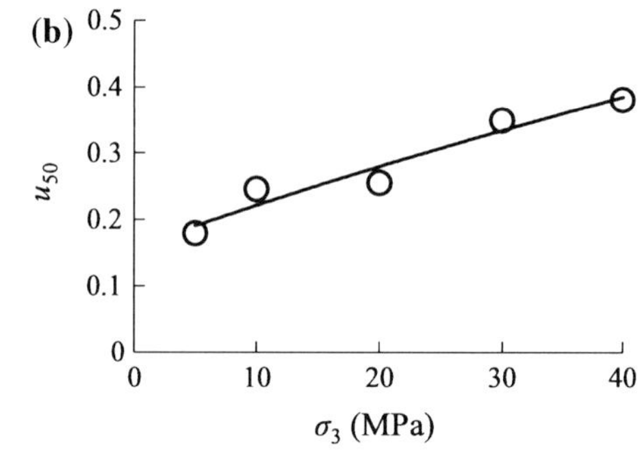

Fig. 149.2 Relationship between the E_{50}, μ_{50} and confining pressure (β = 15°). **a** Elastic modulus. **b** Poisson's ratio

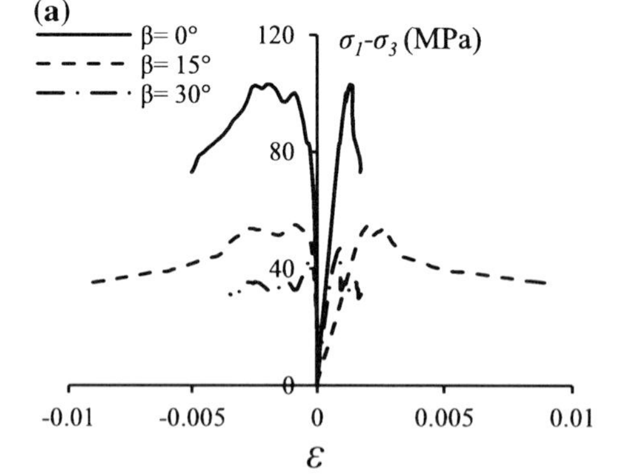

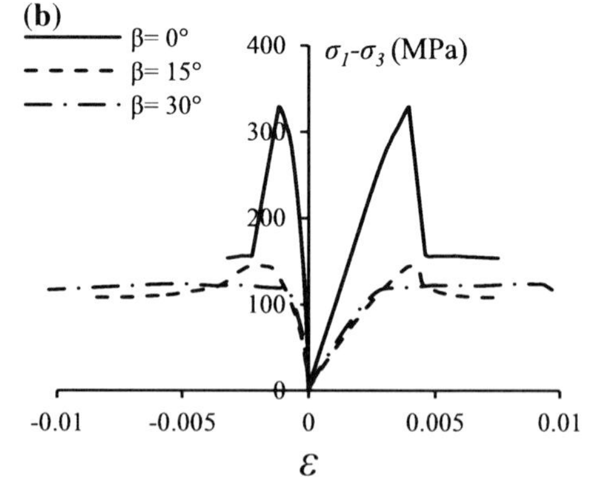

Fig. 149.3 Typical stress-strain curves of phyllite specimens with fixed confining pressure. **a** σ_3 = 5 MPa. **b** σ_3 = 40 MPa

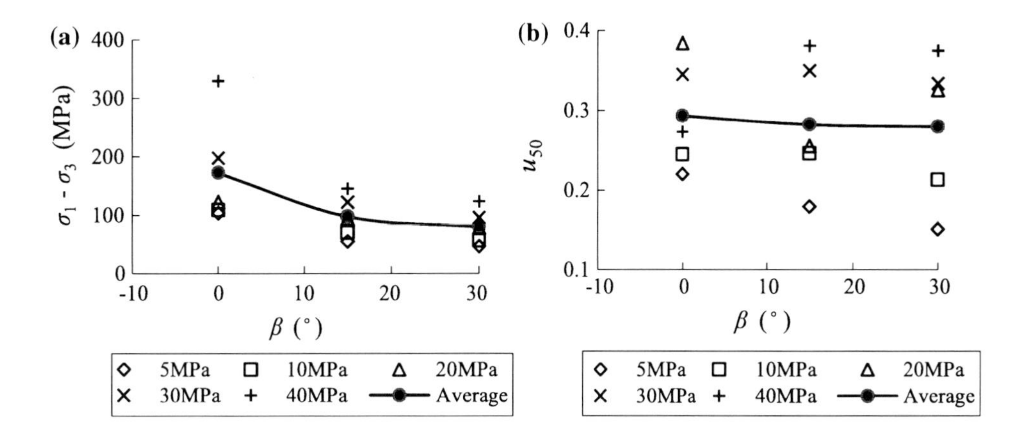

Fig. 149.4 Relationship between the peak strength, Poisson's ratio and orientation angle. **a** Peak strength. **b** Poisson's ratio

where μ_{50} is the Poisson's ratio, E_{50} is the secant modulus of elasticity at half of the peak strength, σ_1 is the axial stress, ε_1 is the axial strain, σ_3 is the confining pressure, and ε_3 is the lateral strain.

The relationships between the E_{50}, μ_{50} and confining pressure are shown in Fig. 149.2. The E_{50} and μ_{50} increase with increasing confining pressure at the same orientation angle. Contrast the E_{50} and μ_{50} at confining pressure of 5 MPa, the E_{50} increased by 40.3, 28, 47.3, 44.5 %, and the μ_{50} increased by 37.2, 42.5, 95.3, 112.8 %, respectively.

The deformation and damage of the specimens are closely related to the confining pressure. The high confining pressure can improve the mechanical properties of phyllite, the strength and stiffness of the phyllite increase with increasing confining pressure, and the residual strength also slightly increases with confining pressure increases. The effect of confining pressure on the elastic modulus is relatively smaller than the peak strength and Poisson's ratio.

149.3.2 Mechanical Features and Anisotropy

The typical stress-strain curves of phyllite specimens with fixed confining pressure are shown in Fig. 149.3. The stress-strain curves indicate that the failure mode of rock with various orientation angles is different. When $\beta = 0°$, rock generates brittle failure; when $\beta = 30°$, rock generates ductile failure.

The relationships between the peak strength, Poisson's ratio and orientation angle are shown in Fig. 149.4. The peak strength decreases with increasing orientation angle at the same confining pressure. The average peak strength of the specimens is 172.2, 97.4, and 80.1 MPa when $\beta = 0, 15, 30°$. The ratio of uniaxial strength perpendicular to strength at orientation of 30° is defined as the degree of anisotropy, the degree of anisotropy of average peak strength is 2.1 ($\beta = 0°$) and 1.2 ($\beta = 15°$).

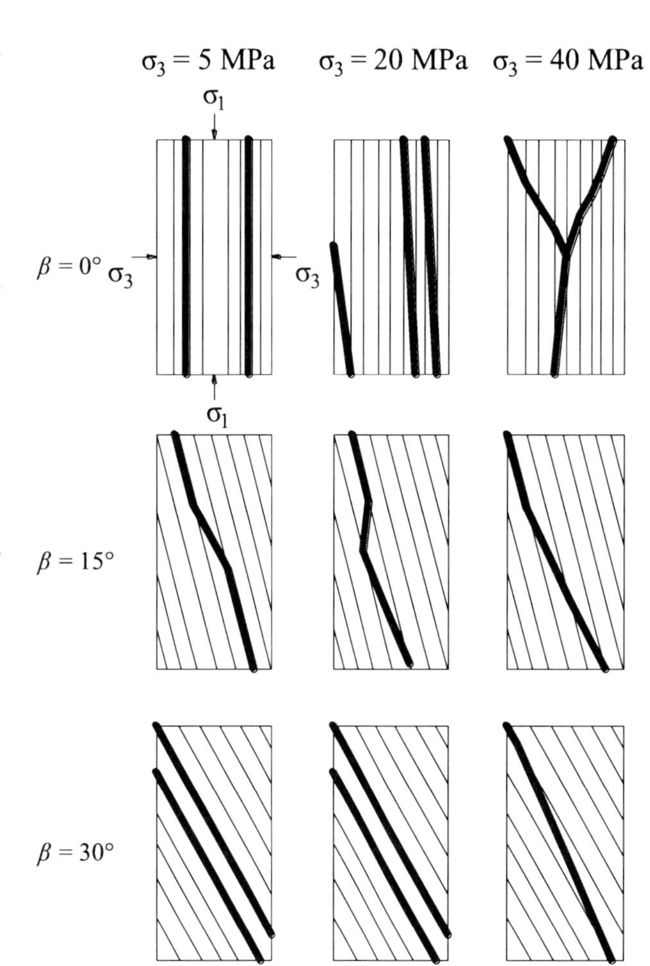

Fig. 149.5 Typical failure patterns

The average Poisson's ratio decreases with increasing orientation angle at the same confining pressure. The average Poisson's ratio of the specimens is 0.293, 0.282, and 0.280 when $\beta = 0, 15, 30°$. The degree of anisotropy on Poisson's ratio is smaller than the peak strength.

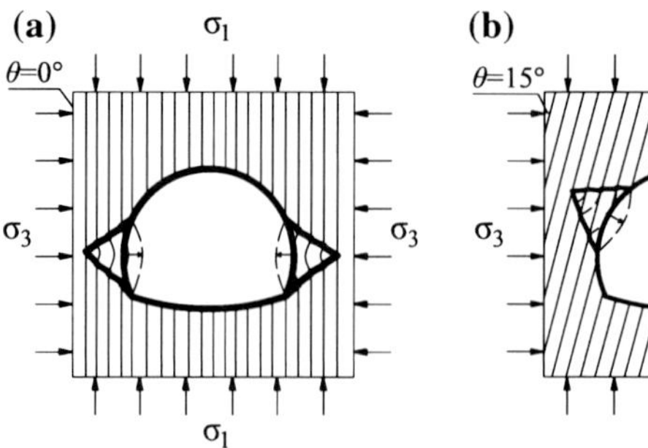

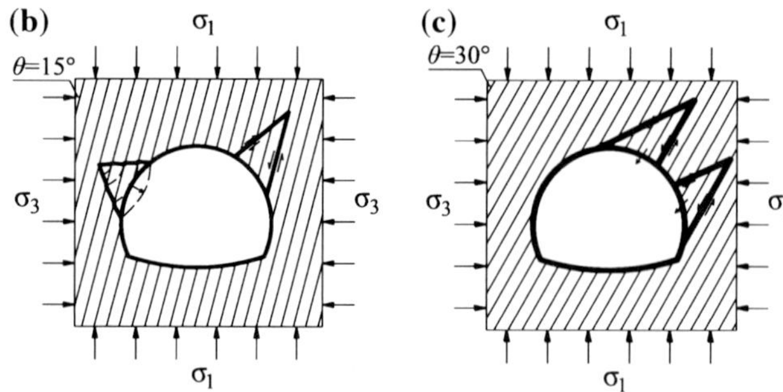

Fig. 149.6 Failure patterns of surrounding rock in deep tunnel. **a** Split and buckling. **b** Split and shear. **c** Shear sliding

149.3.3 Mechanism of Samples Failure

The typical failure patterns are shown in Fig. 149.5. At confining pressure of 5 and 20 MPa (low confining pressure condition), splitting along the foliation planes is common toward low value of β, specifically $\beta = 0°$. At $\beta = 15°$, shearing and splitting failure is observed. At $\beta = 30°$, the phyllites show a pure shear failure across the planes of the foliation.

Under high confining pressure of 40 MPa, at $\beta = 0°$, splitting along the foliation planes and shear failure along a plane oblique to the foliation planes is observed. At $\beta = 15°$ and 30°, the specimens sheared along the weak planes with some development of "kinks", i.e., step-shaped sloping plane, the shearing mechanism controls the behavior of rocks.

The confining pressure of brittle-ductile diversion decreases along with the orientation angle increases; at $\beta = 0°$, the confining pressure of transition is about 30 MPa; at $\beta = 30°$, the confining pressure of transition decreases to 5 MPa.

149.3.4 Modes of Tunnel Surrounding Rock Failure

According to the above test results, the three failure modes of surrounding rock in deep tunnel are inferred, as shown in Fig. 149.6. The angle θ is designated as the angle of the major principal stress and the rock layer plane. when $\theta = 0°$, the thin layer is easily bent and split, the vertical rock layers in sidewalls tend to be bent under the major principal stress and buckles towards the tunnel, as shown in Fig. 149.6a.

When $\theta = 15°$, the thin rock layer in left arch is easily bent and split (Fig. 149.6b), and in right shoulder, the combination of joints leads easily to the development of potential wedges, the shear sliding mechanism controls the behavior of wedges.

When $\theta = 30°$, the shear mechanism controls the behavior of surrounding rocks (Fig. 149.6c), shear sliding of wedge is mainly controlled by discontinuities, shear sliding may appear in the right arch, shoulders and sidewalls of the tunnel.

149.4 Conclusions

1. The peak strength, residual strength, elastic modulus and Poisson's ratio of the phyllite increase with increasing confining pressure. The effect of confining pressure on the elastic modulus is smaller than the peak strength and Poisson's ratio.
2. The peak strength and Poisson's ratio decrease with increasing orientation angle. The effect of anisotropy on the peak strength is larger than Poisson's ratio.
3. The split and shear mechanisms of samples are observed, their failures are brittle and ductile, respectively.
4. The three modes of surrounding rock failure in deep tunnel are derived from the test results, which are split and buckling, split and shear, shear sliding.

Acknowledgements This research is supported by National Natural Science Foundation of China (Grant Nos. 41102189 and 41230635), Program for Research Group of SKLGP (Nos. SKLGP 2013Z004 and SKLGP2009Z002).

References

Gao CY, Xu J, He P et al (2005) Study on mechanical properties of marble under loading and unloading conditions. Chin J Rock Mech Eng 3:456–460
Jaeger JC (1960) Shear failure of anisotropic rock. Geol Mag 97(1):65–72
Li ZC (2008) Experimental study on the uniaxial compression of anisotropic rocks. J Railway Sci Eng 03:69–72
Liu SL, Chen SX, Yu F et al (2012) Anisotropic properties study of chlorite schist. Rock Soil Mech 12:3616–3623
Liu YS, Fu HL, Wu YM et al (2013) Experimental study of elastic parameters and compressive strength for transversely isotropic rocks. J Cent S Univ (Sci Technol) 08:3398–3404
Ramamurthy T, Venkatappa Rao G, Singh J (1993) Engineering behaviour of phyllites. Eng Geol 33:209–225
The Professional Standard Compilation Group of People's Republic of China. SL264-2001, Specifications for rock tests in water conservancy and hydroelectric engineering. China Water Power Press, Beijing

Quantification of Rock Joint Roughness Using Terrestrial Laser Scanning

150

Maja Bitenc, D. Scott Kieffer, Kourosh Khoshelham, and Rok Vezočnik

Abstract

Rock joint roughness characterization is often an important aspect of rock engineering projects. Various methods have been developed to describe the topography of the joint surface, for example Joint Roughness Coefficient (JRC) correlation charts or disc-clinometer measurements. The goal of this research is to evaluate the accuracy, precision and limits of Terrestrial Laser Scanning (TLS) for making remote measurements of large-scale rock joints. In order to find the most appropriate roughness parameterization method for TLS data and to analyse the capability of TLS for roughness estimation, experiments were made with a 20 × 30 cm joint sample. The sample was scanned with TLS and compared to reference measurements made with the Advanced TOpometric Sensor (ATOS) system. Analysis of two roughness parameterization methods, virtual compass and disc-clinometer, and angular threshold method, showed that the latter is less sensitive to noise. Comparative studies of ATOS and TLS roughness parameters indicate that the TLS can adequately quantify surface irregularities with a wavelength greater than 5 mm from a distance of 10 m.

Keywords

Joint roughness • Laser scanning • Rock mass characterization • Rock mechanics

150.1 Introduction

Rock joint roughness is an important factor influencing the potential for shear displacement to occur along an unfilled discontinuity at low normal stress (Patton 1966). Several methods have been developed to measure and parameterize roughness amplitude, anisotropy and scale effects, and to utilize these results in a joint shear strength failure criterion, e.g. (Patton 1966; Barton and Choubey 1977; Grasselli and Egger 2003). Prior research has typically considered small joint samples (<1 m^2) measured in the laboratory environment. Comparatively few studies have investigated larger scale measurements of roughness in the field using shadow profilometry, total station, TLS, photogrammetry and ATOS. Among these, TLS enables fast, accurate and detailed acquisition of distant, inaccessible, large-scale surfaces. TLS data can be used for the extraction of first-order roughness (Sturzenegger and Stead 2009), which is defined by Priest (1993) as "surface irregularities with a wavelength greater than about 10 cm", but the scale and range limitations of TLS measurements have yet to be investigated.

This contribution summarizes an experiment designed to investigate the intrinsic scale and range limitations of TLS. The influence of TLS data resolution and noise on roughness

M. Bitenc (✉)
Eleia iC d.o.o., Dunajska cesta 21, 1000 Ljubljana, Slovenia
e-mail: maja.bitenc@student.tugraz.at; bitenc.m@gmail.com

M. Bitenc · D.S. Kieffer
Institute of Applied Geosciences, Graz University of Technology, Rechbauerstraße 12/EG, Graz, 8010 Austria

K. Khoshelham
ITC Faculty of Geo-information Science and Earth Observation, University of Twente, Hengelosestraat 99, 7514 AE, Enschede, The Netherlands

R. Vezočnik
DFG Consulting, d.o.o., Pivovarniška 8, 1000 Ljubljana, Slovenia

G. Lollino et al. (eds.), *Engineering Geology for Society and Territory – Volume 6*,
DOI: 10.1007/978-3-319-09060-3_150, © Springer International Publishing Switzerland 2015

measurements is studied with ATOS data serving as a reference. Two different roughness parameterization methods are tested to evaluate parameter sensitivity to noise.

150.2 Quality of TLS Point Cloud and Roughness

The quality of a TLS point cloud, namely point accuracy and precision, and resolution, define to what detail roughness amplitude and wavelength (scale) can be observed.

The accuracy and precision of laser point position depends on instrumental errors of laser scanner, environmental conditions (e.g. lightness) and surface features (e.g. reflectivity). If noise is not separated and eliminated from the data, the joint roughness will be overestimated (Khoshelham et al. 2011). In this research, only noise related to the TLS range is considered. Besides, it is assumed that the noise is randomly distributed and that no systematic errors are present. With such assumptions, noise can be reduced by averaging redundant data points. Roughness parameter sensitivity to noise is studied using ATOS data, to which different levels of noise were added. Differences of roughness parameters computed from the noiseless and noise-induced ATOS data indicate the parameter noise sensitivity.

The resolution of TLS points is governed by nominal point spacing set at acquisition and actual footprint size, which depends on scanning geometry and laser beam width. The effective resolution defines the level of detail that can be resolved from a scanned point cloud. Decreasing resolution (i.e. increasing the sampling interval), results in smoothing of the discontinuity surface, indicating that data resolution defines roughness scale. Ignoring the variation of measurement resolution leads to misleading roughness estimation (Tatone and Grasselli 2012). Thus, when comparing joint roughness parameters using different measurement techniques, data should first be resampled to the same resolution.

150.3 Parameterization Methods

In this research, two roughness parameterization methods are applied. To facilitate roughness computations, the coordinate system of the data was aligned with the mean joint plane, with the x- and y-axis coinciding with joint dip and strike, respectively, and the z-axis with roughness amplitude.

The compass and disc-clinometer technique is a traditional, contact-based method of joint roughness measurement (Fecker and Rengers 1971). Discs of different sizes are placed on the joint surface. Dip and dip direction of the disc is measured, which correlate to roughness amplitude and direction, respectively. Roughness scale-dependency can be evaluated using discs of different sizes. The compass and disc-clinometer method is applied to digital data by using orthogonal least squares (OLS) linear regression. For the TLS data having embedded noise, a plane is fitted to all laser points lying within the area covered by a virtual disc. Data redundancy reduces the noise effect on plane calculation. For the reference ATOS data (which are assumed to be free of error) iterative plane fitting is performed.

The angular threshold concept was initially developed to identify potential contact areas during direct shear testing of artificial rock joints (Grasselli 2001). Based on joint surface damage patterns, it was found that only portions of the joint surface that face the shear direction and are steeper than a threshold inclination θ^*provide shear resistance. A higher proportion of steeply inclined facets is indicative of a rougher surface, and is reflected by a larger area under the curve that expresses the potential contact area ratio as a function of θ^*. The area under the curve is taken as the roughness parameter (henceforth referred to as the Grasselli parameter, R). The parameter R depends on shearing direction and the 3D surface representation, but does not consider the scale effect.

150.4 Experiments and Results

Data acquisition. A joint sample of fossil rich limestone was fixed on a wooden plate equipped with eight reference targets (Fig. 150.1a). The smaller circular ATOS targets with radius 7 mm were placed precisely at the center of 10 cm square TLS targets (Fig. 150.1b). The sample and targets were scanned with the Riegl VZ400 laser scanner (Riegl 2013), and imaged with the ATOS I measurement system (Capture3D 2013). Multiple TLS measurements were taken with different nominal resolution in the perpendicular direction and at a distance of 10 m.

Data preparation. The target centers were extracted from corresponding point clouds. The ATOS target centers were processed simultaneously with data acquisition in the ATOS I software. TLS target centers were computed using an image matching algorithm (Kregar et al. 2013). Using target coordinates the TLS and ATOS datasets are co-registered and transformed into a new coordinate system aligned with the joint plane. To eliminate TLS range noise and to enable roughness parameter comparison, both ATOS and TLS point clouds were interpolated into 1 and 5 mm grids. Each grid center was assigned the median height of the points within the grid cell. The triangulated 1 mm ATOS grid is shown in Fig. 150.1c.

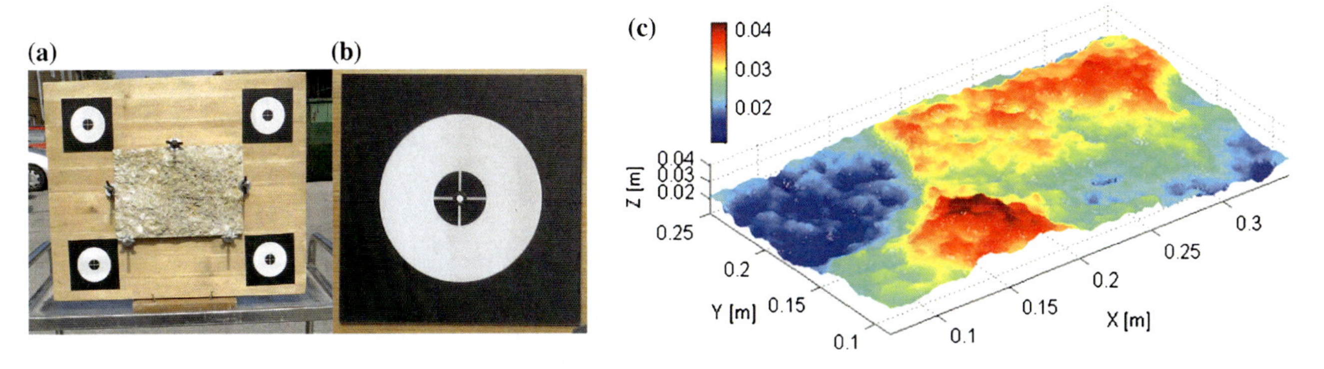

Fig. 150.1 **a** Experimental setup with joint sample and registration targets; **b** target zoom-in; and **c** triangulated surface from 1 mm ATOS grid (**c**)

150.4.1 Roughness Parameter Sensitivity to Noise

For the virtual disc-clinometer and Grasselli parameter algorithm, the dimension of the virtual disc and the grid size are chosen to be 5 mm, respectively. Reference ATOS parameters are then computed based on the described methodologies (Sect. 150.3). Five noise levels (1–5 mm) are added to the ATOS data and roughness parameters are recomputed. The median and 25th–75th percentiles of differences between reference and noisy parameters are shown in Fig. 150.2, where the dip differences are computed for the same disc positions (Fig. 150.2a) and Grasselli parameter differences for the same directions (Fig. 150.2b). Boxes 1–5 in Fig. 150.2 correspond to the five noise levels. Comparing the medians in Fig. 150.2, one can see that the dip is more sensitive to noise (i.e. has bigger median differences) than the Grasselli parameter. The reason might be that dip is computed directly from the data points and that orthogonal least squares results in artificially steep planes, when the noise level is close to disc size.

Reference ATOS dip measurements were also compared to results computed from original TLS data and 1 mm gridded TLS data (Fig. 150.2a, 6th and 7th box, respectively). Comparison of the 6th box to 7th shows that noise reduction by averaging the height within 1 mm grid cells was successful. The reference ATOS Grasselli parameters are compared to parameters computed from 5 mm grid TLS data (Fig. 150.2b, 6th box). Comparison of the 6th box to 1st indicates that TLS data resampled in 5 mm grid contain less than 1 mm noise.

150.4.2 Roughness Parameter Comparison

Based on results summarized in Sect. 150.4.1, TLS roughness parameters are compared to the reference ATOS parameters. A TLS grid of 1 mm and 5 mm are taken as

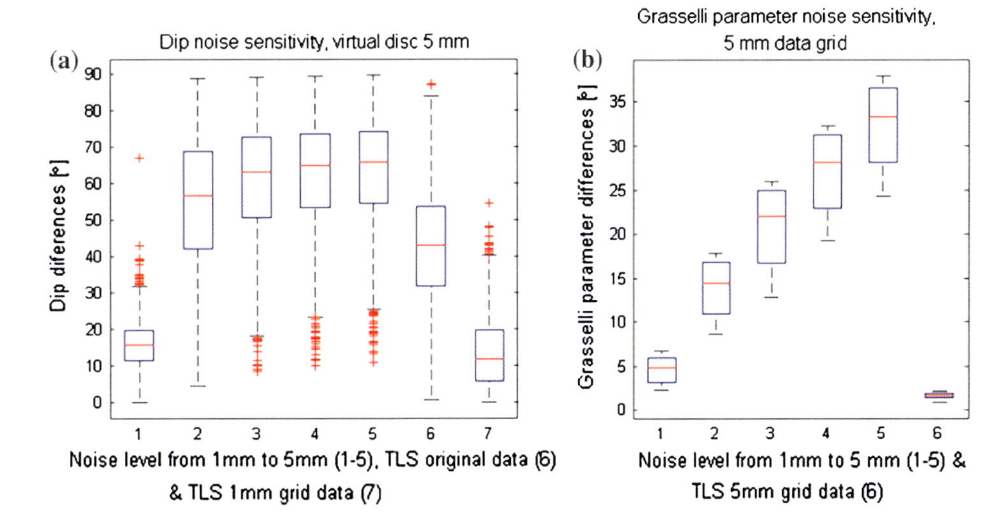

Fig. 150.2 Median (*horizontal lines*) and 25th–75th percentiles (*boxes*) of parameter differences indicate: **a** dip magnitude; and **b** Grasselli parameter noise sensitivity at a scale of 5 mm. Boxes 1–5 correspond to 5 noise levels from 1 mm to 5 mm, boxes 6 and box 7 to TLS data of different noise level

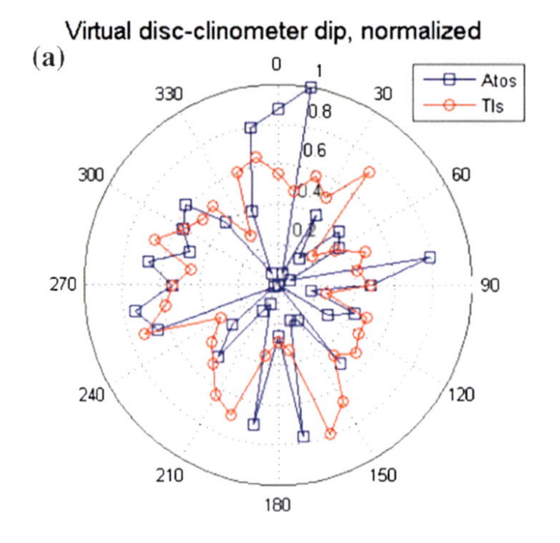

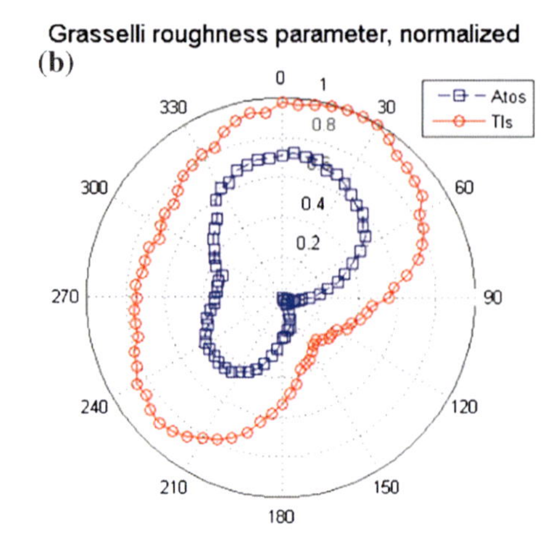

Fig. 150.3 Stereoplots comparing normalized ATOS and TLS roughness parameters computed with **a** virtual disc-clinometer method and **b** Grasselli method

input for the virtual disc-clinometer and Grasselli methods, respectively. Stereoplots in Fig. 150.3 show ATOS and TLS roughness parameters normalized to interval between zero (minimum) and one (maximum), which enables more detailed comparison. Direction 0° corresponds to y-axis (see Fig. 150.1c).

150.5 Conclusions

The use of TLS data for joint roughness computation has been investigated using two empirical roughness parameterization methods (virtual disc-clinometer method and Grasselli angular threshold method). Both methods are capable of representing roughness amplitude and its direction dependency. The sensitivity of roughness parameter measurements to TLS noise indicates that the Grasselli parameter is least sensitive. Dips of discs and the Grasselli parameters computed from gridded TLS data (reduced noise) were compared to the same parameters computed from reference ATOS measurements. Plot of maximum virtual disc dips in all dip directions show relatively poor correlation between ATOS and TLS results. Grasselli parameter plot show a significantly higher correlation between the data sets; however, the TLS surface roughness is systematically overestimated.

Acknowledgements The Slovenian National Building and Civil Engineering Institute kindly provided access to their ATOS measuring system and assisted in data acquisition.

References

Barton N, Choubey V (1977) The shear strength of rock joints in theory and practice. Rock Mech Rock Eng 10(1):1–54

Capture3D (2013) Atos I, configurations. Accessed Oct 2013, from http://www.capture3d.com/products-ATOSI-configuration.html

Fecker E, Rengers N (1971) Measurement of large scale roughness of rock planes by means of profilograph and geological compass. In: Proceedings symposium on rock fracture, Nancy, France

Grasselli G (2001) Shear strength of rock joints based on quantified surface description. École Polytechnique Fédérale de Lausanne. Lausanne, EPFL

Grasselli G, Egger P (2003) Constitutive law for the shear strength of rock joints based on three-dimensional surface parameters. Int J Rock Mech Min Sci 40(1):25–40

Khoshelham K, Altundag D et al (2011) Influence of range measurement noise on roughness characterization of rock surfaces using terrestrial laser scanning. Int J Rock Mech Min Sci 48(8): 1215–1223

Kregar K, Grigillo D, et al (2013) High precision target center determination from a point cloud. ISPRS workshop laser scanning 2013. Antalya, Turkey. ISPRS Annals of the Photogrammetry, Remote Sensing and Spatial Information Sciences, vol II-5/W2

Patton FD (1966) Multiple modes of shear failure in rock. 1st ISRM Congress. Lisbon, Portugal, International Society for Rock Mechanics, pp 509–513

Riegl (2013) Laser Scanner VZ-400, Datasheet. Accessed Oct 2013, from http://www.riegl.com/nc/products/terrestrial-scanning/produktdetail/product/scanner/5/

Sturzenegger M, Stead D (2009) Close-range terrestrial digital photogrammetry and terrestrial laser scanning for discontinuity characterization on rock cuts. Eng Geol 106(3–4):163–182

Tatone B, Grasselli G (2012) An investigation of discontinuity roughness scale dependency using high-resolution surface measurements. Rock Mech Rock Eng 1–25

Elaboration and Interpretation of Ground Investigation Data for the Heterogeneous 'Athens Schist' Formation; from the 'Lithological Type' to the 'Engineering Geological Formation'

151

Georgios Stoumpos and Konstantinos Boronkay

Abstract

Line 4A of the Athens Metro will encounter the Athens Schist, a heterogeneous rock mass comprising a variety of slightly metamorphosed lithological types. Due to the complexity of the rock mass, interpretation of ground investigation data poses a challenge since the direct attribution of laboratory and in situ data to geotechnical design profiles is not appropriate. The methodology that was applied for the evaluation of these data is based on the scheme lithological type—engineering geological unit—engineering geological formation—geological formation, each being a subgroup of the latter. The main aspect of this approach takes into consideration the scale of reference of each assigned category of geomaterials. This elaboration and interpretation methodology, in conjunction with the engineering geological profile, which is compiled in terms of engineering geological formations, provides a comprehensive background for all subsequent geotechnical design needs.

Keywords

Geotechnical interpretation • Intact rock • Rock mass • GSI • Properties

151.1 The Project

Line 4A of the Athens Metro is a new planned line comprising 9 stations and ~8 km of a single ~9.5 m diameter tunnel. The line will connect the central-north suburbs of Athens with the city centre, passing through some of the most densely populated neighbourhoods which, at this moment, are not served by a rapid transit system. This paper refers to the methodology that was adopted for the elaboration of ground investigation data in the framework of the production of the General Final Design of Line 4A.

G. Stoumpos (✉) · K. Boronkay (✉)
Attiko Metro S.A., Messoghion 191, 11525 Athens, Greece
e-mail: gstoumpos@ametro.gr

K. Boronkay
e-mail: kboronkay@ametro.gr

151.2 The Geological Setting

The alignment, of Line 4A runs along the west slopes of Tourkovounia (or Lykovounia) hill range; the tallest and most extensive hill range within the Athens basin which divides it into eastern and western parts.

The geological setting of the project consists of alpine formations and local thin quaternary and man-made deposits. The alpine formations are (from top to bottom) the Crest Limestones, the Sandstone-Marl Sequence and the Athens Schist (Upper and Lower). The quaternary deposits include scree and alluvial deposits. Man-made deposits of small thickness are also locally encountered. Of these geological formations, the one that is being discussed in this paper is the Athens Schist, due to its heterogeneity and complexity.

G. Lollino et al. (eds.), *Engineering Geology for Society and Territory – Volume 6*,
DOI: 10.1007/978-3-319-09060-3_151, © Springer International Publishing Switzerland 2015

Fig. 151.1 Typical transition from the upper to the lower Athens Schist. The transition is marked by the presence of a shear zone between 17.9–19.3 m

151.2.1 Athens Schist

This formation is an upper Cretaceous heterogeneous rock mass that comprises a variety of slightly metamorphosed lithological types (Marinos et al. 1971) which can be distinguished into two sub-formations: the upper and the lower (Fig. 151.1).

The heterogeneity of the Athens Schist is evident in the frequent, yet often unsystematic, alternations of members with significant differences in terms of mineral composition, grain size and strength; differences that result from the stratigraphic and tectonic inherent intricacy of the rock mass.

The upper formation (commonly referred to as the upper unit, Koukis and Sabatakakis 2000) consists mainly of alternations of metasandstones and metasiltstones. Limestones, calcareous sandstones, calcareous siltstones, schists (chlorite quartzitic, chlorite epidote, calcareous chlorite) as well as phyllites and calcareous phyllites are also encountered. Limestones are often karstified. Sporadically, thin layers and lenses of quartz can be found. The alternations comprise beds which may reach several meters in thickness for the hard rock (metasandstones) but usually are of the order of some decimetres. Even where metasandstones prevail, thin interlayers of weaker material (metasiltstones) are present in most cases.

The lower formation (commonly referred to as the lower unit, Koukis and Sabatakakis 2000) comprises shales, metasiltstones and metasandstones. Thin layers of calcareous metasandstones, chlorite schists as well as intercalations and lenses of quartz are also encountered. The lower Athens Schist is considered the poorest rock mass when compared with the upper formation, since its constituents, and especially the prevailing shales, are of a weaker nature.

151.2.2 Tectonics

Athens Schist constitutes a tectonic nappe pile, formed during Upper Cretaceous (alpine) folding and thrusting. As a result, the rock mass usually exhibits intense fracturing even within its relatively competent members and shear zones are frequent in the form of cataclasites or clayey fault gauge. The formations are persistently folded and locally intensely folded with tight to isoclinal rootless folds.

It has to be noted that, due to its weaker nature, the lower formation was more vulnerable to tectonic deformation. As such, plastic deformation and shearing is more intense with shear zones of considerable thickness (of the order of some meters). In these shear zones the Athens Schist exhibits a chaotic structure with persistent metasandstone and quartz lenses "flowing" in a fissured clayey matrix. In many cases, due to the difference in brittleness between the upper and the lower unit, the transition is manifested by the presence of a shear (detachment) zone (Koukis and Sabatakakis 2000; Fig. 151.1).

151.3 The Challenge

Due to the complexity of the rock mass, interpretation of ground investigation data poses significant challenges. Since the rock mass is heterogeneous with alternations of weaker (e.g. metasiltstone) and stronger members (e.g. metasandstone) the direct attribution of laboratory and in situ tests' data to geotechnical profile strata is not appropriate.

Regarding the Attiko Metro projects, it is common practice to utilize the Hoek–Brown failure criterion to derive strength and deformability parameters for design needs. Hoek–Brown failure criterion emphasises on the lithology of the intact rock, through the intact rock strength σ_{ci} and the constant m_i and on the quality of the rock mass through the Geological Strength Index (GSI). GSI is a system that allows for a reduction in the intact rock parameters by evaluating the structure and the discontinuity surface condition. GSI was introduced because '*a system based more heavily on fundamental geological observations and less on "numbers" was needed*' (Hoek and Marinos 2007).

Indeed, this is the case with the Athens Schist. It is apparent that due to the heterogeneity (various alternating lithological types) and the tectonic history (intense shearing and/or fracturing) of the rock mass, the elaboration of given intact rock properties and the attribution of GSI values requires systematic and careful steps as well as geological and engineering geological awareness.

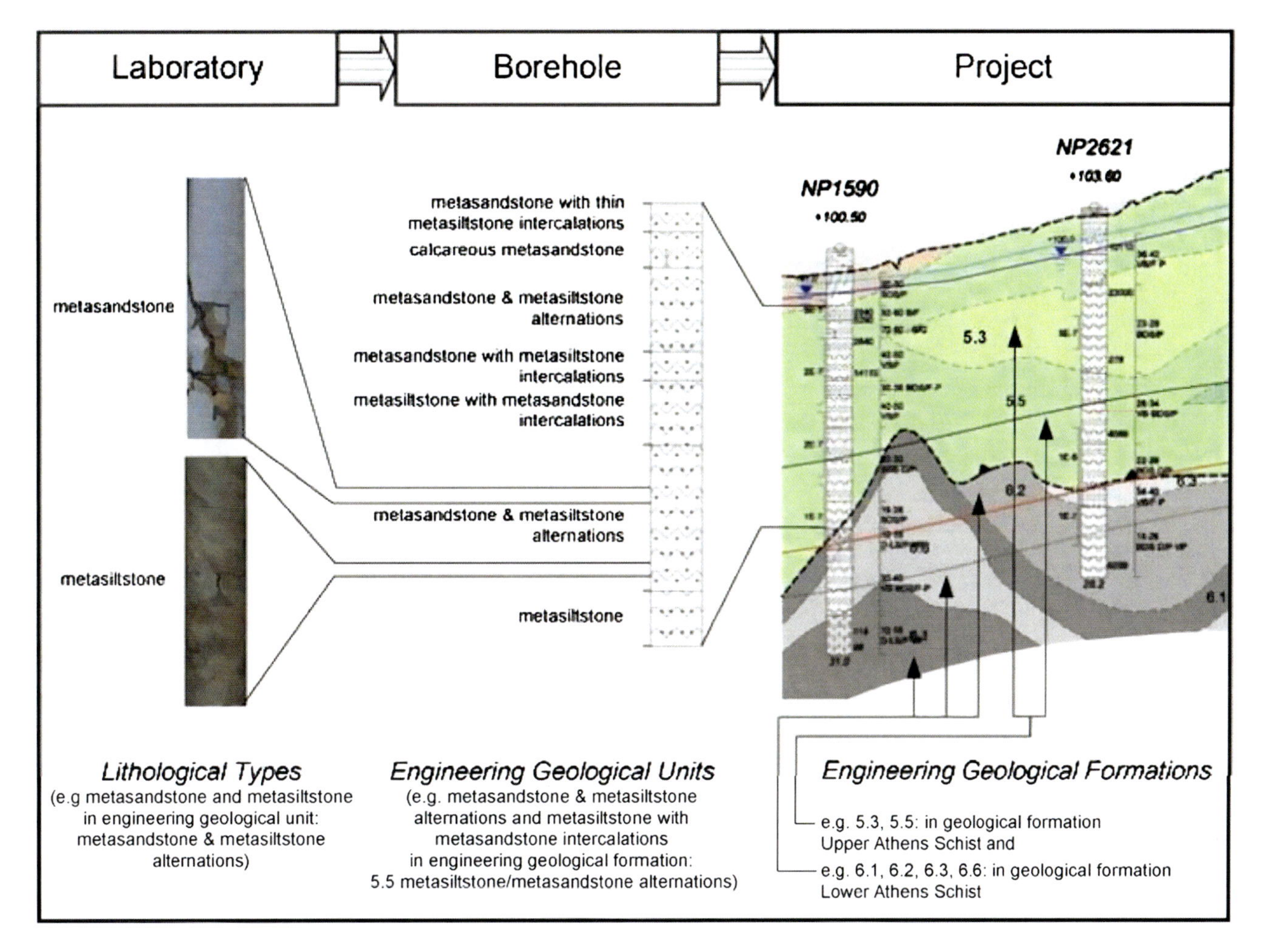

Fig. 151.2 Explanatory graph showing the "flow" of the approach from the laboratory sample scale to the project site scale, namely from the lithological type to the engineering geological formation. In reality, two processes that feed one another, are simultaneously undertaken; synthesis of data from the engineering geological units to the engineering geological formations and analysis of data from the engineering geological units to the lithological types

151.4 The Approach

The methodology that was applied is based on the scheme lithological type—engineering geological unit—engineering geological formation—geological formation, each being a subgroup of the latter (see Fig. 151.2). The main aspect of this approach takes into consideration the scale of reference of each assigned category of geomaterials. Namely, the lithological type refers to the laboratory samples, the engineering geological unit refers to the borehole samples, the engineering geological formation refers to the project site and the geological formation refers to the broader area. In other words, the behaviour of the rock mass during project construction is best described when referring to engineering geological formations. Thus, all available data had to be "translated" in terms of engineering geological formations and attributed to them.

During borehole logging, description and rock mass classification referred to engineering geological units. In this context, the engineering geological unit is the basis of the evaluation and the primary information that was recorded. Obviously, the engineering geological units often comprised more than one lithological type.

These lithological types that were distinguished within the engineering geological units were identified and categorized. Lithological types where the basis for the elaboration of all ground investigation data that refer to the intact rock, namely uniaxial compression tests and point load tests. Moreover, the engineering geological units were grouped

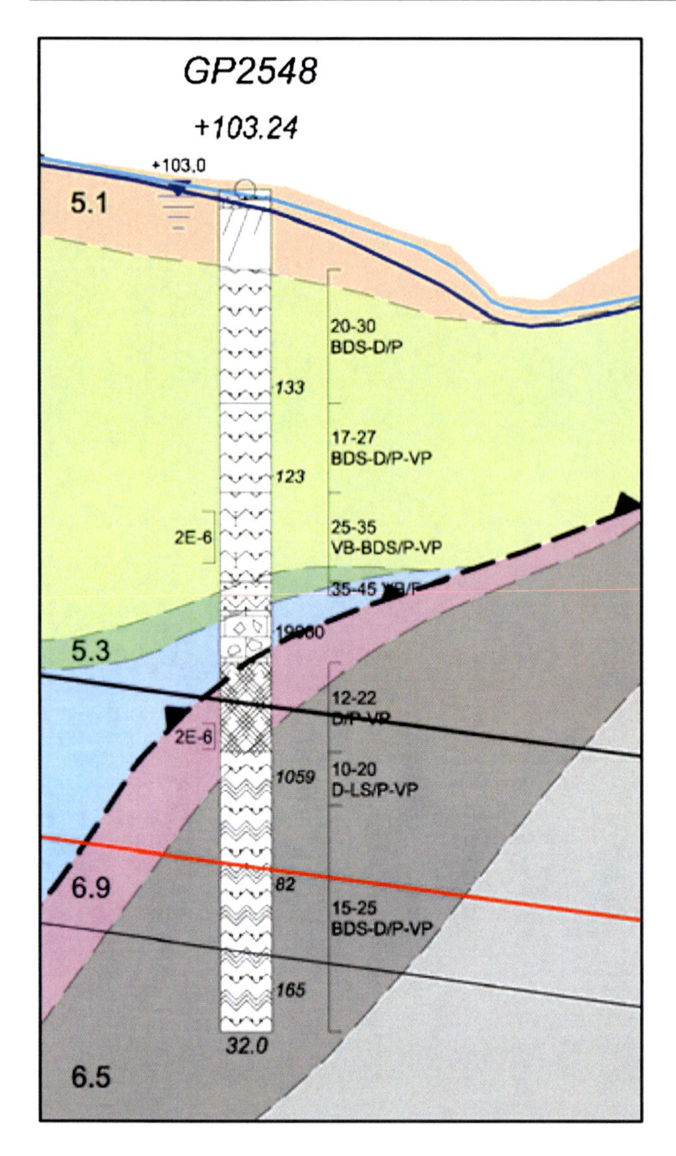

Fig. 151.3 Detail of the geological—hydrogeological—geotechnical longitudinal section. The primary information, namely borehole logging and GSI value as well as structure and surface conditions, are shown in and on the right of the borehole column respectively. q_u and σ_{ci} laboratory tests' results are shown adjacent to the right of the borehole column (in kPa) while in situ permeability tests' results are shown adjacent to the left of the borehole column (in m/s) with the length of the tested section of the borehole also being indicated. The background shows the engineering geological formations (5.1, 5.3, 6.9 etc.). The section is accompanied by a table with proposed σ_{ci}, m_i, GSI and MR values per engineering geological formation (this table is not presented in the present paper since the production of the GFD of the project is still in the pre-tender phase)

into engineering geological formations. These formations are the basis for the engineering geological longitudinal section and the cornerstone of the whole scheme. Rock mass classification values that where assigned to engineering geological units where elaborated to refer to engineering geological formations. This procedure was based on reducing GSI values for the length of the core run on which they have been assigned.

Intact rock properties were elaborated in terms of lithological types per engineering geological formation. Alongside, for each lithological type the constant m_i was considered.

Based on the Marinos et al. (2011) work on heterogeneous rock masses, the percentage of participation of the various lithological types in each engineering geological formation was taken into account and as such, intact rock properties per engineering geological formation were derived.

151.5 The Outcome

This elaboration and interpretation methodology offers a full set of intact rock parameters and rock mass classification values, per engineering geological formation, that can be directly inputted into the Hoek–Brown failure criterion.

In conjunction with the engineering geological profile (geological-hydrogeological-geotechnical longitudinal section), which is compiled in terms of engineering geological formations (see Fig. 151.3), this approach provides a comprehensive background for all subsequent geotechnical design needs.

References

Hoek E, Marinos P (2007) A brief history of the development of the Hoek–Brown failure criterion. Soils and Rocks, vol. 2, Nov 2007

Koukis G, Sabatakakis N (2000) Engineering geological environment of Athens, Greece. Bull Eng Geol Environ 59(2):0127–0135

Marinos G, Katsikatsos G, Georgiadou-Dikeoulia E, Mirkou R (1971) The Athens' Schist formation, I. stratigraphy and structure. Annales Géologiques des Pays Helléniques, 1ere série, tome 23, 183–212 (in Greek)

Marinos V, Fortsakis P, Prountzopoulos G (2011) Estimation of geotechnical properties and classification of geotechnical behaviour in tunnelling for flysch rock masses. In: Anagnostopoulos A et al (eds) Proceedings of the 15th European conference on soil mechanics and geotechnical engineering. Part 1, Athens 2011, pp 435–440

Incorporating Variability and/or Uncertainty of Rock Mass Properties into GSI and RMi Systems Using Monte Carlo Method

152

Mehmet Sari

Abstract

This paper introduces a probability-based methodology that can be used to evaluate alternative approaches on account of the uncertainties associated with predicting the rock mass properties. The use of this methodology is illustrated through its application for two rock mass case studies. In this regard, the probabilistic spreadsheet models are developed for the strength estimation of Kizilkaya ignimbrite and New Zealand greywacke. The frequency histograms and/or the density functions that best describe the data distribution are used as inputs in GSI and RMi systems. This approach allows the variability and/or uncertainty of the available data to be adequately taken into account during simulations. The developed spreadsheet models are also used to quantify the influence of various material and discontinuity characteristics on the resultant strength properties of the studied rock masses. Sensitivity analysis explicitly shows that joint spacing and UCS of intact rock are the most effective parameters on the estimated rock mass strengths.

Keywords

Rock mass • Kizilkaya ignimbrite • Sensitivity analysis • Failure criterion • UCS

152.1 Introduction

A rock mass is a system composed of intact rock pieces separated by a network of discontinuities consisting one or more joint sets, where mechanical behavior and properties are highly variable. Intact rock refers to unfractured blocks between structural discontinuities. A discontinuity is described as "a plane of weakness that has zero or low tensile strength or tensile strength lower than stress levels generally applicable in engineering applications" (Anonymous 1977). Discontinuous rock masses are generally heterogeneous, anisotropic and there is unpredictable spatial variability in both the intact material and discontinuity properties. Due to the heterogeneous properties of the intact rock material and the discontinuity network these masses show a high degree of variability. The intact rock strength indicates the ability of the jointed rock mass to resist shearing failure through the intact pieces of rock. Each discontinuity has a different degree of strength along its length. Therefore, any acceptable solution to a jointed rock mass model should consider both variability of the intact rock and the discontinuities that govern the stability of the rock masses. However, a clear distinction should be made to clarify the difference between the variability of the GSI and σ_{ci} since the first one corresponds to a large volume of rock mass and the second one to a small specimen.

Due to the uncertainty of inherent characteristics that define rock masses, each will exhibit high degree of variability, and thus cannot be evaluated based on deterministic approaches. Empirical models developed for the estimation of rock mass properties are based on the rock mass classification systems such as Q, GSI, RMi. These systems involve a large number of input parameters, each subject to substantial uncertainty. As stated by Riedmuller and Schubert (1999) the complex properties of a rock mass could not sufficiently be described by a single number. Since there are

M. Sari (✉)
Department of Mining Engineering, Aksaray University, 68100 Aksaray, Turkey
e-mail: mehmetsari@aksaray.edu.tr

G. Lollino et al. (eds.), *Engineering Geology for Society and Territory – Volume 6*,
DOI: 10.1007/978-3-319-09060-3_152, © Springer International Publishing Switzerland 2015

so many elements or factors that influence the engineering properties of rock masses and the inherent variability of the value is very large, it is important to describe and characterize rock masses in a way to find representative ratings and values for use in engineering design. An accurate assessment of rock mass strength must involve sensitivity analysis and must attempt to model the inherent variability and uncertainty in these parameter estimates.

The Monte Carlo method may be a useful tool for rock engineers and engineering geologists who study in complex rock mass conditions. This technique already takes into considerations such as anisotropy, heterogeneity and other factors which lead the rock to behave stochastically. Probabilistic sensitivity analysis is one method for performing a sensitivity analysis in which all parameters subject to uncertainty are varied simultaneously by Monte Carlo sampling from the distributions postulated for those parameters.

This study is aimed to develop a methodology for taking uncertainty and/or variability caused by natural phenomenon. For this purpose, intact rock and discontinuity properties of two different rock masses are defined as probability distributions in GSI and RMi systems. The developed stochastic models are used to estimate the probability distributions of rock mass strength and the most effective parameters on the estimated values.

152.2 Previous Studies

In the literature, there are a limited number of studies which considers only the stochastic estimation of strength and deformability characteristics and variations of rock masses. Kim and Gao (1995) presented a probabilistic method of estimating the mechanical characteristics of a rock mass, using the third type asymptotic distribution of the smallest values (extreme value statistics) and MC simulation. They used the chi-square goodness-of-fit test to prove that the distribution reflects the inherent variability of the properties of a basaltic rock. Hoek (1998) applied the same method to estimate variation in the Hoek–Brown properties of a hypothetical rock mass, and assuming that all three input parameters of the criterion can be represented by normal distributions. Sari (2009) and Sari et al. (2010) demonstrated the use of MC simulations to evaluate the strength and deformability of rock masses by including the uncertainties of the intact rock strength and discontinuity parameters. They concluded that the MC method provided a viable means for assessing the variability of rock mass properties.

152.3 Materials and Method

152.3.1 Kizilkaya Ignimbrite

The Kizilkaya ignimbrite well outcrops at Kizilkaya village and in the Ihlara Valley (Fig. 152.1). The study area covers the historical and touristic Ihlara Valley, which is about 14 km in length and covers 52 km^2. The rocks in the study area are basically classified as pyroclastic, called Selime tuffs, Kizilkaya ignimbrites, and Hasandag ashes. With the influence of both water and wind erosion, interesting rock shapes and morphological figures have been formed in the slopes of the valley. There are also joint systems developed on the rock mass as a result of cooling of material deposited during volcanic eruptions.

The engineering geological properties of the exposed Kizilkaya ignimbrite were determined on the basis of field observations/measurements and laboratory tests by the author. The main orientation, spacing, persistence, aperture, filling, weathering, and roughness of the discontinuities were described using the scan-line survey method following the ISRM (2007) description criteria. A total of 260 discontinuities were measured along a straight outcrop surface using a measuring tape and compass. A total of 18 rock blocks were collected from the field, then 112 cube specimens were

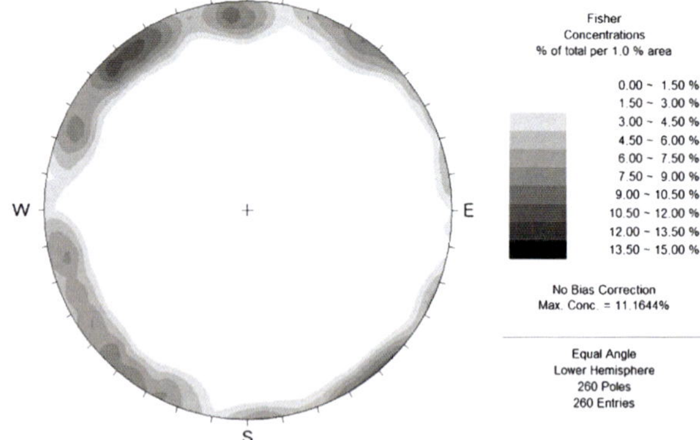

Fig. 152.1 A view of the Kizilkaya ignimbrite and discontinuity orientations

Table 152.1 Discontinuity and intact rock properties of Kizilkaya ignimbrite

Parameter	Distribution type (Mean ± Std. dev.)
UCS (MPa)	Normal (46.06 ± 8.63)
Joint spacing (m)	Lognormal (2.575 ± 1.195)
Joint persistence (m)	Normal (27.09 ± 6.79)
Joint aperture (cm)	N. exponential (3.31 ± 3.26)
Joint roughness	25 % Very rough, 50 % rough, 25 % smooth
Weathering	25 % Fresh, 50 % slightly, 25 % moderately
Infill	50 % Unfilled, 25% sand and gravel, 25 % soft fillings
Groundwater	Dry
m_i	Normal (13.0 ± 2.0)

prepared from the blocks for laboratory testing. The uniaxial compressive strength (UCS) tests were conducted according to Turkish Standards of methods of testing for natural building stones (TS 699 1987). These values are given in Table 152.1.

152.3.2 New Zealand Greywacke

Closely jointed greywacke rock masses are widespread throughout New Zealand and much of New Zealand's infrastructure is constructed upon greywacke rock masses (Fig. 152.2). Cook (2001) aimed to find common physical properties of defects typical of NZ greywacke rock masses, use these data to identify parameters which have a greater effect on rock mass strength. Cook (2001) measured the following rock mass properties: joint orientation, defect spacing, persistence length, type of joint termination, defect aperture, type of infilling material, type of surface roughness, waviness. A database of greywacke properties was also developed by Stewart (2007) based on previous studies upon un-weathered greywacke around New Zealand. The database included descriptions of greywacke defect properties and mechanical properties of the intact rock and joints. Greywacke in the study area is composed of hard sandstones, sandstones inter-bedded with mudstones, and mudstones. The summary information for this dataset is given in Table 152.2.

Table 152.2 Discontinuity and intact rock properties of New Zealand greywacke

Parameter	Distribution type (Mean ± Std. dev.)
UCS (MPa)	Normal (244.7 ± 33.3)
Joint spacing (cm)	Exponential (5.1)
Joint persistence (m)	Lognormal (0.36 ± 0.58)
Joint aperture (mm)	15 % None, 30 % 0.1<, 40 % 0.1–1, 10 % 1–5, 5 % 5>
Joint roughness	5 % Very rough, 50 % rough, 10 % slightly rough, 30 % smooth, 5 % slickensided
Weathering	50 % Fresh, 25 % slightly, 25 % moderately
Infill	30 % Unfilled, 50 % hard filling, 15 % 5 mm< soft filling, 5 % 5 mm> soft filling
Groundwater	Dry
m_i	Normal (17.0 ± 1.5)

152.3.3 GSI and RMi Classification Systems

The Geological Strength Index (GSI) (Hoek et al. 1995) and the Jointing Parameter (JP) of the Rock Mass index (RMi) (Palmstrom 1995) are two of the most known and frequently used indexes. RMi has been developed by Palmstrom (1995)

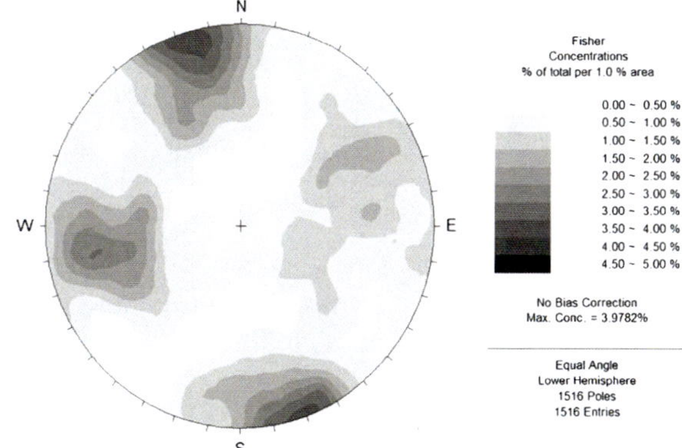

Fig. 152.2 A view of the New Zealand greywacke and discontinuity orientations

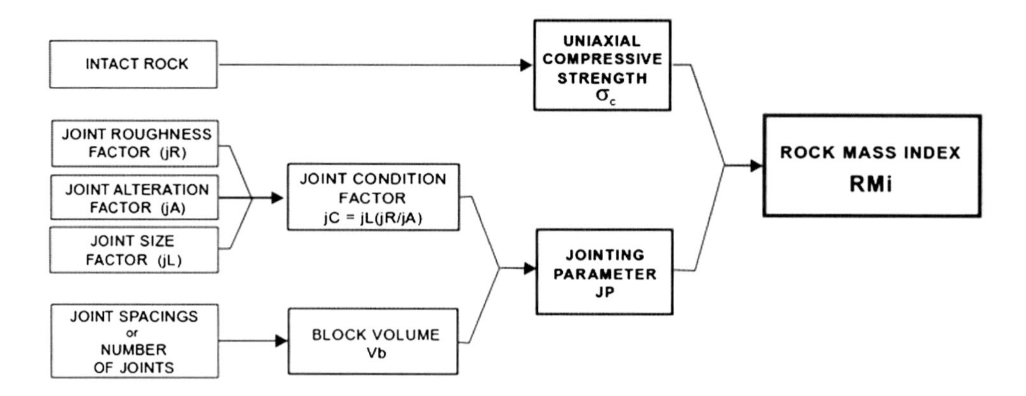

Fig. 152.3 The layout of the rock mass index, RMi (after Palmstrom 1995)

with input of the geological parameters that have the greatest influence on rock mass strength (Fig. 152.3).

RMi is based on the principle that the joints intersecting a rock mass tend to reduce its strength. Consequently, it is expressed as:

$$RMi = \sigma_c \times JP \quad (152.1)$$

Here σ_c = the uniaxial compressive strength of intact rock (in MPa), measured on 50 mm samples. JP = the jointing parameter, expressing the reduction in strength of the intact rock caused by the joints. It incorporates the main joint features in the rock mass. The jointing parameter was found as

$$JP = 0.2 \times jC^{0.5} \times V_b^D \quad (152.2)$$

where $D = 0.37 \times jC^{-0.2}$.

At failure, the generalized Hoek–Brown criterion relates the maximum effective stress, σ_1 to the minimum effective stress σ_3 through the functional relation:

$$\sigma_1 = \sigma_3 + \sigma_{ci}\left(m_b \frac{\sigma_3}{\sigma_{ci}} + s\right)^a \quad (152.3)$$

where m_b extrapolates the intact rock constant m_i to the rock mass:

$$m_b = m_i \exp\left(\frac{GSI - 100}{28 - 14D}\right) \quad (152.4)$$

σ_{ci} is the uniaxial compressive strength of the intact rock and s and a are constants that depend upon the rock mass's characteristics:

$$s = \exp\left(\frac{GSI - 100}{9 - 3D}\right) \quad (152.5)$$

$$a = \frac{1}{2} + \frac{1}{6}\left(e^{-GSI/15} - e^{-20/3}\right) \quad (152.6)$$

The uniaxial compressive strength of the jointed rock masses is calculated from the following equation suggested by Hoek et al. (2002)

$$\sigma_c = \sigma_{ci} s^a \quad (152.7)$$

The concept of a global "rock mass strength" σ_{cm} is proposed by Hoek et al. (2002) and it could be estimated from the Mohr–Coulomb relationship as

$$\sigma_{cm} = \sigma_{ci} \frac{(m_b + 4s - a(m_b - 8s))(m_b/4 + s)^{a-1}}{2(1 + a)(2 + a)} \quad (152.8)$$

The most recent version of the generalized Hoek–Brown failure criterion (Hoek et al. 2002) is employed to estimate the rock mass strengths of the Kizilkaya ignimbrite and New Zealand greywacke. In the Hoek–Brown criterion, the GSI is the most important scaling parameter for allocating the strength and deformation properties determined in the laboratory to the field scale rock mass (Hoek et al. 1995). In earlier versions of this criterion, Bieniawski's RMR was used for this scaling process. Due to the usage of different rating scales for each parameter, the RMR is more suitable for variability determination. It is possible to objectively obtain a frequency distribution of RMR for computing purposes in a probabilistic analysis. However, by using a GSI chart only a subjective estimate of variability to field scale rock mass can be possible. The following relation is suggested by Hoek (1998):

$$GSI = RMR_{89} - 5 \quad (152.9)$$

Since GSI system is mostly descriptive in nature and prone to subjectivity, a linkage between descriptive geological terms and measurable field parameters has been proposed by Sonmez and Ulusay (1999), Cai et al. (2004) and Russo (2009). Cai et al. (2004) proposed a quantitative approach, using block volume and joint condition factor, to utilize the GSI system. Once the block volume (V_b) and the joint condition factor (J_c) are known, the GSI value can be determined from the following equation presented by Cai and Kaiser (2006):

$$GSI = (26.5 + 8.79 InJ_c + 0.79 InV_b)/(1 + 0.0151 InJ_c - 0.0253 InV_b) \tag{152.10}$$

where J_c is a dimensionless factor and V_b is in cm^3.

On the basis of the conceptual similarity of the GSI with the Joint Parameter (JP) used in the RMi, Russo (2009) derived a relationship between the two indexes in order to obtain a reliable, quantitative assessment of the GSI by means of the basic input parameters for the determination of the RMi. On the basis of the above correlations, a quantitative estimation of the GSI is made, by defining the parameters concurrent to the evaluation of JP, i.e. the block volume (V_b) and the joint condition factor (jC) as follows:

$$GSI = 153 - 165/(1 + (JP/0.19))^{0.44} \tag{152.11}$$

152.4 Monte Carlo Simulation and Sensitivity Analysis

The probabilistic spreadsheet models are developed for the estimation of rock mass strengths of Kizilkaya ignimbrite and New Zealand greywacke. The frequency histograms and/or the density functions that best describe the data distribution in Tables 152.1 and in 152.2 are used as inputs in GSI and RMi systems. Excel add-in ModelRisk (Vose Software 2012) program provides a simple and intuitive implementation of a MC simulation together with prepared spreadsheet models. In a MC simulation, a random value is selected for each of the inputs, based on the range of estimates. The model is calculated based on this random value. The result of the model is recorded, and the process is repeated many times. At the end, basic statistics (mean, standard deviation, range, min, max, etc.) for the output are computed from the collected data.

In this study, 5,000 iterations are performed for different rock mass properties using the Latin Hypercube sampling method. This method applies stratified sampling technique to closely resemble the input probability distribution with fewer realizations. Although there are some correlations expected between input parameters, they are assumed to be independent for the sake of simplicity of the problem at hand.

The frequency distributions of calculated GSI values are given in Fig. 152.4 for the Kizilkaya ignimbrite and New Zealand greywacke, respectively. It can be easily discerned that the GSI values estimated from the RMR system (Eq. 152.9) for both rock types illustrate more consistent and less scattered data compared to two equations suggested by Cai and Kaiser (2006) (Eq. 152.10) and Russo (2009) (Eq. 152.11). One of the fallows of Cai and Kaiser (2006) method is that it consistently produces GSI values greater than 100 which is meaningless. In spite of having closer mean values between Cai and Kaiser (2006) and RMR equations, three suggested methods predict dissimilar and highly variable values of GSI for the same rock masses.

Estimated strength values of rock masses are given in Fig. 152.5. Two equations suggested by Hoek et al. (2002) and the one calculated from RMi of Palmstrom (1995) predict completely different strength values. Necessary GSI values for the estimation of σ_c and σ_{cm} values in the Hoek–Brown failure criterion are the ones calculated from the RMR dependent equation (Eq. 152.9). For the Kizilkaya ignimbrite, the estimated rock mass strength of σ_{cm} and RMi are very close to each other, around to 12.5 MPa. For New Zealand greywacke, however, values of σ_c and σ_{cm} are closer than RMi value. It can be said that different rock mass properties of two completely dissimilar rocks are subjected to different rock mass behavior. This is mostly due to attaining various priorities to rock mass characteristics in their empirical equations by the founders.

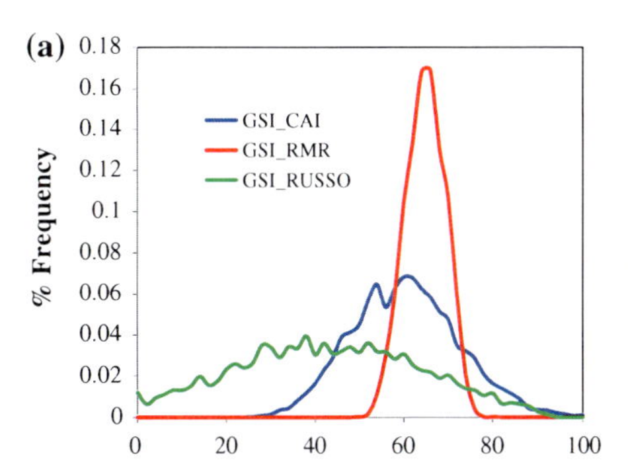

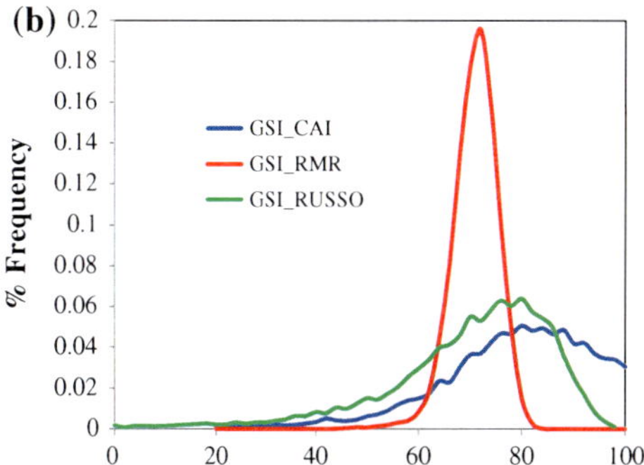

Fig. 152.4 Simulated GSI frequency distributions of **a** Kizilkaya ignimbrite and **b** New Zealand greywacke using three quantitative approaches

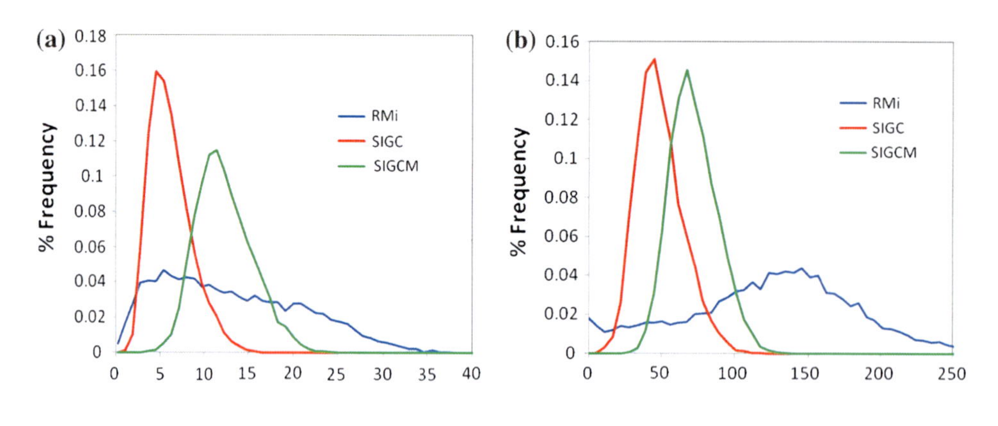

Fig. 152.5 Probabilistic estimates on the rock mass compressive strength (MPa) of **a** Kizilkaya ignimbrite and **b** New Zealand greywacke using GSI and RMi systems

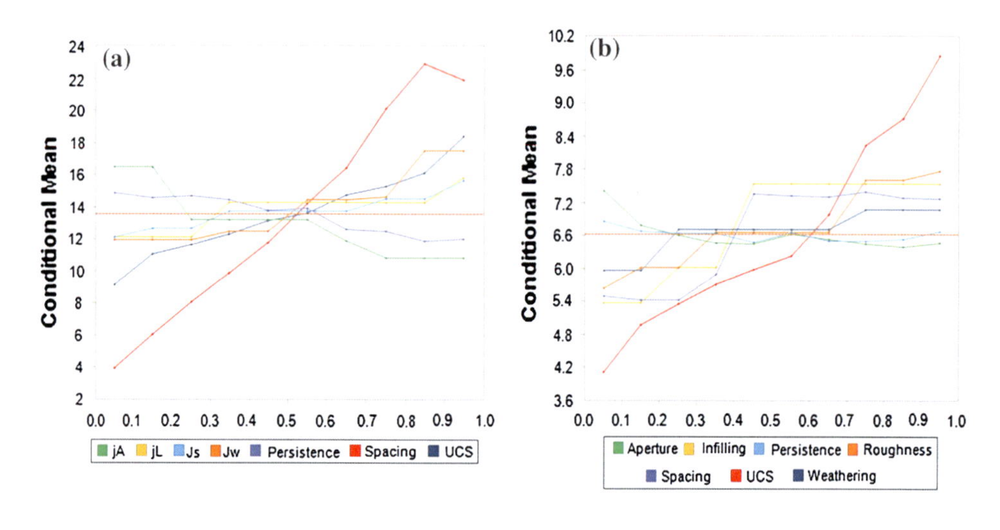

Fig. 152.6 Spider plots for **a** RMi and **b** σ_c of the Kizilkaya ignimbrite

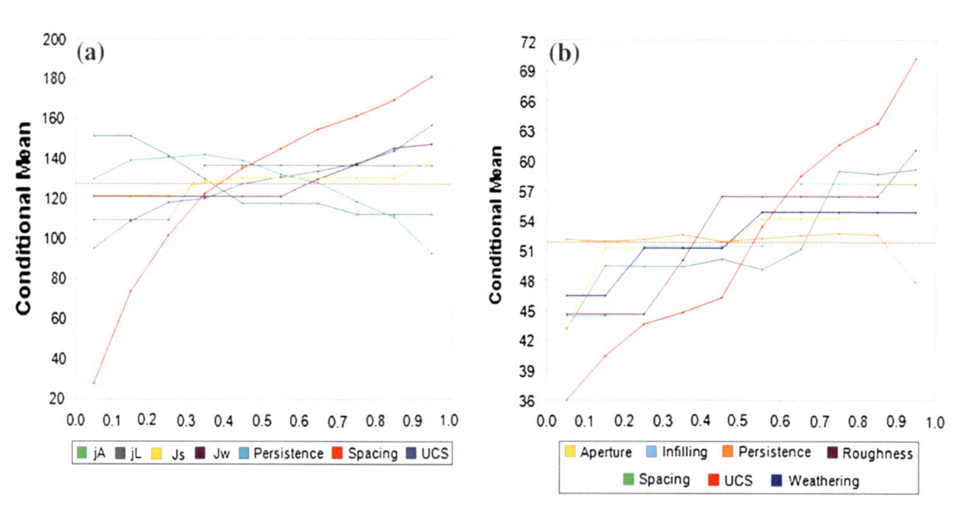

Fig. 152.7 Spider plots for **a** RMi and **b** σ_c of the New Zealand greywacke

Sensitivity analysis is a method employed to find most influential input parameters on the output parameters. If an input parameter changes every time on its full range during simulation while other parameters are kept constant on their mean values, potential effects of each input parameter can be easily discerned on the output results. There are some methods for performing sensitivity analysis on the data, namely, rank order correlations, contribution to variance and spider plots. Spider plots describe graphically how sensitive value of an output variable is to the input variables of a simulation model. The flatter the line for an input variable, the less sensitive it is.

In Figs. 152.6 and 152.7, an analysis has been performed on the sensitivity of the mean of the RMi and σ_c outputs. It has been performed by splitting up simulation data from input distributions into ten groups in terms of their cumulative probability. The simulation data are filtered by software for each of these groups to find the corresponding RMi and σ_c mean values that occurred when the input variable being analyzed lies within each percentile band. The statistic of interest (i.e. mean in this case) is then calculated for the filtered data. Repeating this analysis across each percentile range for each selected input variable produces the spider plots in Figs. 152.6 and 152.7.

In these plots, the horizontal dashed line shows the mean of the unfiltered RMi and σ_c values as a reference. The vertical range that an input line covers reflects the degree of sensitivity that output statistic has to this input value. So, for example, when spacing lies in its 0–10 % range in Fig. 152.6, the RMi mean is approaches to 4 MPa, and when spacing lies in its 90–100 % range, the RMi mean is closer to 22 MPa, a difference of 18 MPa. Reviewing all graphs, it can be easily discerned that while the mean values of RMi is most sensitive to spacing of Kizilkaya ignimbrite and New Zealand greywacke, the mean values of σ_c is most sensitive to UCS of the two rock masses.

152.5 Conclusions

A probability based analysis is performed in the study to account for uncertainty and/or variability reflected by most of the rock masses encountered in large construction projects. Three empirical equations proposed to calculate GSI value quantitatively are investigated for the estimation of strength of two rock masses subject to completely different geologic characteristics. First equation is the generic one proposed by Hoek and Brown (1997) for estimating GSI values from the RMR scores. Second and third ones are proposed by Cai and Kaiser (2006) and Russo (2009) to estimate GSI values quantitatively. It was found that three suggested equations generated completely different values of GSI. It was also found that the most influential parameters were depended on which equations were used in the estimation of rock mass strength. In case of calculated RMi, the most effective input parameter on the rock mass strength was the spacing of discontinuities. On the other hand, in case of σ_c and σ_{cm}, the most influential input parameter was the UCS of intact rocks.

Regarding the estimation of the GSI in the design stage, Cai (2011) is recommended to apply the point estimate method, which reflects rock mass property uncertainties in the numerical analysis of tunnel and cavern stability. Inclusion of more data in the design process may not only provide a more accurate estimation and allow for the recognition of dispersion, but can also alter the ultimate estimate that is made.

References

Anonymous (1977) The description of rock masses for engineering purposes. Report by the Geological Society Engineering Group working party. Quart J Eng Geol 29:67–81

Cai M, Kaiser PK, Uno H, Tasaka Y, Minami M (2004) Estimation of rock mass deformation modulus and strength of jointed hard rock masses using the GSI system. Int J Rock Mech Mining Sci 41:3–19

Cai M, Kaiser PK (2006) Visualization of rock mass classification systems. Geotechnol Geol Eng 24:1089–1102

Cai M (2011) Rock mass characterization and rock property variability considerations for tunnel and cavern design. Rock Mech Rock Eng 44:379–399

Cook GK (2001) Rock mass structure and intact rock strength of New Zealand greywackes. M.Sc thesis, University of Canterbury

Hoek E, Kaiser PK, Bawden WF (1995) Support of underground excavations in Hard Rock. Balkema, Rotterdam 215

Hoek E, Brown ET (1997) Practical estimates or rock mass strength. Int J Rock Mech Min Sci 34:1165–1186

Hoek E (1998) Reliability of the Hoek–Brown estimates of rock mass properties and their impact on design. Int J Rock Mech Min Sci 35:63–68

Hoek E, Carranza-Torres CT, Corkum B (2002) Hoek–Brown failure criterion-2002 edition. In: Proceedings of 5th North American Rock mechanics symposium, Toronto, Canada, pp 267–273

ISRM (2007) The complete ISRM suggested methods for rock characterization, testing and monitoring: 1974–2006 [Ulusay R, Hudson JA (eds)]. ISRM Turkish National Group, Ankara, 628 pp

Kim K, Gao H (1995) Probabilistic approaches to estimating variation in the mechanical properties of rock masses. Int J Rock Mech Min Sci 32:111–120

Palmstrom A (1995) RMi-a rock mass characterization system for rock engineering purposes. Ph.D. thesis, University of Oslo

Riedmuller G, Schubert W (1999). Rock mass modeling in tunneling versus rock mass classification using rating methods. In: Proceedings of the 37th US Rock mechanics symposium, Vail, Colorado

Russo G (2009) A new rational method for calculating the GSI. Tunn Undergr Space Technol 24:103–111

Sari M (2009) The stochastic assessment of strength and deformability characteristics for a pyroclastic rock mass. Int J Rock Mech Min Sci 46:613–626

Sari M, Karpuz C, Ayday C (2010) Estimating rock mass properties using Monte Carlo simulation: Ankara andesites. Comput Geosci 36:959–969

Sonmez H, Ulusay R (1999) Modifications to the geological strength index (GSI) and their applicability to stability of slopes. Int J Rock Mech Min Sci 36:743–760

Stewart SW (2007) Rock mass strength and deformability of unweathered closely jointed New Zealand greywacke. Ph.D. thesis, University of Canterbury, 455 pp

TS 699 (1987) Tabii Yapı Taşları-Muayene ve Deney Metotları. Türk Standartları Enstitüsü, Ankara (in Turkish)

Vose Software (2012) ModelRisk help, Vose Software. www.vosesoftware.com

Using of Multivariate Statistical Analysis in Engineering Geology at the Pest Side of the Metro Line 4 in Budapest, Hungary

153

Nikolett Bodnár, József Kovács, and Ákos Török

Abstract

The geological setting of a part of the new metro line in Budapest, Hungary is very complex and large amount of ambivalent historic data available on the physical properties of rocks. The Late Oligocene–Miocene sediments representing wide ranges of lithologies and mechanical properties. Core logs, drilling reports and records of laboratory analyses were studied for better understanding of the local geology, and to prepare a database on engineering geologic properties of the materials. Using this database, geologic sections were prepared and multivariate statistical analyses were used. Based on it four distinct groups were identified including swelling clays, non-swelling clays, medium clays, sands + silts. The results allowed a better correlation of the strata in the area, and a reconstruction of the geologic evolution. The obtained data sets can be used as input parameters for the design of the tunnel and stations.

Keywords

Metro construction • Bentonite • Clay • Silt • Statistical analysis

153.1 Introduction

The construction of the new metro line (line no 4) of Budapest is nearly completed. The studied section is located in the central part of the line between Kelenföldi Railway Station and Keleti Railway Station (Fig. 153.1).

In the mid-60 to early 80s 500 exploration drillings for the construction of the metro line (Szlabóczky 1988). The physical parameters of these drillings and the later period exploration boreholes were used for the present study. The aims were to outline the engineering geological properties of a mixed sedimentary system including bentonitic clays, silts and sandy deposits. The tunnel system and the stations of the studied area are located in this heterogeneous system. By using multivariate statistical analyses it is possible to estimate the physical parameters of these lithotypes.

N. Bodnár (✉) · Á. Török
Department of Construction Materials and Engineering Geology, University of Technology and Economics, Műegyetem rakpart 3, Budapest, 1111, Hungary
e-mail: bodnar.nikolett@gmail.com

J. Kovács
Department of Applied Geology, Eötvös Loránd University, Pázmány Péter sétány 1/C, Budapest, 1117, Hungary

153.2 Geological Settings

Along the metro line Eocene, Oligocene and Miocene strata are covered with Quaternary sediments (Raincsákné2000) (Figs. 153.2 and 153.3).

The metro line can be divided into three parts on the basis of geological-tectonical settings: Buda part (from the Kelenföld railway station to the Gellért square), the Danube crossing part (from the western part of the Gellért square to the Pest lower quay) and Pest part (from the Fővám square to the Keleti railway station, Dózsa György street).

On the studied area the sediments become younger from the SW to NE on the other riverside in Pest. The line cut through Upper Oligocene and Miocene strata. This part is lithologically more diverse. The Oligocene and Miocene beds are covered by river deposits of the Danube which contains groundwater (Juhász 2000) (Fig. 153.4).

G. Lollino et al. (eds.), *Engineering Geology for Society and Territory – Volume 6*,
DOI: 10.1007/978-3-319-09060-3_153, © Springer International Publishing Switzerland 2015

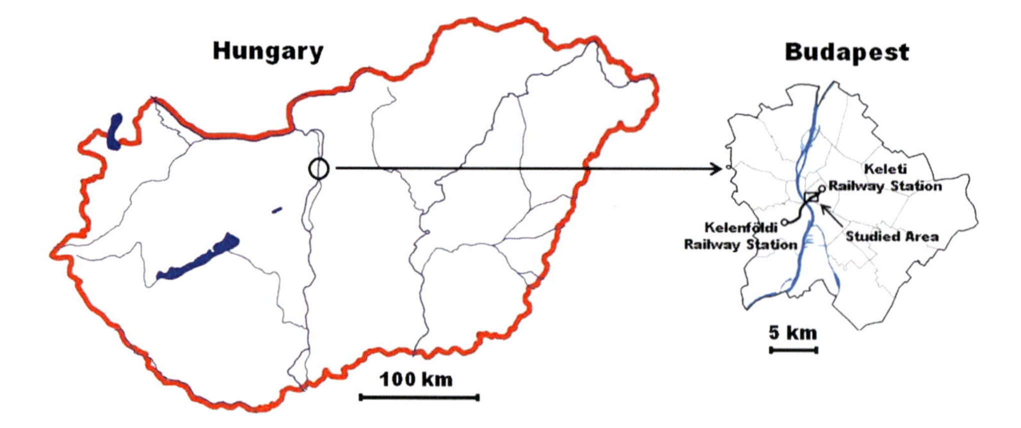

Fig. 153.1 The studied area in Budapest marked by a *rectangle*

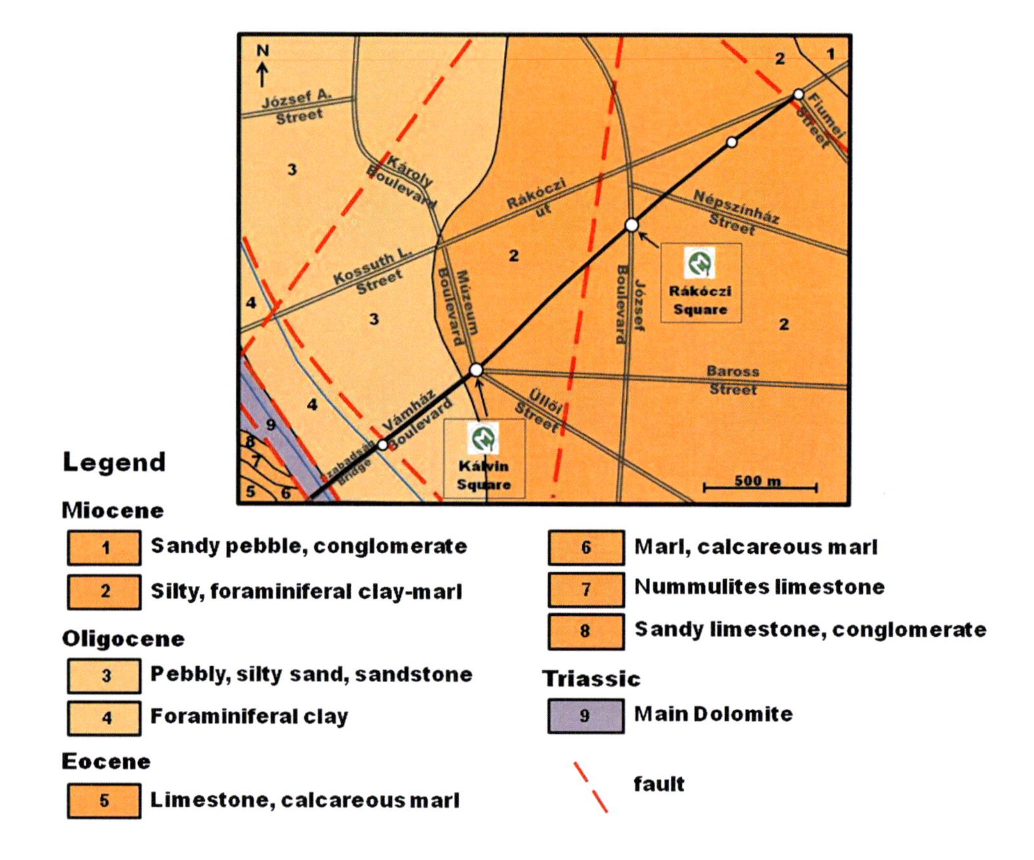

Fig. 153.2 Uncovered geological map

153.3 Methods

The data used in this study was obtained from borehole logs and core descriptions. Data set of the boreholes was used for the statistical analyses. Several of these parameters (e.g. index of plasticity) cannot be defined for such lithologies. From the selected and gathered twelve parameters data filtration showed that there are very poor correlation between several properties, thus these values should not be taken into consideration in multivariate statistical method (Miller and Kahn 1962). After the filtration only five parameters remained including void ratio, dry bulk density, angle of friction, cohesion and compressive strength. 252 samples

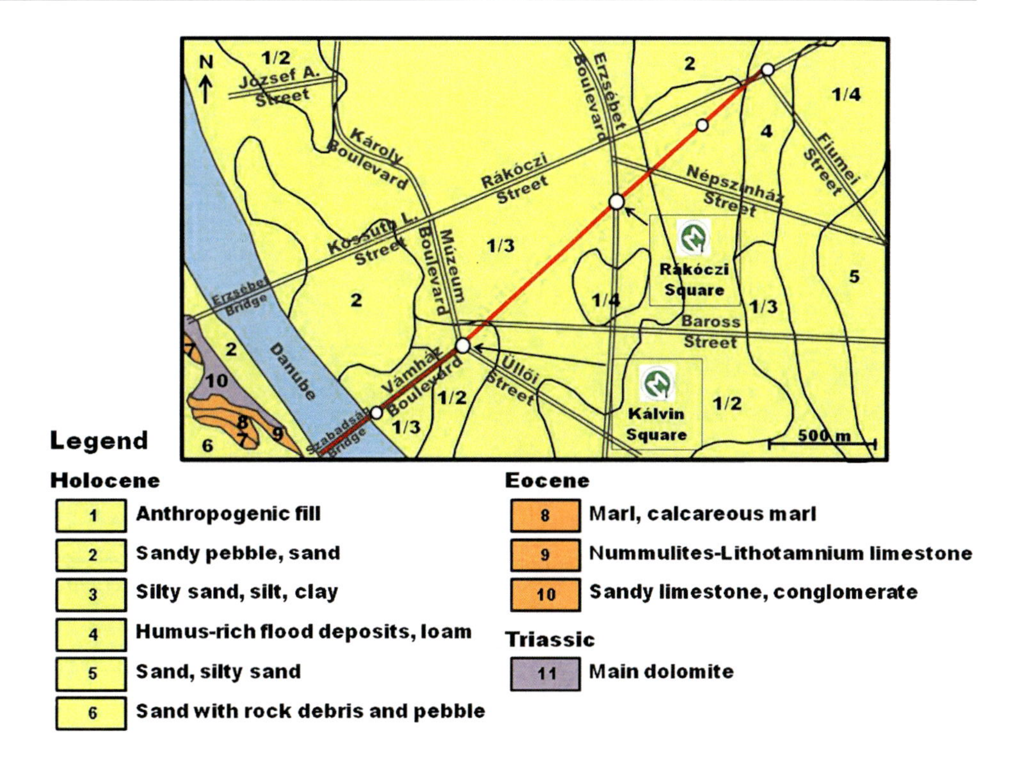

Fig. 153.3 Covered geological map

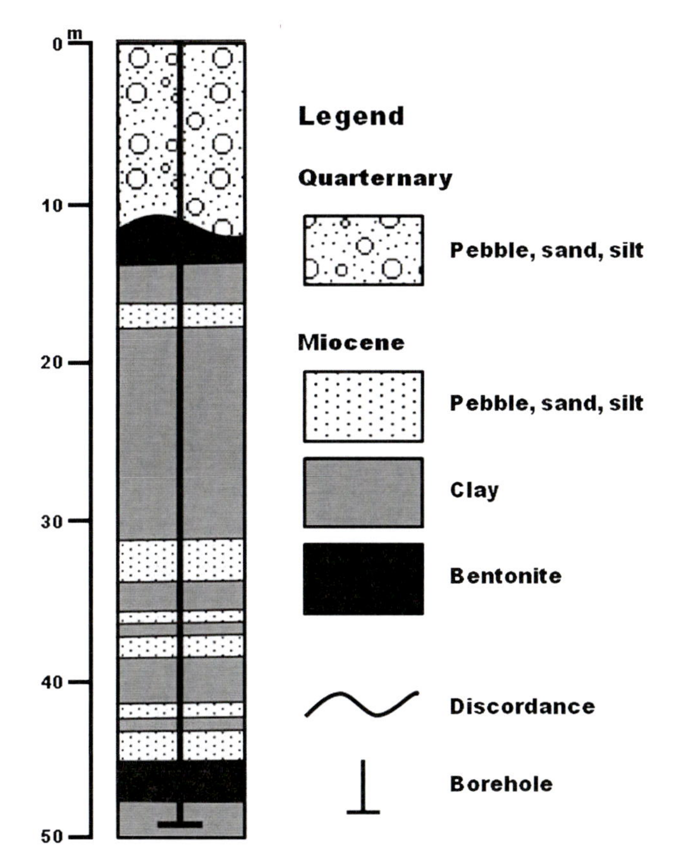

Fig. 153.4 Typical engineering geological borehole at the studied area

which contain 1,260 data were analysed by the means of mathematical statistics.

Cluster analysis was used to form groups from different samples according to their similarity. Additionally discriminant analysis was also performed according to the method listed in (IBM 2010).

153.4 Results

The cluster analysis of 252 samples of the five parameters provided four groups as results. The samples were not evenly assigned into the four groups. In some groups more, while in some other clusters fewer samples were found. The four groups are: swelling clays, non-swelling clays, medium clays, sands + silts. Interestingly, that the bentonites (which are usually among the swelling clays) do not belong to the group of swelling clays in this analysis (Fig. 153.5). Based on these results it has become clear, that these bentonites are more permeable than most bentonites. The average water conductivity of these samples was 10^{-7} m/s, although the mean values of water conductivity of bentonites is 10^{-10} m/s in general.

The groups created with the cluster analysis were verified with discriminant analysis, which proved that the groups were mathematically correct.

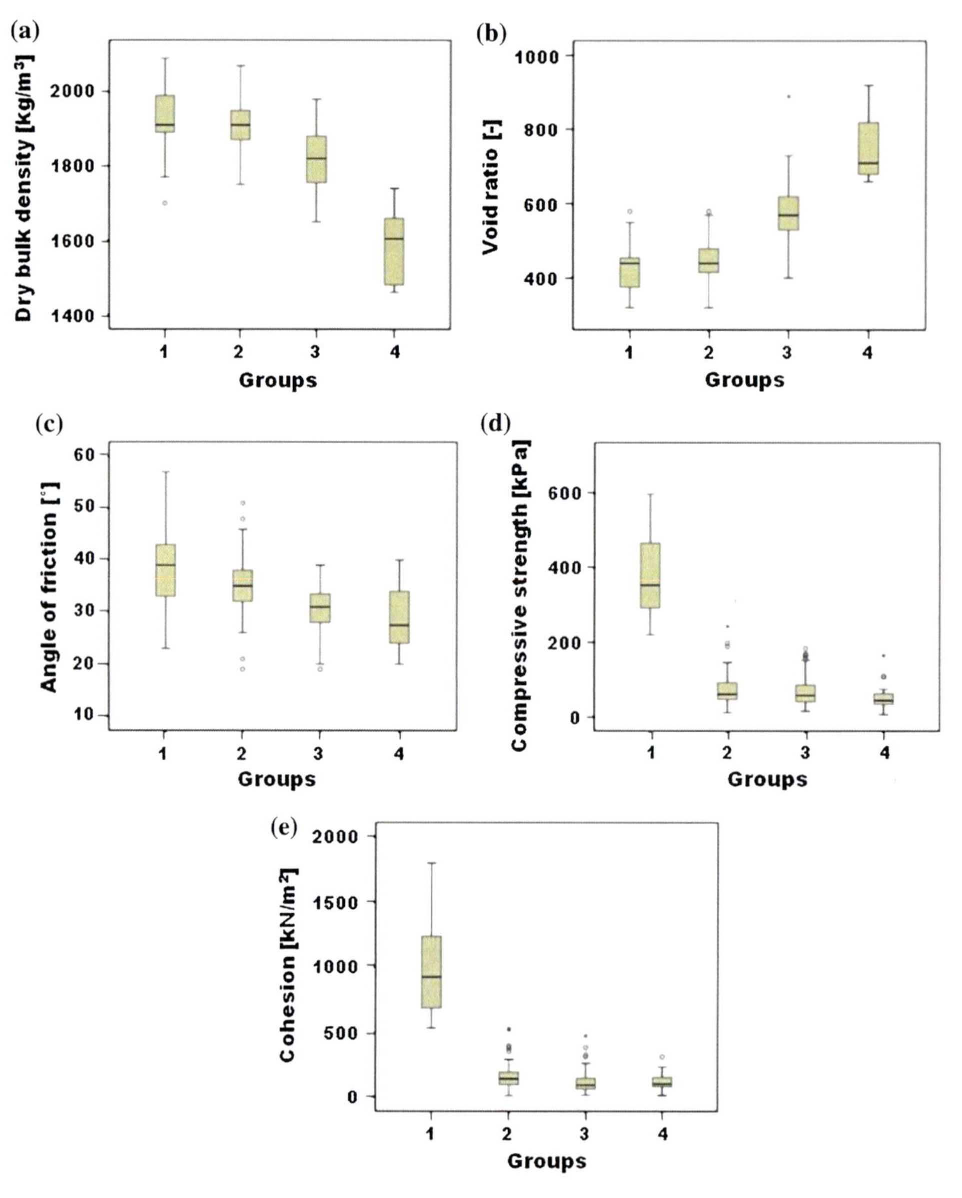

Fig. 153.5 Results of cluster analysis: **a** bulk density, **b** void ratio, **c** angle of friction, **d** compressive strength, **e** cohesion. Ranges are marked by *solid vertical lines*, *boxes* indicate the *upper* and *lower quartile* with *horizontal lines* representing median values. Extreme values are indicated by *dots*

153.5 Conclusions

The studied area has a complex geological structure with thin impermeable strata and lenticular aquifer bodies that makes difficult to use simple engineering geological input parameters for the design of the new metro line in Budapest. This study has proved that the use geomathematical and multivariate statistical methods allow the distinction of poorly described and very mixed laboratory and lithological data sets, that were made in the past. It was possible to identify four groups including swelling clays, non-swelling clays, medium clays, sands + silts. The multivariate statistical analyses provided valuable information for understanding the geomechanical parameters of lithologies and helped in outlining the input values for the design of the structures.

Acknowledgements The presentation of the research has been supported in the framework of the project "Talent care and cultivation in the scientific workshops of BME" project by the grant *TÁMOP—4.2.2. B-10/1—2010-0009.*

References

IBM (2010) Introduction to statistical analysis with PASW Statistics. IBM Company, Chicago IL, USA, 274 p

Juhász J (2000) A 4. sz. metró kutatásának hidrogeológiai eredményei (Results of hydrogeological exploration of metro line)—Földtani Kutatás, 37/2, 25–35 pp (in Hungarian)

Miller RL, Kahn JS (1962) Statistical analysis is the geological sciences. Wiley, New York, 483 p

Raincsákné Gy (2000) A Budapest 4. sz. metróvonal és környezetének földtani viszonyai (Geological setting of metro line 4 and its environment)—Földtani Kutatás, 37/2, pp 4–19 (in Hungarian)

Szlabóczky P (1988) A metrós fúrások földtani eredményeinek átfogó ismertetése (Complex geological evaluation of core drillings of metro)—Földtani Közlöny, 118/1, 61–66 pp (in Hungarian)

Evaluation of the Swelling Pressure of the Corumbatai Formation Materials

154

R.F.C. Souza and O.J. Pejon

Abstract

Materials in which the change in moisture content leads to an increase in their volume are known as expansive soils, and may cause damages in engineering works. Due to lack of soil with better qualities, and the need to use these soils as construction materials, it is necessary to study this behavior. Tests were performed with the inundation of samples with distilled water by the constant volume method. In addition to the mineralogical and geotechnical tests (index properties), the samples were submitted to the methylene blue adsorption test and the scanning electron microscopy. Results of swelling tests showed higher swelling pressure for compacted air dried samples, with reached values about 200 kPa, while 133 kPa for undisturbed air dried samples. Low values of swelling pressures can be justified by the absence of minerals of smectites group. Finally, the results of swelling pressure tests were compared with the values obtained from relations proposed in the literature to predict swelling pressure of expansive soils.

Keywords

Swelling pressure • Swelling clays • Expansive soils

154.1 Introduction

The phenomenon of expansion in soils can intensely affect civil engineering, such as highways, runways, railroads, canals, and buildings, causing damages in their structures. Swelling soils presents volume changes when in contact with water and can mobilizing considerable swelling pressure. Many factors influence the mechanism of swelling, and this phenomenon can be affected by physical soil properties and also state stress (Nelson and Miller 1992).

Numerous studies have been published relating swelling pressure to index properties and predicting methods. The index properties are related such as dry unit weight, initial water content, consistence limits, clay content, and cation exchange capacity (Alonso et al. 1992). To predict and evaluate the swelling pressures it has been proposed laboratory tests. The most commonly used tests are the free swell, constant volume, zero swell and swell-consolidation tests that examining swelling behavior in one direction (Basma et al. 1995). In addition, empirical relations have been developed with the basic soil properties and swelling pressure (Raman 1967; Chen 1988; Pereira and Pejon 1999).

The objective of this paper is to present results of swelling pressure in clay materials through equipment designed to measure the expansion constant volume.

154.2 Materials and Experimental Methods

The studied materials are from Sao Paulo state, southeastern Brazil, they are sedimentary materials (claystones and siltstone) from Corumbatai Formation (Permian) and exhibit geological and geotechnical characteristics leading to expansive behavior, as high clay. These materials do not

R.F.C. Souza (✉) · O.J. Pejon (✉)
University of Sao Paulo, Ave. Trabalhador Sancarlense, 400, Sao Carlos, 13.566-590, Brazil
e-mail: rafaela_faciola@yahoo.com.br

O.J. Pejon
e-mail: pejon@sc.usp.br

G. Lollino et al. (eds.), *Engineering Geology for Society and Territory – Volume 6*,
DOI: 10.1007/978-3-319-09060-3_154, © Springer International Publishing Switzerland 2015

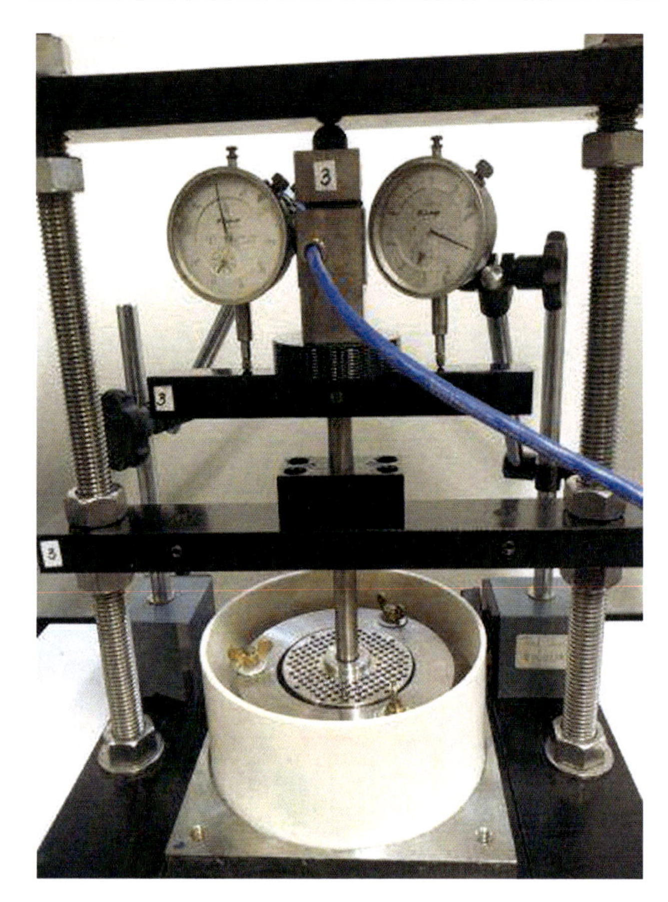

Fig. 154.1 Axial swelling stress in manual press

exhibit in their mineralogical constitution the expansive clay mineral montmorillonite, but have properties that indicate their potential to volume change, such as high water absorption capacity, and plasticity index (IP = 44). Natural soil samples were collected from the side of the highway where it is located the Corumbatai Formation, and were

Table 154.1 Soil geotechnical and mineralogical properties

Soil characteristics	
Percentage of clay content	52
Bulk density (Mg/m^3)	1.81
Specific gravity (Mg/m^3)	2.69
Dry density (Mg/m^3)	1.46
Optimum moisture content (%)	23.7
Void ratio	0.84
Porosity (%)	45.65
Degree of saturation (%)	76.24
LL (%)	76
PL (%)	32
PI (%)	44
CEC (cmol$^+$/kg)	31.05
SS (m^2/g)	242.75

reserved to conduct the tests. The swelling tests are performed in undisturbed and compacted samples and air-dried in atmosphere conditions before testing.

The materials were characterized for their geotechnical properties and mineralogical composition. Samples were submitted in dry unit weight test, moisture content, consistence limits, grain-size distribution, X-ray diffraction of clay fraction, cation-exchange capacity, and methylene blue adsorption.

To determine the swelling pressure has been developed a device that consist a manual press similar to that proposed by ISRM (1989). This equipment allows measuring the expansion of the material at constant volume. Figure 154.1 shows the equipment that consist a set of manual press with an edometric cell and a load cell.

The volume constant method is a technique called the direct method because of its ability to provide the swelling force directly (Kayabali and Demir 2011). This test consists in soil

Fig. 154.2 X-ray diffractometer of the Corumbatai Formation sample

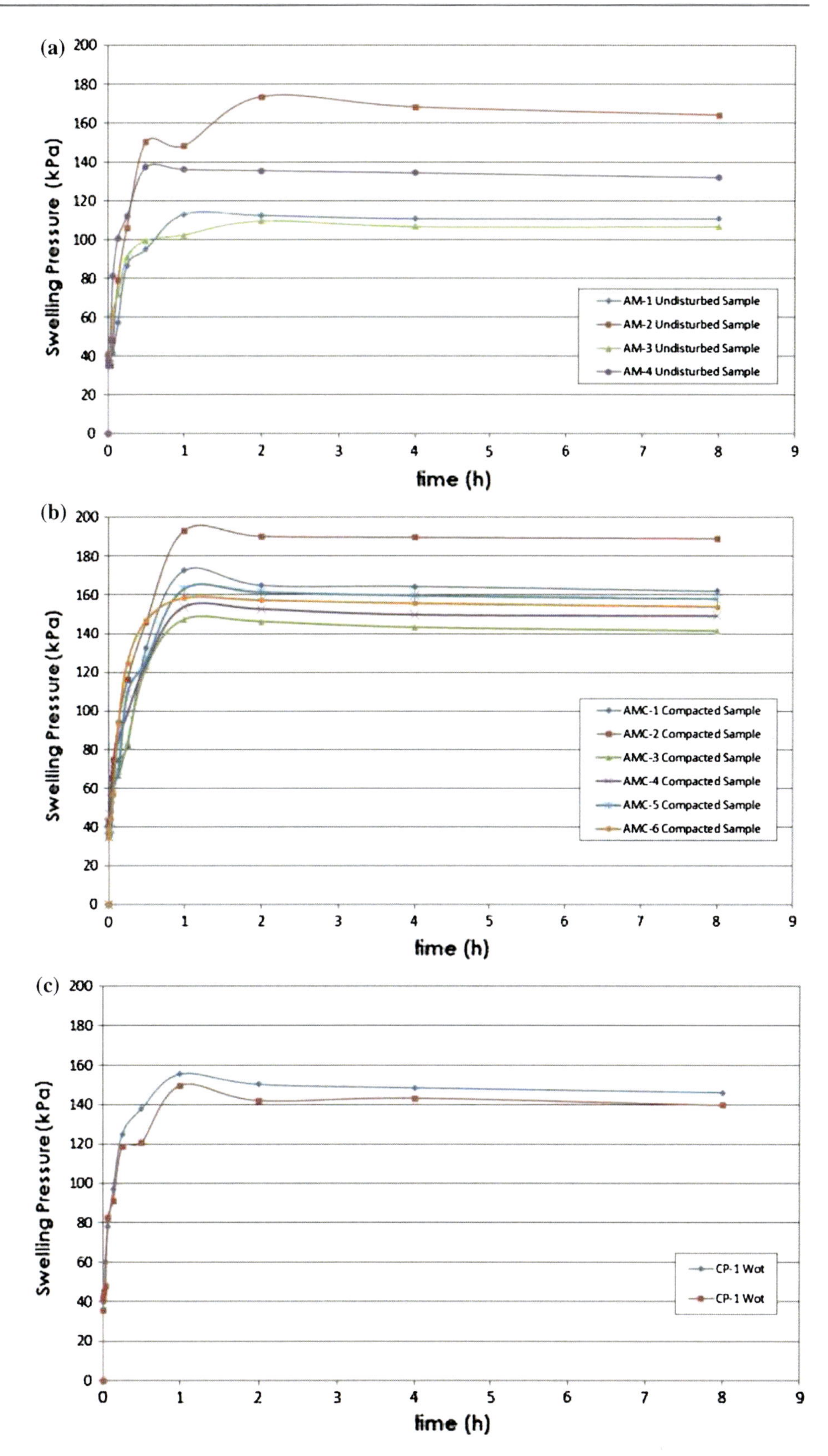

Fig. 154.3 Swelling pressure: **a** Undisturbed samples; **b** Compacted samples; and **c** Compacted at optimum moisture content sample

sample placed in oedometer cell. It is placed in the manual load device, and a seating of 25 kPa is applied, so there is no difference between piston connected to the load cell and the top plate on the sample. In a next step, the soil sample is inundated with water, and increments of vertical stress are applied to prevent swelling. At the time the specimen is

Table 154.2 Medium values of sample properties

Sample condition	Moisture $content_{medium}$ (%)	$Density_{medium}$ (Mg/m^3)
(a)	9.36	1.47
(b)	11.63	1.57
(c)	23.68	1.47

inundated, variations from the deformation are read, and it's preferably kept between 0.005 and 0.01 mm. The specimen remains under pressure and there is no tendency to swell. After 24 h, the vertical stress is recorded as the swelling pressure.

154.3 Results and Analyses

Three conditions of specimen were used to compare swelling pressure. The first series of test included determination of swell pressure using four undisturbed samples, and the second series using six compacted samples in optimum moisture content. In the first two conditions analyzed, the samples were molded and placed to dry in atmosphere air until the moisture content stabilization. Two compacted samples were tested at optimum moisture content in the third condition.

Table 154.1 shows the results of the characterization tests and mineralogical compositions performed with the Corumbatai Formation material. The sample presents fine particle in almost all its composition, varying from 52 % of clay and 44 % of silt. The high porosity and low density can also explain the expansive behavior. The consistence limits values were high which indicates a highly plastic material. The cationic exchange capacity (CEC) and specific surface are indicative of clay minerals with normal activity, and the probable minerals are illite and halloysite.

No smectite clay mineral group was exhibited by X-ray diffraction (Fig. 154.2). Illite and kaolinite were the dominant clay minerals detected. This result was confirmed in thermal differential analysis.

According to the correlations proposed by Raman (1967) and Chen (1988) about consistence limits, this material can be characterized as very expansible. However, about correlation proposed by Pereira and Pejon (1999) considering clay percentage and methylene blue volume, this material present low swell, due to the presence of kaolinite and absence of smectite.

Swelling pressure results are presented in Fig. 154.3 in (a) for the first, (b) second, and (c) to third condition analyzed. The undisturbed sample presented swelling pressure about 150 kPa, and it is clear by the result shown in Fig. 154.3b that compacted samples increased the swelling behavior. Undisturbed sample reached maximum axial swelling pressure in about 2 h with 173 kPa, while compacted sample reached it in about 1 h with 193 kPa of maximum value. The compacted air-dry sample reached maximum axial swelling pressure in about 1 h with 155 kPa. In general, the swelling pressure showed similar values which were not very significant in terms of magnitude. The swelling pressure reached similar values to samples tested at optimum moisture content. Therefore, according to Table 154.2, the moisture content did not influence on the expansion, but only did so on the effect of compaction.

154.4 Conclusions

The Corumbatai Formation materials did not exhibit mineral of smectite group, but present an expansive behavior, showing values of swelling pressure. The geotechnical properties also indicated that the material presents a potential expansive behavior. Results showed that swelling pressure can be more expressive in compacted samples with higher densities. The device presented satisfactory performance and simple handling, and it can be applied to other types of expansion tests.

We conclude that sedimentary materials with high clay content, highly plastic and low percentage of expansive clay minerals also need attention before to deploying large or small buildings. Therefore, careful with this type of material should start from the design phase, extending to all stages of construction, even in the post use, since the expansion can cause damage to the building (cracks and breaks), which can shorten life, or derail in extreme situations.

Acknowledgments The authors thank CNPq, FAPESP and CAPES for the financial support to this research.

References

Alonso EE, Gens A, Josa A (1992) A unified model for expansive soil behavior. In: Proceedings of the 7th international conference on expansive soils, Dallas, Tex., 3–5, Texas Tech University Press, Ludbbock, Tex., pp 24–29

Basma AA, Al-Homoud AS, Malkawi AH (1995) Laboratory assessment of swelling pressure of expansive soils. Appl Clay Sci 9 (5):355–368

Chen FH (1988) Foundation on expansive soils. American Elsevier Science Publication, New York

ISRM: International Society for Rock Mechanics (1989) Commission on swelling rock and working group on swelling rocks of the commission on testing methods. Suggested methods for laboratory testing of argillaceous swelling rocks. Int J Rock Mech Min Sci and Geomech Abstr 26(5):415–426

Kayabali K, Demir S (2011) Measurement of swelling pressure: direct method versus indirect method. Can Geotech J 48:354–364

Nelson JD, Miller DJ (1992) Expansive soils: problems and practice in foundation and pavement engineering. Wiley, 259 p

Pereira EM, Pejon OJ (1999) Estudo do potencial expansivo dos sedimentos ar- gilosos da Formação Guabirotuba na Região de Alto Iguaçú—PR. 9 Congresso Brasileiro de Geologia de Engenharia. ABGE, 16 p, São Pedro–São Paulo

Raman V (1967). Identification of expansive soils from the plasticity index and the shrinkage index data. Indian Eng Calcutta 11(1): 17–22

Classification of Weak Rock Masses in Dam Foundation and Tunnel Excavation

155

V. Marinos, P. Fortsakis, and G. Stoumpos

Abstract

The term weak rock mass is usually associated in design and construction with potential problems or even failures, although this is not always the case. Weak rock masses could be cases with very low intact rock properties, highly tectonized or/and weathered rock masses, rock masses with members of low strength and/or inherent heterogeneity. In this paper, the potential problematic behaviour of weak rock masses in tunnel excavation and dam foundation is discussed. A rock mass could be generally characterized as weak when its potential failure is driven by the inadequacy of its strength (σ_{cm}) as a system to bear the stresses that are imposed upon it and not by local failure of its components (intact rock and discontinuities). On the other hand, anisotropic failures like planar slides in dam abutments or wedge failures in tunnels do not constitute failure patterns that can be directly associated with weak rock masses. A general borderline could be that a weak rock mass: (i) in tunnelling can develop shear failures and deformations even under medium overburden and (ii) in dams can raise serious concerns for the foundation using any other dam type (e.g. arch dam, concrete gravity dam) than an earth/rock fill dam.

Keywords

Weak rock mass • Behaviour • Tunnel • Dam

155.1 Introduction

Typically, the term weak rock mass, if not correctly assessed is associated in design and construction with potential problems or even failures. However, this term usually attempts to describe and quantify the quality of the rock mass via an absolute geotechnical approach, without implying the problematic rock mass behaviour, irrespective of the kind or characteristics of the project. Therefore a weak rock mass could be associated with severe squeezing in tunnelling under considerable overburden and the same weak rock mass could be stable for excavation in smaller depth. On the other side tunnelling through a very competent rock mass could be a stable project or associated with failures such as wedge slide under low overburden or rockburst.

Marinos (1993) described that the weak rock masses could be distinguished to the rock masses that were "born" weak (e.g. mudrocks, siltstones), the ones that became weaker through a retrogression from an original stronger material (weathering, alteration or tectonic deformation) and the rock masses that became weaker, although still strong, since they are associated by genesis with weak rocks at a scale affecting the engineering behaviour of the formation

V. Marinos (✉)
Faculty of Sciences, School of Geology, Laboratory of Engineering Geology and Hydrogeology, Aristotle University of Thessaloniki, 54124 Thessaloniki, Greece
e-mail: marinosv@geo.auth.gr

P. Fortsakis
School of Civil Engineering, Department of Geotechnical Engineering, National Technical University of Athens, Heroon Polytechniou 9, 15780 Zografou, Athens, Greece

G. Stoumpos
Attiko Metro S.A., 191–193 Messoghion Avenue, 11525 Athens, Greece

G. Lollino et al. (eds.), *Engineering Geology for Society and Territory – Volume 6*,
DOI: 10.1007/978-3-319-09060-3_155,

they belong (alternation of competent and incompetent rocks—heterogenous rock masses such as flysch). Thus, it is difficult to draw limits in order to stick the label "weak rock" since other intrinsic or secondary features may impose a weak rock mass behaviour (stress dependent and time dependent behaviour, as creep or swelling or dissolution, change of water content, slaking, inherent anisotropy).

Regarding the intact rock, many different definitions of the weak rock can be found in literature, usually associated with the Uniaxial Compressive Strength (UCS, σ_{ci}). Rocha (in Oliveira 1993) proposed $\sigma_{ci} = 2$ MPa as the limit between the soils and rocks. According to the definition of ISRM (1981) σ_{ci} = 2–6 MPa corresponds to very weak rocks whereas 6–20 MPa to weak rocks. Similar definitions could be set for the rock mass strength (e.g. σ_c, σ_{cm}).

The characterization of the rock mass as weak primarily depends on the lithology of the intact rock and the state of the rock mass. These can all be quantified by the use of e.g. the intact rock strength σ_{ci} and the Geological Strength Index (GSI) as for the case of the Hoek-Brown failure criterion. In that sense, to describe a rock mass as weak or strong should be a straightforward process regardless of a particular project. However, this is not always the case.

Indeed, it is easy to categorize a rock mass with, for example, $\sigma_{ci} < 5$ MPa and GSI < 20 as weak or a rock mass with, for example, $\sigma_{ci} > 100$ MPa and GSI > 70 as strong. Yet, the question that arises has to do with all the rock masses that lay amidst. This is why one has to consider the engineering project type, scale and geometry. After all, the attribution of the term weak or strong to a rock mass "as it stands" is only academically useful. It is the project-rock mass interaction that is important and in particular the types of failure that are associated with each given conditions.

The potential problematic behaviour of a category of weak rock masses in tunnel excavation and dam foundation is discussed in the paper. In general, anisotropic failures such as planar slides in dam abutments, wedge failures in tunnels do not constitute failure patterns that can be associated with weak rock masses. For example in the case of tunnelling the weak rock masses would be related with the tunnel face instability, large displacements or overstress of the support shell and not with gravity driven problems like ravelling ground or wedge failure.

Consequently, a rock mass could be characterized as weak when its potential failure is driven by the inadequacy of its strength (σ_{cm}) to bear the stresses that are imposed upon it; inadequacy that refers to the rock mass as a whole and not as individual constituents (only intact rock or only discontinuities). Moreover, the characterization of the rock mass in an engineering problem does not only depend on the rock mass properties but also to the potential failure pattern. A general borderline could be that a weak rock mass: (i) in tunnelling can develop shear failures and deformations even under relatively low overburden and (ii) in dam foundation raise serious concerns for the foundation using any dam type (e.g. arch dam, concrete gravity dam) apart from an earth fill dam.

155.2 Weak Rock Masses and Tunnels

Conventional excavation of tunnels within weak rock masses can be better approached by identifying the failure modes that could potentially emerge. These failure modes are usually associated with stress-induced phenomena such as face instability and squeezing, manifested by tunnel face extrusion, tunnel closure or shell overstress.

A factor that could be adopted to identify a weak rock mass behaviour in tunnelling is the ratio of the rock mass strength divided by the geostatic stress at the tunnel level (σ_{cm}/p_o, Hoek and Marinos 2000). The low values of this ratio could be a result of high excavation depth or/and low σ_{cm} that is further analysed in low σ_{ci} and/or low GSI. The Tunnel Behaviour Chart (TBC, Marinos 2012) in Fig. 155.1 is a comprehensive tool to indicate combinations of rock mass structure, intact rock strength and overburden height that may develop stress-induced behaviour.

Typical examples of such behaviour include tectonized heterogeneous sedimentary sequences such as flysch, highly tectonized, altered and/or weathered ophiolites such as foliated serpentinites, heavily sheared homogeneous sedimentary rocks of low intact rock strength such as claystones, shales, molassic formations etc.

For example, molassic formations, when confined in depth, have not developed stress relaxation, and exhibit a compact structure and usually correspond to high GSI values (Marinos et al. 2013). Nevertheless, the intact rock strength, especially in the case of siltstones, is usually low, thus leading to a low rock mass strength and potentially to weak rock mass behaviour. On the other hand, there are cases where namely strong rock masses have developed a typical weak rock mass behaviour. One example would be the excavation of the new Gotthard Tunnel in the area of Faido where a yielding support was adopted for the excavation through a tectonically sheared and foliated gneiss formation under the overburden of 1100 m (Personal observation).

155.3 Weak Rock Masses and Dams

Dam foundation in weak rock masses often includes challenges like compressible rock masses, presence of incompetent members and fractured and sheared zones of low to very low strength, poorly to medium cemented rocks, diverse heterogeneity and presence of cavities. These conditions often lead to a single choice selection of the

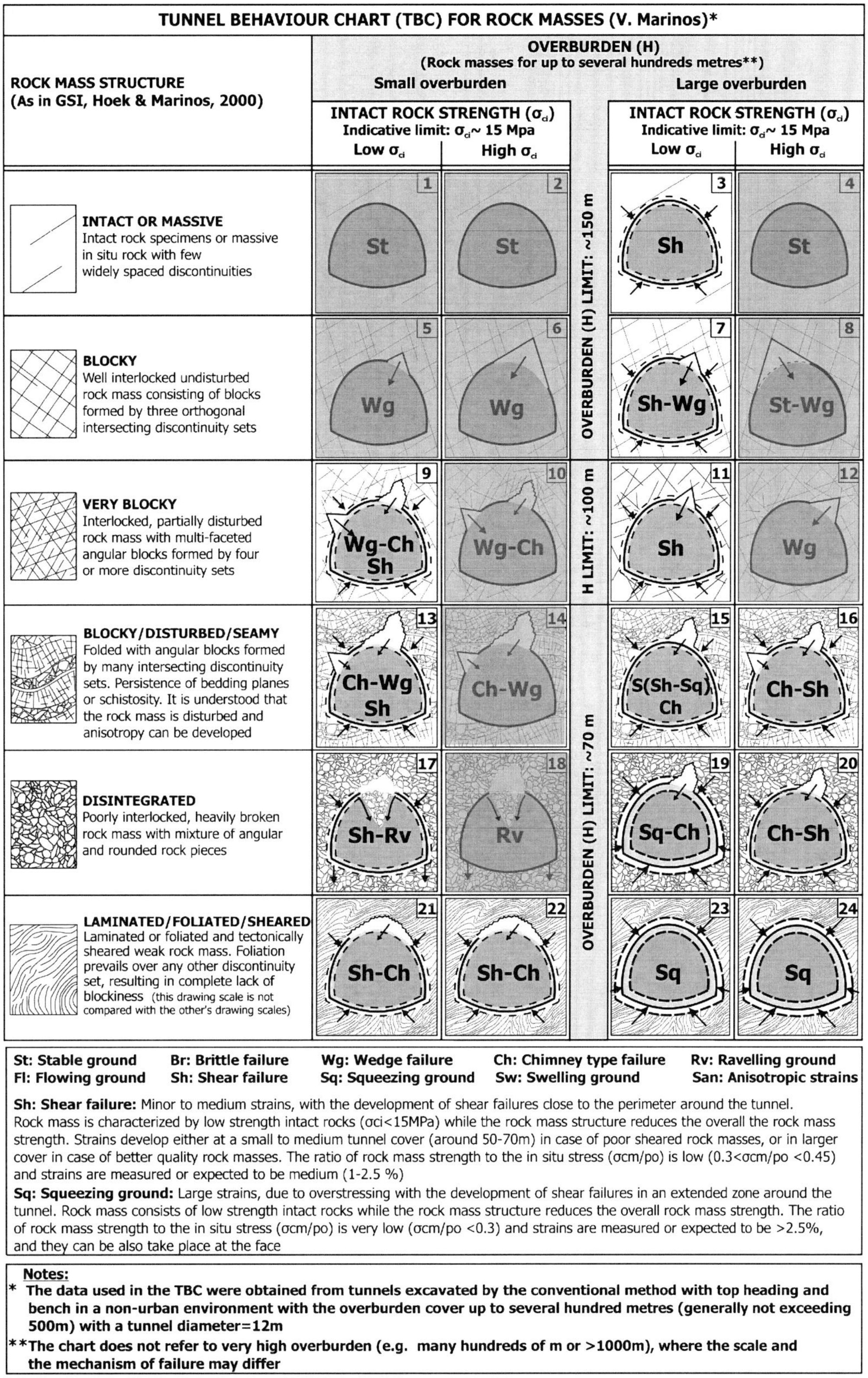

TUNNEL BEHAVIOUR CHART (TBC) FOR ROCK MASSES (V. Marinos)*

ROCK MASS STRUCTURE (As in GSI, Hoek & Marinos, 2000)	OVERBURDEN (H) (Rock masses for up to several hundreds metres**) Small overburden — INTACT ROCK STRENGTH (σ_{ci}) Indicative limit: σ_{ci}~ 15 Mpa — Low σ_{ci}	High σ_{ci}		Large overburden — INTACT ROCK STRENGTH (σ_{ci}) Indicative limit: σ_{ci}~ 15 Mpa — Low σ_{ci}	High σ_{ci}
INTACT OR MASSIVE Intact rock specimens or massive in situ rock with few widely spaced discontinuities	1 St	2 St	OVERBURDEN (H) LIMIT: ~150 m	3 Sh	4 St
BLOCKY Well interlocked undisturbed rock mass consisting of blocks formed by three orthogonal intersecting discontinuity sets	5 Wg	6 Wg		7 Sh-Wg	8 St-Wg
VERY BLOCKY Interlocked, partially disturbed rock mass with multi-faceted angular blocks formed by four or more discontinuity sets	9 Wg-Ch Sh	10 Wg-Ch	H LIMIT: ~100 m	11 Sh	12 Wg
BLOCKY/DISTURBED/SEAMY Folded with angular blocks formed by many intersecting discontinuity sets. Persistence of bedding planes or schistosity. It is understood that the rock mass is disturbed and anisotropy can be developed	13 Ch-Wg Sh	14 Ch-Wg	OVERBURDEN (H) LIMIT: ~70 m	15 S(Sh-Sq) Ch	16 Ch-Sh
DISINTEGRATED Poorly interlocked, heavily broken rock mass with mixture of angular and rounded rock pieces	17 Sh-Rv	18 Rv		19 Sq-Ch	20 Ch-Sh
LAMINATED/FOLIATED/SHEARED Laminated or foliated and tectonically sheared weak rock mass. Foliation prevails over any other discontinuity set, resulting in complete lack of blockiness (this drawing scale is not compared with the other's drawing scales)	21 Sh-Ch	22 Sh-Ch		23 Sq	24 Sq

St: Stable ground **Br: Brittle failure** **Wg: Wedge failure** **Ch: Chimney type failure** **Rv: Ravelling ground**
Fl: Flowing ground **Sh: Shear failure** **Sq: Squeezing ground** **Sw: Swelling ground** **San: Anisotropic strains**

Sh: Shear failure: Minor to medium strains, with the development of shear failures close to the perimeter around the tunnel. Rock mass is characterized by low strength intact rocks (σci<15MPa) while the rock mass structure reduces the overall the rock mass strength. Strains develop either at a small to medium tunnel cover (around 50-70m) in case of poor sheared rock masses, or in larger cover in case of better quality rock masses. The ratio of rock mass strength to the in situ stress (σcm/po) is low (0.3<σcm/po <0.45) and strains are measured or expected to be medium (1-2.5 %)

Sq: Squeezing ground: Large strains, due to overstressing with the development of shear failures in an extended zone around the tunnel. Rock mass consists of low strength intact rocks while the rock mass structure reduces the overall rock mass strength. The ratio of rock mass strength to the in situ stress (σcm/po) is very low (σcm/po <0.3) and strains are measured or expected to be >2.5%, and they can be also take place at the face

Notes:

* **The data used in the TBC were obtained from tunnels excavated by the conventional method with top heading and bench in a non-urban environment with the overburden cover up to several hundred metres (generally not exceeding 500m) with a tunnel diameter=12m**

****The chart does not refer to very high overburden (e.g. many hundreds of m or >1000m), where the scale and the mechanism of failure may differ**

Fig. 155.1 Tunnel behaviour chart (TBC). Stress-controlled problematic behaviour that corresponds to weak rock masses is highlighted (Marinos 2012)

Table 155.1 Rock mass elements that can characterise a rock as "weak" in dams

Rock material	Engineering geological elements enabling, if present, to characterise the rock as "Weak rock mass" in dam foundation
Conglomerate, Sandstone	Poor cementation material (e.g. clayey), low diagenesis. Low strength and deformation modulus. Cases with frequent intercalations with pelitic interlayers
Marl, Siltstone Mudstone, Slate	Presence of clayey minerals, poor cementation. Possible laminated with diagenetic planes. When tectonically disturbed they form very weak rocks masses
Clayshale	Low rock mass strength (foliated structure, low intact rock strength) and deformation modulus -Sheared surfaces, slickensided surfaces. Swelling minerals. Slaking potential
Evaporites	Dissolution phenomena, presence of voids (gypsum, halite). Swelling (anhydrite)
Limestone, Marble	Internal karstic structure, voids empty or soil filled. Cases of intercalations of weak pelitic layers (phyllites in the case of marble)
Molasse	Alternations of sandstones with siltstones (most common). Diverse heterogeneity. Presence of members with low strength. Slaking potential when exposed
Flysch	Alternations of sandstones with siltstone the most common. Diverse heterogeneity. Presence of members with low strength. Unlike molasse tectonically disturbed structures (structural complexity in space due to folding and presence of sheared tectonic zones and layers). Persistence of discontinuities with low strength in depth
Volcanic rocks	Generally strong, but may alternate with pyroclastic compressible, swelling or erodible material. Heterogeneity in strength and deformation modulus. Weathering or alteration with presence of unstable minerals/high plasticity soils. Joints from cooling/lava tunnels
Pyroclastic rocks	Highly erodible, collapsible or swelling. Extreme variability within the fromation
Granite	Generally strong but possibility of extended weathering presence. Irregular weathering profile and bedrock interface (less weathered boulders within completely weathered material). Concealed sheet joints close to surface
Basic-Ultra basic/ Ophiolites	Generally strong but peridotites may present serpentinized zones of low strength in irregular geometry within the mass. Shear zones with altered compressible material
Graphitic-Chlorite schists. Phyllites	Dense schistosity with weak planes. Possible low strength and modulus. Weathered to other clay minerals
Gneiss	Generally strong. Weak in extended weathered/sheared brecciated zones only
Quartzite	Strong but often quartzitic layers alternate with incompetent phyllites
Metamorphic schists	Generally strong. Possible presence of weak zones due to tectonic shears and to slickensided schistosity planes

construction of an embankment, earth/rock fill dam and/or, seldom, the strengthening of the foundation zone or the selection of an alternate dam location. Cases where various rocks can be characterized as "weak" in dam foundation engineering are presented in Table 155.1.

155.4 Conclusions

The term weak rock mass attempts to describe and quantify the quality of the rock mass via an absolute geotechnical approach, irrespective of the kind or characteristics of the project. However, it is the authors' view that the type of engineering project is also important in order to predict the behaviour of the rock mass during construction.

Several definitions and a discussion about the limits of weak rock masses were presented in the paper and a classification of weak rock masses according to the engineering project type was directed. The authors assessed the problematic behaviour of weak rock masses in tunnel excavation and dam foundation in order to achieve this scope.

Characterisation of "weak rock masses" according to the engineering project type was performed based on different approaches. In tunnelling, the combinations of rock mass structure, intact rock strength and overburden height that may lead to the development of stress-induced behaviour were highlighted. In dam foundation the engineering geological elements enabling, if present, to characterise different kinds of rock as "weak rock" were presented.

References

Hoek E, Marinos P (2000) Predicting tunnel squeezing in weak heterogeneous masses. Tunnels and Tunnelling International, Part 1, November Issue, pp 45–51; Part 2, December Issue, pp 34–36

International Society for Rock Mechanics (ISRM) (1981) Commission on classification of rocks and rock masses. Int J Rock Mech Min Abstr. 18:85–110

Marinos P (1993) General report session 1: hard soils-Soft rocks: geological features with emphasis to soft rocks. geotechnical engineering of hard soils-soft rocks. In: Anagnostopoulos et al (eds) Balkema, Rotterdam, ISBN 9054103442, pp 1807–1818

Marinos V (2012) Assessing rock mass behavior for tunnelling. J Environ Eng Geosci XVIII(4):327–341

Marinos V, Fortsakis P, Prountzopoulos G (2013) Tunnel behaviour and support in molassic rocks. The experiences from 12 tunnels in Greece. In: Kwasniewski M, Lydzba D (eds) Rock mechanics for resources, energy and environment (EUROCK2013). CRC Press, Boca Raton, pp 909–914

Oliveira R (1993) Weak rock materials. In: Proceedings of the 26th annual conference of the engineering group of the geological society, Leeds, UK, Sept 1990

Applicability of Weathering Classification to Quartzitic Materials and Relation Between Mechanical Properties and Assigned Weathering Grades: A Comparison with Investigations on Granitic Materials

156

A. Basu

Abstract

The ongoing process of weathering in nature produces progressive but intricate changes in rock microstructure. Evaluating mechanical behaviors of rock materials with reference to weathering grades is, therefore, important for an engineering work encountering weathered rocks. The common 6-fold weathering classification for uniform materials is meant to capture gradational change of rock materials depending on degree of decomposition. Research in this regard has been limited within polymineralic rocks (e.g. granite etc.) where degree of alteration of constituent minerals helps recognize such gradation. Quartz being the most resistant mineral to weathering is the chief mineral constituent of quartzite and therefore, capturing intricate gradational change of quartzite in response to weathering or categorization of weathering grades of quartzitic materials is a challenging task. In line with the author's involvement in three different research topics, this paper presents salient points in categorizing weathering grades of granitic rock materials from Hong Kong and southeastern Brazil that are subsequently compared with the issues in characterizing weathering grades of quartzitic rock materials from eastern India. An overall assessment of mechanical behaviors of these rocks with reference to assessed weathering grades is also outlined.

Keywords

Granite • Quartzite • Weathering classification • Mechanical properties

156.1 Introduction

The ongoing process of weathering in nature produces progressive but intricate changes in rock microstructure at shallow depths where most engineering works are confined, especially in tropical and sub-tropical areas. The most widely used weathering classification system such as ANON (1995) by and large resembles the 6-fold classification scheme developed by Moye (1955) in which Grades I–IV represent rocks whereas higher grades stand for soils. The weathering classification for engineering purposes has been formulated to address the need for a common but simple basis of communication with underlying messages mainly on the possible ranges of mechanical properties (e.g. Dearman and Irfan 1978; Hencher and Martin 1982). Research in this regard has been limited within polymineralic rocks (e.g. granite etc.) where degree of alteration of constituent minerals helps recognize such gradation. Quartz being the most resistant mineral to weathering is the chief mineral constituent of quartzite and therefore, capturing intricate gradational change of quartzite in response to weathering or categorization of weathering grades of quartzitic materials is a challenging task. Subsequently, research on mechanical characterization of quartzitic materials with reference to weathering grades does not seem have gained much attention where quartzite is considered as one of the competent/suitable rocks for various engineering purposes.

In this paper, the author first presents salient points in categorizing weathering grades of granitic rock materials from Hong Kong and southeastern Brazil that are

A. Basu (✉)
Department of Geology and Geophysics, Indian Institute of Technology Kharagpur, Kharagpur 721302, India
e-mail: abasu@gg.iitkgp.ernet.in

G. Lollino et al. (eds.), *Engineering Geology for Society and Territory – Volume 6*,
DOI: 10.1007/978-3-319-09060-3_156,

Table 156.1 Weathering classifications of granitic and quartzitic materials

Hong Kong granite (after Basu 2006)		Granite from southeastern Brazil (after Basu et al. 2009)		Quartzite from eastern India (after Basu et al. 2011)	
Characteristics	WG	Characteristics	WG	Characteristics	WG
No discoloration	I	No discoloration	I	No discoloration/staining	I
Grains have vitreous luster		Grains have vitreous luster		Grains have vitreous luster	
Equigranular texture with intact grain boundaries		Equigranular texture with intact grain boundaries		Intact grain boundaries	
Slight to moderate staining	I–II				
Grains have vitreous to sub-vitreous lustre					
Intact grain boundaries					
High staining	II	Slight to moderate staining	II	Slight to moderate staining	II
Grains have sub-vitreous to dull luster		Grains have vitreous to sub-vitreous lustre		Grains have vitreous to sub-vitreous luster	
Intact grain boundaries		Intact grain boundaries		Intact grain boundaries	
		Grains have sub-vitreous to dull luster	II–III		
		White clay minerals are common			
		Intact grain boundaries			
Moderately decomposed	III	Moderately decomposed	III	Moderately stained	III
Soft white clay minerals can be scratched by nail		Abundant soft white clay minerals can be scratched by nail		Grain boundaries not very intact	
Grain boundaries not very intact		Intact grain boundaries		Can be broken easily by a geological hammer	
Can be broken easily by a geological hammer		Can be broken easily by a geological hammer			
Moderately to highly decomposed (more gritty and clayey appearance of feldspars)	III–IV				
Can be broken more easily by a geological hammer					
Highly decomposed (powdery feldspars) with loose grain boundaries	IV	Highly decomposed (powdery feldspars) with loose grain boundaries	IV	Highly stained with loose grain boundaries	IV
NX core can be broken by hand		Large pieces can be broken by hand		Large pieces can be broken by hand	
Does not slake in water		Does not slake in water		Does not slake in water	

Fig. 156.1 Photomicrographs depicting differences in intactness of quartz grain boundaries and in overall staining with respect to weathering grades of quartzites from eastern India (after Basu et al. 2011)

subsequently compared with the issues in characterizing weathering grades of quartzitic rock materials from eastern India. An overall assessment of mechanical behaviors of these rocks with reference to assessed weathering grades is also outlined.

156.2 Categorization of Weathering Grades of Granitic and Quartzitic Materials

Granitic rocks of various weathering grades from Hong Kong were investigated by Basu (2006). These rocks belong to the intrusive units of Kowloon Granite and Mount Butler Granite (Sewell et al. 2000). A set of six recognition factors (discoloration and staining, grain boundaries, relative strength, decomposition, microcracking, and disintegration) was identified and used by Basu (2006) to describe gradational changes of the granitic rocks over the weathering spectrum. It was proposed that a weathering classification system with intermediate grades (if possible to be recognized) as presented in Table 156.1 (broadly in compliance with the common 6-fold classification) is more efficient or detailed than the 6-fold classification in capturing the weathering induced changes of granitic rocks (up Grade IV). Microscopic observations like increase in chloritization of biotites, sericitization of feldspars and quartz intra-granular crack density with the enhancement of weathering intensity also substantiated this weathering classification (Table 156.1).

A weathering classification (broadly conformable with the common 6-fold material classification scheme) for granitic rocks that belong to the Itu Granitic Complex (IPT 1981) in southeastern Brazil was presented by Basu et al. (2009) (Table 156.1). However, an intermediate class 'Grade II–III' was assigned in this classification (Table 156.1) with its distinct differentiable decompositional characteristics compared to Grades II and III. Microscopic studies also substantiated this classification.

Basu et al. (2011) attempted to apply the 6-fold weathering classification scheme to quartzitic rock materials from Chaibasa Formation (Saha 1994) at Jaduguda in the state of Jharkhand, eastern India. Although individual quartz grains of investigated quartzitic materials do not portray any discoloration as a manifestation of degree of decomposition, macroscopic appearances of these rocks differ noticeably as weathering intensifies because of alteration of other minerals (e.g. biotite and other iron bearing minerals) that constitute less than 5 % of the entire rock volume and loosening of quartz grain boundaries. Consequently, the common 6-fold classification provides a solid guideline even in case of quartzitic materials to categorize weathering grades (Table 156.1 and Fig. 156.1).

156.3 Mechanical Behaviors of Granitic and Quartzitic Materials with Reference to Weathering Grades

Uniaxial compressive strength (UCS) and tangent Young's modulus at 50 % of the peak stress (E) were determined as per stipulations by ASTM (2001) in all three investigations carried out by Basu (2006), Basu et al. (2009, 2011).

Table 156.2 Ranges of mechanical parameters with respect to assessed weathering grades of granites and quartzites

Hong Kong granite (after Basu 2006)						
WG	I	I–II	II	III	III–IV	IV
UCS (MPa)	196.45–116.30	106.34–83.13	68.21–31.14	26.83–13.61	25.14–7.64	6.32
E (GPa)	53.19–42.90	31.79–21.92	25.32–7.02	15.82–5.22	11.83–5.38	4.46
BTS (MPa)	11.36–8.48	7.76–4.73	5.24–3.80	2.07–1.20	1.85–1.68	0.97
Granite from southeastern Brazil (after Basu et al. 2009)						
WG	I	II	II–III	III	IV	
UCS (MPa)	214–153	161–134	137–107	88–73	Undetermined	
E (GPa)	70.17–61.49	61.18–55.23	54.51–47.19	52.00–41.08	Undetermined	
Quartzite from eastern India (after Basu et al. 2011)						
WG	I	II	III	IV		
UCS (MPa)	205.76	114.75–51.71	80.00–42.46	42.44		
E (GPa)	45.56	15.88–6.07	5.22–3.45	1.08		
BTS (MPa)	11.35–8.99	8.50–2.54	4.79–1.87	2.29–1.65		

Brazilian tensile strength (BTS) was determined following ISRM (1978) and ASTM (2001) specifications by Basu (2006) and Basu et al. (2011). The ranges of these parameters with reference to assessed weathering grades (WG) of granites and quartzites are summarized in Table 156.2. Although deterioration of the rock materials is apparent from the decreasing trend of these parameters as weathering intensifies, overlapping of values over adjacent grades is often observed. In case of all three parameters (i.e. UCS, E and BTS) of granites and quartzites, the maximum absolute drop takes place in the early stage of weathering. This can be attributed to sudden augmentation of micro-cracks/flaws of both granitic and quartzitic materials with the onset of weathering. As weathering advances, weakening of the skeletal structure of these rocks also proceeds in a progressive manner. However, at elevated stages of weathering, absolute drops in these parameters due to change in weathering grade do not remain as sensitive to weakening of the skeletal structure or mechanical coherence of granites and quartzites due to induced heterogeneity and flaws as they do at the early weathering stage.

156.4 Conclusions

Based on the study presented in this article, the following conclusions are drawn:

- The conventional 6-fold weathering classification for uniform materials is as capable of capturing gradational weathering induced changes of quartzitic materials as it is for polymineralic crystalline rocks like granite.
- The highest degree of sensitivity of UCS, E and BTS of both granite and quartzite at the early stage of weathering can be attributed to sudden augmentation of micro-cracks/flaws of these materials with the onset of weathering.

References

Anon QJ (1995) Eng Geol 28 207–242
ASTM (2001) ASTM Standards on Disc, 04.08, West Conshohocken, PA
Basu A (2006) PhD thesis, The University of Hong Kong
Basu A, Celestino TB, Bortolucci AA (2009) Rock Mech Rock Eng 42 73–93
Basu A, Ghosh N, Das M (2011) J Eng Geol XXXVII 217–223
Dearman WR, Irfan TY (1978) In: Proceedings of the international symposium on deterioration and protection of stone monuments, UNESCO, Paris, 5–9 June 1978
Hencher SR, Martin RP (1982) In: McFeat-Smith I, Lumb P (eds) Proceedings of the seventh southeast Asian geotechnical conference, Hong Kong
IPT (Instituto de Pesquisas Tecnologicas) (1981) Mapa Geologico do Estado de Sao Paulo
ISRM (1978) Int J Rock Mech Min Sci Geomech Abstr 15:99–103
Moye DG (1955) J Inst Eng 27:287–298
Saha AK (1994) Crustal evolution of Singhbhum North Orissa, Eastern India. Geological Society of India Memoir 27
Sewell RJ, Campbell SDG, Fletcher CJN, Lai KW, Kirk PA (2000) The pre-quaternary geology of Hong Kong. Geotechnical Engineering Office, Hong Kong

Performance of Forepole Support Elements Used in Tunnelling Within Weak Rock Masses

157

J. Oke and N. Vlachopoulos

Abstract

The mechanics associated with forepoling structures explicitly, has never been fully investigated in order to determine the associated support mechanics when installed in isolation and/or in groups as an umbrella arch. Further, numerically, these support structures cannot be modelled using the commonly used, industry standard, two-dimensional (2D) numerical software packages. In numerical software using three-dimensional (3D) codes, these support elements are commonly standardized using pile or rockbolt simplified noded elements that do not truly describe their behaviour when subjected to the true 3D stress conditions that result at face or near the face due to the tunnel or mining excavation process. As such, a deficiency exists with regards to prediction of the interaction of umbrella arch support systems with forepoles element for tunnelling practices within weak rock masses. Methods have been developed in order to predict the behaviour of radial support systems to include temporary support elements such as rock bolts, steel sets, and liners such as the convergence-confinement method. However, these tools do not have the ability to capture the influence of support systems installed longitudinally at the face of tunnel such as forepole, and core reinforcement elements. In an attempt to improve tunnel design strategies, this paper will focus on the mechanical response of the application of the forepole element as part of the umbrella arch method installed in deep and shallow excavations; As well, other issues associated (influences) with the use of forepoles are also highlighted and discussed.

Keywords

Weak rock • Forepoles • Tunnel temporary support • Convergence-confinement • Tunnel support design

157.1 Introduction

This paper defines the concept of utilizing forepole temporary support elements in conjunction with the convergence-confinement method (Carranza-Torres and Fairhurst 2000) for design purposes. It also introduces concepts that are related to the forepole/support arch concept and the gaps which currently exist with their usage. The primary and specific purpose of this portion of the investigation was to capture the support-rock interaction/mechanics associated with the forepole temporary support elements (as seen in Fig. 157.1). The forepole umbrella creates a stable excavation environment. One postulated mechanism of support is the redirection of 3D stress flow around the tunnel (Gibbs et al. 2007). This concept has also been labeled the arch effect by (Lunardi 2000). In order to attempt to determine the true impact of a forepole umbrella on an excavation, an extensive 3D numerical model is required. Such 3D numerical models were developed within FLAC3D (Itasca

J. Oke · N. Vlachopoulos (✉)
GeoEngineering Centre, Queen's-Royal Military College of Canada, Kingston, ON K7K 7B4, Canada
e-mail: vlachopoulos-n@rmc.ca

G. Lollino et al. (eds.), *Engineering Geology for Society and Territory – Volume 6*,
DOI: 10.1007/978-3-319-09060-3_157, © Springer International Publishing Switzerland 2015

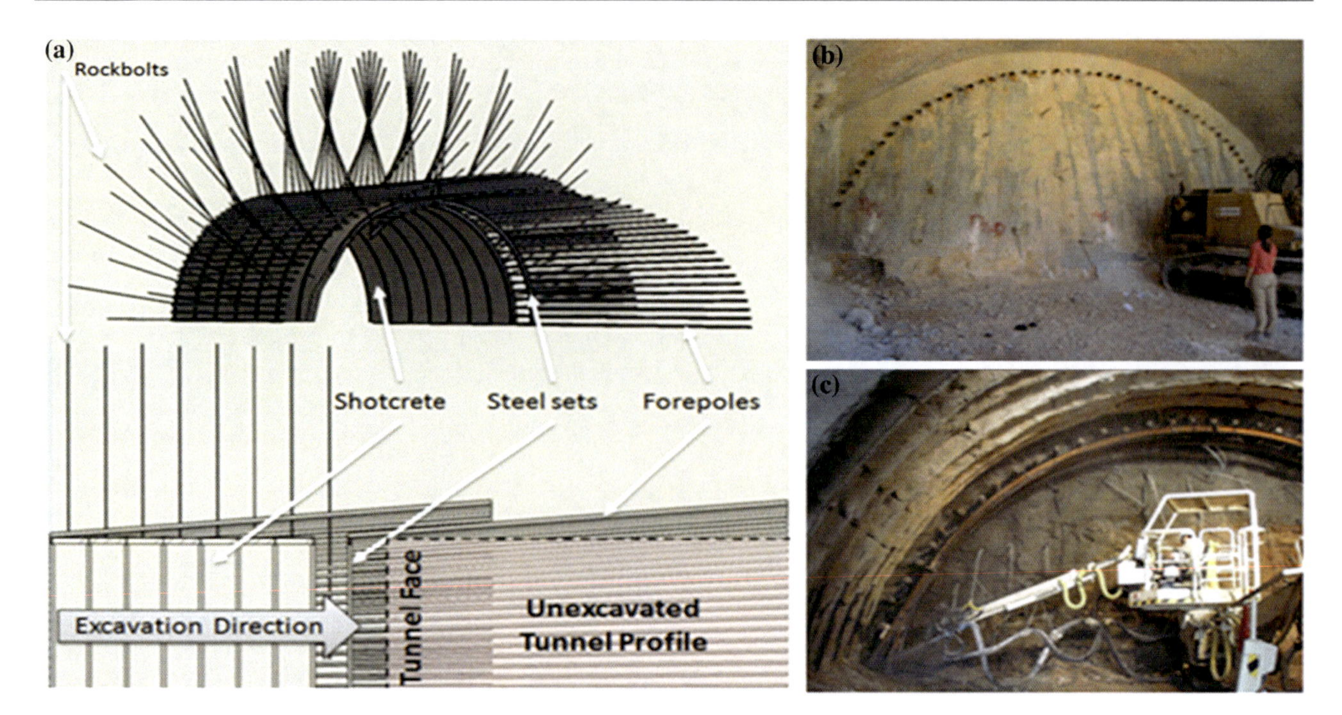

Fig. 157.1 **a** Nominal temporary structural support system used as part of the observational method of support. Elements shown are: forepoles without niches; steel sets (i.e. H-piles); *shotcrete*; and *rockbolts*. Core face reinforcement is not depicted. **b** Forepole installed at the crown of a tunnel excavation. **c** Forepole installed resting on steel sets of a tunnel excavation

2009) for the analysis cited herein. Selected factors of consideration when designing with forepole elements are: spacing between forepoles, positioning, and the size of the forepoles. Other factors that contribute to (but not limited to) the influence and performance of forepole elements is the way that they are simulated numerically, their influence at the design stage (i.e. influence on analytical, unsupported solutions) and their use in the reduction of surface settlements in shallow tunnel environments. These factors affecting the performance of the forepole support element and overall tunnel behavior are highlighted and summarized below.

157.2 Forepole Temporary Support Elements: Nomenclature

Forepoles are temporary support elements employed most often in tunnelling projects associated with weak rock. They are used ahead of the face (in combination with other temporary support), and add stabilization the plastic zone created ahead of the face due to tunnelling effect. They are installed longitudinally to allow for stable excavation underneath the structural umbrella, formed by an arrangement of multiple forepoles (Fig. 157.1). The use of the name forepole as opposed to other support elements has been addressed recently by Oke et al. (2013). The authors proposed specific nomenclature associated with temporary support systems for the design and construction of tunnels. By definition, forepoles have a larger diameter than spiles and their lengths are greater than the height of the tunnel excavation.

157.3 Factors Affecting the Performance of Forepole Support Elements and Tunnel Behaviour

157.3.1 Forepole Spacing, Positioning, and Size

The authors investigated the effects of forepole spacing, positioning and size with respect to tunnel performance (Oke et al. 2012a). A parametric analysis associated with the spacing design parameter was initiated as spacing was increased by 0.05 m until 0.20 m in order to observe the effects of the increased spacing, to cases of 30 cm and beyond. Further points were then investigated to capture the full range of spacing and to ensure the baseline run was capturing the minimal group (or forepole interaction) effect from the influence of adjacent forepoles The results of such an analysis are seen in Fig. 157.2 for a 6 m overhang of the umbrella arch consisting of overlapping forepole elements. The results indicate that when the forepole member is of small diameter (less than 60 cm), it is able to “pass through” the rock mass, allowing the rock mass to flow around it.

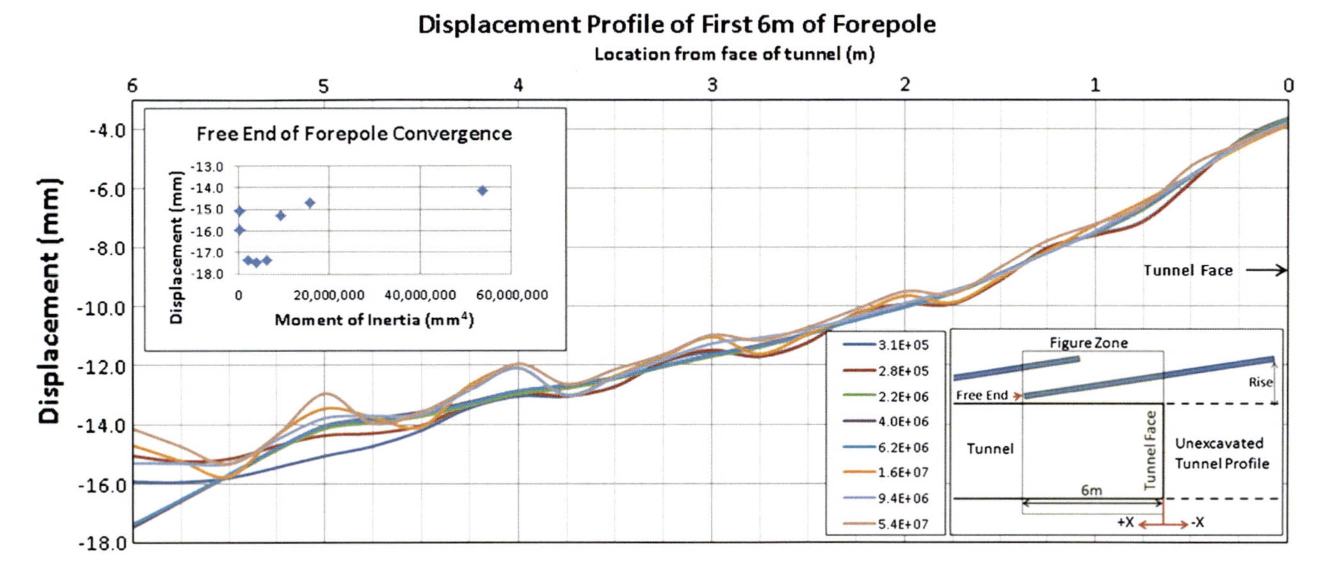

Fig. 157.2 Displacement profile of first 6 m of forepole from numerical analysis of varying size with 6 m of overhang. The plot illustrates the values of displacement of the free end of the forepole

However, if the forepole is too stiff, the rock mass will fail and flow around the structural support member. Having a diameter smaller than 60 cm or modulus of inertia greater than 1.6e7 mm^4 created less tunnel convergence. However, it is necessary to note that the 3D numerical analysis was conducted using a continuum model, and thus, failure modes of overbreak and spalling could not be simulated. It was determined that the optimal range of forepole sizes (within the respective rockmass) should have a modulus of inertia less than 6.2 mm^4, while also having a diameter no less than 101 mm. The reader is referend to the paper cited for further details concerning the findings of the this investigation; conclusions from the research are also summarized in the conclusion section of this paper herein.

157.3.2 Numerical Modelling of Forepole Elements

As with any numerical analysis, there are limitations associated with the use of an idealized element within a numerical software package. Within FLAC3D (Itasca 2009), the forepole was modelled a pile element. The pile element is a straight segment of uniform, bisymmetrical cross-sectional properties lying between two nodal points with six degrees-of-freedom per node. Both a normal-directed (perpendicular to the pile axis) and shear-directed (parallel with the pile axis) frictional interaction occurs between the pile and the finite difference grid. This idealized frictional interaction is illustrated in Fig. 157.3. In addition to skin-friction effects, end-bearing effects can also be captured. Piles may be loaded by point or distributed loads. The element is defined by its geometric, material and coupling-spring properties. The only interaction the forepole has with the rock mass is through other structural members (i.e. forepole is connected to the liner, and the liner is connected to the rock mass), and interaction parameters (coupling-spring properties). These coupling-spring parameters are a function of the rock mass parameters (internal angle of friction for the rock mass φ_{rm}), annulus created during installation, and grout injected into and around the element (if used). These parameters are difficult to obtain without in situ testing (Volkmann and Schubert 2007). The interaction parameters were investigated in the paper by the authors

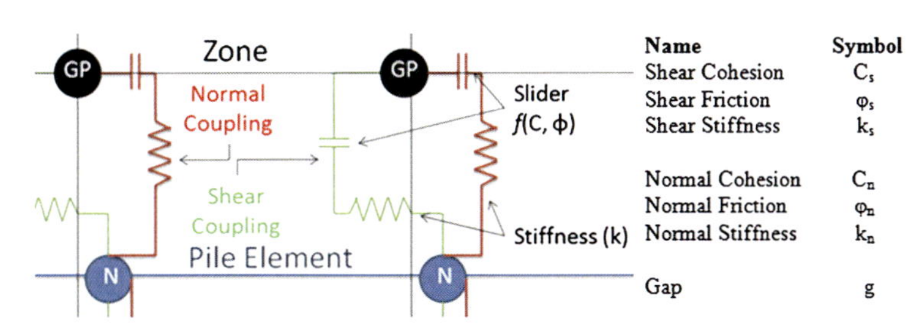

Fig. 157.3 Ideal illustration of normal and shear interaction of the pile element on the zone. Interface elements connecting the forepole (pile element) with the rock (zone)

(Oke et al. 2012b) and it was determined that the magnitude of these input parameters had a significant influence to the overall behaviour of the tunnel under varying conditions and values. As such, one must ensure the proper values or range of values are utilized as inputs to numerical models. Of note is that the influence of interaction stiffness and cohesion on the overall tunnel convergence as well as the longitudinal forepole displacement. For more details associated with this study, the reader is referred to Oke et al. (2012b).

157.3.3 Modified Longituninal Displacement Profile due to Forepole Support

The Convergence-Confinement method (Carranza-Torres and Fairhurst 2000) is a simple engineering design method which has aided tunnel designers with determining preliminary designs (through trial and error) of temporary support arrangements using crude approximations. A summary of the salient components of the method and supporting material can be found within RocSupport (2013). The Longitudinal Displacement Profile (LDP) as part of the Convergence-Confinement Method is an analytical approach which associates the tunnel wall deformations with the actual physical location along the tunnel axis. The LDP consists of Elastic Solutions and several empirical plasticity models of varying intensity (Vlachopoulos and Diederichs 2009). The elastic solutions are reasonable for plastic analysis provided that the plastic radius of the tunnel does not exceed twice the radius of the tunnel. Furthermore, when the plastic radius is greater than twice the radius of the tunnel, two equations are required to capture the profile transitioning at the tunnel face (i.e. one relationship to describe the deformation past the face into the rock mass and another relationship to capture the displacement of the open tunnel cavity). This transitioning is in agreement with other authors such as Unlu and Gercek (2004). However, none of the referenced LDPs take into consideration the effect of temporary support. To improve upon this, the authors have built upon and modified the work of Vlachopoulos and Diederichs (2009). Conceptually, the unsupported and supported (i.e. supported with forepoles) LDP is seen in Fig. 157.4. The authors have determined a series of relationships that allow for the adjustment of the LDP based on the use of temporary support. The method can provide an improved approximation of the stress fields ahead of the tunnel face. This improved approximation can also be used for the simplification of stresses which are applied to pre-support measures (umbrella arches, core-reinforcement, etc.). Additionally, this work highlights the requirement for further modification/improvements to the LDP with respect to curvature when support is installed. This requirement is essential in order to ensure accurate installation of support.

157.3.4 Reduction of Surface Settlement due to Forepole Umbrella Arch Support

The authors have calibrated a three dimensional (3D) numerical model to the documented work of Yasitli (2012), in an effort to capture the reduction of surface settlement when a Forepole Grouted Umbrella Arch (FpGUA) is added to the pre-existing support system. However, the calibration of such support elements is a challenging endeavour due to the interaction parameters used in the numerical analysis. The authors' calibration has also enabled them to further

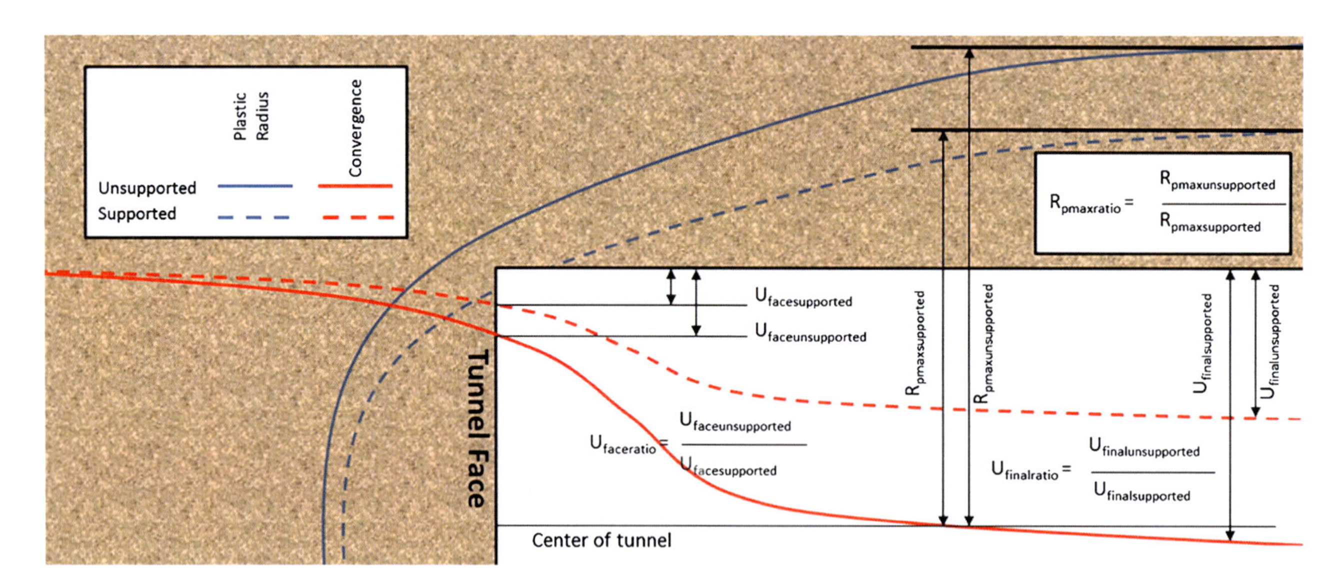

Fig. 157.4 Illustration of the change of the plastic radius, final convergence of the tunnel and displacement of the tunnel face due to support, where *U* displacement and *R* radius)

quantify the Umbrella Arch Selection Chart (UASC) (Oke et al. 2013), a design chart developed in order to aid tunnel engineers and designers in terms of the type of Umbrella Arch to use for certain tunneling conditions. The results of FLAC3D analysis were validated through a comparison of selected analytical solutions for surface settlements as well as cited examples in literature.

157.4 Summary and Conclusions

The authors have undertaken a series of research studies associated with the mechanistic behavior of forepole support elements as part of the overall Umbrella Arch temporary support scheme. Each of the factors investigated have contributed to a better understanding into the behavior and influences of the use of these systems as well as the associated limitations. A significant contribution to the tunnelling industry is the development of the Umbrella Arch Selection Chart (UASC). With more rigorous research in this regard, an optimized use of forepoles within the overall temporary tunnel support scheme can be realized.

References

Carranza-Torres C, Fairhurst C (2000) Application of the convergence-confinement method of tunnel design to rock masses that satisfy the Hoek-Borwn failure criterion. TUST 187–213

Gibbs P, Lowrie J, Keiffer S, McQueen L (2007) M5 East—design of a shallow soft ground shotcrete motorway tunnel. Australas Tunn Soc 1–6

Itasca (2009) FLAC3D v4. Fast lagrangian analysis of continua in 3 dimensions

Lunardi P (2000) The design and construction of tunnels using the approach based on the analysis of controlled deformation in rocks and soils. T&T int 3–30

Oke J, Vlachopoulos N, Marinos V (2013) The pre-support nomenclature and support selection methodology for temporary support systems within weak rock masses. Geotech Geol Eng J 31(5). doi 10.1007/s10706-013-9697-4

Oke J, Vlachopoulos N, Diederichs MS (2012a) Sensitivity analysis of orientations and sizes of forepoles for underground excavations in weak rock. In: American Rock Mechanics Association 46th US Rock Mechanics/Geomechanics Symposium, June 2012, Chicago, IL, USA

Oke J, Vlachopoulos N, Diederichs MS (2012b) Improved input parameters and numerical analysis techniques for temporary support of underground excavations in weak rock. In: Proceedings of the 21st Canadian Rock Mechanics Symposium RockEng12, May 2012. Edmonton, Alberta, Canada

Rocscience (2013 Jan 17) RocSupport v3.011

Unlu T, Gercek H (2004) Effect of Poisson's ratio on the normalized radial displacements occurring around the face of a circular tunnel. Tunn Undergr Space Technol 547–553

Volkmann G, Schubert W (2007) Geotechnical model for pipe roof supports in tunneling. In: Proceedings of the 33rd ITA-AITES World Tunneling Congress, underground space—the 4th dimension of metropolises. Taylor & Francis Group, Prague, ss 755–760

Vlachopoulos N, Diederichs MS (2009) Improved longitudinal displacement profiles for convergence confinement analysis of deep tunnels. Rock Mech Rock Eng 42(2):131–146. doi:10.1007/s00603-009-0176-4

Yasitli N (2012) Numerical modeling of surface settlements at the transition zone excavated by New Austrian Tunneling Method and Umbrella Arch Method in weak rock. Saudi Soc Geosci

158 The Research of Shear Creep Behaviors of Saturated Sericite-Quartz Phyllite

Guang Ming Ren, Xin lei Ma, Bo Wen Ren, and Min Xia

Abstract

Along with the thorough development of underground engineering, people pay more attention on the long-term deformation stability of the tunnels. However, creep characteristic is an important factor to arouse surrounding rock deformation failures and control engineering designs. Phyllite is a common soft rock in engineering practice, when sample is saturated, water will not only impose the softening effect on phyllite, but also it will influence their creep mechanical properties under the long-term loads. In this paper, selecting samples from headrace tunnel GaoPingpu hydropower station in China, we carried out the shear creep test of saturated phyllite along the schistosity surface under different stress levels. Further, we systematically analyzed and compared the test results with the samples in air-dried conditions, and found that water has great effects on the strength of phyllite: the long-term strength and yield stress of saturated phyllite are only 70–78 and 75–89 % of the air-dried ones respectively. At the same time, based on the Burgers creep model, we discussed the impacts of water on the creep parameters' values of phyllite. This result can provide a theoretical basis to calculate the creep deformation of the tunnel surrounding rocks.

Keywords

Phyllite • Shear creep • Softening effect • Creep parameters

158.1 Introduction

Research on rheological test first began in 1901, and its main purpose was to explore the rock deformation and flow properties from a geologic perspective (Liu 1994). As time goes by, numerous rock engineering accidents have happened due to the rock creep, and the significance of rock creep and its aging strength in engineering have been gradually recognized. In recent years, rock rheology has become a hot issue in the study field of rock mechanical properties, and one of the main methods is laboratory testing (Yang and Cheng 2011; Yang et al. 1999; Maranini and Brignoli 1999; Li et al. 2008; Li and Xia 2000; Boitnott 1997; Fabre and Pellet 2006; Tomanovic 2006; Pierre et al. 2005; Chen et al. 2008; Yang et al. 2005; Gao et al. 2010; Liu et al. 2009). The influence of water has been studied in details as the experimental variables for its significance in affecting the creep characteristics of rock and its strength of schistosity surface. Wawersik and Brown (2007) demonstrated experimentally that the creep deformation of granite and sandstone increases with increasing water content. In addition, dry and saturated specimens will differ significantly in their stabilization creep rates. For porous sandstones, Bernabk et al. (1994) found that the presence of water dramatically increased the creep rates near failure compared to dry samples. Based on the creep test of red sandstone, Sun et al. (1999) discovered that the long-term compressive strength of saturated brittle-viscous red sandstone was about

G.M. Ren (✉) · X.l. Ma · B.W. Ren · M. Xia
State Key Lab of Geological Hazard Prevention and Engineering Geological Environment Protection, Chengdu University of Technology, No. 1 Erxianqiao East Road, 610059, Chengdu, China
e-mail: renguangming@cdut.cn

X.l. Ma
e-mail: maxinlei1990@163.com

G. Lollino et al. (eds.), *Engineering Geology for Society and Territory – Volume 6*,
DOI: 10.1007/978-3-319-09060-3_158, © Springer International Publishing Switzerland 2015

46.3 % greater than the air-dried rock samples. Zhu and Ye (2002), through performing the creep test on tuff, have obtained the influence rule of water on the creep properties of rock. Guangting et al. (2004) also proved that the instantaneous deformation modulus of weak conglomerate significantly reduced after being soaked in water and its rheological properties became more obvious. Xie and Shao (2006) revealed that the creep deformation of porous chalks is much higher in water saturated sample than in oil saturated sample (the creep properties of oil saturated samples are similar with dry samples). Chen et al. (2009) used Burgers model to describe the creep properties of red-bed soft rock and also found that the effect of moisture content on creep properties of red-bed soft rock was extremely remarkable. Chen et al. (2010) perfected the influence rules of water on the creep properties of red-bed soft rock through more in-depth research. To sum up, recently, some international scholars have studied a lot about water influences on creep properties of the rock mass, such as granite, sandstone and tuff rocks etc. However, fewer articles have reported the effect of water on the shear creep characteristics of soft rock along the schistosity surface. Hence this paper focus on the study of shear creep characteristics of the soft phyllite under air-dried and saturated conditions. It provides the rheological parameters for the engineering designs, thus having an important practical and theoretical significance.

158.2 Sample Characteristics and the Test Method

Experimental samples are collected from the Gao Pingpu hydropower station diversion tunnel in FuJiang, SiChuan. And they belong to the Maoxian Group of Silurian Strata (Smx) .Through the indoor rock sliced identification, samples are determined to be the slightly weathered sericite-greisen phyllite. And the main minerals of the phyllite are quartz and sericite with a small amount of fine-grained carbonate and pyrite, which has a fine granular blastic texture and phyllitic structure. We have chosen four air-dried samples and four saturated specimens in our test, and divided them into four groups ($A\#_{1,2}$–$D\#_{1,2}$) correspondingly. Since phyllite rock is soft, contains schistosity plane, and is prone to be fragmentized, therefore, we used irregular cubes, sized approximately 70–80 mm × 70-80 mm × 80 mm (length by width by height). The average density of air-dried specimens is 2.78 g/cm^3; Its average water absorptivity is about 61 %; the saturated average water absorptivity is 77 %.And for the saturated simples, in order to maintain their saturated state during the entire test, the samples were placed in a specially made test basin and submerged in the water .The water level was kept 1–2 cm above the vertical top of the specimen. Because the rocks mass are easy to be soften and slacken by water, the saturated specimens were made from air-dried specimens directly before the experiment. The test was performed under relatively constant temperature and humidity conditions on the special rock creep test device which belongs to the State Key Lab of the Chengdu University of Technology. It was conducted along the schistosity of the phyllite, and according to the Tjong-Kie Tan loading method (Tan and Kang 1980) to be applied pressure which is a step wise loading function whereby, in this manner, the creep deformation can be obtained as a function of stress and time on only one sample. In order to reduce the measuring error caused by the heterogeneity of the deformation of the samples, the shear rheometers were installed in different directions, and the mean values of their records were taken as the final values (He et al. 2002). According to quick shear tests, the normal stress was confined as 0.20, 0.40, 0.60, 0.80 MPa respectively. And the shear stress was divided into 6–14 levels with each level sustaining approximately 72 h (under every shear stress step, after 72 h, the shear creep displacement of the test in a day would be less than 0.001 mm, which had met the stability standard of the creep displacements) (Specifications for rock tests in water conservancy and hydroelectric engineering, SL264-2001, in Chinese). We have made sure that each specimen deformed along the schistosity during the last stage of the loading process. This integrated test revealed the shear deformation —time curves of phyllite under the same normal stress, but different shear stresses.

158.3 The Analysis of Creep Experimental Curves of Phyllite Under Air-Dried and Saturated Conditions

According to the above method, we got the typical shear creep curves of air-dried and saturated specimens under multi-level stress, as shown in Fig. 158.1. The shear creep curves of samples at all load levels were extremely similar and the samples did not show any obvious accelerated shear creep. As the test loaded, phyllite instantly showed elastic response followed by creep deformation which grew along with time and the rate of which gradually slowed down as the time getting longer. Finally, both samples went into the stable creep stage. Since then, the creep deformation increased continuously at a very small constant speed over the period of time. In the final loading phase of each specimen, a sudden shear failure appeared in a short time.

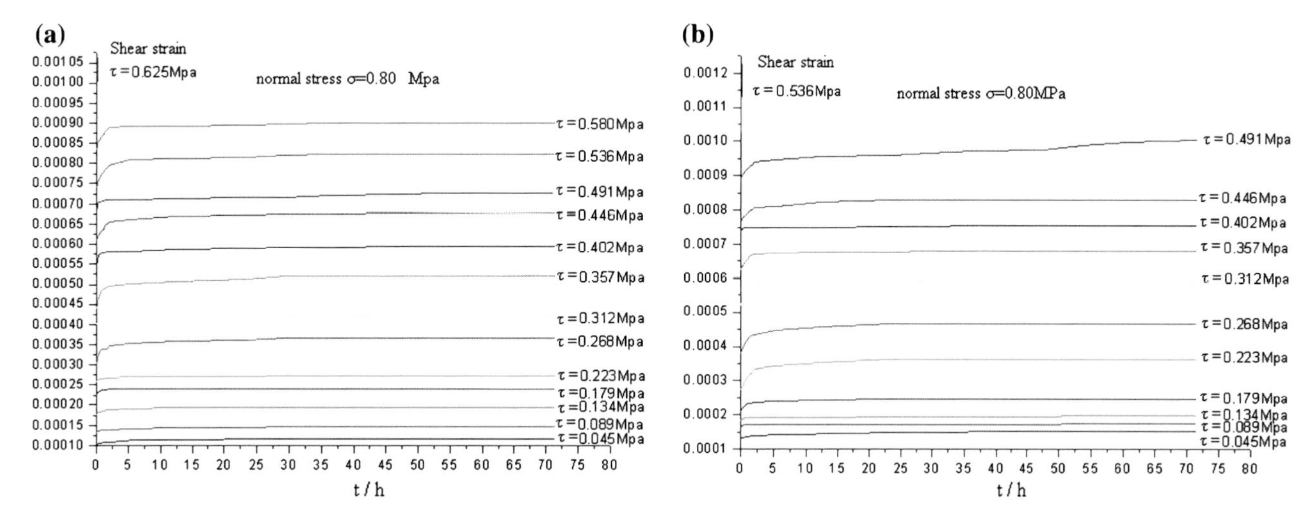

Fig. 158.1 Sheer creep curves under the normal stress of 0.800 Mpa **a** air-dried sample D$\#_1$, **b** saturated sample D$\#_2$

158.4 Analysis of the Effect of Water on Phyllite Rheological Parameters

158.4.1 The Rock Creep Model and Parameters Determination

Based on multiple domestic and foreign studies of rock creep characteristics and the shear creep curves of this test, in this paper, we selected Burgers mechanical creep model to describe the creep characteristics of phyllite. It is connected in series with Maxwell and Kelvin models which is applicable to viscoelastic rock. It can well express creep properties, such as instantaneous deformation, decelerating creep and steady creep etc. (Zhou 1990; Cai 2002; Xu 1997). According to the properties of series, Burgers constitutive equation (Zhou 1990) is:

$$\ddot{\tau} + \left(\frac{E_1}{\eta_2} + \frac{E_2}{\eta_2} + \frac{E_1}{\eta_1}\right)\dot{\tau} + \frac{E_1 E_2}{\eta_1 \eta_2}\tau = E_2\ddot{\varepsilon} + \frac{E_1 E_2}{\eta_2}\dot{\varepsilon} \quad (158.1)$$

where τ is shear stress (MPa); E is the elastic deformation modulus (MPa) and η is viscous fluid viscosity (0.1 Pa·s). Based on the instantaneous superposition principle, the creep equation of Burges model is established as follows:

$$\varepsilon = \frac{\tau_0}{E_1} + \frac{\tau_0}{\eta_1}t + \frac{\tau_0}{E_2}\left(1 - \exp\left(-\frac{E_2 t}{\eta_2}\right)\right) \quad (158.2)$$

Besides, Eq. (158.2) shows that: When t = 0, the model performs instantaneous elastic deformation. Then the strain gradually increases with the time going on. Finally, the creep deformation is increased by the viscous components creeping steadily.

By using the Burgers model and the least square method, we can obtain the creep curves. And then, the creep parameters for Burgers model, such as E_1, E_2, η_1, η_2, can be confirmed.

158.4.2 Analysis of the Effect of Water on Rheological Parameters of Phyllite

158.4.2.1 Analysis of the Effect of Elastic Modulus E_1 and E_2

The value of the instantaneous elastic modulus (E_1) of the air-dried and saturated specimen is similar and its average ratio is 1.280. This illustrates that water has a small impact on the instantaneous elastic modulus of phyllite. In addition, E_1 of each sample went up slightly along with the increase of shear stress, which means that the instantaneous elastic strain of phyllite made almost linearly elastic change with shear stress increase. The instantaneous elastic strain curves are shown in Fig. 158.2.

As shown in Fig. 158.2, under the same water condition, the instantaneous deformation of specimens was closely correlated to the normal stress and shear stress. Namely, when the shear stress was kept at a constant level, the normal stress and the instantaneous shear displacement of rock samples were negatively correlated; In contrast, when the normal stress was kept constant, the shear stress and the instantaneous shear displacement of rock samples showed the positive correlation. Furthermore, it also indicates that the instantaneous elastic strain of saturated specimen is greater than air-dried ones. And when the shear load increases, the increment of instantaneous elastic strain of the water-saturated specimens was greater than the air-dried samples as well. This may be caused by the ability of water

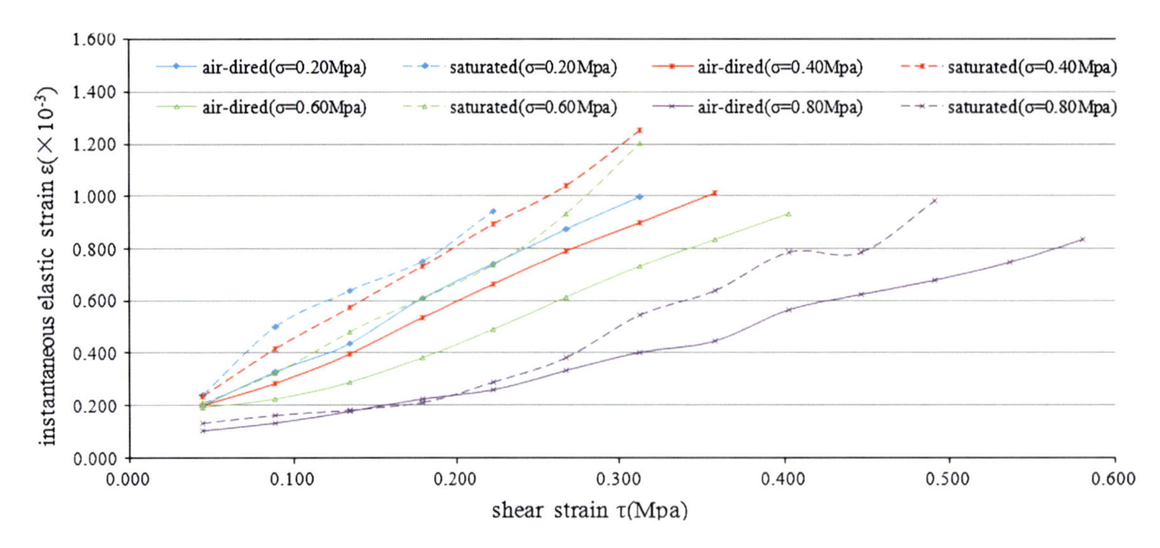

Fig. 158.2 The curves of shear stress—instantaneous strain

to dissolve and soften the rocks. Similarly, water has a comparable influence on E_1 and E_2, and the ratio of E_2 under air-dried condition to that under saturation is 1.326. In other words, the elastic modulus of phyllite E_1, E_2, under the influence of water, will reduce by about 21.9 and 24.6 %, respectively.

158.4.2.2 Coefficient of Viscosity η

The viscous coefficient η is a critical parameter to the viscous characteristics of rock. It varies with the stress level and water content of the test samples. The value of η_1, η_2 are relatively discrete. But in terms of the average, the value of η_1, η_2 under the air-dried condition are 1.275 and 1.724 times of those under water-saturated state, respectively. In other words, the viscous coefficient η_1, η_2 decrease by about 21.6 and 42.0 %, respectively.

158.4.3 Determining the Phyllite Shear Creep Constitutive Equation

As the stress of the test samples is small and little difference amongst them, and if the effect of stress on creep parameters is not considered, the average creep parameters of air-dried and water-saturated samples can be obtained, then the creep parameters were applied into Eq. (158.2), so we can get the phyllite creep constitutive equation:

$$\text{Air} - \text{dried} \quad \varepsilon(\text{t}) = \frac{\tau}{0.48} + \frac{\tau}{764.859}t + \frac{\tau}{6.580} \times \left(1 - e^{-0.185t}\right)$$

$$\text{Saturated: } \varepsilon(\text{t}) = \frac{\tau}{0.357} + \frac{\tau}{432.974}t + \frac{\tau}{4.961} \times \left(1 - e^{-0.240t}\right)$$

158.4.4 The Phyllite Shear Strength

The long-term strength is one of the important parameters reflecting the creep properties of rock. Through the isochronous curves, a most commonly method, we can get the value of strength (Sun 1999; Zhou 1990). And it is based on some related specification of rock mechanics test. Here, we joined the turning points by smooth curve, so the corresponding stress of the arc is the long-term strength of the sample. Then we listed every sample's long-term strength in Table 158.1. And through the tests, we presented the yield stress of each sample (τ_f) in Table 158.1 together.

Comparing the long-term strength and yield stress of the air-dried samples with the saturated ones (in Table 158.1.), these obvious conclusions are easy to be indicated: under the same normal stress, the long-term strength of saturated phyllites is only 0.70–0.78 times of the air-dried ones and the mean value is 0.747 times; to the yield stress of the saturated phyllite, it is 0.75–0.89 times of the air-dried samples and the average is 0.830. These results demonstrate

Table 158.1 The long-term strength and yield stress of phyllite under different conditions

Group NO.	A$\#_1$	A$\#_2$	B$\#_1$	B$\#_2$	C$\#_1$	C$\#_2$	D$\#_1$	D$\#_2$
Normal stress/Mpa	0.20		0.40		0.60		0.80	
Conditions	Air-dried	Saturated	Air-dried	Saturated	Air-dried	Saturated	Air-dried	Saturated
the long-term strength/Mpa	0.245	0.171	0.31	0.242	0.345	0.253	0.48	0.365
Yield stress/Mpa	0.357	0.268	0.402	0.357	0.446	0.357	0.625	0.536

Table 158.2 The long-term shear strength parameters of the phyllite under air-dired and saturated conditions

Parameters conditions	C_∞/Mpa	Φ_∞/°
Air-dried	0.175	19.34
Saturated	0.121	16.01

that phyllite is easy softened by water, resulting in the decrease of its strength. Based on the normal stress and the long-term strength listed in Table 158.1, we determined the parameters of the long-term shear strength and is presented in Table 158.2.

As shown in Table 158.2, the strength parameters of phyllite decrease because of the softening effect of water. And the reduction of C_∞ and Φ_∞ is 0.054 Mpa and 3.33°, respectively.

158.5 Conclusions

The experiment was conducted along the schistosity surfaces. The shear creep curves of phyllite, under the air-dried and saturated conditions, perform the similar form. All of the tested phyllite has the instantaneous and steady deformation behavior under the action of shear stress. And finally the shear failure happened suddenly. At the same time, the instantaneous deformation of specimen, if under the same water content, is closely correlated to the normal stress and shear stress. Namely, as the shear stress was constant, normal stress and the instantaneous shear deformation of rock samples were negatively correlated; and when the normal stress was steady, the shear stress showed a positive correlation with the instantaneous shear deformation of rock samples.

In accordance with Burgers model, comparing the creep property of air dried samples with water-saturated ones, the conclusion emerged with stark clarity: the average values of the elastic modulus E_1, E_2 and the viscous coefficient η_1, η_2 of phyllite, under the action of water, will reduce by about 21.9, 24.6, 21.6 and 42.0 %, respectively.

For the air dried and water saturated samples, the analysis of the test results about the long-term strength and yield stress demonstrate that: under the same normal stress, the long-term strength of saturated phyllite is 0.70–0.78 times of the air-dried ones (the average is 0.747); the ratio of the yield stress of the saturated samples is 0.75–0.89 times of the air-dried specimens (average value is 0.830); and the value of strength parameters (C_∞ and Φ_∞) of the saturated samples reduced 0.054 MPa and 3.33°, respectively. All of above properly proved that water can soften phyllite and decrease its strength. This result will be useful to determine the long-term strength of the similar rock-mass in other future engineering projects.

Acknowledgements The authors would like to acknowledge financial support by the National Natural Science Foundation Project of China (No. 41072229).

References

Liu X (1994) The conspectus of rock rheology. Geological Publishing House, Beijing (in Chinese)

Yang S-Q, Cheng L (2011) Non-stationary and nonlinear visco-elastic shear creep model for shale. Int J Rock Mech Min Sci 48(6):1011–1020

Yang C, Daemen JJK, Jian-hua YIN (1999) Experimental investigation of creep behavior of salt rock. Int J Rock Mech Min Sci 36(3):233–242

Maranini E, Brignoli M (1999) Creep behaviour of a weak rock: experimental characterization. Int J Rock Mech Min Sci 36(1):127–138

Li Y, Wang Z, Tang M et al (2008) Relations of complete creep processes and triaxial stress-strain curves of rock. J Cent South Univ Technol 15(s1):311–315

Li Y, Xia C (2000) Time-dependent tests on intact rocks in uniaxial compression. Int J Rock Mech Min Sci Geomech Abstr 37(3):467–475

Boitnott GN (1997) Experimental characterization of the nonlinear rheology of rock. Int J Rock Mech Min Sci 4

Fabre Géraldine, Pellet Frédéric (2006) Creep and time-dependent damage in argillaceous rocks. Int J Rock Mech Min Sci 43(6):950–960

Tomanovic Z (2006) Rheological model of soft rock creep based on the tests on marl. Mech Time-Depend Mater 10:135–154

Béresta P, Antoine Blumb P, Pierre Charpentiera P, Gharbia H, Valés F (2004) Very slow creep tests on rock samples. Int J Rock Mech Min Sci 42:569–576

Chen Y-j, Wu C, Fu Y-M (2008) Rheological characteristics of soft rock structural surface. J Cen S Univ Technol 15(s1):374–380

Yang S, Su C, Xu W (2005) Experimental investigation on strength and deformation properties of marble under conventional triaxial compression. Rock Soil Mech 26(3):475–478 (in Chinese)

Gao YN, Gao F, Zhang ZZ et al (2010) Visco-elastic-plastic model of deep underground rock affected by temperature and humidity. Min Sci Technol 20(2):183–187

Liu X, Su J, Wang X (2009) Experimental investigation on shear rheological properties of tuff lava with different grades. Chin J Rock Mech Eng 28(1):190–197 (in Chinese)

Sun J (2007) Rock rheological mechanics and its advance in engineering applications. Chin J Rock Mech Eng 26(6):1081–1106 (in Chinese)

Bernabk Y, Fryer DT, Shively RM (1994) Experimental observations of the elastic and inelastic behaviour of porous sandstones. Geophys J Int 117:403–418

Sun J (1999) Rheological behavior of geomaterials and its engineering applications. China Architecture & Building Press, Beijing (in Chinese)

Zhu H, Ye B (2002) Experimental study on mechanical properties of rock creep instauration. Chin J Rock Mech Eng 21(12):1791–1796 (in Chinese)

Guangting L, Hu Y, Fengqi C et al (2004) Rheological property of soft rock under multiaxial compression and its effect on design of arch dam. Chin J Rock Mech Eng 23(8):1237–1241 (in Chinese)

Xie SY, Shao JF (2006) Elastoplastic deformation of a porous rock and water interaction. Int J Plast 22(12):2195–2225

Chen W, Yuan P,Liu X (2009) Study on creep properties of red-bed soft rock under step load.Chin J Rock Mech Eng 28(s1):3076–3081 (in Chinese)

Chen C, Lu H, Yuan C et al (2010) Experimental research on deformation properties red-bed soft rock. Chin J Rock Mech Eng 29 (2):261–270 (in Chinese)

Tan TK, Kang WF (1980) Locked in stresses, creep and dilatancy of rocks, and constitutive equations. Rock Mech 13(1):5–22

He M, Jing H, Sun X (2007) Engineering mechanics of soft rock. Science Press, Beijing (in Chinese)

SL264-2001 (2001) Specifications for rock tests in water conservancy and hydroelectric engineering (in Chinese)

Zhou W (1990) Advanced rock mechanics. Hydraulic and Hydroelectricity Press, Beijing (in Chinese)

Cai M (2002) Rock mechanics and engineering. Science Press, Beijing (in Chinese)

Xu Z (1997) Rock mechanics. Hydraulic and Hydroelectricity Press, Beijing (in Chinese)

Part XV

Radioactive Waste Disposal: An Engineering Geological and Rock Mechanical Approach

Convener Prof. Ákos Török—*Co-convener* Jian Zhao

Handling of radioactive waste represents a major environmental concern in modern society due to existing nuclear power plants and is also related to the management of contaminated land after major nuclear catastrophes. The rationale of this session is to provide a forum to papers dealing with various aspects of radioactive waste disposal on surface, near surface and in subsurface storage facilities. It covers aspects of engineering geology, rock mechanics, environmental geology and hydrogeology of site selection, site operation and post-closure periods. The results of in situ and laboratory analyses, including engineering geological and rock mechanical measurements at small and large scales are also included with special emphasis on the existing experimental sites, planned and operating depositories. Computer modelling of the geological environment (soft and hard rock) in combination with hydrogeological and heat transfer models will also form parts of the session.

159 Investigation of Mineral Deformation and Dissolution Problems Under Various Temperature Conditions

J.H. Choi, B.G. Chae, C.M. Jeon, and Y.S. Seo

Abstract

To understand the effects of temperature on the pressure dissolution of mineral, we conducted some experiments using monocrystalline quartz samples. The first of these was a flow thorough experiment to investigate temperature effects for dissolution mechanism. The samples were stressed mechanically by pressing one sample against the other. The flow through experiments was conducted at two different temperatures (35 and 70 °C) at the same pH (pH 11.7) level. The value of the applied stress was 7.32 and 25.27 MPa. During each of these dissolution tests, the solution was sampled regularly and analyzed by an Inductively Coupled Plasma-Atomic Emission Spectrometry (ICP-AES) technique to measure Si-concentration. With the measured Si-concentration, a dissolution rate constant was computed for variety of stress and temperature conditions. It is therefore shown that the rate constant is proportional to the temperature, as expected and as indicated in the literature. It should be noted that the rate constant for the highly stressed case (25.50 MPa) and highly temperature case were considerably greater than for the mildly stressed cases and lower temperature cases. Also, island-channel patterns characterized by micro-cracks a few nanometers in length were seen on the dissolved parts of the samples. The findings and the measured data in this research may be useful for the future development of theoretical models for pressure dissolution and its validation.

Keywords

Quartz dissolution • ICP-AES • CLSM • Dissolution rate

159.1 Dissolution Experiment

The samples for the dissolution experiments were taken from a block of monocrystalline quartz, which may have involved some crystalline lattice defects. The contact area of specimen for flow-through experiments, it was 7.068 mm^2.

A flow through setup was used for the second series of experiments to keep the pH level constant (11.7). Reagent grade NaOH was used to adjust pH of the solution. After the dissolution experiments, the contact areas of the quartz specimens were observed by the confocal laser scanning microscope (CLSM) to identify the type of the dissolution mode (Choi et al. 2008). For quantitative evaluations of the dissolution speed, the Si concentration was measured by Inductively Coupled Plasma Atomic Emission Spectomerty (ICP-AES) to obtain the dissolution rate.

J.H. Choi (✉) · B.G. Chae · C.M. Jeon
Korea Institute of Geoscience and Mineral Resources, 124 Gwahang-no, Yuseong-gu, Daejeon, 305-350, Korea
e-mail: jhchoi@kigam.re.kr

Y.S. Seo
Chungbook National University, 52 Naesudong-ro, Heungdeok-gu, Cheongju Chungbuk, 361-763, Korea

G. Lollino et al. (eds.), *Engineering Geology for Society and Territory – Volume 6*,
DOI: 10.1007/978-3-319-09060-3_159,

159.1.1 Flow-Through Experiment

A schematic diagram of the flow through experiment is shown in Fig. 159.1. The experimental setup consisted of two subsystems: a loading subsystem in a temperature-controlled oven and a CSLM observation subsystem.

The quartz sample was held in a liquid chamber filled with a NaOH solution. The pH level of the solution was 11.7, and the flow speed was 0.15 ± 0.05 ml/h. An iron weight was mounted on the liquid chamber to apply mechanical stress. The solution was sampled every three or four days to measure Si-concentration and to calculate the actual dissolution rate as a function of elapsed time.

159.1.2 Surface Observation and Analysis

After the dissolution test, the surface conditions of the samples were observed by CLSM. The CLSM was an Olympus OLS1100, which uses an Ar laser with a wavelength of 488 nm. The scanning method was light polarization using two resonant galvano mirrors, which enable high-speed, high resolution imaging of a wide area. The size of the image data (the number of pixels) given by the CLSM is usually chosen as 1,024 × 1,024. With a 100× lens, the resolution (length per pixel) is about 125 nm for an imaging area of 0.128 × 0.128 mm^2. A higher magnification may be achieved by digitally zooming from a factor of 1–6. To obtain the dissolution rates, the Si concentrations were measured using an ICP-AES (SPS1500R), which uses an Ar gas to create high-energy plasma. Following a standard procedure for the ICP-AES, the equipment first was calibrated with a standard solution of known Si-concentrations, and then liquid in the current experiments were tested. The overall and actual dissolution rates were calculated from the time variations of the measured Si-concentrations.

159.2 Results and Discussions

159.2.1 Surface deformation

In this study, the contact surfaces of the quartz samples were observed after the dissolution test by CLSM. In Fig. 159.2a–d, the CLSM photos show the elevation of the scanned area of the sample. The small photos in (a), (b) and (d) are the upper contact area. The whitish area indicates a lower elevation, therefore indicating dissolved parts. As seen in these images, the dissolved parts are found only on the contact area.

159.2.2 Temperature Effect

To observe the temperature effect on dissolution, experiments were carried out at three different temperatures (35 and 70 °C). The pH level was kept at 11.7 using the flow through setup. The applied stress for these series of tests was either 7.32 or 25.27 MPa. The levels of stress and pH were thus determined to emulate chemio-mechanical conditions that could be encountered in deep underground nuclear waste disposal facilities. The Si-concentration was measured with the ICP-AES by sampling the solution every three or four days to observe the temporal evolution.

The experimental results are shown in Table 159.1 as Si concentrations and as the dissolution rates at each sampling interval. Note that the overall dissolution rates are shown at the bottom row of this table. The data given in this table are plotted in Fig. 159.3 as a function of time. The plots show

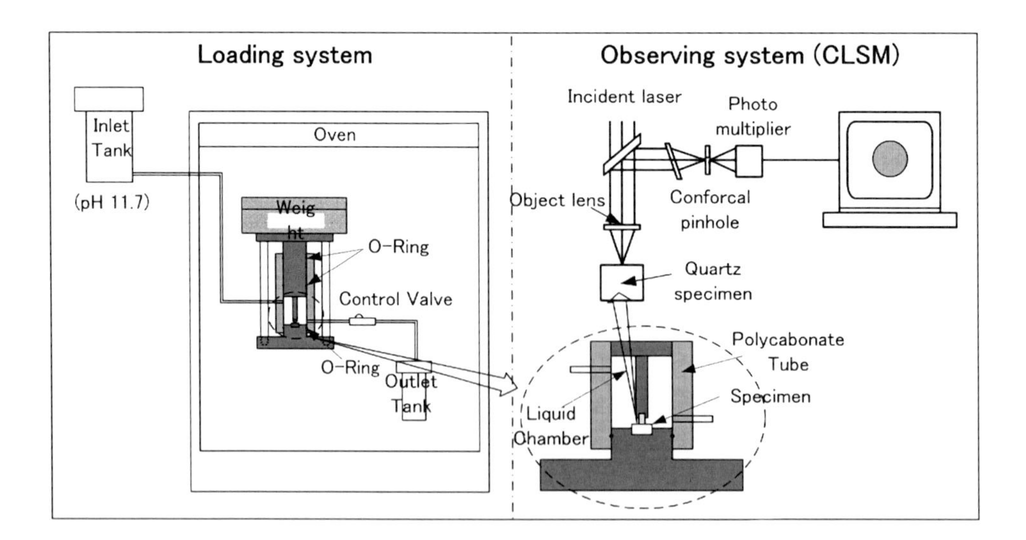

Fig. 159.1 Schematic diagram of the flow-through experimental setup

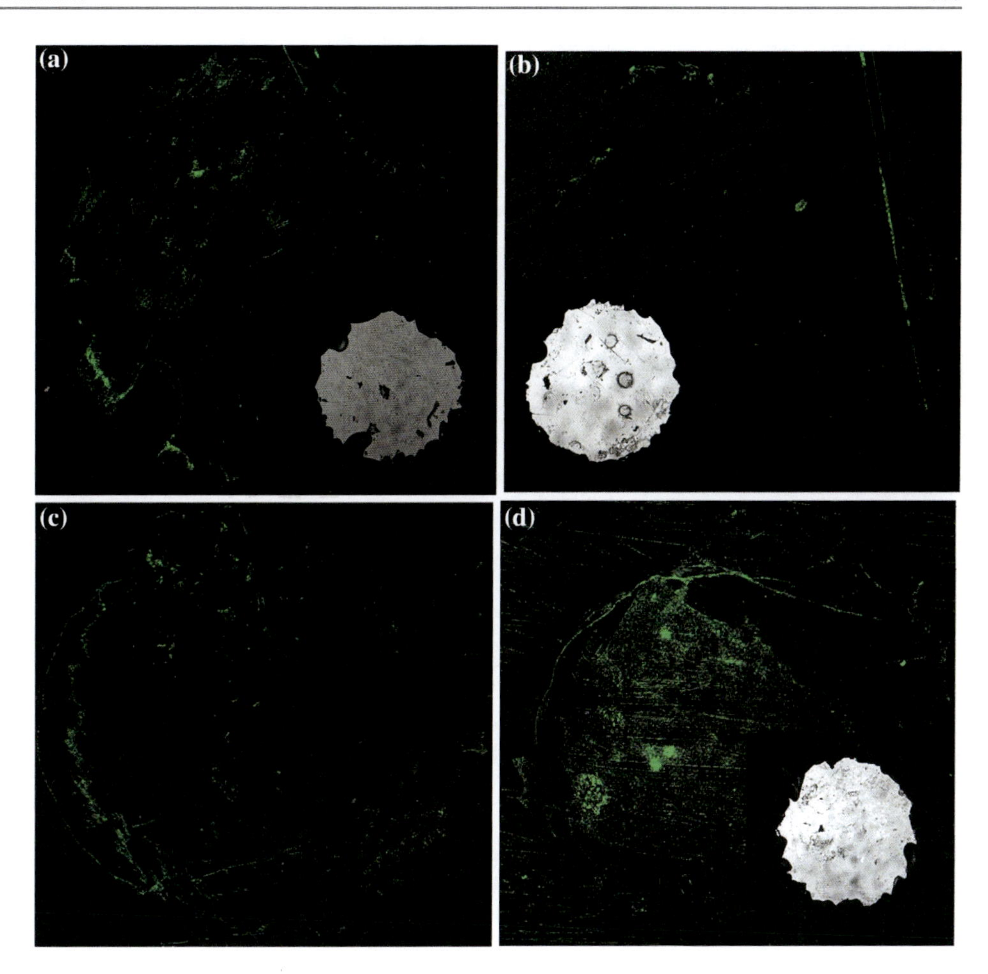

Fig. 159.2 CLSM photos of the dissolved surfaces of the samples

Table 159.1 Time variation of the actual dissolution rate and the overall dissolution rate (v)

Days	7.32 MPa			25.27 MPa		
	35 °C	50 °C	70 °C	35 °C	50 °C	70 °C
2		5.64E−11				
3	8.95E−12			2.95E−10	3.49E−11	
4			3.28E−11			1.15E−10
5		3.44E−11				
6	9.98E−12			1.40E−11		
7					5.53E−11	
8		2.47E−11	6.65E−11			1.67E−10
9	1.02E−11					
10				4.61E−11		
11		2.22E−11			6.82E−11	
12	1.06E−11		7.99E−11			1.27E-10
13				1.25E−11		
14		2.33E−11				
15	1.23E−11				7.46E−11	
16			8.59E−11	1.15E−11		1.13E−10
17						
18		1.74E−11				
20					5.50E−11	
v	1.04E−11	2.76E−11	6.63E−11	1.27E−11	5.86E−11	1.31E−10

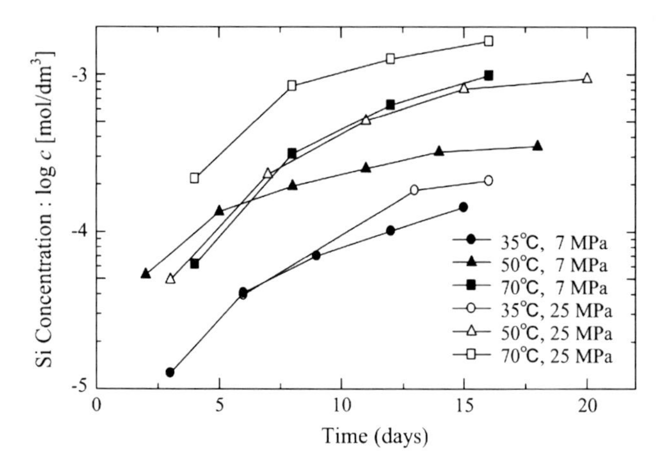

Fig. 159.3 Time variation of Si-concentration

that the dissolution rate is proportional to the temperature at every applied stress level.

The observed temperature dependence may be explained by the adsorption mechanism. According to a sodium sorption experiment Berger et al. (1994), the interaction between the cation and surface was dramatically changed by increasing the temperature from 25 to 150 °C. They explained that sodium has a weak interaction with the surface at a low temperature, but it undergoes specific adsorption at a higher temperature. Hence, the dissolution rate increases with increased temperature in this study. The isotherms for these results are as follows:

$$\upsilon = 1.62 \times 10^{-12}T - 4.87 \times 10^{-11} \quad \text{at } 7.32\text{ MPa} \tag{159.1}$$

$$\upsilon = 3.38 \times 10^{-12}T - 1.08 \times 10^{-10} \quad \text{at } 25.27\text{ MPa} \tag{159.2}$$

159.3 Conclusions

In this work, quartz dissolution rates were measured by flow through experiments using monocrystalline quartz specimens. The experiments were carried out at various temperature levels to develop dissolution rate isotherms. It was found that the dissolution rate increases with an increase in each of the experimental parameters, i.e., temperature, as reported in the literature. It should be noted, however, that our results show that the stress accelerates quartz dissolution so significantly that the rate constants for the samples with applied stress were four to five orders of magnitude greater than those for the samples with no stress. Because the effect of mechanical stress on the dissolution is of great importance, the activation energy E_a was calculated and presented as a function of stress to show that E_a is inversely proportional to the applied stress. With regard to the dissolution mechanism, CLSM observation revealed that the island-channel model best describes the condition of the dissolved surface of the specimens. This means that the dissolution may be viewed as a growth of surface breaking cracks. Hence, the significance of stress in quartz dissolution may be partly attributed to stress concentration at the crack tip, which could accelerate crack growth.

References

Berger G, Cadore E, Schott J, Dove PM (1994) Dissolution rate of quartz in lead and sodium electrolyte solutions between 25 and 300 °C : Effect of the nature of surface complexes and reaction affinity. Geochim et Cosmochim Acta 58(2):541–551

Choi JH, Anwar AHMF, Ichikawa Y (2008) Observation of time-dependent local deformation of crystalline rocks using a confocal laser scanning microscope. Int J Rock Mech Minig Sci 45:431–441

160 Analysis of Permeability Coefficient Along a Rough Fractures Using a Homogenization Method

Chae Byung-Gon, Choi Jung Hae, Seo Yong-Seok, and Woo Ik

Abstract

To compute a permeability coefficient along a rough fracture that takes into account the fracture geometries, this study performed detailed measurements of fracture roughness using a confocal laser scanning microscope, a quantitative analysis of roughness using a spectral analysis, and a homogenization analysis to calculate a permeability coefficient at the micro- and macro-scale. The homogenization analysis is a type of perturbation theory that characterizes the behavior of microscopically inhomogeneous material with a periodic boundary condition in microstructure. Therefore, it is possible to analyze accurate permeability characteristics that are represented by the local effect of the facture geometry. The C-permeability coefficients that are calculated using the homogenization analysis for each rough fracture model exhibit an irregular distribution and do not follow the relationship of the cubic law. This distribution suggests that the permeability characteristics strongly depend on the geometric conditions of fractures, such as the roughness and the aperture variation. The homogenization analysis may allow to produce more accurate results than the preexisting equations for calculating permeability.

Keywords

Permeability coefficient • Rough fracture • Confocal laser scanning microscope • Homogenization analysis • Multi scale

160.1 Introduction

It is important to evaluate a long-term safety of natural barrier for radioactive waste disposal in rock masses. Among the characteristics of natural barrier, fluid flow along the rock fractures is one of the most important factors to migrate the nuclide. Therefore, it is necessary to evaluate the permeability of rocks, especially permeability along rock fractures. Many researches that take the fracture geometries into consideration have contributed to the ability to characterize the fluid flow in a fractured rock mass. Initially, a parallel plate model with an impermeable matrix was used (Kranz et al. 1979; Renshaw 1986; Snow 1965). This model is based on a cubic law that implies that the flow rate is proportional to the cube of the fracture aperture. A model composed of a rough fracture's surfaces with an impermeable matrix has also been suggested (Renshaw 1986; Brown 1987; Cook et al. 1990; Neuzil and Tracy 1981; Pyrak-Nolte et al. 1988; Witherspoon et al. 1980; Zimmerman et al. 1992). This model takes into consideration the partial contact areas along the fracture walls that may result from roughness of the fracture. Recently, a realistic fracture model has been suggested that uses the real geometry of a fracture that simulates the

C. Byung-Gon (✉) · C.J. Hae
Korea Institute of Geoscience and Mineral Resources, 124, Gwahang-no, Yuseong-Gu, Daejeon, 305-350 Korea
e-mail: bgchae@kigam.re.kr

S. Yong-Seok
Chungbuk National University, 52, Naesudong-no, Heundeok-Gu, Cheongju, 361-763 Korea

W. Ik
Kunsan National University, 558, Daehangno, Kunsan, 573-701 Korea

G. Lollino et al. (eds.), *Engineering Geology for Society and Territory – Volume 6*,
DOI: 10.1007/978-3-319-09060-3_160, © Springer International Publishing Switzerland 2015

permeability characteristics (Kranz et al. 1979; Cook et al. 1990; Olsson 1992; Raven and Gale 1985; Trimmer et al. 1980; Zimmerman and Bodvarsson 1996). The results of the previous studies that were mentioned above have shown that the permeability characteristics change irregularly because of the fracture geometries. Particularly in a rough rock fracture, the fluid flow occurs as a selective flow, such as a channel flow that is dependent on the fracture roughness. Therefore, a numerical analysis of the permeability characteristics should include the fracture geometry when possible.

The purpose of this study was to calculate the permeability of a single fracture while taking the true fracture geometry into consideration. The fracture roughness was measured using a confocal laser scanning microscope (CLSM). These data were used to reconstruct a fracture model for numerical analysis using a homogenization analysis (HA) method. The HA method is a perturbation theory that was developed to characterize the behavior of a microscopically inhomogeneous material that involves periodic microstructures. The HA permeability coefficient was calculated based on the local fracture geometry and the local material properties (the water viscosity).

160.2 Analysis of the Fracture Roughness

The specimens in the study were composed of Jurassic coarse-grained granite. These specimens were collected from the drilled cores in the Iksan area in the mid-western part of Korea. The six core specimens, which contained a single, fresh fracture that was parallel or subparallel to the long axis of the core, were studied in detail. They were referred to as GRA, GRB, GRC, GRD, GRE, and GRF and were 5.5 × 11.0 cm in size (Fig. 160.1).

Fig. 160.1 Core specimens used for the measurements of roughness

The fracture roughness was measured using a confocal laser scanning microscope (CLSM; Olympus OLS 1,100) to collect high resolution digital data on the surface roughness. The CLSM has a high level of resolution and contrast in the direction of the light axis because of its confocal optics. In this study, the sampling spacing was 2.5 μm for the x and y directions. To collect the 2-D roughness data, one scan line was placed on the fracture's surface along the long axis of the fracture. The resolution for the z-direction was 10 μm in the study because the difference in the roughness from the highest peak to the lowest peak was large.

After the Fourier spectral analysis and the noise reduction procedures were done for all of the data for each specimen, a reconstruction procedure of the roughness geometry was performed using the influential frequencies of all of the components of frequency. The reconstructed roughness was in agreement with the measured roughness data. The reconstructed roughness features were used to develop the fracture models for the analyses of the fluid flow along a fracture (Fig. 160.2).

160.3 Computation of the Permeability Coefficients of the Rough Fracture Models Using a Homogenization Analysis

For the calculation of the permeability coefficients with the consideration of the detailed fracture geometries, this study introduced the homogenization analysis (HA) method. The HA method is a perturbation theory that was developed to characterize the behavior of a microscopically inhomogeneous material that involves periodic microstructures (Ichikawa et al. 1999; Sanchez-Palencia 1980). The HA permeability coefficient was calculated based on the local fracture geometry and the local material properties (the water viscosity). HA method considers flow characteristics in both micro scale and macro scale by introducing the incompressible Navier-Stokes equation. The method calculates characteristic velocity and characteristic pressure in a unit cell using a micro scale equation, and then, calculates HA-permeability coefficient. It draws a macro scale equation, namely HA-flow equation, by averaged material velocity and HA-permeability coefficient.

The C-permeability coefficients were computed using several rough fracture models. The important objective of the homogenization analysis was to understand the changes of the permeability characteristics that are dependent on the roughness patterns. The fracture models were constructed with a consideration of the fracture roughness that was analyzed using the fast Fourier transform.

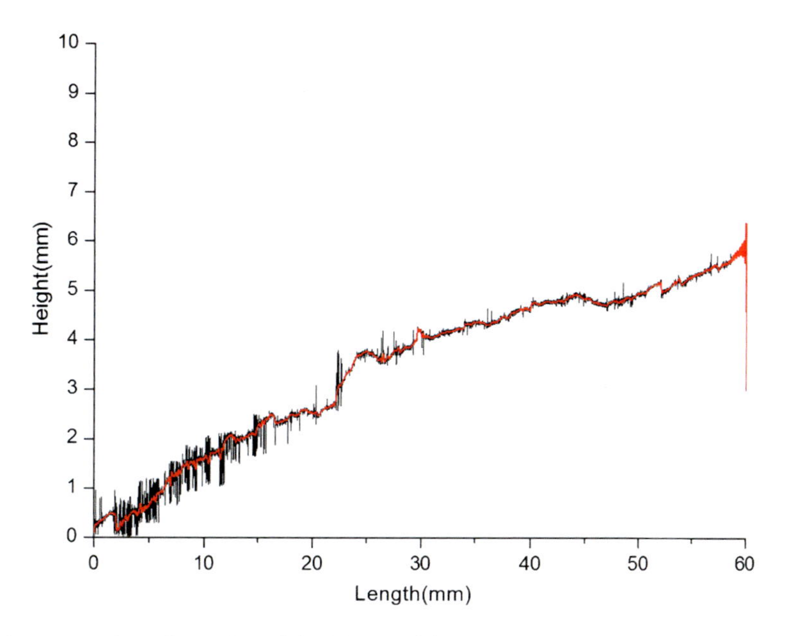

Fig. 160.2 Comparison of the measured roughness data of the *left* part of GRA (*black*) with the smoothed roughness data (*red*)

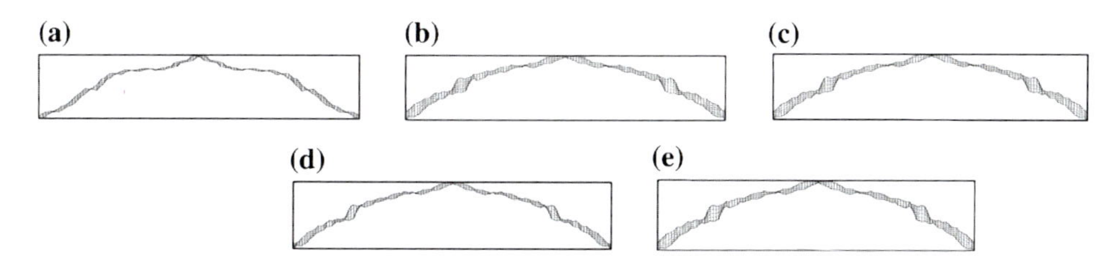

Fig. 160.3 Fracture models of GRA showing various roughness features and apertures at each stage. Exaggerated 100 times in the *vertical* direction. **a** 1st stage **b** 2nd stage, **c** 3rd stage, **d** 4th stage, and **e** 5th stage

The HA permeability characteristics were determined for various roughness conditions. For several types of roughness features that were measured using the CLSM, the upper fracture wall was displaced at intervals of 1 mm in the shearing direction. This shear displacement was introduced for five stages, which resulted in various aperture values along the fracture. The permeability coefficient was calculated at every stage of this displacement. The fracture models of each specimen are shown in Fig. 160.3.

The results of the C-permeability coefficients at each stage of the shear displacement are shown in Table 160.1. It was determined that the permeability coefficients were irregularly ranged from 10^{-4} to $10{-}^{1}$ cm/sec, whereas the coefficients of the previous parallel plate models were uniformly distributed within certain range. This difference of permeability between this study and the parallel plate models was due to the complicated change in the roughness and the aperture values, which increased the shear displacement in the current models.

Table 160.1 Permeability coefficients of the five stages of shear displacement for each specimen

Shear disp	C-permeability coefficient (cm/sec)					
(mm)	GRA	GRB	GRC	GRD	GRE	GRF
1.0	7.46E-03	1.67E-03	1.68E-01	1.43E-01	5.05E-01	5.31E-03
2.0	2.28E-03	2.00E-03	4.67E-04	4.45E-01	2.35E-02	5.31E-03
3.0	7.18E-03	7.79E-03	2.32E-04	7.60E-02	1.16E-01	4.16E-04
4.0	1.79E-02	1.79E-03	1.91E-04	1.48E-03	5.09E-02	3.22E-03
5.0	1.55E-01	1.67E-03	1.83E-04	5.94E-04	6.83E-03	1.76E-03

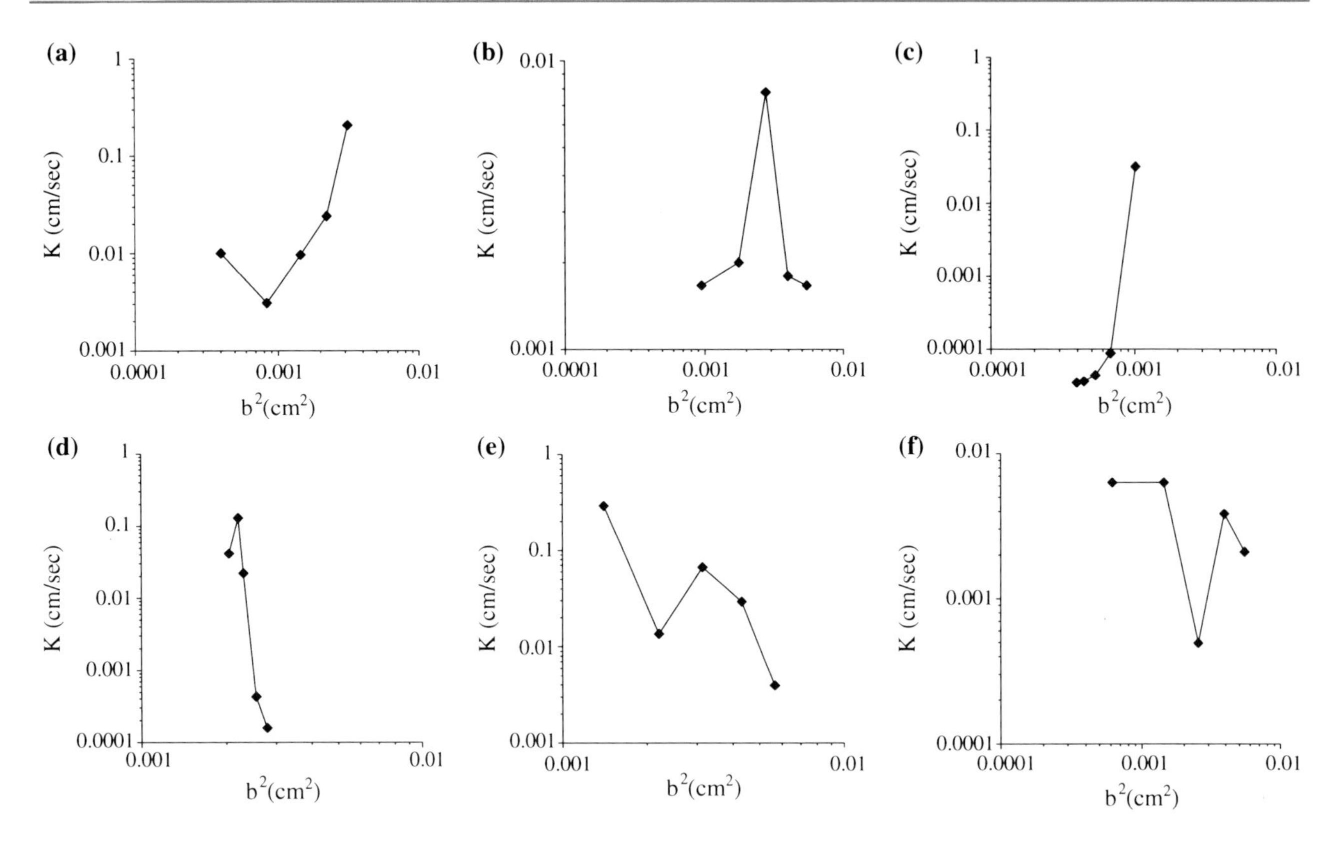

Fig. 160.4 Relationship between the C-permeability coefficients and the aperture square. **a** GRA, **b** GRB, **c** GRC, **d** GRD, **e** GRE, and **f** GRF

The relationships between the square of the mean aperture, *b*2, and the calculated permeability are shown in Fig. 160.4. The nonlinear relationship indicates that the cubic law was not suitable for the rough fracture case.

160.4 Conclusion

The fracture roughness was measured in detail using the CLSM in this study. The measured data were analyzed to identify the effective geometrical characteristics using a Fourier spectral analysis and low pass filtering. To characterize the hydraulic conductivity, the permeability along a single fracture was computed using the HA method and the measured fracture geometry data. Taking into consideration the change of the roughness pattern along a fracture, the permeability coefficient was calculated using the rough fracture models. The calculation results revealed various changes in the permeability that depended on the roughness patterns and the aperture values. The irregular distribution of the roughness and the aperture along a fracture may introduce a negative proportional relationship between the aperture and the permeability coefficient, even though the mean aperture becomes larger. This relationship clearly demonstrates that the permeability characteristics are very sensitive to the geometry of the roughness and the aperture.

On the basis of the above results, the homogenization analysis method is suggested as an appropriate numerical method for the analysis of fracture permeability. Because the HA method considers the fracture geometries on the micro- and macro-scale, this method can be used to understand the permeability characteristics, such as the local effect and the overall influence, of a fracture. The understanding of the permeability characteristics of each rock fracture by the HA method can contribute to the evaluation of overall hydraulic conductivity of rock masses and fracture networks in deep disposal of radioactive wastes.

Acknowledgment This work was supported by the Radioactive Waste Management of the Korea Institute of Energy Technology Evaluation and Planning (KETEP) grant funded by the Korea Government Ministry of Knowledge Economy (2012171020001B).

References

Brown S (1987) Fluid flow through rock joints; the effect of surface roughness. J Geophys Res 92:1337–1347

Cook A et al (1990) The effects of tortuosity on flow through a natural fracture. In: Proceedings of the 31st US symposium on Rock mechanics. pp 371–378

Ichikawa Y et al (1999) Unified molecular dynamics and homogenization analysis for bentonite behavior: current results and future possibilities. Eng Geol 54:21–31

Kranz R, Frankel A, Engelder T (1979) The permeability of whole and jointed Barre Granite. EOS Trans AGU 58:1229

Neuzil C, Tracy J (1981) Flow through fractures. Water Resour Res 17:191–199

Olsson W (1992) The effect of slip on the flow of fluid through a fracture. Geophys Res Lett 19:541–543

Pyrak-Nolte L, Cook N, Nolte D (1988) Fluid percolation through single fractures. Geophys Res Lett 15:1247–1250

Raven K, Gale J (1985) Water flow in a natural rock fracture as a function of stress and sample size. Int J Rock Mech Min Sci Geomech Abstr 22:251–261

Renshaw C (1986) On the relationship between mechanical and hydraulic apertures in rough-walled fractures. J Geophys Res 100:24629–24636

Sanchez-Palencia E (1980) Non-homogeneous media and vibration theory. Springer-Verlag, Berlin

Snow D (1965) A parallel plate model of fractured permeable media. PhD thesis, University of Califonia, Berkeley, USA

Trimmer D et al (1980) Effect of pressure and stress on water transport in intact and fractured gabbro and granite. J Geophys Res 85:7059–7071

Witherspoon P et al (1980) Validity of cubic law for fluid flow in a deformable rock fracture. Water Resour Res 16:1016–1024

Zimmerman R, Bodvarsson G (1996) Hydraulic conductivity of rock fractures. Transp Porous Media 23:1–30

Zimmerman R, Chen D, Cook N (1992) The effect of contact area on the permeability of fractures. J Hydrol 139:79–96

161 In Situ Quantification of Hydrocarbon in an Underground Facility in Tight Salt Rock

Benjamin Paul, Hua Shao, Jürgen Hesser, and Christian Lege

Abstract

Within the framework of hydraulic characterisation of deep geological formations for the disposal of high level radioactive waste, a long-term investigation program has been carried out to quantify local hydrocarbon occurrences (HCO) regarding the exploration of tight salt rock. In twenty boreholes, each with a length of 6 m and a diameter of 46 mm, the volume of the exposed hydrocarbons (HC) and the influenced area in the rock mass have been determined. These boreholes were plugged with mechanical single packers. For a few years the pressure development in these boreholes has been recorded automatically and the amount of HC flowed into the test-interval was measured. The pressure development within the twenty boreholes differs significantly in dependence on the borehole locations. In terms of pressure increase, four groups can be distinguished: almost no, low, intensive and very intensive. Based on the equation of state for ideal gas and considering the gas composition as determined by laboratory analysis, the calculated gas amount agrees very well with the measured data. The chemical analysis of exposed gas and fluid phase hasn't shown significantly changes of the chemical composition in time. The evaluation of long-term monitored pressure development and the chemical analysis provide a better understanding concerning the distribution of HC in tight sedimentary rock mass.

Keywords

HLW disposal • Salt rock • Hydraulic characterisation • Quantification • Hydrocarbon

161.1 Introduction

A repository for high level radioactive waste has to encase the radioactive material for a long time to prevent contact with fluids and a release of any radioactive material. The prediction for a safe deposition of high level radioactive waste in an underground repository for about 1 million years requires detailed knowledge of chemical and physical properties of the host rock. One of the most significant issue is ascribed to the hydraulic integrity in dependence of mechanical, chemical and thermal influences. Another important issue, which may affect the safety and the assessment of the suitability, especially related to the disposal of heat producing radioactive waste, is the occurrence of HC. HC were counted to the natural components of salt rock (Bornemann et al. 2008; Gerling et al. 1991). If HC are detected, their distribution and possible interconnectivities have to be determined and the volume of these occurrences has to be estimated with regard to the long-term performance of the potential repository.

161.2 Investigation Layout and Instrumentation

A long-term investigation program concerning the volume of HCO and potential interconnectivities has been carried out in a salt dome in northern Germany in a depth of about

B. Paul (✉) · H. Shao · J. Hesser · C. Lege
Federel Institute for Geosciences and Natural Resources (BGR), Stilleweg 2, 30655 Hannover, Germany
e-mail: benjamin.paul@bgr.de

G. Lollino et al. (eds.), *Engineering Geology for Society and Territory – Volume 6*,
DOI: 10.1007/978-3-319-09060-3_161, © Springer International Publishing Switzerland 2015

840 m bsl. The HC appear in the "Staßfurt" series (z2) almost exclusively within the "Knäuelsalz" (z2HS1). To quantify the HCO in the near field of galleries 20 boreholes, each with a length of 6 m and a diameter of 46 mm, were drilled and afterwards plugged with mechanical single packers to measure the pressure development and the inflow volume of gaseous and liquid HC. The boreholes were drilled at locations where HC were visible either with common artificial light or with ultraviolet light. For reference purpose few additional boreholes were drilled and instrumented in areas were no HC were detected. The (detailed) localisation of the measuring boreholes in HC impregnated areas is based on the mapping results of Amelung and Schubert (2000) and an additional visual survey. Additional results from core analysis of adjacent boreholes were considered. The HCO were classified regarding the intensity of fluorescence (Amelung and Schubert 2000). The boreholes should engage an area with a high amount of HC but a leakage through the Excavation Damaged Zone (EDZ) should be avoided. To achieve this, the single packers were set in a borehole depth of about 3 m. Depending on inclination and orientation of the boreholes the minimal distance between packer and side wall is about 2–3 m. All boreholes were drilled upwards in order that gravitation could be used for sampling. Two pipes with an inner diameter of 4 mm enable the separate sampling of the liquid and the gaseous phase of the HC. The chemical composition of the liquid and gaseous phase has been determined in regular intervals.

161.2.1 Equipment

Due to the investigation of HC, the used equipment is explosion protected. The measuring system consists of a mechanical single packer system, pressure sensors, a measuring and control system (MCC) as well as pipes for sampling and cables (Fig. 161.1). The volume of each test interval has been reduced using a dummy made of high quality polytetraflourethylene (PTFE) in order to detect minor pressures.

161.2.2 Realisation of the Measurements

After drilling the packer system was installed immediately to reduce the primarily degassing of the HCO. In the first time after installation every 30 s pressure values were recorded automatically. Depending on the pressure development the frequency was adapted for the specific borehole conditions. Due to safety reasons the maximum pressure build up has been restricted to 1 MPa by the local mining authority. The MCC has the possibility to send an alert if the pressure comes close to the maximum pressure. When the alert has been triggered, staff from the operating company got samples of liquid and gaseous HC and reduced the pressure in the test-interval. The amount of the sampled gases and liquids were measured and documented periodically. In some cases, especially shortly after the installation, some boreholes had to be opened during the weekend because of the rapid pressure increase and the lack of manpower. The loss of gas, however, was approximated based on the measured gas flow rate before and after opening.

161.3 Data

The 20 boreholes cover a wide range of different HCO. From clearly under common artificial light visible HC over slightly only under ultraviolet light visible HC to locations without HC (referential boreholes) all transitions are included in the investigation program. Due to pressure development and amount of HC, four main categories can be distinguished:

Category 1: almost no pressure build up (<0.03 MPa; no venting needed)

Category 2: low pressure build up (<1 MPa; no venting needed)

Category 3: intensive pressure build up (initially frequent pressure build up; venting needed)

Category 4: very intensive pressure build up (permanent pressure build up; venting needed)

Eleven boreholes, more than a half of all boreholes, belong to category 1, with almost no pressure build up. Four of these boreholes are reference boreholes, which were drilled in areas where no HC were suggested. Two boreholes show a low pressure build up and can be allocated to category 2. One of them was drilled to determine the pressure development of a HCO which is connected to a thin anhydrite layer. Four boreholes belong to category 3 with an intensive pressure build up. Only three boreholes have an extraordinary position because of their very intensive pressure build up and the amount of inflowing gaseous and liquid HC (Table 161.1).

The two boreholes 6 and 6a were drilled to investigate the extent of a visible HCO and the interconnectivity between these boreholes. The two boreholes cross each other with a minimal distance of about 0.6 m. Borehole 6 showed a very intensive and long lasting pressure build up with an amount of more than 2,000 l under standard conditions, while borehole 6a showed no evidence of HC occurrences at all.

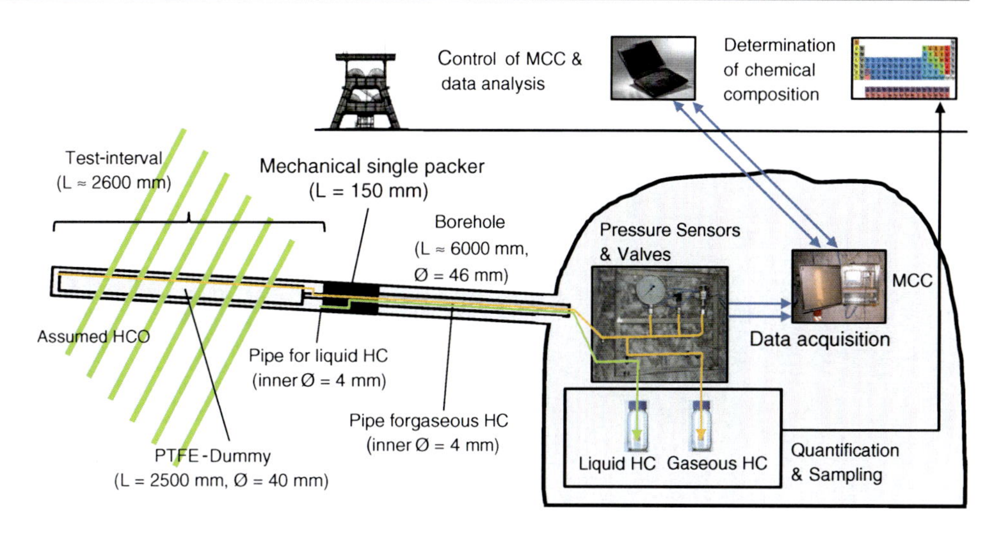

Fig. 161.1 Schematic sketch of the measurement setup

Table 161.1 Summary of boreholes without category 1, except borehole 6a as at December 2012

Borehole	Period (d)	Total sampled gas (stdl)	Total sampled liquids (ml)	Total samples gaseous/ liquid	Pressure build up	Category
1	537	121.0	1,350	14/6	Intensive	3
2	514	1,331.7	1,883	99/69	Very intensive	4
3	517	287.3	1,443	19/15	Very intensive	3
4	516	80.4	93	9/5	Very intensive	3
5	515	3,271.6	16,015	176/159	Very intensive	4
6	511	2,383.3	1,664	150/112	Very intensive	4
6a	510	–	–	–	Almost no	1
7	252	148.2	575	9/4	Intensive	3
8	314	3.7	–	1/0	Low	2
9	70	–	–		Low	2

161.3.1 Data Analysis

Based on the measured pressure development the volume of the inflowing gas was approximated by using the mass balance equation. The inflowing gas is assumed to be an ideal gas so that the following state equation is valid:

$$m = P * \frac{V}{R_S * T} = P * \frac{V}{(\frac{R}{M} * T)} \tag{161.1}$$

P interval pressure (Pa)
R_s spec. gas constant (J/kg K)
M molecular weight (kg/mol)
V interval volume (m^3)
R universal gas constant 8.314472 (J/kg K)
T absolute temperature (K)

Under isothermal conditions the mass flow can be calculated with:

$$m = \frac{\partial m}{\partial t} = \frac{V}{R_S * T} * \frac{\partial P}{\partial t} \tag{161.2}$$

The specific gas constant depends on the gas composition. A synthetic gas constant value was calculated according to a composition of about 40 % methane, 40 % nitrogen and 20 % other gases (carbon dioxide, propane and oxygen), given by the chemical analysis of samples. The measured and calculated gas volume correlates very well. Figure 161.2 shows exemplarily the correlation of the calculated and measured gas volume of borehole 2.

For calculation purposes the gas volume at the beginning was set to a constant volume between 1.4 and 2 l in dependence on the pipe length and the installation depth of the packer. This volume should be reduced by the inflow of liquid HC during measuring. The reduction has been considered in the calculation in the way that the inflowed volume of liquid HC is subtracted from the entire volume. The inflow of

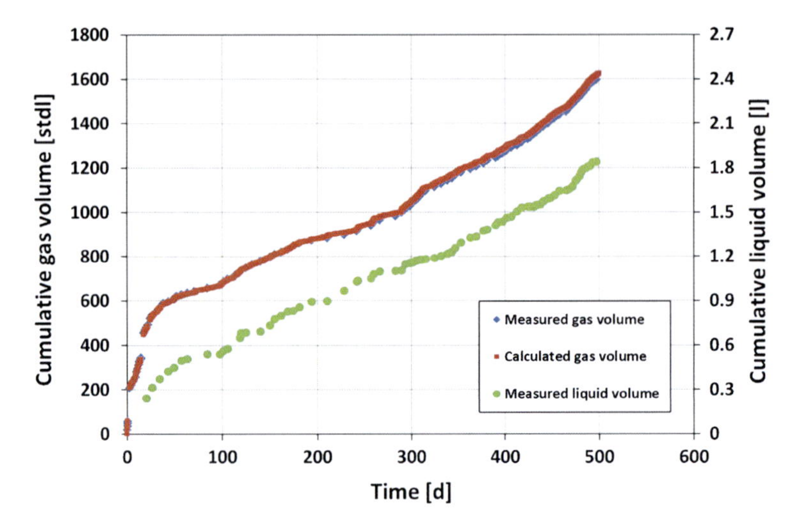

Fig. 161.2 Correlation of calculated and measured gas volume on the basis of data of borehole 2

liquids is assumed to be nearly linear within a sampling interval. By a known mass flow the volume of gas under normal conditions can be calculated at a certain time with:

$$V_t = m * (t - t_1)/\rho_{syn} \quad (161.3)$$

t point of time of last pressure determination

ρ_{syn} synthetic density of gases

Hence the cumulative gas volume can be calculated by summarising V_t.

161.4 Conclusions

Generally, liquid and gaseous HC occurred simultaneously, based on the data measured up to now. If a drilling encounters a HCO, gaseous HC were released due to pressure decrease. To determine the amount of gaseous and liquid HC the inflowing HC were measured separately. These measured data were used to verify the calculations, which are based on the pressure data. The excellent accordance of calculated and measured results confirms the suitability of the presented calculation method for determination of inflowing HC in relation to the measured pressure development. Because of the influence of the liquid phase on the entire volume, a measurement of the liquid phase is still necessary. The investigation showed that adjacent boreholes, even within a distance of less than 1 m to each other can behave completely different. Reason for that is the very low permeability in tight salt rock beyond the excavation disturbed zone, which prevents the mobilisation of HC. The estimation of a hydraulic relation between two boreholes depending on category 4 is not yet possible because of the continuing intensive pressure development.

References

Amelung P, Schubert J (2000) Dokumentation der Kondensatvorkommen im Hauptsalz der Staßfurt-Folge, unpubl. report, DBE, Gorleben/Peine

Bornemann O, Behlau J, Fischbeck R, Hammer J, Jaritz W, Keller S, Mingerzahn G, Schramm M (2008) Description of the Gorleben site Part 3: results of the geological surface and underground exploration of the salt formation structure. BGR, Hannover

Gerling P, Beer W, Bornemann O (1991) Gasförmige Kohlenwasserstoffe in Evaporiten des deutschen Zechsteins. Kali und Steinsalz 11:376–383

162 Relationship Between the Fractal Dimension and the Rock Mass Classification Parameters in the Bátaapáti Radioactive Waste Repository

Rita Kamera, Balázs Vásárhelyi, László Kovács, and Tivadar M. Tóth

Abstract

A fractured rock mass is divided by the fracture sets into mixed shapes of blocks and wedges. The geotechnical parameters of the rock mass are obviously dominated by the presence and the spatial distribution of fracture sets. In rock engineering, the determination of geotechnical parameters of rock masses is a key issue for the prediction of expected behaviour of the fractured rock mass. Several measurements and empirical methods can be used to investigate the characteristics of the fractures including its spacing and orientation. The goal of this paper is to find the possible connection between the calculated fractal dimension (D) and the derived values originating from several rock mass classification systems (e.g. RQD, RMR, Q, GSI). Principally, the derivation of fractal dimensions and the rock mass classification systems are based on empirical methods. For our present investigation the fractures visible in scores of pictures taken by automatic photo robot of the tunnel faces were used at the low and intermediate radioactive waste repository located at the Bátaapáti site in Southern Hungary. The BENOIT software was used to calculate the single fractal dimension according to the box-dimension methodology and the rock mass classification values were evaluated with the JointMetriX3D (ShapeMetriX3D) software. Our results show that some connection between geotechnical parameters and fractal dimension values can be obtained by the applied methodology. Using these empirical relationships, it is possible to carry out more safety construction and calculations in the radioactive waste repository in Bátaapáti.

Keywords

Fracture network • Damage model • Fractal • Radioactive waste • Depository

162.1 Introduction

It is a basic importance to get reliable information of fracture networks of different geological formations, in order to understand fractured hydrocarbon reservoirs, to optimize radioactive waste depository or even to exploit geothermal energy. Although, there are numerous simulations methods for fracture network modeling, input data calculations are still completely depend on subjective decisions.

R. Kamera (✉)
Golder Associates Ltd., Budapest, Hungary
e-mail: Rita_Kamera@golder.com

B. Vásárhelyi
Department of Structural Engineering, University of Pécs, Pécs, Hungary

L. Kovács
Rockstudy Ltd., Pécs, Hungary

T.M. Tóth
Department of Mineralogy, Geochemistry and Petrology, The Faculty of Science and Informatics, University of Szeged, Szeged, Hungary

G. Lollino et al. (eds.), *Engineering Geology for Society and Territory – Volume 6*,
DOI: 10.1007/978-3-319-09060-3_162, © Springer International Publishing Switzerland 2015

The aim of this study is to show the relationship between the fractal dimension and the geotechnical parameter based on the Bátaapáti low-, and intermediate radioactive waste depository of Hungary.

162.2 Study Area

The investigated area is situated in the Mórágy granite body (Fig. 162.1), which is the host rock of the radioactive waste disposal in Hungary. A significant part of the Carboniferous intrusion (Buda et al. 1999) is covered by young sediments and only a small portion of it is avaliable for outcrop survey. Petrographically, the igneous body is composed of diverse granitoid subtypes, such as monzonite and monzogranite, which due to a polymetamorphic evolution under greenschist facies conditions, exhibit a slightly foliated structure (Király and Koroknai 2004). During the subsequent post-metamorphic deformation events, a mutual fracture network developed. Based on evaluation of the BHTV data of over 60.000 single fractures representing 20 wells, two main groups of faults can be emphasized. One cluster shows very high dips (70–80°), with a NW-SE strike; whilst another shows a strike of ENE-WSW and a similar dip (Maros et al. 2004).

162.3 Method

162.3.1 Fracture Network Digitalization

Due to different physical circumstances rock bodies are deformed and complex fault systems may be developed. The rocks brake along defined planes and the fracture network is set up from these discrete fractures (Fig. 162.2).

In the first part of the investigation the geometric parameter such as the fractal dimension values and in the second step the different rock mass classification systems (i.e. RQD, RMR, GSI, Q) values were developed. During this study 140 samples were analyzed to determine the parameters using the JointmetriX3D code.

During the digitization there may be some problems with identification of the fractures:

- Breccia zone: Due to the intensive deformation, breccia zones of different-scale may develop along the shear plane. Discrete fractures can be hardly identified in these zones that consist of angular rock pieces. In this case only those fractures should be considered, which can be defined unambiguously;
- Fracture wall: In those cases when fractures are almost parallel with the image plane they cannot be taken into consideration;

Fig. 162.1 Geographical position of the research area

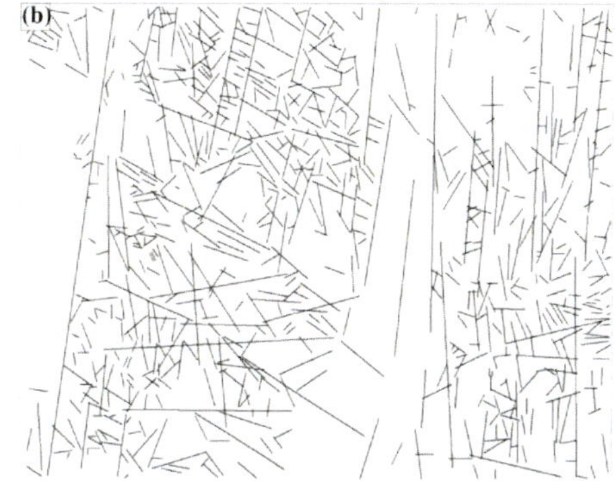

Fig. 162.2 **a** A tunnel end photo from Bátaapáti site; **b** The digitized fractures

- Numerous fractures: A fracture network consists of strongly and less fragmented zones. The determination of discrete fractures can be difficult inside those intensively fragmented zones. In this case only those fractures should be digitized that can be identified unambiguously at the magnification in question;
- Sheared fractures: Through subsequent tectonic events any discrete fracture may be sheared. Portions of a sheared fracture are interpreted as two individual fractures; and
- Unfinished and covered with each other fractures were ignored during the digitalization process (La Pointe et al. 1993).

Unfinished and covered with each other fractures were ignored during the digitalization process (La Pointe et al. 1993).

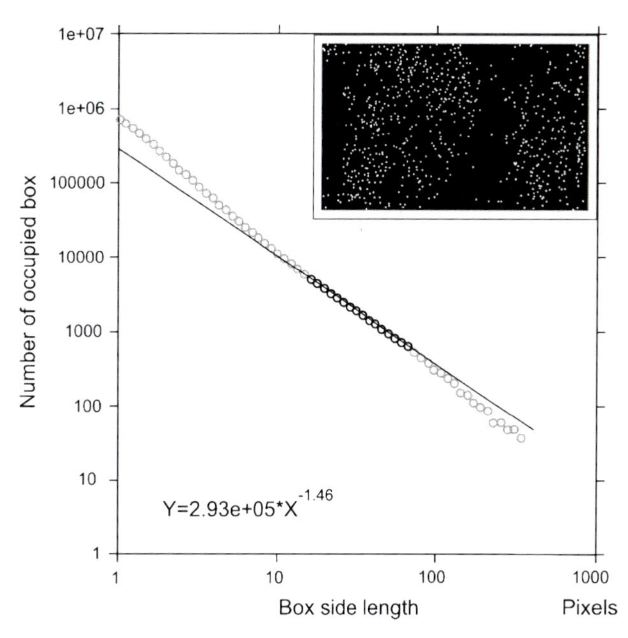

Fig. 162.3 The fractal dimension calculation based on the midpoints of the digitized fractures

162.3.2 Fractal Dimension Calculation

In the second step of the evaluation was the calculation of the fractal dimension (D) using the box-counting method (Mandelbrot 1985). This approach is based on the fracture midpoints distribution. The box dimension relates to the result of counting the box number (N) of size r required to cover the fracture pattern. This value can be calculated with the following algorithm:

$$N(r) \sim r^{-D} \qquad (162.1)$$

A black-and-white image was used as an input file that shows the midpoints distribution of the fractures. A point diagram was generated with the "box-counting" method, from which the dimension value was calculated by using linear regression slope. The two sections of the line after and before the breakpoints were deleted to determine the D value. Before the first breakpoint, the size of the points was calculated instead of the complexity of the pattern. After the second breakpoint the values are not representative because the box sizes are too large. The fractal dimension of the fracture pattern was generated from the middle section of the linear regression line (Fig. 162.3).

162.4 Rock Mass Characterization as Damage

We relate the empirical fracture and disturbance measures of rock mechanics (RQD, Q, RMR, GSI, etc.) to a damage measure D. According to the physical interpretation of damage, the value $D = 0$ characterizes the intact rock and $D = D_{cr}$ stands for the fractured rock mass at the edge of failure. As the rock mass quality measures are zero at maximal possible damage and are one at the undamaged state, we suggest the simplest linear relationship interpreting them as damage measures

$$D_{RM} = 100\left(1 - \frac{D}{D_{cr}}\right) \text{ therefore } D = D_{cr}\left(1 - \frac{D_{RM}}{100}\right). \qquad (2)$$

Here $D_{\rm RM}$ is the damage of the rock mass. RM would be one of the rock mass classification systems (RQD, RMR or GSI value, i.e. between 0 and 100). As a simplification, hereafter we assume that $D_{cr} = 1$. In our case this is not a restriction because we do not associate a direct physical meaning (e.g. crack density, fractal dimension of the crack system) to damage and we accept the normalization and measurement methods of the rock mass quality measures as a proper characterization.

We assume that:

- The mechanical parameters of the rock mass in case of $D = 0$ are equal to the intact rock parameters
- The relationships can be modeled by an empirical function.

162.4.1 Damage Variable and Deformation of Rock Mass

We can transform the functions above, introducing the damage as an independent variable, into the form

$$\frac{E_{rm}}{E_i} = \exp(-AD). \quad (3)$$

The values of the material parameter A corresponding to the published equations are between 2.7 and 4.2 according to the results of Ván and Vásárhelyi (2010).

162.4.2 Damage Variable and Unconfined Compressive Strength of the Rock Mass

Similarly to the deformation module of the rock mass, were recalculate the empirical equations of the unconfirmed compressive strength (σ_{cm}) for the following form, using the compressive strength of the intact rock (σ_i):

$$\frac{\sigma_{cm}}{\sigma_c} = \exp(-BD). \quad (4)$$

The average value of B is 5.43 (between 4.17 and 6.56) and 8.61 (7.65–9.84) in case of undisturbed and disturbed rock mass, respectively (Ván and Vásárhelyi 2010). The purpose of the paper was to present an explicit mathematical relationship between fractal dimensions and rock mass classification parameters. As a result of our investigations, no mathematical formula could be deduced which describes the inherent connection between these parameters with adequate correlation coefficient. Although this report does not necessarily represent any connection between the applied data, it can stimulate some interests and thoughts for further investigations in association with this problem.

References

Buda Gy, Lovas Gy, Klötzli U, Cousen BI (1999) Variscan granitoids of the Mórágy Hills, South Hungary. Beihefte zur European J Miner 11(2):21–32

Király E, Koroknai B (2004) The magmatic and metamorphic evolution of the north-eastern part of the Mórágy Block. Annu Rep Geol Inst Hung 2003:299–310

La Pointe P, Wallman PC, Derschowitz WS (1993) Stochasticestimation of fracture size from simulated sampling. Int J Rock Mech Mining Sci Geomech Abs 30:1611–1617

Mandelbrot BB (1985) Self-affine fractal dimension. Phys Scr 32:257–260

Maros G, Koroknai B, Palotás K, Fodor L, Dudko A, Forián-Szabó M, Zilahi-Sebess L, Bán-György E (2004) Tectoic analysis and structural evolution of the north-eastern Mórágy Block. Annu Rep Geol Inst Hung 2003:371–380

Min KB, Jing L, Stephansson O (2004) Determining the equivalent permeability tensor for fractured rock masses using a stochastic REV approach: method and application to the field data from Sellafield, UK. Hydrogeol J 12(5):497–510

Ván P, Vásárhelyi B (2010) Relation of rock mass characterization and damage. In: Vrkljan I, Balkema (ed) Rock Engineering in Difficult Ground Conditions (Soft Rock and Karst), pp 399–404

Direct Shear Strength Test on Opalinus Clay, a Possible Host Rock for Radioactive Waste

163

Buocz Ildikó, Török Ákos, Zhao Jian, and Rozgonyi-Boissinot Nikoletta

Abstract

In the past decades the long term safe storage of highly active radioactive waste has been one of the greatest challenges for engineers and scientists to be solved. It is internationally agreed that the radioactive waste should be placed into deep geological formations, namely into granite, salt or clay stone. The latter type, in particular the Opalinus Clay stone, is the potential host rock for such repository in Switzerland. The aim of the paper is to broaden the knowledge on this rock type. Direct shear strength tests were carried out, to provide data on the maximal shear strength, friction coefficient and the connection between the plane of the bedding and the plane of the natural fault. The samples were derived from the major fault zone. The specimens were sheared along natural open joints under 1 MPa constant normal load (CNL) condition. Another set of experiments were carried out on intact samples disturbed by a soft clay vein. The shear plane was the same as the plane of the intrusion. On these specimens different normal stresses were applied, between 0.5 and 2 MPa. Apart from the maximal shear strength and the friction coefficient, the cohesion and the consolidation settlement (vertical displacement during the application of the initial normal stress, before the shear test started) were determined as well.

Keywords

Opalinus clay • Direct shear • Open joints • Clay vein

163.1 Introduction

The final placement of radioactive waste is a complex engineering task, which by today became an urgent problem to be solved. All around the world several laboratories carry out countless experiments on the host rocks of potentially highly active radioactive waste disposal sites—mainly clay stone, granite and salts. These experiments aim at gaining relevant information in order to prove that the facility is able to function safe for several hundreds and thousands of years. The Mont Terri Rock Laboratory, situated in the north-western part of Switzerland, has operated already for more than 15 years, and has performed more than 110 small- and large-scaled experiments on Opalinus Clay stone, out of which 43 are still running. Besides the Laboratory, several Universities and Institutes carry out further experiments on the Opalinus Clay, deriving from Mont Terri.

Direct shear strength tests were carried out at the Laboratory for Rock Mechanics of the École Politechnique Fédérale de Lausanne on Mont Terri rocks that were collected during the excavation of the FE Gallery in 2012. The

B. Ildikó (✉) · T. Ákos · R.-B. Nikoletta
Department of Construction Materials and Engineering Geology, Budapest University of Technology and Economics, Műegyetem rkp. 3, Budapest 1111, Hungary
e-mail: ildikobuocz@yahoo.com

Z. Jian
Department of Civil Engineering, Monash University, Building 60, Melbourne, VIC 3800 Australia

G. Lollino et al. (eds.), *Engineering Geology for Society and Territory – Volume 6*,
DOI: 10.1007/978-3-319-09060-3_163, © Springer International Publishing Switzerland 2015

investigated shaly facies Opalinus Clay rocks are homogeneous, barely visible laminated clay stones with low sand content, derived from the shear fault zone of the site.

The paper summarizes the first stage evaluation of the direct shear strength test results of these Oplainus Clay stones following the guidelines of the International Society for Rock Mechanics (1974). The second stage evaluation will take in consideration extra strength value influencing parameters, such as surface roughness, and the elevation of the shear plane.

163.2 Test Methodology

The test methodology as well as the evaluation of the data were both carried out according to the suggestions of the ISRM et al. (1974).

Two types of direct shear strength tests were performed along natural discontinuities of the Opalinus Clay stone. One was along open joints, where the bedding of the samples enclosed an angle with the shear plane, and the other was on closed fractures infilled with a softer clay material. The fracture was parallel to the bedding in all cases.

163.2.1 Shear Apparatus

The shear machine operated with an upper box fixed against movements in the shear direction, and a lower box moving with constant speed of 0.8 mm/min in the shear direction. The upper box could rotate around the point where the normal load was applied, and vertical movements were allowed as well. During the tests, displacements in vertical and horizontal directions were measured on the upper box by 3–3 linear variable differential transformers (LVDT), allowing to follow all the movements. One LVDT measured the shear displacement on the lower box (Fig. 163.1).

163.2.2 Test Procedure

Tests were carried out under constant normal load (CNL) conditions. Before the shear test started, a 1 MPa normal stress was applied for the samples with open joints. In the same way, for the healed samples four normal stresses with different magnitude were used, i.e. 0.5, 1.0, 1.5 and 2.0 MPa. For all tests, subsequent to the application of the normal stress, the consolidation settlement was measured. The samples had a size of 50 × 50 mm and, as the shear displacement cannot exceed the 10 % of the length of the sample according to the ISRM et al. (1974), the shear displacement was restricted to 5 mm.

163.3 Test Results and Evaluation

The two types of tests were evaluated in a different way. For the open joints the main purpose was to find a connection between the bedding plane and the plane of the natural joint. For the samples with clay-healed fractures, where the bedding plane and the plane of the fracture were parallel, the goal was to determine the cohesion and the rate of the consolidation by applying different normal loads.

163.3.1 Test Results for Open Joints

For a limited number of samples, the plane of the natural joints enclosed an angle with the bedding plane. This angle varied between 0 to 49°. Tests were carried out under 1 MPa CNL. The evaluation focused on samples where the angle was greater than 0 degrees and less than 50. Figure 163.2 shows that the shear stress after 4 mm shear displacement, i. e. in the residual range, increases with the increase of the angle.

Figure 163.3 shows the maximal shear strength values at 1 MPa normal stress. The average of the values is 0.48 MPa with a deviation of 0.1.

The friction coefficient ranges between 0.33 and 0.73, with an average value of 0.45.

163.3.2 Test Results for Closed Fractures

As introduced before, four magnitudes of normal stresses were applied on the specimens, i.e. 0.5, 1, 1.5, 2 MPa. The friction coefficient decreased and the maximal shear strength increased with the increase of the normal load. The vertical displacement (consolidation) was measured from the beginning of the normal loading until the start of the shear test. Table 163.1 gives the average values of the above mentioned properties for each normal stress.

The cohesion was determined by approximating the maximal shear strength—normal stress results with a linear approach. The value obtained for this parameter is 0.42 MPa.

163.4 Conclusion

Direct shear strength tests were carried out on two types of Mont Terri Opalinus Clay stone, i.e. with open joints and clay healed fractures.

Specimens with open joints did not always have their plane of discontinuity parallel to the bedding, therefore the effect of the magnitude of the angle on the test results was investigated. In the range of angles examined ($0° < \alpha < 50°$)

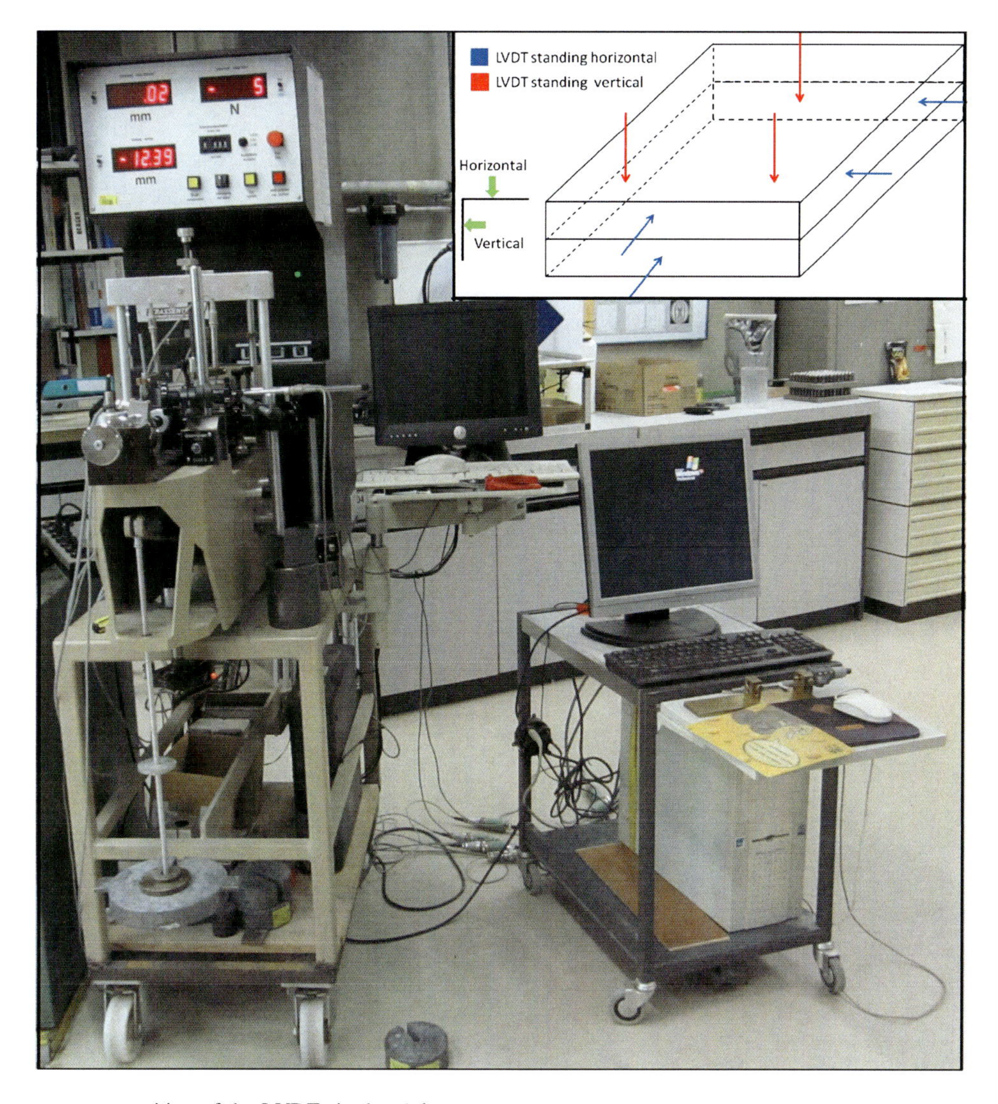

Fig. 163.1 Shear apparatus; position of the LVDTs in the *right upper corner*

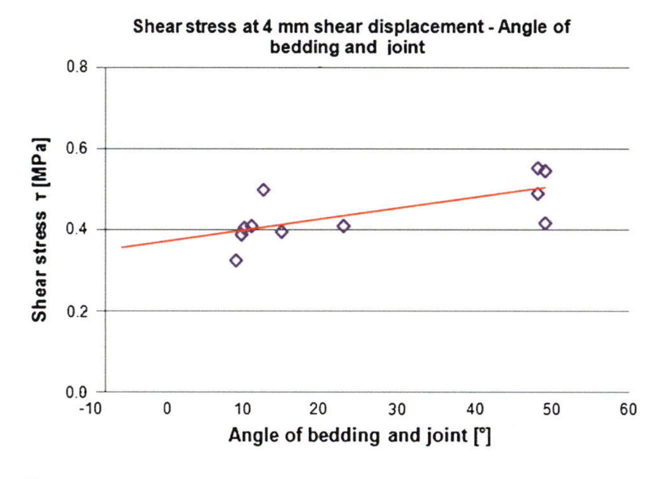

Fig. 163.2 Shear stress at 4 mm shear displacement for samples in which the plane of the bedding encloses an angle with the plane of the natural joint (angle greater than 0° and less than 50°)

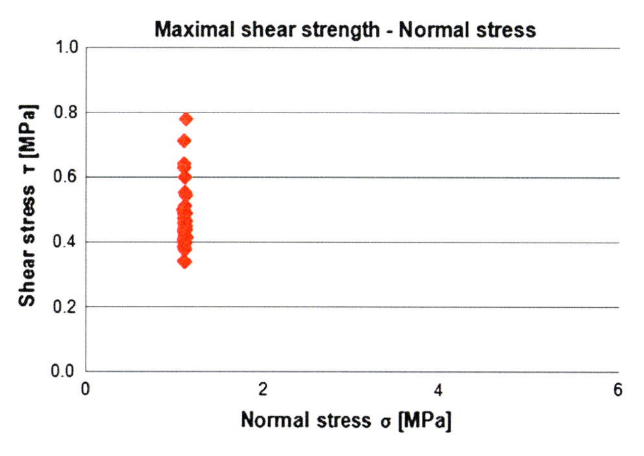

Fig. 163.3 Maximal shear strength—normal stress diagram

Table 163.1 Average values for maximal shear strength, friction coefficient and consolidation settlement for every normal stress condition

Normal stress (MPa)	Average maximal shear strength (MPa)	Average friction coefficient (–)	Average consolidation settlement (mm)
2.0	0.58	0.28	1.08
1.5	0.66	0.45	0.56
1.0	0.52	0.51	0.28
0.5	0.44	0.85	0.19

the greater the angle was, the higher the maximal shear strength became.

The average maximal shear strength of samples with open joints was 0.48 MPa, and the average friction coefficient 0.45.

The closed joints contained a softer clay vein that had a cohesion of 0.42 MPa. Four normal stresses were used during the experiments, the greater this stress was, the lower the friction coefficient became. The opposite effect was observed on the value of the maximal shear strength.

Acknowledgments The Authors would like to thank Nagra and Swisstopo, who allowed and helped in collecting samples from the Mont Terri Underground Test Laboratory, and the staff of the Laboratory for Rock Mechanics at EPFL. They would also like to thank the Department of Construction Materials and Engineering Geology at BUTE, for the support provided during the experiments, both under the technical and the scientific points of view. The financial support of SCIEX Grant no. 11.062 is highly appreciated.

Reference

ISRM, Franklin JA, Kanji MA, Herget G, és Ladanyi B, Drozd K, és Dvorak A, Egger P, Kutter H, és Rummel F, Rengers N, Nose M, Thiel K, Peres Rodrigues F, és Serafim JL, Bieniawski ZT, és Stacey TR, Muzas F, Gibson RE, és Hobbs NB, Coulson JH, Deere DU, Dodds RK, Dutro HB, Kuhn AK, és Underwood LB (1974) Suggested methods for determining shear strength. In: International Society for rock mechanics commission on standardization of laboratory and field tests, vol 1, pp 131–140

Significance of Joint Pattern on Modelling of a Drill and Blast Tunnel in Crystalline Rock

164

Dániel Borbély, Tamás Megyeri, and Péter Görög

Abstract

The first underground radioactive waste repository for low and intermediate level waste in Hungary is being built in the outskirts of the village of Bátaapáti. The total length of tunnels driven to date is over 5,200 m including two inclined access tunnels, the base tunnels and the first two emplacement chambers. The tunnels were driven in fractured granitic rocks. Based on the discontinuities the host rock of the repository can be considered as an assembly of blocks, therefore the discrete element modelling approach can be used to provide representative results of its behaviour. A hybrid continuum-discrete model is presented, where the near-field is modelled as a blocky rock mass, and the far-field is modelled as a continuum using the built in deformable blocks of the 3DEC software. Convergence monitoring was carried out in the modelled section that offers the ability to check the validity of analysis results. This paper focuses on the effect of joint pattern on the prediction capability of the discrete model. Different possible representations of the measured joint pattern were examined to assess the effect of joint pattern on the results, particularly on internal forces in rock support.

Keywords

Distinct element modelling • Joint pattern • Hybrid continuum-discrete modelling • Radioactive waste repository

164.1 Introduction

Design in fractured rock can be challenging especially, when the rock mass behaviour is governed by the block movement. The blocky nature of rock can be considered with distinct element modelling approach. The National Radioactive Waste Repository Project (NRWRP) is a good example to research the numerical modelling of blocky rocks. Due to the project radioactive safety first approach and its public recognition all the required date with good quality available for investigation. This paper is focuses on two fundamental issues about advanced numerical modelling of the repository. The first goal is the reduction or optimization of the computational effort of three dimensional modelling. The second aim is to assess the effect of joint pattern representation on results.

164.2 National Radioactive Waste Repository Project (NWRP)

The repository facilitates the low and medium activity radioactive waste of Nuclear Power plant of Hungary. The area of the repository is close to the village of Bátaapáti, at South Hungary that is part of the geological unit of Mórágy Block that is composed of granitic rocks (Kovács et al. 2012).

D. Borbély (✉) · P. Görög
Budapest University of Technology and Economics, Budapest, Hungary
e-mail: borbely.daniel@epito.bme.hudaniel.borbely@mottmac.comdaniborbely@gmail.com

P. Görög
e-mail: gorog.peter@epito.bme.hu

T. Megyeri
Mott Macdonald Hungary Ltd., Budapest, Hungary
e-mail: tamas.megyeri@mottmac.com

G. Lollino et al. (eds.), *Engineering Geology for Society and Territory – Volume 6*,
DOI: 10.1007/978-3-319-09060-3_164, © Springer International Publishing Switzerland 2015

Construction of the first two emplacement chambers was finished in 2012. The facility consists of tunnels, shafts, caverns, access roads and portals. The access tunnels arrive to the chambers at the reference base level of 240 m below ground level. The rock support in the investigated section was 150 mm thick steel fibre reinforced sprayed concrete, supplemented with systematically rock bolting in a 1.5 m × 1.0 m raster. Full-face excavation was used for the tunnel with diameter of app. 7 m, the length of the applied bolts is 3 m.

Regarding that the NWRP is in focus, carefully conducted, comprehensive geotechnical investigations were made. During the excavation face mapping was performed. The joint pattern applied in the paper is based on the face mapping. Systematic rock sampling and laboratory test were carried out during construction to determine the properties of intact rock and the rock mass (Kovács et al. 2012). The shear strength of the joints had been measured by laboratory tests (Buocz et al. 2010). The validity of the laboratory measurement was checked by plain strain distinct element model of a monitoring section (Horváth et al. 2012). Convergence measurement arrays were installed in 11 sections so far. In these sections relative displacements of the rock mass surrounding the excavation has been measured continuously in 6 radial directions.

164.3 Distinct Element Model, Representation of Rock Mass and Rock Support

The distinct element approach is most suitable for moderately fractured rock masses where the number of fractures is too large for the continuum-with-fracture-elements approach. The key concept of distinct element modelling (DEM) is that the domain of interest is treated as an assemblage of deformable blocks and the contacts among them need to be identified and continuously updated during the entire deformation process (Jing and Stephansson 2007). Three type of model with different purpose were used. First a continuum based finite element model was set up using Phase2 software to validate the properties of rock mass and rock support. Second type: small scale numerical tests were made to determine joint stiffness according the description of Jing and Stephansson (2007) using the three-dimensional distinct element code 3DEC 4.1. Several models with different joint stiffness were tested, and the one with the best fitting results to the validated continuum properties were selected. Using the input parameters applied in the finite elements model and the best fitted properties (determined in the small scale tests) a hybrid continuum-discrete model was built. In case of the hybrid model, the near-field is modelled as a blocky rock mass (using the joint properties and the intact rock properties), and the far-field is modelled as a continuum using the built in deformable blocks of the 3DEC with the rock mass properties. The resulted convergence and the internal forces were compared with field data. The prediction capabilities of using different joint patterns are compared in this study.

Mohr-Coulomb constitutive model with tension cut-off is assigned to the deformable blocks assuming that the intact rock is a linear elastic-perfectly plastic material. Plain strain distinct element modelling confirmed that the joints can be represented with Mohr-Coulomb constitutive model (Horváth et al. 2012).

Four different joint pattern (Pattern A to Pattern D) based on the same measurement were examined in the paper (Fig. 164.1). The same measurements are interpreted in different ways. In case of Pattern A, three joint sets from the measured six were selected. It is assumed that the joint persistence is 100 %, i.e. the joints are continues across the entire model. The joints are not follows a random distribution, but the spacing, dip and dip direction are a given as deterministic value. This is the simplest approach, but this kind of representation might be very helpful on the early stage of a project, when limited information about joint pattern is available. Pattern B follows the actual distribution of joints and all six joint sets are considered. The joint spacing is equal for all six join set and based on the average joint set of the rock mass.

Pattern C is examining the effect of spacing. It is the same as Pattern A but the spacing is 4 m instead of 2 m. In case of Pattern D, the persistence of the joints is 80 % and the spacing is 3.2 m, giving an average block volume equal to Pattern C. The longitudinal redistribution of stresses around the advancing tunnel face was considered with stress relaxation method. The relaxation factor (proportion of stress relief before any support is installed) was measured on site. Linear elastic behaviour of the tunnel lining is a valid assumption since no sign of plastic deformation of the lining was observed at the analysed section. The early age properties of the concrete were determined according to Chang and Stille (1993). The support provided by rock bolt is taken into account with cable elements (considering axial loads).

164.4 Discussion of the Results

According to the Phase2 models, the displacement shows good agreement with the field measurements. The internal forces in the bolts and the liner are similar with the expectations. The liner is below the plastic limit in the model and it is in accordance with the field observation. Hence the set of rock mass parameters are considered valid for the given section.

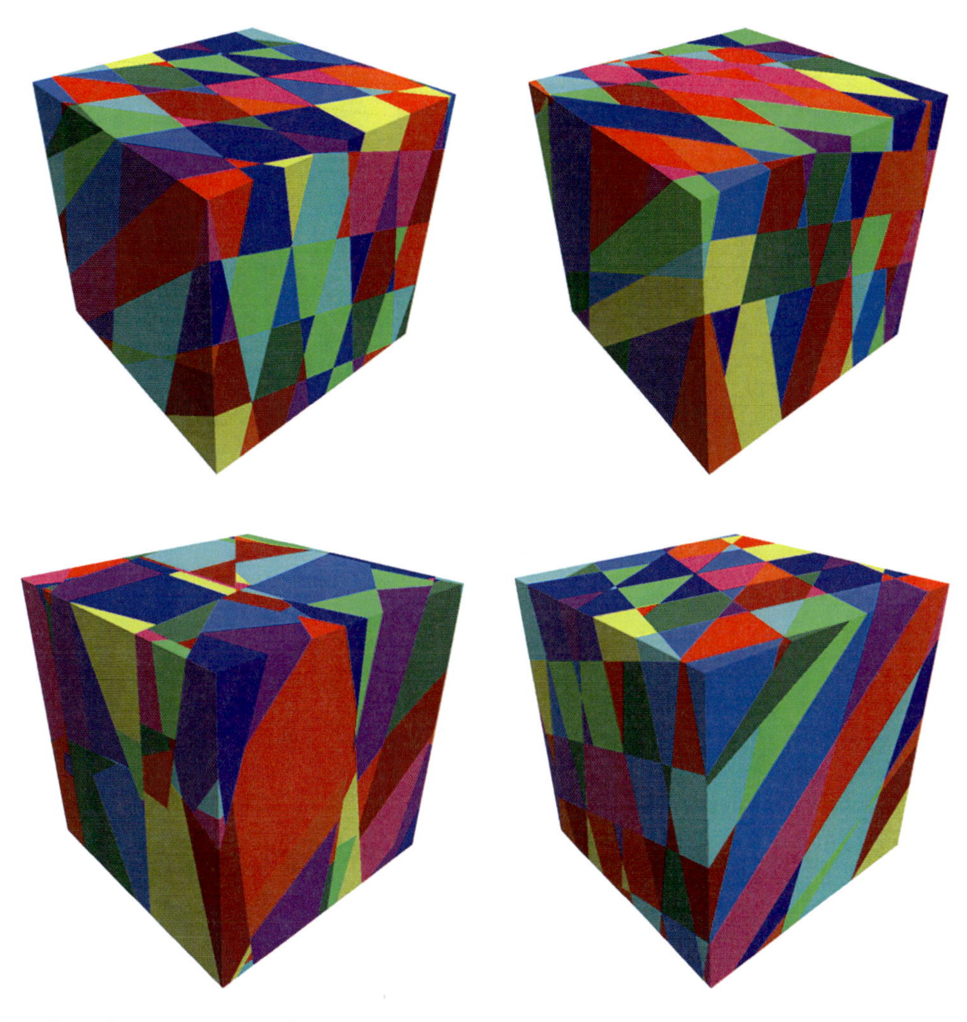

Fig. 164.1 Pattern A (*top left*), B (*top right*), C (*bottom left*) and D (*bottom right*)

The joint stiffness was calibrated to the measured rock mass stiffness using the small scale numerical models, with short run time (couple of minutes). The displacements in the hybrid model (with the calibrated joint stiffness) showed good agreement with the monitoring results (Fig. 164.2). It should be noted, that in some cases the hybrid model was re-run with a slightly different joint stiffness (to finalize the calibration) but the calibration of a DEM model is less time consuming using the small scale numerical tests.

As it can be seen in Fig. 164.3, the internal forces are similar in case of the four models, the point cloud represents the magnitude of the bending moments and axial forces do not show significant difference. In fact the difference between models with the same joint pattern and slightly different joint stiffness are higher than the difference between models with different joint pattern but with app. same displacements. According to this model, it can be concluded, that the effect of the joint pattern is significantly lower than the effect of the joint stiffness (and the displacements).

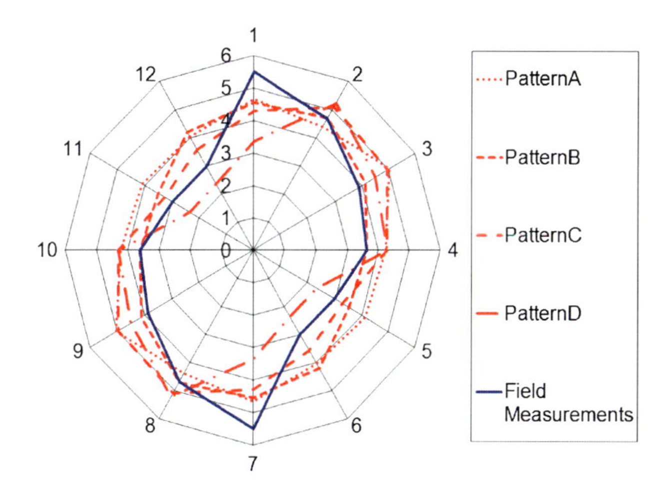

Fig. 164.2 Measured and calculated displacements in mm

The internal forces in the DEM model are significantly higher than it was found in the VEM model made with Phase2. One of the possible reasons of this phenomenon is that the rock mass induced forces, and the wedge movement induced forces can be examined in one model (it is not the same case if a continuous model is supplemented with a wedge analysis) therefore the internal forces with different cause are superimposed.

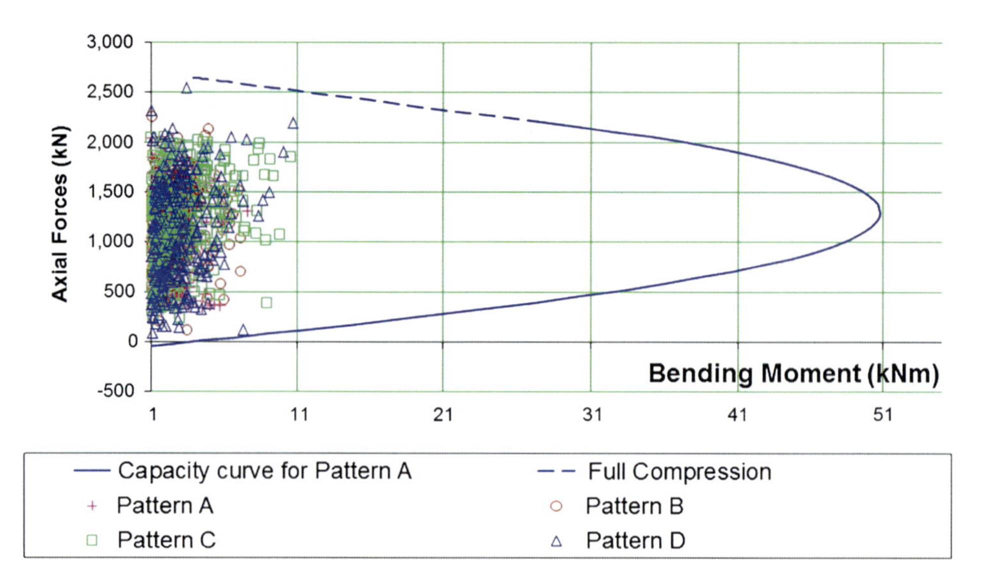

Fig. 164.3 Comparison of internal forces in the four models

One of the biggest advantages of the DEM is that the rock support can be optimized. In case of an anisotropic rock mass, the internal forces can be significantly different from the ones calculated with an isotropic continuum model. In these cases the DEM gives good alternative to use the rock support (such as bolts or additional reinforcement) in the places where they are really needed.

164.5 Conclusions

The aim of this paper was to explore the effect of joint pattern on the internal forces in linings. Four different joint pattern based on the same field measurement are presented here. Back analysis was carried out to calibrate the displacements. The calculated values showed a great agreement with the monitoring results. The calibration was performed with small scale model, to reduce the computational effort of the models. Hybrid continuum-discrete modelling was carried out. In line with expectations, the number of elements and the run time were reduced significantly, whilst the behaviour of the model was in agreement with the discontinuous model. Application of continuous representation of far-field is recommended in three-dimensional distinct element modelling. The biggest disadvantage of the DEM is the waste amount of additional input parameters. The joint pattern, joint strength and stiffness are to be determined. Probably the most important result of these models, that the joint pattern had smaller effect to the results than the joint stiffness.

According to this the following modelling procedure can be suggested: firstly, set up a reasonable continuum model (it can be considered as a common engineering task). Determine a reasonable estimation of the joint pattern and joint strength. It can be done with relativity simple field measurements, such as face mapping and joint characterisation. Then the displacements of the distinct model are to be calibrated to the continuous one. According to the results of the discrete model, the rock support can be optimized. With this approach the required additional input parameters can be determined as an estimation of the joint pattern and the joint strength.

Acknowledgments I want to express my gratitude to RHK Ltd. and Mecsekérc Ltd. for granting me permission to collect and use data related to the NRWR project. I would like to express my special thanks to Mott MacDonald Magyarország Ltd. for granting me access to the numerical modelling code.

References

Buocz I, Rozgonyi-Boissinot N, Görög P, Török Á (2010) Laboratory determination of direct shear strength of granitoid rocks; examples from the host rock of the nuclear waste storage facility of Bátaapáti (Hungary. Cent Eur Geol 53(4):405–417

Chang Y, Stille H (1993) Influence of early age properties of shotcrete on tunnel construction sequences. In: Shotcrete for underground support VI. American Society of Civil Engineers, pp 110–117

Horváth Z, Megyeri T, Váró Á, Görög P (2012) Discrete element modelling of the Mórágy granite formation in Southern Hungary. In: Horváth T (ed) 1st Eastern European tunneling conference. Veszprém, Budapest. Paper 22. ISBN:978-963-89638-0-2

Jing L, Stephansson O (2007) Fundamentals of discrete element methods for rock engineering: theory and applications'. Elsevier, Oxford

Kovács L, Deák F, Somodi G, Mészáros E, Máté K, Jakab A, Vásárhelyi B, Geiger J, Dankó G, Korpai F, Mező G, Darvas K, Ván P, Fülöp T, Asszonyi C (2012) Geotechnical investigation report, RHK-K-032/12

Special Requirements for Geotechnical Characterization of Host Rocks and Designing of a Radioactive Waste Repository

165

László Kovács and Balázs Vásárhelyi

Abstract

The safe management of radioactive wastes and spent fuels is one of the greatest technical-scientific challenges of 21st century. Radioactive wastes and spent fuels must be isolated safely from the biosphere until they become harmless as a result of decay process of radioisotopes. It takes even a few hundred thousands or one million years in some cases. Recently the final disposal in a stable geological environment seems to be the only feasible way for that. It is an unusual and multidisciplinary task to select an appropriate host rock, characterize that, design and construct an underground facility (including the engineering barriers), which fits to the features of geological barrier completely. Numerous special requirements to be taken into account both for earth-scientist and geo-engineers have been conceived worldwide during the last few decades. This paper tries to summarize the most important radwaste-specific issues of the geotechnical characterization programmes and the functional and static design of a repository, by introducing some example from the Hungarian L/ILW repository (National Radioactive Waste Disposal—NRWR at Bátaapáti) and the characterisation programme preparing the Hungarian HLW/SF (Boda).

Keywords

Radioactive waste • Repository • Geotechnical characterization • Design • NRWR—bátaapáti

165.1 Introduction

By the classification systems applied in most European countries radioactive wastes can be divided according to their activity (low, intermediate and high activity level wastes—LLW, ILW and HLW) and characteristic lifetime (short-lived and long-lived—SL and LL). Spent nuclear fuels (SF) represent a separate category in legal viewpoint (most of the countries do not classify SF as waste, including Hungary) but technically they have to be handled together with the HLW. Radioactive wastes and SF must be isolated safely from the biosphere until they become harmless as a result of decay process of radioisotopes. Particularly HLW and spent fuels are very dangerous substances with significant heat production, extremely high radiation level and radiotoxicity, so their danger remains during even a few million years. Consequently the safe management of radioactive wastes and spent fuels is one of the greatest technical-scientific challenges of 21st century.

L. Kovács (✉)
Kőmérő (RockStudy) Ltd., Pécs, Hungary
e-mail: kovacslaszlo@komero.hu

B. Vásárhelyi
Department of Structural Engineering, University of Pécs, Pécs, Hungary

G. Lollino et al. (eds.), *Engineering Geology for Society and Territory – Volume 6*,
DOI: 10.1007/978-3-319-09060-3_165, © Springer International Publishing Switzerland 2015

Wide international consensus is developed that the only possible and ethical solution for the long-term isolation of radioactive wastes is the final disposal based on a complex protective system of engineered and natural barriers (multi-barrier system). The key element of this system is the suitable natural (geological) barrier, but the waste form and the engineered barriers also have to be selected and designed according to the long-term stability and safety of the overall system.

A certain part of HLW and SF (depending on the selected back-end option) has to be disposed in deep geological repositories anyway. The Hungarian Public Limited Company for Radioactive Waste Management (PURAM) considers as reference scenario the direct disposal of SF. L/ILW can be disposed both in near-surface facilities or deep geological repositories. Hungary has developed both types of repositories: non-NPP origin radwastes are disposed in a surface facility at Püspökszilágy (hosted by loess and argillaceous sediments), while the NPP-origin L/ILW gets to the National Radioactive Waste Disposal (NRWR at Bátaapáti) in a granitic host rock (Mórágy Granite Formation).

Although the scientific and ethical bases of final disposal have been already given, lots of technical and scientific efforts have to be taken before licensing and operation of a repository. The reliable characterization of a potential host rock, the development of safety concept of final disposal, the confirmation of its long-term stability and safety, the detailed design and the construction generally take a few decades even in the case of L/ILW. The nationwide screening preparing the NRWR started in the beginning of 1993 and the very first underground disposal chamber was put into operation at the end of 2012. This relatively short implementation period was made to be possible by the extremely high level of public acceptance of local communities and the wide political consensus.

The council directive No. 2011/70 of European Union declares that each member state shall elaborate and implement its own national program with clear milestones and timeframes for the management of spent fuel and radioactive waste, covering all types of waste and all stages from generation to disposal. In conformity with this prescription PURAM develops mid and long term plans, which is updated in each year and accepted by the government.

The siting, characterization and design processes preparing a repository do not need to be uniform in every country. They depend mostly on the type of potential host rock, the geological, hydrogeological and climatic situation, the applied waste classification system, the technical, scientific, social and economical level, the internal regulations of the given country, etc. Only a limited number of general (mostly methodological) prescriptions and liabilities are defined in the relevant Safety Standards issued by International Atomic Energy Agency (IAEA) available in three categories (Safety Fundamentals, Safety Requirements and Safety Guides. The most important prescription is that the suitability of selected geological environment and the whole disposal system have to be verified by integrated Performance/Safety Assessments (PA/SA) based on the FEP-approach (Features, Events and Processes) and uncertainty-analyses in each of the project.

165.2 Key Questions of the Geotechnical Characterization and Design

The Eurocode 7 standard also has to be taken into account by the geotechnical characterization and design of a repository. Unfortunately, it does not consist of any commitments to radioactive waste repository up to now (see Alejano et al. 2013). Those prescriptions, nevertheless, are related only to the main rules, basic definitions and requirements of geotechnical works, but they do not specify the professional details of a program to be implemented. Since the radioactive waste disposal facilities are classified always into the Geotechnical Category 3, the elements of geotechnical characterization and design program have to be customized according to the features of host rock and the special character of the facility.

165.2.1 Elements and Implementing of the Geotechnical Investigation Programme

As the consequence of the above mentioned facts the main aim of the geotechnical characterization programme must be to understand and forecast the real behaviour of rock mass and reduce the uncertainties of input data for design and PA/SA. For fulfilling this requirement it is essential to determine the variability of the parameters/processes/phenomena (as a function of rock types, weathering level, parting, lateral and vertical position, anisotropy, etc.). A comprehensive investigation and design programme is required for achieving this aim including field tests and laboratory measurements, advanced geomathematical evaluation and extended numerical methods, fully-coupled models (Hudson and Feng 2007).

In general, the following geotechnical-rock mechanical parameters/processes/phenomena should be investigated in the framework of each radwaste projects:

- Virgin and tunnelling-induced (primary and secondary) stress field, including stress-dependent hydro-mechanical processes and phenomena;
- Strength and deformability of intact rock matrix and various discontinuities (by laboratory tests, special in situ tests and/or numerical back-calculations);
- Spatial variability of geotechnical features (by geophysical methods and continuous application of empirical geotechnical classifications);
- Transient (tunnelling-induced) and long-term (time-dependent or rheological) deformation processes (by various in situ deformation tests);
- The extension, distribution and the complex, coupled thermo-hydro-mechanical-chemical (THMC) processes of the Excavation Damaged/disturbed Zone (EDZ/EdZ);

Of course, the investigation programme has to include some waste specific elements, too: e.g. in the case of direct disposal of HLW and SF the study of heat- and radiation-sensibility of the potential host rock and the pre-selected elements of engineering barriers plays extremely important role. The special thermo-mechanical behaviour of Boda Claystone Formation—BCF (the potential host rock of Hungarian HLW/SF) could be studied very effectively in the deepest URL of the world (more than 1,000 m), where the virgin rock temperature exceeded 50 °C. Site (or host rock) specific issues have to be also taken into account: the creep and after all the long-term stability are also a critical question of unconsolidated argillaceous sediments or salty host rocks. In the case of overconsolidated BCF this question is not determinant not even under 1,000 m depth (Kovács 2001). Special geotechnical task can be the determination of mechanical properties of a clayey fault zones inside a granitic host rock (like in NRWR) not only for preventing geotechnical problems, but for protecting the integrity of the most important confining elements of site (Kovács et al. 2012).

Further requirements for the adequate implementation of geotechnical-rock mechanical investigations:

- Representativity (we should measure there, that and so regularly, which is enough for sufficient understanding of features, events and processes);
- Determining the scale effects; development of appropriate up- and downscaling procedures;
- Consistent, systematic applications of unified methods for the comparability of different rock zones;
- Generally it is unsatisfactory to evaluate the geotechnical —rock mechanical parameters separately. The consideration of coupled thermo-, hydro and chemical processes is also required;
- Investigations should not affect the isolation capabilities of host rock;
- Quality assurance/quality control requirements (reconstructability, objectiveness, documenting, long-term data preservation, etc.);
- Avoiding the quick aging of the results (liability for applying the best, up-to-date available technologies).

165.2.2 Key Questions of Functional and Static Design

Most important tasks for the functional and static design are to optimize the main elements of the overall final disposal system and harmonize them to the safety concept and to each other. The suitability and safety of disposal facility have to be proved in a very detailed and complex licensing procedure. Long-term radiological safety determined by PA/SA must be the primary consideration, so some usual engineering aspects remain in the background, e.g. limitations of applicable materials and supporting systems (e.g. plastics and metals with high gas formation risk and/or corrosion rate cannot be inbuilt). All of the inbuilt materials and devices have to be carefully selected and strictly documented. The construction works and the final disposal are generally parallel activities, so the tunnels and infrastructures have to be divided and separated into nuclear controlled zone and building zone. To fulfil the legal requirement for the retrievability of radioactive wastes is one of the greatest challenges for the designers.

Engineered barriers, the required tunnels and chambers, the surface and underground infrastructural background have to be constructed and operated generally in very complex geological, hydrogeological, geotechnical environment. The aggregate construction and operation time of a repository generally exceeds 100–150 years. Due to the final disposal activities, the possibility for the routine maintenance of tunnels and chamber in the controlled zone is quite limited. Beside of geotechnical investigations only the most advanced methodologies static and functional design can guarantee the stability and the operability of the whole system during such a long period. The strong interactions between the engineered and natural elements of the repository system require deep understanding the real long-time behaviour of rock mass not only for earth-scientists but for the designers, too. The complex, coupled THMC processes of the EDZ/EdZ have to be also taken into account at the optimization of the layout and other parameters of disposal chambers (see e.g. Deák et al. 2013). Whereas the mentioned parameters/processes/phenomena can be recognized gradually during the construction, the design and implementation have to be also remained flexible until the end of final operational licensing procedure. That principle is called "DESIGN AS YOU GO-approach" in the international practice. In the case of HLW and SF that principle is particularly important, so the application of a two-stage design and excavation process is compulsory. In the first step an underground research laboratory (URL) is constructed inside the host rock and operated during 15–20 years. Investigations and operation of URL should not affect the isolation capabilities of host rock but the results must be representative also for the repository area. The URL operated in Hungary in 1990s was connected to the former uranium mine, far from the potential zone of final disposal, so it did not affect the long-term radiological safety. The new URL under preparation recently is going to fulfil the mentioned requirement.

There are some other task-specific difficulties during the design of radioactive waste disposal and the repository:

- Demand for detailed constructional and operational risk management;
- Demand for environmentally friendly implementation and operation;
- Demand for the application of complex, up-to-date technologies and logistics;
- Considerable effort is needed to handle the political and social aspects;
- Project-approaches: generally strict quality requirements, time and financial constraints.

165.3 Conclusion

The accomplishment of safe final disposal of radioactive wastes and spent fuels is a non-usual and multidisciplinary task, which is not solvable only by applying the professional routine methods and approaches of the "traditional" engineering projects. It can be only based on the most updated results of earth-sciences and continuous cooperation of scientists, geo-engineers and designers during each stage of the process (the site selection, characterization of host rock, iterative development of safety concept, the design and implementation of repository), which altogether are lasting even one or two centuries until the closure and the end of the institutional control of the facility. The principles listed in this article are entirely taken into account in both Hungarian NPP-origin radwaste projects being in progress recently: NRWR at Bátaapáti (L/ILW) and Boda-project (HLW/SF).

References

Alejano LR, Bedi A, Bond A, Ferrero AM, Harrison JP, Lamas L, Migliazza MR, Olsson R, Perucho Á, Sofianos A, Stille H, Virely D (2013) Rock engineering design and the evolution of Eurocode 7. In: Rock mechanics for resources, energy and environment. Eurock 2013, pp 777–782

Deák F, Kovács L, Vásárhelyi B (2013) Modeling the excavation damage zones in the Bátaapáti radioactive waste repository. In: Rock mechanics for resources, energy and environment. Eurock 2013, pp 603–608

Frank R, Bauduin C, Driscoll R, Kavvadas M, Krebs N, Ovesen N, Orr T, Schuppener B (2004) Designer's guide to EN 1997-1 Eurocode-7: geotechnical design—part 1: general rules

Hudson JA, Feng XT (2007) Updated flowcharts for rock mechanics modelling and rock engineer design. Int J Rock Mech Min Sci 44:174–195

Kovács L (2001) Partial self-healing effects of a highly indurated claystone formation (BCF) discovered by in situ measurements. Self-healing topical session proceedings organised by OECD NEA/RWM/Clay Club. Proceedings, pp 47–57

Kovács L, Mészáros E, Deák F, Somodi G, Máté K, Jakab A, Vásárhelyi B, Geiger J, Dankó G, Korpai F, Mező G, Darvas K, Ván P, Fülöp T, Asszonyi C (2012) Updated version of the geotechnical integrated report of NRWR (available only in Hungarian). Manuscript. PURAM's Archives, RHK-K-032/12

Public Limited Company for Radioactive Waste Management—PURAM (2013) The 13th intermediate and long-term plan for the activities to be financed from the central nuclear financial found. Manuscript. http://www.rhk.hu/docs/publications/PURAM_plan_13.pdf

166 Rock Mechanical and Geotechnical Characterization of a Granitic Formation Hosting the Hungarian National Radioactive Waste Repository at Bátaapáti

László Kovács, Eszter Mészáros, and Gábor Somodi

Abstract

Construction of a radioactive waste disposal should have a wide-range investigation and design program. It is important to suit the design, construction and monitoring tasks of the underground facilities to the rock mechanical-geotechnical character of the host rock in many aspects, such as the long-term environmental and radiological safety, the safety requirements at the workplace and economical reasons. The gained information enables the optimization of the advance and support systems. Getting to know the behaviour of the surrounding rock mass has a great importance in case of a radioactive waste repository, as the geological barrage is one of the main elements of the safety system. So the in situ measurement is not just important for the static safety during the work phase, but it is the element of the long-term safety system, too. It helps to recognize the changes in the environment and interfere if needed.

Keywords

Radioactive waste repository • Granite • Rock mechanical data • Tunnel face mapping • Monitoring system

166.1 Introduction

For final disposal of Hungarian operational and decommissioning LLW and short-lived ILW produced by Paks Nuclear Power Plant a new facility (National Radioactive Waste Repository—NRWR) is under construction at Bátaapáti in a granitic host rock. After a 12-year-long preparation process two inclined shafts were constructed using drill and blast technology to reach the repository depth. Having obtained the required licences and the acceptance of the Parliament and the local community the construction of repository began in September 2008. The structures of underground infrastructural background (water pumping plants separately for construction and final disposal activities, compressed air and electricity plants, etc.) and the loop tunnel system hosting the disposal chambers had been constructed until 2010. The first two underground disposal chambers were completed in September 2011. The customer of the project is the Public Agency of Radioactive Waste Management (PURAM), the main contractor is MEC-SEKÉRC Co. KŐMÉRŐ (RockStudy) Ltd. was charged with performing and evaluating the geotechnical and rock-mechanical in situ and laboratory measuring program and geotechnical documentation.

L. Kovács (✉) · E. Mészáros · G. Somodi
Kőmérő (RockStudy) Ltd, Pécs, Esztergár L. Str. 19, 7633, Hungary
e-mail: kovacslaszlo@komero.hu

166.2 Geotechnical Research Program

Construction of a radioactive waste disposal should have a wide-range investigation and design program. Information gained from geotechnical research is basic data for the design, construction and operational phases of the facility.

DOI: 10.1007/978-3-319-09060-3_166,

166.2.1 Geotechnical Classification of the Rock Mass

The regular geotechnical rock mass classifications were performed on cores of exploratory drillings and of each tunnel faces. A detailed forecast of the rock mass properties was compiled on the basis of pilot holes. These were the basis to the everyday decisions on the advances and applicable support systems.

166.2.1.1 Core Documentation

Before the tunnel driving pilot holes were drilled and investigated in detail. Because of the large (~ 100 m^2) section of the chambers, three holes had to be drilled along the whole chamber. Two of them were in the roof area, as it plays the biggest role in the stability and one in the floor area. The verification of the forecast made from three boreholes was investigated after the construction. It seems that the three boreholes with the mentioned arrangement gave a good prognosis of the whole sections.

166.2.1.2 Tunnel Face Mapping

The geotechnical classification of each faces was performed using the RMR (Bieniawsky 1989), the Q (Barton et al. 1974) and the GSI (Hoek et al. 1995, with modification published by Cai et. al 2004) method. The verifying of the results by each other was essential as this was the first research programme which gave such data of the Mórágy Granite Formation. The parallel use of the three methods helped to derive easier design parameters and remark subjectivities, accidental mistakes also.

The documentation of the tunnel faces was based on onsite observations and 3D optical surface mapping (Gai et. al 2007 and Gai and Pötsch 2008). The orientations of main characteristic discontinuities were defined in the 3D models (Fig. 166.1). This information was applied in surface roughness determination and fracture system modelling also (Krupa et al. 2013).

The digital documentation method allows studying the faces and the tunnel wall not just on site but later for further investigation tasks, too. It has a great importance at Bátaapáti, as the tunnel wall is supported immediately by a shotcrete layer. With the digital documentation the whole chamber can be visualized which can be a basis of a 3D database of geological features.

166.2.2 Laboratory Investigations

The cores of every borehole were systematically sampled for standard rock mechanical laboratory tests, which were performed according to the recommendations of ISRM. By the laboratory tests the main rock mechanical parameters of the intact rock (i.e. uniaxial compressive strength, Young's modulus, Poisson's ratio, tensile and shear strength, Hoek-Brown and Mohr-Coulumb parameters) could be determined which gave the basic information for the static design. Elastic properties were determined during uniaxial compressive tests carried out with deformation measurements using bonded strain gauges. CT-scanning was performed on selected UCS samples to filter out the effect of micro-cracks. Varifying these results we recalculate parameters taking into consideration Martin and Chandler (1994), Eberhardt et al. (1998), Diedrichs and Martin (2010) theories and suggestions.

We used multiple failure test (Kovari et al. 1983) also for precising peak strength and determining residual strength of the granite.

The numerous data enabled geostatistical evaluation which made the geotechnical characterization more specified. According to discriminant analysis the six main rock types of the formation were not recognizable in point of laboratory measurement types. Although histograms and distributions may differ, they can be handled separately in two groups, as monzonite-type and monzogranite-type. Monzonite type group has higher strength and elastic laboratory values. It is squarely statable that there are differences between values of laboratory investigations on the strength of depth and location. Differences seem to follow petrologic zones.

166.2.3 In Situ Measurements and Monitoring System

A comprehensive geotechnical monitoring system was applied in the first two chambers (Fig. 166.2). In addition to the numerous optical convergence and load indicator sections, 4 radial MPBX-extensometers (section Ext-10 in Fig. 166.2) were installed to measure the radial displacement of the surrounding rock mass caused by the tunnel driving and in the long-term monitoring, 8 load cells (LC-01-04 and LC-05-08 in Fig. 166.2.) for controlling rock bolts, 6 two-directional gauges (LB-01-06 in Fig. 166.2.) for measuring the deformation of shotcrete have been installed and continuously measured. 6 CSIRO HI-cells (Bkc-7-12 in Fig. 166.2.) were installed for determining 3D distribution and magnitudes of stress changes around the chambers during tunnelling.

The measurements provided important data not just for the short- and long-term stability of the chambers, but for the optimization of the further chambers, too.

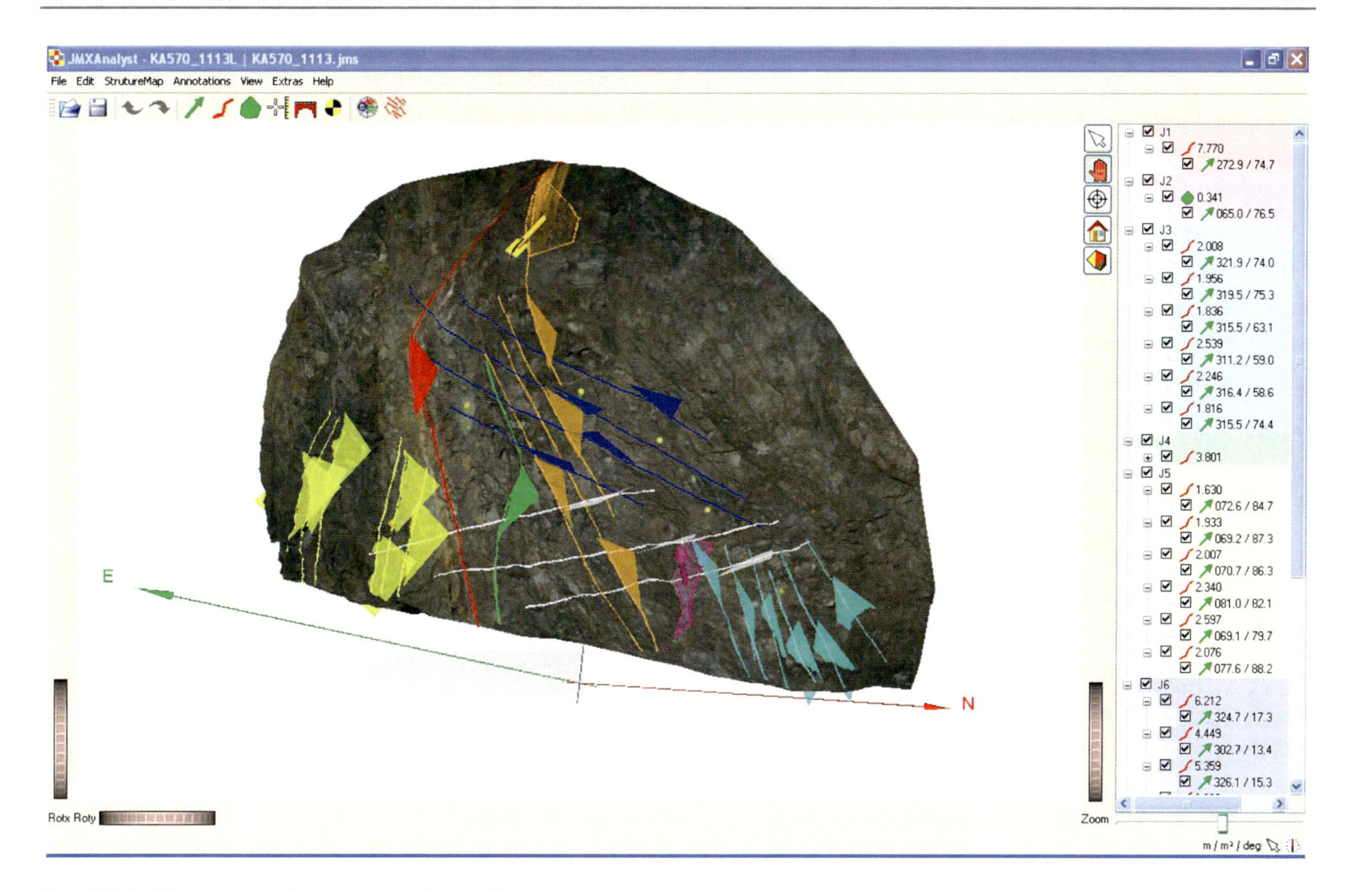

Fig. 166.1 3D documentation of a tunnel face with the structure map

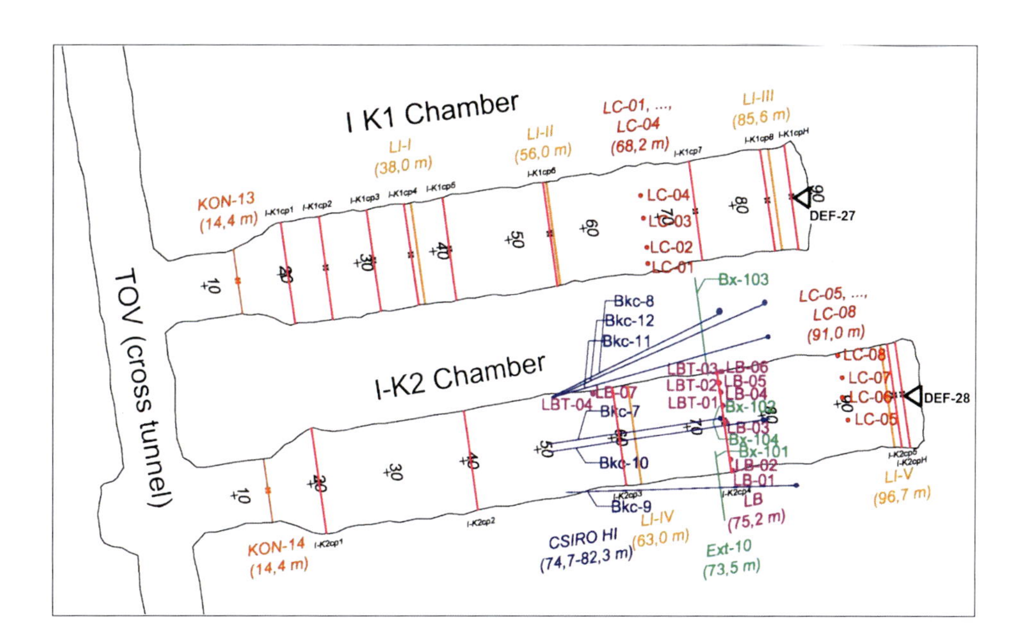

Fig. 166.2 The in situ measurement and monitoring system in the first two chambers at Bátaapáti NRWR

166.3 Conclusions

The results have provided important data for verification of the design and for revealing the behaviour of the surrounding rock mass in short and long term monitoring period. Beside the geotechnical data gathering the investigation programme also helps the long-term safety assessment.

Acknowledgment The authors gratefully acknowledge the permission of PURAM to publish this paper.

References

Bieniawski ZT (1989) Engineering rock mass classifications. John Wiley and Sons, Pennsylvania

Barton N, Lien R, Lunde J (1974) Engineering classification of rock masses for the design of the tunnel support. Int J Rock Mech Min Sci 6(4):189–236

Cai M, Kaiser PK, Uno H, Tasaka Y, Minami M (2004) Estimation of rock mass deformation modulus and strength of jointed hard rock masses using GSI system. Int J Rock Mech Min Sci 41:3–19

Diederich MS, Martin CD (2010) Measurement of spalling parameters from laboratory testing. In: Rock mechanics in civil and environmental Engineering. Zhao, Labiouse, Drudt, Mathier (eds). EUROCK2010, ISRM, Taylor & Francis Group, London, pp 323–326

Eberhardt E, Stead D, Stimpson B, Read RS (1998) Identifying crack initiation and propagation thresholds in brittle rock. NRC Can Can Geotech J 35:222–223

Gaich M, Pötsch W, Schubert (2007) High resolution 3D imaging for site characterisation of a nuclear waste repository, 1st Canada—U. S. rock mechanics symposium 2007, Vancouver, Canada, pp 69–75

Gaich A, Pötsch M (2008) Computer vision for rock mass characterization in underground excavations. ARMA, 2nd U.S.—Canada rock mechanics aymposium, San Francisco, p 14

Hoek E, Kaiser PK, Bawden WF (1995) Support of underground excavations in hard rock. In: Balkema AA, Rotterdam, Reprinted in 2005, p 215

Kovari K, Tisa A, Einstein HH, Franklin JA (1983) Suggested methods for determining the strength of rock materials in triaxial compression: revised version. Int J Rock Mech Min Sci Geomech Abstr 20 (6):283–290

Krupa Á, Deák F, Görög P, Buocz I, Török Á (2013) Quantitative roughness profiling of fracture surfaces of a granitic host rock at a radioactive waste disposal site. In.: Marek Kwasniewski, Dariusz Lydzba (ed) Rock mechanics for resources, energy and environment. ISBN 9781138000803

Martin CD, Chandler NA (1994) The progressive fracture of Lac du Bonnet Granite. Int J Rock Mech Min Sci Geomech Abstr 31 (6):643–659

Part XVI

Subsurface Water in Tunnels: Prediction, Estimation, Management

Convener Mr. Yogendra Deva—*Co-conveners* Dr. Y.P. Sharda, S.N.C. Lavlin and Prof. S.K. Singh, Delhi University, A. Bishwapriya, Patna University and Shazia Tabassum, Delhi University

Introduction: Subsurface water remains the most significant factor in tunneling problems. In association with crushed rock, it often leads to flowing ground conditions—a nightmare for tunneling personnel. Numerous tunneling projects the world over have suffered time and cost escalations due to subsurface water in large quantities or extreme flowing ground conditions. The subject still remains enigmatic and calls for state-of-the-art investigation and construction methodologies for demarcating and managing such problematic zones. Session Concept: The session will have papers and invited lectures on research, and case studies on investigations and handling of subsurface water/ flowing ground conditions in tunnels. Objective: The paper contents and discussions during the session would be compiled and summarized to outline state-of-the-art approach for the prediction, estimation and management of subsurface water/ flowing ground conditions.

Ground Water Management for Large Under-Ground Storage Caverns

167

Saikat Pal, G. Kannan, Vijay Shahri, and A. Nanda

Abstract

Underground unlined (or mined) rock caverns is one of the economical alternative for buffer storage of crude oil to ensure energy security of import dependent countries. The principle of such storage employs hydrodynamic containment of the product. As the large rock caverns are excavated by conventional drill and blast method, the groundwater management is of utmost importance to conserve ground water and avoid inadvertent de-saturation of rock mass. This is added to the constraints due to inflow of water during excavation of rock mass like other underground projects. Consequently, hydrogeology forms an important aspect during planning, investigation and subsequent construction stages. The present paper focuses on groundwater management during construction for an ongoing storage cavern in India. In the process it highlights early identification of permeable structural features, the adopted grouting philosophy, the water curtain system and the required hydro-monitoring.

Keywords

Hydrodynamic containment • Water curtain • Hydro-monitoring

167.1 Introduction

The principle of storage of crude oil in large unlined mined rock cavern ensures tightness of product by directing ground water gradients towards the storage caverns (Amantini et al. 2005). This is known as hydrodynamic containment. The unlined storage caverns are constructed below natural ground water table. The ground water level during construction and operation stage of the cavern is maintained by uninterrupted artificial charging of water curtains so as to rejuvenate the ground water regime (Usmani et al. 2010). The water curtains comprise of water curtain galleries (WCG) and horizontal and vertical water curtain boreholes (WCBH) drilled from these galleries. These WCBH are charged with water and encases the storage caverns. The project under discussion comprises of four large caverns (900 mL × 20 mW × 30 mH), three (6.5 × 6.5 m) water curtain galleries, two circular shafts (8.2 m diameter) for pumping of crude and seepage water and access tunnel (12 × 8 m) for facilitating construction.

S. Pal (✉) · G. Kannan · V. Shahri · A. Nanda
Engineers India Limited (EIL), New Delhi, India
e-mail: saikatindia@yahoo.co.in

G. Kannan
e-mail: gopi.kannan@eil.co.in

V. Shahri
e-mail: vijayshahri@gmail.com

A. Nanda
e-mail: ananda@eil.co.in

167.2 Hydro-Geological Model

During excavation, a minimum hydraulic head equivalent to 20 m of water above the horizontal water curtain level to be maintained in order to ensure hydraulic gradient >1 (Aberg 1977). This is to prevent de-saturation of rock mass surrounding the cavern. Thus impediment of uncontrolled inflow of groundwater in tunnels as well as conservation of groundwater is necessitated. A hydro-geological model of the project is prepared based on the project geological model

G. Lollino et al. (eds.), *Engineering Geology for Society and Territory – Volume 6*,
DOI: 10.1007/978-3-319-09060-3_167,

and various explorations during investigation and pre-construction stage (Usmani et al. 2010).

The rock type in the project area is granitic gneiss belonging to the Peninsular Gneissic complex of India. The project is seated in a hilly terrain with thick laterites and lateritic soil at the top followed by weathered and fresh granitic gneiss. The permeability of the soil and lateritic portion is high of the order of 10^{-5}–10^{-6} m/sec where as the permeability of the massive gneissic rock is very low of the order of 10^{-9} m/sec. However, 3–4 sets of prominent discontinuities were observed including a sub-horizontal joint which shows permeability in the range from 10^{-8} to 10^{-6} m/sec (some horizontal joints show very high permeability of the order of 10^{-4} m/sec locally). Sub-vertical and sub-horizontal dolerite dykes are also encountered with the permeability of the order of 10^{-6}–10^{-7} m/sec. The hydrogeological model also includes of groundwater level contouring and residual seepage evaluation apart from the permeability distribution.

During excavation, the hydro-geological model was constantly updated by: • Structural projections of permeable features • Updating probing and grouting detail • updating the seepage points and permeability values of all WCBH and manometer holes drilled from underground; and • Correlating all above data.

167.3 Probing and Grouting

Locating water ingress features ahead of cavern excavation face helps to plan judicious treatment. Thus continuous systematic probing was envisaged during design and planning stage with provision of probing kept for each alternate faces. Systematic probing with original frequency was performed during excavation of WCG, access tunnel and top heading of caverns. Probe holes, 10–12 m long destructive drill holes, were drilled ahead of excavation faces, 2–3 in numbers depending on the surface area of excavated face. For all seeping probe holes, depth, rate and pressure of water inflow were recorded to take decision on grouting, magnitude of grouting and parameters of grouting. The hydrogeological model was updated by significant features identified during excavation of small tunnels and were confirmed during top heading excavation.

The WCG (20 m above caverns) and access tunnel will be cut off from the caverns and will be completely flooded with water during operation. So, there were no design seepage limits for this tunnel and grouting was taken up only if the ingress affects the construction works or if the ground water level is affected. Incase of caverns 2 other criteria also influenced grouting. They were intake quantity of interfering WCBH and designed residual seepage. Based on these, the trigger value for pre-grouting was decided as 3 L/min/m/bar. If the seepage from probe holes was more, fan pre-grouting was performed on the face ahead.

Pre-grouting was preferred and carried out from top heading by modifying the grout fan as suited to disposition of feature. In case of persistent features overlapping grout fans were constructed from alternate faces.

Once the disposition of major hydrogeological features were finalized on the basis of excavation data, the probing were optimized for subsequent bench levels and grouting were concentrated in the zones where the features are anticipated to be negotiated in the respective elevations (Fig. 167.1). Accordingly, pre-grouting plan of all benches was made. Side wall pre-grouting form higher bench were carried out in the identified zone with sub vertical grout holes directed to intersect the feature and constitute grout curtain to cutoff wall seepage. Invert pre-grouting from last bench was carried out with target to cutoff seepage up to depth of 5 m from invert.

This reduced the time and efforts at each level and helped to expedite excavation. No major grouting was required apart from predicted areas. However, provisions were kept for probing and grouting with respect to sub horizontal features revealed at each level.

Post grouting was necessary for one or more of the following reasons: •Pre-grouting insufficient to maintain ground water level, •Residual seepage more than design seepage limit (30 L/min/100 m section), •Increase in water intake of WCBH, and • Hindrance to rock support.

167.4 Water Curtain System

The final patterns of WCBH were directed with aim to cater water to the identified high permeable features:

- The horizontal holes were directed to 70° w.r.t tunnel axis considering the orientation of prominent joints.
- Peripheral vertical curtains (Fig. 167.2) were introduced to counteract the drawdown along low dipping highly permeable features (permeability 10^{-5} m/sec) and to isolate from upcoming facilities in adjacent area.
- 40 m ahead of cavern excavation, horizontal and vertical WCBH were pressurized with water at 3–4 bars. The sequence of cavern excavations was thus guided by availability of water charged boreholes.

In order to check the efficiency of designed water curtain, hydro-tests were carried out to ensure their ability to maintain the desired hydrostatic potential. As per original design, the efficiency of entire horizontal curtain was to be tested in a single test after heading excavation of entire cavern system. However, the test was distributed in different sections. The sections were selected for testing as per excavation schedule so that, excavation works can be continued parallel in other part of caverns. Additional holes were

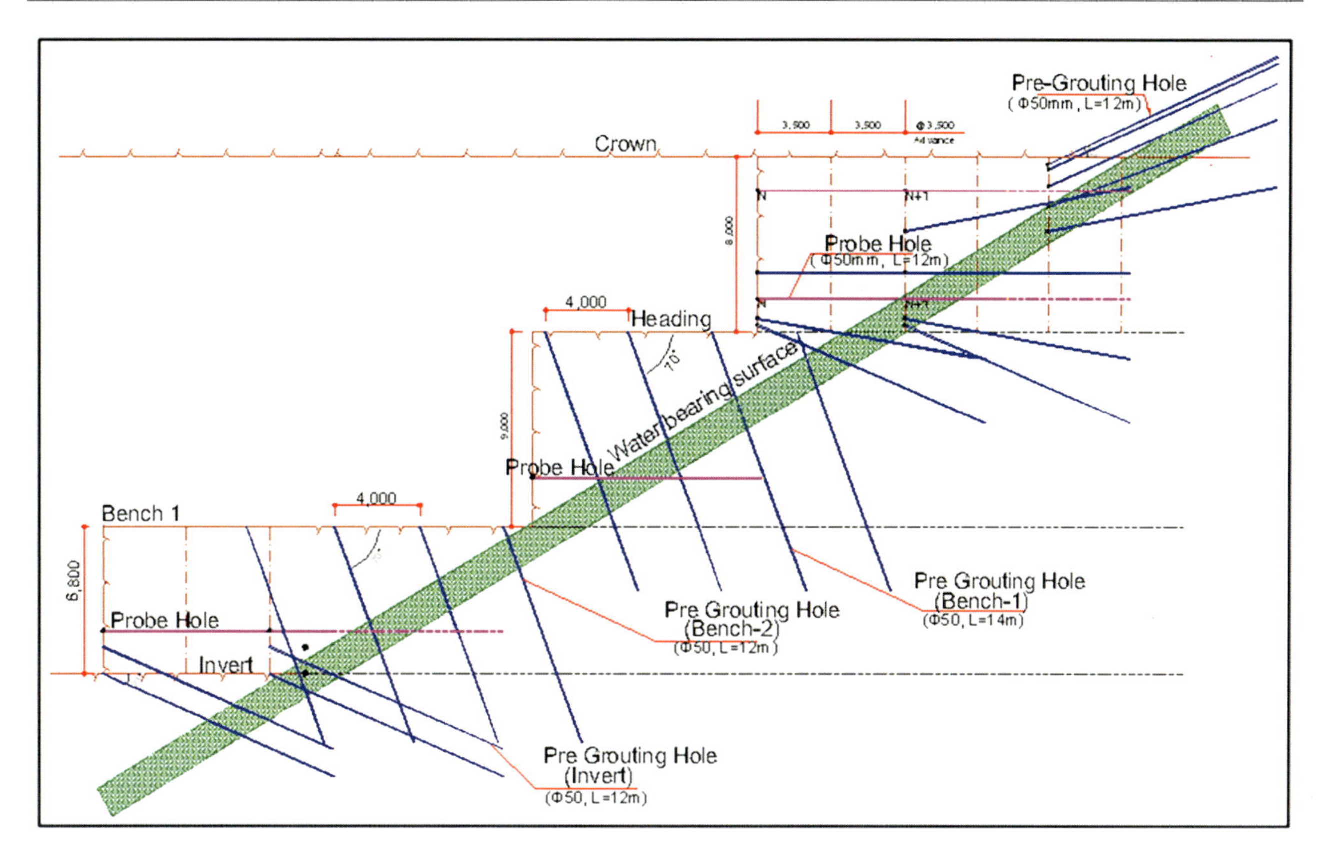

Fig. 167.1 Scheme for grouting

Fig. 167.2 WCBH distribution

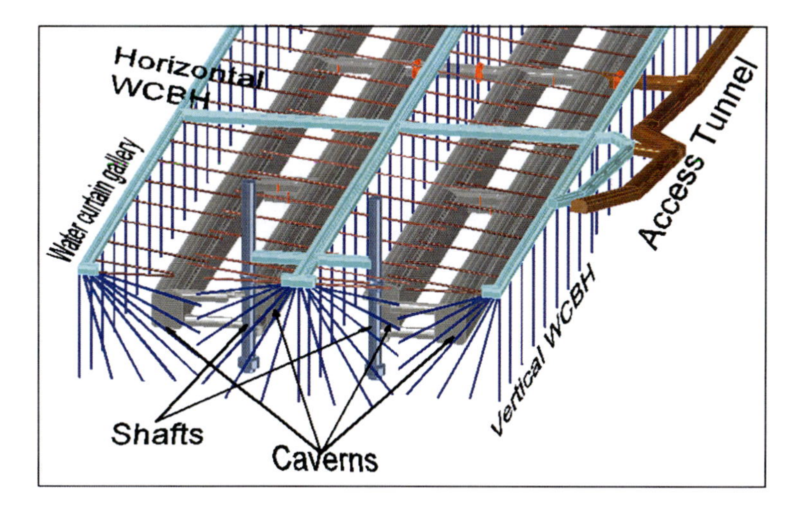

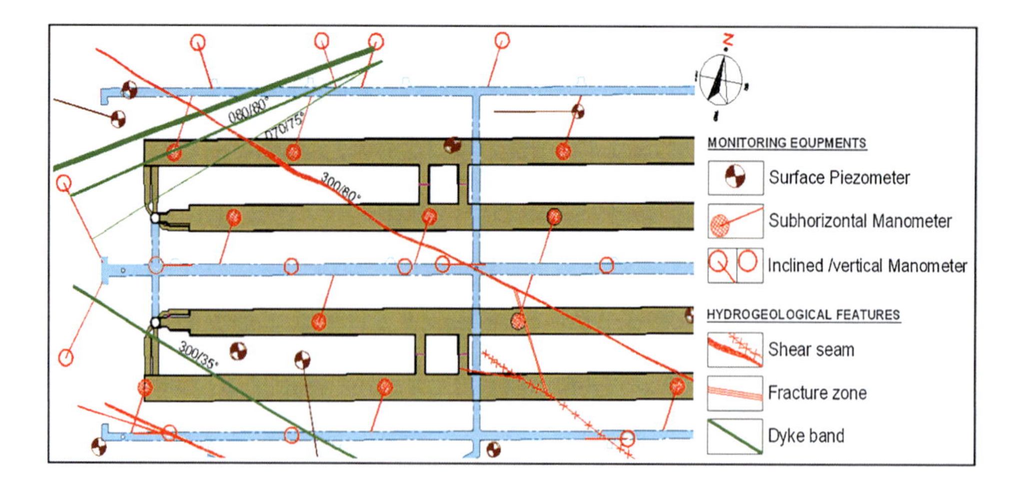

Fig. 167.3 Hydro-geological monitoring plan of manometers and piezometers

recommended and drilled wherever the hole spacing of 20 m was insufficient to maintain the desired potential.

Management of WCBH during grouting was an important aspect. The water pressure of the boreholes in the affected area was lowered to facilitate rock bolting and grouting. However constant flow at optimum pressure was required and pressure was monitored during gout injection so that the holes do not get clogged with grout mix.

167.5 Monitoring and Seepage Measurement

Hydro-monitoring and seepage measurements help to understand the groundwater balance of seepage versus recharge (natural and artificial). The monitoring/measurements are essential to judge requirement of grouting as well as effectiveness of grouting. It is also mandatory to evaluate residual seepage and compare it with the designed seepage levels to confirm the adequacy of design for seepage pump in shaft.

The daily hydro-geological monitoring comprised mainly • Ground water level monitoring through surface piezometric wells • Hydraulic potential measurement from underground WCG by manometer and pressure cells. These were installed adapted to monitor identified major water bearing features (Fig. 167.3). • Pressure and water intake measurement of all WCBH.

All the measurements were integrated to ensure overall well being of the hydrogeological conditions.

In the project, seepage measurements were carried out through following methods:-

• **Indirect seepage measurement**—the daily difference of outgoing and incoming water assessed by using flow meters. • **Measurements from individual seepage points**—the total seepage from individual seepage points on crown and walls were mapped and measured monthly to have an idea about change in locality as well as quantity of seepage. • **Direct seepage measurement**—the total seepage measurement in isolated sections. Isolation was done by constructing concrete/clay weirs across the gallery.

The residual seepage measured in the storage caverns was between 20 and 25 L/min/100 m (Usmani et al. 2012). Almost equal amount of water was being recharged through WCBH to maintain the natural ground water level. The ground water level was in between 30 and 80 m above horizontal water curtain level.

167.6 Conclusion

The dynamic approach of continuously updating hydro-geological model was of immense help in correct and timely anticipation of hydro-geological features. This aided to readiness of addressing situations. The stipulated hydro-geological guidelines were adapted to construction friendly procedures. This also helped in expediting the work pace by optimization of activities like *probing*. The ground water management of the project was completed successfully by judicious combination of *grouting and groundwater recharging*. The basic criterion of saturation of rock mass was maintained and at the same time the overall seepage was controlled within the designed capacity.

Acknowledgment The authors wish to acknowledge the support rendered by the subsurface projects team and wish to place on record their thanks to the management of EIL for granting permission to publish the paper.

References

Amantini E, Francois C, Anne M (2005) Groundwater management during the construction of underground hydrocarbon storage in rock caverns. In: 9th international mine water congress (IMWA 2005)

Usmani A, Nanda A, Kannan G, Jain SK (2010) Hydraulic confinement of hydrocarbons in unlined rock caverns. ISRM, Delhi

Usmani A, Kannan G, Nanda A, Jain SK (2012) Seepage assessment in tunnels under different field conditions. In: Proceedings of Indian geotechnical conference, Dec 13–15, 2012, New Delhi

Usmani A, Nanda A, Kannan G, Jain SK Analysis of groundwater flow in underground storage caverns. In: 6th ICEG (international conference on environmental geotechnology), Nov 2010, New Delhi

Aberg B (1977) Prevention of gas leakage from unlined reservoirs in rock. In: Proceedings rock store, pp 399–413

Experience from Investigation of Tectonically Extremely Deteriorated Rock Mass for the Highway Tunnel Višňové, Slovakia

168

Rudolf Ondrášik, Antonín Matejček, and Tatiana Durmeková

Abstract

The designed highway tunnel Višňové will cross the mountain ridge Lúčanská Fatra in Slovakia in its length of 7.48 km. The selection of the tunnel alignment was preceded by a preliminary geological investigation, which indicated a complex geological structure. Consequently, a pilot tunnel was included into the planned geologic investigation. Two tunnelling methods were tested. Complicated geologic structure, intensive faulting and fracturing of rocks and intricate hydrogeological condition were proved by the pilot tunnel. Excavation of the pilot tunnel was variable, all rock mass behaviour types occurred. Enormous problems were linked with negotiation of sudden concentrated inflows of groundwater from fault zones reaching up to 160 l/s, followed by groundwater erosion.

Keywords

Engineering geological investigation • Pilot tunnel • Groundwater inflow

168.1 Introduction

The proposed highway tunnel Višňové will be an integral part of a passage of the Slovakian D1 Highway between Žilina and Ružomberok (Fig. 168.1). The tunnel will cross the Lúčanská Fatra Mountains. The Lúčanská Fatra horst is in tectonic contact with the Žilinská kotlina basin from the west and the Turčianska kotlina basin from the east (Fig. 168.2).

A tunnel line was selected from several variants on the basis of a preliminary investigation and environmental impact assessment (EIA). Geological investigation took place in 1995–1998 and consisted of geological mapping, geophysical investigation and several boreholes to a depth 120 m. The pilot tunnel in the designed investigation was excavated using the Tunnelling Boring Machine (TBM) from the east side (4,293 m) and the New Austrian Tunnelling Method (NATM) predominantly from the west side (3,187 m). The highway tunnel realization was supposed to start in 2003. Due to various reasons this was postponed to 2014.

R. Ondrášik (✉) · T. Durmeková
Faculty of Natural Sciences, Department of Engineering Geology, Comenius University, Mlynská dolina, Bratislava 842 15, Slovakia
e-mail: ondrasik@fns.uniba.sk

T. Durmeková
e-mail: durmekova@fns.uniba.sk

A. Matejček
Geofos, Ltd, Veľký Diel 3323, Žilina 010 08, Slovakia
e-mail: antonin.matejcek@geofos.sk

168.2 Preliminary Geological Investigation

The most important data resulting from the preliminary investigation are:

(a) two exploratory boreholes to a depth 120 m;
(b) engineering geological map of the designed tunnel on a scale of 1:5,000;
(c) regional geological study with a geological heterogeneity map (Fig. 168.2);
(d) several geophysical profiles and a longitudinal geological profile of the designed tunnel (Fig. 168.3).

DOI: 10.1007/978-3-319-09060-3_168,

Fig. 168.1 The highway D1 with the designed tunnel Višňové location

The geological heterogeneity map helped to understand the geological condition of the area. Its compilation was based on regional geology evolution analysis, data from investigation of other projects in the area, proper field research and analysis of aerial photos. Overthrusting in the Upper Mesozoic and differential tectonic movements during Upper Tertiary evolution influenced the designed tunnel's geological condition. Downward movement along listric faults resulted in rock stress release in the upper parts of a massif and was reflected in joint widening and ground water saturation (Fig. 168.4).

168.3 Geological Investigation by a Pilot Tunnel

A pilot tunnel was driven along the south tunnel line from November 1998 to August 2002 (Matejček et al. 2002, Matejček and Bohyník 2006). In addition to the geological investigation, it will drain groundwater ahead of tunnel construction starts and it will serve as the temporary emergency corridor for the northern tunnel line which will be constructed and used for traffic as the first.

168.3.1 Western Pilot Tunnel Section Driven Using NATM

The west gallery section was driven into Paleogene flysch and Mesozoic strata, and a short passage in altered granitic rock (mylonites) tectonically intensively deformed. Tectonic contact between Paleogene and Mesozoic rock consists of breccia in a zone up to 20 m thick, steeply declined to the Žilinská kotlina basin. Mesozoic limestone—dolomite complex contains marl, shale, schistose and tectonically crushed carbonates. This lithological heterogeneity is reflected in complicated hydrogeology. Permeable disrupted

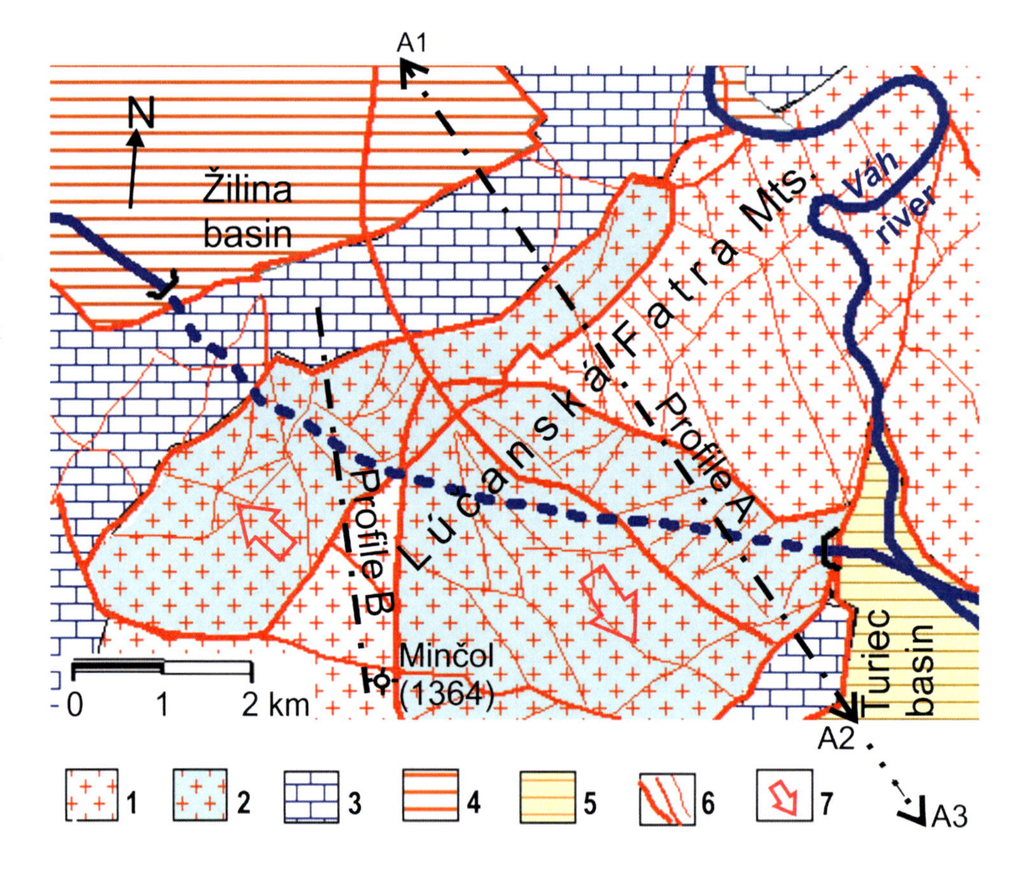

Fig. 168.2 Scheme of tectonic inhomogeneity (modified after Ondrášik et al. 2000). *1* Crystalline granitic rock, *2* Crystalline granitic rock with open joints and saturated, *3* Mesozoic rock (carbonates predominate) locally disrupted and karstified, *4* lower Tertiary flysch strata, *5* upper Tertiary shale and conglomerates, *6* faults, *7* direction of down movements and stress release

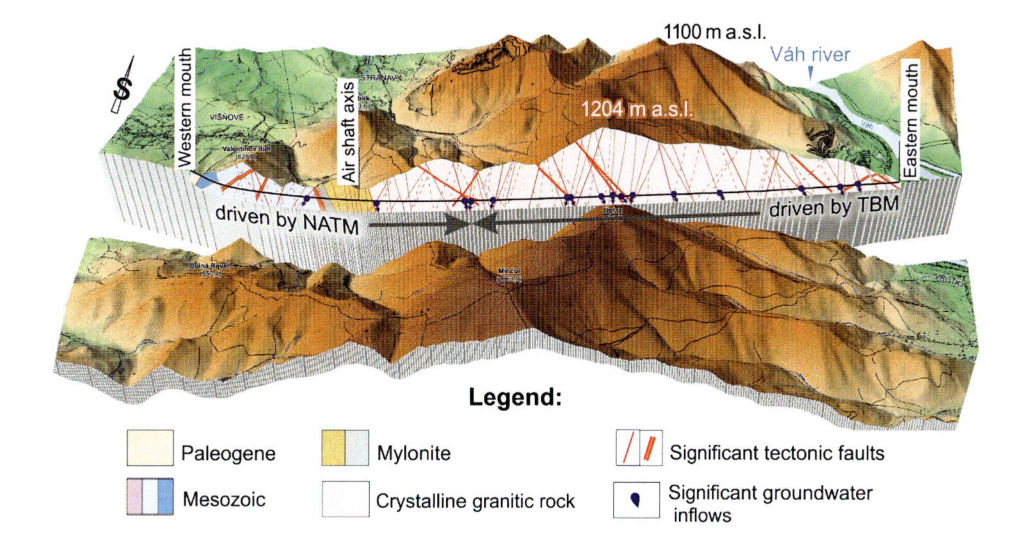

Fig. 168.3 A three dimensional geological profile on the axis of the pilot tunnel Višňové

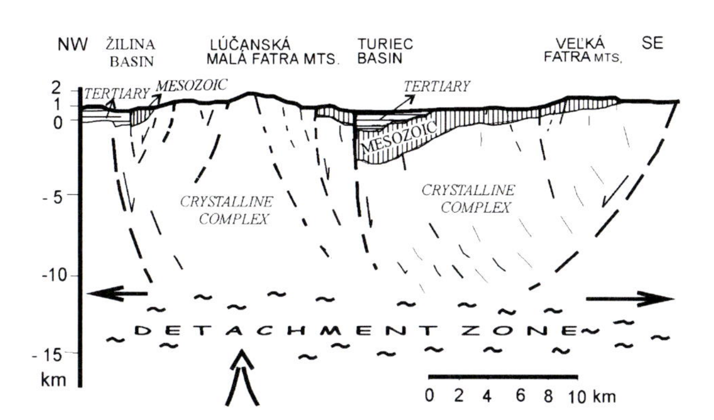

Fig. 168.4 A kinematic model of the Lúčanská Fatra horst and the Turčianska kotlina basin—*left* part of the section is delineated on the map in the Fig. 168.2 as the profile A, *right* part is out of the map (Ondrášik et al. 2000)

and karstified carbonates alternate with impermeable shales, marls and faults filled with clay. Shales and marls are swelling in contact with water.

A tectonic fault with crushed rock up to 70 m thick occurs at the contact between Mesozoic suit and granitic rock. Marginal part of granitic rock mass in a zone 300 m thick consists of crushed mylonites.

The most problematic activity was downward driving up to a distance of 1,300 m from the pilot tunnel's mouth. Concentrated groundwater inflow exceeding a standard limit 30 l/s. Outflows died out some 80–100 m from the gallery face. The total groundwater inflow reached more than 140 l/s. Water from the west section of the pilot tunnel concentrated at the face and was pumped out until perforation to the east section, which declined towards the east mouth. Groundwater inflow from the gallery face was reduced by 35 m long drainage wells declined at 17 % with perforated steel tubing of 156 mm in diameter. Groundwater runoff into tubes was initially gravitational under high pressure, later by pumping by seed drill into sump. There were 7 sumps altogether. Dewatering of the rock mass was successful by pumping about 60 l/s from drainage wells and numerous outflows. Polluted water near the pilot face was separated and pumped at about 30 l/s into sump tanks. Average advance of the tunnelling was 75 m per month.

168.3.2 Eastern Pilot Tunnel Section Driven Using TBM

The eastern pilot tunnel section was driven into granitic rock, predominantly tonalite, using TBM. The first part up to 70 m was driven by NATM. Weathered rock was up to 300 m. Fault zones of three main systems predominated in the rock mass. Average total groundwater inflow reached up to 10 l/s. Monthly digging advance reached up to 200 m.

Water bearing fault zone up to 12 m wide of oblique position to the pilot tunnel occurred at a distance of 442 m from the mouth. Concentrated groundwater inflow reached up to 80 l/s. Suffosion and caving-in took place. Concrete floor was executed in front of the cutting head. Negotiating this fault in the length of 50 m took 96 days.

Various fault zones were cut onwards up to the distance 2,810 m from the mouth and 180 m/month was driven, which doubled the advance in comparison to NATM. Groundwater inflow gradually increased up to 120 l/s at the pilot tunnel face. Total outflow from the eastern passage of the pilot tunnel reached up to over 420 l/s. The gallery floor deformed under TBM in fault zones and stabilization measures took place in front of the cutting head, often under water shower from the roof of the gallery, and caused mining retardation.

Wide groundwater bearing fault zone oblique to the gallery was intersected at the distance of 2,892 m from the adit. Ground water inflow up to 100 to 120 l/s under pressure 3.1 MPa occurred after cutting its marginal part. Due to caving-in a large cavity of 8 m long, 3 m wide and 9–12 m high size was forming.

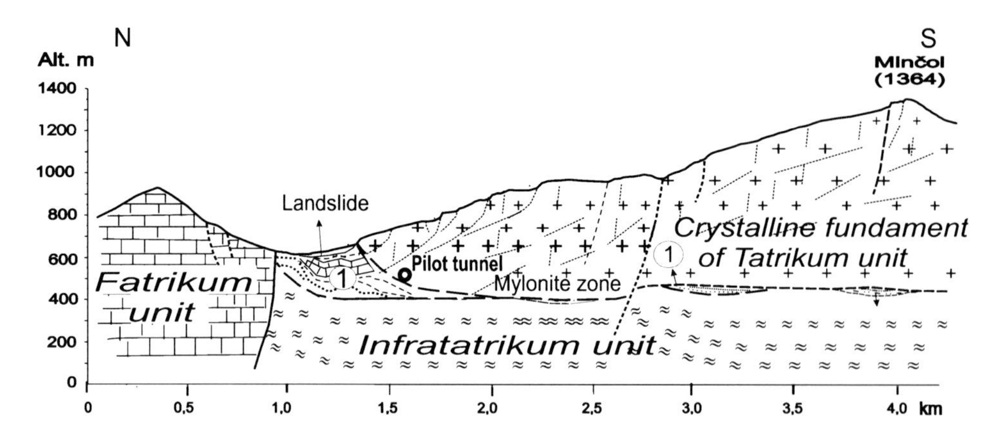

Fig. 168.5 Geological profile across the tunnel—profile B in the Fig. 168.2 (Ondrášik et al. 2009)

A concrete barrier was built in front of the collapsed earth and a set of drainage wells were drilled to enable stabilization of the cavern roof by ascending cement injection trough 14 m long tubes placed in horizontal boreholes. Cement was washed out in critical passages with concentrated water streams. These critical parts were tightened by polyuretan resin before cementing. Polyuretan resin mixed with cement proved to be strong enough to stabilize the unstable rock roof of the gallery.

Pilot tunnel drainage affected a reduction in the discharge of creeks on the surface and some of them even dried out in the surface zone of about 7 km width. It is supposed that rock mass dewatering will minimize tunnelling risks. As well there will be time enough to decide if water from the tunnel is used for water supply or measures will be taken to protect tunnel walls against ground water inflow under pressure, and reverse the hydrogeological regime near to the initial state.

Tectonically crushed rock represented about 40 % of excavated rock in the gallery, which was four times more than it was supposed. Geologically conditioned overbreaks exceeded 300 m^3, and this means about 0.3 % of total excavated rock. Rock mass was drained before the highway tunnel is constructed. Groundwater chemical analyses indicated only low hydrocarbonate aggression in some zones. Sulphur aggression was proved in a passage from 300 to 800 m of the west pilot tunnel mouth.

Mylonites found in the gallery stimulated to supplementary research concerning a thrusting of Paleozoic granitic rock mass (Tatricum) over older basement (Infratatricum) (Ondrášik et al. 2009). It is supposed that mylonites were generated on the zone of overthrusting during Mesozoic cycle of Alpine orogenesis at the depth of 10–15 km. Geologic profile across the tunnel was constructed (Fig. 168.5) to illustrate enormous disruption of overthrusted granitic rock.

168.4 Conclusion

Rock mass in the pilot tunnel was found tectonically intensively fractured in a range 40 %. Except of first category, all categories of tunnelling rock classification were found. Groundwater inflow was extremely high, particularly along the fault zones. Maximum total water outflow from the granitic rock part of the gallery reached more than 420 l/s when the face was 4,600 m from the adit and decreased to 200 l/s after several months. Concentrated inflows reached up to 120 l/s under pressure 3.1 MPa and created suffosion with incavitation up to some tenth m^3, 0.3 % in total of excavated rock. Maximum groundwater inflow occurred at a gallery face and gradually reduced to zero at a distance 50–100 m. Decrease of water discharge occurred in creeks on the surface. Groundwater inflows were also from the west gallery section; however these were supposed to happen. Negotiating of wide fault zones filled with clay required particular nontraditional excavation techniques for each fault zone.

The pilot tunnel construction brought invaluable knowledge and experience to be used in the construction of the highway tunnel.

Acknowledgment Authors appreciate generosity of the directory of the highway company to use data from the geological investigation. We acknowledge the Ministry of Education, Science, Research and Sport of the Slovak Republic for funding the project VEGA No. 1/0828/13.

References

Matejček A, Bohyník J, Hyčko M (2002) Diaľnica D1 Višňové—Martin, tunel Višňové. Final report (in Slovak). Geotest, Geofos and Geohyco companies. Geofond Bratislava

Matejček A, Bohyník J (2006) The importance of the exploration gallery driven for the Višňové tunnel. Tunel 15, 4, GRAFTOP Prague, pp 12–17

Ondrášik R, Matejček A, Holeša Š, Vrábľová K (2000) Gravitational tectonics along the tunnel Višňové in the Lúčanská Fatra Mts. Slovakia Mineralia Slovaca 32:429–438

Ondrášik R, Putiš M, Gajdoš V, Sulák M, Durmeková T (2009) Blastomylonitic-cataclastic zones and their influence on slope deformations in the Lúčanská Fatra Mountains. Mineralia Slovaca 41:395–406

Verification and Validation of Hydraulic Packer Test Results in a Deep Lying Tunnel Project

169

Ulrich Burger, Paolo Perello, Sacha Reinhardt, and Riccardo Torri

Abstract

For the deep lying, trans boundary Brenner Base Tunnel several packer tests have been carried out in deep boreholes for a hydrogeological characterisation of metamorphic rocks. For permeability testing (expected range from $K = 10^{-7}$ to 10^{-12}m/s) Pulse-, Slug-, Drill stem and active flow tests were applied. Due to the importance of the hydraulic parameters for the technical design of the tunnel system and the environmental risk assessment, the testing results were methodologically verified. The verification method and the final result of the evaluation are shown. Calibrate numerical models of built tunnel sections are used to validate the hydraulic testing results, first experiences and results in a granite tunnel section are shown.

Keywords

Tunnel • Packer tests • Verification • Hydraulic conductivity • Metamorphic rocks

169.1 Overview BBT Project

The trans-boundary Brenner Base Tunnel is a 55.6 km long railway base tunnel project with a max. overburden of 1.850 m. As part of the Transeuropean network axes 1 (TEN1 Helsinki-Valetta) the base tunnel will connect Austria (Innsbruck) with Italy (Franzensfeste/Fortezza). The procedure project has been worked out from 2004 until 2008, since 2008 the tunnel is under construction. Until 2013 approximately 28 km of tunnels (3 access tunnels, exploration tunnels and first main tunnel-sections) have been built. It is planned that the excavations of the tunnel system (total length: 200 km due to the 3-tube tunnel system: 1 exploration tunnel, 2 tubes for the main tunnels) will be finished in 2021.

U. Burger (✉)
Galleria di Base del Brennero—Brenner Basistunnel BBT SE, Amraser Straße 8, 6020, Innsbruck, Austria
e-mail: ulrich.burger@bbt-se.com

P. Perello
GDP Consultants (I-Torino), Saint-christophe, Italy
e-mail: perello@gdpconsultants.eu

S. Reinhardt
Solexperts (CH-Mönchaltorf), Mönchaltorf, Switzerland
e-mail: sacha.reinhardt@solexperts.com

R. Torri
SEA consulting srl (I-Torino), Turin, Italy
e-mail: torri@seaconsult.eu

169.2 Hydraulic Testing

169.2.1 General Procedure

The hydraulic tests were carried out after finalizing the deep core borings (200–1.320 m). On the basis of the core logging and results of the borehole geophysics the borehole sections to be tested and the corresponding depth and length of the packed-off test intervals were defined. The hydraulic tests were carried out from borehole bottom upwards and covered usually the whole borehole. The test interval length of the double packer tests ranged from 5 to 35 m, the single packer tests were made usually in the deepest testing interval or as long testing section (even >100 m).

G. Lollino et al. (eds.), *Engineering Geology for Society and Territory – Volume 6*,
DOI: 10.1007/978-3-319-09060-3_169, © Springer International Publishing Switzerland 2015

169.2.2 Methodology of Testing

Hydraulic tests were carried out using a straddle packer system, which consisted of an upper and lower inflatable packer in order to confine a test interval section in the borehole. The test section between the packers comprises a perforated rod allowing formation water to enter the riser pipe or, conversely, allowing injecting water through the test rods into the formation. Three pressure transducers were measuring the pressures below, within (interval pressure) and above the test interval (annulus pressure). The test system was installed to the specific depth by means of a test tubing (pipe). A downhole shut-in valve, mounted between the system and the test tubing, enabled to close and open the connection between the test interval and the test tubing instantaneously. All parameters including pressure (pressures down-hole, atmospheric pressure), temperature in the borehole, flow rate and other parameters were recorded by means of an automated data acquisition system and displayed real-time on the PC-screen of the test engineer.

The reliability of estimated formation parameters is increased by carrying out several test procedures (methods) for the same test interval (Quinn 2012). After packer positioning and inflation, a test series starts with an initial pressure recovery phase with closed downhole shut in valve, which allows the test zone pressure to recover toward the static formation pressure.

The following test sequences depend on the estimated transmissivity and the pressure potential of the specific test zone. With a pulse test or a slug test a first transmissivity estimate of the test zone can be provided. A pulse test is conducted by exposing the test interval to a short under or overpressure and monitoring the pressure response as it recovers toward the formation pressure. The over- or under pressure is produced by emptying or filling the test tubing with closed downhole shut-in valve. The pulse is transmitted to the test zone by opening and closing the downhole shut-in valve for around 3–10 s. The penetration depth into the formation of a pulse test is considerably small (in the range of dm–m). Pulse tests are also used to determine the compressibility of the test zone. The implementation of a slug test is similar to a pulse test, but the imposed pressure pulse in a slug test recovers towards formation pressure with open downhole shut-in valve. During the open shut-in valve-period, the water level in the test tubing corresponds to the current pressure in the test interval and active in- or outflow from the formation to the test tubing (or vice versa) takes place. This phase is also referred to as slug flow phase.

In tight formations, the slug flow phase is normally interrupted by closing the downhole shut-in valve. The following accelerated pressure recovery with closed shut-in valve may also be analysed, if the pressure recovery during the previous slug flow phase is small in relation to the initial pressure pulse. The Drill stem test derives from the petroleum industry and consists roughly of two consecutive slug withdrawal tests, which are interrupted after a certain time by closing the downhole shut-in valve.

The recovery behaviour of pulse and slug tests are influenced by skin effects (diminished or enhanced permeability in the vicinity of the borehole). Therefore, a test sequence should contain an active flow test as a constant head injection test or a constant rate test, if feasible. Flow tests are less sensitive to skin effects. Constant head and constant rate tests are performed by extracting or injecting water for a certain time span (usually 20–30 min) and by maintaining constant either (i) the injected/extracted flow rate, with a consequent variations of the head, or (ii) the head with a consequent variation of the injected/extracted flow rate. These tests provide a larger scale permeability value (larger penetration depth into the formation), depending on the transmissivity and the storage coefficient of the formation. In low permeable formations, the radius of investigation ranges to a few meters and is still relatively small. Constant rate and recovery phases after shut-in (downhole valve closed) allow the use of transient pressure analysis methods (Bourdet et al. 1989) based on the analysis of the derivative of pressure versus the appropriate time function (natural logarithmic or Agarwal/Horner superposition time) in a diagnostic plot. The method facilitates the diagnostic of the different flow phases of a test which supports the correct use of the straight-line analysis method and provides information of flow behaviour in the formation.

Test analysis was conducted using type curve, straight-line analysis methods together with diagnostic log-log pressure plots. The Cooper-Bredehoeft-Papadopulos type-curves were used to analyse both slug and pulse tests (Cooper et al. 1967; Bredehoeft and Papadopulos 1980). Constant head injection tests were analysed according to Jacob and Lohman (1952) and Doe and Geier (1990), recovery tests after Agarwal (1980) and Horner (1951).

169.2.3 Overview Output of Testing

All the tests were focused on fractured metamorphic rocks of various nature, where the primary permeability related to porosity is very low. It was considered that in these rocks, the permeability is mainly governed by fractures and faults and their aperture, spacing and infilling. The tested rocks are all characterised by low to high grade metamorphic conditions (greenschist to amphibolite facies) and have been grouped in the following classes: 1—Phyllites (30 tests); 2—Metabasites and serpentinites (2 tests); 3—Calcschists with prevalent carbonatic composition (8 tests); 4—Calcschists with prevalent phyllosilicatic composition (37 tests); 5—Quartzites (3 tests); 6—Gneissic rocks (7 tests); 7—Granites

(14 tests); 8—Marbles (8 tests); 9—Mixed successions including interlayered anhydrites, anhydritic schists, phyllites and quartzo-micaschists (17 tests). A total of 126 tests have been executed; 78 of these have been done in boreholes located in Austria and 48 in Italy. A general overview of the executed tests is shown in the two diagrams of Fig. 169.1, irrespective of the tested rock-type. The same diagrams also show the general results of the verification works that will be discussed in detail in the following paragraphs.

169.3 Verification of the Testing Results

The verification analysis has been carried out with the aim to evaluate the reliability degree of the hydraulic test results in relation to the geological and structural setting of the test interval. The study consists of a comparative analysis of the rock sampled at the depth of the hydraulic test and the result of the hydraulic testing. It is necessary to specify that a test with low or very low degree of reliability does not indicate that the test is to be rejected but the output seems to be anomalous with respect to the geological setting.

169.3.1 Rock Mass Classification

As a first step, the completeness of the input data has been verified in order to give a first evaluation of the verification process quality. Then, the core drill sample analysis has been carried out regarding (i) lithotype, (ii) depth of the test and (iii) rock mass conditions. The rock mass of each tested interval has been geologically classified on the basis of visual inspections of the drill cores, BHTV analysis and RQD values. The following categories have been defined:

1. rock mass without tectonisation characterized by standard fracturing degree;
2. fault zones distinct in cataclasite or tectonic breccias; it is also reported the presence of "gouge " (core zone) and/or lateral damage zone;
3. rock mass affected by dissolution phenomena with the development of karst morphologies; this concerns mainly carbonatic and evaporitic rocks;
4. rock mass with enlarged fractures owing to gravitational collapse phenomena regarding the shallow portion of the rock mass (until about 300 m deep).

169.3.2 Evaluation Sheet Format

Data have been stored in a database management system specifically designed, from which an "Evaluation sheet" could be extracted directly. The sheet includes several sections as (i) general framework, (ii) characteristics and results of the hydraulic test, (iii) description of the geological features of the tested section, (iv) representation of the borehole logs, core drill photographs and hydrogeological map of the

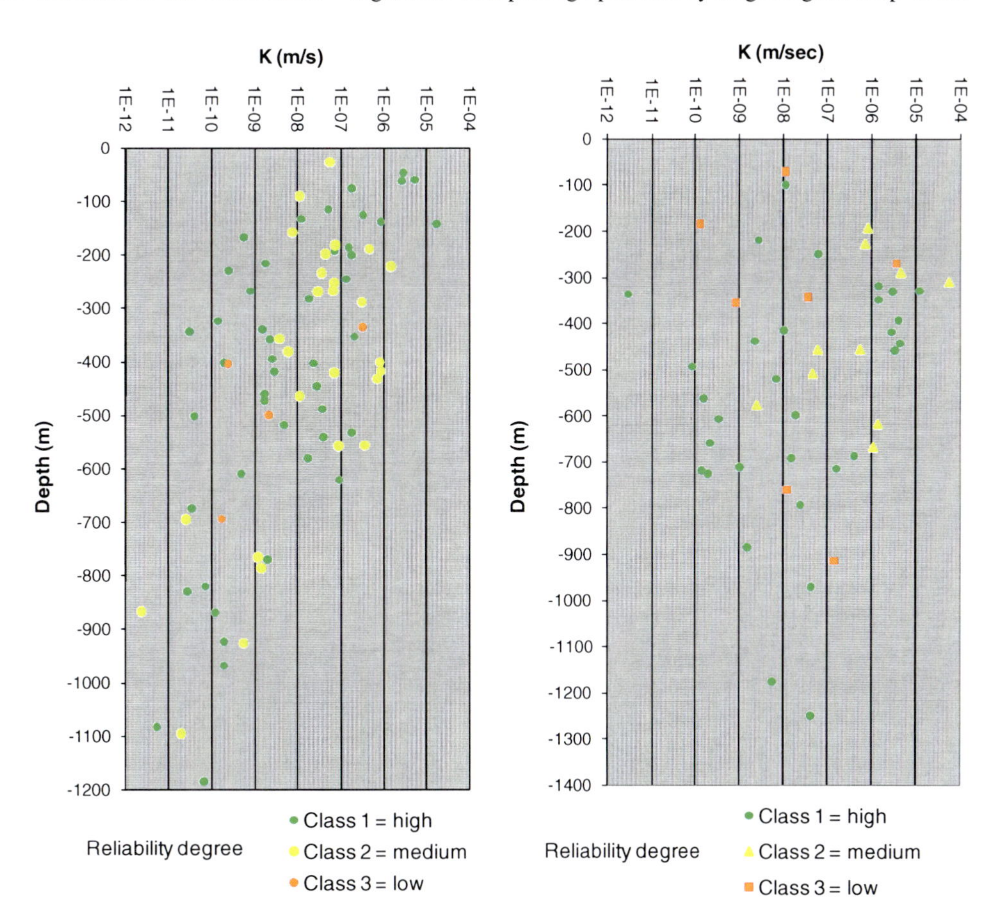

Fig. 169.1 Total distribution of hydraulic conductivity versus depth for the tests executed in Austria (*left*) and Italy (*right*), irrespective of the tested rock-type

test site and (v) the results of the verification process. The results of the study are formulated in terms of (a) "overall evaluation" which consists in a qualitative description and (b) "reliability degree" defined as:

- Class 1 = High
- Class 2 = Medium
- Class 3 = Low
- Class 4 = Very low

The results of the reliability analysis have been also presented in tables and graphs that allow visualizing the distribution and the range of variation of the data.

169.3.3 Verification Results

The evaluation of 126 hydraulic tests carried out in testing intervals lying from 180 to 1320 m depths and with K-values ranging from 10^{-6} to 10^{-12} m/s shows that:

- more than 60 % of the testing results belongs to the class 1, 30 % to the class 2, 10 % to the class 3;
- no test result showed a very low reliability (class 4);
- the reliability of the test results doesn't depend from the rock types. All different rock types are even part of class 1.

Methodological verification analysis shows that borehole hydraulic testing, even applied in deep boreholes and for different rock types provide valuable data for rock mass characterization. It can be observed that homogeneous and low permeability rock mass conditions have a greater degree of reliability. Regarding the granite lithotype, more than 75 % of the hydraulic tests have been evaluated with high and medium reliability degree (classes 1 and 2). The permeability ranges from very low degree ($<1 \times 10^{-8}$ m/s) and medium to high degree (between 1×10^{-6} and 1×10^{-8} m/s) in relation to the fracturing degree. A low degree of reliability has been recognized for a high degree of fracturing: in these cases the tests results gave a low permeability values, probably due to a reduced connectivity of the fractures network.

169.4 Validation

169.4.1 Methodology

Using calibrated numerical models for excavated tunnel sections back analysis are performed. The hydraulic conductivity for the different rock masses obtained by the numerical models is compared to the verified values derived from the hydraulic packer tests.

169.4.2 Result of the Validation in the Granite Area

The Aica-Mules exploration tunnel was excavated in granitic rocks. In this case feedbacks on rock permeability have been obtained from hydrogeological numerical modelling (Feflow 6.0; DHI-WASY GmbH). In the granite not affected by faulting and characterised by a simple fracture network, the hydraulic conductivities obtained by borehole tests ranges from 1×10^{-7} to 1×10^{-8} m/s. These test results were classified as "Class 2: medium reliability". The permeability as output from the numerical hydrogeological model is approximately 1×10^{-9} m/s, therefore 1–2 orders of magnitude lower than the conductivity obtained by in situ testing. This incongruence is probably related to the fact that small scale hydraulic conductivities with locally high importance for the hydraulic condition are not relevant for the conductivity at a large, pluri-decametric scale, due to the poor hydraulic connectivity among fractures related to a low fracturation density.

As far as fault zones are concerned, a large variability has to be taken into account for the hydraulic conductivity, due to the presence of products deriving from cataclastic fragmentation having mechanical characteristics and granulometric composition ranging over a wide range. The in situ hydraulic testings demonstrate these hydraulic conductivity variations along fault zones, with values spanning over the range from 5×10^{-6} to 5×10^{-9} m/s. The lower value has been observed in a single case and is related to a core zone where the granite is reduced to a fine grained fault gouge, comparable to a silty clay. The higher values are representative for fault damage zones where the fracture density is high to very high (indicatively 5–30 joint/m^3) and where decimetric to metric layers of tectonic breccias are locally present. The packer tests, despite their number is rather limited, seem to indicate that in fault zones high hydraulic conductivities are by far the most common condition. The prevalent large hydraulic conductivity observed at the small scale seems to be mostly confirmed at the great scale by the numerical models with conductivities ranging from 1×10^{-6} to 8×10^{-7} to m/s.

References

Agarwal RG (1980) A new method to account for producing time effects when drawdown type curves are used to analyze pressure buildup and other test data. Society of petroleum engineers, SPE Paper 9289, presented at SPE-AIME Meeting, Dallas, Texas, Sept 21–24

Bourdet D, Ayoub JA, Pirard YM (1989) Use of pressure derivative in well-test interpretation. SPE Formation Eval 4:293–302

Bredehoeft JD, Papadopulos SS (1980) A method for determining the hydraulic properties of tight formations. Water Resour Res 16 (1):233–238

Cooper HHJR, Bredehoeft JD, Papadopulos SS (1967) Response of a finite-diameter well to an instantaneous charge of water. Water Resour Res First Quarter 1967:263–269

Doe TW, Geier JE (1990) Interpretation of fracture system geometry using well test data, technical report 91-03. Swedish Nuclear Fuel and Waste Management Co. (STRIPA), Stockholm, Sweden

Horner DR (1951) Pressure build-up in wells, third world petroleum Congress. E. J. Brill, Leiden II, pp 503–521

Jacob CE, Lohman SW (1952) Nonsteady flow to a well of constant drawdown in an extensive aquifer. Trans Am Geophys Union 33 (4):559–569

Quinn P, Cherry JA, Parker BL (2012) Hydraulic testing using a versatile straddle packer system for improved transmissivity estimation in fractured-rock boreholes. Hydrogeol J 20:1529–1547

Change in Hydraulic Properties of Rock Mass Due to Tunnelling

170

Bernard Millen, Giorgio Höfer-Öllinger, and Johann Brandl

Abstract

Presented and discussed are aspects of changes to the hydraulic properties of rock mass due to tunnelling, but in particularly the change due to the opening or making of new fractures and voids in the rock mass as a result of deformation or relaxation of the tunnel opening and hence an associated increase in the permeability of the rock mass surrounding the tunnel. This process principally occurs in tunnels were high stress fields persist effecting rapid and large scale deformation or relaxation (>100 mm) in the presence of (heavily) jointed rock mass and fault zones. Examples and observations from water and ground inflow events, associated with hydraulic property changes which have occurred during the mining of the Tapovan-Vishnugad Head Race Tunnel, India, are given and which have caused trapping of the Tunnel Boring Machine (TBM) resulting in time delays and cost increases.

Keywords

Ground deformation • Increased permeability • Water inflow • Ground inflow

170.1 Introduction

A change in the hydraulic properties of rock mass due to tunnelling has been observed in the Garhwal Himalaya, Uttarakhand, India during construction of the deep seated 12.1 km long head race tunnel (HRT) for the NTPC Ltd 520 MW Tapovan-Vishnugad Hydro Electric Project (HEP). The project area lies within the Dhauliganga and Alaknanda Valleys and consists of high strength medium to high grade metamorphic rocks belonging to the Central Himalayan Crystalline Series (Heim and Gansser 1939; for a recent review see Yin 2006).

Since October 2008, 8.6 km of the HRT has been under construction by a Herrenknecht Double Shield—Tunnel Boring Machine (DS-TBM). The rest of the HRT is been constructed by drill and blast methods. During TBM driving, a steel reinforced 0.3 m thick concrete hexagonal segmental lining is inserted behind the machine, pea ballasted and grouted making the internal finished diameter of the TBM-HRT 5.64 m; the excavation diameter is 6.57 m.

To date three TBM trapping events—all associated with subsurface water inflow—have severely hampered the HRT excavation resulting in time delays and cost increases. The first trapping event occurred in December 2009 at chainage (Ch) 3,016 m at a depth of some 900 m in a heterogeneous fault zone (Brandl et al. 2010; Millen and Brandl 2011). During trapping the front and telescopic shields were jammed in and dented by major wedge slides. Approximately 24 h later, massive surges of high pressure subsurface water, containing faulted rock material, broke two crown segments of the segmental lining immediately behind the tailskin with the initial flow rates reaching circa 700 L/s compounding the trapping problem.

The second and third trapping events—which to date have not been reported on—happened in February and October 2012 at Ch 5,840 and 5,859 m respectively in the same circa 20 m wide fault zone at a depth of some 700 m. This fault zone lies at a very acute angle to the tunnel axis meaning the TBM had (will have) to drive through this zone

B. Millen (✉) · G. Höfer-Öllinger · J. Brandl
Geoconsult ZT GmbH, Hoelzlstraße 5, 5071 Wals bei Salzburg, Austria
e-mail: bernard.millen@geoconsult.eu

G. Lollino et al. (eds.), *Engineering Geology for Society and Territory – Volume 6*,
DOI: 10.1007/978-3-319-09060-3_170,

for at least 35 m. At the time of writing the TBM was still trapped at Ch 5,859 m.

When the second event occurred, the face and surrounding conditions were initially dry and due to over excavation and collapse a cavity of several cubic metres had developed around the cutter head and front shield of the machine in the soil-like stiff clay-rich fault gouge and breccia of which the fault zone consists of. As in the first trapping event, water inflow (1–2 L/s) started some 20 h later. The situation then greatly deteriorated as the water rapidly eroded the water sensitive fault gouge and breccia causing further cavity development, ground creep and ground inflow through the cutter head and shield openings trapping the TBM.

Interestingly, in both described events, the water inflow started approximately 1 day after the initial trappings. The events caused subtle and rapid major changes to the hydraulic properties—particularly increased permeability—of the rock mass within several metres and probably many tens of metres of the tunnel walls.

170.2 Hydraulic Properties of Rock Mass

The hydraulic behaviour of rock mass is much more complicated than in granular soil and there can be huge differences in properties over short distances caused by the distribution of rock types, their tectonic history and distribution in relation to a drainage basin or system.

Subsurface water flow in rock mass is generally governed by (1) the (unaltered) rock porosity and permeability and (2) the rock mass porosity and permeability. In tunnelling geology the usual concern is with the flow of subsurface water through a granular soil or rock mass as is the case in this paper.

Porosity is the measure of the voids in a material and is given in a percentage. However, the voids in a rock must be in contact otherwise the rock is impervious. Permeability is a general term which describes the ability of a porous medium to allow the flow of fluid through it. Permeability is given in K (the coefficient of permeability) after experiments by Darcy (1856) and is also called *hydraulic conductivity* and has the unit m/s. For flow to take place through a saturated material there must be a difference in total head across the medium.

The hydraulic permeability of rocks is generally very low and ranges commonly from 10^{-13} to 10^{-10} m/s. Exceptions are for example coarse grained porous sandstones.

Hydraulic conductivity or permeability and subsurface water flow in rock mass is mainly governed by the interconnection of fractures (bedding, schistosity, joints, faults, etc.) in the rock mass or *fracture permeability*. The hydraulic

Fracture Width	k [m/s]	≙ (Soil)
0,1 mm	0,6 E-06	Silt
	0,3 E-06	
0,2 mm	0,5 E-05	Sand
	0,2 E-05	
0,4 mm	0,4 E-04	
	0,2 E-04	
0,7 mm	0,2 E-03	Gravel
	0,1 E-03	
1,0 mm	0,6 E-03	
	0,3 E-03	

Fig. 170.1 The hydraulic permeability of rock mass with a fracture system of defined width and an average distance between fractures of 1 m in comparison to the hydraulic permeability of granular soils from (Wittke 1984) taken from Prinz (1997)

permeability of rock mass is not only inhomogeneous but also highly anisotropic.

In Fig. 170.1 is given an idealised and homogeneous rock mass model in comparison to the hydraulic permeability of granular soils.

The rows of values show that the permeability in the direction of wider fractures is the 3rd–4th negative power (10^{-3}–10^{-4}) of the mean fracture width in relation to the cubic law for fluid flow in rock fractures (Witherspoon et al. 1980). This means that a single wide fracture or narrow distances between multiple fractures, for example in joint and fault zones, can or on the whole control the permeability of the rock mass.

The subsurface water flow is also extremely dominated by the geometry of the fractures and their extension. Fracture walls are also generally rough and the fracture width can change over short distances.

Where there are sequences of low and high permeability in a rock mass there is not only fracture anisotropy to consider in the horizontal direction but also the bedding or schistosity anisotropy of the rock mass to be considered in the vertical direction.

In dissimilarity to granular soil aquifers, where homogeneous and isotropic characteristics are not fully given but a quasi homogeneity and isotropy, fractured rock mass aquifers, with very different fracture widths and grades of opening, have very anisotropic and turbulent flow regimes (Prinz 1997).

170.3 Change in Hydraulic Properties of Rock Mass

Changes to the hydraulic properties of a rock mass are achieved principally by the following: (1) chemical processes such as precipitation, dissolution and weathering; (2) erosion processes; (3) closure of fractures and voids by physical processes (4) opening or making new fractures and voids by physical processes. Process (1) can either decrease or increase the hydraulic permeability, process (3) causes a decrease and processes (2) and (4) cause an increase.

Long term erosion and dissolution and weathering of rock mass causing hydraulic changes (or increased permeability) has been examined before particularly in relation to problems with the long term operation of water transport tunnels (e.g. Gysel 2002; Lipponen et al. 2005) and will not be discussed further.

There are plenty of examples of large scale water and related ground inflows and hence the development of cavities caused by short term erosion events (e.g. Schwarz et al. 2006; Wenner and Wannenmacher 2009). However the recognition that such events radically change the hydraulic properties and framework of the rock mass surrounding the tunnel up to several tens of metres from the tunnel walls has not been considered to date in discussion of the same.

As stated in the introduction (Chap. 1), events surrounding the development of water and ground inflow during the excavation of the Tapovan-Vishnugad HRT led to changes to the hydraulic properties of the rock mass around the tunnel. Not only the processes (1) and (2) played major roles but also process (4), the opening or making of new fractures and voids. The changes occurred in high stress fields and effected rapid and large scale deformation or relaxation (>100 mm) in sections of (heavily) jointed rock mass and fault zones surrounding the TBM.

As presented in Chap. 2, hydraulic permeability in rock mass is principally governed by fracture permeability. If say some 10 mm of deformation occurs and this affects, for discussion purposes, some 10 m of jointed rock mass evenly away from the tunnel walls and this deformation is taken up evenly by the opening of 10 existing fractures from 0.1 to 1. 1 mm at 1 m spacing, then there will be an automatic increase in the hydraulic permeability of several orders of magnitude according to the model given in Fig. 170.1. Of course in reality, at such low levels of deformation, this will not be the case and the deformation will be taken up within the first 1–2 m of the tunnel walls and be much more anisotropic. However, when >100 mm of deformation or relaxation is registered then the range of impact will be much greater reaching at least several if not tens of metres into the tunnel walls.

In the case of the TBM trapping event at Ch 3,016 m (see Chap. 1), the water inflow reaching circa 700 L/s started approximately 24 h after the initial trapping. The TBM over cuts the shields by a maximum of 200 mm at the crown and it is known from observations during the recovery of the TBM that the gap or over cut between the bore and shields closed completely in the upper arch during the event therefore large scale deformation of the jointed and faulted rock mass took place. In this case the processes of change included (1) opening of new and/or existing fractures leading to increased hydraulic permeability around the tunnel and (2) high pressure subsurface water wash out or erosion of joints and the fault core zone, which contained clay and other soil-like materials, into the tunnel leading to development of a conduit-like structures and associated water and ground inflow.

The conduit-like structures were quickly generated in the fault zone i.e. the opening of water bearing joints in connection with the fault core zone containing soil-like material led to rapid erosion of the fault core and the development of an extensive interconnected 3-D network of erosion channels and hence a radically changed hydraulic framework. The fault core was also found to be in direct connection with a sequence of pervious water bearing heavily jointed quartzitic gneiss and quartzite some 40 m above Ch 3,016 m. This sequence of rocks was later encountered between Ch 3,110 and 3,250 m where the combined flow-rate reached up to 60 L/s and an immediate decrease in the water inflow at Ch 3,016 m was observed confirming the water inflow model as given in Millen and Brandl (2011). At the time of TBM restart at Ch 3,016 m in March 2011 the water inflow at Ch 3,016 m had reduced to circa 120 L/s. By the time the TBM had past Ch 3,250 m the water inflow at Ch 3,016 m had reduced to circa 60 L/s.

In the event at Ch 3,016 m the dilemma was not the fact that relaxation and an increase in permeability took place close to the tunnel, but that a nearby pervious water bearing aquifer was short circuited through a fault zone due to the relaxation and hence the large high pressure water and ground inflow developed. If no water bearing pervious sequence had been tapped, then there would have been little water inflow, however a deterioration and increase in permeability of the rock mass surrounding the tunnel would still have taken place.

In the case of the TBM trapping event at Ch 5,840 m, the water inflow of some 1–2 L/s started 20 h after the event. The undisturbed soil-like stiff clay-rich fault gouge and breccia which make up the 20 m wide fault zone are impervious and were acting as an aquitard to the above lying semi-pervious but highly pressurized saturated jointed gneiss and schists sequences. In one of the exploratory bore holes carried out, a water inflow was recorded during drilling of circa 1 L/s emerging out of jointed augen gneiss and a pressure of 14 bar was measured after installation of a packer and manometer after completion of the bore hole. As at Ch 3,016 m, large scale deformation took place resulting in complete closure of the gap between the bore and shields and the development of new cracks or fractures in the stiff soil-like fault gouge and breccia hence allowing water from the overlying jointed rock sequences to penetrate through into the TBM causing erosion of the same and a rapid deterioration of the ground conditions surrounding the machine (see Chap. 1).

170.4 Consequences and Conclusions

To date the authors have found no specific literature on changes to hydraulic properties related to large scale rock mass deformation or relaxation around tunnels and the related increase in permeability due to the opening or making of new fractures and voids but believe it to be important aspect of water inflow into tunnels and understanding its control.

The increase in permeability of the ground by one or two orders of magnitude due to loosening created by blasting and the delay before the installation of temporary support has been recognised for some time (Howard 1991). It is noted here that the observations of increased permeability occurred during DS-TBM mining where the segment support is first installed and finalised (pea ballast and grouting) behind the 12.5 m long shields.

Generally it is important to control water inflow but it is even more important to try and stop erosion by the same and the generation of ground inflow as this has a more detrimental effect on the rock mass surrounding the tunnel and the tunnel advance itself. Preventative measures to stiffen up ground include grouting, ground freezing or the installation of grouted pipe roofs and the like ahead of the face into suspected poor water bearing geology. However, such measures are expensive and often met with scepticism about whether or not they will work and or reduce costs in the long term.

Remedial measures to fill cavities and the like are costly and cause time delays. All three areas where the Tapovan-Vishnugad TBM has become trapped to date will require extensive grouting and other works to be carried out such as replacement of the segmental lining with a more robust lining as the HRT will be a pressure tunnel during operation.

The measures carried out to achieve TBM recovery in the three cases quoted included (1) Ch 3,016 m: the building of a 143 m long by-pass tunnel, a 25 m long water diversion drift and an extensive drilling and grouting campaign as described in Millen and Brandl (2011); (2) Ch 5,840 m: installation of an extensive 2 layer chemically grouted pipe roof above the machine and then over mining under the same using a steel girder, shotcrete and rock anchor support systems by access through the telescopic shield; (3) Ch 5,859 m: over mining from behind the tailskin using a fore poling, steel girder, shotcrete and rock anchor support systems.

The information on hydraulic change presented here should also be considered during the implementation of grouting schemes and during the determining of the behaviour of subsurface water inflow into underground structures in the short, middle and long term i.e. prior, during and after construction.

Further thought has to be given as to where these changes occur i.e. in what types of rock or rock mass can they be expected and whether such changes are more dependent on the existence of competent (brittle deformation) or incompetent (plastic deformation) rock mass and the prevailing hydrogeological framework.

Acknowledgements The authors are indebted to various staff of NTPC Ltd., Larsen & Toubro Ltd., Alpine—Mayreder GmbH, Herrenknecht AG and Geoconsult ZT GmbH for providing data for this paper.

References

Brandl J, Gupta VK, Millen B (2010) Tapovan-Vishnugad hydroelectric power project—experience with TBM excavation under high rock cover. Geomech Tunn 3(5):501–509

Darcy H (1856) Les Fontaines Publiques de la Ville de Dijon (in French) [The Public Fountains of the City of Dijon], Dalmont, Paris

Gysel M (2002) Anhydrite dissolution phenomena: three case histories of anhydrite karst caused by water tunnel operation. Rock Mech Rock Eng 35(1):1–21

Heim A, Gansser A (1939) Central Himalaya. Soc Helv Sci Nat Zurich 73:1–245

Howard AJ (1991) ITA working group on maintenance and repair of underground structures. Report on the damaging effects of water on tunnels during their working life, vol 6, no I, pp 11–76. Tunnelling and Underground Space Technology

Lipponen A, Manninen S, Niini H, Rönkä E (2005) Effect of water and geological factors on the long-term stability of fracture zones in the Päijänne Tunnel, Finland: a case study. Int J Rock Mech Min Sci 42:3–12

Millen B, Brandl J (2011) TBM recovery under high cover and extreme water-inflow, Himalayas, India. In: Kolić D (ed) 1st international Congress on tunnels and underground structures in South-East Europe "using underground space" April 7–9. Dubrovnik, Croatia, p 10. PDF on USB-Stick on back cover of book, ISBN 978-953-55728-6-2

Prinz H (1997) Abriß der Ingenieurgeologie (in German) [synopsis of engineering geology], 3rd edn, p 545. Ferdinand Enke Verlag, Stuttgart

Schwarz L, Reichl I, Kirschner H, Robl KP (2006) Risks and hazards caused by groundwater during tunnelling: geotechnical solutions used as demonstrated by recent examples from Tyrol, Austria. Environ Geol 49:858–864

Wenner D, Wannenmacher H (2009) Alborz service tunnel in Iran: TBM tunnelling in difficult ground conditions and its solutions: 1st regional and 8th Iranian tunneling Conference, vol 18, 20 May, Tehran, Iran

Witherspoon PA, Wang JSY, Iwail K, Gale JE (1980) Validity of cubic law for fluid flow in a deformable rock fracture. Water Resour Res 16:1016–1024

Wittke W (1984) Felsmechanik: Grundlagen für wirtschaftliches Bauen im Fels (in German) [rock mechanics: principles for economic construction in rock], p 1050. Springer, Berlin

Yin A (2006) Cenozoic tectonic evolution of the Himalayan orogen as constrained by along-strike variation of structural geometry, exhumation history, and foreland sedimentation. Earth-Sci Rev 76:1–131

Groundwater Ingress in Head Race Tunnel of Tapovan: Vishnugad Hydroelectric Project in Higher Himalaya, India

171

P.C. Nawani

Abstract

In underground projects vast uncertainties are confronted in terms of hydrogeology and geology, more often in mountainous terrains like Himalaya. Groundwater is often the main source of problems in tunneling projects and it is more troublesome during construction. Groundwater problems like sudden water ingress or continuous seepages are considered to be the most difficult conditions to predict. The problem is often due to ingress of water in higher volumes than predicted, or worse, the unanticipated encounter of water during tunneling. In Tapovan—Vishnugad Hydro electric project (520 MW), a run of the river scheme, the Head Race Tunnel (HRT) is being driven by Tunnel Boring Machine (TBM). Tunnelling is through the rocks of Central Crystallines which are heavily stressed due to—the presence of Main Central Thrust (MCT) in close proximity and presence high rock cover (>1 Km.). The Tunnel Boring Machine (TBM) was stuck up at RD 3016 m (rock cover ± 990 m) due to major rock fall on the shield near cutter head, in December, 2009. This led to sudden ingress of water (600–700 L/s) in HRT, damaging the pre-cast lining in the crown portion. The water ingress was from quartzite rock through some wide open joints, fractures or fault which were progressively widened due to ground water movement and caused instability problems. The major problems associated with the heavy ingress of water were found to be—difficult working conditions, damage to the structures, stability problems and environmental impact to the ground water resources. The work was called off for more than 10 months in the year 2010. A bypass tunnel was driven to reach the location where TBM was stuck up and damaged so as to carryout repairs. The ingress of water was diverted from HRT through a drift to make the HRT workable.

Keywords

Higher Himalaya • Central crystalline • Heavily stressed • Bypass tunnel

171.1 Introduction

Underground projects located in Himalaya are often confronted with uncertainties and complexities in terms of geology, hydrogeology and in-situ stress because of the prevailing compressional environment in Himalayan region. Groundwater is often the main source of problems in tunnel construction. The occurrence of water is difficult to predict accurately, hence it is prudent to be prepared for large variations both with respect to locations and volumes. Adverse situations like sudden ingress of groundwater are the most difficult conditions to assess and also quite challenging in handling. This ultimately affects productivity and thus cost.

Tapovan—Vishnugad Hydroelectric Project (520 MW), a run-of-the-river scheme, is presently under construction by NTPC Ltd., in the Higher Himalaya of Uttarakhand State in India . The project envisages harnessing of hydro-potential of Dhauliganga—a tributary of Alaknanda river, by utilizing

P.C. Nawani (✉)
Jindal Power Ltd, Gurgaon, India
e-mail: drnawanipc@jindalsteel.com; drnawanipc@gmail.com

G. Lollino et al. (eds.), *Engineering Geology for Society and Territory – Volume 6*,
DOI: 10.1007/978-3-319-09060-3_171, © Springer International Publishing Switzerland 2015

a head of 518 m, with a barrage near Tapovan across Dhauliganga, and a 11.69 km long head race tunnel of 4.8 m dia. and an underground powerhouse (4 × 130 MW) on the left bank of Alaknanda. The reservoir pondage will have a maximum depth of 22 m and have live storage of 13 m, between FRL at EL 1803.5 m and MDDL EL 1790.5 m, with a capacity of 0.57 Million m^3. It will have a small submergence area confined to a 10 ha area providing a short term daily storage to allow peaking.

In this project, tunneling of head race tunnel (HRT) is being done by tunnel boring machine (TBM) in about 9 km length of HRT and about 6 km tunneling by TBM has been achieved so far (Fig. 171.1). The rocks of Central Crystalline Group which are hard, brittle and heavily stressed, due to the prevailing compressional environment and presence of Main Central Thrust (MCT) near the project site and also high vertical cover (upto >1 km) occurring at various points along the tunnel alignment, form the tunneling medium for the entire 11.69 km long HRT.

In December 2009, when HRT was being driven under a high cover zone, there was a rock mass failure followed by a sudden heavy ingress of groundwater from the tunnel face. The TBM shield was damaged and stuck up at RD ± 3,016 m. This incident led to immediate suspension of tunneling which continued for months in the year 2010–11. A bypass tunnel (BPT) was driven to access the HRT reach where TBM was stuck up with the aim to assess the damages suffered to channelize water inflow away from HRT through a drift size opening. Tunneling was resumed in the year 2011.

171.2 Geological Setting

The project area is located in the Higher Himalayan region where the Central Crystallines composed of medium to high grade metamorphic rocks of Tapovan—Helong Formation and Joshimath Formation are thrusted over the Lesser Himalayan rocks along the Main Central Thrust (MCT), near Helong about 2 km downstream of the location of underground powerhouse. The rocks exposed in the area are quartzites, mica schists, banded gneiss, augen gneiss, amphibolites of Tapovan—Helong formation and coarse grained, garnet—mica gneiss, garnet Kyanite gneiss of Joshimath formation.

Fig. 171.1 Excavation of HRT by using TBM

171.2.1 Geology Along Head Race Tunnel (HRT)

The HRT alignment passes through a rough and rugged terrain on the left bank of Dhauliganga and Alaknanda rivers. HRT is being driven through the rocks of Tapovan—Helong Formation dipping 20°–40°/NNE-NE overlain by rocks of Joshimath Formation dipping due NE with considerable variations forming a broad syncline (Fig. 171.2). The synformal structure is favourable for groundwater storage. The fine to medium grained quartz mica gneiss intercalated with schist and two quartzite bands (25–55 m thick) of Tapovan Formation are overlain by the coarse grained garnet—biotite—Kyanite gneiss of Joshimath Formation.

The quartzites are traversed by three prominent joint sets dipping at 35°–40°/N 290–020, 80°/N 360 and vertical N10°E (strike). Three prominent joint sets in quartz mica gneiss are dipping at 30°–40°/N 285–25, 45°–55°/N 070–110 and vertical/N 60° E (strike).

171.2.2 Rock Mass Characteristics

The rock masses of Tapovan—Helong formation are characterized by rock mass values—Q = 7–14, RMR = 60–75 and Q = 6–10 and RMR = 55–65 for augen gneiss and quartzite respectively. The Q and RMR values of 6–10 and 60–65 respectively have been estimated for the coarse grained biotite—Kyanite gneiss of Joshimath Formation (Naithani and Krishna Murthy 2006).

The vertical cover above the HRT grade ranges from 300 m in the initial reaches near intake which gradually increases to 900—1,000 m in the middle and then gradually decreases to 200 m near the surge shaft location. High stress condition exists in the high vertical cover reaches (about 1 km).

171.3 Groundwater Ingress in HRT

On 24 Dec. 2009, while driving the HRT by a double shield TBM, under a high cover zone of ±990 m comprising hard and brittle rocks of Tapovan—Helong Formation overlain

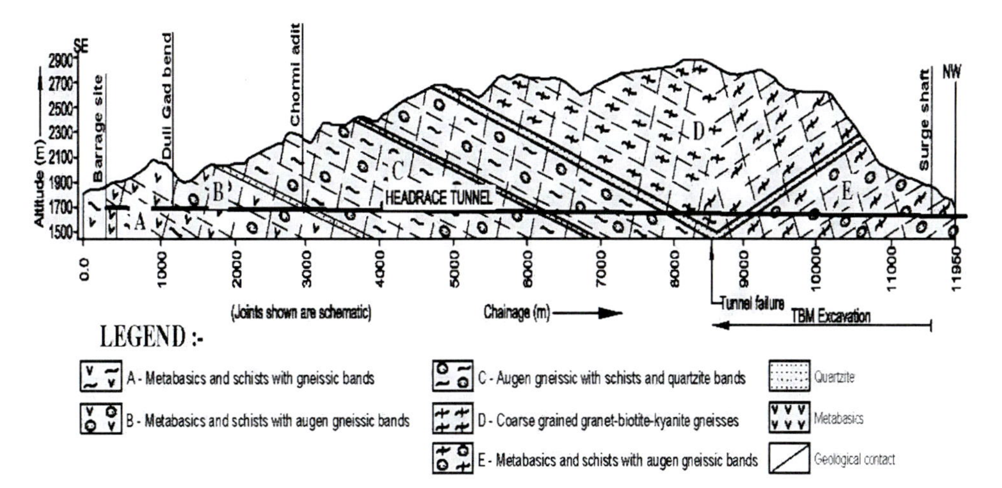

Fig. 171.2 Longitudinal section along headrace tunnel (Modified after Naithani and Krishna Murthy 2006)

by rocks of Joshimath Formation and thick overburden materials, rock mass failure occurred which hit the front shield and immediately stopped TBM. Subsequent attempts to free the machine with very high thrust forces failed. By the following day, a massive surge of ground water started entering the tunnel under considerable pressure into the annulus of the completed but not yet grouted segmental lining and caused failure of two roof segments, allowing a sudden heavy ingress of groundwater (600–700 L/s) from the tunnel face at RD ± 3,016 m. The heavy groundwater ingress from the quartzites rocks of Tapovan—Helong Formation, forming a synformal structure, was considered to be through the pathways—open joints/fractures or fault zone which were further widened due to stress relief and removal of infilling material by continuous movement of water (Nawani 2010) (Fig. 171.3). Initially the water was clean and clear but later it became muddy and milky in colour. This incident led to complete suspension of tunneling operation at the site.

171.3.1 Bypass Tunnel

For sometime wait and watch strategy was adopted but when there was no significant change in the groundwater ingress, it was considered essential to construct an access tunnel to the location of TBM at RD ± 3,016 m in order to assess the ground condition and repair the TBM. In the meantime, with passage of time, there was an appreciable reduction in water ingress and discharge was reported to be 200–250 L/s. A 140 m long, D-shaped (4.2 m dia.) bypass tunnel was excavated, from the left side, beginning from concrete lining segment No. 1837 (ch. 9945.6 m from the downstream and of HRT) and running parallel to HRT, to reach TBM cutter head location (Fig. 171.4). At this location, it was observed that rock mass failure has occurred above the concrete lining segment No. 1905 and 1906, and TBM shield was damaged near cutter head. The segments 1905 and 1906 were partly removed to examine the cavity formed due to failure of rock blocks above the crown portion of HRT.

Fig. 171.3 Water ingress from segment No. 1905

Fig. 171.4 View of By-pass Tunnel

The water inflow was channelized through a drift—size opening excavated between BPT and HRT.

171.4 Geotechnical Assessment of Rock Mass Behaviour

Rock mass behaviour around the Collapse Zone where cavity has formed, above the crown of the HRT, was examined and geotechnically assessed. The banded quartzite rock mass is mostly blocky and the rock mass failures are mainly structurally controlled due to interplay of two joints J_1 (foliation) and J_2 (cross-joint), triggered by high stress relief and high water inflow under pressure. The rock mass above the concrete lining segment No. 1904 was found to be intact and it, however, failed above segment No. 1905 and 1906. The cavity size further increased due to heavy ingress of groundwater Fig. 171.5). The pea gravel grout around few pre-cast lining segments was also washed out under high pressure of water. The washed out grout material choked the pea gravel holes and as a result water gushed in under high pressure from pea gravel holes in crown portion of HRT. The vertical extension of the cavity was estimated to be 8–9 m in NE direction and the width of this zone is more than 12 m. The detailed geotechnical assessment indicated that the cavity was formed above TBM cutter head between RD ± 3,016 m and RD ± 3,000 m.

Supportive measures, repair work and treatment

- Removal of damaged segments from the collapsed zone (segment No. 1906–1908), followed by cavity backfilling.
- Water ingress has been channelized through drift by inserting the flexible pipe in to the cavity.
- Installation of steel arch to support the pre-cast segment lining from No. 1857–1908.

Fig. 171.5 Close view of cavity and water ingress (which was later on diverted in to drift) above segment No. 1905 in HRT (TBM)

- Rock bolting (Swellex 4 m long, dia 25 mm) and SFRS in the cavity zone above segment no. 1905 and 1906.
- Gaps on the left and right side walls have been filled with the thick shotcrete.
- Consolidation grouting above HRT crown, area between collapsed zone and BPT starting point, was done from the right wall of BPT.
- Replacement of damaged telescopic shield, after movement of the gripper shield.
- Dismantling of backup units of TBM (21 Nos. of gantries).
- Grouting in the cavity zone from segment no. 1904 onwards and for this purpose pipes of 104 mm dia and 1215.18 m length were used.

171.5 Other Significant Observations

- Surface water bodies—lakes, springs and streams were studied by project geologists in and around Auli area, along the HRT alignment, to monitor if there is any change in water level or discharge. No change in water level of lakes (Chatrakund, Sunilkund) and no decrease in water discharge in streams and springs were found.
- No sign of subsidence was noticed in the ground near and above the HRT alignment.
- The water samples were analysed by NIH (National Institute of Hydrology, Roorkee) and the results suggest that the water bearing strata are being recharged at higher elevation (±3,000 m) and it is a mixture of snow/glacier melt water and rain water.
- Discharge of water from the cavity is having a decreasing trend and present discharge is ±120 L/s. Water is clean and clear and flows along southwest dipping joints.

171.6 Conclusion

Groundwater ingress is the key issue which impacts design and construction of tunnel projects. Heavy water inflow under high pressure at times causes unexpected delays in construction schedule of underground projects. In Tapovan—Vishnugad Hydroelectric project, tunneling along the head race tunnel (HRT) alignment by double shield TBM suffered a setback when rock mass failure followed by heavy ingress of groundwater, with 600–700 L/s discharge, occurred at RD ± 3016 m where the rock cover was as high as ±990 m. The water ingress was from the water bearing quartzite bands of Tapovan—Helong Formation, disposed as synformal wide open fold, through structural discontinuities—open joints/fracture or fault zone which were further widened in the form of a cavity due to stress relief and movement of water. The problem was addressed through excavation of a bypass tunnel of about 140 m length and 4.2 m diameter, parallel to HRT, to reach the location of incident at RD ± 3,016 m. The groundwater inflow was channelized through a drift—size opening driven through the rock column between BPT and HRT. The cavity formed in the crown portion of HRT was supported and made up, and the double shield TBM was repaired. The tunneling was resumed in 2011.

Prediction of critical hydrogeological condition along tunnel alignment is very essential and an advance planning to tackle the potential risks associated with water ingress is very important for ensuring the tunneling work in safe and controllable manner. Probe drilling to know rock mass condition and groundwater pressure in advance followed by pre-excavation grouting(using Micro cement) with pressure in order to make the ground water-tight, i.e, pre-injection of rock mass, is considered to be the best option. The main purpose of pre-injection in tunneling is to reduce the leakage of water into the underground excavation and also to improve the stability of rock mass. Post-excavation grouting must be avoided as it is difficult and also very expensive—20–50 times more.

Acknowledgements The author is thankful to the project authorities of NTPC and Resident Geologists of GSI for sharing their views during the author's inspection visits to the project site as Director, National Institute of Rock Mechanics (NIRM) in April 2010 and June 2011.

References

Naithani AK, Krishna Murthy KS (2006) Geological and geotechnical investigations of Tapovan—Vishnugad Hydroelectric Project, Chamoli District, Uttarakhand, India. Published in J Nepal Geol Soc 34 (spl. Issue):1–16

Nawani PC (2010) A note on inspection visit to Tapovan—Vishnugad Hydroelectric Project on 1–2 April 2010. Unpublished inspection note submitted to NTPC

Investigation Constraints in Subsurface Water Aspect of Hydropower Development in the Indian Himalaya

172

Y.P. Sharda and Yogendra Deva

Abstract

Keeping pace with the overall growth, the infrastructure in India—a rapidly growing nation—developed many fold during the recent past. In particular, the Himalayan region witnessed a boom in hydro power and, in view of rugged topography and consequent lack of suitable open spaces, depended more and more on subsurface spaces for housing project components including long water conductor systems. The Himalaya, due to its location in the inter plate region, is a highly seismic region with a very complex geological setup. Combined with very rugged topography with high ridges and narrow and deep valleys, construction of subsurface components invariably encounter various geological adversities amongst which sudden ingress of subsurface water in large quantities, often with flowing ground conditions, happens to pose one of the most critical problems during construction. This is one of the common causes for delays in construction work for long durations that, in turn, leads to severe time and cost overruns. Long head race tunnels of hydroelectric projects like Dul Hasti, Parbati Stage-II, Tapovan Vishnugad, etc. are some of the better known examples that suffered long delays due to this problem. While, other subsurface structures like surge shafts, desanders and powerhouse caverns are constructed in limited spaces with lesser variations in ground conditions, the long tunnels encounter frequently changing ground conditions and face serious constraints in investigations due to inaccessibility and high superincumbent cover that pose constraints in dependable estimations. Besides innovations in standard investigation techniques, possible solution lies in systematic advance probing during construction.

Keywords

Himalaya • Tunnel • Investigation • Water • Delays • Advance probing

172.1 Introduction

Along with other infrastructure facilities, availability of energy plays a critical role in the sustainable development of the economy of a nation. For sustainable growth, electricity, gas and water supply sectors should also grow at the same rate as the economy. Recent economic liberalization policy in India has had a significant impact on all sectors of development, with infrastructure and power, in particular, getting prominence. Rapid industrialization, coupled with

Y.P. Sharda (✉)
SNC Lavalin Engineering India Limited, 3rd Floor, Redisson Commercial Plaza, Mahipal Pur, New Delhi, 110 037, India
e-mail: yp.sharda@snclavalin.in

Y. Deva (✉)
Indo Canadian Consultancy Services Ltd., Geological Survey of India, Citadel 1502, Eldeco Green Meadows, Sector-Pi, Greater Noida, India
e-mail: yogendradeva@gmail.com

G. Lollino et al. (eds.), *Engineering Geology for Society and Territory – Volume 6*,
DOI: 10.1007/978-3-319-09060-3_172,

the expansion of the agricultural sector, and overall improvement in the quality of life and urbanization of rural India, has pushed up the demand for energy requirements. In today's scenario, the generating capacity is far below the current demand resulting in chronic power shortages in almost all parts of India. The problem becomes acute during peak hours. It is true that it may be possible to meet base load demand through other fossil fuel based or nuclear power plants, but, hydro power- the perennial energy is the only answer for peak demand through which it is also possible to achieve low carbon and environment friendly growth as compared to energy from other sources. In addition, development of hydro power projects also provides the added advantage of opening up avenues for development of remote and backward areas of the country. Despite being recognised as a renewable source of energy, the share of hydro power in the overall generation capacity of the country has been steadily declining since 1963—from 44 % in 1970 to about 21.18 % today. This could be attributed to long gestation periods in case of hydroelectric projects, capital investment, remote location of potential sites, lack of infrastructure in the desired areas and above all the environmental concerns.

Keeping in view the importance of hydropower, the Central Electricity Authority (CEA) carried out re-assessment studies of hydroelectric potential of the country. These studies indicated availability of economically exploitable hydroelectric potential of the country as 148,701 MW of which only 23.34 % has been harnessed so far. It is also evident from the reassessment study that 70 % of this economically exploitable hydro potential is concentrated in the Himalayan part of the three major river basins, viz. Indus, Ganga and Brahmaputra. Therefore, in order to fulfil the demand for power and to even go on to be a power surplus state, India has no option but to harness Himalayan Rivers for meeting the ever growing demand for energy. In the quest of harnessing more and more hydropower, it has become necessary to move into remote and difficult areas that offer geologically and geotechnically challenging sites, harsh climatic conditions and almost non- existing infrastructure facilities. The target is tough and calls for all out efforts in achieving self sufficiency in hydropower development.

172.2 Complexities of the Himalaya

Himalaya, the world's largest mountain chain with some of the highest peaks extends for over 2,500 km along the northern borders of the country. It is characterised by very rugged topography dissected by deep valleys resulting in high to very high relief and snow covered peaks. The altitude in the Himalayan terrain attains the maximum of 8,848 m in the Mount Everest—the world's highest mountain peak and every mountaineer's dream. The climate varies from alpine snow to sub-tropical hot and humid. By virtue of its birth related to the collision of the Indian and the Eurasian Plates, the Himalayan region is home to geological and tectonic complexities and high seismicity. The combined effect of various complexities results in major geological and geotechnical problems both over ground as well as underground. For the underground structures, excessive mountain/overburden stresses with problems like squeezing, rock bursting, infrequent incompetent and problematic zones across faults, shears and thrusts, huge quantities of subsurface water, flowing ground conditions and geothermal pockets with hot springs or dry heat, etc. are just a few examples of the innumerable challenges encountered while driving tunnels. The deepness of a tunnel makes it rather difficult and expansive to get a reliable picture of geological and hydrological situation through which a tunnel has to be built (Sanders 2011). On the other hand, surface problems like unstable slopes, landslides, debris flows, snow avalanches, earthquakes, liquefaction etc. are some of the problems that plague many a surface components of hydro projects during investigation as well as construction. In general, serious problems due to large quantities of subsurface water, particularly in combination with crushed rock leading to flowing ground conditions, is the foremost and nearly common to any long tunnel in the Himalayan terrain.

172.3 Investigation Constraints for Tunnels in Himalaya

During the hydropower development in the Himalaya, relatively easier sites were developed earlier and the difficulty level increased as development moved into difficult remote sites in higher reaches due to rugged nature of terrain leading to poor accessibility, low population density, and hostile climate. With the greater thrust on the development of run-of-the-river schemes due to environmental and rehabilitation concerns, more and more power projects are now planned as run-of-the-river schemes that, by implication, involve long water conductor systems, viz. head race tunnels as given in following Table.

Most of these tunnels were constructed, or were taken up for construction, after very limited investigations that were restricted due to various constraints discussed in the paper.

Sr. no	Name of project	Feature	Length (km)	Maximum discharge (l/m)	Generation capacity (MW)
1.	Nathpa Jhakri HEP, Himachal Pradesh, India	HRT	27		1,500
2.	Karcham Wangtu HEP	HRT	17		1,000
3.	Dul Hasti HEP	HRT	10.63	72,000	360
4.	Rampur HEP	HRT	13		
5.	Parbati HEP stage-II	HRT	31	7,000	800
6.	Maneri Bhali HEP stage-II	HRT	16		
7.	Kameng HEP	HRT	14.5		
8.	Luhri HEP	HRT	38		
9.	Tapovan—Vishnugad HEP	HRT	11.5		

These constraints make it very difficult to carry out intensive investigations along the tunnel routes, including even detailed engineering geological mapping, leading to serious constraints in building a reliable geological model and in predicting the problems precisely that are likely to be encountered during construction stage. The data acquisition, therefore, tends to depend more upon projections from long distances, even from the opposite banks of the rivers and, consequently, adds to the errors on large scale. Subsurface investigations through drilling, drifts, geophysical surveying, etc. face extreme problems of equipment/material transportation including drilling fluids. In particular, advanced technology for deep drill holes, that is not available easily and even if available, it is almost impossible to transport bigger drilling machines and drilling fluid to remote sites. Even remote sensing techniques have not been of much help in most of the cases due to thick overburden cover supporting dense vegetation.

In view of the general investigation constraints discussed above, the project planning depends more on the construction stage advance probing either through advance drilling or other indirect methods like TSP that, invariably, get sacrificed in the name of progress and many a time lead to disastrous consequences.

172.4 Implications of Investigation Constraints in the Himalaya

Shortfalls in investigation due to various constraints have far reaching consequences in the execution of long tunnels compared to other localised components like caverns for powerhouse, desanders, surge shafts/chambers etc. This is attributable to long tunnels being more exposed to far greater variations in ground conditions that are difficult to be investigated due to accessibility problems and high superincumbent cover. The problems assume greater dimensions due to complex geological and tectonic settings. The other localised components, on the other hand, are constructed in confined or limited space with very little or no variation in ground conditions.

Huge quantities of subsurface water connected to perched water tables is a common feature during tunnel execution and is nearly impossible to investigate. While, discharges from such sources disappear or reduce significantly in due course of time depending upon the magnitude of the source, interconnection with perennial sources is not uncommon and may turn out to be a major construction problem. In good rock condition, the subsurface water on its own may not be as big a problem as it is when it punctures through poor rock condition like fractured and crushed rock in shear or fault zones. The resultant flowing ground conditions turn out to be nightmares for the executing personnel and, many a time, are capable of stalling construction activity for months, or, in extreme cases, for years.

172.5 Serious Subsurface Water Problems in Himalayan Tunnels

Some of the projects with long tunnels as water conductor systems in the Himalaya, that have suffered due to sudden ingress of large quantities of water without any warning are discussed in the following paragraphs.

172.5.1 Dul Hasti Hydroelectric Project

The 53 m^2, partly TBM and partly DBM driven circular head race tunnel of the 390 MW Dul Hasti Project in northwest Himalaya encountered several artesian blowouts with peak discharge at 72 m^3/min at RD 1,194 m that declined with time and stabilised at 7.5 m^3/min after about 6 months (Deva et al. 1994) is of particular importance. The "pseudo fluvial" outwash material amounted to 4,000 m^3 and led to partial burial of the TBM. The TBM was buried in another such blow out with huge flowing ground condition, and same had to be abandoned. The tunneling for the remaining

60 % of the tunnel length could continue by DBM after over 14 months, by constructing a by-pass tunnel. A number of fracture/crushed zones of 1–3 m width charged with water cut across the tunnel alignment. These when intercepted in the tunnel effected blowout under high hydrostatic head of +200 m and drained the water into the tunnel with mud/slush and angular rock fragments and flowing ground conditions. These took several weeks to months to drains, till then halting of tunneling was the only solution. The tunneling was accomplished with many challenges/difficulties in design and construction of support system and lining in the weak zones due to selective under excavations and supporting in the weak zones and time and cost over runs in 2005, with a construction history of over +15 years.

The artesian conditions have been linked to confinement of fractured quartzite beneath an impervious barrier of phyllite with a major tectonic lineament (fracture/fault?) in the vicinity responsible for aquifer porosity and groundwater channelization (Fig. 172.1). The HRT had already been detoured as the "loop alignment" for bypassing the problems due to an exceptionally thick 'buried valley' of the Chenab River (Fig. 172.2). The incidence, that took 186 days for resumption of tunnelling, was handled basically through the dewatering arrangements under gravity with pumping assistance (60–80 m^3/min capacity), followed by mucking and limited probing through destructive drilling. The unconventional design of upstream slope of the HRT at the intake helped in drainage under gravity and saved the tunnel excavation from many problems and consequent long delays.

According to Winter et al. (1994), the geological features of the project area are of exceptional nature and the tunnel layout, the tunnel design, the investigation and construction methods have to be continuously adapted to these new and exceptional conditions. According to Sengupta et al. (2008), abrupt interception of an aquifer resulted in sudden ingress of water with sediments, pebbles and gravel ranging in size from 0.5 to 1,000 mm resulting in formation of a cavity and accumulation of the muck near the mouth of the cavity. Again a blowout at three closely spaced locations at invert sprang a surprise. The blowouts carried slushy discharge of about 700 L/s in the beginning and subsequently increase to 1,100 L/s. This later stabilised to 50–70 L/s. The water carried and deposited 2,500–3,000 m^3 of muck comprising sand, silt and pebbles. The blowout caused extensive damage to machinery and delayed the project by several years and further tunnelling was done by drilling and blasting. The inflows reduced to 150 L/s within 100 days and flows of reduced to 100 L/s continued for a period of over 5 years (Mcfeat-Smith 2008).

172.5.2 Parbati Hydro Electric Project

This 800 MW installed capacity project involving trans-basin transfer of water is under construction in Kullu District of Himachal Pradesh. The project involves the construction of a 91 m high concrete diversion dam on the Parbati River, a 31.23 km long headrace tunnel on the left bank of the Parbati River to convey the water to a surface power house located on the right bank of Sainj River, a tributary of the Beas River. The ground cover over the HRT exceeds 1 km for considerable length. In conjunction with high relief, the rugged topography, steep slopes, dense forest cover and complete lack of accessibility led to very limited investigations and consequent decision to rely on advance probing. On 18th November 2006, at Ch. 2,700 m, water along with slush started coming out from a probe hole and soon the discharge rose to about 2,500 lpm with 120 m^3 slush comprising sand and silt that deposited in the tunnel. The situation became out of control after about 8 h when the discharge increased from 2,500 lpm to about 7,000 lpm and all the machinery including the TBM were drowned in water and slush (Fig. 172.3). The tunnel excavation came to a standstill and, even after seven years, still continues to be suspended. The tunnel is still discharging at the rate of 4,000 lpm.

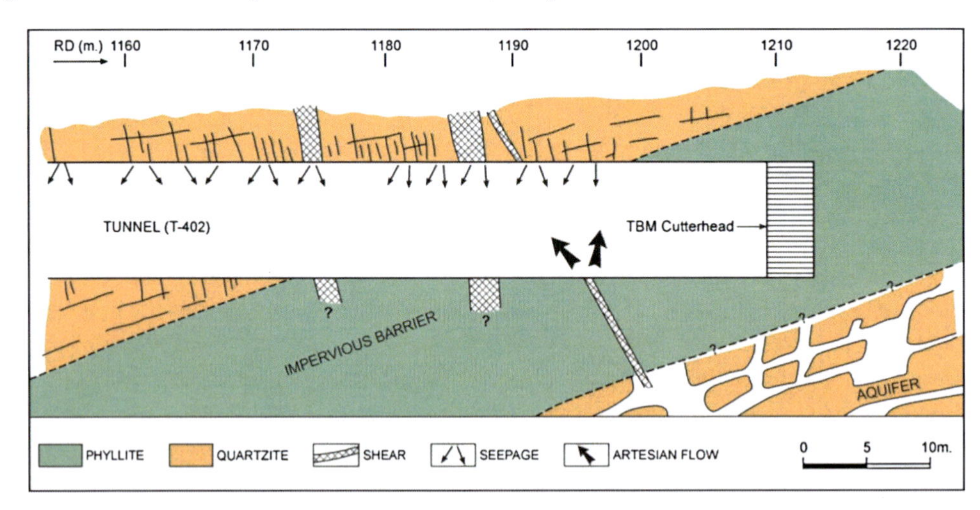

Fig. 172.1 The artesian blowout setting at Dul Hasti HEP

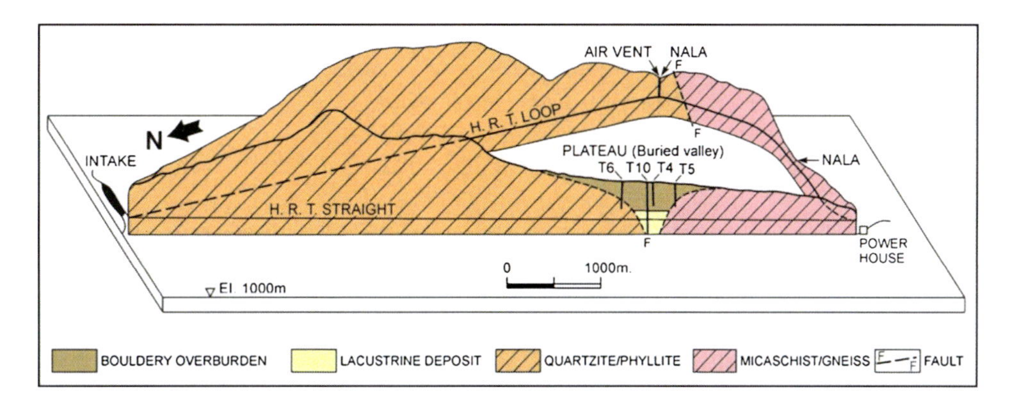

Fig. 172.2 Bypassing the deep buried valley along Dul Hasti HRT alignment

Fig. 172.3 Heavy discharge with slush in Parbati HRT

172.5.3 Tapovan Vishnugad Hydroelectric Project

The TBM driven 11.5 km long HRT of the 520 MW hydroelectric project is located in the state of Uttrakhand is being constructed through three construction adits aggregating 1.25 km length. The ground cover over the tunnel exceeds 1 km in the middle reaches. The TBM got struck as large quantities of water suddenly flooded the tunnel on 25th December 2009 at RD 3,016 m. The rate of inflow increased to about 700 lpm and resulted in complete stoppage of work for about 15 months. Work could be resumed only after the excavation of a bypass tunnel for dewatering at the face and freeing of the TBM cutter head which had got stuck. The water inflow continues at 100 lpm and is being drained through the bypass tunnel.

172.5.4 Kameng Hydroelectric Project

Located in North Eastern Himalaya, the project with 14.5 km long HRT is being excavated through four construction adits. According to Kalita (2011), large quantities of slush along with heavy ingress of water measuring up to 400 l/m was encountered suddenly at RD 422.60 m from the Face-2. The slush comprised loose sub-angular fragments of quartzite measuring 60–80 mm with quartz crystals and clay. Further advancement of the HRT was possible only after 8 months of all out efforts through poly urethane and micro cement grouting, and fore-poling with MS heavy duty pipes.

172.5.5 Karcham Wangtu Hydroelectric Project

Located in western part of Himalaya, this 1,000 MW installed capacity project has an 11.28 m diameter and 17 km long head race tunnel as part of the water conductor system. The HRT encountered extremely poor and flowing ground conditions for a length of about 200 m. The tunnel advancement was handled through DRESS technology. The HRT also encountered about 350 m long high temperature zone with temperature going up to 98 °C. This difficult ground condition was negotiated by resorting to multiple drifting, continuous injection of cold water through drill holes and reducing the charging time to within 40 min to avoid deflagration and immature detonation. In addition to the above, problems of rock bursting and spalling were also experienced in reaches with cover exceeding 700 m (Sati et al. 2011).

172.5.6 Maneri Bhali Hydroelectric Project

Located in Himalayan State of Uttrakhand, this run-of-the-river scheme has water conductor system comprising a 6.0 m diameter and 16 km long head race tunnel (HRT). According to Gajbhiye and Bhattacharjee (2011), the HRT below Dhanari Gad was being driven through highly jointed and frequently sheared whitish cream quartzite of Proterozoic Garhwal Group when it abruptly encountered river borne sediments accompanied by ingress of large quantities of water under high head. The incidence delayed the progress of the HRT considerably and further advance could be made by bypassing the zone by realigning the HRT and resorting to intensive pre-grouted umbrella arch over the crown.

In addition to the above, several other projects suffered badly due to abrupt ingress of large quantities of water flooding the tunnels. Some of the better known incidences are from projects like Beas—Satluj Link Project with HRTs in two stretches of 13 km length each and Nathpa Jhakri Hydroelectric Project with 27 km long HRT. In case of Nathpa Jhakri HEP, the HRT also encountered hot water with temperature going up to 68 °C.

172.6 Conclusions

With its fast developing economy and its ever increasing demand for energy, India has no other option but to harness the vast hydroelectric potential of the great Himalayan chain. These young mountains, in general, are characterized by rugged topography, high relief, deep valleys with steep and high slopes, thick soil supporting dense vegetation in middle and lower reaches, thick glacial deposits in higher reaches, complex geology and high seismicity. These factors result in retarding the development of infrastructure facilities in many remote areas. The hydropower developers, therefore, face serious constraints in carrying out desired investigations for the project components. In particular, due to wide variations in ground conditions and high superincumbent cover, long tunnels face the maximum brunt and usually remain the most poorly investigated components of the projects. The excavation of such tunnels, therefore, springs up problems that are not fully appreciated during project planning. Huge quantities of subsurface water, particularly the flowing ground conditions, are the most common and dreaded tunneling problems. These are capable of stalling construction activities in a big way and result in over runs of cost and time in project execution. The losses to the developer sometimes assume colossal proportions.

It is, therefore, imperative that all out efforts are made during the investigation stage so that a dependable geological model along the tunnel route is developed and the construction methodology defined precisely with dependable estimation. Based on project specific geological and hydrological conditions, the standard investigations through geological mapping, remote sensing, drilling, drifting, geophysical surveying, etc. are to be blended and optimized. Since, the investigation results are not expected to be fully satisfactory, construction stage investigations attain great importance. Over and above the standard investigation approach, the following investigation strategy may help achieve greater information. It would also be advisable to supplement the investigations with other state-of-the-art global technology. It is reiterated that an extra expenditure in investigations, both during investigation stage as well as during construction stages, will give ample rewards during implementation stage. Along with the investigations, it is also necessary to assess the risk factors and keep remedial plans ready for managing them efficiently (Mcfeat-Smith 2008).

172.6.1 Investigation Stage

Detailed geological mapping, the backbone of any investigation campaign, with a good blend of remote sensing, regional geology, and review of existing information from neighboring projects, if any, is the most effective mode of investigation, both technically and commercially. For reasons of accessibility constraints, the data collection would be nearest to the tunnel alignment, followed by data projection and development of geological model. Absolute judicious decisions are to be exercised in the observed ground condition, i.e. lithological and structural variations.

The geo-hydrological conditions, particularly the locations and discharge details of surface water sources like springs and nalas are to be recorded and integrated in to the geological model.

Subsurface investigations by drilling, drifts and geophysical surveying may even be carried out in geological test sections away from the tunnel alignment, but the conditions must be fairly similar to the projected conditions along the tunnel alignment.

172.6.2 Construction Stage

Advance probing by percussion drilling combined with occasional core drilling, if required, for tunnel forecast is a well established construction stage technique, but, is often sacrificed for much less important aspects. Elaborate observations and interpretation during the drilling operation by an expert geologist can provide vital geotechnical and engineering geological inputs that can lead to planning of timely and adequate remedial measures averting disastrous tunneling problems. Such religiously followed approach during the construction of 14 km long Banihal railway tunnel in Jammu and Kashmir through most complex geological setting proved a boon and helped complete the project in time.

Other indirect advance probing techniques like Tunnel Seismic Profiling (TSP) may be adopted depending upon the ground conditions.

On site preparations for dealing with heavy subsurface water ingress, e.g. pipe roofing, drilling and grouting, adequate arrangement for pumping, etc. need to be kept in readiness for use whenever the requirement arises.

Ambiguity in tunneling contracts, particularly for less practiced TBM driven tunnels, led to lack of clarity in dealing with emergent situations and has resulted in unprecedented delay of many projects by several years. The contracts for such projects must have provisions for handling the emergent situations in clear terms with roles of each agency or individuals clearly defined so that decisions are taken quickly and implemented on war footing.

References

Deva Y, Dayal HM, Mehrotra A (1994) Artesian blowout in a TBM driven water conductor tunnel in North-West Himalaya, India. In: Proceedings of 7th international congress IAEG, Portugal, Balkema, Rotterdam, pp 4347–4354

Gajbhiye PK, Bhattacharjee NR (2011) Tunnelling through weak and water charged strata by ground improvement—a case from Maneri Bhali HE Project-II, Uttrakhand. Abs. vol., international conference on underground space technology and 8th Asian regional conference of IAEG, Bangalore, India

Kalita GC (2011) Tackling of a critical zone encountered in Face-II of HRT, Kameng HE Project—a case study. Abs. vol., international conference on underground space technology and 8th Asian regional conference of IAEG, Bangalore, India

Mcfeat-Smith I (2008) Tunnelling in the Himalayas: risk assessment and management for tunnelling in extreme geological conditions. In: Proceedings of world tunnel Congress 2008- underground facilities for better environment and safety- India, pp 1748–1760

Sanders MR (2011) Tunnelling in high overburden with reference to deep tunnels in Switzerland, Spain and Iran. Abs. vol., international Conference on underground space technology and 8th Asian regional conference of IAEG, Bangalore, India

Sati KK, Sood KK, Singh P, Goyal DP (2011) Excavation of 17 km long Head Race Tunnel for 1000 MW Karcham Wangtu hydroelectric project in South West Himalaya. Abs. vol., international conference on underground space technology and 8th Asian regional conference of IAEG, Bangalore, India

Sengupta SK, Saxena PK, Bakshi SS (2008) Tunnel boring machine in India- experience, prospects and challenges. In: Proceedings of world tunnelling Congress 2008- underground facilities for better environment and safety

Winter Th, Binquet J, Szendroi A, Colombet G, Armijo R, Tapponnier P (1994) From plate tectonics to the design of Dul Hasti hydroelectric project in Kashmir (India). Engg Geol 36:211–241

173 Prediction and Management of Ground Water for Underground Works in Himalayas

Akhila Nath Mishra and S. Kannan

Abstract

The Himalayan geomorphology guides important hydrologic systems recharging ground water occurrences. Ground water conditions mainly depend on origin and occurrence of precipitation. Prediction and managing ground water occurrences in mountainous regions is difficult due to inadequacy of data. From ground water levels in the wells and the natural spring discharge levels it is possible to construct the groundwater contour maps, showing the form and elevation of water table to some extent. Tunneling in complex geological set up exhibiting highly heterogeneous rock mass conditions can be problematic, which may lead to disastrous and significant delay if not adequately foreseen in mountainous regions. High displacements, stability problems and ground water inflows are common phenomena observed during tunneling through fault zones. Advanced methods of prediction and investigation/probing ahead of the face during construction are useful to minimize tunneling difficulties and apply requisite supporting methods. The influence of ground water during tunneling through high mountainous terrain poses problems during construction by any of the TBM, DBM or NATM methods. The paper addresses ground water investigation methods and targets for obtaining the key parameters for estimating the rock mass behavior, rock mass characteristics and criteria for the selection of appropriate construction methods, based on estimations.

173.1 Introduction

Development of a region demands harnessing water resources for irrigation, power generation and to mitigate frequent floods. The importance of river valley projects in India has increased enormously during the recent years, which pose challenges to engineering skills, more so when these resources are located in the Himalayan region having complex geological conditions. Here tunnels have to encounter incompetent rock mass, with quasi-elastic folds, faults and thrusts of various magnitudes with high ingress of water .The chances of existence of the sub-terrainian water with high head is much more than that in other regions. It is very difficult to design a suitable layout of project components in such terrain with mixed lithology, varying tectonic behavior and trapped water reservoirs with considerably high head and may become a very costly and hazardous operation. Tunnel lengths increase when the alignment is through ideal geological conditions to reduce the impact of squeezing, swelling, running ground, ingress of water with a poor state of rock, excessive temperature and gases in rocks. The past experiences gained in combating such problems have provided the technical skill to the designers and engineers for the future tunnels in the Himalayan region.

173.2 Physiographic Zones of Himalayas

The raising of Himalaya has resulted in repeated deformation of the sedimentary successions accumulated in the Tethys Sea. The Himalayas is characterized by its distinctive structural architecture and unique sedimentary and tectonic history. Himalayan mountains system stands as a series of

A.N. Mishra (✉) · S. Kannan
GMR Consulting Services Pvt. Ltd., New Delhi, India
e-mail: mishra.akhil1974@gmail.com

S. Kannan
e-mail: sknnhpc@rediffmail.com

G. Lollino et al. (eds.), *Engineering Geology for Society and Territory – Volume 6*,
DOI: 10.1007/978-3-319-09060-3_173, © Springer International Publishing Switzerland 2015

ranges in the form of largest individual geosyncline ranges 1500 miles long and from 150 to 250 miles broad with varied structure, composition and stratigraphy from Kashmir to Assam. The long alignment of Himalayan system has been divided into : the Punjab Himalayas from the Indus to the Sutlej (563 km long); Kumaun Himalayas from the Sutlej to the Kali (322 km), Nepal Himalayas from the kali to the Tista (805 km) and the Assam Himalayas from the Tista to the Brahmaputra (1165 km long). The following four major litho-tectonic physiographic zones have been recognized, which are separated from each other by major faults and thrusts systems. (i) Outer Himalayan Siwalik belt, (ii) Lesser Himalaya, (iii) Great Himalaya, (iv) Tethys Himalaya.

The Lesser Himalayas is characterized by deep valleys and steep slopes. The majority of the hydroelectric projects are located in the Lesser Himalaya due to deep dissected valleys favorable for high dams and big reservoirs posing high water heads for power generation. The lesser Himalaya, with elevations ranging from 1500 to 3000 m, has strongly folded and thrusted sedimentary and metamorphic rocks housing huge plutonic and volcanic bodies (of Paleazoic and Precambrain age) outcropping in the dissected valleys.

173.3 Ground Water Investigation Methods

173.3.1 Field Investigative Methods

Investigation of ground water involves many of the typical challenges faced when defining groundwater flow within fractured bedrock. Experiences gained during the process emphasize combining a variety of exploratory techniques to complete the investigation. The investigation includes geological mapping, monitoring of wells (by installing and gauging); analyzing groundwater samples; conducting fracture-trace analysis, surface geophysical survey; subsurface exploration by boreholes, permeability test in bore holes and observations of discharges in probe holes during tunneling.

173.3.1.1 Fracture-Trace Analysis

Fracture-trace analysis to identify potential fracture zones down gradient of the dam site and to determine potential migration pathways to assist in locating monitoring wells. Electrical resistivity surveys are conducted to get the ground truth results of the fracture-trace analysis. Pumping tests are also conducted to evaluate the hydraulic properties of the bedrock and the fracture system in the bedrock.

173.3.1.2 Bore Hole Exploration

Subsurface field investigations for ground water needs detail study of geological and geotechnical parameters to locate bore holes at various locations to establish ground water level and hydraulic gradients. Geological information is recorded during exploratory drilling to prepare a systematic geological bore hole log.

173.3.1.3 Electrical Resistivity Survey

An electrical resistivity geophysical survey is conducted to determine whether there is an anomaly indicative of water bearing fractured or weathered zone .The electrical resistivity (direct current) shows variations among different geological materials. Within crystalline rock, variations in resistivity are primarily due to water content within fractures and weathered zones. Therefore, a fractured or weathered zone in the bedrock is expected to show a zone of relatively low resistivity within a background of relatively high resistivity. This helps in correlating with exploratory bore hole data and other geophysical explorations, interpretations to predict ground water conditions in the area of interest.

173.3.1.4 Seismic Geophysical Survey

Two types of seismic geophysical tests are widely used in the exploration program (1) Seismic reflection testing and (2) MASW and REMI testing. The objective of the geophysical studies was to characterize subsurface geology, structure and geotechnical conditions up to a target depth of approximately 90–120 m bgs, at interested areas. Seismic reflection testing aims in collecting compression wave velocity (P-wave) information at identified locations. P-wave reflection data acquired along selected profiles, using instruments like IVITM. Mini Buggy vibratory source helps in evaluating weak zones or any fault below ground surface.

173.3.1.5 Hydraulic Conductivity Packer-Pressure Testing

Packer pressure testing is performed at 3 m intervals/segments in the boreholes to determine the hydraulic conductivity at each interval. From the recorded data of flow rates and the pressure the average hydraulic conductivity is calculated using the following formula:

$$\mathbf{K} = (\mathbf{q}/2\mathbf{nLH})\,\mathbf{ln}(\mathbf{L}/\mathbf{r}) \quad \mathbf{when\,L} > 10\mathbf{r}$$

where: k = hydraulic conductivity, cm/sec, q = constant rate of flow into test interval, cm^3/sec, L = length of test interval, cm, H = differential head of water at test interval, cm, r = radius of hole, cm, ln = natural logarithm Taking all information of Ground water level from bore hole data in the area of interest, the water table contouring can be done (Fig. 173.1)

173.3.1.6 Probe Hole

The probe holes are taken from the headings in advance for a detailed geological study of the advancing face, to predict and estimate the advancing tunneling conditions. Drilling

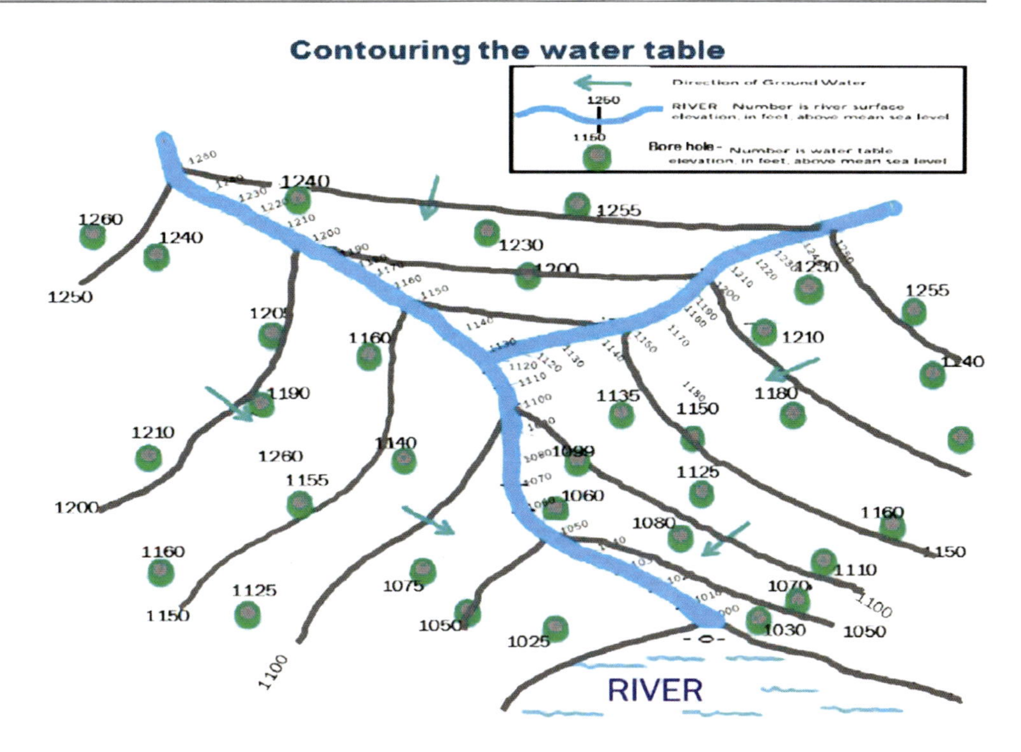

Fig. 173.1 Flow of ground water and contouring of water table

information like hammering sound, color of drilling water, nature of fine materials, penetration rate and measurements related to leakages, water loss and flowing status recorded during advancing of probe holes gives prior information of tunnel media before boring or excavation of tunnel. Problematic zones like highly permeable strata, fractured or crushed zones, fault or sheared zones detected by probe holes and its valuable interpretation of data often saves tunnel structures from driving into geological uncertainties. Information gathered on the basis of the probe holes is then employed in the tunnel excavation planning and execution of tunnel excavation which may be significant during construction. A number of probe holes can be designed to evaluate the actual difficulties if any and explore tunneling risks at predetermined locations, and plan for geological, geophysical, hydro geological or hydro chemical studies.

173.4 Tunneling Problems in Himalayas

The tunneling problems mainly squeezing, swelling and running ground conditions etc. encountered in the Himalayan region basically pertain to the stability of tunnels. The experience with different approaches for supporting the underground openings in the region reveals that the conventional rock-mass classification system (empirical approach) will help in developing realistic assessment of the design parameters. Underground works can be executed smoothly if adequate knowledge of appropriate method for computation of the parameters with expertise is timely implemented. The analytical approach has been tried upon as an alternative to classification systems. Some simple and reliable relationships out of the studies have been driven. The results from these derivations are found nearer to the results obtained through observational approach. However, the accuracy suffers due to lack of knowledge of rock-mass behavior under the state of in situ stress. The pressures generated under such conditions are capable of causing failure of heavy supports and containment/confinement is the major problem. The Himalayan region is within major earthquake region owing to the continuous collision of the plates (Plate Tectonics), and hence is subjected to so high pressures. It is an intricate problem to understand the effects of rock stresses whenever any underground excavation is planned. The problems of excessive dewatering, very high temperatures and gases in tunnels are also reported from the projects in the Himalayan region.

173.4.1 Running Ground Condition

The running ground condition is water soaked fragmented rocks often containing larger rock fragments embedded in a matrix of fine grained material that has almost no strength and flow into tunnel as slurry. The running ground is quite a common feature found associated with faults, thick shear zone, thrusts and buried fossil valleys in the Himalayan region. The initial pressures generated by running ground may be high to cause failure of very heavy supports. The typical examples are the power tunnels of Beas Sutlej Link and Maneri Bhali Stage-II in the Central Himalayas (Table 173.1).

Table 173.1 Summary of water inflow problems of some Himalayan tunnels

S. no	Name of project, tunnel and size	Rock type	Tunneling problems
1	Yamuna hydel project Ichari Chibro tunnel 6.2 km long, 7 m dia.	Quartzite, slate and limestone	High over-break, Running and squeezing condition
2	Pandoh Baggl tunnel Beas Sutlej link 13 km long & 7.62 dia.	Phyllite and granite with schistose band and kaolinised pockets	Heavy water inflow, running ground, swelling and squeezing condition
3	Maneri Bhali stage-I HRT16.8 km, 6 m dia.	Quartzite, basic rocks, lime stone and slate	Heavy water inflow, squeezing condition and high overbreak
4	Maneri Bhali stage-II HRT16 km, 6 m dia.	Metavolcanic, quartzite, lime stone, phyllite and graywacke	Running ground conditions, high squeezing and roof collapse condition
5	Nathpa Jhakri HEP HRT 30 km, 5 m dia.	Gneiss, granite, schist and amphibolite	Mild squeezing and excessive running ground conditions
6	Tenga hydel project HRT 16.5 km, 5 m dia.	Granite gneiss, schist, phyllite, quartzite, shale and sandstone	Running ground condition, water inflow, Mild squeezing condition and gases in rock
7	Loktak hydel project power tunnel 6.25 km, 3.65 m dia.	Sandstone and shale	Heavy water inflow, excessive pressures and gases in rock
8	Uri-II hydro project tail race tunnel	Sandstone, siltstone, shale and fluvio-glacial material	Flowing and frequent flooding condition and excessive dewatering
9	Dul Hasti HEP HRT 9.5 km, 8.3 dia.	Quartzite, phyllite and weak mylonite	Heavy water inflow, cavity formation and TBM buried
10	Tapovan Vishnugad HEP HRT 11.97 km, 5.4 dia.	Gneiss, quartzite, schist and meta-basics	Heavy water inflow, Rock fall, TBM stuck

173.4.2 Hot Springs and Geothermal Zone

Problems of excessive temperature (39–40 °C) are reported in the power tunnel, of 3.2 m excavated diameter, of Bhabha Hydro Electric Project in Central Himalaya which lies within the zone of lesser and Greater Himalayas. Most of hot springs are located along this zone. Underground temperature of water, rock and humid tunnel atmosphere when exceed about 32 °C make working conditions very difficult. An extensive system of refrigeration or rapid circulation of tunnel air is required to work in tunnel having temperature beyond 32 °C. The problem of this project was tackled by supplying fresh air through efficient ventilation, reducing the pull per blast to 1.0-1.2 m per day and by introducing the alternative rotation of workmen for short period, in order to take care of exhaustion. Similar thermal problems were anticipated in the tunnel of Nathpa Jhakri Project since two geothermal zones, Nathpa-Sholding (2.5 km long) and in Wadhal-Manglad (3.37 km long) sections, were encountered along the tunnel route. The HRT was expected to pass through folded metamorphic rocks comprising gneiss, quartz-mica schist, amphibolite and some granite and pegmatite. In the Nathpa–Sholding section (Ch. 1,600–4,100 m) hot water (34.5 ºC–50.7 ºC) was encountered locally at some places. The Wadhal—Manglad geothermal zone (Ch. 17,067 m and 20,440 m) began with a major hot water blowout (54 °C; 90–100 l/s) that flooded the tunnel to about 300 m from the heading as the tunnel gradient did not permit natural drainage.

173.5 Construction Method

Generally construction of tunnel by conventional drill and blast, NATM or by TBM method requires considering the all relevant aspects of tunneling like size of project, geological set up, equipment deployment, de-watering techniques, timely support & required skilled manpower to deliver a project as per schedule economically (Table 173.2). Conventional excavation is the excavation performed by conventional tunneling and shaft driving methods using drilling and blasting or by manual means. Multi-drift /multi-segmental excavation is defined as excavation of tunnel heading in segments. TBM excavation provides considerable advantages with high progress rates and some cost savings from the elimination of temporary construction of structures, in comparison to conventional Drill and Blast.

173.6 Ground Water Management

When ingress of water is too high causing hindrances in tunneling activity, normal practice of tunneling halts and generally pre injection grout method and use of polyurethane

Table 173.2 Summary of construction methods and difficulties in Himalayan tunneling

Feature	Drill and blast	TBM
Stability	Lossening of ground rockmass, wedge failure and roof collapse Increases water inflow	Mechanical solutions for temporary stabilization of the face area, and the rear zone
		Naturally stable
		Ideal for mass transit, pilot tunneling, unlined hydro and water conveyance tunnels
		Superior flow characteristics may eliminate lining requirement
Shape	Irregular excavated tunnel profile due to presence of over/under break	Smooth bored tunnel profile
Overbreak	Overbreak increases	Over-break negligible
Support	Demands additional rock supports Costly filling with concrete	Tunnel support may be reduced by 90 %
Operating	All skills required very cyclic, dangerous and unpleasant working environment	Continuous (non cyclic),repetitive operation safer and more pleasant working environment than in drill and blast
Blasting	Reduces stability	Eliminated
Crews	All skills required under high risks	Mechanical solution available for stability and temporary support at the face, work area and permanently behind the excavation operation
Access structure	Shafts and adits necessary to open multiple headings	Can eliminate all temporary access structures, particularly if the project is well laid out

grout are considered to manage the ground water. Such advanced ground water stabilization methods minimize tunneling risks.

173.6.1 Pre Injection Method

Modern pre-injection method of grouting ahead of tunnel heading offer significant advantages in avoiding mishaps and hindrances during situations like water ingress or mechanically poor ground. The basic idea of pre-injection is to treat the ground prior to the excavation by injecting a grout into the ground, surrounding the tunnel, through drill holes by the pumping under pressure. Modern cost effective methods and material technology for pre-injections with rapid setting micro cement in combination with colloidal silica in underground construction aims at strengthening the ground conditions, hence improving the stand-up time as much as possible during construction. Exploratory drilling, layout of grout holes, injection method, type of grout and grout mix designs are the main features for pre-injection grouting. Evaluation or control of the injection needs site specific decisions regarding repeated injections or to commence excavation through the treated ground after grout injection till the termination criteria is reached. Deployment of drilling machine and drilling method needs input of ground properties and its limitation for length of drill, cycle of probe drilling, injection fan drilling and grouting packers to suit the required hole/pipe diameter. Wastage of grout can be controlled by close monitoring of all processes involved to avoid hydro-fracturing or unwanted injection of grout far away from the structure.

173.6.2 Polyurethane Grout

Polyurethane grout can be extremely useful which reacts with water and expand due to production of CO_2 and it has proven successful when grouting with cement suspension does not work in sealing rock openings and post grouting has to be performed. The joint openings may be too narrow for the particles in the cement to enter, or flowing water in the rock transports the suspension away and/or dilutes it. Advantages of the process are (i) polyurethane grouts have better penetrability, because they don't contain particles and (ii) they can be used to control the cement propagation, since their gel time can be programmed. The viscosity of polyurethane is higher than that of water; the polyurethane front is stable when it is pumped into ground water. A resolute or absolute grouting pressure should be used when polyurethane grout is pumped into flowing water i.e. sufficiently high to cut off the water flow and facilitate grout penetration upstream, thus preventing flushing out of the grout.

173.7 Conclusion

Managing adverse ground water occurrence properly during the execution of underground works is the key to accelerated implementation of tunneling projects. Adequate geological investigation during project report preparation and its interpretation, reliability of the predicted geology and the predictions on the ground water conditions likely to be encountered is very important for the designers and executing agency. This enables the associated agency to plan accordingly and make adequate provisions in the contract documents to handle such situations. Continuous geological monitoring of the rock mass and predictions by project geologist during execution, inputs from expert committees helps the designers to go in for mid-way corrections in the designs. Prompt decision making by the implementing agency is vital in managing adverse ground water occurrences.

References

Bajaj AK, Sing TP, Chandra shekhar Iyer J Managing adverse geological occurrences-key to accelerated implementation of underground hydropower projects. Central Water Commission, New Delhi, India World Congress 2008-underground facilities for better environment and safety-India

Geotechnical site investigations for underground projects, volume1 and volume2 (1984) National Academy Press, Washington D. C

Hoek E, Palmeiri A (1998) Geotechnical risks on large civil engineering projects. International Association of Engineering Geologists Congress, Vancouver, Canada, 21–25 Sept 1998 (Keynote address for theme-I)

Schubert W, Fasching A, Goricki A Tunneling in fault zone-state of the art Lee I, Mo In, Yoo C, You K (eds) Safety in underground space, proceedings of the ITA-AITES2006 World Tunnel Congress

Sharma HK, Chauhan RS, Kuthiala S Challenges in design and construction of HRT of Nathpa Jhakri hydroelectric project (1500 mw)-A case study, Satulej Jal Vidyut Nigam limited, Shimla, India. World Tunnel Congress2008-underground facilities for better environment and safety, India

U. S. Army Corps of Engineers (1984) Geotechnical investigations, EM1110-1-1804, Feb.1984

Part XVII

Sustainable Water Management in Tunnels

Convener Dr. Antonio Dematteis—*Co-conveners* Joe Gurrieri, Alessandro Gargini, Giona Preisig

The concept of sustainability of tunnels needs the achievement of correct management of groundwater in tunnel projects. Equally, attention should be given to environmental and construction impacts and to opportunities that can be obtained exploiting drained waters, including the heat that it is often associated with. This session is directed to all operators in the field of design, supervision and construction of tunnels, such as universities and research institutes, designers, construction companies, control authorities, contracting authorities. Interesting topics for this section are the following: past experience; water inflow rates; examples of exploitation of water and heat; methods of forecasting; technologies for drainage, waterproofing and collecting; impact on Groundwater Dependent Ecosystems; risk and mitigation analysis; communication to decision makers and concerned peoples; laws, rules and policy framework.

Methodological Approach for the Valorisation of the Geothermal Energy Potential of Water Inflows Within Tunnels

174

Riccardo Torri, Nathalie Monin, Laudo Glarey, Antonio Dematteis, Lorenzo Brino, and Elena Maria Parisi

Abstract

Recent progress in underground excavation technology allows the construction of deep tunnels for the complementary exploitation of road, civil and hydroelectric infrastructures, e.g., the uptake of renewable hydroelectric and geothermal energy sources. This approach is employed to assess the positive effects of a civil work on a social and manufacturing community. The article describes the methods adopted for the study and design of the Lyon–Turin railway Tunnel, and identifies scenarios for exploitation of the drained hot water. Hydrogeological and geothermal models show the presence of a localized water inflow, with temperatures over 40 °C within tunnel sectors under the highest topographic covers (around 2,000 m). A standard for the water intake structure in the tunnel has been defined. Potential users of the resource have also been identified. An evaluation of energy facilities programs for domestic and industrial purposes has been carried in order to favoring local communities involved by the project. Characteristic elements are used to define the energy potential and ways of employment of the warm water drained by the tunnels and available at their portals. Potential consumers of the resource are also indicated.

Keywords

Tunneling • Hydrogeology • Geothermal energy potential • Geothermal model • Groundwater

174.1 Overview of the Lyon–Turin Project

The 57 km length basis tunnel between St-Jean-de-Maurienne (Savoy, France) and Susa (Piedmont, Italy) is a part of the future new railway link which will connect 2 main European cities: Lyon and Turin. This new railway belongs to the Trans-European Transport Core Network (Mediterranean corridor) which will link Algeciras in Spain to Budapest and the EU border. Preliminary works on the French side, involving the construction of 3 access galleries in the Maurienne Valley, are now completed with the excavation of around 9 km of tunnels in various kind of rocks. All these access galleries allowed a better understanding of the geological structures. On the Italian side, a new preliminary tunnel is in progress: the Maddalena survey gallery.

In this article we present a hydrogeological and geothermal reference model with some suggestions for deriving the best advantage from the water drained by underground works. The evaluation has been carried out for the main tunnel from its highest point to the eastern portal and for the accessory tunnels (Maddalena adit and Clarea ventilation

R. Torri (✉) · A. Dematteis
SEA Consulting Srl, C.so Bolzano 14, 10100, Turin, Italy
e-mail: torri@egteam.eu

N. Monin · L. Brino · E.M. Parisi
LTF SAS, Chambery, France

L. Glarey
Lombardi SA, Minusio, Switzerland

G. Lollino et al. (eds.), *Engineering Geology for Society and Territory – Volume 6*,
DOI: 10.1007/978-3-319-09060-3_174, © Springer International Publishing Switzerland 2015

well) in Italy; a similar analysis for the French part of the project is still in progress.

A main principle of the 'New Lyon–Turin rail-link' project is to minimize water drainage induced by the underground works. This is thought in order to reduce (i) the impacts on the water resources and (ii) problems created by water inflow during the excavation and utilization of the tunnel.

174.2 Hydrogeological and Geothermal Model

The tunnel crosses a variety of paleogeographic and tectonic geological units.

For the deep-lying trans frontier Lyon–Turin basis Tunnel several borehole temperature measurements have revealed the local geothermal gradient. A geothermal numerical model carried out during the previous project phases (Brino et al. 2008) assigned a specific thermal conductivity and geothermal gradient to each geological unit crossed by the tunnel; moreover, the project has been differentiated thermally in homogeneous sections.

According to such a model, specific temperature values have been assigned to the water inflows predicted along the tunnel to evaluate their thermal potential. The final temperature of the mixed water expected at the eastern portal was estimated to assess the possibility of their exploitation. In the Table 174.1, the geothermal gradient and the temperature range expected at the basis tunnel elevation (from the high point to the Italian portal) are given.

The temperatures and the geothermal gradients decrease owing to the presence of water flowing along the shear zones from the surface to the deep portion of the rock mass.

The central sector of the tunnel alignment is of thermal significance: it will be excavated under about 2,000 m of overburden through the metamorphic rocks of the Ambin Massif. Water inflows are expected mainly at the interception of faults and deformation zones; discharge rate, chemistry and water inflow temperature depend on the connection degree with their recharge areas at the surface.

Table 174.1 Geothermal gradients and their variation for the tectonic units involved by the excavation of the basis tunnel, from its highest point to the Italian portal. The temperatures range expected at the basis tunnel elevation is also given

Tectonic Unit	Geothermal gradient (°C/km)	Temperature at tunnel elevation (°C)
Gypsum Nappe	20–30	22–31
Ambin Massif	25–33	21–47
Tectonic fault zone	16	12–22
Piemontese Nappe	17–25	10–20

However, the geothermal model proposed for this sector provides temperatures higher than 40 °C (about 47 °C) related to the normal geothermal gradient measured in the boreholes perforated in the external belt of the Ambin Massif.

174.2.1 Evaluation of the Potentially Exploitable Water Inflow

Exploitation of the water inflow has been evaluated in function of the hydro-geochemical parameters. These parameters permitted to distinguish water aggressiveness and potability (sulphates and chlorides); by reconstructing the global and local geothermal gradient, potable waters with temperatures up to 25 °C and warmer waters available for thermal exploitation have been separated.

In keeping with the current French (Arrêté du 11 janvier 2007) and Italian regulations (D.Lgs. 152/2006, D.Lgs. 30/2009), the criteria for evaluation of the exploitable water potential have set the threshold at 250 mg/l for the contents of sulphates and chlorides and at 25 °C for the water temperatures; regarding to the inflow type an intermittent discharge has been distinguished by a continuous.

A cross analysis between data from deep boreholes (i.e. hydraulic packer test, temperature logs and hydrochemistry analysis) has led to hydrogeological characterization of metamorphic rocks lying at the tunnel elevation.

The central part of the Ambin Massif meets all water exploitation criteria: high topographic covers (>2,000 m), a normal geothermal gradient (30 °C/km) and a medium to medium-low permeability degree of the fissured rock (K ≅ 1E-07 m/s). The discharge of drained water in this sector is about 40 l/s with a temperature of 46 °C. A low sulphide (about 50 mg/l) and chloride content (about 100 mg/l) is expected.

Assessment of the water exploitation must take account of the possible contribution of the Clarea and Maddalena tunnels that drain into the basis tunnel since they intersect its alignment at the deepest sectors of the Ambin Massif. The data collected during the excavation of the Maddalena survey gallery will provide very useful information for evaluation of the current water exploitation potential.

The total flow, coming from the water available along the main tunnel and at the intersection with the Clarea and Maddalena adits, could range between about 40 (basis tunnel only) and 100 l/s (basis tunnel plus the adits).

The real exploitation potential of the waters drained in the tunnel, of course, can only be evaluated in steady-state conditions, and once the works are finished. Depending on the hydrogeological conditions encountered during excavation, the characteristics of the water drained within the tunnel and collected for exploitation must remain stable over time. Depending on the most fruitful exploitation scenario, therefore implementation of the monitoring protocols established by the current legislation and good-practice design is essential.

174.3 Possible Exploitation Scenarios

174.3.1 A Standard Operational Approach

In general terms, punctual water inflow possesses the best exploitation characteristics. From an operational point of view, the following steps must be foreseen and considered during the excavation phase:

1. construction of a temporary uptake structure;
2. definition of a monitoring program of the punctual water inflow for measurements of discharge rate, electrical conductivity, temperature and pH (for at least one year);
3. characterization of water quality by hydro-geochemical, bacteriological, isotopic and radiometric laboratory tests (for at least one year and according to regulations, depending on the type of use expected).

Determination of the final water intake system requires the definition of the following parameters: (i) identification and localization of the main water inflows; (ii) Discharge rate of water inflow; (iii) hydraulic head of water inflow; (iv) water quality; (v) administrative concessions and current legislation; (vi) technical specifications of construction materials and definition of the excavation technique(s), depending on the hydrogeological context.

174.3.2 Energy Potential of the Water Drained by the Basis Tunnel

The energy potential of warm water has been preliminarily evaluated with the relation:

$$Pot = P * C * D * dT$$

where, *Pot* is the thermal power [w], *P* is the discharge rate [m^3/s], *C* is the specific heat of water [4,186 J/kg/K], *D* is the water density [997 kg/m^3] and *dT* is the thermal gradient [K].

The analysis gave the temperature of the waters released from the energy plant. This is an important factor to evaluate the environmental impact subsequent to their introduction into the natural hydrographic network at the surface. Assuming the use of a simple heat exchanger, the following energy potential values were defined, based on the thermal gradient and upon a discharge rate of 100 l/s and a water temperature of 36 °C issued from the design (Table 174.2).

Table 174.2 Energy potential defined in function of the design input data and thermal gradient

Input water temperature (°C)	Output water temperature (°C)	Thermal Gradient (°C/step)	Energy Potential (kW)
36	31	5	2,000
36	26	10	4,000
36	16	20	8,000

174.3.3 Evaluation and Identification of the Potential Users of the Resource

Some scenarios for an energy development strategy for domestic and manufacturing purposes, in favor of the local community initiatives in the region interested by the basis tunnel project, are depicted in the following paragraphs. A study on the energy needs and outcomes of such a development strategy on the local social and cultural context must be part of the project's objectives. For this reason the following proposals have been developed:

1. Heating of the building and water of a public swimming pool;
2. Heating of the buildings and domestic hot water production in the international railway station and safety station (solution internal to the railway line project);
3. District heating system serving the local Municipality.

The heating of swimming pool water, commonly at temperatures as high as ~28 °C, would be possible using a simple heat exchanger (without heat pump); however, the installation of a heat pump must be considered in the case of the building and/or district heating system, in order to attain temperatures around 36 °C.

An operating heat pump for such purposes requires the availability of a certain electric power (see next tables). (Table 174.3).

It must be noted that the thermal potential naturally available is much higher than the demand for the two cases of the public swimming pool and the international railway station (2,301 kW compared to 400 kW in the case of the pool and 830 kW in the case of buildings of the station and safety area).

The construction of a district heating plant is a realistic option. Its size could be similar to those implemented in cities such as Sondalo and Tirano (Central Alps, Northern Italy), as the latters are very similar to Susa both in terms of demographic characteristics and logistics (as in altitude and geo-morphological context). The water output temperature is about 16 °C. The thermal power capacity is equal to about 4*2,500 kW (10 MW). The heating capacity of water leaving the Basis Tunnel would thus be fully exploited.

Table 174.3 Main characteristics of the heating systems for the developed proposals

Proposal	Heat demand rated (kW)	Thermal capacity (kW)	Preliminary characteristics of the system
Communal swimming pool (building included)	400	2,301 (equivalent of 20 apartments)	Nr of heat pump: 1
			Heat pump type: water/water
Building of the International railway station	830		Electric power demand: 403 kW
			COP: 5.71
			Output water temperature: 31 °C
District heating plant	10,000	10,082	Nr of heat pump: 4
			Heat pump type: water/water
			Electric power demand: 1,838 kW
			COP: 5.49
			Output water temperature: 16 °C

174.4 General Considerations

The water exploitation study includes, as a first step, subdivision of tunnel alignment, in order to exploit the sectors with the most significant geothermal features (e.g. T > 25 °C) via special pipes, depending on the intended purpose.

Possible use of warm waters from the Basis Tunnel can be confirmed, as well as feasibility of the proposals depicted in this paper, such as:

- heating of a public swimming pool;
- district heating system;
- heating/cooling of the railway station buildings;
- other possible initiatives near to the portal (fish farms, nurseries, etc.)

Other possible ways to exploit the warm waters drained by the Basis Tunnel have been considered; they could be:

- Realization of glasshouses to grow exotic fruit and vegetable species;
- Realization of water catchments to farm sturgeons and/or produce caviar.

These options have been already realized at the Tropenhaus in Frutigen (Switzerland), at the northern portal of the Lötschberg Tunnel (water discharge rate 110 l/s, temperature 17–20 °C) (Hufschmied and Brunner 2010).

All the above solutions are viable thanks to the characteristics and thermal potential of the waters drained by the Basis Tunnel.

References

Arrêté du 11 janvier 2007, Annexe I, Partie II, Tableau B relatif aux limites et références de qualité des eaux brutes et des eaux destinées à la consommation humaine mentionnées aux articles R. 1321-2, R. 1321-3, R. 1321-7 et R. 1321-38 du code de la santé publique

Brino L., Monin N., Poti P., Piraud J. and Buscarlet E. (2008) Modélisation géothermique et système de refroidissement pour le tunnel de base de la nouvelle liaison ferroviaire Lyon-Turin, Congrès International AFTES 2008, Monaco, 6–8 ottobre 2008, pp 31–37, Edition Spécifique

Decreto Legislativo 3 aprile 2006, n°152—Norme in material ambientale. Pubblicato nella Gazzetta Ufficiale n. 88 del 14 aprile 2006 —Supplemento Ordinario n. 96

Decreto Legislativo 16 marzo 2009, n°30, Allegati 3 e 5—Attuazione della direttiva 2006/118/CE, relativa alla protezione elle acque sotterranee dall'inquinamento e dal deterioramento. (09G0038). pubblicato nella Gazzetta Ufficiale n. 79 del 4 aprile 2009

Hufschmied P. and Brunner A. (2010) The exploitation of warm tunnel water through the example of the Lotschberg Base Tunnel in Switzerland. Geomech Tunn 3(5). doi:10.1002/geot.201000045

Impacts on Groundwater Flow Due to the Excavation of Artificial Railway Tunnels in Soils

175

Gabriele Bernagozzi, Gianluca Benedetti, Francesca Continelli, Cristiano Guerra, Renato Briganti, Santo Polimeni, Giuseppe Riggi, and Fabio Romano

Abstract

The engineering design for tunnel excavation in soils may produces an hydrogeological barrier to the natural groundwater flow. Upstream of the tunnel the groundwater table may rise while downstream may be lowered. This paper examines three case studies, all in a urban context (Turin, Parma and Venice—Italy). For each case a brief geological framework is provided, as well as performed Modflow models for investigating this problematic are presented. The quantified perturbation induced by the hydraulic barriers shows the classical behavior on the natural groundwater hydrodynamic leading to rising/lowering water table.

Keywords

Railway artificial tunnels • Hydrogeological barrier • Modflow models

175.1 Foreword

Realization of civil engineering works below the groundwater table generates an interference to the natural groundwater flow. In general, for excavation in soils, a sheetpile network or diaphragms walls is implemented, which may be an obstacle to groundwater flow leading to an increase and a decrease of the water table upstream and downstream of the barrier, respectively (Fig. 175.1). This effect cannot be neglected in the design phase as changes in groundwater level, may induce flooding events in basements of structures located upstream of the excavation and settlements of buildings foundations located downstream. This issue is investigated for three case studies. Changes in groundwater level due to hydraulic barriers, may also have environmental impacts in rural areas that always have to be evaluated in design phases.

175.2 Case Studies 1: Artificial Tunnel at Nichelino Railway Station Near Turin

The study was carried out in the context of the railway line doubling between Turin and Pinerolo. All the railway path is built at the surface, except for a part of about 2.5 km at the Nichelino station, which is an artificial tunnel. Excavations are supported by diaphragms walls at depth of about 18 m. Soils were characterized by an extensive survey campaign mainly performing boreholes. All available data were used to process a reference geological model of the site. In this area from ground level up to about 20 m of depth there are Quaternary alluvial and fluvial-glacial deposits overlaid on Astiane sands, Lugagnano clays and Villafranchiano silty sands at greater depths (Bondesan et al. 2008). The Modflow model has been divided in two layers: (1) the top layer (20 m thick) with horizontal permeability of 8×10^{-5} m/s corresponding to Quaternary alluvial deposits and (2) the lower layer, with

G. Bernagozzi
Free lance, Bologna, Italy

G. Benedetti (✉) · F. Continelli
ENSER S.R.L., Faenza, Italy
e-mail: gianluca.benedetti@enser.it
URL: http://www.enser.it

C. Guerra
Free lance, San Marino, Republic of San Marino

R. Briganti · S. Polimeni · F. Romano
ITALFERR, Gruppo Ferrovie dello Stato Italiane, Roma, Italy

G. Riggi
ITALFERR, Gruppo Ferrovie dello Stato Italiane, Milano, Italy

G. Lollino et al. (eds.), *Engineering Geology for Society and Territory – Volume 6*,
DOI: 10.1007/978-3-319-09060-3_175, © Springer International Publishing Switzerland 2015

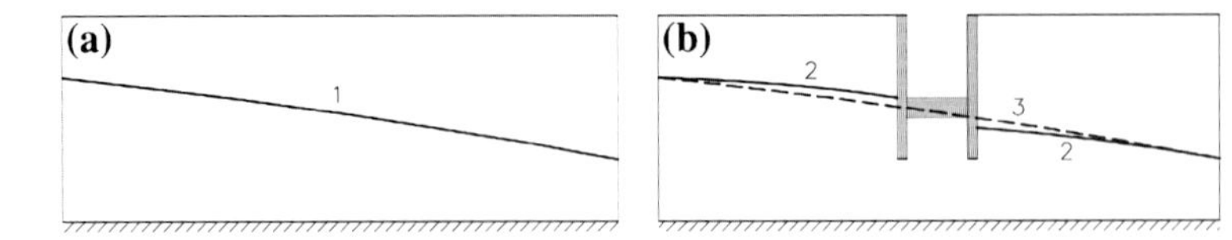

Fig. 175.1 **a** Natural groundwater table before the construction; **b** situation after the construction. *1* natural groundwater table; *2* effect of the hydraulic barrier on the groundwater table; *3* natural groundwater table before the excavation

variable permeability depending on the encountered lithology. Boundary conditions match surface hydraulic with constant hydraulic heads for Sangone river at north, Po river at east, Chisola river at south (Fig. 175.2).

Model calibration has been performed by multiple trial runs varying permeability of layers in a reasonable range of values in order to obtain the best match between measured and simulated data.

Perturbation produced by the realization of the tunnel has been simulated with no-flow cells conditions. For a conservative approach diaphragm walls reach the Quaternary base. Maximum variations between before and after excavation perturbation detected by the model are of 40 cm close the tunnel, with a rapid attenuation away from the path. In Fig. 175.1 is shown with a dashed line the area with variations of 15 cm. Positive changes are indicated with a plus sign (+), while negative ones with a minus sign (−).

175.3 Case Studies 2: Natural Tunnel Between Mestre and Venice Airport

The project design connect the railway station of Mestre with the Venezia Marco Polo airport. The path of about 8 km long, will be developed almost entirely in natural tunnel with the exception of a short part close to the Mestre railway station. The first part of the path will be realized in an artificial way: excavations will be supported by diaphragm walls driven to a depth of about 20 m from the ground level. Soils affected by the excavation are of alluvial type, always with a high fraction of fine materials. More permeable lenses, which correspond to riverbed fill material deposits, have an average permeability of about 10^{-5} m/s while deposits between alluvial cones have a lower permeability of about 5×10^{-8} m/s. The area has

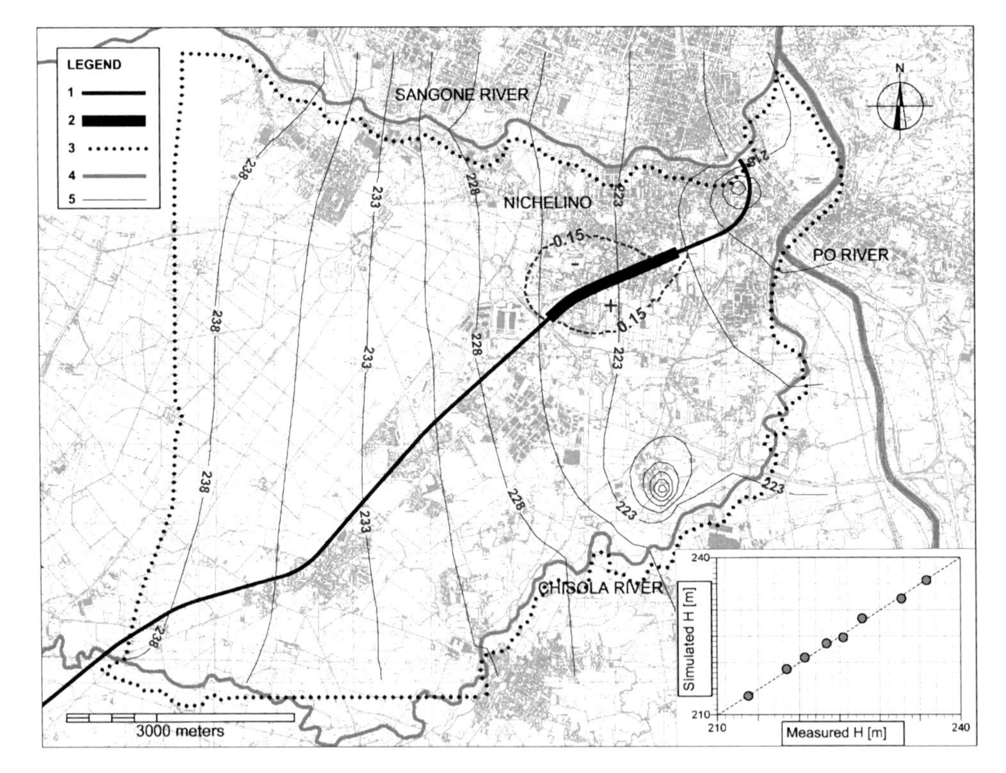

Fig. 175.2 Hydrography, boundary conditions and simulated hydraulic heads (*H*) for Turin Pinerolo railway close to Nichelino underground station. *1* Turin Pinerolo railway line; *2* Nichelino underground station; *3* limits of the Modflow model; *4* Sangone, Chisola and Po rivers; *5* groundwater table before excavation. *Bottom right corner*: matching between simulated and measured hydraulic head before tunnel excavation

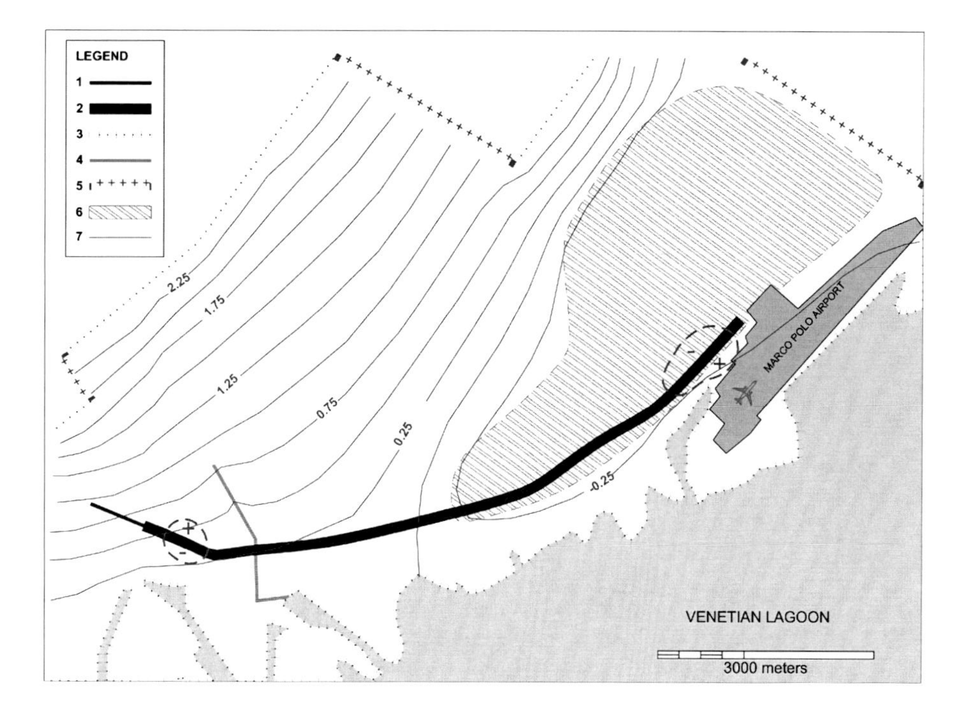

Fig. 175.3 Schematic representation of the Modflow model between the railway station of Mestre and Venezia Marco Polo airport. *1* Artificial tunnel designed with diaphragm walls; *2* design natural tunnel; *3* boundary conditions at hydraulic constant head; *4* artificial channel; *5* model limits; *6* reclaimed land drained by water pumps and channels; *7* groundwater table before the tunnel excavation

been intensively investigated over the years, because of the characterization and remediation campaigns carried out at the nearby industrial complex of Marghera. The reference hydrogeological model was constructed via the interpolation of hundreds of boreholes and water wells logs. The final result is a Modflow model with 9 layers of about 5 m thick. Each layer is obtained by the spatial interpolation in the horizontal direction of the available permeability data.

The model has dimensions of 11.5 km in the East–West direction and 8 km in the North–South direction.

The boundary condition set downstream was the Venice lagoon, modeled as a hydraulic constant head at 0 m. Given the shallowness of the lagoon this condition is limited to the upper layer of the model. Upstream the hydraulic head is constant and identified on the basis of the reference isophreatic lines. Monitoring data and available hydrogeological maps show that in the North West area of Venezia Marco Polo airport there is a portion of land with groundwater level below the sea level. This phenomenon is due to the drainage of water from a network of channels for land reclamation purposes. In the southern part of the model there is also an artificial channel that extends inland up to intercept the tunnel path. This channel has been simulated using the River function of the software.

The model was calibrated by successive approximations primarily by changing the drainage level and the conductance in the area affected by the land reclamation and secondly by acting on the artificial channel features. The result, which demonstrates a good correlation with the reference hydrogeological map, is shown in Fig. 175.3.

Fluctuation of natural groundwater flow are extremely modest and always contained within a few tens of centimeters. The explanation is to be found in the hydraulic gradient, which is very low, and in the fact that the tunnel thickness is small in comparison to the aquifer system thickness. In the airport area after the tunnel construction the water tends to be depressed North of the construction and raised South of the construction. This effect is explained by the drainage action which results in a groundwater flow from the lagoon to the land.

175.4 Case Studies 3: Artificial Tunnel Along the Parma Osteriazza Railway Line

The study area is located between the rivers Taro and Parma. Excavation works will affect the shallow aquifer, located from 35 to 10 m from ground level. Top of the aquifer there is a clayey layer of about 10 m of thickness which can be

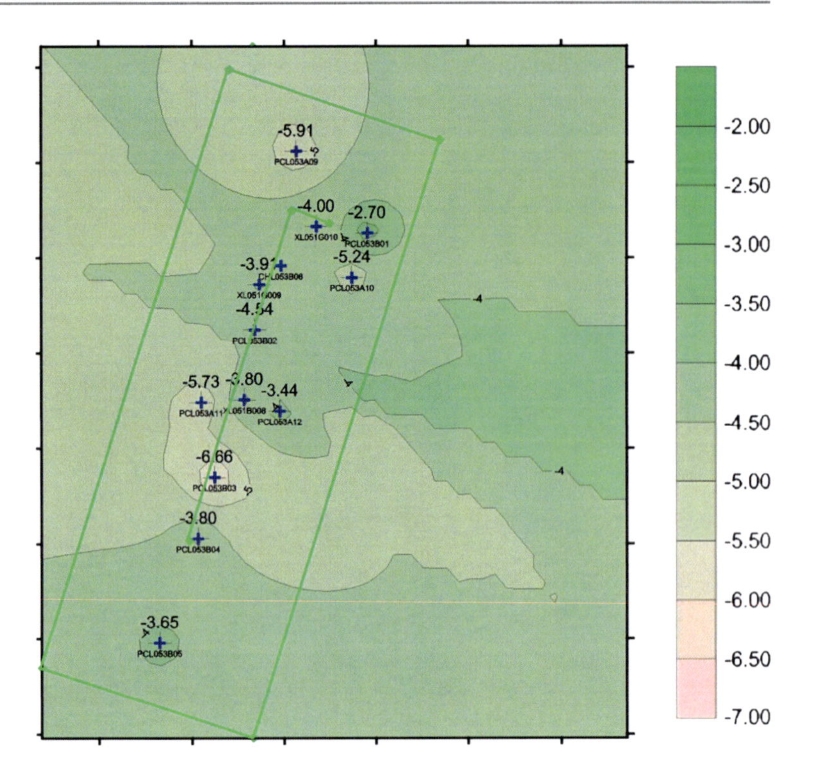

Fig. 175.4 Coefficient of permeability [unit log(10)k where k is m/s] contour map. Points indicate the location of the Lefranc tests along the artificial railway tunnel

considered substantially impermeable. The aquifer is composed mainly of sands and gravels with varying percentage of silt and clayey discontinuous lenses inside. The hydraulic gradient is about 0.001, with flow from North to South in confined conditions. The aquifer permeability was investigated through a series of Lefranc tests and the average values is equal to 3×10^{-4} m/s. (Fig. 175.4)

A Modflow model and an analytical analysis based on the method of fragments (Pavlovsky 1956; Harr 1962; USACE 1993), have shown that the interference increases in relation with hydraulic gradient changes. Acting with a conservative approach, the model was created imposing a gradient of approximately 5 times higher than that observed. Also in these conditions, changes in hydraulic head were always contained within a few tens of centimeters. The reason for the low interference is probably due to the fact that the tunnel direction is almost parallel to groundwater flow.

175.5 Conclusions

All 3 case studies indicate that deeper is the hydraulic barrier and higher is the natural hydraulic gradient, more significant will be the barrier effect. For the investigated sites impacts are always modest and range within a few tens of centimeters. This because of the low natural hydraulic gradients for all areas.

However, the impact of hydraulic barriers on groundwater flow should never be neglected. Engineering designs including partial occlusions of an aquifer should always be accompanied by hydrogeological investigations quantifying impacts on groundwater.

References

Bondesan A, Primon S, Bassan V, Vitturi A (2008) Le unità geologiche della Provincia di Venezia, Provincia di Venezia, Università di Padova

Calda N, Valloni R, Bedulli F (2007) Three-dimensional representation of permeability barriers and aquifers recharge in the pleistocene deposits of the parma alluvial plain. Mem Descr Carta Geol D'It LXXVI:97–108

Carraro F (1996) Il Quaternario—Revisione del Villafranchiano nell'area-tipo di Villafranca d'Asti. Ital J Quat Sci Aiqua 9:fasc.1

Harbaugh A (2005) MODFLOW-2005, the U.S. geological survey modular ground-water model—the ground-water flow process. Techniques and Methods 6-A16, U.S. Geological Survey, http://pubs.usgs.gov/tm/2005/tm6A16/

Harr ME (1962) Groundwater and seepage. McGraw-Hill Book Company, New York

Provincia di Venezia—Servizio Geologico, Difesa del Suolo e Tutela del Territorio (2011), Atlante geologico della Provincia di Venezia

U.S.A.C.E (1993) Seepage analysis and control for dams, U.S. Army Corps of Engineers, Washington, http://140.194.76.129/publications/

Chemical and Isotope Composition of Waters from Firenzuola Railway Tunnel, Italy

176

L. Ranfagni, F. Gherardi, and S. Rossi

Abstract

Forty-six water points from stream waters, springs, wells and tunnel seeps have been monitored for major reactive aqueous components and isotope composition of water over the period 2004–2008 in an area (Mugello basin, Northern Apennines, Italy) impacted by the tunneling works for the Bologna-Firenze high-speed railway. Based on geological, hydrogeological and geochemical information, an integrated conceptual model has been elaborated which considers the drainage of waters of different age in the Firenzuola railway tunnel. Precipitations and waters from small perched aquifers rapidly percolate in the tunnel through well-defined fractures zones and mix with more than 100 years old waters. Dilute, near neutral pH waters of Ca–HCO_3 composition are generated through limited interaction of meteoric waters with local lithologies, whereas more saline, alkaline pH waters of Na–CO_3–HCO_3 composition originate through prolonged interaction of meteoric waters with flysch rocks.

Keywords

Chemical composition • Isotope composition • Railway tunnel • Northern Apennines • Mugello, Italy

176.1 Introduction

Opened in 2009, the new Bologna-Firenze high-speed railway has a length of 78.5 km, and includes 73.3 km of tunnels. The Tuscan stretch of the new railway includes 6 mainline tunnels (three of which more than 15 km in length). The Bologna-Firenze railway is one of the most important links in the Italian rail network, and connects the Emilia Romagna region with Tuscany region under the Apennines.

The Firenzuola tunnel, approximately 44.5–59.6 km southwards of Bologna, in the Mugello basin, was excavated at an elevation of 300–350 m a.s.l, below the main apenninic watershed, located at an elevation of 1,000–1,100 m a.s.l. (Rodolfi et al. 2004). The excavation of the Firenzuola tunnel caused a significant drop of groundwater levels regionally, due to the seepage of surrounding groundwater into the tunnel. About 250 and 600 L/s of groundwater currently discharge into this tunnel during the low- and high-flow season, respectively, but a maximum inflow of 900 L/s has been estimated during the period of maximum crisis. A number of springs and creeks have dried out as a consequence of the excavation works of the tunnel (AA.VV. 2008).

The main purpose of this study was to elaborate an integrated hydrogeological and geochemical model of the area impacted by the tunnel, and possibly identify the main structures connecting the tunnel with streams and surficial aquifers.

L. Ranfagni (✉) · S. Rossi
Environmental Protection Agency of Tuscany (ARPAT), Via Porpora 22, 50144, Florence, Italy
e-mail: luca.ranfagni@arpat.toscana.it

F. Gherardi
Consiglio Nazionale delle Ricerche (CNR), Istituto di Geoscienze e Georisorse (IGG), Via Moruzzi 1, 56124, Pisa, Italy

G. Lollino et al. (eds.), *Engineering Geology for Society and Territory – Volume 6*,
DOI: 10.1007/978-3-319-09060-3_176,

176.2 Geological and Hydrogeological Setting

The Northern Apennines represent the exposed portion of a large orogenic wedge generated by a process of crustal shortening and thickening associated with the subduction of the Adriatic plate under the Iberian plate (Martini et al. 2001). The most important geological units in the area are represented by the turbidites of Oligo-Miocenic age of the Marnoso-Arenacea formation (Vincenzi et al. 2014).

The Firenzuola tunnel has been prevalently bored throughout sandstones (greywackes), marls and, subordinately, siltstones and claystones of Lower-Middle Miocene age. These rocks show large secondary permeability solely in correspondence of major fault and fractured zones, mainly created by the post-orogenic tectonic detensioning of Apennines. In general, fractures develop in arenaceous banks, and, occasionally, fractures with a persistence exceeding that of the individual layers may be recognized. Large-scale water circulation develops in these features, mostly WNW-ESE trending. Fractures are more frequent near anticline/syncline axes and/or faults oriented in the Apennine direction (WNW-ESE). Associated with these structures, large tabular sub-vertical aquifers (up to 200 m wide, and up to some kilometers length), perennial springs and streams occur.

176.3 Methods

A total of 46 water sampling points, were repeatedly collected during the period 2004–2009. All water samples were analyzed for major (Ca, Mg, Na, K, Cl, HCO_3, SO_4) chemical constituents, and for the isotopic composition of water ($\delta^{18}O$, δ^2H, ^3H). Temperature, pH, electric conductivity were measured in the field. The chemical analyses were performed in the laboratories of ARPAT, Firenze, as follows: Cl and SO_4 by ionic chromatography, Ca, Mg, Na and K by atomic adsorption spectroscopy. The percent charge-balance error of samples discussed in this study is typically below 5 %. The $^2H/^1H$, $^{18}O/^{16}O$ isotope ratios and tritium (^3H) concentrations were determined in the laboratories of CNR-IGG, Pisa. Analytical accuracies are typically ±1 ‰ for δ^2H, ±0.05 ‰ for $\delta^{18}O$ and ±0.5 UT for ^3H.

176.4 Results and Discussion

Three main types of waters have been considered for this study: (i) creeks and streams (group A), (ii) springs and wells (group B); (iii) tunnel seeps (group C). All the sampled waters have pH values between 6.7 and 10, with the most alkaline values associated to tunnel seeps. Waters inflowing in the Firenzuola tunnel are generally more saline (330–848 mg/L) than waters from springs (238–508 mg/L) and streams (314–574 mg/L). Surficial waters have a Ca–HCO_3 signature, whereas the chemical composition of tunnel inflows varies from Ca–HCO_3 to Na–CO_3–HCO_3 with increasing salinity. The prolonged interaction of meteoric waters with host rocks causes C_{TOT} (Fig. 176.1a) and Na_{TOT} (Fig. 176.1b) concentrations to increase in tunnel seeps. Decreasing tritium contents (^3H < 4 UT; Fig. 176.1) are good tracers of (i) prolonged underground residence times, and (ii) of the lack of mixing with meteoric waters (^3H > 5 UT) rapidly infiltrating in the Marnoso Arenacea flysch.

From 2004 to 2008, only for some tunnel points we observed a decreasing of prolonged residence water contribution. In the same lapse of time, the sampling point with more "mature" waters didn't show significant changes in composition. Latest chemical analysis, carried out in 2012 at the southern entrance of the tunnel, shows that a significant contribution of prolonged circulation waters remains.

Reaction path modeling has been applied to investigate the processes governing the chemistry of the tunnel seeps (Fig. 176.2). A flow-through configuration has been used to

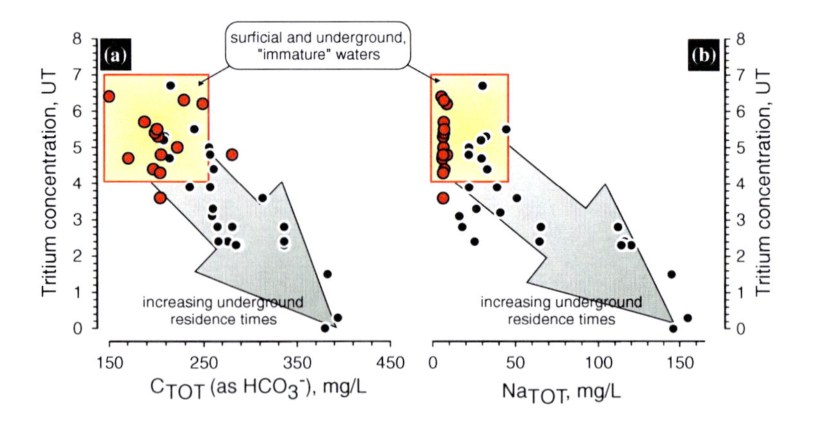

Fig. 176.1 Tritium versus C_{TOT} (*A*) and tritium versus Na_{TOT} (*B*) diagrams for Firenzuola Tunnel waters; *red dots* springs and streams; *black dots*: tunnel points

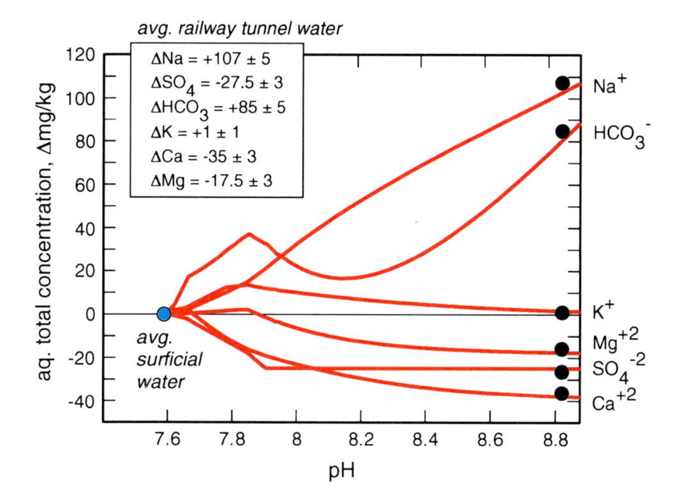

Fig. 176.2 Geochemical modeling of water-rock interactions occurring in the Marnoso Arenacea flysch. Groundwater total aqueous concentrations are computed as delta mg/kg-H_2O, with respect to the average composition of surficial waters (streams) infiltrating in the study area. pH is used as the master reaction progress variable. Consistently with their longer underground residence times, tunnel waters (*black dots*) plot at the end of the reactive path, and represent a sort of "geochemically mature end-member"

simulate the evolution of a packet of fluid traversing the Marnoso Arenacea aquifer, under the assumption that the interacting fluid has an initial chemical composition comparable to waters usually recognized in small, perched aquifers. This zero-dimensional, forward model considers that quartz, albite, K-feldspar, chlorite and carbonatic and/or dolomitic cements are the main reactive phases in the system, according to petrographic and mineralogical evidences (Gandolfi et al. 1993). Numerical calculations (Fig. 176.2) indicate that the hydrolysis of the Marnoso Arenacea flysch is a possible way to generate the most saline, and alkaline waters sampled in the central part of the Firenzuola tunnel (Southern sector—zone I of Fig. 176.3).

The stable isotope composition of the sampled waters ($\delta^{18}O$ and $\delta^{2}H$ values between −8.7 and −7.0, and between −55 and −42 ‰, respectively) is adequately represented by a meteoric line having a deuterium excess, d, of 15, similar to precipitations falling in Central Italy. The distribution of points indicates that tunnel groundwaters have a meteoric origin, and that the observed differences in the isotopic ratios are mostly due to seasonal effects. The meteoric recharge

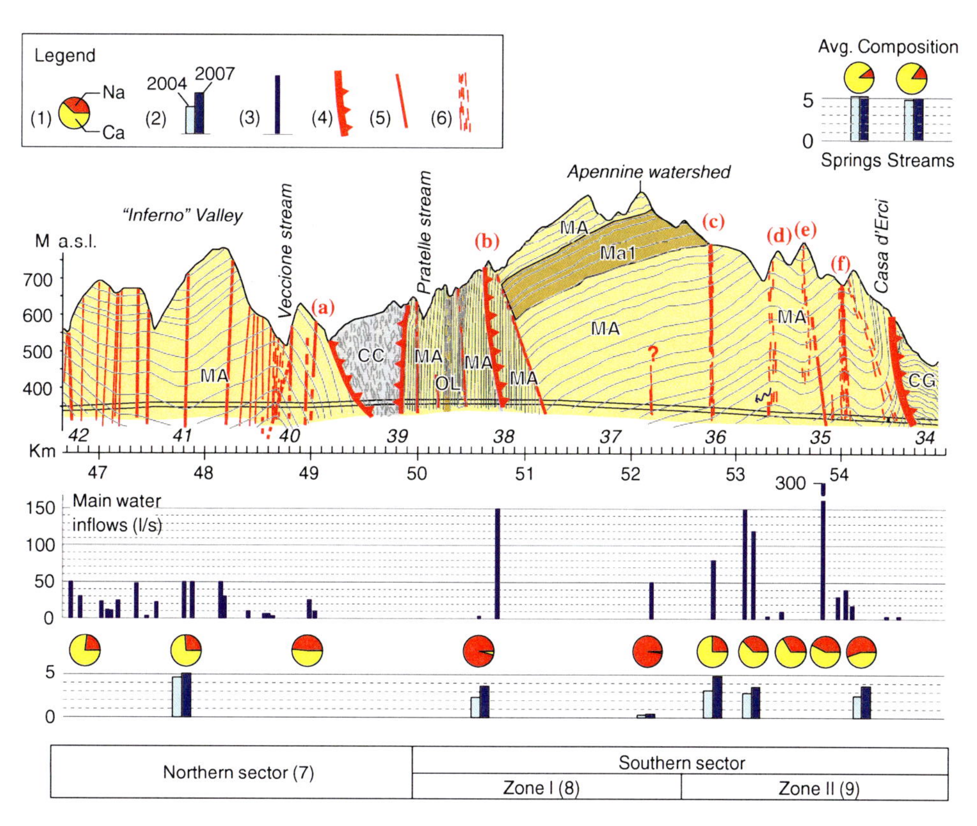

Fig. 176.3 Geostructural and hydrogeological cross-section of the Firenzuola Tunnel. Information on the location and on selected chemical parameters of the main water inflows is also provided: (*1*) Ca/Na ratio; (*2*) tritium content (UT) in 2004 and 2007; (*3*) inflow rates (l/s) during the excavation of the tunnel. Major structural features are: (*4*) thrusts; (*5*) faults or master joints; (*6*) fractured belts. The Northern sector (*7*) of the tunnel, dominated by the rapid infiltration of meteoric waters, is hydraulically set apart from the Southern sector (*8* and *9*, see text) by the permeability barrier CC (Chaotic Complex). The most important geological formations (*fm*) are: Marnoso-Arenacea fm. (*MA*); Marnoso-Arenacea fm., marl-rich facies (*MA1*); Olistostrome (*OL*); Caothic Complex (*CC*); Castel Guerrino fm. (*CG*). Vertical circulation of groundwater is supposed to prevalently occur through the following tectonic discontinuities (*red labels*): (*A*) Isola tectonic line; (*B*) Fognano overthrusting (*C*); Ronta fault; (*D*) Frassineta Nord fractured belt; (*E*) Casa d'Erci Nord fractured belt; (*F*) Casa d'Erci Sud fault and fractured belt

occurs at elevations roughly comprised between 700 and 1,100 m a.s.l.

By combining hydrogeological and geological information with data on chemical and isotope composition of waters, we derived a model of underground water circulation which considers two main hydrogeological sectors in the tunnel (Fig. 176.3). These sectors are separated by a chaotic assemblage of blocks of preexisting rocks (olistostrome), which acts as an impermeable barrier. The northern sector is dominated by the rapid infiltration of meteoric waters, and then, by the occurrence of low-salinity, high-^{3}H, near-neutral pH waters, of Ca-HCO_3 composition.

The southern sector comprises two sub-zones: the first one (Zone I), adjacent to the olistostrome, is dominated by relatively ancient waters (likely older than 100 years; ^{3}H generally <2 UT), having high pH values, and Na–CO_3–HCO_3 composition. In this zone, the occurrence of marl-rich, relatively impermeable strata of the Marnoso-Arenacea fm. above the tunnel (MA1 in Fig. 176.3) prevents meteoric precipitation from infiltrating in the excavated area. The remaining zone (Zone II), is characterized by the occurrence of the same "old" waters of Zone I, locally mixed with surficial (^{3}H generally >5 UT) or meteoric (^{3}H up to 10 UT) waters rapidly percolated in the aquifer. Low-salinity, high-^{3}H waters are expected to percolate through well-defined fractures zones, such as the Fognano overthrusting, and the Ronta and Casa d'Erci faults (Fig. 176.3). The occurrence of long residence water circuits, is further supported by the occurrence of waters with extremely low ^{3}H concentrations in the Allocchi tunnel, a railway tunnel excavated in 1890 at about 550 m a.s.l. in the same flysch formation of the Firenzuola tunnel, not far (8 km) from the area of study.

176.5 Conclusions

Combined with geological and hydrogeological information, a detailed chemical and isotope survey of 46 water points (streams, springs, wells and tunnel seeps) has contributed to the definition of an integrated hydrogeochemical model of an area impacted by the tunneling works for the Bologna-Firenze high-speed railway. Water isotopes reveal that waters of different age are drained by the Firenzuola tunnel, with the oldest groundwater component likely being older than 100 years. The chemical characteristics of the Firenzuola tunnel waters are controlled by the mineralogy of the interacting rocks (Marnoso Arenacea flysch) and the extent of water-rock reaction. More prolonged residence times and/or effective water-rock interactions lead to more saline, pH alkaline waters of Na–CO_3–HCO_3 composition.

References

AA.VV. (2008) Relazione sullo stato dell'Ambiente in Toscana 2008. Regione Toscana—ARPAT, 2008, pp 221–224

Gandolfi G, Paganelli L, Zuffa GG (1993) Petrology and dispersal pattern in the Marnoso-Arenacea Formation (Miocene, northern Apennines). J Sedim Petrol 53:493–507

Martini IP, Sagri M, Colella A (2001) Neogene-Quaternary basins of the inner Apennines and Calabrian arc. In: Vai GB, Martini IP (eds) Anatomy of an Orogen: the Apennines and adjacent Mediterranean basins. Kluwer, Dordrecht, pp 375–400

Rodolfi G, Rossi S, Doni A, Ranfagni L (2004) Apennine tunneling works: impacts on the surface and underground water resources. D08 Field trip Guidebook, 32nd international geological congress, 20–28 Aug 2004, Firenze, Italy

Vincenzi V, Gargini A, Goldscheider N, Piccinini L (2014) Differential hydrogeological effects of draining tunnels through the Northern Apennines, Italy. Rock Mech Rock Eng 47(3):947–965

Hydrogeological Modeling Applications in Tunnel Excavations: Examples from Tunnel Excavations in Granitic Rocks

177

Baietto Alessandro, Burger Ulrich, and Perello Paolo

Abstract

Numerical modeling of groundwater circulation has been applied in the Aica-Mules tunnel, an exploration adit that has been used for geological investigations before the excavation of the main Brenner Base Tunnel (high speed railway link between Italy and Austria). Conditions faced by the Aica-Mules tunnel allowed to improve the general knowledge of the geological context and of hydrogeological impacts on groundwater circulation in a granitic massif where the two main tubes of the Brenner Base Tunnel will be constructed. In this perspective, numerical modeling has been applied, both in back- and forward-analysis for the following purposes: (i) in order to get insights on the reliability of the conceptual model relative to the hydrogeological setting during the excavation; (ii) to estimate the risk that the tunnel might interfere with spring catchment areas with possible negative feedbacks on the natural discharge regime; (iii) to forecast the inflow rate expected in the tunnel in the most critical geological sectors related with the presence of fault and fracture systems.

Keywords

Numerical modeling • Tunnel design • Groundwater circulation • Risk management

177.1 Introduction

In engineering geology, hydrogeological numerical modeling is used extensively for multiple purposes (Meiri 1985; Marechal et al. 1999; François et al. 2007). Whenever based on a background of hydrogeological observations and on a solid geological model, numerical modeling is able to provide appropriate solutions related to (i) project design decisions and (ii) environmental impacts on shallow manifestations of groundwater flow such as springs and wells.

B. Alessandro (✉) · P. Paolo
GDP Consultants, Località Grand Chemin, 22, 11020 Saint Christophe, AO, Italy
e-mail: baietto@gdpconsultants.eu

B. Ulrich
Galleria di Base del Brennero, Brenner Basistunnel BBT SE, Piazza Stazione 1, Bolzano, Italy

In relevant tunneling projects, consisting of large underground excavations, the prediction of how groundwater will interact with the constructions is paramount for economic efficiency and for minimization of environmental effects.

In the framework of the Brenner Base Tunnel (BBT), a planned 55 km long railway tunnel running from Innsbruck (Austria) to Fortezza (Italy), currently under design and construction, many hydrogeological analyses and investigations have been carried out during the last 20 years (Brandner and Dal Piaz 2005; Barla et al. 2010; Perello et al. 2013). Results of these studies were used to reconstruct conceptual models for groundwater circulation in sectors interested by planned excavation works. In the southernmost section of the BBT project, this conceptualization constituted the base for numerical models aiming at: (i) validating the hydrogeological conceptual model of the area interested by the Aica-Mules tunnel, (ii) estimating impacts that the tunnel had on the spring catchment areas with possible negative feedbacks on the natural discharge regime, and (iii) forecasting the groundwater inflows in the most critical geological sectors of the main tubes. Models set-up was

G. Lollino et al. (eds.), *Engineering Geology for Society and Territory – Volume 6*,
DOI: 10.1007/978-3-319-09060-3_177, © Springer International Publishing Switzerland 2015

different according to purposes and applications. Detailed 3D modeling taking into account most of main geological and hydrogeological elements was performed in the above mentioned case (i) and (ii), while broad conceptualization and simplification of the hydrogeological context is assumed for the setup of 2D and 3D models referring to case (iii). Models are subject to a strict calibration procedure that involves a comparison between measured and modeled data before and after tunnel perturbation. Forward analysis is based on sensitivity analysis of parameters in order to provide ranges of expected values of groundwater inflow into the tunnel. All models where produced by means of the software Feflow (Diersch 2009).

177.2 Geological Setting of the Aica-Mules Tunnel

The Aica-Mules tunnel, already constructed, constitutes an exploratory excavation of 6.7 m of diameter lying about 12 m below the main tubes of the railway link (currently under design) providing information on the rock mass for reducing construction costs and times of the BBT project. The Aica-Mules tunnel is approximately 10 km long and has been drilled by means of a double shield Tunnel Boring Machine (TBM) starting from the Aica portal in Italy. No rock-mass impermeabilization operations were performed and groundwater free flows into the tunnel.

The Aica-Mules tunnel is entirely located in crystalline rocks pertaining to the South-Alpine domain of the Alpine chain, the so-called Brixen Granite, mainly composed by granites and granodiorites variously crosscut by aplitic and pegmatitic sills. To the North, the Brixen Granite is bounded by a main regional structure, the Periadriatic Line (Figs. 177.1 and 177.2). This structure is marked by a steeply dipping E-W striking mylonitic shear zone affecting a tonalitic body that separates the Brixen Granite and the Austroalpine units. The Brixen Granite has been affected by brittle deformation with structures spanning from simple fracture networks, to cataclastic fault zones with up to a pluri-km-scale persistence.

In the tunnel area the most important tectonic feature is represented by the Riobianco Fault Zone, a sub-vertical N–S striking fault. All these structures affect in a different manner the permeability of the granite and play an important role in influencing the groundwater circulation pattern and the discharges that occurred during the tunnel excavation. Insights into the hydraulic conductivity of the rock-mass for the area are available from hydraulic packer tests that were performed in deep drill holes. The tests have shown that hydraulic conductivity is high only along fracture systems. The fault zone permeability is generally high, despite it

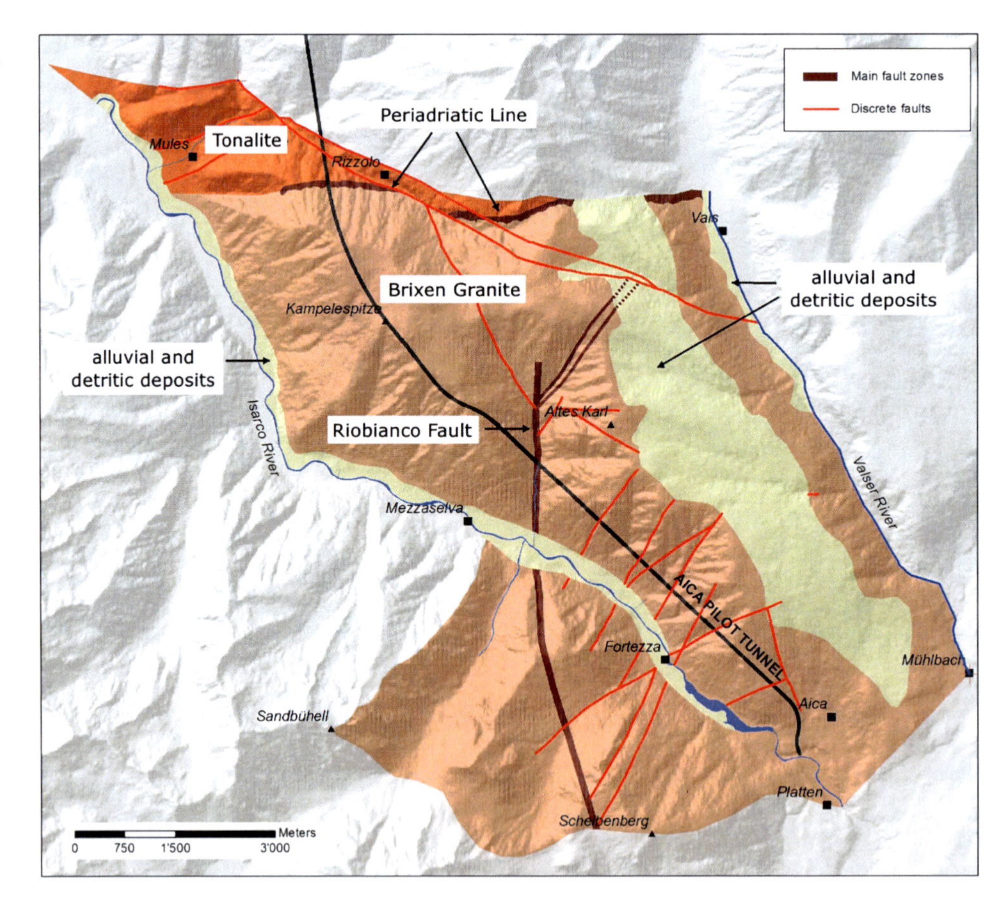

Fig. 177.1 Geological patterns integrated in numerical models of the Aica Tunnel

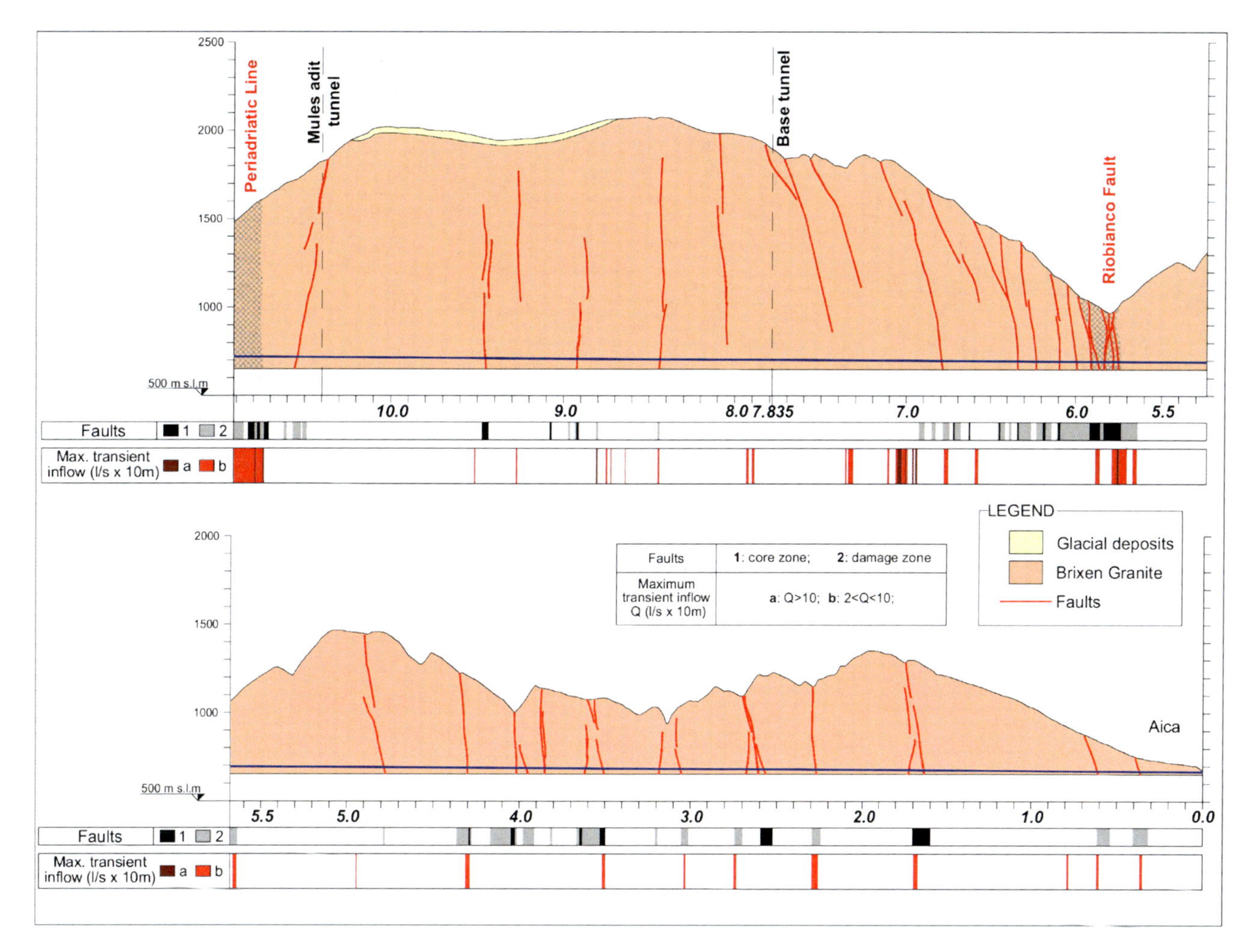

Fig. 177.2 Geological cross-section along the Aica Tunnel with the main water inflows occurred during the excavation

shows a strong variability due to the presence of products deriving from cataclastic fragmentation with a size ranging from fine-grain to coarse-grain. Hydraulic tests indicate values spanning over three orders of magnitude, in the range of 5×10^{-9}–5×10^{-6} m/s.

177.3 Numerical Models

177.3.1 Model of the Aica-Mules Exploration Tunnel

During the excavation of the Aica-Mules tunnel water flow was continuously monitored allowing to reconstruct the hydrographs relative to water discharges into the tunnel. Various inflow events have been recorded that can be correlated with fault or fracture zones intersection. The most relevant tunnel inflow was recorded at the intersection with the Riobianco Fault (around pk 5 + 700), with a peak of net discharge of 220 l/s, progressively decreasing to less than one third of this value. A second major inflow zone with peak discharges similar to those of the Riobiancofault was intersected (around pk 7 + 000), but this time without any connection with fault zones, since only a small fracture zone was detected.

A 3D finite element model (Fig. 177.3) covering entirely the sector above the excavated section of the Aica-Mules tunnel has been implemented in order to gain insights on the hydrogeologic behavior shown by the aquifer during the tunnel drilling. This model was calibrated in steady-state conditions and was then used for transient simulations relative to the different excavation phases of the tunnel; boundary conditions accounted either for hydraulic heads along the Isarco River and the main streams, and for the effective infiltration calculated for the top boundary. This model allowed a back-analysis based on the comparison of the numerical results with the hydrogeological data collected during the monitoring. While main fault zones (i.e. RiobiancoFault and Periadriatic Line) have been treated as equivalent porous media, minor faults have been handled as discrete features with distinctive hydrogeological properties.

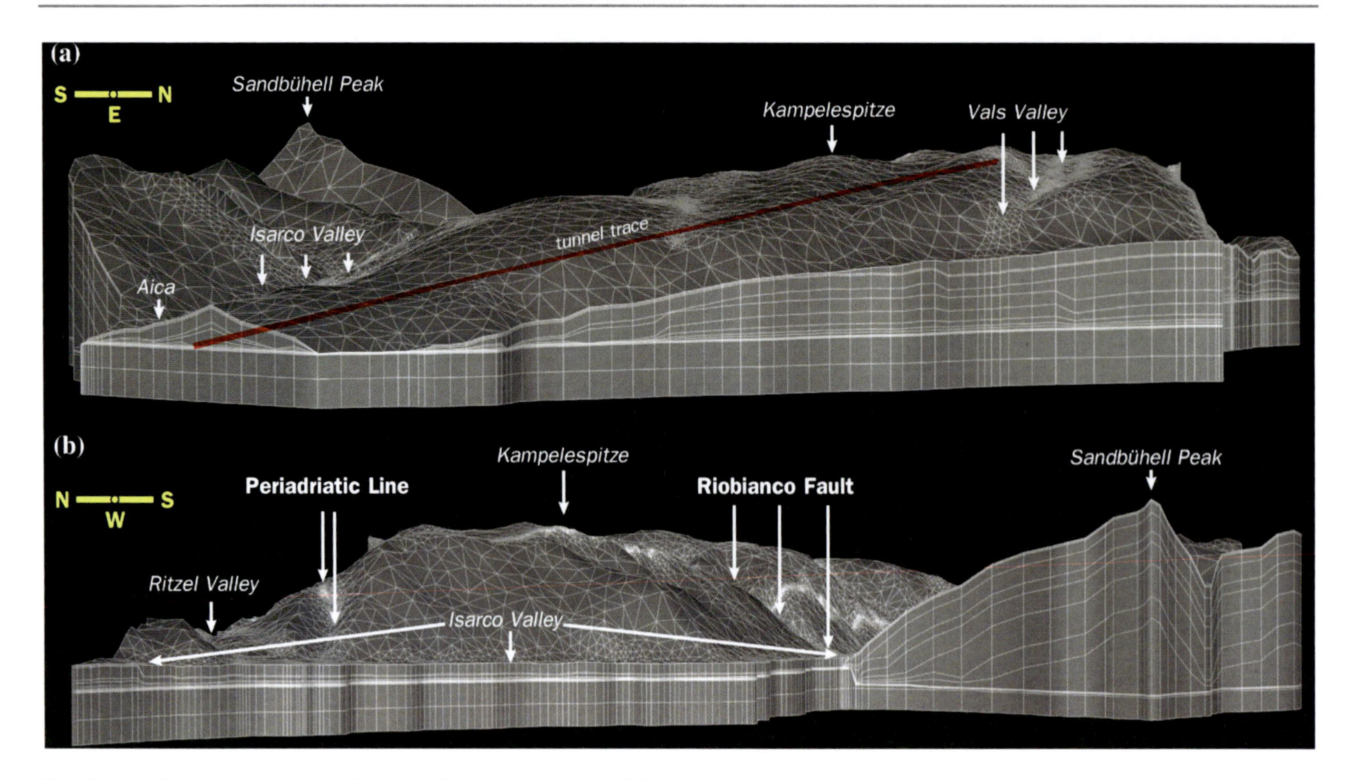

Fig. 177.3 3D geometry and meshes used for the numerical model of the Aica-Mules tunnel

A satisfying calibration was obtained only for some portions of the aquifer while other discrepancies with the observational data couldn't be smoothed out unless assuming very different hypotheses with respect to the ones that were initially considered. The main inconsistency between the measures made on site and model results was relative to the very rapid pressure pulse that was recorded by a piezometer (Mu-B-04/05), located 5-km away from the Riobianco Fault and measuring the water level at depth in the Bressanone Granite. Quite surprisingly, the piezometer recorded a decrease in the water level after less than 3 days from the fault interception by the tunnel. Since that moment, the groundwater level in the Mu-B-04/05 started showing a slow but constant decrease that kept on for 5 months without any restoration, until getting to a final drastic decrease that likely corresponded to the interception by the water table of an impervious limit.

To justify the very fast response of the piezometer that followed the fault interception by the tunnel, it has been assumed a conceptual model considering that the piezometer was hydraulically connected with the Riobianco Fault through a set of permeable faults crosscutting the Bressanone Granite. After several trials it was apparent that a satisfying calibration was impossible to achieve only by modifying hydraulic boundary conditions and/or rock properties. This observation led to the perception of a different condition lying behind the real nature of the aquifer. So far, according to the available geological data, the aquifer was considered as an unconfined aquifer with a free groundwater surface and with direct connection with surface infiltration. But, to obtain satisfying calibration the aquifer hosted in the fault system had to be treated as a confined aquifer. Sealing phenomena along the fault, not evident from field and drilling data, could explain this particular behavior.

177.3.2 Model for Impacts Evaluation of Spring Catchment Areas

Drainage of groundwater systems and drying out of existing spring catchments and groundwater wells is recognized as one of the most important issues to take into account when forecasting environmental impacts connected with the excavation of long and deep tunnels. In this context, numerical models have been applied in order to forecast impacts on the groundwater reservoir feeding some springs. Since natural variations in the spring discharge regime occur, it is important to distinguish whether a decrease in the discharge rate is due to natural reasons or to the tunnel influence. Analyses carried on model outputs allowed to identify zones with different expected drawdown. This computation helps constraining and differentiating those springs that are more likely to be subject to negative feedbacks due to the tunnel drainage from those that are less affected by this effect.

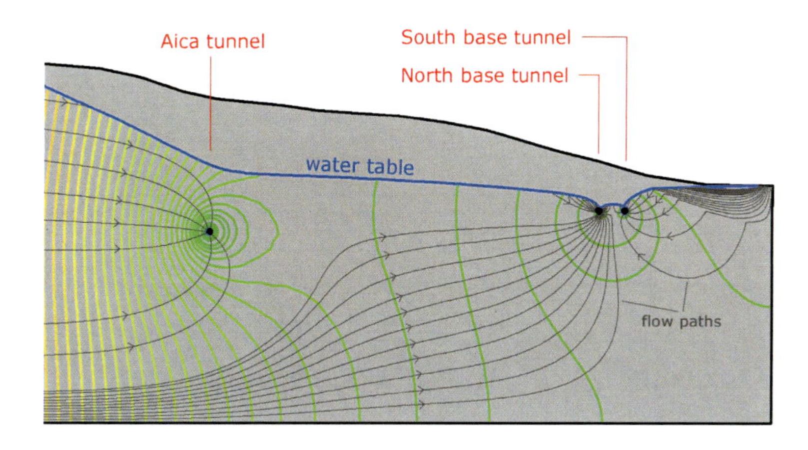

Fig. 177.4 2D model used for inflow rate estimation within the Aica tunnel and the Brenner Basis Tunnel

177.3.3 Model for Inflow Rates

After back analysis, new models have been implemented for forecasting the hydrogeological conditions during the future construction of the main BBT tubes. A set of 2D and 3D numerical models have been set up in zones where main inflows are expected in relationship to the occurrence of main faults and fracture systems (Fig. 177.4). These models aimed at forecasting the water discharges into the BBT tunnel tubes under the hypothesis of different operational conditions (excavation of the first tube, of the second tube, etc.).

Simulations were carried out via a parametrical approach accounting for uncertainties related with the geological model. For example different permeability and fault distributions were considered in the models. The obtained results were used to identify possible ranges of expected tunnel inflow values, as well as for identifying local variations in the flow systems (Fig. 177.4).

177.4 Conclusions

This study shows that, whenever constrained by adequate data and consistent interpretation, groundwater numerical modeling applied to tunneling can serve as a valuable tool for an appropriate design and environmental risk assessment. In the frame of the Aica-Mules Tunnel, numerical modeling has been applied for validating the hydrogeological conceptual model, for estimating the potential impact of the tunnel on spring catchments and for forecasting the inflow rate expected in the tunnel.

References

Barla G, Ceriani S, Fasanella M, Lombardi A, Malucelli G, Martinotti G, Oliva F, Perello P, Pizzarotti EM, Polazzo F, Rabagliati U, Skuk S, Zurlo R (2010) Problemi di stabilita‘ al fronte durante lo scavo del cunicolo esplorativo Aica—Mules della Galleria di base del Brennero. In: Barla G, Barla M (eds) XIII ciclo di conferenze di meccanica e ingegneria delle rocce MIR 2010. Torino 30 Novembre–1 Dicembre 2010, pp 175–213

Brandner R, Dal Piaz GV (2005) Brenner Basis Tunnel/Tunnel di Base del Brennero: Strukturgeologische Kartierung und erganzende geologische Studien/Rilevamento geologico-strutturale e studi geologici integrative, BBT SE unpublished internal report, contract D0104, 1011 p

Diersch HJG (2009) FEFLOW finite element subsurface flow & transport simulation system—reference manual. DHI-WASY GmbH ed

François B, Tacher L, Bonnard Ch, Laloui L, Triguero V (2007) Numerical modelling of the hydrogeological and geomechanical behaviour of a large slope movement: the Triesenberg landslide (Liechtenstein). Can Geotech J 44(7):840–857

Marechal JC, Perrochet P, Tacher L (1999) Long-term simulations of thermal and hydraulic characteristics in a mountain massif: The Mont Blanc case study, French and Italian Alps. Hydrogeol J 7 (4):341–354

Meiri D (1985) Unconfined groundwater flow calculation into a tunnel. J Hydrol 82:69–75

Perello P, Baietto A, Burger U, Skuk S (2013) Excavation of the Aica-Mules pilot tunnel for the Brenner base tunnel: information gained on water inflows in tunnels in granitic massifs. Rock Mech Rock Eng. doi:10.1007/s00603-013-0480

Effects on the Aquifer During the Realization of Underground Railway Works in Turin

178

Stefano Ciufegni, Fabrizio Bianco, Adriano Fiorucci, Barbara Moitre, Massimiliano Oppizzio, and Francesco Sacchi

Abstract

The "Underground Railway Link" of Turin (Piedmont, Italy), in the section between Porta Susa and Corso Grosseto stations, represented a work with no small difficulties in planning and execution, in fact it has entailed the difficult problem of crossing through an underpass the Dora Riparia River with a double artificial tunnels. The potential interference of this work with unconfined aquifer has necessitated a complete hydrogeological investigation aimed at the design of a monitoring plan for the entire line of action, compatible with the presence of road and railway surface network as well as sub-systems directly interconnected to these. Monitoring, still in place, started from August 2005, with a series of campaigns of detection given by manual acquisition and since November 2006 has been implemented with the installation of data logger. The monitoring system has provided for the control of the variation of the groundwater level ante operam, during construction and post operam in order to control in detail the development and the process of hydrodynamic adaptation of the aquifer system to the new structure in the succession of the different implementation phases.

Keywords

Underground railway • Groundwater level • Monitoring • Turin

178.1 Introduction

The "Underground Railway Link" of Turin (Piedmont, Italy), in the section between Porta Susa and Corso Grosseto stations, represented a work with no small difficulties in planning and execution, in fact it has entailed the difficult problem of crossing through an underpass the Dora Riparia River with a double artificial tunnels. For the realization of the work has been required the construction of a system of bulkheads and a bottom plug, that penetrate in the subjacent unconfined aquifer flowing in the almost orthogonal direction to the work.

The potential interference of this work with unconfined aquifer has necessitated a complete hydrogeological investigation aimed at the design of a monitoring plan for the entire line of action, compatible with the presence of road and railway surface network as well as sub-systems directly interconnected to these (Civita et al. 2002; Civita and Pizzo 2003).

S. Ciufegni · F. Sacchi
Italferr S.p.A., Rome, Italy

F. Bianco · A. Fiorucci (✉) · B. Moitre
Department of Environment, Land and Infrastructure Engineering (DIATI), Politecnico di Torino, Corso Duca degli Abruzzi 24, 10129 Turin, Italy
e-mail: adriano.fiorucci@polito.it

B. Moitre
e-mail: barbara.moitre@tin.it

M. Oppizzio
ATI: Astaldi S.p.A., Rome, Italy

M. Oppizzio
Vianini S.p.A, Rome, Italy

M. Oppizzio
Di Vincenzo S.p.A., San Giovanni Teatino, Italy

M. Oppizzio
Impresa Rosso S.p.A., Turin, Italy

G. Lollino et al. (eds.), *Engineering Geology for Society and Territory – Volume 6*,
DOI: 10.1007/978-3-319-09060-3_178,

178.2 Hydrogeological Conceptual Model of the Study Area

In the study area there are three hydrogeological complexes, from the oldest to the most recent:

- silty-sandy complex: it consists of sands in silty matrix with permeability from low to medium, host the deep aquifer of the Turin foothills and affects the central and central-eastern portion of the area.
- alternations complex: it consists of layers of clay alternating with horizons of gravel and sand, does not surface in the study area, but in the western sector is located at a depth of about 30 m.
- gravelly-sandy complex: it consists of gravel and sand with weakly silty ma-trix, includes the alluvial deposits of the fundamental level of the Turin plain and terraced alluvial systems affecting the river areas of the waterways of the study area.

The study area is characterized by two aquifers more or less interconnected among themselves and with the Dora Riparia River. It is possible to distinguish an unconfined aquifer and a multi aquifer. The unconfined aquifer is located mainly in the gravelly-sandy complex. The Dora Riparia River regulates the flow field in the investigated area, in fact imposes extreme dominance on the aquifer with its meandering in the western part of the area and its irregular oscillations of level, related to the considerable extension of its catchment area.

In the eastern sector of the study area, below the level of the sandy-gravelly aquifer is present a sandy layer which is directly interconnected with the above complex. This area is therefore characterized by a multilevel aquifer (Civita et al. 2007).

Considering a wider area subtended by the Dora Riparia, the morphology of the piezometric surface is radial complex with just accentuated concavity downstream and a gradual increase of the distance between the piezometric curves (Civita et al. 2007). From the northwestern limit of the study area, the distance between the piezometric curves turns out to be quite smooth. Proceeding towards S, it is clear the influence of the Dora Riparia on groundwater flow. The general direction of groundwater flow is from W to E. In the southern sector, the aquifer system affected by the draining action of the Dora Riparia: the directions of the flow, in fact, initially from NW to SE, veer towards the right bank of the river highlighting a draining action, while at the side left, the piezometric values are slightly lower and impose the outflow from the Dora the aquifer (Fig. 178.1, Civita et al. 2007).

178.3 The Monitoring Network

Monitoring, which started with a series of campaigns of measures through manual acquisition by phreatimeter since August 2005, has been implemented, starting in November 2006, with the installation of automatic acquisition instrumentation. The aim was to update, with constant frequency, the time evolution of the groundwater level fluctuations related to the aquifer interested by the work.

In consideration of the fact that the work could interfere directly, as well as with the subjacent aquifer, even with the surface watercourse which is hydro-geologically interconnected, the monitoring system has provided the installation of two hydro-metric stations on river course over which the continuous measurement of the groundwater levels on different equipped piezometers.

The monitoring network was made up initially of twenty control stations of piezometric levels of the aquifer and two hydrometric stations on the river Dora Riparia (Fig. 178.1).

The monitoring system has provided the variation of the groundwater level ante operam, during construction and post operam in order to control in detail the development and the process of hydrodynamic adaptation of the aquifer system to the new structure in the succession of the different implementation phases (Civita and Pizzo 2003).

178.4 Analysis of the Data

The piezometric trend recorded in the last 7 years, in various monitoring points, outlines the progression of the work in the gallery. In this work we analyzed the data, inherent to the piezometric changes induced from the realization of the underground railway link, often measurement points (Fig. 178.2). It must always be kept in consideration the fact that the data analyzed may be affected by medium seasonal natural fluctuations, other than ongoing works.

The piezometers were divided into three groups, representing three distinct areas located along the course of the work, two located in the left and one on the right bank of the Dora Riparia. In the first group includes the measuring stations on the orographic right SA17 and SA18 upstream and SI5 downstream of the work. The second group includes piezometers SA3 and SA12 upstream and piezometers SA4 PC2 downstream. The third group consists of the measurement points more distant from the river, the DA7 and SA7 piezometers upstream and the SA5 piezometer downstream.

The piezometric level, in the period between May 2006 and April 2012, outlines the progression of the work in the gallery. In general, until the autumn of 2008, we can see the reaction of the aquifer to the impacting presence of the work just performed without bypass. The piezometric levels trends of the monitoring points upstream and downstream diverge. There are increases upstream and decreases downstream of the system of bulkheads, realized to secure the path to the dangers of flooding. Since September 2008, there were changes to the piezometric surface area due to groundwater dewatering

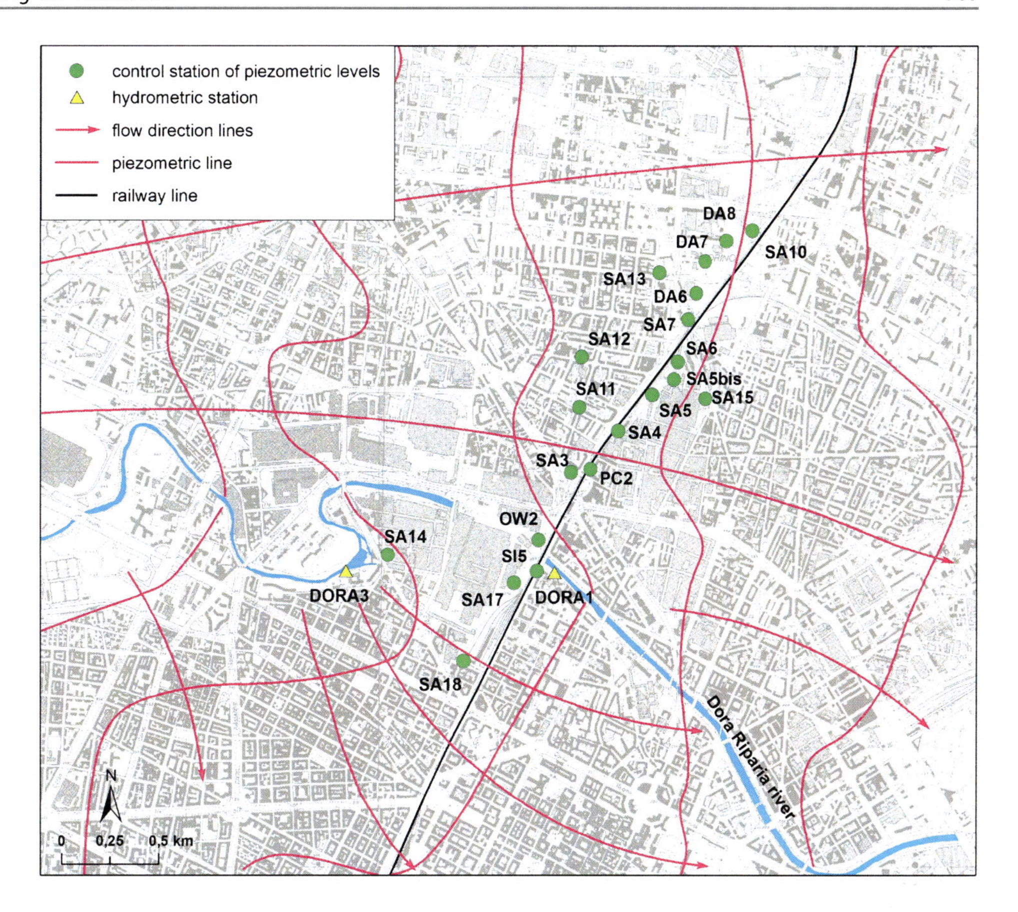

Fig. 178.1 Flow field and monitoring network

operations during the execution of the work. From May, 2012, after a serious of operations aimed at stabilizing the unconfined aquifer and compensation of dam-effect, the configuration of the levels is gradually and significantly improved, tending to the natural recovery of the initial conditions. In fact, the delta of the levels measured upstream–downstream are considerably less than the forecast calculated using a mathematical model, between raising to upstream and lowering to downstream.

In recent months considered is evidence of greater tendency towards stabilization of the levels to a state of balance and greater mitigation of dam-effect compared to the previous period. In fact, in February 2013 was carried out the final phase of the groundwater bypass, which led to significant and positive effects on the conformation of the aquifer in the downstream sector, with imported recoveries of the piezometric levels in the monitoring points taken into account. The decrease of the difference between the piezometric level upstream and downstream of the work is evident between the measuring points SA17 and SI5 and between SA3 and PC2.

In conclusion levels upstream work are in line with the pre-construction measures with a modest superelevation that wards off eventual known environmental problems. The levels downstream work, instead, show a significant recovery in conjunction with the final commissioning of the entire by-pass system.

178.5 Conclusion

The underpass of the Dora Riparia of the "Underground Railway Link" of Turin caused a dam effect on the unconfined aquifer which can be considered not relevant on the hydrogeological structure taken into account. In fact, from the data collected can be seen as the commissioning of the technician bypass has caused a mitigating effect of the disturbance induced by the work on the natural flow of the aquifer. From the data obtained, it highlights the fundamental importance of monitoring to be performed during the realization of works that may impact with groundwater.

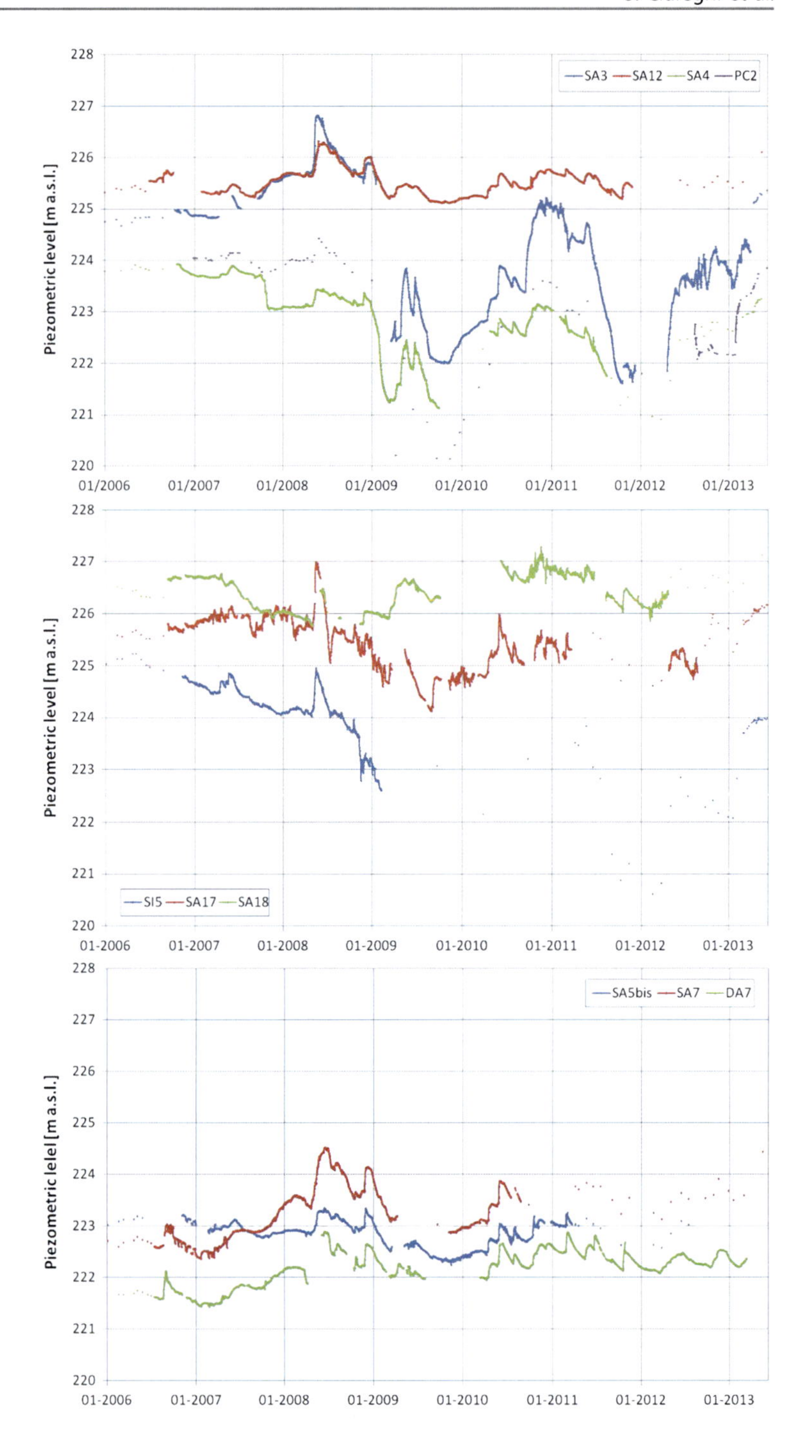

Fig. 178.2 Piezometric levels

References

Civita M, Pizzo S (2003) Hydrogeological interference areas: the importance of preliminary prognosis in the designing of a high velocity railway line. In: Proceedings of AVR 03 "First International Workshop on Aquifer Vulnerability and Risk", Salamanca, Messico, pp 182–193

Civita M, De Maio M, Fiorucci A, Pizzo S, Vigna B (2002) L'ambiente idro-geologico: l'importanza della prognosi preliminare. Atti MIR 2002 "Le opere in sotterraneo e il rapporto con l'ambiente", Torino, Ed. Patron, Bologna, pp 73–106

Civita M, Danese A, Nicola E, Pizzo S (2007) Previsione e monitoraggio degli effetti sull'acquifero nel corso della realizzazione di un'importante opera ferrovia-ria in sotterraneo nella città di Torino. GEAM. Geoingegneria Ambientale e Mine-raria, vol 44.2, pp 49–68

Proposal for Guidelines on Sustainable Water Management in Tunnels

179

Antonio Dematteis

Abstract

This paper illustrates the progress of the working group named GESTAG, from the Italian chapter of the International Association of Hydrogeologists, elaborating a guideline for hydrogeological activities in tunneling design and construction. The aim is to improve sustainable Water Management in tunnels. The impact on drilling production of water drainage during excavation, the valorization of groundwater resources drained during excavations, and the associated heat, and the environmental impacts of drainage on Groundwater Dependent Ecosystems and the aquifers are considered. GESTAG is composed by experts from different backgrounds (universities, designers, public authorities managers). The guideline is expected to be published in late 2014.

Keywords

Tunnel • Water management • Environmental sustainability • Monitoring • Italy

179.1 Introduction

Tunneling and big urban excavation works sprawl rapidly as meet the ever increasing requirements of a fast and sustainable mobility either between cities or inside cities' transportation network. At the same time, the rapid growth of renewable energy is sourcing the development of underground hydroelectric power plants. Such a growing interest in exploiting underground represents a great opportunity to limit the environmental impact from large infrastructure projects.

However, underground excavation has also the greatest impact on groundwaters. Severe consequences, e.g. hydraulic head drawdown, base-flow impoverishment, subsidence and groundwater contamination, may occur during and after excavation, making more difficult local people acceptance of ongoing projects. Groundwater inflow in tunnels should not be simply considered as a "geological accident" and a geotechnical obstacle for tunnel excavation and stability. Attention should be paid to the interaction between tunnel and groundwater flow systems. In the stages of both environmental impact assessment, and hydrogeological modelling and monitoring, the expected modifications of local groundwater budget terms (recharge-discharge) along with ecological effects against groundwater dependent ecosystems must be carefully taken into account. The hydrogeological reference model assumes central importance in this context, being the basis of any impacts prediction. Investigations, surveys, studies and calculations that are necessarily carried out to define the model must be encoded and described in detail. This allow to quantify the model reliability, which is an input to the risk analysis of time, cost and environmental and social impacts of the work.

179.2 Need of a Guideline

Experience put in evidence the need of guidelines for coordinating all hydrogeological activities in tunneling design, in order to improve environmental sustainability. This should help hydrogeologists and decision makers to

A. Dematteis (✉)
SEA Consulting s.r.l., Corso Bolzano 14, 10121 Turin, Italy
e-mail: dematteis@seaconsult.eu

G. Lollino et al. (eds.), *Engineering Geology for Society and Territory – Volume 6*,
DOI: 10.1007/978-3-319-09060-3_179, © Springer International Publishing Switzerland 2015

improve the water resources management in tunneling. A hydrogeology-based approach, founded upon a more or less common protocol, can surely help to improve the management and interaction of "drilling production" and geotechnical excavation constrains with the environmental impacts. At the same time, the water resources drained by underground excavations, and the associated heat, should also be valorized. Based on of this experience came the idea to establish guidelines with the group of experts from different backgrounds (universities, designers, public authorities managers) listed above. The guideline is expected to be published in late 2014.

179.3 The GESTAG Working Group of the IAH

This paper illustrates the progress of the working group named GESTAG, established on 20/06/2012 by the Italian Chapter of the International Association of hydrogeologists (IAH). The working group is constituted by the following persons: U. Burger (BBT SE, Brenner Basis Tunnel); F. Capozucca (ANAS S.p.A.); P. Cerutti (ECOTER CPA s.r.l.); A. Dematteis (SEA Consulting s.r.l.) chairman of the WG; A. Gargini (Università di Bologna Alma Mater Studiorum); M. Governa (Regione Piemonte), F. Grosso (Hydrodata S.p. A.); F. Marchionatti (Politecnico di Torino) M.E. Parisi (LTF SAS, Lyon Turin Ferroviaire); P. Perello (GDP Consultant s.r.l.); M. Petitta (Università di Roma La Sapienza); M. Petricig (Regione Piemonte) M. Polemio (CNR IRPI); G. Preisig (Université de Neuchatel); L. Ranfagni (ARPA Toscana- VIA/VAS Sector); G. Ricci (Geodata Engineering S.p.A.); S. Skuk (BBT SE, Galleria di Base del Brennero); M. Tallini (Università dell'Aquila); R. Torri (SEA Consulting s.r.l.); V. Vincenzi (freelance geologist); A. Geuna (GDTest s.r.l.).

The aim of GESTAG working group is to propose a guideline on Sustainable Water Management in Tunnels. GESTAG is subdivided into five sub-groups, analyzing five different topics: (i) experience feedback on tunneling hydrogeology, based on verified review of tunnels already realized; (ii) geological and hydrogeological reference models, and methods of forecasting and management of the relationship between water and tunnel; (iii) impact on the Groundwater Dependent Ecosystems (GDE) and the environment, i.e. springs, streams, and wetlands; risk analysis and mitigation; (iv) communication, training, support to decision makers, advertising, education on the perception of possible environmental impacts from tunneling; (v) existing laws and policy framework, perspectives and development of new guidelines.

179.3.1 Experience Feedback

A data base of different field experiences on tunneling hydrogeology, inflow rates, examples of exploitation of water and heat represents a starting point and a useful calibration of modeling based forecasting. Experiences should include verified review of case-studies. A synthesis of different case-studies, in Italy and abroad, where either a professional best-practice and/or a research-focused approach have contributed to treat groundwater flow as a most significant target of analysis, is considered as the basis for the guidelines proposal.

179.3.2 Geological and Hydrogeological Reference Model; Methods of Forecasting and Management Water-Tunnels Relationships

The evaluation tool of the interferences induced by the works with the underground waters, starting from the different options of project up to the definitive solution, is the Hydrogeological Reference Model (HRM). A HRM is constituted by a conceptual and numerical schematization of the aquifers able to represent, both to a qualitative and quantitative level, the actual state of the underground circulation, and the impacts evaluation by overlaying the project alternatives. The HRM is developed in its main structure within the project phase, but it is subject to revisions and refinements on the base of the data-flow deriving from the monitoring activities, during the work and *post-operam* phases, with the aim to support an objective quantification of the interferences. The HRM must be the tool of analysis to be used for the prediction of impacts. These are the forecasting of the inflow into the tunnel, the forecasting of drawdown hazard that can affect the depletion of springs, wells and watercourses, and the design of technologies for drainage and/or waterproofing.

179.3.3 Valorization of Groundwater Resources and Heat

The water drained from the tunnel is a resource that cannot be wasted and which must imperatively be recovered. And this also applies to the heat associated with water in deep tunnels such as those in the Alps. The valorization of groundwater resources drained during excavations, and the associated heat, can be made if planned in the design phase, by means of appropriate catchment and housing of evacuation pipes in the tunnel. This approach must include an analysis of the external demand of water and heat, near the portal of the tunnel where the resource can be delivered. Good examples that demonstrate the feasibility of these solutions are few but do exist, and should be studied, analyzed and reproposed systematically.

179.3.4 Impact on Groundwater Dependent Ecosystems and the Environment; Risk Analysis and Mitigation of Impacts

Groundwater dependent ecosystems (GDEs) are habitats that must have access to groundwater to maintain their ecological structure and function and are critical components in the conservation of the earth's aquatic biodiversity. These comprise a complex and often biodiverse subset of the world's ecosystems, and can be found in marine, coastal, terrestrial, cave, and aquifer environments (ref. GDE Network of the International Association of Hydrogeologists, http://iah.org). In addition to the GDE should obviously be included potential impacts on humans water uses. This require the analysis of springs, streams, and wetlands. Through the activities of modeling and monitoring it is possible to achieve a defined layout of the effects induced by the infrastructure on the underground water environment (GDE, springs, streams, etc.).

The impact layout described above is to be shared with the stakeholders, including environmental control Authorities and the inhabitants, in order to find shared solutions, and limit possible litigations. This process help to identify and activate alternative supply plans where interferences that cannot be eliminated.

179.3.5 Communication and Education

The communication and education is one of the most important aspects to be managed to the rapid progress of the project. The experience shows that where these aspects are neglected, the project progresses slowly, and the costs increase. All of the following tasks must be overseen and carried out in parallel with the design and implementation of the work: training, advertising, education on the perception of possible environmental impacts from tunneling, and understanding of opportunities to valorize water and heat resources.

179.3.6 Existing Laws and Policy Framework

The analysis of existing laws and policy frameworks is one of the tasks of the GESTAG group. All suggestions included in these guidelines have to comply with the national rules. However, close examination of the rules, that by the time we made just for those in Europe, indicates a general lack of specific standards for tunneling.

Part XVIII

Uncertainty and Risk in Engineering Geology

Convener Dr. Luca Soldo—*Co-convener* Jean Piraud

The Geological and Geotechnical Model (GGM) is the cornerstone of the design of major civil works, from feasibility evaluation to design during construction. The GGM visualizes, describes and quantifies the geological and geotechnical features and their characteristics in-depth, their estimated reliability, with the potential related hazards.The GGM is affected by uncertainties arising from the accuracy and completeness with which the subsurface conditions are known. Its limits span among events with a known probability (sometimes referred to as statistical uncertainty), and true uncertainties, which are events with unknown probability (sometimes referred to as indeterminacy) or, worst still, completely unforeseen events. Uncertainty may be thought of as a continuum ranging from zero for certain information to intermediate levels for information with statistical uncertainty and known probabilities to high levels for information with a "true uncertainty". This session aims to be a round table to collect practical guidelines for the understanding and quantification of uncertainties and reliability of GGM for large civil engineering works as a base for the identification of the "best practices" to face them (ref. Also IAEG Commission C 28—Reliability quantification of the geological model in large civil engineering projects).

The Design Geological and Geotechnical Model (DGGM) for Long and Deep Tunnels

180

Alessandro Riella, Mirko Vendramini, Attilio Eusebio, and Luca Soldo

Abstract

Design errors frequently stay behind construction cost overruns and delays. A relevant part of these errors may arise from an inadequate knowledge of the geological and geotechnical conditions: several geological, hydrogeological and geotechnical aspects can remain unknown, partially or completely, prior to actual construction of engineering project. These unknowns usually exist in inverse proportion to the effectiveness of the geological and geotechnical investigations. On the other hand it is experienced that, because of the intrinsic complexity of the geological context, also rigorous investigation approaches are subject to several sources of uncertainty and random geological features or events remain difficult to be predicted and characterized. In particular for long and deep tunnels, these uncertainties and the resulting risks in construction could be especially severe. The Design Geological and Geotechnical Model (DGGM) has been intended from several authors in the last years as a conceptual framework inside which, progressively, searching adequate answers to the above mentioned limits. Inside the DGGM all the collected data are comprehensively stored and interpreted, anticipating and characterising the ground conditions with their related risks. Because of unavoidable limits rising from accuracy and completeness with which the subsurface conditions may be known, it is also a mean for identifying the variability and uncertainties of the data and of the derived geological context, with the related hazards and risks, providing the basis for a rationale design procedure.

Keywords

Tunnel design • Geology • Geotechnics • Uncertainties • Risks

180.1 Geological and Geotechnical Unknowns and Risks

The IMIA-ITIG has recently published (2011) the analysis of several cases of major tunnel failures (for projects from 1994 to 2010), identifying among the causes an "insufficient ground investigation" level (12 % of the cases) and for 41 % some "design errors", that often can derive from inadequate countermeasures for facing geological and geotechnical hazards.

As known several geological, hydrogeological and geotechnical aspects can remain unknown, partially or completely, prior to actual construction of engineering project, mostly due to intrinsic investigation ("diagnosis") difficulties.

A. Riella (✉) · M. Vendramini · A. Eusebio · L. Soldo
Geodata Engineering S.p.A., Corso Duca degli Abruzzi 48/e, Turin, Italy
e-mail: ari@geodata.it

M. Vendramini
e-mail: mve@geodata.it

A. Eusebio
e-mail: aeu@geodata.it

G. Lollino et al. (eds.), *Engineering Geology for Society and Territory – Volume 6*,
DOI: 10.1007/978-3-319-09060-3_180,

In particular for long and deep tunnels, these uncertainties and the resulting risks in construction could be especially severe: "…. *the deeper the tunnel, the larger the uncertainties; the higher the probability of encountering adverse or unforeseen conditions for tunnelling, the greater the effort and the cost for site investigations to reduce the uncertainties*" (International Tunnelling Association ITA Report no 4—Long tunnels at great depth 2010).

Also the effectiveness of the modern procedures of "flexible design" and "Risk Management"—today integrated inside the best practices of geotechnical design (Chiriotti and Grasso 2002; AFTES GT32.R2A1 2012)—is enhanced when based on a sound preliminary "diagnosis" phase. The Risk Management must be applied in the very early stages of the project, and then up-dated along its completion. To understand the geological and geotechnical environment with which the project will face, identify the related hazards, evaluating their probability of occurrence and impact is of unparalleled importance.

These unknowns usually exist in inverse proportional to the effectiveness (amount, nature and quality) of the geological and geotechnical investigations (U.S. National Committee on Tunnelling Technology, USNCTT 1984; Consiglio Nazionale delle Ricerche 1997; Site Investigation Steering Group 2007).

Many literature references and rules of procedure underline the importance of a complete and proper investigation campaign (USNCTT 1984; U.S. UTRC 1996; ITA Working Group no. 17, 2010; AFTES GT32.R2A1 2012). Based on an analysis of 89 underground projects the USNCTT observed that in more than 85 % of the cases the inadequate level of the investigation leaded to claims and time/cost overruns. The USNCTT publication made recommendations as to minimum requirements for any project, especially considering the different order of magnitude (as percentage of capital cost) between investigation levels (<1 %) and claims levels (12–20 % and upwards).

The recent study "Analysing International Tunnel Costs" (Worcester Polytechnic Institute 2012) emphasises, again, the key role of the preliminary site investigations: "*We recognize the issues with convincing Clients to spend more money in the early stages of a project, when the overall viability, constructability and financing is still unknown, but all of our research subjects described a direct correlation between the amount of SI and cost savings*".

It isn't worthless to observe that the expenditure on site investigation as a percentage of total project cost is often low, not rarely ranging from a mere 0.1 to 0.3 %. Over the past years ground investigation prices seem to be forced down in real terms and investigation today is often based upon "minimum cost and maximum speed".

On the other hand it is experienced that sometimes, because of the intrinsic complexity of the geological context, also rigorous investigation approaches are subject to several possible sources of uncertainty that could arise from the need for simplifications, heterogeneity, inherent randomness, imperfect interpretative concepts and hypotheses, measurement inaccuracies, sampling limitations, insufficient sample numbers, and others. Random or partially random geological features or events such as e.g. karst, gas and water circuits remain difficult to be predicted and characterized (e.g. Fig. 180.1).

If the reliability of the geological model largely depends on the effectiveness (amount, nature and quality) and reliability of the initial investigation phase the "core" of a sound and comprehensive understanding of whatever is necessary for the design requires "something more", as it will be described in the following.

180.2 The Design Geological and Geotechnical Reference Model

In the last years several authors (Soldo 1997 in *Progetto Strategico Gallerie*; Venturini 2001; Knill 2002; IAEG Commission C25 and C28) have progressively proposed the concept of Geological and Geotechnical Model (GGM or DGGM, Design GGM, underlying the bi-univocal relationships between itself and the project, then its importance inside the design procedure) a conceptual framework where the collected data are comprehensively stored (factual data) and interpreted, anticipating and characterising the ground conditions with their related risks.

A model is conceived as a tool (built on the base of the available data; the model itself could then be defined as a tool for store and process the input data) to understand, define, quantify, visualize, or simulate a certain aspect of the nature. It requires a selection and identification of the relevant aspects of a situation in the real world and then using different types of models for different aims:

- conceptual models are intended as tools to better understand the reality,
- graphical models to visualize the reality,
- operational models to define something (e.g. a variable, term, or object) in terms of a process (or set of validation tests) needed to determine its existence, duration, and quantity,
- mathematical models are intended as tools to quantify (objects or processes).

The model is built following some main steps:

- do assumptions that simplify the system to its essential aspects,
- identify initial and/or boundary conditions,
- identify and quantify operating processes,

Fig. 180.1 Random, heavy water inflow through massive basalts (*Kárahnjúkar Hydroelectric Project, Iceland. Headrace Tunnel*)

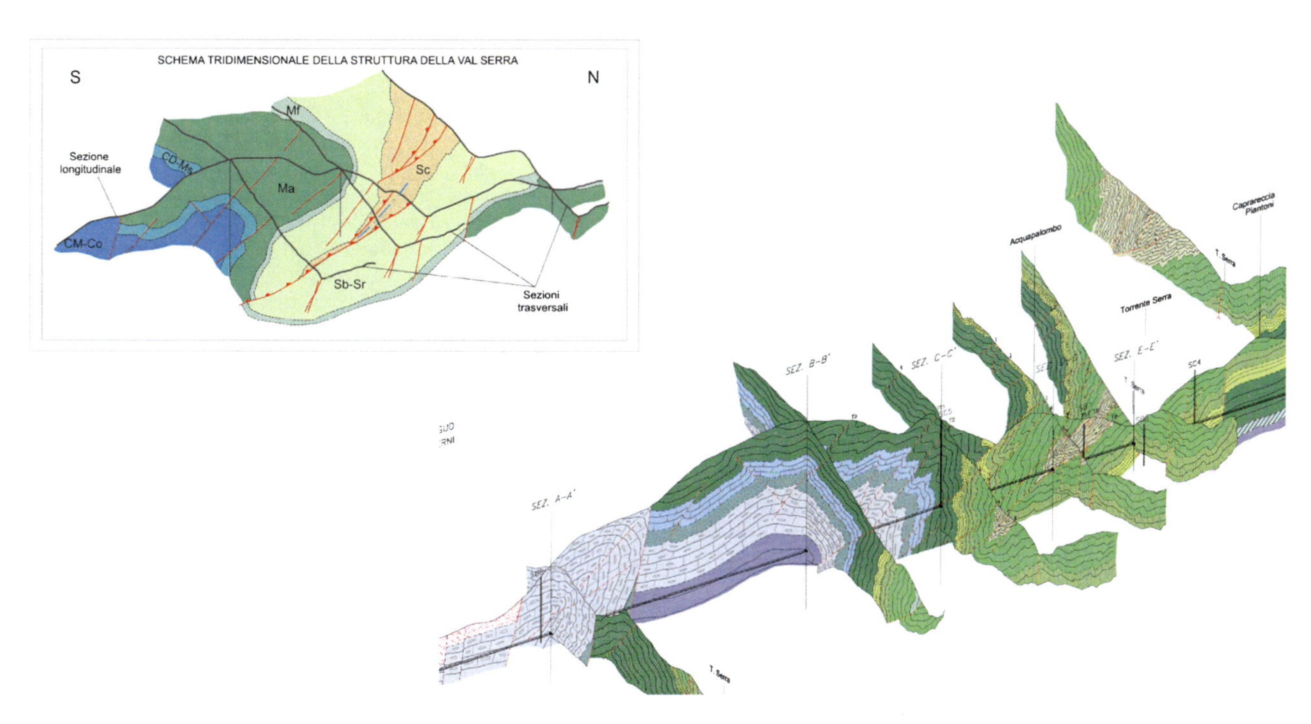

Fig. 180.2 Terni—Spoleto railway base tunnel, offprint of the basic geological model (*CM-Co* Calcare Massiccio-Corniola; *CD-MS* Calcari Diasprigni-Marne del Serrone; *Ma* Maiolica; *MF* Marne a Fucoidi; *SB-Sr* Scaglia Bianca-Scaglia Rossa; *Sc* Scaglia Cinerea)

- identify and quantify any changes to the system being considered, and
- define, or at least understand, the applicability of the model.

The DGGM is eventually an "engineering geological" model and this influence its nature. "*The same geological setting will interact with different engineering projects in different ways and will require different questions to be asked, different models to be developed and different investigations carried out. Certain engineering geological parameters may be more critical than others and some projects, by their very nature or setting, are exposed to more geological risk*" (IAEG Commission C25, Draft Report). Nevertheless the authors emphasise that its roots (and finally its overall effectiveness) remain deeply into the background of the Geological and Hydrogeological Model. The result of the geological studies primarily consists in the evaluation of the geometry of geological bodies and characteristics at depth, at various scales. From this it can be derived a design Model that can be focused on some particular aspects such as lithology, groundwater, geomorphology, or rock structure and properties (Geological and Hydrogeological Model). On these first, fundamental leaps, can be built the Geotechnical Reference Model (GGM, e.g. IAEG C28 "Reliability quantification of the geological and geotechnical model in large civil engineering projects") (Fig. 180.2).

The Geotechnical Model is built on the Geological Model describing the range of engineering parameters and ground conditions (with their variation and reliability) that must be considered in the design (Knill 2002). Simultaneously the Geotechnical Model eventually simplifies (also to meet the requirements of the selected method of mathematical and physical analysis) the Geological Model by defining and characterizing volumes of ground with similar engineering properties, and identifies boundaries (with their variability, see e.g. AFTES GT32.R2A1 2012) at which changes of geotechnical conditions may occur.

Because of unavoidable limits rising from accuracy and completeness with which subsurface conditions may be known it is also a mean for identifying their related variability and uncertainties, the related hazards and risks, providing the basis for plan eventual additional site investigation and for a correct design procedure. The DGGM can be then finally described as the framework in which the expected risks are recognized and characterized.

Finally, not in order of importance, the DGGM will include the assessment of its effectiveness and reliability (with the evaluation of the associated uncertainties). Some approaches consider the quality of the investigation procedures (e.g. in term of extension of the geological mapped area, number and length of boreholes). Others consider the quality of the input data, still others the entire model itself.

The DGGM is conceived to be built of two main parts:

- the first, where all the collected data will be stored,
- the second, with the complete model, as derived from the input data interpretation.

Both of them, even if coming from interpretation, are intended to be as objective as possible in order to reduce biased interpretations of results. Another basic expectation is to document, archive and share all data and methodology, making them available for any possible reason of scrutiny by other specialists.

The DGGM must be focused on the engineering needs of the project. The provided information must be disclosed and comprehensible for all the specialists inside the design team, and eventually to the *non-specialists* (ideally for all the stakeholders interested to the project) as much as possible.

The DGGM must be suited and fulfil the current laws, norms, standards and procedures, together, in case, with requirements of the Owner or Third Parties. Because the DGGM must be suited with the project itself isn't useless to point out that there isn't a universal protocol for its construction. The reliability of the model can be high, for example, because supported by a good field mapping, also without many boreholes. It must also be considered the complexity of the geological context: monotonous sequence of horizontal, homogeneous, alluvial layers can be effectively studied also with few boreholes.

References

AFTES Recommendation GT32.R2A1 (2012) Recommendations on the characterisation of geological, hydrogeological and geotechnical uncertainties and risks

Bistacchi A, Massironi M, Menegon L (2010) Three-dimensional characterization of a crustal-scale fault zone: the Pusteria and Sprechenstein fault system (Eastern Alps). J Struct Geol 32:2022–2041

Consiglio Nazionale delle Ricerche (1997) Progetto Strategico Gallerie

Efron N, Read M (2012) Analysing international tunnel costs. Worcester Polytechnic Institute

Eusebio A, Soldo L, Pelizza S, Pettinau D (2005) Preliminary geological studies for the design of Mercantour base tunnel. AFTES Tunnelling for a sustainable Europe, Chambery

Grasso P, Chiriotti E, Xu S, Kazillis N (2007) Use of risk management plan for urban mechanized tunnelling projects: from the establishment of the method to the successful practice. In: Underground space, the 4th dimension of metropolises. Taylor & Francis, London

IAEG Commission C25. Use of engineering geological models. Part 1: introduction to engineering geological models

IAEG Commission C28—Reliability quantification of the geological and geotechnical model in large civil engineering projects. Terms of Reference

IMIA-ITIG (Reiner H) (2011) International Tunneling Insurance Group. Developments in the tunneling industry following introduction of the tunneling code of practice. IMIA annual conference, Amsterdam

ITA WG 17 (2010) Long tunnels at great depth

ITA Working Group no. 2 (2004) Guidelines for tunnelling risk management. Tunn Undergr Space Technol 19

Knill, J. (2002). First Hans Cloos Lecture', 9th congress of the International Association for Engineering Geology and the Environment. Geol Environ 62:1–34. Durban, South Africa

Mallet JL (2002) Geo-modeling. Oxford University Press, New York

Perello P, Venturini G, Dematteis A, Bianchi GW, Delle Piane L, Damiano A (2005) Determination of reliability in geological forecasting for linear underground structures: the method of the R-index. Géoline 2005—Lyon

Soldo L (1998) L'importanza delle indagini preliminari nella progettazione di opere in sotterraneo. PhD Thesis, Politecnico di Torino

The British Tunnelling Society (2003) The joint code of practice for risk management of tunnel works in the UK. Published by The British Tunnelling Society

US National Committee on Tunnelling Technology (1984) Geotechnical site investigations for underground projects

Research on Overall Risk Assessment and Its Application in High Slope Engineering Construction

181

Tao Lianjin, An Junhai, Li Jidong, and Cai Dongming

Abstract

Due to the special landform and physiognomy in the mountain highway construction, it's often necessary to excavate of road cutting in a complex geological environment, including high and steep slope. And the security of cut slope in construction has a deep influence on the normal development of the whole project. Whether the construction of high slope engineering is safe or not depends on many uncertainty factors. Generally, the sole safety factor can't deal with these uncertainties. Therefore, in the course of construction operation, analyzing and assessing the risk of slope engineering to avoiding and reducing risk is substantial. According to the characteristics of the slope engineering projects, considering to the experience of this kind of project management, the engineering geological conditions, site conditions, the height of slope and climate conditions were chosen as the overall risk assessment indicators of high slope engineering in the construction safety, and demonstrate the rationality of selecting the indicators in detail. Afterwards, based on the knowledge of probability theory and mathematical statistics, the scores of overall safety risk of construction can be divided into four levels, then evaluate the criticality of slope construction quantitatively and thereby provide scientific decision-making basis for different risk rank of slope engineering to take corresponding risk control measures. Furthermore, the special risk assessment of construction safety is suggested to the high slope engineering whose overall risk rank is level III and over. At last, based on Hubei Yunshi highway which is being constructed, the overall risk assessment on the construction of cut slope engineering was made and suitable evaluation result was obtained. Then compared with the actual construction situation, found evaluation results and actual situation are keep in coincide. So the effectiveness of the risk assessment system is verified.

Keywords

Cut slope • Overall risk assessment • Evaluation index system • Risk rank

181.1 Introduction

With the rapid development of high grade highway in mountainous areas, a large amount of cut slope engineering emerged and many slope deformation and failure occurred at the same time. The security issues of cut slope are also increasingly prominent, which have caused large numbers of casualties and huge economic losses in mountainous areas of the world. The total number of deaths due to all kinds of landslides activities is more than 25 annually which exceeding the average losses due to earthquakes (Krohn and Slosson 1976). The most disastrous landslides have claimed as many as 100,000 lives (Li and Wang 1992). Factors that affect the safety of cut slope are numerous and complex and there is no effective risk analysis and evaluation theory on multi-factor system at present. In recent years, risk analysis

T. Lianjin · A. Junhai (✉) · L. Jidong · C. Dongming
College of Architecture and Civil Engineering, Beijing University of Technology, Beijing 100124, China
e-mail: tsanjunhai@126.com

G. Lollino et al. (eds.), *Engineering Geology for Society and Territory – Volume 6*,
DOI: 10.1007/978-3-319-09060-3_181, © Springer International Publishing Switzerland 2015

and assessment, as an important tool, was used by increasing experts and scholars to address uncertainty inherent in slope hazards. Einstein proposed the landslide risk evaluation framework: Natural datum—Dangerous figure—Hazard figure—Risk figure—Management figure, which realizes the change of the risk assessment from the qualitative to the quantitative (Einsten 1988). Zhang Yong-xing established the risk assessment of slope system, which can analysis the risk of slope based on the physical and mechanical parameters. Huang Run-qiu proposed a geological disaster risk assessment method which based on the GIS system (Huang 2000).

Even though study on slope risk have achieved a lot of achievements, however, cutting slope engineering, as an important part of the highway engineering, is still a blank area of the study about the risk evaluation during its construction.

181.1.1 The Failure Modes

The destruction of the rocky slope types roughly included: (1) Collapse failure mode; (2) Translational sliding failure mode; (3) Wedge failure mode; (4) Toppling failure mode.

181.1.2 The Established Process of Cut Slope Risk Evaluation Index System

The established process of highway cut slope risk evaluation index system divided into the following steps and the established processes of system were shown in Fig. 181.1.

181.1.3 Overall Risk Evaluation Index Classification Assignment and Its Standard

How to define the Slope risk criteria value is not only a technical problem, but also a comprehensive indicator including political, economic, social, engineering technology, natural environment, cultural background and other factors. The classification of evaluation index and value standard of cut slope engineering construction safety risk overall assessment was shown in Table 181.1.

The formula to calculate overall risk of cut slope construction safety:

$$R = A1 + A2 + A3 + A4$$

A1 The scores assigned by engineering geological conditions;
A2 The scores assigned by Site environment;
A3 The scores assigned by slope height;
A4 The scores assigned by climate conditions.

181.1.4 The Overall Risk Classification Standard

The high slope construction safety risk overall assessment classification standard can be gotten based on "Independent identically distributed central limit theorem" and slope expert experience which was shown in Table 181.2.

To the high slope whose risk level is the level III (high risk) and over, the dynamic assessment on the whole construction process was proposed.

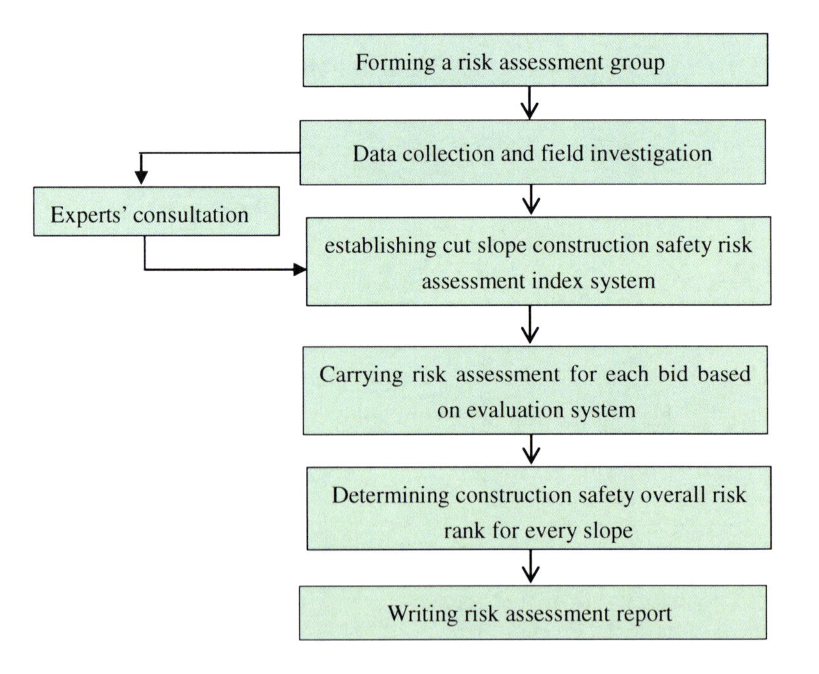

Fig. 181.1 The cut slope construction safety risk assessment program flow chart

Table 181.1 Cut slope construction safety overall risk assessment index system

Evaluation index	Classification			Score	Explain	Evaluation score
Engineering geological conditions $A1 = (a + b + c + d + e)$	Landform and physiognomy (*a*)	Steep slope, canyon and other dangerous area		3–4	Relative height difference is up to 130 m, the mountain is round in shape, the average slope is 35°	2
		General mountain area		1–2		
	Formation lithology (*b*)	Expansive Soil, frozen soil, collapsible loess and other special soils		4	The quaternary system diluvial silty clay and lower Triassic marl and limestone	2
		Silty clay, gravel soil and so on		3		
		Soft rock (siltstone, clay shale and so on)		2		
		Medium hard and hard rocks (dolomitic limestone, quartz schist and so on)		1		
	Geological structure = ($c1$ + $c2$ + $c3$)	Fault ($c1$)	Yes	1–2	No fault	0
			No	0		
		Rock mass structure ($c2$)	Granular structure	3	Thick layer structure	1
			Cataclastic texture	2		
			Layered structure	1		
		Main structural plane direction ($c3$)	Identical or close (consequent slope)	3–7	The right of slope was a consequent slope after excavation, slope: 215° ∠45°, joint: 280° ∠13°	6
			opposite (reverse slope)	1–2		
	Groundwater (*d*)	Plentiful groundwater		2–3	Karst water in majority, the water was not uniform	2
		Poor groundwater		1		
	Unfavorable geological condition (*e*)	Slide, dangerous Rock mass, collapse and rock pile and other unfavorable geology		1–2	Karst collapse may occur	2
		There are no unfavorable geological condition		0		
Site environment *A2*	There are buildings and structure in construction site			1	A village near slope farmland, rural road access	1
	There are no buildings and structure in construction site			0		
Slope height (m) *A3*	Rock slope: ≥50 m or soil slope: ≥40 m			7	Left slope is about 44.8 m high, right slope is about 34.0 m high	5
	Rock slope: 40 ~ 50 m or soil slope: 30 ~ 40 m			5–6		
	Rock slope: 30 ~ 40 m or soil slope: 40 ~ 50 m			3–4		
	Rock slope: <30 m or soil slope: <20 m			1–2		
Climate conditions *A4*	Severe climate event happens frequently in construction area (strong winds, heavy rains and snows)			1	Damp and rainy, precipitation concentration and intensity	0
	Climate environment in good condition, it doesn't affect the construction safety			0		

Fig. 181.2 A certain cut slope of Yunshi highways. **a** Bedding landslide caused by excavation, **b** slope with mass broken rock

Table 181.2 The cut slope construction risk overall assessment classification standard

Risk level	Calculated score (R)
Level IV(extremely risk)	$R \geq 25$
Level III (high risk)	$20 \leq R \leq 24$
Level II (moderate risk)	$16 \leq R \leq 19$
Level I (low risk)	$6 \leq R \leq 15$

181.1.5 Project Case

This paper analyses the rationality of the risk assessment method using somewhere in a 2012 survey Yunshi highway excavation slope as an example (As shown in Fig. 181.2). The detailed descriptions of this slope are shown in Table 181.1.

The concrete evaluation analysis is shown in Table 181.1.

The overall risk of this cut slope construction safety:

$$R = A1 + A2 + A3 + A4 = 21$$

The conclusions which shown as follows can be gotten based on the cut slope construction risk overall assessment classification standard. The overall risk rank of this cut slope are level III (high risk). To ensure construction safety, construction safety management department should carry controlling work based on the point of construction technical measures and site safety management. The risk assessment results were identical with the actual situation, which verified the correctness of the construction risk assessment method witch this paper proposed.

181.1.6 Conclusions

(1) This paper mainly describes five failure mode and mechanism of the mountainous area highway rocky slope, points out the damage types that may occur in the process of its construction, advances a slope construction safety impact index and the weighting influenced by rock slope based on a large number of survey data of high slope somewhere in yunshi expressway rock slope, then determines the size of the overall risk classification standard combined with the empirical method and the central limit theorem, and established a more scientific, reasonable and systematic risk evaluation index system of high rocky slope.

(2) The geological conditions and height of high slope are the two most important factors affecting the safety of construction, directly determines the level of the overall risk level, construction should be mastered before slope stability and the weathering degree. Don't construct blindly.

(3) This paper analyses and evaluates Yunshi highway rocky slope construction safety risk based on some engineering cases, the evaluation result is consistent with the engineering practice. The high slope overall risk evaluation index system has a certain reference value for building and construction and predicting and avoiding high mountain area highway rocky slope construction risk.

References

Champman CB, Cooper DF (1987) Risk analysis for large project: models, methods and cases, Wiley, Hoboken

Chunlong N (2012) Risk analysis theory and application for slope engineering. Central South University, Changsha

Einsten HH (1988) Special lecture: landslide risk assessment procedure. In: Proceedings of 5th international symposium. Lausanne, Landslide, pp 1075–1090

Krohn JP, Slosson JE (1976) Landslide potential in the United States. Nat Emerg Train Cent 30(1):30–33

Li T, Wang S (1992) Land slide Hazard sand their Mitigation in China. Science Press, Beijing 84 pp

Wang S-C, Zhe X-S, Zhou Z-J et al (2010) Application of engineering fuzzy set theory analysis on AHP to highway slope stability evaluation. J Saf Environ 10(5):189–192

Xia Q-T (2008) Engineering construction risk assessment and management. Fujian Archit Constr 10(5):97–99

Zhang Y-Xg, Chen Y, Wen H-J et al (2008) The research of landslide hazards assessment system. J Chongqing Jianzhu Univ 30(1):30–33

The Research of Geological Forecast Based on Muti-source Information Fusion

182

S. Cui, B. Zhang, F. Feng, and L. Xie

Abstract

In recent years, a large number of mountain tunnels, which have long distance and large buried depth, have been built. But China is a mountainous country with complicated geological conditions. And due to the complicated geological conditions, different kinds of geological disasters often occur during the construction. Therefore, the geological forecast is an important approach to reduce geological disasters effectively. This paper is based on the geological forecast conducted in Jucheng tunnel, Yangquan, Shanxi province. In this paper, a new method of geological forecast is put forward, which is based on the tunnel geological hazards assessment and combine with the stress-strain analysis by using FLAC^{3D} and tunnel seismic tomography (TST). Through risk analysis to identify hazard sections, then select the typical tunnel face of high-risk section and analyze the stress-strain of surrounding rock. Therefore, the stability properties of tunnel section can be obtained. After that, the geological forecast, based on the TST, is conducted on the typical tunnel face. Over all, these new method can forecast the construction geological disasters of different scales of sections (such as the tunnel section or specific tunnel face). Moreover, the new method has obtained the ideal effect in the geological forecast of Jucheng tunnel.

Keywords

Geological forecast • Tunnel geological disasters • Analytic hierarchy process • FLAC^{3D} • Tunnel seismic tomography

182.1 Introduction

Along with China's infrastructure construction, tunnel construction technology has developed rapidly. Tunnel engineering construction speed is faster, and the tunnel depth is also growing. Because of the unclear understanding of geological conditions ahead of the face, tunnel construction will bring great blindness, and unexpected landslides, cave, gushing and other accidents may happen. How to solve the above problem tunnel engineering has become a growing concern.

Using the method of land sonar in the Yangzhai tunnel and Lvchang tunnel, Zhong et al. (1995) successfully detected the cave of 40–80 m distance range ahead of the face. In Sichuan Tieshan Tunnel, good results have been achieved to forecast the geological situation within 50 m ahead of the face by horizontal sonic profiles (Yang et al. 1997). Dai et al. (2005) used TSP203 to probe BAJIAOXI tunnel, forecasting rock geological nature, location and size about 100 m ahead of face to ensure safety and quality. Wang (2008) recognized more accurate position ahead of the face in Taizhou tunnel and the distribution of rock mass on the top of the shallow buried tunnel by using GPR.

S. Cui · B. Zhang (✉) · F. Feng
China University of Geosciences, School of Engineering and Technology, Beijing, China
e-mail: sc_zhb@cugb.edu.cn

L. Xie
Beijing Urban Construction Exploration & Surveying Design Research Institute Co., Ltd., Beijing, China

G. Lollino et al. (eds.), *Engineering Geology for Society and Territory – Volume 6*,
DOI: 10.1007/978-3-319-09060-3_182, © Springer International Publishing Switzerland 2015

In this paper, relying on the Giant City Tunnel project of Yangquan-Niangziguan highway, risk checklists and analytic hierarchy process method have been applied to access the risk of geological disasters in the construction of the tunnel, and analyze the surrounding rock stability of the tunnel after excavation situation by FLAC^{3D} software. In the tunnel construction process, TST geological advanced prediction technology has been adopted for advance geological forecast of each section.

182.2 Engineering Situation and Geological Conditions

Giant City Tunnel of Yangquan-Niangziguan highway is located in the northeast of the Huge Town, Yangquan City, Shanxi Province. The line is roughly east-west, and cross two high mountains. The top of the mountain maximum elevation is about 724 m, the maximum depth of the tunnel is 167 m. Tunnel site area belongs to the tectonic denudation accumulation and tectonic karst landform. It is in Shanxi upfaulted zone of Sino Korean paraplatform. Second-level structure includes the Qinshui platform concavity and the broken arch of Taihang Mountains. The simple stratum structure has no obvious signs of a new tectonic movement. Based on regional data from 1:500,000 geological map and engineering detailed investigation results, the upper strata of the tunnel site area is the Quaternary Aeolian deposits, diluvial layer coverage, underlying bedrock is Carboniferous sandstone and Ordovician limestone.

182.3 Geological Forecast Based on Multi-source Information Fusion

182.3.1 Risk Assessment of Tunnel Geological Disaster

In this paper, the risk assessment is before the stage of tunnel excavation. Geological disaster risk assessment is mainly based on tunnel engineering design data and geological data. Main objective is various geological disasters may occur in the process of tunnel construction. Combined with hierarchical analysis method of expert surveys, geological factors of the risk index system were established: eccentric compression (C1), lithology and weathering degree (C2), normal faults (C3), development and erosion of karst fissure water (C4), karst (C5). Huge City tunnel was two-way separated tunnel, geological and structural condition were different in the direction of the line. According to the regional geology and survey report, the assessment of the tunnel was divided into three sections, namely K3+290–K3+920, K3+920–K4+380, K4+380–K4+740.

Risk factors and the impact between risk events were respectively analyzed through the checklist method. The judgment matrix of risk index system has been obtained. By solving the theoretical solution of the judgment matrix, the comprehensive weights were calculated and sorted, which showed a clear understanding of five underlying risk factors that affects the target. Eccentric compression of tunnel terrain, lithology and weathering degree and normal faults had very important influence in Table 182.1. Karst and fissure water also need to be carefully considered.

182.3.2 Numerical Simulation of the Stability

In the results of tunnel geological disaster risk assessment, the three risk zones were finally selected. According to the classification of surrounding rock, characteristics of rock mass and boundary conditions in the survey report, three sections of typical rock were elected, namely K3+600–K3+604–K4+260–K4+264 and K4+480–K4+484.

The statistical results of the model calculations were shown in Table 182.2. The stability of surrounding rock of each section in excavation without supporting was worse, characterized by large plastic zone area. The minimum principal stress of surrounding rock was larger, so was the vertical and horizontal displacement. After adding the supporting measures, the stability of surrounding rock was improved. The plastic zone area decreased, the minimum principal stress and horizontal and vertical displacement was reduced.

182.3.3 Tunnel Geological Prediction During Construction

After the pre-tunnel geological disaster risk assessment and stability analysis of tunnel surrounding rock, in the specific tunnel excavation process, a variety of unexpected geological problems will be encountered. Therefore, during the construction geological prediction is particularly important, especially for tunnel construction process, forecasting geological conditions ahead of the face. The tunnel geological prediction is a TST geological prediction technique.

TST technical is short for tunnel seismic tomography. Its observation system uses spatial layout, and receive and excitation system layout on both sides of the tunnel wall rock. Seismic waves are stimulated by the bursting of small doses, received by the geophones. TST can get the accurate distribution of wave velocity of the surrounding rock in front of face and the location images of the geological body.

As shown in Fig. 182.1, TST observing system is arranged as follows: ① 12 detectors(S1–S12), Arranged on both sides of the wall, each side six, spacing of 4.0 m; ② 6

Table 182.1 Results of geological disaster risk assessment

No.	Mileage		Risk factors	Risk events
	Starting	Termination		
1	K3+290	K3+920	Eccentric compression	Landslide
2	K3+920	K4+380	Normal faults	Landslide, water inrush
3	K4+380	K4+740	Lithology and weathering degree	Landslide, water inrush

Table 182.2 Results of numerical simulation calculation

Model	Working condition	Plastic zone	Minimum principal stress (MPa)	Maximum horizontal displacement (cm)	Maximum vertical displacement (cm)	Stability
K3+600–K3+604	No supporting	Larger	−1.1	−4.8	−11.8	Poor
	With supporting	Smaller	−1	−1.1	−3.3	Better
K4+260–K4+264	No supporting	Larger	−0.68	−0.7	−4.2	Poor
	With supporting	Smaller	−0.65	−0.5	1.4	Better
K4+480–K4+484	No supporting	Larger	−0.88	−2.4	−7.5	Poor
	With supporting	Smaller	−0.83	−0.7	−2.4	Better

explosion sources(P1–P6), arranged in two side, three on each side. The first hole of each side is apart from the detector 4, 24 m space of the rest.

The tunnel was segmented for forecasting, each prediction distance was 100–150 m. The early stage of the tunnel geological disaster risk assessment and surrounding rock stability analysis have found three Issues outstanding location, so the geological advanced prediction results of these three location were displayed.

(1) Geological prediction probe results ahead of the ZK4 +597 face:

The observations were obtained from effective records of 36 channels, getting geological migrated image and seismic wave velocity distribution curve of surrounding rock within 100 m ahead of the ZK4+597 face of left lane tunnel exit, shown in Figs. 182.2 and 182.3.

In geological migrated image, abscissa was tunnel mileage and working face coordinates was 0; Vertical axis is horizontal transverse width of the tunnel. The blue stripes represented rock interface changes from hard to soft, red indicated the interface from soft to hard and blue-red indicates the presence of rock fracture zones. Figures 182.5, 182.6 and 182.7 showed the rock velocity distribution curve, which reflected the distribution of rock mechanical properties. Intensive strips reflected complex structure and structural development, corresponding high velocity zone in the wave velocity distribution. Less strips area reflected uniform density of surrounding rock, corresponding high velocity zone.

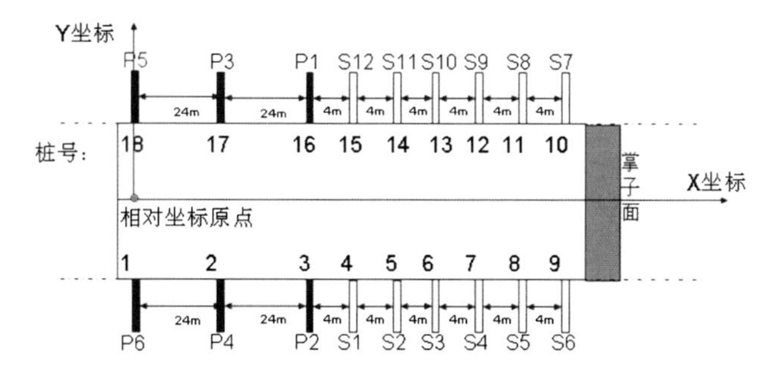

Fig. 182.1 TST excitation and reception mode

Using the above results, geological conditions within 100 m ahead of the ZK4+597 face was divided into two sections:

Section 1: ZK4+597–ZK4+570, 27 m length, lower rock velocity (P-wave velocity of 2,500 m/s), development structure plane and crack, intermediary weathered-strong weathered, poor stability, grade IV rock.

Section 2: ZK4+570–ZK4+497, 73 m length, velocity of 2,600 m/s, better stability and integrity, intermediary weathered, grade IV rock.

(2) Geological prediction probe results ahead of the ZK4 +292 face:

Geological conditions within 100 m ahead of the ZK4 +292 face was divided into four sections (shown in Figs. 182.4 and 182.5):

Section 1: ZK4+292–ZK4+272, 20 m length, velocity of 2,400 m/s, development structure plane and crack, intermediary weathered-strong weathered, poor stability, grade V rock.

Section 2: ZK4+272–ZK4+252, 20 m length, velocity of 2,400 m/s, better stability and integrity, intermediary weathered, grade IV rock.

Section 3: ZK4+252–ZK4+242, 10 m length, velocity of 2,400 m/s, fault or fracture zone may exist, grade V rock.

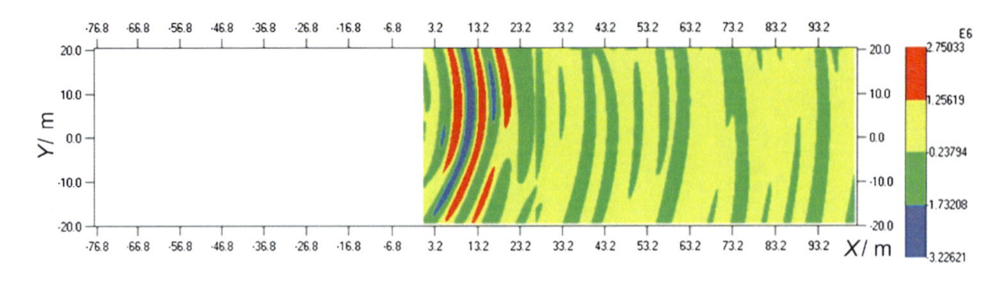

Fig. 182.2 Geological migrated image ahead of the face

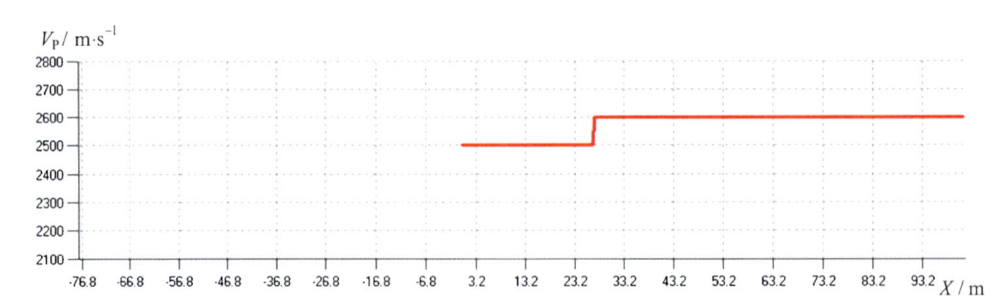

Fig. 182.3 Distribution curve of seismic wave velocity

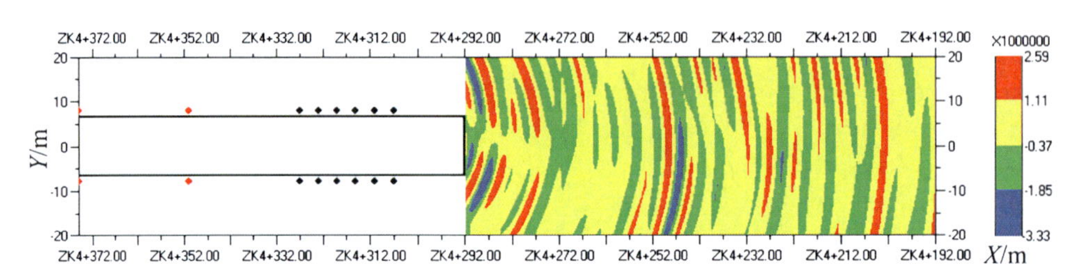

Fig. 182.4 Geological migrated image ahead of the face

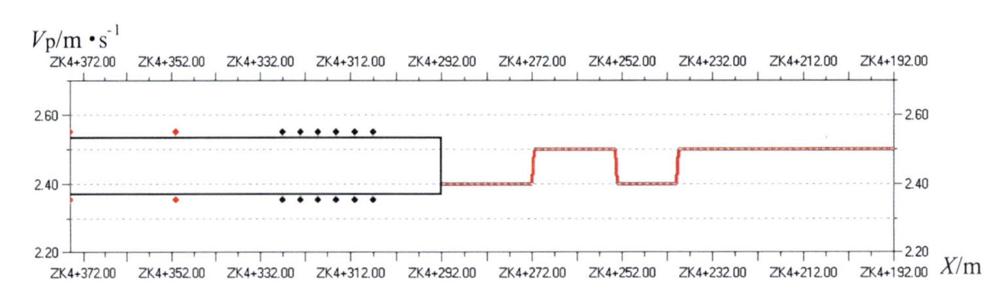

Fig. 182.5 Distribution curve of seismic wave velocity

Section 4: ZK4+242–ZK4+192, 50 m length, velocity of 2,500 m/s, better stability and integrity, intermediary weathered, grade IV rock.

(3) Geological prediction probe results ahead of the YK3 +480 face:

Geological conditions within 100 m ahead of the YK3 +480 face was divided into three sections (shown in Figs. 182.6 and 182.7):

Section 1: YK3+480–YK3+504, 24 m length, low velocity of 2,000 m/s, fault or fracture zone may exist, grade V rock.

Section 2: YK3+504–YK3+537, 33 m length, velocity of 2,200 m/s, better stability and integrity, intermediary weathered, grade V rock.

Section 3: YK3+537–YK3+580, 43 m length, velocity of 2,400 m/s, development structure plane and crack, intermediary weathered-strong weathered, poor stability, grade V rock.

Compared detailed forecast results with the actual situation after excavation, these two cases were almost identical except a little difference on the determination of rock mass.

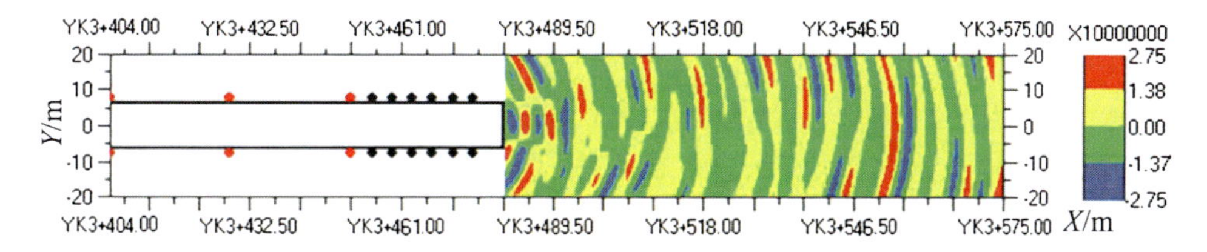

Fig. 182.6 Geological migrated image ahead of the face

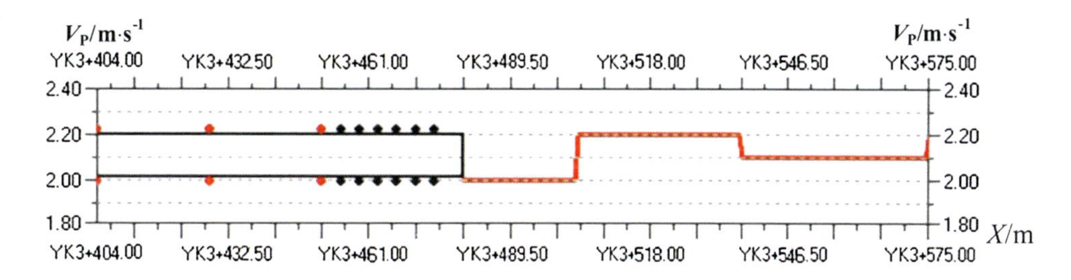

Fig. 182.7 Distribution curve of seismic wave velocity

182.4 Conclusions

In this paper, relying on the huge city tunnel engineering, geological forecast based on multi-source information fusion was studied. The main conclusions and results are as follows:

(1) Using hierarchical analysis method, with regional geological data and survey data of tunnel site, a tunnel geological disaster risk assessment index system was established. Analyzing the underlying risk factors of the index system, three risk zones were obtained, and landslide is the most possible risk events, the second is the water inrush.

(2) Using FLAC^{3D} numerical analysis software, by analysis for plastic zone, stress and displacement characteristics, the stability of surrounding rock during excavation without supporting was worse.

(3) Using TST geological prediction method to predict the situation of the tunnel, the scale situation of unstable rock mass ahead of the face was otained, which were good construction guidance. It came to predict the situation the first two methods cannot.

(4) Tunnel geological disaster risk assessment, rock stability analysis and tunnel geological prediction were layer upon layer progressives. The research process was from the macro to the specific. Tunnel geological disaster risk assessment drew risk zoning of geological disasters; tunnel surrounding rock stability which were no supported and supported after excavation of each risk partition were analyzed by numerical analysis software; Tunnel geological prediction predicted body size of unstable rock mass in specific problematic partition. Three results are complementary and have good guiding significance on the construction.

Acknowledgement This project is supported by the Fundamental Research Funds for the Central Universities of China (2010ZY14) and The National Natural Science Foundation of China (NSFC) (40902086).

References

Black K, Kopac P (1992) The application of ground penetrating radar in highway engineering. Public Roads 56(3)

Dai QW, He G, Feng DS (2005) Application of the TSP-203 system in geological advanced prediction of tunnel. Prog Geophys 20(2):460–464

Flemming J, Holger LA, Peter S, Esben A, Egon N (2003) Geophysical investigations of buried valleys in Denmark: an integrated application of transient electromagnetic and feflection seismic surveys and exploratory well data. J Appl Geophys 53–56

Johnston MJS (1997) Review of electric and magnetic fields accompanying seismic and volcanic activity. Surv Geophys 18:441–476

Otto R, Button E, Bretterebner H, Schwab P (2002) The application of TRT-Ture Reflection Tomography-at the Unterwald Tunnel. Geophysics 2:51–56

Scheers B (2001) Ultra-wideband ground penetrating radar with application to the detection of anti personnel landmines. Royal Military Academy, Department of Electrical Engieering and Telecommunication. pp 258–360

Song XH, Gu HM, Xiao BX (2006) Overview of tunnel geological advanced prediction in China. Prog Geophys 21(2):605–613

Wang RQ (2008) The study of advanced detecting methods on tunnel geological hazards. Central South University, Changsha

183 Presentation of the Activity of the AFTES WG 32: Considerations Concerning the Characterization of Geotechnical Uncertainties and Risks for Underground Projects

G.W. Bianchi, J. Piraud, A.A. Robert, E. Egal, and L. Brino

183.1 Introduction

The feedback derived from completed underground projects is still far from being satisfactory for all aspects relating to the characterisation of uncertainties, unforeseen circumstances and risks relating to underground space:

- The graphical representation of these uncertainties on geological cross-sections is often incomplete, ambiguous or completely lacking;
- In reports, the description of uncertainties is often insufficient, whether they relate to geotechnical properties, the position of events, the frequency of unpredictable phenomena;
- There is no recognised, unequivocal methodology for taking these uncertainties into account in so-called "Risk analysis" reports;

Faced with these findings, in 2009, AFTES reactivated working group GT32, with a view to establishing a methodology for properly identifying and representing uncertainties, analysing and managing the risks arising from these uncertainties.

To contribute to this objective, the GT32 recommendation aims mainly:

SEA Consulting Srl, Turin, Italy

Present Address:
G.W. Bianchi
EG-TEAM STA, Turin, Italy

J. Piraud
ANTEA Group, Orléans, France

A.A. Robert
CETU, Lyon, France

E. Egal
Egis, Lyon, France

L. Brino
LTTF, Turin, Italy

- Specifying terminology in terms of uncertainties and risks;
- Establishing a methodology for examining risks;
- Making proposals with a view to improving certain practices and tools, such as analysis of the reliability of investigations, graphical representation of uncertainties on geological cross-sections, etc.

A detailed description of the proposed methodology is found in the text of the recommendation (AFTES WG 32.2 2012).

183.2 Risk Management Methodology

The geotechnical risk management methodology which AFTES recommends should be applied for studies comprises three major phases:

1. Compiling a review of geotechnical knowledge and uncertainties covering geological, hydrogeological and geotechnical data;
2. Geotechnical risk assessment based on the summary of data; this phase in turn comprises three stages: risk identification, analysis and evaluation;
3. Geotechnical risk treatment.

183.2.1 Review of Geotechnical Data and Uncertainties

The proposed steps for performing a review of geotechnical data and uncertainties are as follows:

- Presentation of the raw data available;
- Assessment of its reliability;
- Summary and interpretation;
- Summary of uncertainties and in particular of gaps in knowledge (Register of Uncertainties).

G. Lollino et al. (eds.), *Engineering Geology for Society and Territory – Volume 6*,
DOI: 10.1007/978-3-319-09060-3_183, © Springer International Publishing Switzerland 2015

183.2.1.1 Presentation of the Raw Data Available

During this first stage, a complete list must be drawn up comprising all documentary data and relating to worksites conducted in similar terrain; the results of specific investigations conducted for the project should be added to these data.

The nature and quantity of available data, their distribution, source and date of acquisition must be clearly stated.

183.2.1.2 Reliability of Data

The second stage corresponds to a critical evaluation of the quality of data. This stage is highly recommended to correctly define the contribution of these data to drawing up the geological model. It is also appropriate to evaluate the extent of gaps in knowledge, i.e. "what is not known". Among the factors to be taken into account to evaluate reliability, the complexity of the local geological context, the nature of investigation works as well as the physical distribution of this work and its spatial "density" may be mentioned.

183.2.1.3 Summary and Interpretation

The third stage consists in drawing up a geological model displaying the designer's idea of the geological context and expected construction environment. The presentation of this model includes two types of documents:

- a report, detailing the hypotheses deemed the most likely based on the analysis of all the data;
- graphical documents: geological and hydrological models and especially provisional longitudinal geotechnical profiles, along with as many cross-sections as necessary and a horizontal cross-section of the project.

It is in these documents that uncertainties with respect to interpretation should be pointed out, in particular on graphical elements. This longitudinal profile shall also include information about the variability of the parameters within each sub-section, such as the following:

- the dispersion of parameters, to allow the finalisation of methods (excavation, mucking, temporary support, etc.);
- the characteristic values chosen for the various geotechnical magnitudes;
- the limits within which the main parameters vary.

183.2.1.4 Register of Geotechnical Uncertainties

The fourth stage consists of summarising the uncertainties identified at the end of the previous operation and compiling a "Register of Uncertainties". This Register of Uncertainties should be limited to a list of the identified uncertainties, without analysing the consequences. On completion of this first phase of "Review of Geotechnical Knowledge and Uncertainties", the elements drawn up during the four stages described above are brought together in a single document including both a report and diagrams, as well as the Register of Uncertainties.

183.2.2 Geotechnical Risk Assessment

For each of the risks under consideration, the risk assessment phase includes three distinct phases:

- risk identification,
- risk analysis (in the strict sense of the term),
- risk evaluation.

183.2.2.1 Risk Identification

Risk identification requires the analysis of uncertainties with respect to their effects on expected results. Normally all uncertainties are a source of risk, but some of them may have virtually no effect at all. Only uncertainties for which the deviations induced with respect to the geological model are sufficiently significant to cause notable consequences, need to be identified as risks.

For each of the uncertainties identified, several hypotheses may be formed:

- for a given event: a variable number of occurrences, different locations or more or less serious consequences;
- for a "lack of geological knowledge": various configuration hypotheses for the geological context.

183.2.2.2 Risk Analysis

The Risk analysis stage includes three operations:

- quantification of the consequences arising from an event identified as a risk;
- quantification of the likelihood of this event and/or consequences;
- determination of the level of risk (significance of the risk) by combining the consequences and likelihood.

Quantification of the Consequences Arising from an Event

To proceed with risk assessment, the designer must draw up one or more hypotheses for each event identified, describing the circumstances caused by the occurrence of the event. The consequences of the same event may affect several objectives and each of these objectives in a different way. For each event, an analysis of its consequences on each of the objectives should therefore be conducted. Practically speaking, for geotechnical risks, the most relevant general objectives are site safety, cost, lead time, performance and the environment. The consequence is usually estimated as being the additional costs and/or extra time required to treat the event encountered.

Quantification of the Likelihood of an Event

The following stage consists of determining the "likelihood" of the identified event and/or its consequences. The likelihood of the event itself depends on a number of factors characterising the level of knowledge:

- the amount of investigation works carried out, its relevance and its quality of execution;
- the geographical proximity of investigation works to the structure;
- the complexity of the geological context.

Determining the Level of Risk (Significance of the Risk)
The "level of risk" qualifies the significance of the risk and is usually expressed by combining the likelihood with the consequence, both of which are evaluated by the designer. The combination of the likelihood and consequences may be "qualitative, semi-quantitative, quantitative or a combination of the three, depending on the circumstances."

183.2.2.3 Risk Evaluation

The designer then proceeds with risk evaluation by comparing their estimated level of risk to the risk criteria expressed by the project owner. For each of the risks, the project owner may take two attitudes:

(1) Refuse the risk and request that the designer:
(2) Accept the risk, with or without treatment:

183.2.2.4 Risk Treatment

Risk treatment aims at reducing the importance of risk or eliminating it. Possible actions may include the following:

- eliminating the risk source, e.g. by performing a specific investigation enabling uncertainty to be locally eliminated;
- altering likelihood, also by means of additional investigation enabling the geological model to be further clarified;
- reducing the consequences of an event on the circumstances of completion, through the implementation of preventive technical provisions and altering construction methods
- implementation of an early detection method for the occurrence of an event and early definition of remedial technical measures.

Following application of these measures, a new evaluation of each risk is conducted. If, despite the treatment measures, the risk remains unacceptable, a new "risk treatment" process is launched.

183.3 Graphical Representation of Geological Uncertainties

Maps and geological cross-sections are designed to provide a continuous representation of the geological nature of underground space based on discontinuous observation and data available in varying degrees of abundance and density. They are therefore interpreted "models" providing a two-dimensional representation of the most likely geology and reflect the author's understanding of the geology in question, in line with available data. The abundance and relevance of data will have a primary influence on the reliability of the document. With regard to geological cross-sections for civil engineering, unlike more conceptual "academic" cross-sections, it is particularly important to be meticulous and accurate regarding the geometry of layers (thickness, incline, folds, etc.), the location of contacts and faults, and the uncertainty of these locations. Indeed, the consequences of these uncertainties may be highly significant when it comes to design of the structure, its mode of construction, and so on.

GT32 has formulated a number of recommendations on the way to represent geology and the related uncertainties on documents used for civil engineering. The aim is that ultimately, a graphical representation should be achieved that makes it possible to see the extent of both knowledge and lack of knowledge regarding the terrain that may be crossed by an underground structure. In general, GT32 recommends the following:

- Drawing a clear distinction between the factual data that enabled the geologist to draw a map or cross-section and the interpretations;
- Ensuring that maps and cross-sections feature only unambiguous figures and symbols;
- Representing the uncertainty with regard to the existence and/or geometry of the geological object shown as well as possible on cross-sections, particularly adjacent to the projected structure.

183.3.1 Representation of Data

The geological map constitutes the foundation document for any geological study. Ideally, any geological map should be accompanied by an outcrop map, either in the form of a separate document, or on the geological map itself, with outcrop zones distinguished by darker or closer shading or with a specific outline.

Geological profiles are established using both surface and underground data:

- On the surface, the geological map makes it possible to locate contact points, faults etc. with the related degree of uncertainty;
- At depth, direct observations may be made using boreholes, and in some cases existing underground works or exploratory adits.

Moreover, observations made in boreholes are not always exactly located on the cross-section. The further away the borehole is, the greater the degree of uncertainty of the projection on the cross-sectional plane. Consequently, it is recommended that boreholes should be indicated on cross-sections by distinguishing those that are "close" to the profile plane from those that are farther away with respect to the profile plane.

183.3.2 Representation of Uncertainties Related to the Position of Contacts

The degree of uncertainty relating to the location of each geological object should be represented in detail along the entire length of a cross-section. Four possible ways of representing these extreme positions are described below.

183.3.2.1 Representation N.1

The uncertainty range is shown on the whole of the longitudinal profile for each contact or fault (Fig. 183.1). The resulting uncertainty range may be shown as a line, both on the surface (outcrops) and at depth (for instance, at a borehole which has passed through a clear contact point between formations A and B). This type of representation is appropriate if it only concerns a few contact points, but can become illegible in the event of multiple contact points, with overlapping uncertainty ranges.

183.3.2.2 Representation N.2

Representing the uncertain position of contact points or faults should be done only at the tunnel depth, on a specific profile located beneath the principal cross-section and confined to a narrow vertical area along the tunnel axis (Fig. 183.2). The uncertainty is expressed by a strip of variable width, corresponding to the zone where formations in contact may be encountered.

183.3.2.3 Representation N.3

The extreme locations of the contact points are not shown by their actual geometry on the vertical longitudinal profile, but by standard symbols indicated on a strip located beneath the principal cross-section. Two types of symbol may be used:

- **Type 3a—the uncertainty bar**. The strip features a bar centred on the most probable location of the contact point. In this case, only one uncertainty bar is shown for the entire stratigraphic series (Fig. 183.3);
- **Type 3b—oblique line**. At the top and bottom of the strip, the extreme positions of the contact point are shown for the project, connected by an oblique line: the steeper its gradient, the lower the degree of uncertainty. The advantage of this method is that it clearly visualises the contrasting uncertainty along the cross-section, and it can be applied to successive geological contact points even when these are very close together. Type 3b

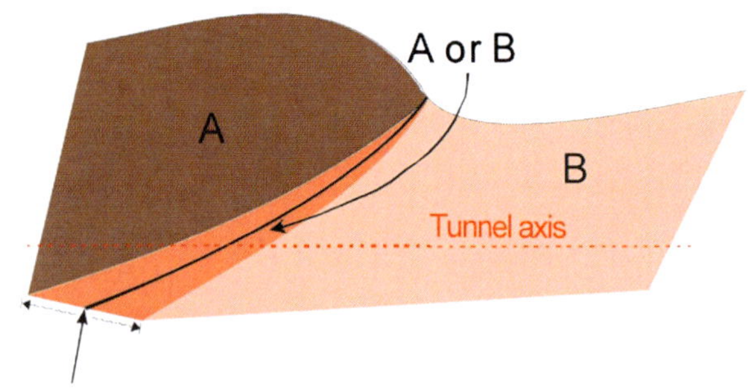

Fig. 183.1 Representation 1: geological longitudinal profile with uncertainty range for a contact location

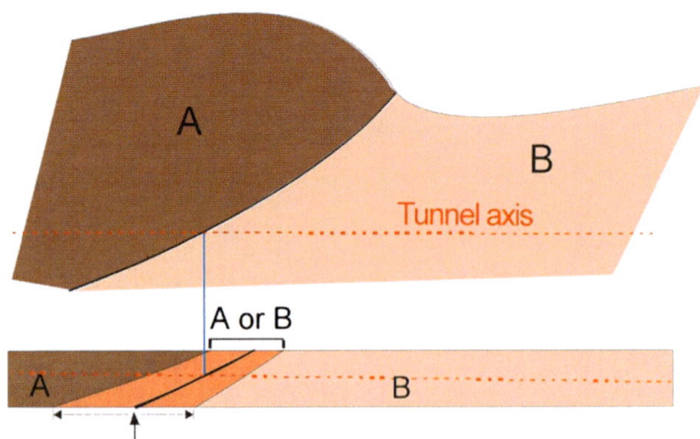

Fig. 183.2 Representation 2: vertical geological cross-section and "mini-profile" at the elevation of the project with an uncertainty range

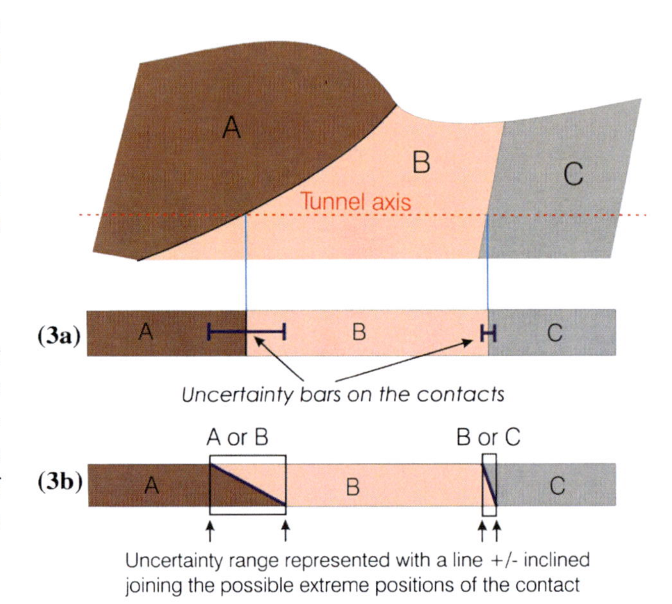

Fig. 183.3 Representation 3. Geological longitudinal profile and strips showing the location of contact points for the elevation of the project, with an uncertainty bar (type 3a) or oblique line (type 3b)

representations must however be clearly explained in the legend, because they are less intuitive than 3a. The uninitiated often confuse the uncertainty range with a horizontal geological cross-section at the tunnel depth, which is not the case.

Reference

AFTES WG 32.2 (2012) Recommendations about characterisation of geological, hydrogeological and geotechnical uncertainties and risk. Tunnels & Ouvrages Souterrains, n. 232

Development of 3D Models for Determining Geotechnical-Geological Risk Sharing in Contracts—Dores de Guanhães/MG/Brazil Hydroelectric Powerplant Case Study

184

Isabella Figueira, Laurenn Castro, Luiz Alkimin de Lacerda, Amanda Jarek, Rodrigo Moraes da Silveira, and Priscila Capanema

Abstract

Discussions concerning the type of contract to be adopted, the responsibilities on risks and possibilities of risk sharing in construction works are time consuming during the formatting of the contracts and, often, last until its closure. Within the main difficulties encountered are the lack of data and subjectivity in the interpretation of available information. In order to reduce these uncertainties in models and the sharing definitions, a research project was proposed focusing on the development of a methodology based on an integrated three-dimensional model to support the risk reports and the definition of risk limits assumed by each stakeholder. Available information from, Dores de Guanhães Hydroelectric Power Plant, was integrated in such environment. From geo-structural point of view the main features found are thrust faults of NNE-SSW direction, which are truncated or displaced by a less intense system in NW-SE direction, affecting granite-gneiss rocks of Açucena Granite belonging to the Mantiqueira Complex (Poente and Guanhães Energia, Basic Project Studies HHP Dores de Guanhães-MG, 2008). Data provided by the construction owner have been uploaded in the 3D modelling software, along with the regional images and structural and field data. Geophysical analyses (dipole-dipole array) were performed and included in the model with a thorough discussion of its possible benefits in general interpretation. Finally, the use of the consolidated 3D model is discussed within a tunnel excavation risk assessment methodology.

Keywords

3D Models • Geotechnical-Geological risk • Geophysical analyses

184.1 Introduction

Discussions concerning the type of contract to be adopted, the responsibilities on risks and possibilities of risk sharing in construction works are time consuming during the formatting of the contracts and, often, last until its closure. Settings involved in these discussions should consider the characteristics of entrepreneurs and builders, the site of the work, the level of uncertainty and safety of the developed models and design features.

Within the main difficulties encountered are the lack of data and subjectivity in the interpretation of available information. When it comes to construction of Hydroelectric Power Plants (HPP), the excavation of underground structures such as the case of diversion tunnels or adduction, the

I. Figueira (✉) · L.A. de Lacerda · A. Jarek · R.M. da Silveira
LACTEC—Institute of Technology for Development, Curitiba, Brazil
e-mail: isabella.figueira@lactec.org.br

L.A. de Lacerda
e-mail: alkimin@lactec.org.br

A. Jarek
e-mail: amanda.jarek@lactec.org.br

R.M. da Silveira
e-mail: rodrigo.silveira@lactec.org.br

L. Castro · P. Capanema
CEMIG, Belo Horizonte, Brazil
e-mail: laurenn@cemig.com.br

P. Capanema
e-mail: pic@cemig.com.br

G. Lollino et al. (eds.), *Engineering Geology for Society and Territory – Volume 6*,
DOI: 10.1007/978-3-319-09060-3_184, © Springer International Publishing Switzerland 2015

main focus of this work, for the preparation of a local geological model, qualification and quantification of geological and geotechnical risks various methodologies have been adopted, but normally not considering the integration of these different databases.

Aiming to reduce uncertainties and clarifying the risk sharing conditions, a Geological-Geotechnical Risk Report is compiled seeking to identify in the adopted models marked characteristics of foundation, including necessary treatments and observed uncertainties. This report is discussed and consolidated between the parties, becoming part of the contract. Although the consolidation of the geological-geotechnical model and the risk report involve a team with participation of geologists and engineers from involved companies in a series of meetings prior to signing the contract, the majority of encountered difficulties are due to lack of multidisciplinary integration and, mainly, the interpretative and subjective character of the elaborated models.

In order to reduce these uncertainties in models and the sharing definitions, a research project was proposed focusing on the development of a methodology based on an integrated three-dimensional model to support the risk reports and the definition of risk limits assumed by each stakeholder. The proposed model seeks a combined analysis of data obtained in preliminary surveys and field investigations, such as regional analysis using satellite images, topographical surveys, geomechanical data of the rock mass from rotating surveys, geophysical surveys from the ground surface, geological mapping focusing on the structuring of the rock mass. Initially, it was carried out a search and selection of software with the ability of integrating 3D data, such as vectors in formats dxf, dwg, shp; rasters from remote sensing data and profiles; and 3D solids.

The main object of this paper is to present modeling data of the region surrounding the intake tunnel of HHP Dores Guanhães, as a case study, which is currently under construction.

184.1.1 Localization and Technical Features

The HPP is located at Guanhães River, a tributary of Santo Antonio River in the State of Minas Gerais State–Brazil (Fig. 184.1).

In this HPP, the water supply circuit, positioned on the left bank, is composed of water intake, intake tunnel, surge tank, penstock and powerhouse. The water intake is located just upstream of the earth dam on left side, corresponding to the beginning of the inlet tunnel, excavated in rock with a nominal diameter of 6.0 m and length of 1,185 m, with a reduced slope.

184.2 Geological Features

Rocky outcropping in the region of the dam are gneisses, whitish gray color with striking foliation in the direction N70 W. The mineralogical composition is quartz, feldspar and biotite, amphibole and garnet being the most abundant accessory minerals. Bedrock is present with signs of mild decomposition, moderately fractured and very consistent, and it is common the occurrence of quartz and pegmatite veins.

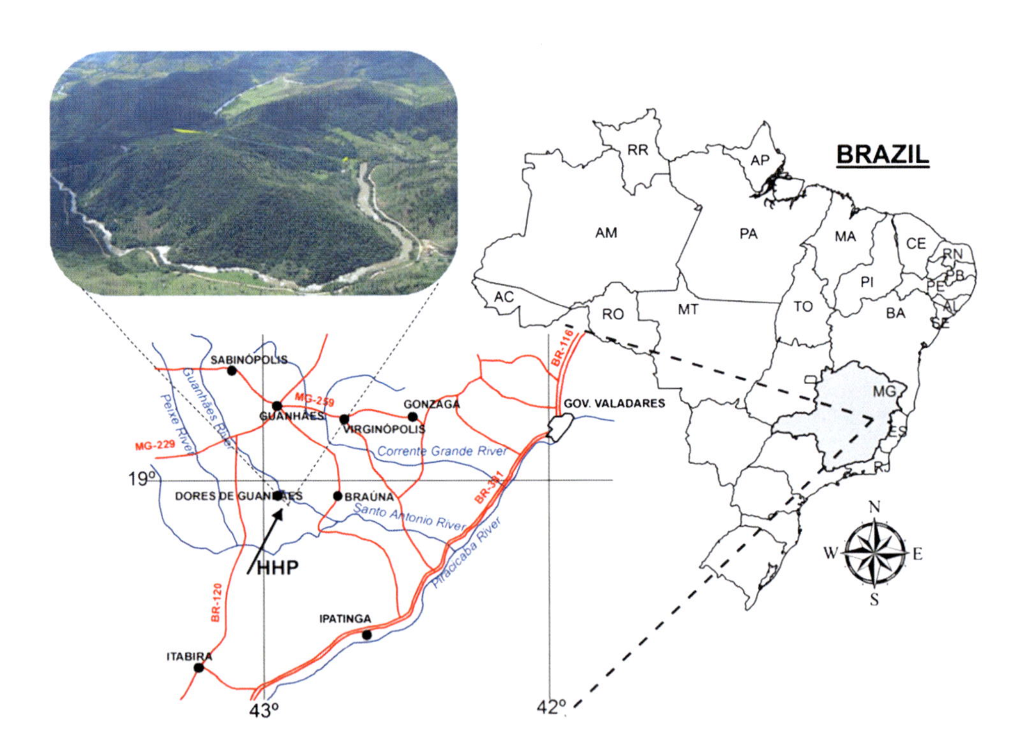

Fig. 184.1 HHP localization

Fig. 184.2 Erosive levels of biotite and amphibole

These veins represent tabular bodies, post-tectonic intrusives, usually with centimetric thicknesses, subparallel to the foliation. In the same direction occur dark gray levels, formed by the merging of biotite and amphibole that are easily weathered and eroded when outcropping in drainages (Fig. 184.2).

184.3 Models

The 3D model was generated from the topographic data which have been integrated into the structural data obtained from geological mapping and geoelectrical survey towards the tunnel that led to the recognition of the structures at depth.

184.3.1 Geophysical Survey

The geophysical survey was conducted by a geoelectrical method that used an artificial source to introduce an electric current in the basement for determining the resistivity of different geological materials in the subsurface.

The survey used an array with 60 channels, with 10 m spaced electrodes at the ground level. Thus, it was possible to achieve at great part of the survey depths very close to the axis of the tunnel, as indicated by a dashed line in Fig. 184.3.

This survey identified a wide range of resistivity values. In general, the passages through soil and regolithe have resistivities less than 5,000 Ω/m. Up to these values are founded rocks that can reach values higher than 50,000 Ω/m. On the subsurface level (initial 10 m) it is possible to identify

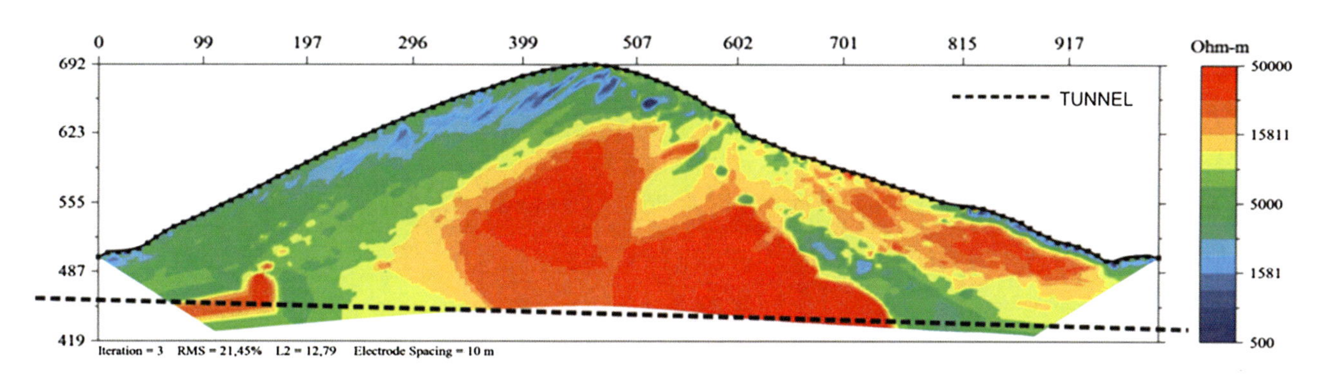

Fig. 184.3 Geophysics results and the tunnel axis indicated by the *dashed line*

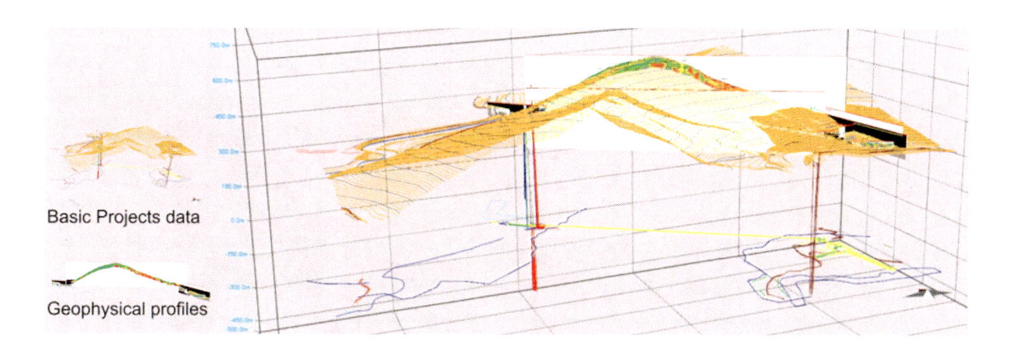

Fig. 184.4 Basic project data and geophysical profiles

wet zones represented by the colors in the range 500–1,500 Ω/m. Yet, these resistivity values are very high, indicating little presence of water.

The discontinuities observed in the rock also exhibit very high resistivity, probably consisting of fractures that are currently dry or with little water.

184.3.2 Modelling

One of the first problems encountered is due to the majority of the data produced during the basic design work are 2D in various cartographic basis (see Fig. 184.4), which makes the integration and interpretation of the data more laborious. Currently these data are in the process of adjustment and transformation to a 3D environment.

To input the data into the MOVE©, it is necessary to have all with the same coordinate system (x, y, z). The data are separated into categories where the direction and dip plans fractures, obtained from the geological mapping (Fig. 184.5), are input into the software. It is also input images of geophysical survey, topography information results by the Kriging interpolate method.

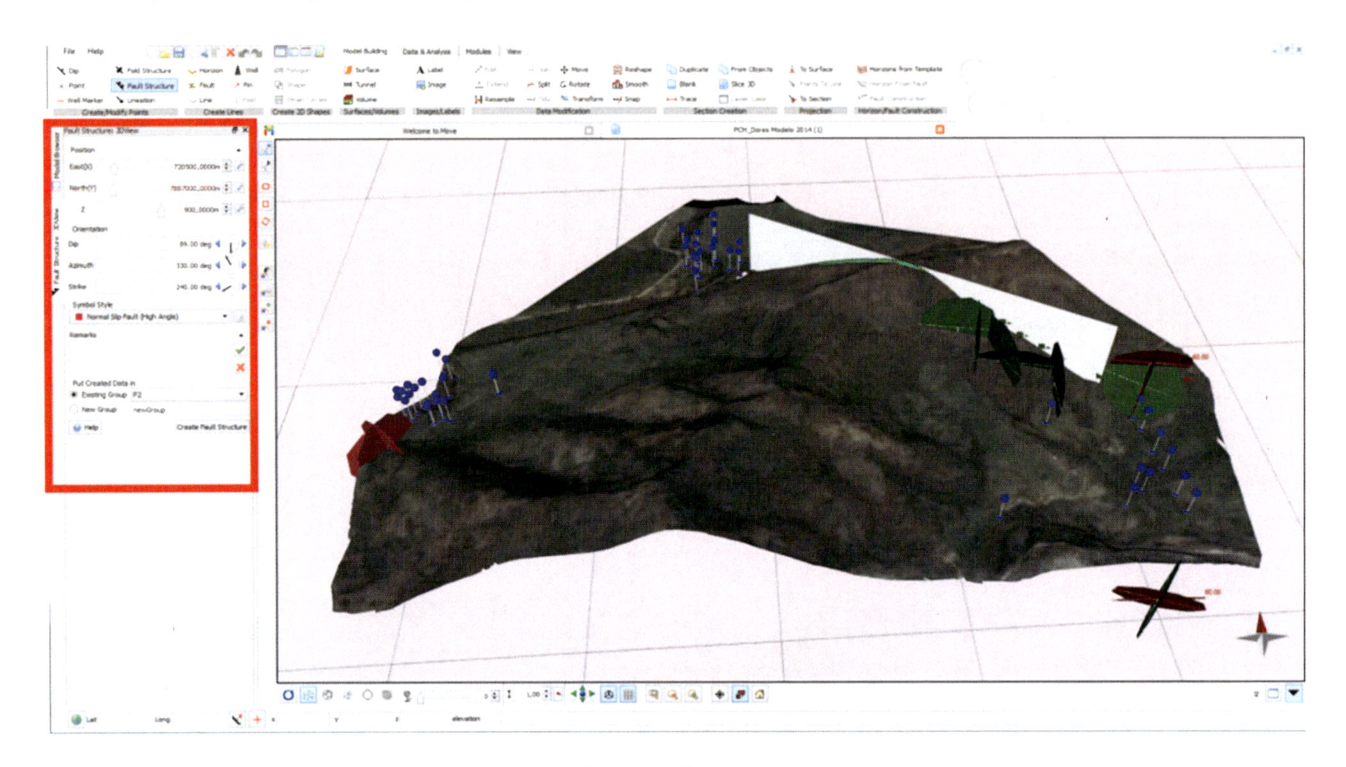

Fig. 184.5 Software Image showing the input information of the plans fractures, in *red box*

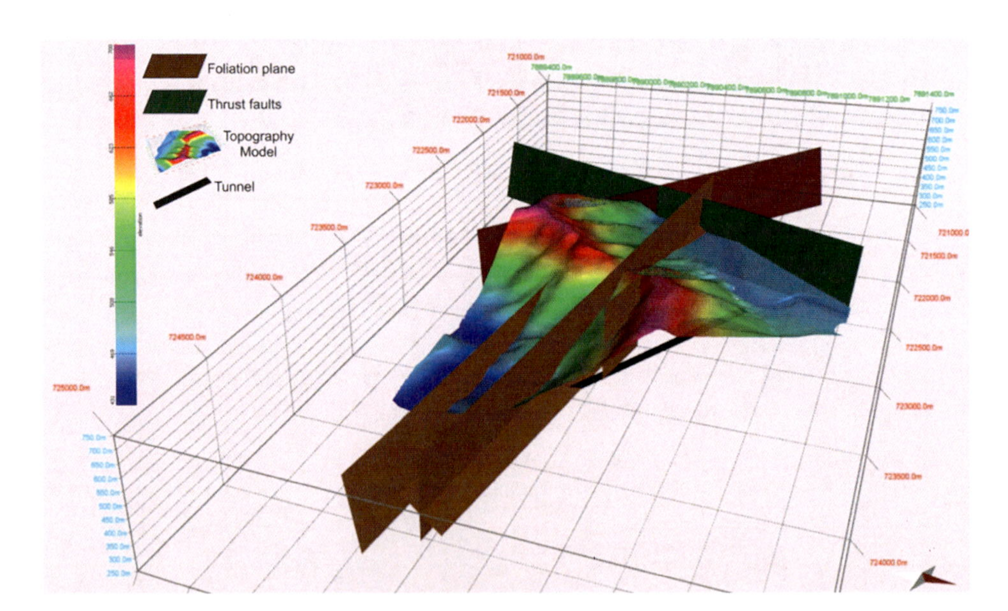

Fig. 184.6 3D modelling with the regional images and structural and field data

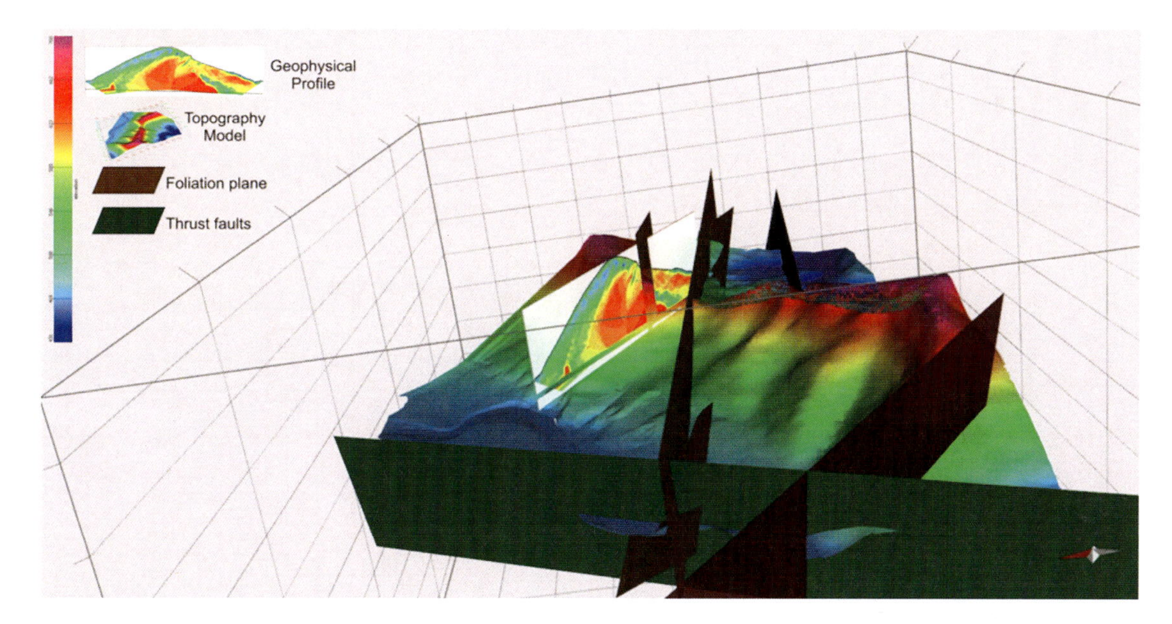

Fig. 184.7 Geophysical survey with structural data

Data provided by the construction owner have been uploaded in the 3D modelling software, along with structural and field data, as seen in Fig. 184.6.

Geophysical analyses (dipole-dipole array) were performed and included in the model with a thorough discussion of its possible benefits in general interpretation (Fig. 184.7). In this simulation it was possible to see that the model with the presence of geophysical section presents a robust result when comparing to the model without this information, it allows to infer the mapping surface into the rock mass.

184.4 Discussions

The integration of three-dimensional models allowed us to visualize the fracture and continuity plans crossing tunnel, and predict directions which fractures and biotite veins occur, helped with geophysical interpretation data.

The geological mapping and MOVE 3D model interpretation, helped to understand the geological features to reduce these uncertainties in models and the sharing definitions.

To enhance the data of 3D modeling, together with data obtained from geological mapping, data should be integrated geomechanics characteristics of the rock mass obtained in geotechnical surveys.

New geological surveys will be undertaken to improve the model and about these risk analysis will be performed with the software Decisions Tools.

References

MOVE© (2013) Midland Valley Exploration Ltd

Poente and Guanhães Energia (2008) Basic project studies HHP Dores de Guanhães-MG

185 Use of Rock Mass Fabric Index in Fuzzy Environment for TBM Performance Prediction

Mansour Hedayatzadeh and Jafar Khademi Hamidi

Abstract

Precise prediction of TBM performance is one of the most crucial issues in mechanized tunneling projects. The main objective of this study is to estimate the TBM rate of penetration by constructing a fuzzy inference system analysis. A database consisting of fabric index of Q rock mass classification system, rock material properties and machine characteristics and performance along the tunnel alignment was compiled. In order to verify the validity of the developed model, the predicted and field measured penetration rates were compared. Results picked out from this predictor model revealed that the model has a strong capability for estimation of TBM performance with a correlation coefficient of 81.5 %.

Keywords

TBM performance • Fabric index • Rate of penetration • Fuzzy logic

185.1 Introduction

Performance prediction of tunnel boring machine is a complex and ambiguous engineering geological problem. This issue is crucial because a precise estimation of machine performance can considerably influence the cost and duration analysis during the planning stage of a mechanized tunneling project. A comprehensive review of the literature on the subject can be found in Khademi Hamidi et al. (2010b) and Farrokh et al. (2012). Among the most recently developed empirical prediction models, some ones have used intelligent systems such as artificial neural network (ANN), fuzzy logic and Neuro-Fuzzy hybrid techniques. Taking into consideration the nature of the problem, the main purpose of the present study is to develop a model by utilizing the fuzzy logic for predicting TBM performance. In order to achieve this aim, a database composed of rock mass properties including fabric indices of four rock mass classification and the angle between plane of weakness and tunnel axis, intact rock properties including uniaxial compressive strength, machine specifications including net thrust per cutter together with actual measured TBM rate of penetration(ROP) is compiled along the tunnel alignment. Alborz service tunnel situated in Tehran-Shomal freeway is chosen as a case study. The Tehran-Shomal freeway project with about 120 km in length is a new freeway to connect the capital Tehran with the city of Chalus at the Caspian Sea in the north of Iran. The freeway alignment has more than 30 twin tunnels for double lanes. The Alborz tunnel is the longest of these with a length of 6,400 m at an altitude of 2,400 m.

At this stage of the Alborz tunnel construction, an exploratory pilot tunnel is being driven in Kandovan region. This tunnel, 5.2 m in boring diameter, will be used for exploratory purposes prior to the start of the construction of main tunnels and will also serve as an emergency access and ventilation way and drainage during the life of the tunnels. Besides, it will be used for transportation and other services during the construction of main tunnels. Longitudinal profile of the geological sections is also shown in Fig. 185.1. A Wirth open TBM having diameter of 5.2 m was used for

M. Hedayatzadeh
Politecnico di Torino, Corso Duca degli Abruzzi. 24, 10129 Turin, Italy
e-mail: mansour.hedayatzadeh@polito.it

J. Khademi Hamidi (✉)
Tarbiat Modares University, Jalal Ale Ahmad Highway, P.O. Box 14115-111 Tehran, Iran
e-mail: jafarkhademi@modares.ac.ir

G. Lollino et al. (eds.), *Engineering Geology for Society and Territory – Volume 6*,
DOI: 10.1007/978-3-319-09060-3_185, © Springer International Publishing Switzerland 2015

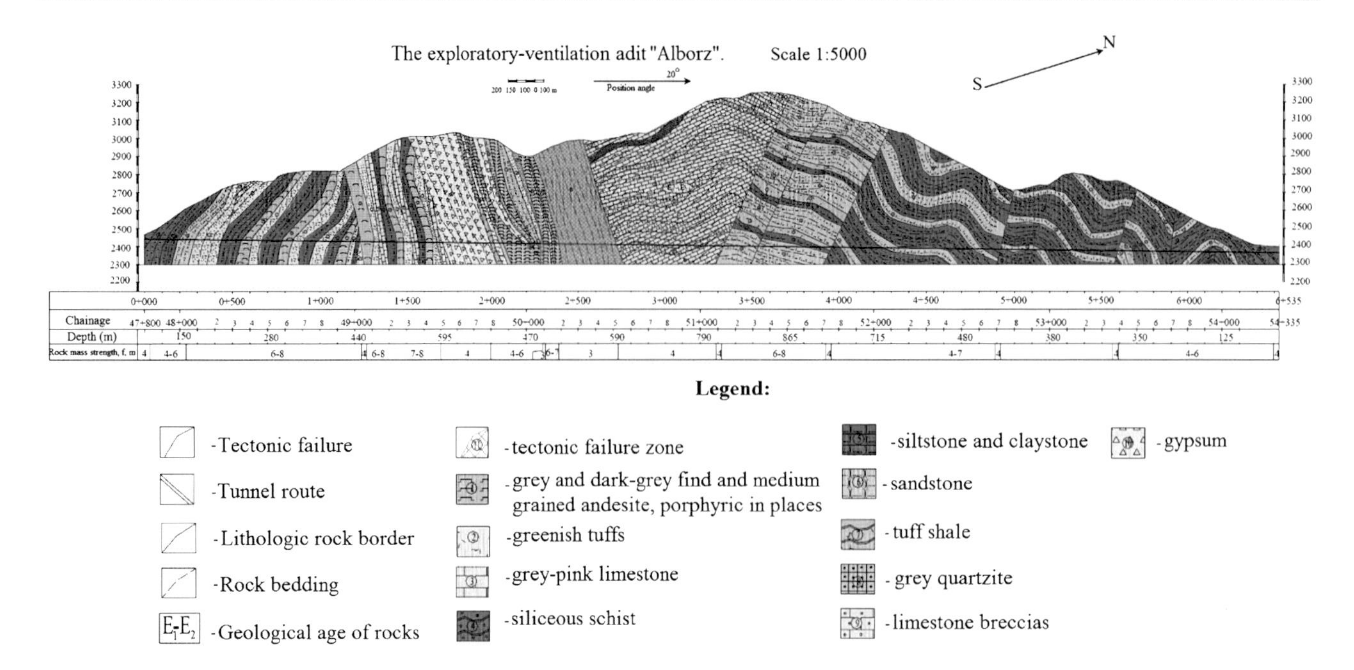

Fig. 185.1 Longitudinal geological profile of Alborz service tunnel (Technical Office Tehran–Shomal Highway 2009)

Table 185.1 Geomechanical parameters for different engineering geological units (Technical Office Tehran–Shomal Highway 2009)

Rock type	Density (kg/m^3)	UCS (MPa)	RQD (%)	RMR
Sandstone	2,700	120	50–75	45–55
Limestone	2,650	85	45–55	45–50
Tuff	2,650	70	50–60	45–50
Gypsum	2,300	65	70–80	55–60
Shale	2,500	55	55–70	50–60

tunnel excavation. Site investigation for the service tunnel included a geological surface mapping, a geoelectrical resistivity survey along the alignment from the surface and some index laboratory tests on rock samples. Geological conditions are complex and overall heterogeneous.

Based on the results of geological site investigations, the main lithological units through which the tunnel was driven consist of sandstone, tuff, gypsum, shale and limestone layers. The main geomechanical characteristics of major lithotypes are summarized in Table 185.1.

185.2 Rock Mass Fabric Index in TBM Performance Prediction

There are many factors influencing the TBM performance such as rock material, rock mass parameters, machine characteristics and operational parameters. The rock mass properties including the orientation, condition and frequency of discontinuities in rock mass, and also intact rock properties such as strength, hardness, toughness and brittleness are crucial ground parameters for analysis of hard rock TBMs. This information along with machine specifications such as thrust and power allow the appraisal and prediction of machine penetration.

Tzamos and Sofianos (2006) correlate four rock mass classifications including RMR, Q, GSI and RMi by introducing rock mass fabric (denoted as F index). The common parameters of these systems, which concern and characterize solely the rock mass (excluding boundary conditions such as stress regime and water pressure), are those used for rating the rock structure and the joint surface conditions. Rock structure is quantified by the block size or the discontinuity spacing ratings (BS) and the joint surface conditions are quantified by the joint conditions ratings (JC). For instance, in the RMR system, the parameters concerning rock structure are the RQD and the spacing of discontinuities, denoted as parameters R_2 and R_3. Their summation, $R_2 + R_3$, defines the BS component. The JC component, which represents condition of discontinuities, is defined by the parameter denoted as R_4.

The fabric indices of four commonly used systems (F_Q, F_{GSI}, F_{RMR} and F_{RMi}) are given in Table 185.2. In this study, the rock mass fabric index of Q system is used.

The orientation of rock mass discontinuities is another influencing parameter on TBM penetration rate and has been widely considered in many prediction models (e.g. Bruland 1998; Gong et al. 2005; Yagiz 2008; Khademi Hamidi et al. 2010b). This parameter is usually determined with regard to the tunnel axis and denoted as Alpha angle (α). To calculate the α angle, orientation of discontinuities and driven direction of TBM have been measured in the field. In this study,

Table 185.2 F index of four rock mass classifications (Tzamos and Sofianos 2006)

$F_Q = (RQD/J_n\ J_r/J_a)$	$BS = (RQD/J_n),\ JC = (J_r/J_a)$	(6.2)
$F_{RMR} = R_2 + R_3 + R_4$	$BS = (R_2 + R_3),\ JC = R_4$	(6.3)
$F_{GSI} = GSI$	$BS = SR,\ JC = SCR$	(6.4)
$F_{RMi} = JP$	$BS = (V_b),\ JC = (jC)$	(6.5)

J_n, J_r and J_a input parameters of Q system; *SR*, *SCR* structural rating and surface condition rating of GSI; V_b, *jC* block volume and joint coefficient factor in RMi

the α angle (F_α), computed for the critical joint set is included in the model.

Intact rock compressive strength, toughness, hardness, brittleness and abrasiveness are some of rock resistivity factors which are usually employed in predicting TBM penetration rate. In several TBM performance studies, it has been indicated that the uniaxial compressive strength (UCS) is the single most important rock parameter controlling TBM ROP (e.g. Cassinelli et al. 1982; Innaurato et al. 1991; Hassanpour et al. 2010). In this study, the UCS normalized by cutter load (F_f) is used in the model. The advantage of the normalized UCS compared to UCS is the elimination of the effect of machine cutterhead thrust.

A database consisting of 34 records from 34 sections along the 6.4 km bored tunnel and containing three independent variables including fabric index of Q system (F_Q), the alpha angle (F_α) and UCS of rock material normalized by cutter load (F_f) and the measured TBM ROPs (i.e. dependent variable) was compiled and subjected to fuzzy logic in order to drive a TBM performance prediction model. Knowledge about the relationship of these factors and their effect on the TBM performance are available from the subject literature. These experiences were the primary source of information for designing the rule bases of the fuzzy model. The reasoning behind the choice of the most related factors and the translation of the expert knowledge into the fuzzy if-then rules is described in detail in Sect. 185.3.

185.3 The Fuzzy Model for TBM Performance Prediction

Construction of the fuzzy model includes several steps: selection of the related input variables, design of the membership functions, translation of the expert knowledge into if-then rules and determination of defuzzification method.

The advantage of fuzzy logic in comparison with traditional prediction models such as statistical methods is its capability to describe the complex and nonlinear behavior which commonly exists in engineering geological problems (Khademi Hamidi et al. 2010a). There are several fuzzy inference systems (FISs) that have been employed in various applications, such as the Mamdani fuzzy model, Takagi–

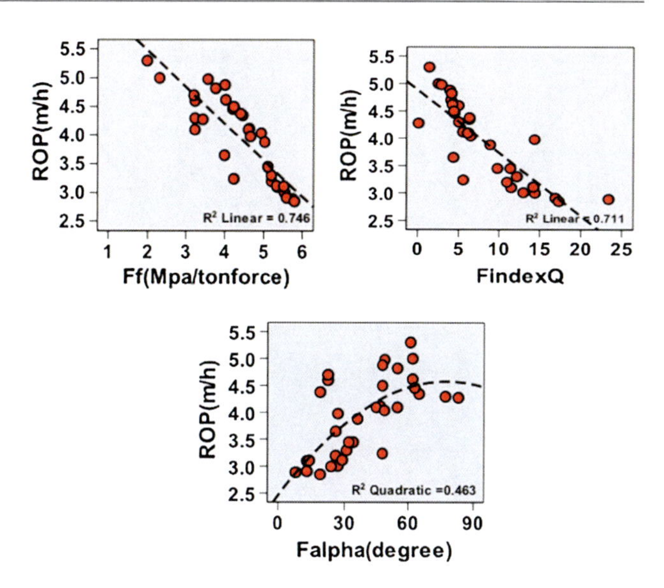

Fig. 185.2 Relationship between measured ROP and three input parameters

Sugeno–Kang (TSK) fuzzy model, Tsukamoto fuzzy model, and Singleton fuzzy model. However, the Mamdani fuzzy algorithm is the most commonly used FIS in engineering geological problems.

Figure 185.2 illustrates the correlations between the individual independent variables and the actual measured ROP. The figure also, includes the coefficients of determination (R^2) which is an indicator of correlation strength.

In construction of a fuzzy model, the use of the proper membership function (MF) is a crucial issue because MFs express the fuzziness of the model's variables. The shape of the membership function of fuzzy sets can be either linear (trapezoidal or triangular) or various forms of non-linear, depending on the nature of the system being studied. In this paper, the triangular membership function is employed because of its simplicity. For instance, the graphical illustration of the membership function of fabric index of Q system is given in Fig. 185.3.

Input–output relationship by fuzzy conditional rules is a significant concept in fuzzy logic. A fuzzy conditional rule is generally composed of a premise and a consequent part (IF premise, THEN consequent). For example, 'if the rock strength is high, then the ROP is low', where the terms high and low can be represented by fuzzy sets or more specifically by membership functions.

The last stage of a FIS is to select the defuzzification method. The aggregation of two or more fuzzy output sets gives a new fuzzy set in the basic fuzzy algorithm. In most cases, the result in the form of a fuzzy set is converted into a crisp result by the defuzzification process. The "centroid of area" (COA) method is very popular and applied for the defuzzification process in this study.

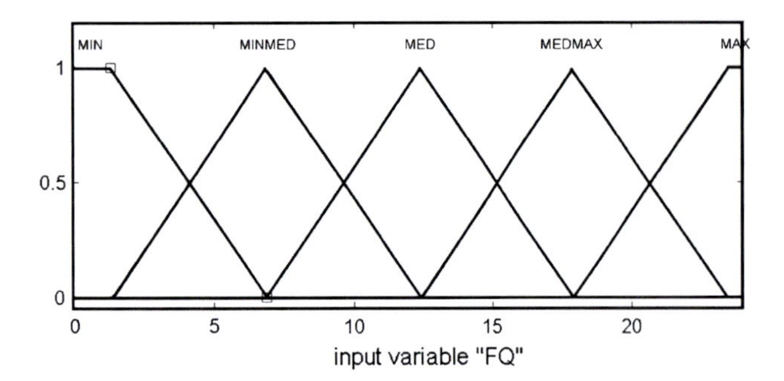

Fig. 185.3 Membership function of the fabric index of Q system, FQ

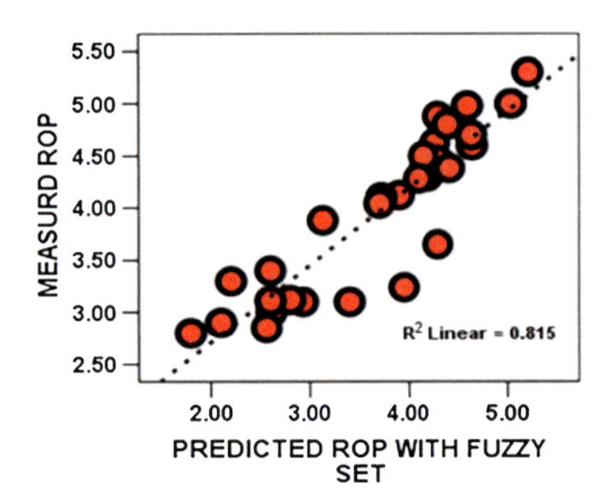

Fig. 185.4 Correlation between measured and predicted ROPs

To validate the model accuracy, the actual field ROPs are compared with the predicted values from the fuzzy model as illustrated in Fig. 185.4. As seen in the figure, the predicted data are in good agreement with the measured ones in database.

185.4 Conclusions

A fuzzy model was developed for the prediction of hard rock TBM penetration rate based on expert knowledge, experience, and data obtained from 34 sections along the route of Alborz service tunnel. In order to predict TBM ROPs, three input variables including fabric index of Q classification system, uniaxial compressive strength of intact rock, cutter load and the angle between tunnel axis and discontinuity planes were utilized. Results obtained from the fuzzy model showed that it has a strong capability to predict the TBM penetration rate, with correlation coefficient of 0.815. However, the ranges of the input data used for development of the proposed prediction model were very limited and as such, the results cannot be considered to be universal and more in depth study is required to extend the finding of this study to develop a universal model.

References

Bruland A (1998) Hard rock tunnel boring. PhD Thesis, Norwegian University of Science and Technology, Trondheim

Cassinelli F, Cina S, Innaurato N, Mancini R, Sampaolo A (1982) Power consumption and metal wear in tunnel-boring machines: analysis of tunnel boring operation in hard rock. In: Tunnelling '82, London. Inst Min Metall 73–81

Farrokh E, Rostami J, Laughton C (2012) Study of various models for estimation of penetration rate of hard rock TBMs. Tunn Undergr Space Technol 30:110–123

Gong QM, Zhao J, Jiao YY (2005) Numerical modeling of the effects of joint orientation on rock fragmentation by TBM cutters. Tunn Undergr Space Technol 20:183–191

Hassanpour J, Rostami J, Khamehchiyan M, Bruland A, Tavakoli HR (2010) TBM performance analysis in pyroclastic rocks: a case history of Karaj water conveyance tunnel. Rock Mech Rock Eng 43 (4):427–445

Innaurato N, Mancini R, Rondena E, Zaninetti A (1991) Forecasting and effective TBM performances in a rapid excavation of a tunnel in Italy. In: Proceedings of the seventh international congress ISRM, Aachen, pp 1009–1014

Khademi Hamidi J, Shahriar K, Rezai B, Bejari H (2010a) Application of fuzzy set theory to rock engineering classification systems: an illustration of the Rock Mass Excavability Index. Rock Mech Rock Eng 43(3):335–350

Khademi Hamidi J, Shahriar K, Rezai B, Rostami J (2010b) Performance prediction of hard rock TBM using Rock Mass Rating (RMR) system. Tunn Undergr Space Technol 25(4):333–345

Technical Office Tehran–Shomal Highway (2009) Technical report

Tzamos S, Sofianos AI (2006) A correlation of four rock mass classification systems through their fabric indices. Int J Rock Mech Min Sci 477–495

Yagiz S (2008) Utilizing rock mass properties for predicting TBM performance in hard rock condition. Tunnell Undergr Space Technol 23(3), 326–39

The Risk Analysis Applied to Deep Tunnels Design—El Teniente New Mine Level Access Tunnels, Chile

186

Lorenzo Paolo Verzani, Giordano Russo, Piergiorgio Grasso, and Agustín Cabañas

Abstract

El Teniente Mine, with 2,400 km of tunnels excavated since the beginning of the last Century is the largest underground copper mine in the world. El Teniente Mine production plan has a thin overlap between the exhaustion of the current production level and the activation of the New Mine Level, located at almost 1,000 m depth, planned for 2017. The infrastructure system involves the construction of 24 km of access tunnels, consisting of two adits, a tunnel for vehicular access of personnel and a twin conveyor tunnel for the transport of the ore. The definition of geological and geomechanical scenarios, as predicted on the basis of the reference models, and the related hazards identification and mitigation (following a risk analysis based design), are cornerstones along the production chain. Tunnel alignment intersects a complex geological environment characterized by rock variability: from igneous (effusive and intrusive) to sedimentary volcanoclastic rocks, with sectors of intense hydrothermal alteration. Due to high overburden and variability of rock mass properties, geomechanical hazards such as squeezing and rockburst are expected, together with caving and flowing-ground conditions crossing fault sectors associated with high hydraulic pressures. This paper synthesizes the design methodology, focused on risk management (Risk Analysis-driven Design, Geodata 2009). The construction of the tunnel is actually in process and then also a preliminary comparison "predicted versus observed" is anticipated.

Keywords

Risk management plan • Risk analysis-driven design • Probabilistic reference scenarios • Hazards mitigation measures

186.1 Introduction

El Teniente Mine, located in the Libertador General Bernardo O'Higgins Region 80 km southeast of Chile's capital Santiago, is the largest underground copper mine in the world, with 2,400 km of deep tunnels producing more than 400,000 tons per year of fine copper.

El Teniente Mine production plan has a thin overlap between the exhaustion of the current production level and the activation of the New Mine Level (NML), located at almost 1,000 m depth, planned for 2017. The underground infrastructures are under construction; the NML will extend life of the mine by 50 years; a deeper level has been investigated at 1,400 m depth (Fig. 186.1).

L.P. Verzani (✉) · G. Russo · P. Grasso
Geodata Engineering S.p.A., Turin, Italy
e-mail: lpv@geodata.it

G. Russo
e-mail: grs@geodata.it

P. Grasso
e-mail: pgr@geodata.it

A. Cabañas
División El Teniente, Corporación Nacional del Cobre de Chile (Codelco), Santiago, Chile
e-mail: acabanas@codelco.cl

G. Lollino et al. (eds.), *Engineering Geology for Society and Territory – Volume 6*,
DOI: 10.1007/978-3-319-09060-3_186,

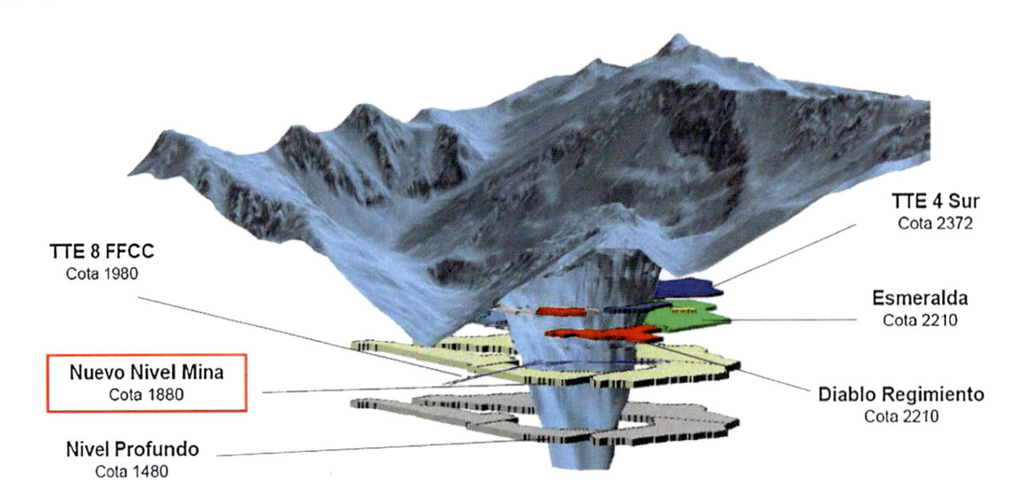

Fig. 186.1 Mine levels (by Codelco)

The NML project foreseen the construction of 24 km of access tunnels, consisting of two adits (Ltot = 6 km) and two main tunnels (Ltot = 9 + 9 km): a tunnel for vehicular access of personnel and a twin conveyor tunnel for the transport of the ore. Geodata Engineering (GDE), as a consultant of Codelco and tunnels Contractor's counterpart on geotechnical issues, has been present on site since April 2012.

Prior to the construction phase, Geodata Engineering (GDE in association with Ingeroc) developed for Codelco a design for the two main tunnels, based on the risk analysis (Risk Analysis-driven Design, RAdD) as reference for the owner about engineering solutions and construction costs and time assessment. Conventional and mechanized excavation methodologies were analyzed.

Constructora de Túneles Mineros–joint venture between Soletanche Bachy and Vinci (CTMSA) won proposing the conventional method (D&B), with two additional adits to increase the number of parallel advances along the main access tunnels (Fig. 186.2). Actually the constructions of both the main access tunnels and the adits are in process.

186.2 Risk Analysis-Driven Design (RAdD)

The design and construction of long tunnels particularly those at great depth, is generally associated with a high level of risks due to a whole series of uncertainties involved. The risk management approach consists in identifying and listing

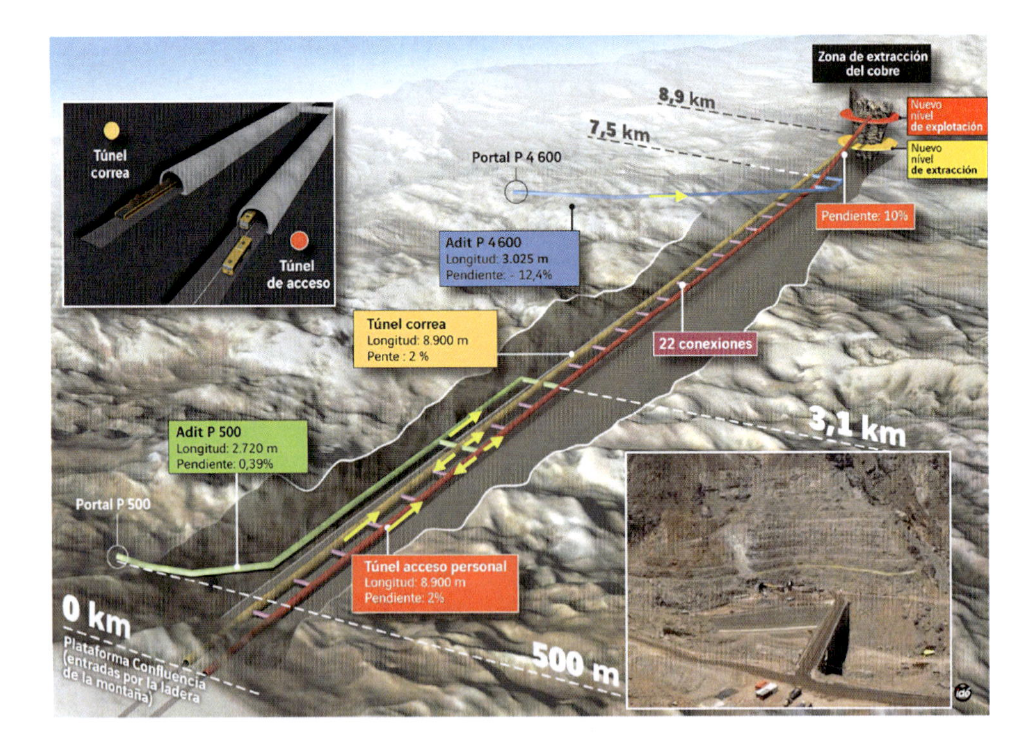

Fig. 186.2 NML tunnel access system (by CTMSA)

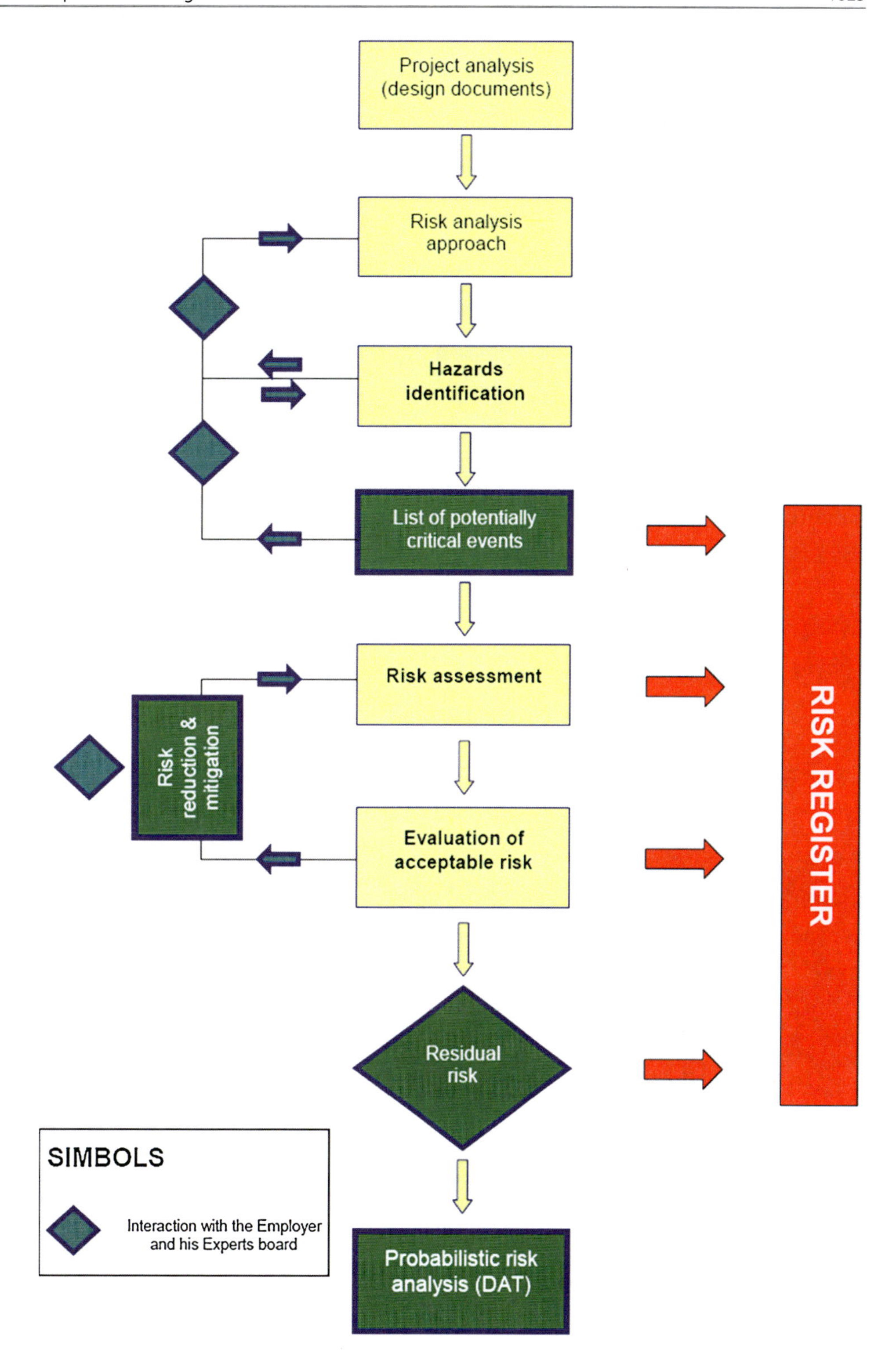

Fig. 186.3 RAdD flow chart

the potential hazards associated with the tunneling activities, assigning a probability of occurrence to each hazard, and allocating an index of severity to the consequence (impact). Two main categories of hazard events are identified in connection to geological and geomechanical issues, namely: (Fig. 186.3)

- Hazard phenomena associated with unfavorable *geological conditions* (fault, water acidity, etc.)

- Geomechanical hazard related to *rock mass behaviour* upon excavation (squeezing, rockburst, etc.).

The risk (R) is defined as the product of the probability of occurrence of the hazard (P) and the related impact (I): R = P*I. In cases where the initial risk (i.e. the risk to which the project is exposed in absence of any mitigation measure) level is not acceptable, the relevant mitigating measures should be identified and designed.

After application of the mitigation measures, an analysis should be performed to reassess the remaining risk level, obtaining an updated risk level, which is called the "residual risk level". It should be examined for acceptance as the maximum risk level that is to be confronted with its "global cost", necessary for reducing or completely eliminating the risk itself.

All the relevant information about the hazards, the associated risks and counter measures are filled and regularly reviewed in a risk register.

186.3 Geological Setting and Related Risks

The regional geology of El Teniente area is characterized by volcanic rocks and sedimentary volcanoclastic deposits, with felsic to intermediate intrusive. As shown in the Geological Reference Model, proceeding from West (portal) to East (mine), the following lithological formations and Rock Mass Unit (RMU) would be crossed:

- Farellones Formation lower and undifferentiated members (FFm, RMU.V1-V2)
- Agua Amarga Hydrotermal Alteration and Breccia (RMU.AA)
- Sewell intrusive Complex (CSW, RMU.i1-CQ-i2)
- El Teniente Mafic Complex (CMET, RMU.i3)
- Braden Breccia (RMU.BB).

The tunnel axis crosses three major faults (F1, F2) and a large number of minor faults (F4). Moreover the El Teniente shear zone (F3) is foreseen along CSW and CMET formations. On the basis of the Geological Reference Report (GRR Codelco-Hatch 2009), some potential geological hazards were identified. Among them, the main ones in terms of impact are: geological structures, hydraulic load and water pH, natural stress field and anisotropy, rock weathering and hydrothermal alteration (Fig. 186.4).

Moreover some additional hazards were analyzed for the mechanized method (TBM): rock hardness-abrasiveness and heterogeneity. The main geological hazards are probabilistically quantified and the risk register is compiled, both for D&B and TBM, considering the required mitigation measures for each potential risk. Since March 2012, the following geomechanical units have been excavated: RMU.V1 and RMU.V2 in FFm (main tunnels and Adit 1); the structural contact RMU. V2/AA (Adit 1) and RMU.I2 in CSW (Adit 2).

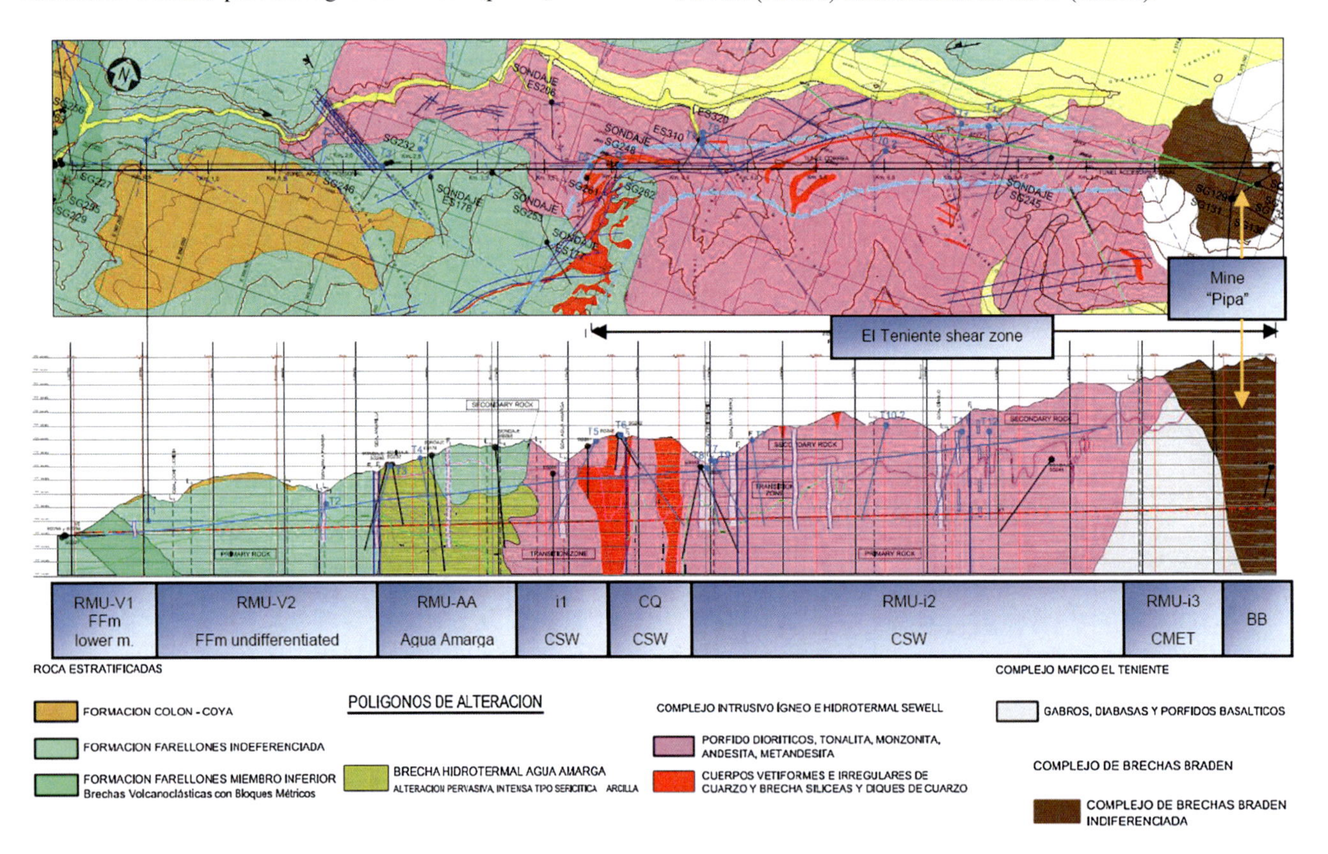

Fig. 186.4 Geological setting

Fig. 186.5 GD Classification (*Notes* Russo and Grasso (2007); δ_0 = radial deformation at the face; R_{pl}/R_0 = palstic radius/ radius of the cavity; σ_θ = max tangential stress; σ_{cm} = rock mass strength. The limits of shadow zones are just indiactive) of the excavation behaviour

↓ ANALYSIS →		Geostructural →		Rock mass: Continuous ↔ Discontinuous ↔ Equivalent C.				
Tensional↓				RMR				
Deformational response ↓	δ_0 (%)	R_{pl}/R_0	Behavioural category ↓	I	II	III	IV	V
Elastic ($\sigma_\theta<\sigma_{cm}$)	negligible	-	a	STABLE				
			b	INSTABLE				CAVING
Elastic - Plastic ($\sigma_\theta\geq\sigma_{cm}$)	<0.5	1-2	c	SPALLING/ ROCKBURST	WEDGES			
	0.5-1.0	2-4	d					
	>1.0	>4	e				SQUEEZING	
			(f)	→ Immediate collapse of tunnel face ↑				

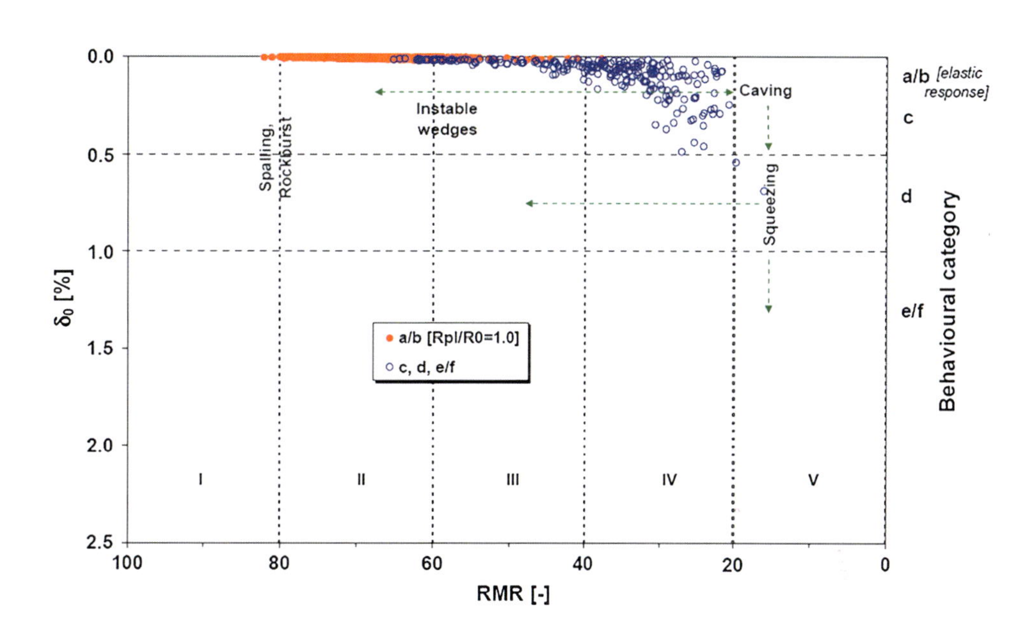

Fig. 186.6 Design scatter diagrams (RMU.V1)

Fig. 186.7 Hazard probability/ intensity (RMU.V1-V2)

RMU Rock Mass Unit (Overburden)	Scenario ↓	Hazard [%]									
		Wedge instability / Rockfall			Caving		Squeezing		Spalling/Rockurst		
	Intensity →	s1	s2	s3	s2	s3	s2	s3	s1	s2	s3
V1 (40<H<290m)	H_LIK	0.3%	67.1%	16.1%	15.0%	0.0%	0.20%	0.0%	1.3%	0.0%	0.0%
	GD_FAV	28.0%	35.1%	29.4%	7.4%	0.0%	0.0%	0.0%	0.1%	0.0%	0.0%
	GD_UNF	0.6%	38.6%	45.3%	15.0%	0.3%	0.0%	0.0%	0.2%	0.0%	0.0%
V2 (250<H<420m)	H_LIK	12.3%	33.1%	38.8%	8.1%	0.7%	1.5%	0.5%	4.7%	0.3%	0.0%
	GD_FAV	15.2%	39.0%	32.6%	5.5%	0.0%	1.5%	0.2%	5.2%	0.8%	0.0%
	GD_UNF	8.2%	25.6%	40.1%	11.3%	2.2%	3.7%	2.0%	6.2%	0.7%	0.0%

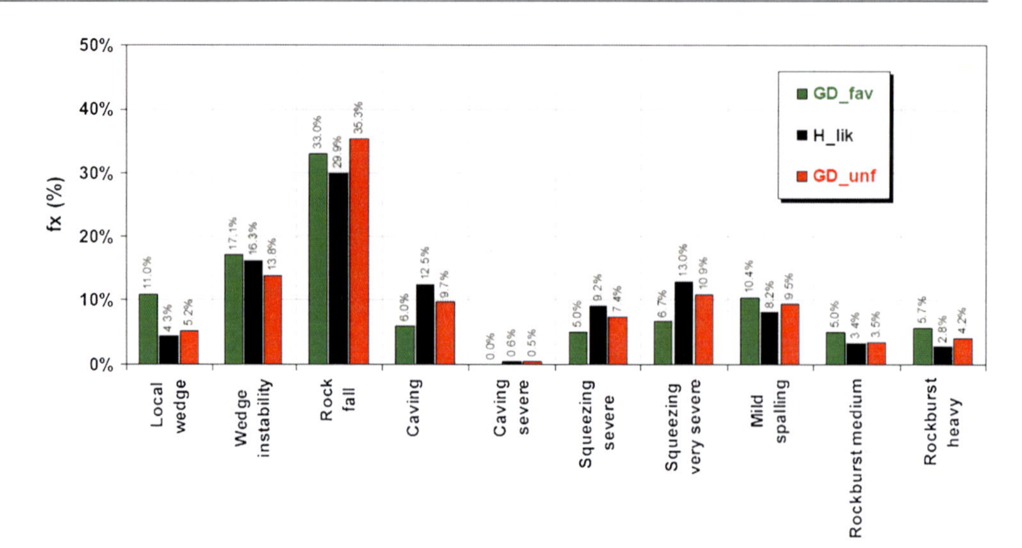

Fig. 186.8 Hazards frequency: probabilistic results for the n.3 scenarios of reference (9 km in personnel tunnel)

Fig. 186.9 Risk register (RMU. V2)

Hazard identification: CATEGORY / Sub-category / TYPE / Sub-type / HAZARD		Hazard Probab. [P]	Primary risk: D&B Impact [I]	D&B Risk [R=PxI]	TBM Impact [I]	TBM Risk [R=PxI]	Mitigation measures* (*or cross-reference to geomechanical hazard [→]): D&B	TBM
GEOMECHANICAL HAZARDS (EXCAVATION BEHAVIOUR AND LOADING CONDITION RELATED)								
Gravity driven instability								
B1	ROCK BLOCK FALL (→ OVERBREAKS)	5	2	10	1	5	M01,M02,M23,M24	M01,M22,M23,M24
B2	CAVING (→FACE / CAVITY COLLAPSE)	4	3	12	2	8	M01,M02,M03,M06,M07,M08,M24	M01,M08,M22,M24,M25,M27
Stress induced instability								
B3	ROCKBURST	2	3	6	2	4	M1,M2,M21,M23,M26,M27	M01,M22,M23,M26
B4	SQUEEZING, FACE EXTRUSION	2	3	6	4	8	M01,M02,M05,M07,M21,M24,M25,M27	M01,M22,M25,M27
Mainly water influenced (fault zone)								
B5	FLOWING GROUND	5	5	25	5	25	M01,M02,M06,M07,M08,M24	M01,M08,M22,M25,M27
B6	WATER INRUSH	5	5	25	5	25	M01,M02,M06,M07,M08,M24	M01,M08,M22,M25,M27
B7	PIPING	5	5	25	5	25	M01,M02,M06	

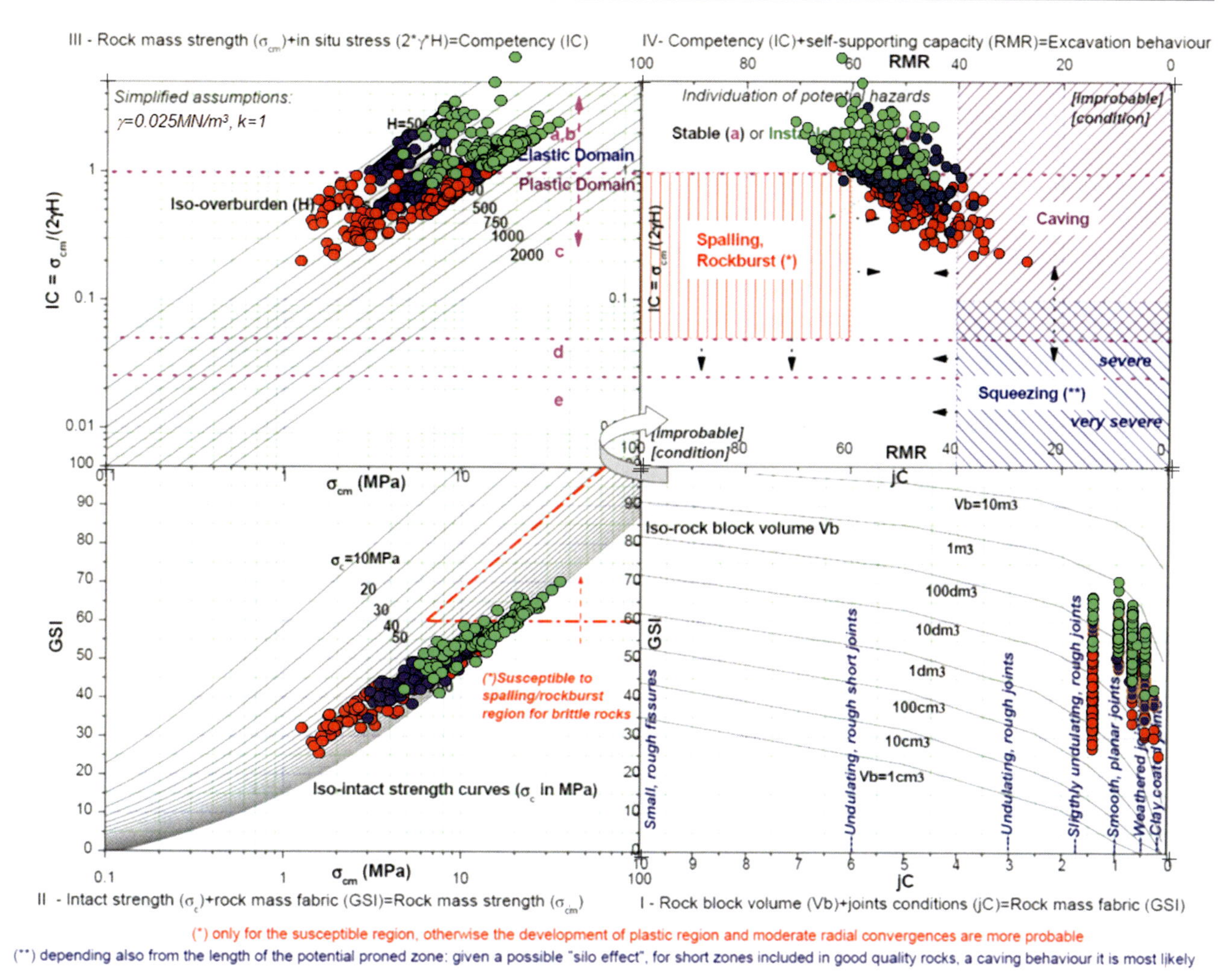

Fig. 186.10 RMU-V1, results of the geomechanical classification at tunnel face by the method of GDE multiple graph (Russo 2009)

186.4 Geomechanical and Residual Risks

The first step for RAdD is the geotechnical characterization of the different RMU. Related to the available information, the statistical analysis has adopted different approaches in order to define, in a probabilistic way, the three reference scenarios as geomechanical inputs for design: the most-likely scenario (from previous studies: H_lik), the favorable and the unfavorable ones (by data processing: GDE_fav/unf). Geomechanical hazards are mainly related to ground behaviour upon excavation, thus taking into account the intrinsic properties of rock masses and the associated stress conditions.

The reference classification of the excavation behaviour is based on both stress and geo-structural type analysis (matrix in Fig. 186.5), in the theoretical hypothesis of absence of any design interventions (→primary risk).

An example of the resulting design scatter diagrams is presented for one of the rock mass unit excavated in the personnel tunnel (RMU-V1, Fig. 186.6). The assessment of the geomechanical risks is obtained with reference to the occurrence probability and intensity of the related hazards. Along RMU V1 and V2, mainly geomechanical hazards due to gravity, as wedge instability, are expected (Figs. 186.7 and 186.8). The risk analysis proceeds with the initial risk assessment. Its evaluation involves the estimate of the potential impact (consequence) deriving from the damages related to the identified hazards. The Risk Register (Fig. 186.9) is consequently compiled for each RMU, both for the D&B and the TBM methods. The type and the dimensioning of the stabilization measures will be directly related to the hazards and their potential impact on tunneling (→primary risk). The adequate mitigation measures (design solutions) are consequently individuated, concurring to the composition of the different Section Types, dimensioned and probabilistically distributed along the tunnels. The last step for risk analysis process is the assessment of the new risk level obtained after the application of the design (→residual risk). The risk has been managed and reduced from its initial (primary) level to a lower (residual) value. If all the initial risks have been mitigated and the tunnels construction is not more exposed to unacceptable risks but the residual risk level remains classified as unwanted, some counter-measures are consequently defined.

186.5 Construction

The construction of the NMN tunnel access system started on March 2012, with the Adit 1. Currently, 18 months after the beginning, almost the 35 % of the 24 km totals has been excavated. The experience along the Adit 1 and the two main tunnels (RMU V1-V2 and contact zone RMU.V2/AA), permits to have a comparison with RAdD-design expected conditions. Outside from gully influence areas, along ordinary rock mass sections in RMU.V1-V2, the instabilities mainly related to gravity (wedge instability, rock fall with a lower probability of caving) were expected by GDE risk analysis. By the comparison among data collected during the advancement in RMU.V1, summarized by the method of the "GDE Multiple Graph" (Russo 2009; Fig. 186.10), and the probability of occurrence of the hazards expected by the design (Figs. 186.6 and 186.7, referred to RMU.V1), the reliability and effectiveness of the adopted risk analysis approach is confirmed.

186.6 Conclusions

Eighteen months of advancements in the NMN access tunnel, allow to obtain a first positive feedback on RAdD results. The Risk Analysis is a process that should support and follow a project, from the conceptual up to the construction stage.

The risk should be managed through the implementation of a specific Risk Management Plan (RPM, Grasso et al. 2002), fully integrated in each part of the design study, in accordance to a real development of a "Risk Analysis-driven Design".

References

Degn Eskesen S, Tengborg P, Kampmann J, Holst Veicherts T (2004) Guidelines for tunnelling risk management International tunnelling association, working group no 2. Tunn Undergr Space Technol 19:217–237 (ITA/AITES)

Grasso P. Mahtab M A, Kalamaras G, Einstein H H (2002) On the development of a risk management plan for tunnelling. In: Proceedings of world tunnel congress, Sydney

Russo G. Grasso P (2007) On the classification of the rock mass excavation behaviour tunneling. In: Proceedings of the 11th congress of international society of rock mechanics, Lisbon, 9–13 July 2007

Russo G (2008) A simplified rational approach for the preliminary assessment of the excavation behaviour in rock tunneling. Tunnels et Ouvrages Souterrains 207

Russo G (2009) A new rational method for calculating the GSI. Tunn Undergr Space Technol 24:103–111

187 Development of Measurement System of Seismic Wave Generated by the Excavation Blasting for Evaluating Geological Condition Around Tunnel Face

Masashi Nakaya, Kazuhiro Onuma, Hiroyuki Yamamoto, Shinji Utsuki, and Hiroaki Niitsuma

Abstract

Mountain tunneling in Japan is commonly performed according to the new Austrian tunneling method (NATM), and preliminary investigation for understanding the geological conditions of the mountain are performed from the surface by borehole sampling and seismic exploration using the refraction method. Exploratory boring is performed in only select locations with specific terrain or geological features, whereas seismic exploration using the refraction method is generally performed over the area in which the tunnel will be excavated. The certainty of such seismic exploration is reduced according to depth, making it difficult to ascertain the details of the geological conditions. The results of preliminary investigations must therefore be constantly compared with the actual geological conditions as revealed during construction. Seismic velocity is a significant factor to determine the design of tunnel support. Evaluation of geological condition requires the seismic velocity a round tunnel face, so there is a need for seismic exploration techniques that do not affect the excavation work cycle of tunnels under construction. We have therefore developed the tunnel face tester (TFT), which is a seismic exploration system that uses excavation blasting as wave source. This paper describes the developed system, and the results of verification experiments performed at a tunnel construction site.

Keywords

Mountain tunneling excavation blasting • Seismic evaluation geological condition exploration

187.1 System Components

Developing a system is aimed at exploration in daily tunneling cycle, and easy system to use for tunneling engineers (Nakaya et al. 2013). The system uses a seismic sensor fixed on tunnel walls behind the tunnel face, uses excavation blasting as a wave source, and measures seismic wave automatically.

The system (tunnel face tester unit, Figs. 187.1 and 187.2) is composed portable detector (a) and peripheral devices (b)–(d). The DC sensor (b) has a noncontact connection with the blasting leading wire, and converts the ignition current into a signal. The geophone (c) is of a type commonly used for seismic exploration (OYO GS-20DH; natural frequency: 28 Hz). The geophone is mechanically fixed to the head of a rock bolt (L = 3–4 m) on the tunnel wall, which serves as a waveguide. The ignition signal and seismic waveform data (are recorded on the same time scale) pass through the portable detector, and are recorded using a stereo IC recorder (TASCAM DR-05 up to 96 kHz/24bits resolution WAV recording) (d).

The following equation gives the velocity of seismic waves arrived to the geophone from the distance between the geophone and the tunnel face (L_i) and the arrival time of P wave (t_i).

M. Nakaya (✉) · K. Onuma · H. Yamamoto · S. Utsuki
Hazama Ando Corporation, 6-1-20, Akasaka Minatoku, Tokyo, 107-8658, Japan
e-mail: nakaya.masashi@ad-hzm.co.jp

H. Niitsuma
Tohoku University, Sendai, Japan

G. Lollino et al. (eds.), *Engineering Geology for Society and Territory – Volume 6*,
DOI: 10.1007/978-3-319-09060-3_187, © Springer International Publishing Switzerland 2015

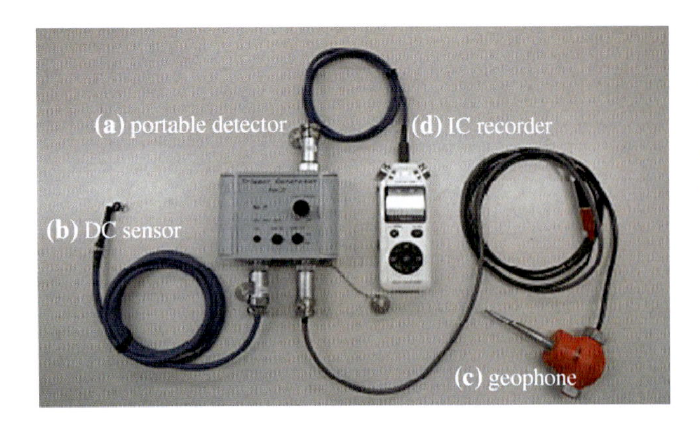

Fig. 187.1 Components of the tunnel face tester

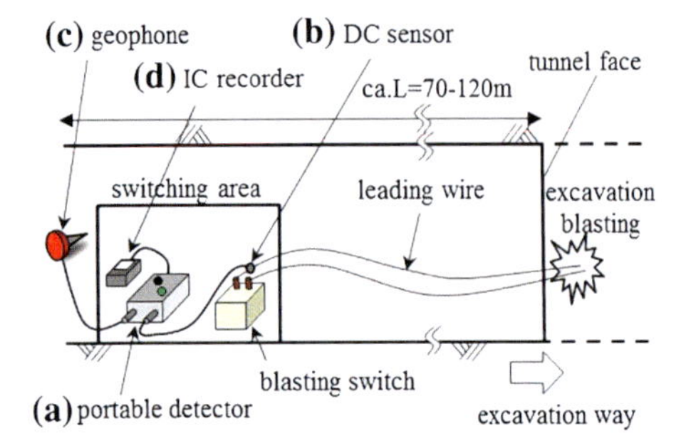

Fig. 187.2 Conceptual diagram of the measurement

$$Vp_i = L_i/t_i \quad (187.1)$$

When seismic waves arriving along the same straight line are measured without changing the location of the geophone, the wave path should be the same as in a previous measurement interval. The change of seismic velocity is thus a combination of the velocities in the previous measurement interval and the advance interval, so the previous measurement interval can be used to calculate the interval seismic velocity V_p' for moreover drilling, by the following equation.

$$Vp_{i\sim i+1}' = (L_{i+1} - L_i)/(t_{i+1} - t_i) \quad (187.2)$$

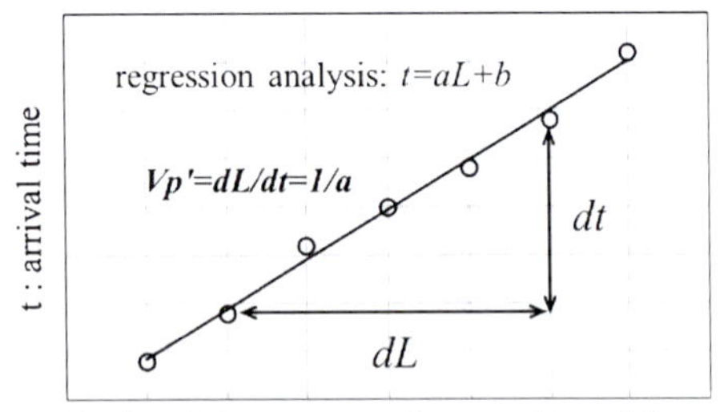

Fig. 187.3 Travel time curve

When the geophone position is unchanged, V_p' in the face advance interval can be determined from travel time curve (t: arrival time of P wave, L: length between geophone and tunnel face) (Fig. 187.3). The travel time curve and V_p' are determined by the following equations.

$$t = aL + b \quad (187.3)$$

$$Vp' = dL/dt = 1/a \quad (187.4)$$

187.2 Verification Testing

187.2.1 Experiment Overview

The Okanyo Tunnel (L = 2,736 m; cross-sectional area: 62.7 m^2) in Iwate prefecture of the northeastern Japan is a road tunnel excavated by the NATM. The geology in this area is characterized relatively homogeneous Granodiorite, which is formed in Cretaceous. We used this site to perform an experiment to evaluate the properties of developed measurement system. We furthermore compared measurement results with seismic velocities measured from the surface by the refraction method. Figure 187.4 shows a geological section of the interval used for the verification experiment. Seismic velocities measured by the refraction method in the interval were expected to be above 3.75 km/sec.

187.2.2 Geological Properties of Cutting Advance Interval

187.2.2.1 Measurement Method

The measurement position, which is during excavation of the examined tunnel were situated within the tunnel approximately 70–120 m behind the tunnel face. A geophone was fixed to an existing rock bolt on the tunnel wall. Each blast excavated 1.0–1.5 m of rock, and measurement sets were established until the length of the face advance interval reached approximately 50 m, at which time the measurement system was moved along with the measurement position in the direction of face advance and measurement was resumed. P wave of first blasting is measured, and measurements of 14 sets are performed over 900 m span.

187.2.2.2 Measurement Results

Figure 187.5 shows the measurement results (travel time curves) for 14 measurement sets performed during excavation over the interval TD 459.9–1,287.5 m, plotting the time required for seismic from each blast to reach the geophone. Geophone positioning for each measurement is shown as

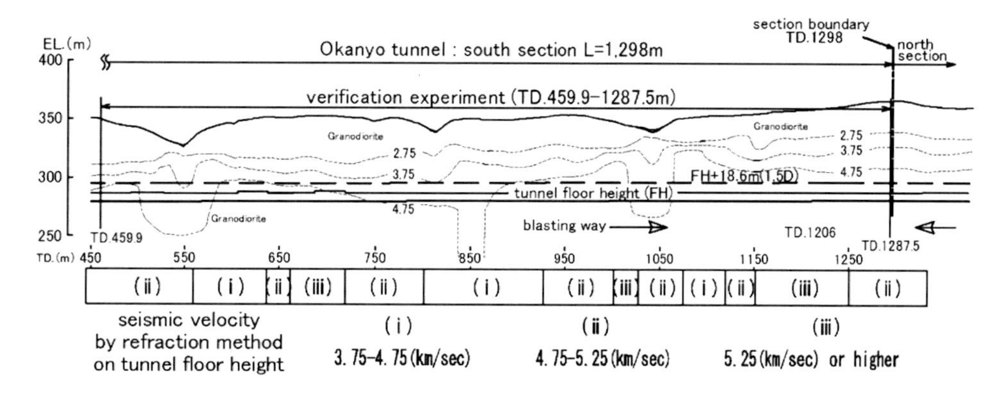

Fig. 187.4 Geological profile of Okanyo tunnel

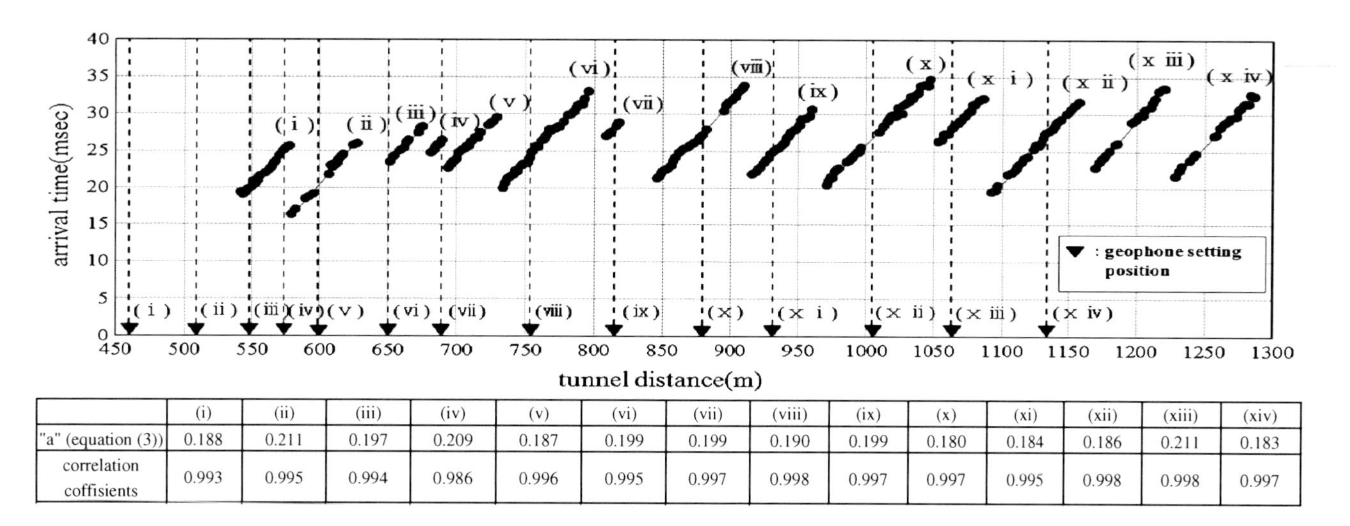

	(i)	(ii)	(iii)	(iv)	(v)	(vi)	(vii)	(viii)	(ix)	(x)	(xi)	(xii)	(xiii)	(xiv)
"a" (equation (3))	0.188	0.211	0.197	0.209	0.187	0.199	0.199	0.190	0.199	0.180	0.184	0.186	0.211	0.183
correlation coffisients	0.993	0.995	0.994	0.986	0.996	0.995	0.997	0.998	0.997	0.997	0.995	0.998	0.998	0.997

Fig. 187.5 Travel time curve in Okanyo tunnel

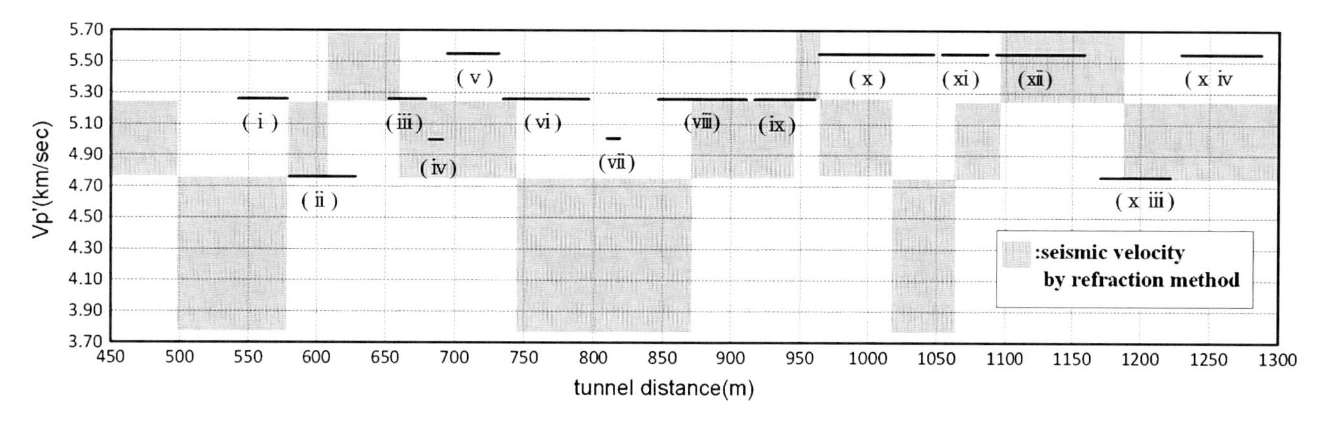

Fig. 187.6 Interval seismic velocity

dotted lines in the figure. These results indicate that obtained measurement data have correlation coefficients of $r = 0.986$–0.998, indicating that exploration was performed with high precision and low variance. And V_p' is calculable from Eq. (187.4).

Figure 187.6 shows V_p' values for the face advance interval as calculated by Eq. (187.4), indicating that the proposed method allows detailed changes in geological properties to be ascertained.

For comparison of refraction method V_p and V_p', Fig. 187.7 shows a reorganization of the data. The figure

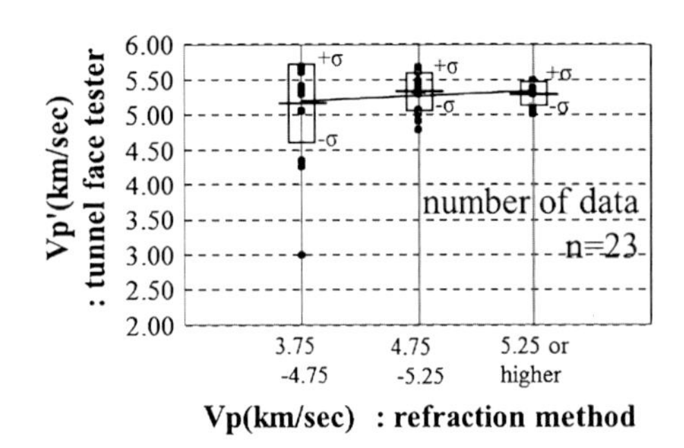

Fig. 187.7 Reorganization for Vp and Vp′

shows extraction method V_p divided into three groups, 3.75–4.75 km/s, 4.75–5.25 km/s, and 5.25 km/s or higher, and V_p' is plotted correspond *to* V_p of same interval. For refraction method V_p of 5.25 km/s or higher, V_p' values were 5.01–5.49 km/s, approximately the same values with a small variation range of 0.5 km/s. For refraction method V_p of 3.75–4.75 km/s, V_p' values were approximately the same, 3.00–5.67 km/s, but slightly faster. These results are likely averaged values over a somewhat wide-ranging area of seismic exploration by the refraction method, indicating the utility of V_p' for determining detailed changes in geological properties.

187.2.3 Discussion

The low variance in the results obtained from the proposed system indicates its potential for precise exploration. Capturing geological properties of cutting advance sections via V_p' is furthermore found to allow more detailed evaluation of geological properties than exploration from the surface.

187.3 Conclusion

We have developed a system for geological property evaluation of tunnel faces that utilizes excavation blasts. The correlation coefficient for travel time data obtained was approximately $r = 0.986$–0.998, indicating that measurements were performed with high condition.

Continuous exploration using the proposed system during excavation should allow calculation of V_p'. It is furthermore effective for obtaining a detailed understanding of changes in geological properties.

Further topics for research include that proposed system can be use seismic reflection method, geological condition ahead the face can be predicted. So we will continue verification experiments to further improve the proposed system.

Reference

Nakaya M, Yamamoto H, Utsuki S, Onuma K, Suzuki M (2013) Development and illustration of system which evaluating geological condition around tunnel cutting face by the excavation blasting(in Japanese). In: Proceeding Tunnel Technology Conference of JSCE, vol 23, pp 209–215

Multidisciplinary Methodology Used to Detect and Evaluate the Occurrence of Methane During Tunnel Design and Excavation: An Example from Calabria (Southern Italy)

188

S. Lombardi, S. Bigi, S. Serangeli, M.C. Tartarello, L. Ruggiero, S.E. Beaubien, P. Sacco, and D. De Angelis

Abstract

The occurrence of high volumes of methane during tunnelling operations is a critical safety factor that can influence the choice of different technical approaches for tunnel design and construction. Moreover, gas accumulation can be influenced by fluid migration along spatially focused preferential pathways (i.e. points along faults and fracture zones) that can result in highly variable gas concentrations along the tunnel trace. This paper proposes a methodological approach to minimise the risks, and costs, related to tunnel construction in rocks with potentially high methane concentrations. This approach combines soil gas geochemistry and structural geology surveys along and across the main faults and fracture systems that occur in the study area. The procedure is based on near-surface sampling and consists of a two-pronged approach: the measurement of fault zone gas emissions and their classification as barrier or conduit zones. Moreover, it is illustrated the importance of measuring a wide spectrum of different gas species, not just methane, for a more accurate interpretation of the geological, geochemical, and structural system. This is due to the potential for multiple gas origins, different gas associations, and various alteration and oxidation processes (e.g., CH_4 oxidation into CO_2) that can modify the geochemical signal along the flow path as gas migrates towards the surface.

Keywords

Soil gas • Faults • Tunnel • Methane

188.1 Introduction

The discovery of significant volumes of methane during the preliminary phases of tunnel design increases the risk of a potential explosion during or after the excavation work. From a strictly design point of view, the excavation of tunnels in areas characterized by documented gas occurrences can be problematic. This is due to perturbation of the rock mass and gas distribution caused by the excavation itself, as tunneling can result in significant changes to the stress field and the pre-existing hydrological and hydraulic equilibrium.

This problem has generally been dealt with during an advanced phase of the project, when the most important choices have already been taken, or directly during the course of the work, imposing the adoption of specific techniques to detect the presence of gas during excavation. In the case of the discovery of gas during the early stages of drilling or excavation it is necessary to adjust, as a consequence, the procedures and digging equipment.

The methodological approach presented in this paper was developed, via a close collaboration between ANAS and the CERI research institute, to address and overcome these problems. This methodology, tested in a geological setting

S. Lombardi · S. Bigi (✉) · M.C. Tartarello · L. Ruggiero · S.E. Beaubien · P. Sacco · D. De Angelis
Dipartimento di Scienze della Terra, Università Sapienza di Roma, P.le A. Moro, 5-00183, Rome, Italy
e-mail: sabina.bigi@uniroma1.it

M.C. Tartarello
e-mail: mariachiara.tartarello@uniroma1.it

S. Serangeli
Anas SpA, Via L. Pianciani, 16, 00185, Rome, Italy

G. Lollino et al. (eds.), *Engineering Geology for Society and Territory – Volume 6*,
DOI: 10.1007/978-3-319-09060-3_188, © Springer International Publishing Switzerland 2015

that is potentially favorable to natural gas accumulation, is based on the identification of the geological conditions that can control and/or contribute to the genesis and migration of the gases.

Starting from the study of the spatial distribution of soil gas anomalies, measured over the zone that will be crossed by a proposed road tunnel, it is possible to develop the tunnel design as a function of gas emission distribution and geological conditions. Adopting this methodology during an early stage of the project, and in a targeted manner, can aid in a more correct prediction of the costs and timeline of the road infrastructure construction.

188.2 Methodology

Near surface gas geochemistry surveys include the measurement of gas concentrations in the shallow soil ("soil gas"), deep gases in boreholes, and the flux of gases from the soil to the atmosphere ("gas flux"). These measurements provide useful information on the geological processes that control the production and migration of gas. Over the years, this methodology has been refined and applied in many different research fields, addressing various environmental, geological, and engineering issues (Annunziatellis et al. 2008; Beaubien et al. 2013 among many others). The structural setting of the area and the classification of the fault deformation characteristics, integrated with the gas concentration distribution in the soil, are essential elements for the reconstruction of a model of fluid migration towards the surface in a given area. Fault and fracture zones are known to be belts of enhanced permeability which can readily transmit fluids, including gases, to the surface. Field data have shown the usefulness of the soil gas method for detecting faults and discontinuities even when faults are buried or cut non cohesive clastic rocks, conditions which make it difficult to recognize their surface expression by traditional means (Lombardi et al. 1996; Ciotoli et al. 1998).

In the case presented here, located in the Ionian part of Calabria, soil gas surveys were conducted along a 1 km wide strip ("buffer") that follows the planned trace of a future road. The obtained values were elaborated using spatial analysis statistical techniques to define threshold values and to localize the main anomalies for each gas species.

188.3 Geological Setting

The study area belongs to the southern Apennine orogenic system; the trace of the future road crosses the easternmost tectonic units belonging to the chain and the more eastern foredeep deposits. The stratigraphy in this area consists of: (a) the Albidona and Saraceno Formations of the Liguride and Sicilide units, which consist of marls, turbiditic sandstones and thick micro polygenic conglomerates, and of calcareous mudstones, grainstone and clays, respectively; (b) the Varicolori Clays Formation, which is tectonically placed on the previous units and which consists of multicolored, over-consolidated clays containing olistoliths of limestone and interbedded mudstone levels; (c) Plio-Pleistocene marine sands and clays; and (d) alluvial deposits. The Sicilidi and Liguridi units, as well as the syn-compressional Oligo-Miocene deposits, were involved in the Apennine orogenic compression from the Upper Cretaceous to the Lower Miocene. The Plio-Pleistocene succession and the carbonate bedrock are still involved in compressional deformation, mainly by left lateral strike slip faults connected to the southward migration of the Calabrian Arc (Monaco et al. 2001). Along the coast, several orders of terraced Quaternary marine deposits (up to 7) are considered evidence of a rapid uplift, due to the activity of southeastern dipping normal faults (Ferranti et al. 2009; Cucci 2004).

188.3.1 Structural Analysis and Soil Gas Survey

A structural analysis of deformation styles has been carried out in the investigated area. The main thrust cropping out in the northern sector is the Argille Varicolori thrust, which juxtaposes this Unit onto the already deformed Cretaceous—Paleogene formations. This hanging wall unit has a chaotic organization with numerous thrust planes within the same Varicolori formation. All these thrust planes, including the basal one, consist of strongly foliated clay-rich deformation bands. The reverse faults and small thrusts that crop out in the area, involving the Albidona and Saraceno Formations, are instead characterized by a brittle and localized deformation, associated with wide, metre-scale fracture zones, succession repetition, and series thickening. The last episode of compressional deformation in the area was due to NW-SE trending strike-slip faults that cut and dissect the previous faults and folds. In many cases strike slip deformation is developed on the limb of a previous fold, along vertical bedding, which can create open fractures that provide a pathway for vertical fluid migration.

In a second phase a soil gas survey was performed along the path of the future road, from Trebisacce in Roseto Capo Spulico, to localize and quantify soil gas anomalies. These anomalies can potentially be correlated with the presence of permeable faults and/or highly fractured zones. In the case of methane, anomalies may also be correlated with the occurrence of specific lithologies or geological settings that are able to generate this gas.

The soil gas survey presented here involved the collection of 422 samples; the analysis has been focused on three gas species (He, CO2 and CH4) due to their geochemical

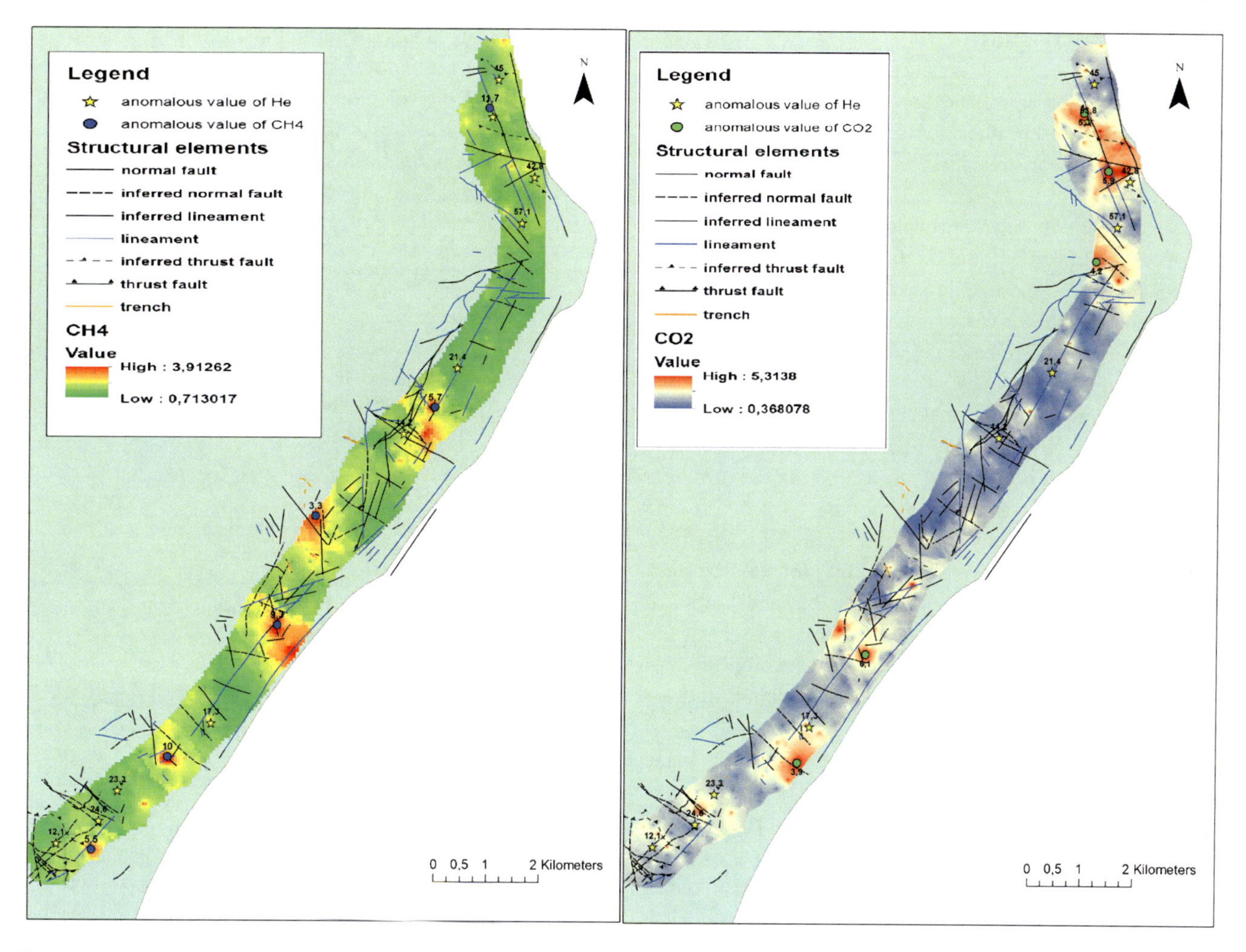

Fig. 188.1 Soil gas concentrations: CO2 (on the *right*) and CH4 (on the *left*)

characteristics, both for discontinuity detection and for the goals of the research itself (i.e., tunnel construction).

Carbon dioxide has a double importance, as it can act as a carrier gas of trace gases (when it is the dominant species) and can be formed by methane oxidation. Helium is recognized as a good tracer of deep structures (e.g. Oliver et al. 1984), because it is a highly mobile and inert gas and thus it can easily migrate through the underground with little attenuation. Finally, CH4 is obviously the focus of this study; it too is highly mobile because of its small molecular size, however it is more soluble than He, can be formed by a number of biogenic and geogenic processes, and can be consumed along the flow path via microbially mediated oxidation processes.

Soil gas contour maps were created for each gas species based on the defined threshold values, with results shown in Fig. 188.1 for CH_4 (left) and CO_2 (right). Each of them highlights areas where an increased circulation of fluids is present. Points of anomalous He concentrations (i.e. higher than atmospheric) have been added to the maps of the other gases. The methane concentration map shows large anomalous zones in the central-southern part of the area, while in the north, around Roseto, the anomalies are very localized (Fig. 188.1).

The same trend can be observed in the CO2 concentration map, where maximum values (c. 6 %) were measured in the central-southern part. Instead, in the north near Roseto and Roseto Castle, an area of anomalies was found in correspondence with the previously described Argille Varicolori thrust. Helium anoma- lies correspond to CO2 anomalies in the north, confirming the occurrence of deep fluid circulation, whereas in the southern sector the anomalous values are aligned along the coast in correspondence with the normal fault system responsible for the deformation of the Pliocene marine terraces.

188.4 Discussion

Based on the analysis of the described dataset, integrated with geological observations, it is possible to describe different modes of gas migration in different sectors of the study area.

The main correlation is between gas concentrations and the outcropping lithologies: the highest anomaly values are present where the Albidona and Saraceno Formations crop out (as in the area of Trebisacce), especially for CO2 and He. In the northern sector, in the area of the Plio—Pleistocene deposits and Quaternary terraces, the values are under the threshold, except for along the valleys where anomalous values are located, as in the case of the Straface and Ferro rivers.

The different concentrations measured in the different geological settings suggest that the gas generation and migration is controlled by two main factors: the structural control of faults and fractures and the lithological and morphological control. In the first case, where the sandstones and clays of the Albidona and Saraceno Formations crop out (deformed by faults associated with densely fractured bands) it is hypothesized that migration is mainly controlled by fractures associated with faults, acting as migration pathways that may feed the accumulation of methane at depth. In this case the soil gas composition is continuously fed from depth, as can be deduced by the high concentration values measured there.

Low permeability Pliocene clays and Quaternary marine terraces outcrop between the Avena and Ferro rivers. Locally, in the Pliocene clay, strata of unconsolidated, highly porous sandstones can constitute small methane reservoirs, even under pressure. Moreover, the Pliocene clays are suspected to be sites of biogenic methane production themselves, as largely documented by studies into petroleum generation in the area (Casero 2004).

The lack of soil gas anomalies in this area can be interpreted as due to the barrier effect of these rocks, which prevent communication between the surface and the fracture systems at depth. This lithological control is supported by the fact that the gas anomalies are found in the valleys, where the Quaternary terraces are interrupted and the Pliocene clays crop out.

In conclusion, this study has shown that integrated soil gas and structural geology surveys are a powerful combination for the study of gas migration and associated pathways, and can be a useful tool during the early phases of design and construction of road works involving tunneling in areas suspected of having potentially dangerous methane accumulations.

References

Annunziatellis A, Beaubien SE, Bigi S, Ciotoli G, Coltella M, Lombardi S (2008) Gas migration along fault systems and through the vadose zone in the latera caldera (central Italy): implication for CO2 geological storage. Int J Greenhouse Gas Control 2(3):353–372. http://dx.doi.org/10.1016/j.ijggc.2008.02.003

Beaubien SE, Jones DG, Gal F, Barkwith AKAP, Braibant G, Baubron JC, Ciotoli G, Graziani S, Lister TR, Lombardi S, Michel K, Quattrocchi F, Strutt MH (2013) Monitoring of near-surface gas geochemistry at the Weyburn, Canada, C2-EOR site, 2001–2011. Int J Greenhouse Gas Control 16(Supplement 1):S236–S262. http://dx.doi.org/10.1016/j.ijggc.2013.01.013

Casero P (2004) Structural setting of petroleum exploration plays in Italy. Spec Volume Italian Geol Soc IGC 32:189–199

Catalano S, Monaco C, Tortorici L, Paltrinieri W, Steel N (2004) Neogene-quaternary tectonic evolution of the southern Apennines. Tectonics 23(2), TC2003. doi:10.1029/2003TC001512

Cucci L (2004) Raised marine terraces in the northern Calabrian arc (southern Italy): a ~600 kyr-long geological record of regional uplift. Ann Geophys 47(4). doi: 10.4401/ag-3350

Ferranti L, Santoro E, Mazzella ME, Monaco C, Morelli D (2009) Active transpression in the northern Calabria Apennines, southern Italy. Tectonophysiscs 476(1):226–251. http://dx.doi.org/10.1016/j.tecto.2008.11.010

Lombardi S, Etiope G, Guerra M, Ciotoli G, Graubger P, Duddridge GA, Gera F, Chiantore V, Pensieri R, Grindrop P, Impey M (1996) The refinement of soil gas analysis as a geological investigative technique. Commission of the European Communities, nuclear science and technology. Final report, EUR 16926 EN

Monaco C, Tortorici L, Catalano S, Paltrinieri W, Steel N (2001) The role of Pleistocene stike-slip tectonics in the neogene-quaternary evolution of the southern Apennines orogenic belt: implication for oil trap development. J Petroleum Geol 24:339–359. doi:10.1111/j.1747-5457.2001.tb00678.x

Oliver BM, Bradley JG, Farra H (1984) Helium concentration in the earth's lower atmosphere. Geochimica et Cosmochimica Acta 48:1759–1469. http://dx.doi.org/10.1016/0016-7037(84)90030-9

Combined Geophysical Survey at the A2 Tunnel Maastricht

189

O. Brenner and D. Orlowsky

Abstract

In advance of the construction of the A2 tunnel in Maastricht a geophysical survey was necessary in order to investigate the position of layers, faults, flint layers, occurrence of rock deposits and potential cavities within the limestone layer. For this purpose three different geophysical methods were combined: CMP refraction seismic, refraction tomography, reflection seismic. The whole survey was performed with three different sources to get the best results for each depth: air sound source, sledgehammer and accelerated weight drop source. After processing and during the interpretation process the results of 18 boreholes were included. Finally all relevant layers could have been detected and corresponded to the results of drilling. Furthermore minor and major faulting in the survey area could be detected. Potential cavities in the limestone were not detected.

Keywords

CMP refraction seismic • Refraction tomography • Reflection seismics • Tunnel construction • A2 Maastricht

189.1 Introduction and Background

The city of Maastricht, The Netherlands, is split by the national motorway A2 into two parts (Fig. 189.1). The A2 Maastricht project organisers developed a plan to reconnect both parts and to reduce air and noise pollution by tunnelling the A2. Cars, lorries and motorcycles shall disappear under the ground. Above the tunnel a green park avenue will be set up.

The Consortium Avenue 2 planned to develop the underground for the construction of the A2-tunnel and to redevelop the adjacent areas along the A2. The purpose of the geophysical investigation was to provide information about the geological condition of the substratum. In detail, the following objectives were specified:

- Profile variation of the layer boundary between gravel and limestone,
- location, pattern and vertical separation of the geological fault in the Maastrichtian and Campanian,
- any occurrence of flint layers and
- any occurrence of rock deposits within the gravel layer.

Furthermore, potential cavities within the limestone should be detected.

For the investigation of the underground the following geophysical methods were applied:

- CMP seismic refraction
- Refraction tomography
- Seismic reflection

For these methods three profile locations have been specified by the client. The total length of the survey line was 2 km. 200 active single geophone stations were used with a geophone spacing of 1 m. The distance between

O. Brenner (✉) · D. Orlowsky
DMT GmbH & Co. KG, Essen, Germany
e-mail: olaf.brenner@dmt.de

D. Orlowsky
e-mail: dirk.orlowsky@dmt.de

G. Lollino et al. (eds.), *Engineering Geology for Society and Territory – Volume 6*,
DOI: 10.1007/978-3-319-09060-3_189, © Springer International Publishing Switzerland 2015

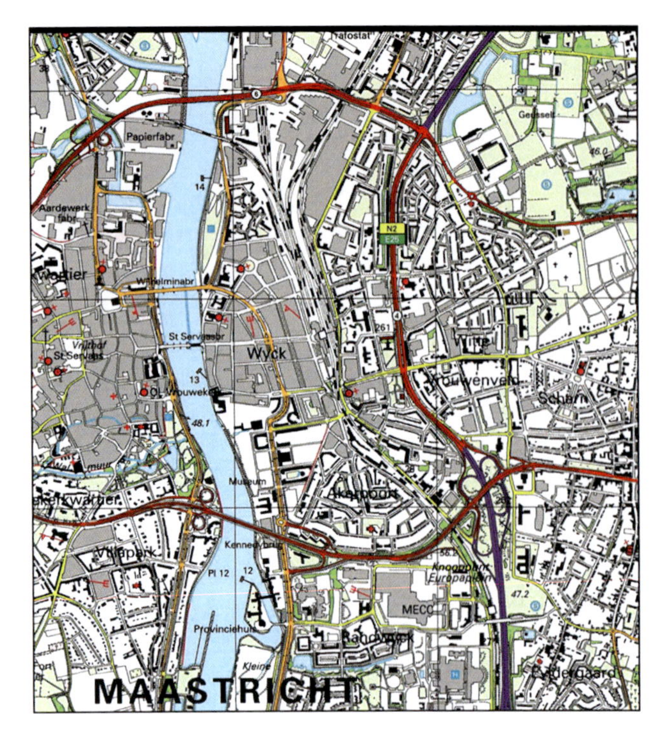

Fig. 189.1 A2 in Maastricht

source locations was 2 m. This layout was selected, so that the ground conditions down to a depth of 50 m below ground level can be determined. The seismic energy was generated by three different sources:

DMT started with a pre-evaluation in 2009 and continued with the final survey in 2010. Both seismic investigation campaigns were carried out during night sessions to minimize interfering noise generated by traffic. Occasionally, the traffic on the A2 was interrupted for the single measurements.

189.2 Methods

189.2.1 General

At the beginning of the survey a source test with three different sources was performed:

- Accelerated weight drop "Mjoelnir"
- Sledgehammer
- Air sound source

This test was necessary to find the best source for different geophysical methods (reflection, CMP refraction and refraction tomography) (Fig. 189.2). Finally for the interpretation the results of the weight drop and the air sound source were used.

The refracted wave could be best detected in the data of the weight drop source. Although in the raw data the air sound source shows clearly a reflector and higher frequencies for higher resolution the stack of the weight drop data gives more information and reflections over the complete section especially for increasing depth, see Fig. 189.3. The data of the sledgehammer gave only poor results on reflections and for far offsets.

189.2.2 Reflection Seismics

Reflection seismic processing was applied to all datasets of the weight drop, sledgehammer and the air sound source. The data of the air sound source contain higher frequencies but also the natural air sound was dominating the results. Compared to the air sound data the results of the weight drop source contain stronger reflections but lower frequencies (Fig. 189.3). The sledgehammer didn't generate reflections in the stacked section.

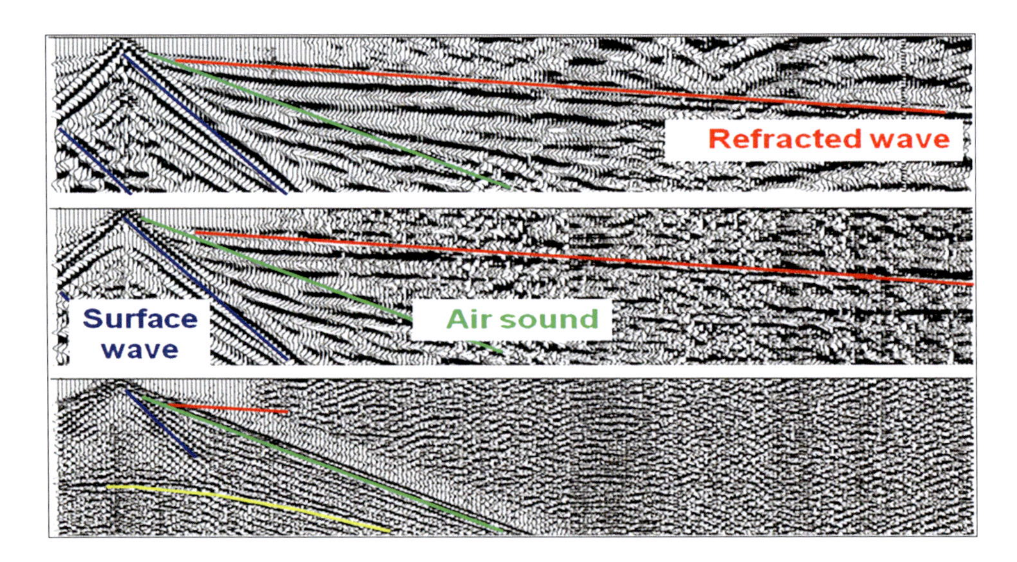

Fig. 189.2 Raw data of the different tested sources *top* accelerated weight drop, *middle* sledgehammer, *bottom* air sound

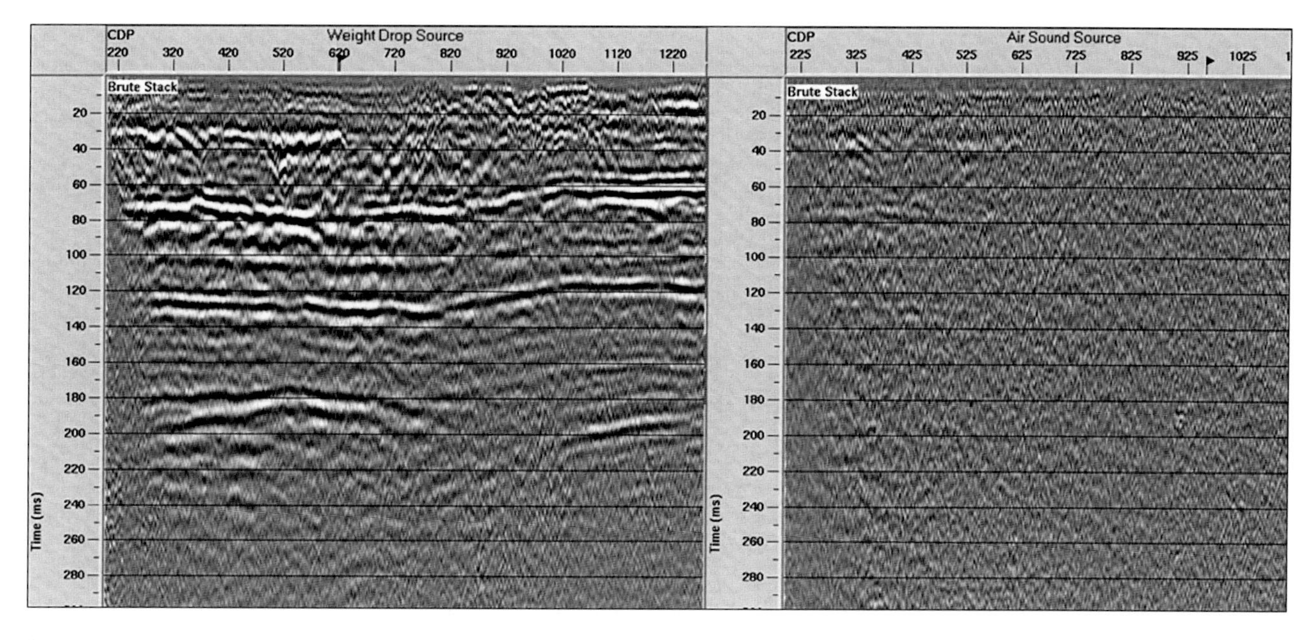

Fig. 189.3 Comparison of Brute stacks for weight drop (*left*) and air sound source (*right*)

The first clear reflector could be detected at a depth of 15 m. Overall 4 reflectors could be determined and correlated to the results of boreholes and known geological structure.

189.2.3 CMP Refraction Seismics

Common-Midpoint (CMP) seismic refraction is a method to improve the signal/noise ratio of the refracted waves. Applying this technique, the shallow underground is described using all information (amplitude, frequency, phase characteristics) of the wave train, following the first break (first-break phase). Thus, the layering can be determined and, additionally, locations of disturbances such as faults, weak zones, and clefts can be detected.

The results of CMP seismic refraction data processing are merged with the seismic reflection data. The refractors are displayed with the complete refracted wave fields and the according depths. Disturbances and anomalies (e.g. cavities) within the refracted waves can be detected. Figure 189.4 shows the difference in the upper 30 m by combining the datasets. With the CMP seismic refraction in the upper part the underground image is more clearly and better for interpretation in this part.

For interpretation all lines were merged to one long line. The merging was done in an interactive way using the refractor depth in a range of 10–30 m with a region of overlapping. For the merging of the lines to one single line a triangle taper function was used to get a smooth result in the overlapping regions of the lines.

189.2.4 Refraction Tomography

Refraction Tomography is a method to produce a model of the distribution of seismic propagation velocity in the shallow underground. It utilizes the first arrivals (first breaks) of seismic wave energy. The method is restricted to shallow depths and a positive velocity gradient (higher velocities in greater depth) is generally assumed.

Depending on the actual velocity distribution a penetration depth of approx. 1/6th through 1/4th of the maximum source-receiver offset can be obtained. For the current survey, a maximum depth of 20 m through 30 m below surface was covered.

Figure 189.5 shows one example result of the refraction tomography in the middle of the survey area. The velocity distribution increases with depth, different layers can be detected. Furthermore areas with low velocities in depth can be detected. Depending on the combined interpretation these variations can be interpreted as weak zones, faults or extension of the weathering zone.

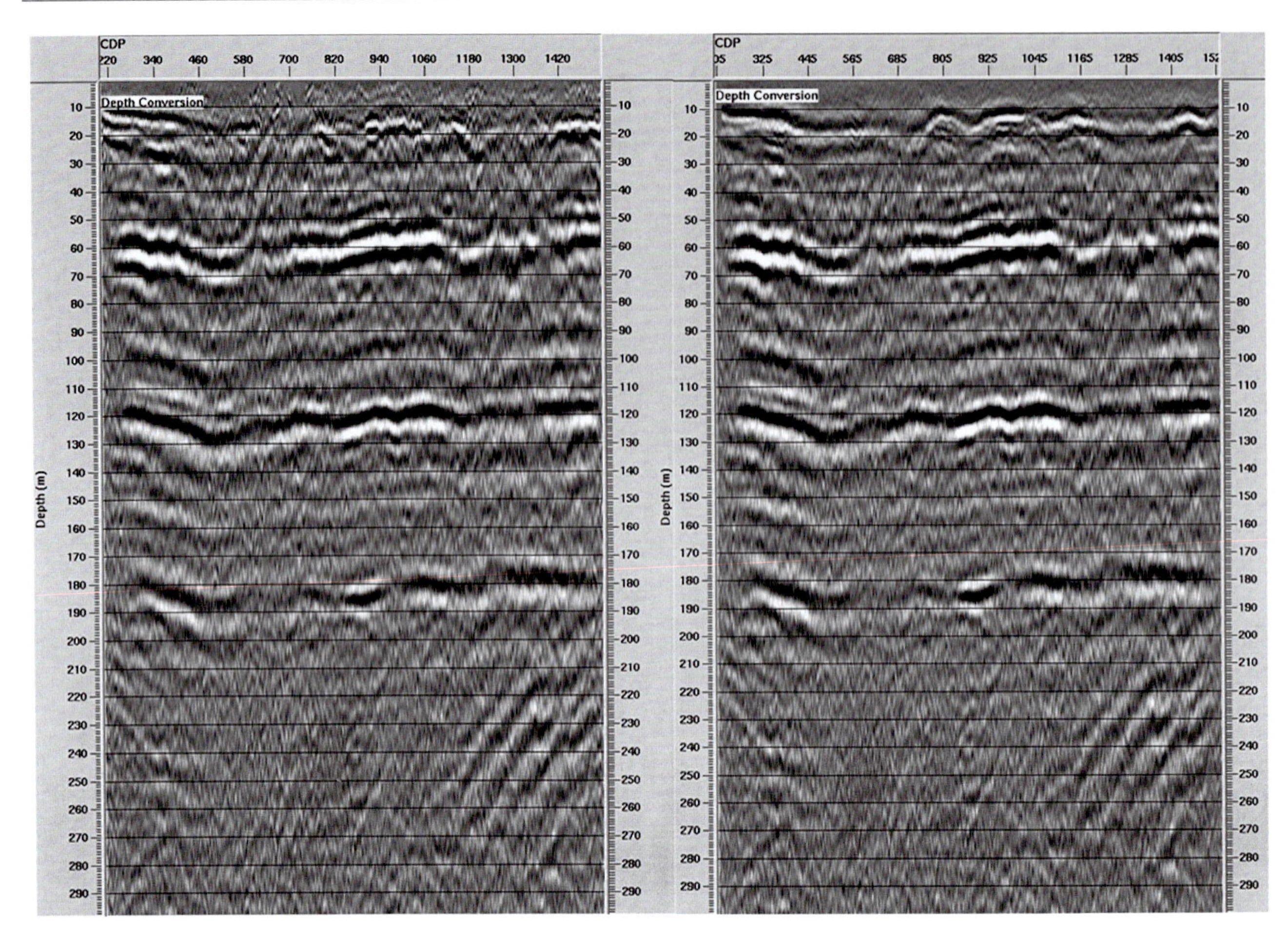

Fig. 189.4 *Left* result from seismic reflection processing, *right* seismic reflection combined with CMP-Refraction processed data

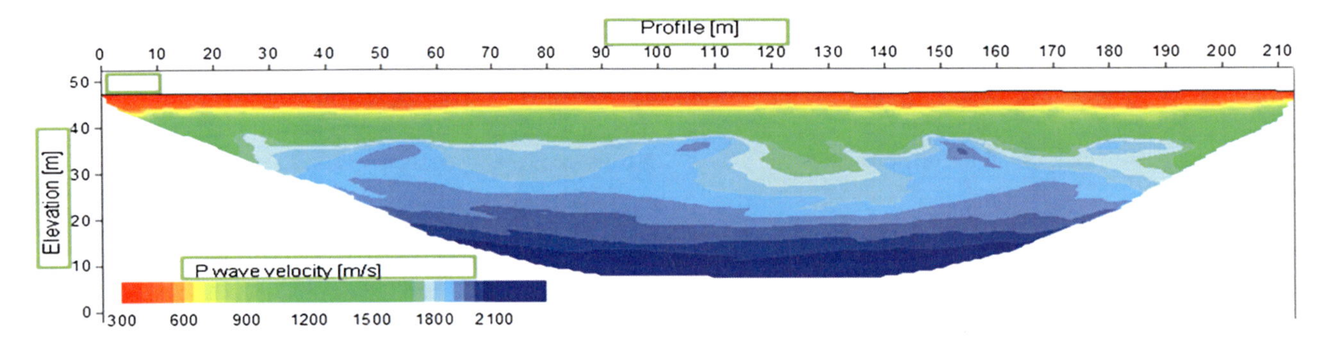

Fig. 189.5 Result of refraction tomography (example)

189.3 Joint Interpretation

After processing of each individual method a joint interpretation by a geologist was performed with the 3D interpretation and modelling software PETREL 2010 (Schlumberger Information Solutions). Overall 18 boreholes were integrated in the final results used to correlate the various layers visible in the seismic refraction and reflection data with the geological boundaries.

The top of the limestone layer including variations could be interpreted and in the middle of the profile several faults and one flintstone layer were detected (Fig. 189.6).

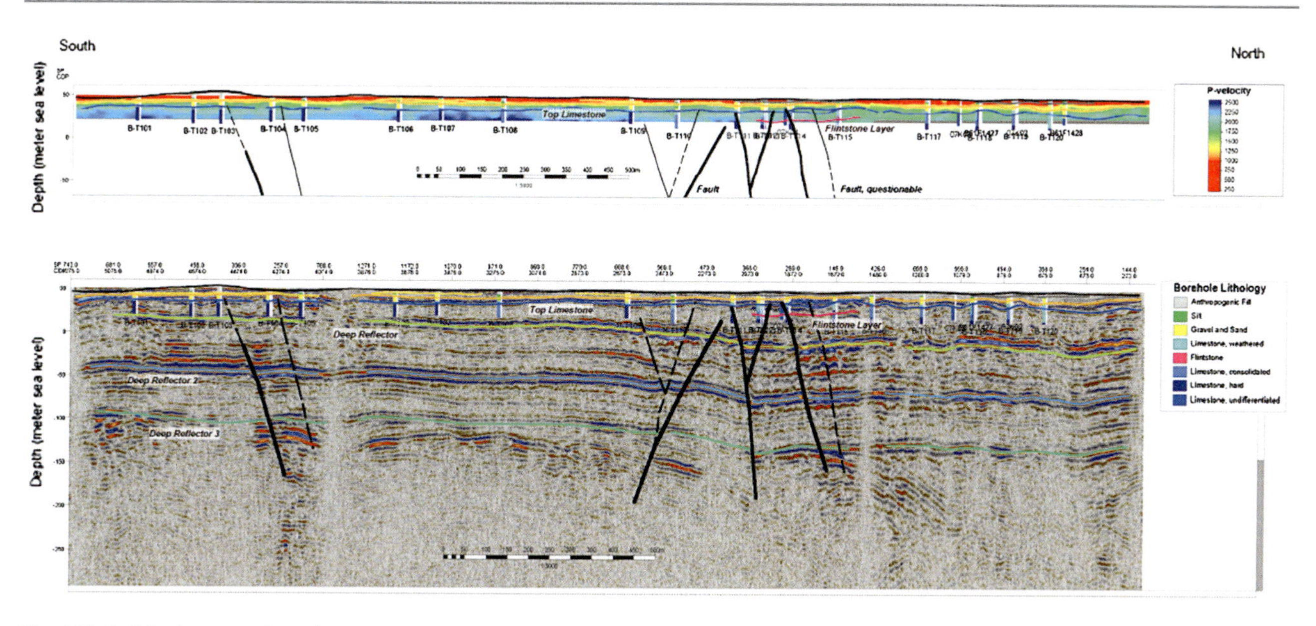

Fig. 189.6 Joint interpretation of geophysical data with integration of borehole results. *Top* refraction tomography, *bottom* combination of cmp refraction and reflection seismic

189.4 Conclusions

The combination of different geophysical processing methods and different kind of sources was very successful for getting a good image of the underground and to achieve all targets of the survey. Two refractors could be detected. The upper refractor correlates to the boundary between the anthropogenetic fine grained fill and the Pleistocene gravels. The second refractor is close to the hangingwall boundary of the weathered limestones of the Maastrichtian. Also seismic reflectors could be determined and correlated with the main lithostratigraphic units. The data also reveals that intensive faulting may be present. Velocity variations derived from refraction tomography analysis indicate a potential high degree of weathering of the ground along the Northern part of the survey area.

Potential cavities in the limestone were not detected.

The Combined Use of Different Near Surface Geophysics Techniques and Geotechnical Analysis in Two Case Histories for the Advanced Design of Underground Works in Urban Environment: Rome Metro B and Torino-Ceres Railway

190

Riccardo Enrione, Simone Cocchi, and Mario Naldi

Abstract

In the urban context of the Italian towns, both historical remnants (archaeological items, man-made cavities, buried walls, etc.) and underground geotechnical "anomalies" (cemented layers, pockets of low-density soils, presence of big boulders, etc.) can strongly increase the hazard risk associated to underground excavation in urban context (as collapse, surface subsidence, interference) or inappropriate choice of the excavation method. This consciousness has led the writers to work out a site-specific multi-investigation method, hereinafter explained through a couple of case histories. The first one considers the reconstruction of a man-made cavities map distribution along the anticipated alignment of the extension of Rome Metro B Line, with the use of GeoRadar and geoelectrical methods, calibrated through boreholes. The second case history is related to the recognition of poor-quality soils and possible cemented layers along the alignment of a new urban railway line in Turin, using a MASW survey and a seismic refraction tomography survey, both coupled with geotechnical investigations. In both cases, the combined use of geophysical and geotechnical investigation has guaranteed the identification of the possible main geohazards, fundamental basis for an appropriate tunnelling design and assessment of the related risk.

Keywords

Geophysics • MASW • Geotechnical investigations • Geohazards • Risk

190.1 Introduction

Too often project main goal and time-related concerns may distract designer and client from the boundary value problems, potentially leading to an underestimation of the so-called minor issues that in the end may turn out in major ones. In this sense, the knowledge of the geological–technical context must be intended as a key of success of the Design phase and not as an annoying formal step only because required by contract specifications or rules. As a consequence, the definition of the Design Geological and Geotechnical Model (DGGM) is the first and inevitable design goal, by which it is possible to achieve high level of sensibility about the underground space and to propose mitigation measures for the risks associated, both during the construction phase and during exploitation. Actually, geological uncertainties and the resulting risks in construction of great civil infrastructures are well understood. However, the use of a step-wise investigation approach in reducing these risks not always receives the deserved emphasis. In particular, a well-planned preliminary investigation can identify the possible hazards and thus deliver a high ratio of benefit to cost. In many designed works, which met severe problems during construction, it was found that often an inadequate model of ground conditions had been used, either because some geological features had been missed or overlooked during investigation or because its significance had been underestimated. This is especially true in case of

R. Enrione (✉) · S. Cocchi
Geodata Engineering, Corso Duca degli Abruzzi 48/e, 10129 Torino, Italy
e-mail: ren@geodata.it

M. Naldi
Techgea Srl, Via Modigliani 26/a, 10137 Torino, Italy

G. Lollino et al. (eds.), *Engineering Geology for Society and Territory – Volume 6*,
DOI: 10.1007/978-3-319-09060-3_190,

underground works in urban areas where the interaction between the geological and geotechnical context with the excavation is dramatically critical and therefore more detailed studies are necessary.

Besides, it must be noted that common geo-investigation methods usually deal with single survey technique at once, both for practical reasons and costs control. In our experience, this approach has turned out to achieve benefits only in the short-term, as immediate saving of time and money. On the other side, in the long-term, this is not a pay-back winning strategy, because of the uncertainties left behind and possible unidentified geohazards that could lead to additional cost and time consumption higher than early savings.

For this reason, a multi-approach investigation method has been worked out, refined in the course of time, on the account of several past experiences, with the aim to achieve the best compromise among all the above listed aspects. Current results show, in short, that coupling geotechnical investigations and geophysics surveys, conveniently integrated in terms of mutual placing in space and time allows gaining high satisfactory level of geological-technical knowledge and time/cost optimization.

The illustrated approach has been consolidated from the experience of many successful completed projects, and has become a standard practice, as well explained by the two following case histories.

190.2 Case A—Rome Metro B Extension

The case history A is related to the definitive design of underground Metro Line B extension, in Rome, between Rebibbia and Casal Monastero. The metro line extension has involved the realization of three new underground stations and the C&C excavation of a double tracks single tunnel, 3,000 m length, between diaphragm retaining walls.

190.2.1 Geological Settings and Geotechnical Investigations

From a geological-technical point of view, superficial deposit are represented by recent alluvial and anthropic soils, with heterogeneous grain size distributions and maximum thickness of 24 m, overlying pyroclastic deposits mainly composed by aches and lapilli of volcanic origin. Into all the pyroclastic sequence gas pockets (also radon), as well man-made cavities are possible; specifically, Roman Age underground pits for building materials extraction can be expected in the *Tufo Lionato* and *Pozzolana* units (Funiciello et al. 2005; Ventriglia 2002).

Beside this, presence of archaeological ruins must be taken into account when excavating in Rome subsoil; previous bibliographic information highlighted from the beginning the possible presence of Roman ruins near the alignment sector, such as old road network, underground services (ancient sewers), cemeteries and thermal baths (Funiciello 1995).

Due to these main anticipated geohazards, all the entire design process has been eventually driven by the associated knowledge and consequent management, conferring to the investigations-based DGGM a key-role in case of high-stakes choices.

Geotechnical investigations, which the DGGM has been based on, consisted in boreholes, CPTUs (Seismic Piezocone penetration test) and lab tests, assuring complete geotechnical characterization and punctual stratigraphical recognition; on the other side, spatial continuity was missing, as well the possibility to identify cavities or ancient remains if not directly intercepted by boreholes. This is where integration with geophysics came in handy.

190.2.2 Geophysical Survey and Discussion

Geophysical survey have been carried out with the aim to map the subsoil in terms of anomalies that might be led back to possible cavities or buried remains, using the previous boreholes and CPTUs as calibration network, thus avoiding false positives, on one hand, and assuring continuity in stratigraphical reconstruction, on the other one. Two techniques have been adopted: Ground Probing Radar (GPR), with different antennas (with frequency of 25, 100, 200 and 600 MHz), specifically oriented to ancient remains detection within the very first meters below ground level (4–5 m), and Electrical Resistivity Tomography (ERT), more oriented toward cavities identification. In order to best address the ERT survey, a preliminary calibration test was conducted in a similar area in presence of a known cavity (archaeological park in Rome). The test was performed with the aim of get the typical resistivity values of the void in the tuff materials (the so-called digital signature) in order to easily recognize similar anomalies along the Project alignment. The test pointed out a strong contrast between the resistivity value of the tuffs (ranging from 100 to 200 ohm.m), and the resistivity values of the test cavity (>800 ohm.m,).

Speaking of specifications, GPR survey was conducted on the account of a regular grid all along the metro anticipated alignment, using three different antennas (100, 200 and 600 MHz). The clayey shallow deposits have slightly conditioned the survey results (strong attenuation on the radar signal), but on the whole the survey was able to reveal a lot of structures down to a depth of 4–5 m with the lower frequency antenna (i.e. 100 MHz). All the founded structures (detected in two directions) have been mapped and cross check with known underground utilities or with opening of nearby gutters. After this comparison, all the underground structures have been mapped as underground utilities or underground tanks and no archaeological remnants have been detected. These results have been critical in defining the final boreholes location to avoid unwanted interference with underlying utilities.

ERT survey was carried out all along the future metro path, with a longitudinal line of 3 m-spaced electrodes and overlapping among consequential lines (roll-along technique). The obtained results have always revealed very low resistivity values (lower than 200 ohm.m.) that can't be referred to cavities presence. The only relevant "anomalies" have been identified in alluvial deposit, in terms of high contrast resistivity (80–100 ohm.m spot compared with the surrounding 10–30 ohm.m average). Again, the geophysical survey has proved to be unique for preliminary assessment of the specific areas to be directly investigated through boreholes; in this sense, the resistivity anomalies turned out to be soil pockets of very loose silty sand (SPT = 1) then correlated to piping phenomena in alluvial soil underlying the main gullies.

190.3 Case B—Torino-Ceres Railway

190.3.1 Geological Settings and Geotechnical Investigations

The case history B concerns the Definitive Design of a metropolitan underground rail junction between Turin Rebaudengo existing station and Caselle Airport located in the North city area. The Project involves a new 2.5 km long tunnel, single tube, double tracks, with a section area of 54 m^2, to be excavated in C&C between diaphragms hydromill-excavated. From the geological point of view, the Project area is underlain by glaciofluvial deposits of *Rissian* age (Quaternary), mainly composed by sandy gravels, cobbles and blocks, occasionally with clayey or silty sand layers.

Even if apparently plain, yet Turin subsoil can present frequent anomalies, as highlighted in past experiences such as *Metro Torino Line 1* or the *Passante Ferrioviario*, namely:

- the cementation of glaciofluvial deposits, and its variability, relevant to excavation technique choice and cutters wearing in case of TBM or hydromill excavations;
- the presence of big blocks (diameter >1 m), hence difficulties during excavation;
- presence of very loose sandy, supposedly affecting the stability of nearing existing building and road, as well the safety conditions of the site equipment, machinery and vehicles, during the excavation.

Considering the possible geohazards listed above, geotechnical investigation have been planned consequently; specifically, common boreholes have been coupled with DAC tests, that is the automatic measurement of drilling parameters to evaluate the mechanical properties of soils, and namely the level of cementation, to be correlated to the soil shear resistance.

Again, in order to better address the investigation plan, final location of tests have been derived from the geophysical results, as showed below, with the aim to limit the number of drillings and placing them right where most effective.

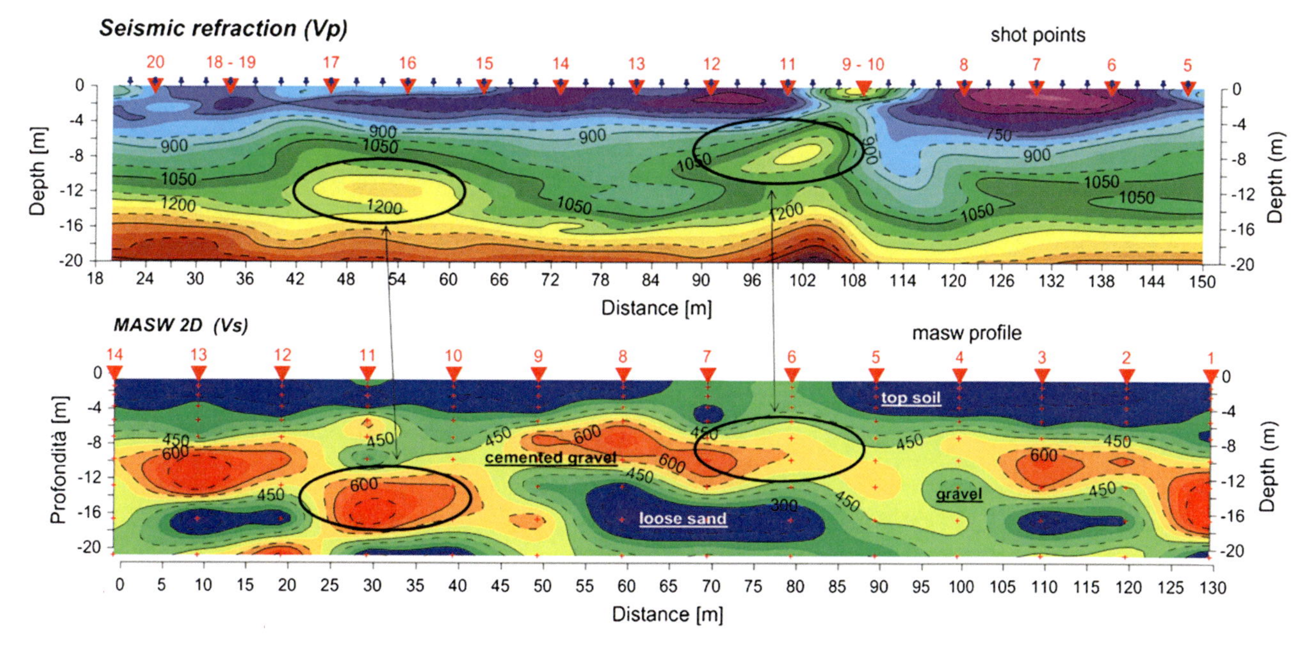

Fig. 190.1 Comparison between seismic refraction and MASW 2D section

190.3.2 Geophysical Survey

According to the objectives of the survey (subsoil profiling to locate cemented layers and loose sand pockets), two different geophysical methods have been applied: P-wave shallow seismic refraction and Multi-channel Analysis of Surface Waves (MASW) continuous profiling (2D section-shear wave velocities) (Park et al. 1999). The combined use of two different seismic methods provides both P-waves and S-waves velocities to estimate the main elastic properties of the soil and a cross-check to reduce the interpretative ambiguity for the DGGM. The results of the survey have provided a very detailed "image" of the subsoil, with some differences between refraction seismic and MASW 2D.

For the irregular distribution of low velocity and high velocity materials, the refraction seismic tomography method does not reveal correctly the geometry and morphology of the sedimentary lenses, while MASW2D methods provides a better resolution of the geometry of the alluvial sedimentary sequence (Fig. 190.1).

190.4 Conclusion

Owing to the anticipated subsoil peculiarities, neither single investigation tests nor separate geophysical surveys would have solved efficiently the risk associated to the possible geohazards affecting the designing of the urban underground works here described. For this reason, an integrated investigation survey has been worked out, coupling traditional geotechnical boring and tests with advanced geophysical surveys, compensating the reciprocal limits, in a complementary way: boreholes and CPTUs has reached the desired investigation depth assuring direct visual recognition and soil geotechnical testing; geophysical survey have fixed the gap between single site tests, which have act meanwhile as calibration point, improving the overall continuity and reliability of the DGGM.

This is how it works: reducing costs while increasing design quality and achieving final Project goals: a methodological approach that has become, for us, common practice.

References

Funiciello R (1995) La geologia di Roma. Il centro storico. Mem. Descr. della Carta Geol. d'Italia, vol. L - Servizio Geologico Nazionale, Roma, p 550

Funiciello R et al (2005) Carta Geologica di Roma alla scala 1:10000, vol. 1. Dipartimento Scienze Geologiche Università Roma Tre – Comune Di Roma – APAT

Park CB et al (1999) Multichannel analysis of surface waves. Geophysics 64(3)

Ventriglia U (2002) Geologia del territorio del Comune di Roma. Servizio Geologico, Difesa del Suolo, Amm. Provinciale Roma

Landslides Induced by Intense Rainfall and Human Interventions—Case Studies in Algeria

191

Ramdane Bahar, Omar Sadaoui, and Samir Sadaoui

Abstract

This paper describes two landslide case histories triggered by rainfall, unfavourable geological, hydrogeological and geotechnical conditions and human interventions. The first one is located about 250 kms South-West of Algiers, at place named Bordj Bounaama in Tissemsilt region. The landslides caused a collapse of a stadium and surrounding wall of the gymnasium and tilting of retaining wall under the earth pressure. The second landslide is one of the most spectacular among the 120 sites affected by landslides since 2000 in Béjaia department, located in Kabylia region. It was occurred at PK 226 + 300 of the National Road RN 24. The landslide caused the collapse of a section of the road of about 120 m, with a main scarp of about 8 m. During the firstly time of the field observation, the landslide mass and the road moved 10 m. Based on extensive geological, geotechnical and hydro-geological investigations, comprehensive of the landslide analyses were made using limit equilibrium and finite element method.

Keywords

Rainfall • Human interventions • Landslides

191.1 Introduction

Every year, particularly in the winter, mountainous and coastline region of Algeria, are affected by landslides displacing an important volume of detritic materials, and causing damages to infrastructures, housing and public facilities. Most disasters of mountain road damage are similar. The landslides take place in geological formations that are particularly favourable to this type of movement, because of the heterogeneity of their facies and the impermeability of some layers. Experience shows that the main cause of the landslides is the combination of several passive and active factors, as geology, morphology, hydrology, climate and anthropic activities. This paper describes two case histories of landslides triggered by rainfall, unfavourable geological, hydrogeological and geotechnical conditions and human interventions.

R. Bahar (✉)
LEEGO Laboratory, LGEA-UMMTO, University of Sciences and Technology Houari Boumediene, BP 32 El Alia, 16111 Algiers, Algeria
e-mail: rbahar@usthb.dz

O. Sadaoui
Department of Civil Engineering, University Mira Abderrahmane of Béjaia, 06000 Béjaia, Algeria

S. Sadaoui
Bureau de Conseils et d'Engineering, 06000, Béjaia, Algeria

191.2 Case Historie of Bordj Bounaâma Landslide

Bordj Bounaama town is in a broader context of the northern fringe of Tellian countrie, and it takes place in a physical-climatic set, consisting generally of a mountain range running parallel to the coast. Schematically the town is surrounded in a monoclinic area surrounded by mountains in all directions. The region is recorded a harsh winter with snow and heavy rains and a hot summer. According to the data, 90 days of frost, 62 days of rain and 10 days of snow are

G. Lollino et al. (eds.), *Engineering Geology for Society and Territory – Volume 6*,
DOI: 10.1007/978-3-319-09060-3_191, © Springer International Publishing Switzerland 2015

Fig. 191.1 Gymnasium view and earthworks

recorded per year. The site, approximately 11 ha in area, has a topographic line whose slope oriented in North-South direction. The largest slopes are found in the North West and South East part of the site, they are on the order of 15–20 %.

During the site investigations, two areas where landslides have already occurred are observed. The first one is located in the southern boundary of the gymnasium, which led to the degradation of the stadium and the surrounding wall (Fig. 191.1). According to the site investigation and local citizens, this landslide was caused by earth works made for a housing project located downstream of the gymnasium (Fig. 191.2). The terraced ground was left unprotected to the weather for two months, and was backfilled after landslide. The second landslide was observed upstream of the third block of 60 housing under construction, where the gabion retaining wall, tilted under the earth pressure (Fig. 191.3). Figure 191.3 shows also the presence of water in the excavations.

The geotechnical investigation consisted on core drilling and pressuremeter boreholes, dynamic penetration tests, seismic refraction profiles, profiles of electrical resistivity and piezometers. According to the logs of core drilling, the local lithology consists of a layer of colluviums of variable thickness overlying a shale layer. The shale layer is altered on surface and becoming hard in depth. The greatest thickness of colluviums was noted at the core drilling 3. It is about 18 m. Piezometer installed in the landslide area has detected water in January, 20th 2008 between 0.50 and 5 m from the ground surface. The soils are saturated. The natural water content w_n varies between 8 and 38 %. The dry density of the encountered soils varies between 14 and 22.6 kN/m^3. The Liquid limit varies between 41 and 63 % and the plasticity index between 18 and 31 %. The shear strength parameters derived from unconsolidated undrained triaxial tests range from 5° to 18° for the friction angle and from 29

Fig. 191.2 Landslide causing disorders to the stadium and surrounding wall

Fig. 191.3 Titling and moving retaining wall

to 182 kPa for the cohesion. A conventional limit pressure ranging from 500 to 4,000 kPa characterizes the soils.

The seismic refraction tests show four layers characterizing by different shear velocity. A surface layer has an average thickness ranging from 0.7 to 2.2 m. The second layer has a thickness between 3.4 and 8.1 m. The third layer has a thickness between 4.8 and 6.8 m. The fourth layer is encountered in average at 10 m from ground surface. These four layers are characterized by a shear wave velocity of 200, 410, 680 and 830 m/s respectively. The resistivity results are low ranging from 5.4 to 99.6 Ωm corresponding to highly conductive ground. The results show a layer about 3 to 8 m can be attributed to the sliding layer and resistivity measurements show a loose ground with water flow at depth.

For this purpose, two profiles, where the various factors of instability are present, were analyzed. In the first zone, the slope is already stabilized and has not changed since the first visit by the passive resistance of downstream buildings. Stability calculations are given for the critical slip circle a safety factor F = 1.295 found in the layer of colluviums and can reach 7.0 m deep. Although the safety factor is >1.2, the risk of instability cannot be ruled out. The slope created by the gabion wall was not taken into account in the calculations. The deformation of the wall reflects this instability. The second area includes the important disorders at the stadium. Stability calculations give a safety factor <1.2, the critical circle has a coefficient of 0.77.

The different results obtained by observations, insitu investigations and water level monitoring show that at the gymnasium and playground whose floors collapsed, the observed instability is due to the discharge of upstream surface water and downstream earthworks made to launch a project of collective housing. The landslide started at the gabion retaining wall can be explained by two factors: the design in the wall, the overload slope has not been considered and the hydraulic system has been modified; the infiltration of runoff through the gabions changed the mechanical properties under the wall and caused a flow of the foundation layer. Hydraulic parameter (runoff and groundwater) was the trigger. Anarchic earthworks, undersized gabion retaining wall and fragile layers of saturated colluviums also contributed to the destabilization of the slope.

191.3 Case Historie of Boulimat Landslide

The exceptional rains of winter 2012 triggered a series of landslides in the Kabylia region, among which the landslide occurred at PK 226 + 300 of the National Road RN 24, is one of the most spectacular (Fig. 191.4). The RN24 traverses the mountainous areas on the lower side. The landslide occurred just of 100 m coastline of the Mediterranean Sea on 17th April 2012 and caused the collapse of a section of the road of about 120 m, with a main scarp about 6–8 m (Fig. 191.4). After 18 h, the landslide mass and the road moved 10 m in the south-north direction. In this section, the road is clamped between the Mediterranean Sea and the mountain with very steep dip (Fig. 191.4). All drainage systems were damaged. The road is closed to traffic between Bejaia and Tizi-Ouzou cities for more ten days, disrupting road communication which is lifeline to the two states and many villages of Kabylia. The disruption experienced by local traffic was substantial. However, the real impacts of the events were economic and social. According to early preliminary investigation carried out on site, the perimeter of the landslide impact area is in order of 2.20 ha. The volume of material movement is estimated to 450,000 m^3. The potential slide surface is deep, located about 15–20 m in depth, which is the interface between the saturated colluviums and the weathered schistose marl. It is mainly caused by the circulation of groundwater drained by the catchments

Fig. 191.4 Assoumeth RN24 landslide, PK 226 + 300

area and the effect of marine transgression. The exceptional rainfall in the region of Béjaia during the month of April 2012 has amplified the phenomenon.

The records covering periods from 1923 to 2010 (ANRH 2011) show that the average rainfall along the cycles recorded is 831 mm. The last two decades are characterized by an average rainfall of 750 mm. The representation of rainfall is characterized by irregularities inter-seasonal and inter-annual. As a general, the relative distribution of 28.94 % from September to November, 39.96 % from December to February, 27.30 % from March to May and 3.8 % from June to August. Almost all rainfall is concentrated over a period ranging from November to March, not exceeding five (05) months: two-thirds (2/3) of the total rainfall is recorded during this period. In addition, according to the data, a peak rainfall was recorded in April with a maximum monthly rainfall of 278 mm, which justifies the initiation of the majority of landslides in the Kabylia region during the spring period. In April, the hillsides were in a saturated condition following a relatively wet spell during the preceding weeks.

According to the geological map of Bejaia, most of the formations encountered along the coastal road RN 24 are composed of flysch and old quaternary layers with dips varying from 30 to 60°. The affected section is formed of a syncline schistose marl of dip 42° S-N, it is topped with a thick layer of saturated colluviums, slipping at the interface with the marl under the effect of a steep slope, hydrostatic pressures and hydraulic gradients. This lithological facies widespread along the coastal region, it consists of a conglomerate of pebbles and blocks decimetric to metric dimensions, impregnated with a clay matrix which is very sensitive to the action of water. These layers of colluviums are very permeable and attract a large volume of water infiltrating the lower slopes as resurgences or sources. In general, training colluviums overcome dips direction (North-South) of schistose marl and very hard and impermeable marly limestone. The interface between the scree slope and marl consists of greenish clay and plastic very sensitive to water; it loses the mechanical characteristics and becomes slippery.

Based on site investigations, the slip surface was encountered at the very soft layer, 13–15 m deep. The emergency measures were to restore immediately traffic between East and Center regions and maintain the connection with the outside for neighboring village people. The temporary reinforcement solution possible technically and economically within an acceptable time is the partial substitution of the slipped ground by an appropriate compacted granular material. Then, a geotechnical investigation was conducted. Based on this investigation results, stability calculations were conducted for two stages; the state at landslide triggering and the state with nailing reinforcement taking into account the residual soil parameters considered for the moved mass (colluviums) are $c = 20$ kPa, $\varphi = 7$ °, $E = 10$ MPa and $\nu = 0.35$). The sliding mechanism are confirmed the observed on site. The results show the instability of the slope, the safety factor F is equal to 0.85. Large displacements, about 1.4 m, of the basin are noted. The results indicate clearly that the state remains unstable after sliding, hence, the importance of the proposed emergency works to stabilize the ground and the road. The classical methods give a safety factor $F < 1.0$. A final design of rehabilitation works was proposed, it consists on the nailing reinforced concrete piles connected by a capping reinforced concrete beam and a proper drainage (Bahar et al. 2013; Cartier and Morbois 1986).

191.4 Conclusion

The two case histories of landslides Bordj Bonaama and Boulimat are triggered by combination of rainfall, unfavourable geological, hydrogeological and geotechnical conditions and human interventions. Hydraulic parameters, runoff and groundwater, were the trigger. Anarchic earthworks, undersized retaining wall and weak layers of saturated colluviums also contributed to the destabilization of the slope.

References

ANRH (2011) Données pluviométriques de la station de Tifra, Béjaia. ANRH, Algérie

Bahar R, Sadaoui O, Sadaoui S (2013) Case study of landslides in Kabylia region, Algeria. In: CDROM seventh international conference on case histories in geotechnical engineering. Chicago, USA

Cartier G, Morbois A (1986) Expérimentation d'un clouage de remblai construit sur un versant instable à Boussy- Saint- Antoine. Bulletin de liaison des Ponts et Chaussées, Paris, France, pp 29–36

Part XIX

Physical Impacts to the Environment of Infrastructure Development Projects – Engineering Geology Data for Environmental Management

Convener Dr. Francisco Nogueira de Jorge M.Sc., Sofia J.A. Macedo Campos

Infrastructure development projects are usually associated with the risk of causing environmental impacts of various types and magnitudes. Proper assessment of physical impacts to the environment in order to design adequate monitoring programmes and establish effective preventive and mitigation measures usually require comprehensive understanding of geological features and dynamics of physical processes. Engineering geology may provide the necessary background concepts, technical resources, and appropriate methods and tools to evaluate, address and manage physical impacts to the environment.

The purpose of this session is to address and discuss the role of engineering geology in the current practices of both environmental management and environmental performance evaluation for the development of new projects as well as during projects' lifetime—design, construction, operation, decommissioning and site rehabilitation.

Using of Man-Made Massives in Russian Mining (Engineering: Geological Aspects)

192

Galperin Anatoly

Abstract

Essential negative impacts of man-made mining massives on the environment and the directions toward their ecology safe utilization has been demonstrated, considering the characteristics of mining regions in Russia and Germany. Based on an engineering-geological, hydro-geo-mechanical and geologic-geomechanical scientific observation over a long term period, a complex approach to the ecologically safe utilization of man-made massives has been developed. Applying conventional interactive monitoring methods, the presented scientific results respond to diverse questions regarding the development of assessment methods for the state in massives emerged from hydraulic fill or dumping, the hydromechanical excavation of tailing dams and hydraulic dumps in deposits with high water content, the removal of hydraulic dumps during the preparation of coal deposits to mining with hydromechanical technologies and the use of residual pits of former open-cast mines for the hydromechanical deposition of mine spoils or for the development of recreation areas.

Keywords

Man-made massives • Tailings dam • Bearing capacity • Stability • Monitoring

192.1 Introduction

Great territories of mining-industrial regions in the Russian Federation, namely with fertile soils, are covered with dumps and hydraulic dumps, storages with the useful materials' wastes. These mining-engineering constructions occupy large areas. They are composed by sediments with the low strength and bearing capacity. As a result it entails development of dangerous geological processes and following-up environmental contamination.

As a result of deposits exploitation and the accompanying redistribution of useful materials there are formed technogenic massives-man-made in natural landscape geological bodies from the fragments of rocks, tailings, ashes, slag, slime etc. Areas of land allotments of the biggest mining enterprises are measured by 1,000 ha of disposing and destroying lands. Deposits exploitation is linking with the lost of valuable components in subsoil during the non-complex processing of extracting mineral materials.

In dumps of Russian mining enterprises great volumes of overburden operations measuring by tens of milliard cubic meters are located. Only in hydraulic dumps and tailing dams nearly 5 milliard m^3 of tailings and more than 1.5 milliard m^3 of overburden operations are located. These constructions are objects of high ecological danger causing contamination of air, underground and surface waters, soils, changing of biocenose on large territories. Carrying out a complex of engineering-geological, hydro-geo-mechanical and geological-geochemical researches of man-made massives is necessary for supporting measures to provide their stability, saving water and soil resources, refilling of mineral base of mining enterprises.

G. Anatoly (✉)
Moscow State Mining University, Moscow, Russia
e-mail: galperin_a@mail.ru

G. Lollino et al. (eds.), *Engineering Geology for Society and Territory – Volume 6*,
DOI: 10.1007/978-3-319-09060-3_192, © Springer International Publishing Switzerland 2015

Professors Mironenko and Shestakov (1974) created scientific direction –hydrogeomechanics within the scope of which taking equal methodological positions it is created "joint and mutual consideration of soil mechanics and filtration of underground waters conformity…with regard to the tasks of hydrogeology and engineering geology in frame of one and the same scientific discipline".

To estimate stability of watered slopes Mironenko suggested principle of bringing large hydrodynamic forces to equivalent contour ones due to powerful effect of underground waters is taken into account very simply-through the meaning of piezometric head on contour of sliding block. This effective estimated way is widely used to estimate stability of slope constructions of hydraulic dumps and tailing dams.

192.2 Main Text

The objects of hydro-geomechanical studies of the Geology Department of MSMU (Moscow State Mining University) were embankment dumps of mixed rocks on weak water-saturated foundations (KMA (Kursk Magnetic Anomaly), Kuzbass (Kuznezk Coal basin) etc.) which were subjected by the depression consolidation as a result of deep water lowering, overlying beds of KMA deposits and Zaporozhsky ore-iron, filling structures of hydraulic dumps and tailing dams in different mining industrial regions (Fig. 192.1) (Galperin 2003; Galperin et.al. 1993a).

New methods of man-made massives' monitoring and means of their realization have been introduced (Galperin et.al. 2012).

Method of complex sounding by means of devices MSMU-DIGES (Fig. 192.2), penetration-logging station VSEGINGEO and type-set of the original probes (Galperin et al. 1994).

Method of remote control condition of hardly accessible filling structures using aerophotogrammetry survey to define settlement of filling structures and its bearing capacity (Galperin et al. 1993b):

- Method of interactive remote control of the slope dumps of filling structures' stability and dumping embankment on the weak foundations according to the relevant hydro-geomechanical criteria of safety by means of mobile communication standard GSM;
- Technical means to measure deformation of slopes of dry rocks dumps on the weak foundations providing setting of critical deformation's measures with the aim to control the process of dump formation.

Engineering-geological zoning allows to compare sites of filling structures and define appearance and limits of their future using. Zoning provides singling out of uniformed in

Fig. 192.1 Space photo of the objects of the Staro-Oskolysky iron-ore distrikt: *1*-open Pit LGOK; *2*-open Pit SGOK; *3*, *4*-overburden dumps LGOK; *5*-overburden dumps SGOK; *6*-reclaiming hydraulic dump "Berezovy Log"(LGOK); *7*-tailings dam LGOK; *8*-active hydraulic dump LGOK; *9*-tailings dam SGOK; GK1–4-profiles of hydrogeomechanical monitoring

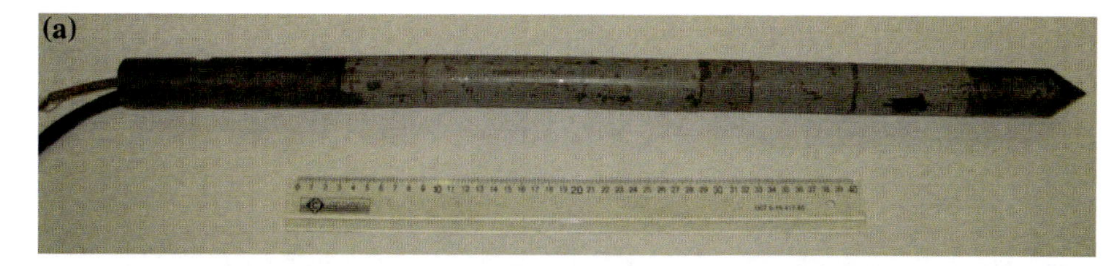

Fig. 192.2 Composite string probes **a** two-parametric (q_s, P_w); **b** two-parametric (τ,P_w); **c** three-parametric (τ,q_s,P_w)

engineering-geological respect taxonomic units of the definite level in limits of the investigating territories.

Materials of the engineering-geological zoning of the filling territories have to provide the decision of the following main tasks:

- Capacity's increasing of the operating constructions on the base of forcing the consolidation of finely-dispersed rocks;
- preparing of filling territories for their following using (reclamation, placement of dry dumps, construction).

For the effective using of hydraulic dumps' territories it is necessary to possess reliable information concerning the state of man-made massif and foresee the behavior of filling foundations in time. In particular, using the hydraulic dumps territories for the dry dumps or reclamation the bearing capacity of filling masses and changing it in time is of great importance.

In the biggest Russian mining-industrial regions KMA and Kuzbass there were made complex investigations of the man-made massives (dumps and filling structures) including natural and laboratory experiments with the using of above mentioned methods of monitoring and defining the mechanical properties of man-made deposits in devices of triaxial compression. The examples of results of engineering-geological zoning of filling territories hydraulic dumps KMA for two biggest open-pits on the base of iron-ore deposits KMA are given on the Fig. 192.3a–c. On the base of these materials there were elaborated recommendations concerning the filling of two levels dumping filling at this hydraulic dump because of lack of dumping areas. At the Fig. 192.3b, c data of zoning of two territories of hydraulic dumps on bearing capacity are given (P_{dop}).

Interactive monitoring of hydraulic dumps and tailing dams allows efficiently to correct constructions and conditions of buildings and estimate possibilities of their future growing. The principal scheme to monitor constructions of industrial hydraulic structures with the using sensors of porous pressure, devices to collect and save monitoring characteristics and their transmission through mobile communication are given at Fig. 192.4.

Nowadays monitoring and operation devices of the objects through the mobile communication standard GSM in several regimes (GPRS, SMS ets.) are widely spread. Device for the remote monitoring of the filling dams where vibrating wire transducers of water pressure ("Gidroproekt", Moscow) was created in VNIMI (Russian Scientific Research Surveying Institute) and a company "Karbon" (St. Petersbourg) (Kutepov et al. 2009). Monitoring of porous pressure in hydraulic engineering constructions is provided by recommendations of ICOLD (International Commission on Large Dams 1996).

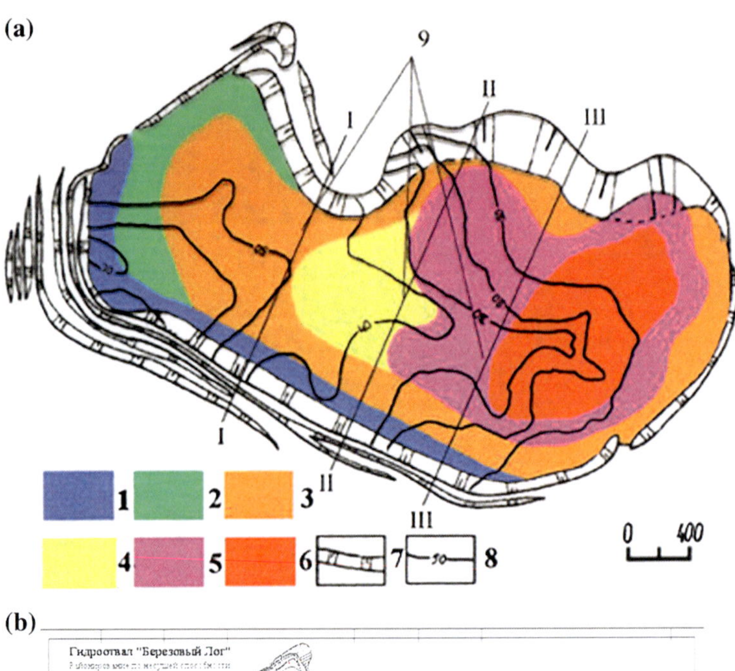

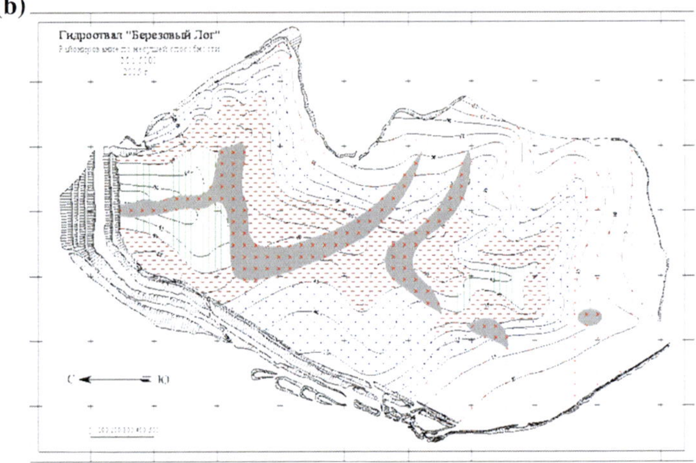

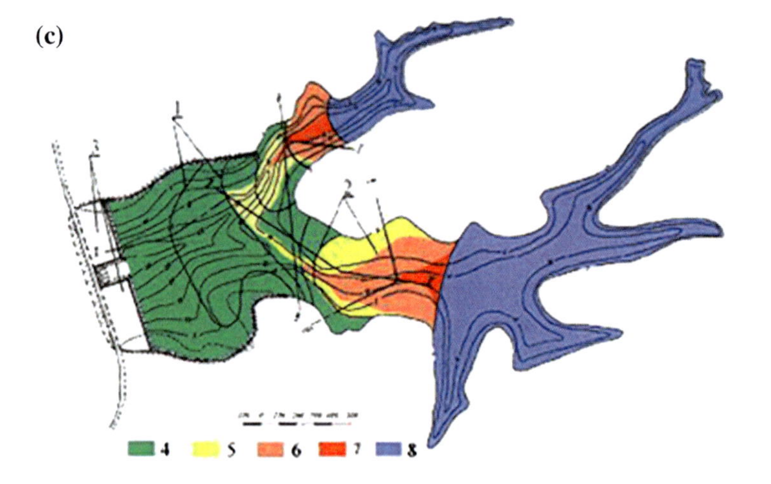

Fig. 192.3 Engineering-geological zoning. **a** A map of engineering-geological zoning of hydraulic dumps "Berezovy Log" (KMA): 1,2,3,4,5,6-engineering-geological sites; 7-slope constructions of hydraulic dumps; 8-izolines offilling structures' thicknese; 9-lines of engineering-geological sections according to the axis 1,3 and 2 drainage dividing prisms. **b** A map of zoning of the hydraulic dump "Berezovy Log" territory on the bearing capacity (Pdop). **c** A map of zoning of the hydraulic dump "Log Shamarovsky" (MGOK) on bearing capacity of the filling massif: 1-borders of the engineering-geological zones; 2-borders of sites with the different bearing capacity of filling soils; 3-reclaiming sites of hydraulic dump; 4-sites of hydraulic dump with $P_{dop} > 0.15$ MPa; 5-sites of hydraulic dump with $P_{dop} = 0.1$–0.15 MPa; 6-sites of hydraulic dump with $P_{dop} = 0.05$–0.1 MPa; 7-sites of hydraulic dump with $P_{dop} < 0.05$ MPa; 8-water surface.

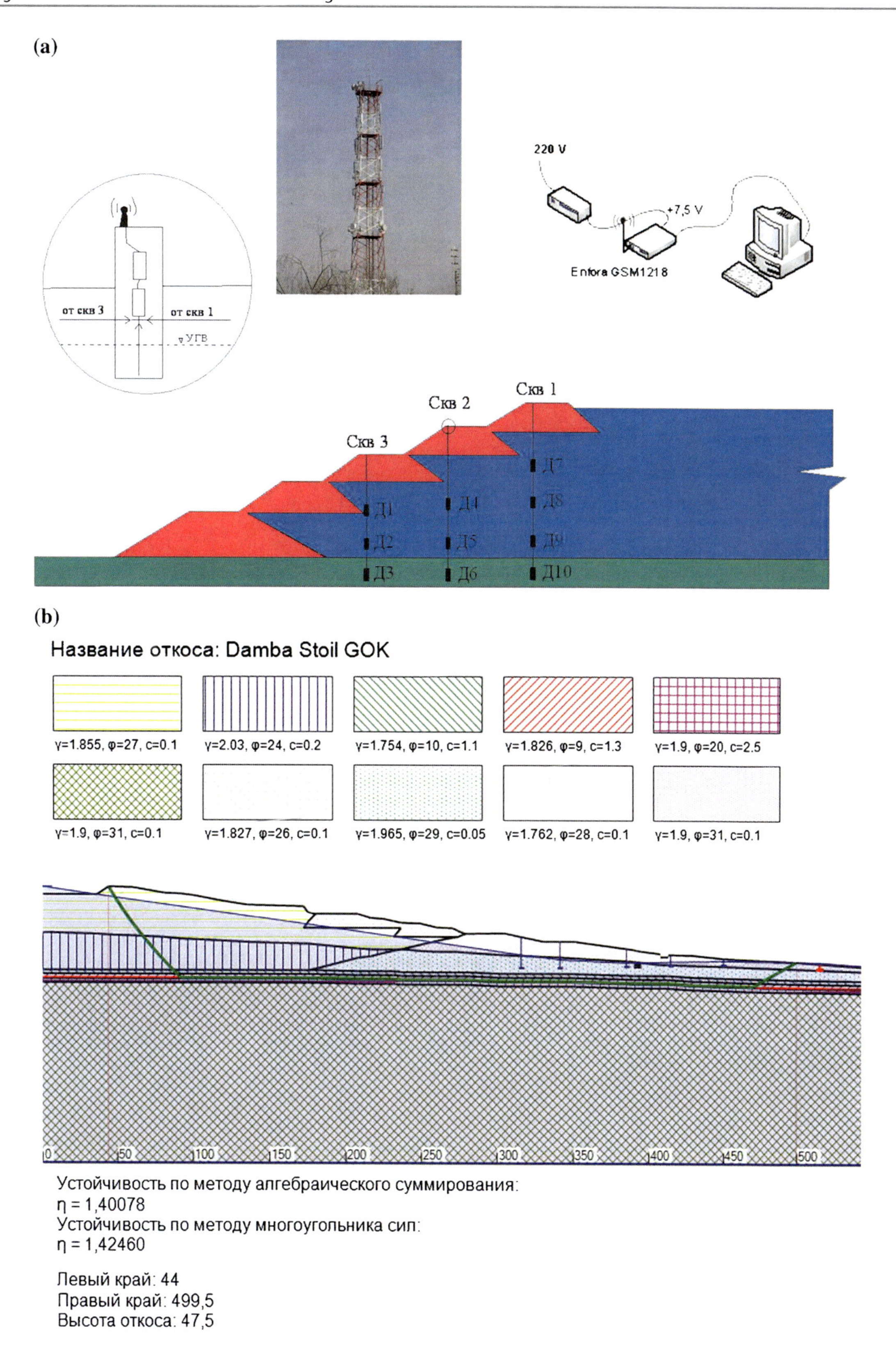

Fig. 192.4 Organization scheme of system of the remote control: **a** data collection, storage and transmission; **b** processing of the receiving information for conditions of the tailing dams SGOK

192.3 Conclusion

Introduction of work's results in the regions Kuzbass and KMA provided operative monitoring of dams' stability and bearing capacity of weak foundations, location on the filling foundations higher than 0.5 milliard m^3 of the "dry" overburden, prevention of fertile soil destroying higher than 5,000 ha, reducing to decades dumps' reclamation time.

Acknowledgments Calling your attention work has been made for the last 20 years by the composite author including staff of MSMU, VNIMI, SPMI (State Petersbourg Mining Institute), VSEGINGEO (Moscow), VIOGEM (Belgorod). We express our thanks for the many years effective cooperation to Professors Y.I.Kutepov and V.S. Krupoderov, Doctor A.V. Kiyants, representatives of mining enterprises KMA and Kuzbass but also colleagues of the Freiberg Mining Academy Prof. V. Forster, Doc. H.Iu. Shef, Prof. H. Klapperih and Doc. N.Tamashkovich.

References

Galperin AM (2003) Geomechanic in surface mining. MSMU

Galperin AM, Zaytsev VS, Norvatov YA (1993a) Hydrogeology and engineering geology. Balkema publ., Rotterdam, Brookfield

Galperin AM, Zaytsev VS, Streltsov VI et al (1993b) Method of monitoring the condition of filling structures. Patent RF no 1188322

Galperin AM, Zaytsev VS, Kirichenko YV et al (1994) Device for the complex sounding of soils. Patent RF no 2025559

Galperin AM, Kutepov YI, Kirichenko YV et al (2012) Using of man-made massives at mining enterprises. In: Mining book

Kutepov YI, Kutepova NA, Milman GL (2009) Methods and technical means of hydrogeomechanical monitoring of industrial hydrotechnical structures safety//Engineering Researches, vol 5

Mironenko VA, Shestakov VM (1974) Basics of Hydrogeomechanics. Nedra

Mironenko VA, Strelsky FP (1989) Practical use of hydrogeomechanics principles with the aim to increase industrial and ecological safety of mining works//Engineering geology, vol 5

Monitoring of tailings dams, ICOLD, Bulletin 104 (1996)

Modeling Optimized UCG Gas Qualities and Related Tar Pollutant Production Under Different Field Boundary Conditions

193

Stefan Klebingat, Rafig Azzam, Marc Schulten, Thomas Kempka, Ralph Schlüter, and Tomás M. Fernández-Steeger

Abstract

The process of Underground Coal Gasification (UCG) bears the potential to produce medium to high calorific syngas for several industrial applications, e.g. electricity generation in the frame of the Integrated Gasification Combined Cycle (IGCC) concept; or Coal-To-Liquid (CTL) technologies as the Fischer-Tropsch synthesis. In view of preferred environmentally sound operations and stable gas qualities for these applications previous global UCG research led to considerable process experience. Despite this knowledge background however UCG still remains a challenging technology as many physical and chemical sub processes are not sufficiently traceable by aboveground instrumentation, in turn hampering enhancement of overall process efficiency and engineering performance. In this context equilibrium modeling becomes a useful strategy to gain a better process understanding of coal gasification at different depths and its related engineering geological boundary conditions (i.e. coal type, p/T conditions and overburden water influx). The recent CO2SINUS project thus investigated sensitivities of various boundary conditions on establishing optimized gas qualities at simultaneous minimum tar production rates during active operation by using a new self-developed thermodynamic model. The main potential of this model approach is seen in the pre-assessment of individual field boundary condition effects, amongst other criteria indicating coal type related gas qualities as well as tar related long-term groundwater pollution risks.

Keywords

UCG • CO2SINUS • Field boundary conditions • Gas qualities • Tar production

193.1 UCG Model Development Within the CO_2SINUS Project

In view of optimized UCG gas qualities the CO2SINUS project analyzed the influence of selected field boundary conditions on gas quality and related tar pollutant production by using a new self-developed steady state equilibrium thermodynamic model approach (cf. Fig. 193.1), based on the principal chemical processes within a fixed bed reactor (cf. Fig. 193.2). This fixed bed engineering approach was chosen as a first starting point to the extent of former reviewed UCG process conditions by Min and Edgar (1987).

Within this thermodynamic model (cf. Fig. 193.1) selected Aspen Plus® software modules (e.g. material streams, mixer- splitter-, reactor- and valve types) represent

S. Klebingat (✉) · R. Azzam · T.M. Fernández-Steeger
Department of Engineering Geology and Hydrogeology, RWTH Aachen University, Lochnerstr. 4-20, 52064 Aachen, Germany
e-mail: klebingat@lih.rwth-aachen.de

M. Schulten
Unit of Technologies of Fuels (TEER), RWTH Aachen University, Wüllnerstr. 2, 52062 Aachen, Germany

T. Kempka
Section 5.3 Hydrogeology, Helmholtz Centre Potsdam, GFZ German Research Centre for Geosciences, Telegrafenberg, 14473 Potsdam, Germany

R. Schlüter
DMT GmbH & Co. KG, Am Technologiepark 1, 45307 Essen, Germany

G. Lollino et al. (eds.), *Engineering Geology for Society and Territory – Volume 6*,
DOI: 10.1007/978-3-319-09060-3_193, © Springer International Publishing Switzerland 2015

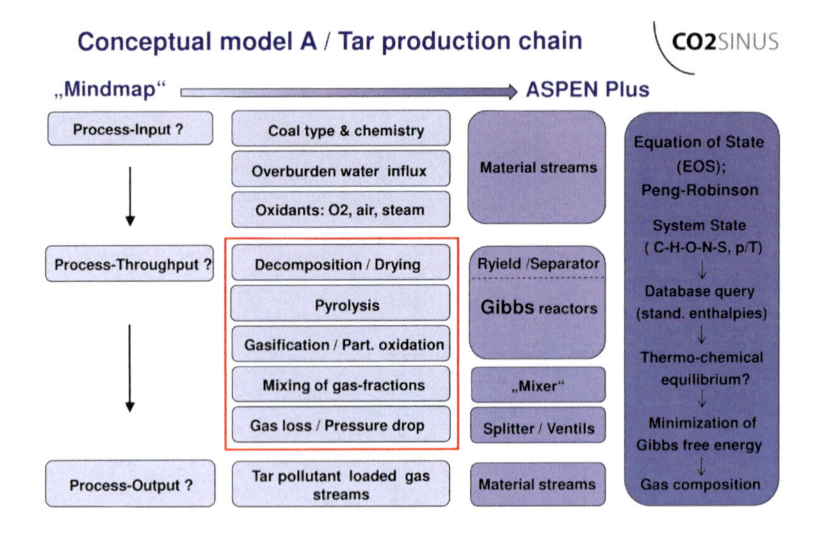

Fig. 193.1 Overview of conceptual steps and mathematical foundations in the frame of thermodynamic UCG tar pollutant production model development and calculation (Klebingat et al. 2013)

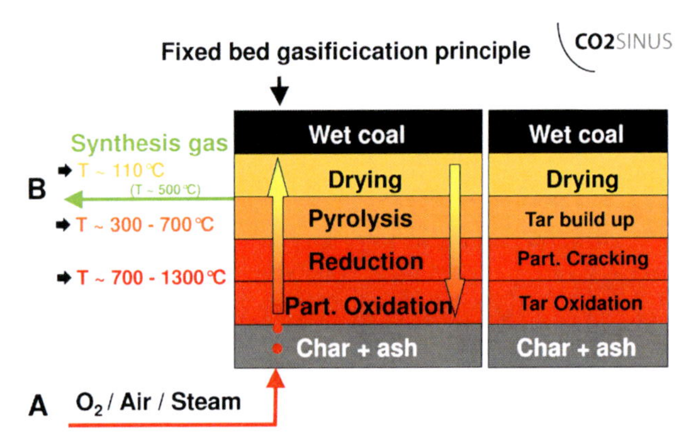

Fig. 193.2 Main UCG chemical processes and related tar behaviour during fixed bed gasification (Klebingat et al. 2013)

the main gas- and tar altering processes of pyrolysis, reduction/partial oxidation and gas mixing to syngas, besides important other UCG boundary conditions (e.g. gasification agent flux/composition, pyrolysis temperature, overburden water influx). A separate model flowsheet governs the hierarchical order of the single unit operations. Gas qualities and related tar pollutant yields for the abovementioned main processes are calculated according to the main principle of Gibbs free energy minimization by solving the Peng-Robinson Equation of State (Peng and Robinson 1976), as well as an Aspen Plus® internal Lower Heating Value (LHV) subroutine.

193.2 Model Data and Selected Sensitivity Studies According to Field Boundary Conditions

The thermodynamic model used within the CO2SINUS project referred to technical data reports from sub-bituminous and bituminous coals of the former US-UCG trial sites at Hanna, Centralia and Pricetown (amongst others Campbell et al. 1974; Bartke and Gunn 1983; Hill et al. 1983, 1984; Moskowtschuk 1997). Each of these coals has been tested with regard to quantitative gas quality and tar-load altering effects for the main gas types of UCG (Pyrolysis gas, Reduction/partial Oxidation gas and mixed syngas fractions). In this context the following field boundary conditions were investigated:

- Pressure
- Temperature
- Water influx
- Gasification agent flux/composition

193.3 Selected Preliminary Results

In view of the examined status of results to date, water influx appears to be one of the most sensitive parameters with regard to gas quality and tar pollutant production changes. Figures 193.3, 193.4 and 193.5 give an example of the Centralia trial's simulation gas quality and tar production trends within the different main UCG gas zones. In this context increased water influx rates lead to desired decreasing tar yields for most gas fractions—the main tar associated pyrolysis gas fraction as well as the wet syngas fraction, respectively. Simultaneously however a drop in gas qualities can also be observed. In agreement with literature our model results thus indicate that increased water influxes cause lower gas temperatures, which in turn generate lower gas heating values. Due to these results water influx remains a difficult field boundary parameter to judge in terms of best practices: in tendency lower water influx rates—favouring the existence of higher temperature fields—will be positive with regard to gas qualities, i.e. higher economic benefit. At the same time this will be at the cost of higher tar production rates.

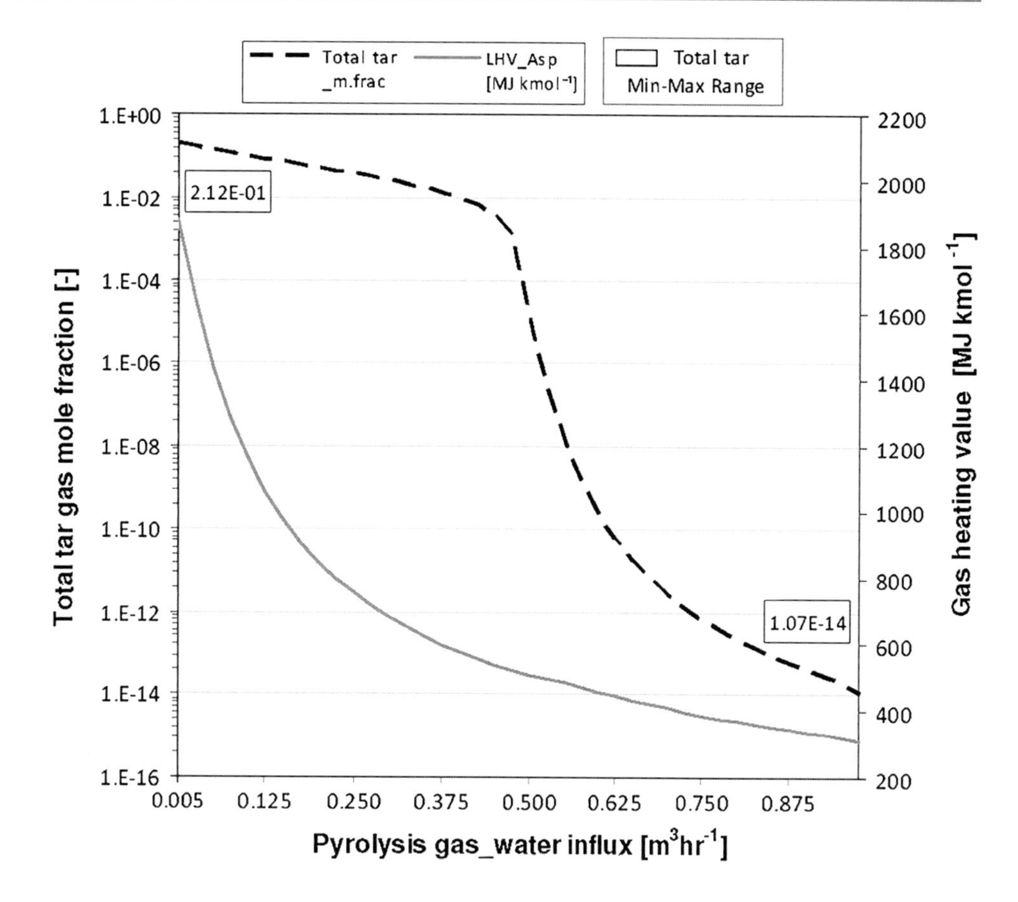

Fig. 193.3 Centralia trial simulation: total tar gas mole fractions and lower gas heating values (LHV) as a function of water influx: Pyrolysis gas

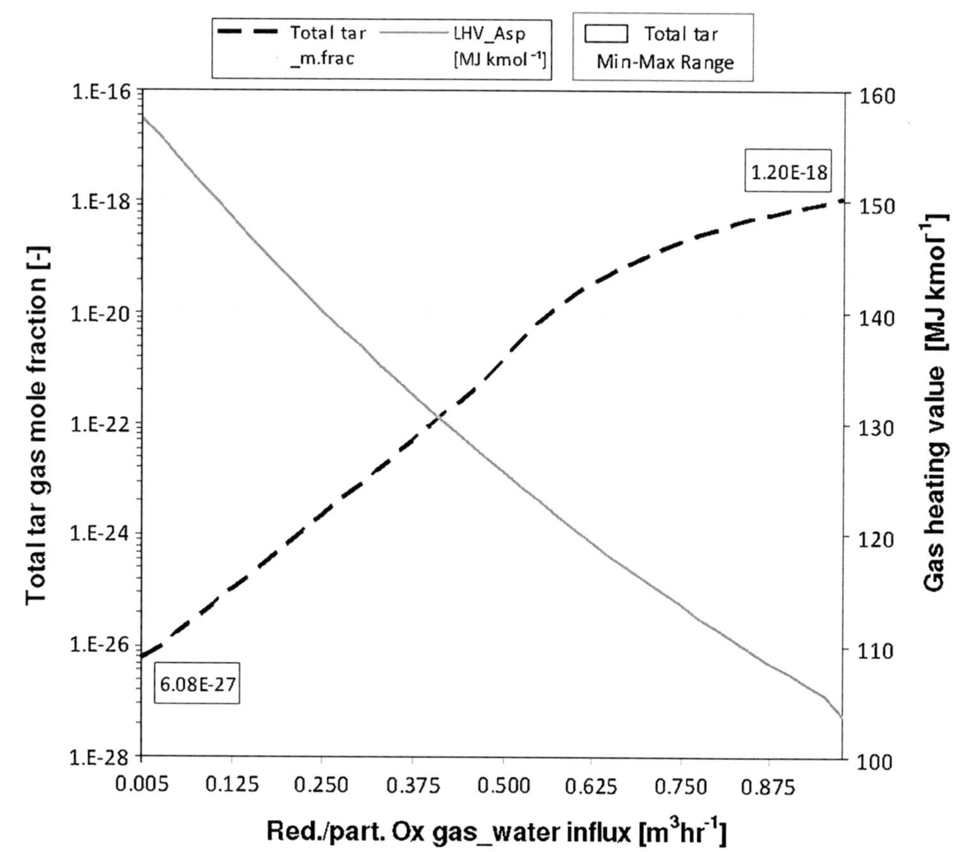

Fig. 193.4 Centralia trial simulation: total tar gas mole fractions and lower gas heating values (LHV) as a function of water influx: Reduction/partial Oxidation gas

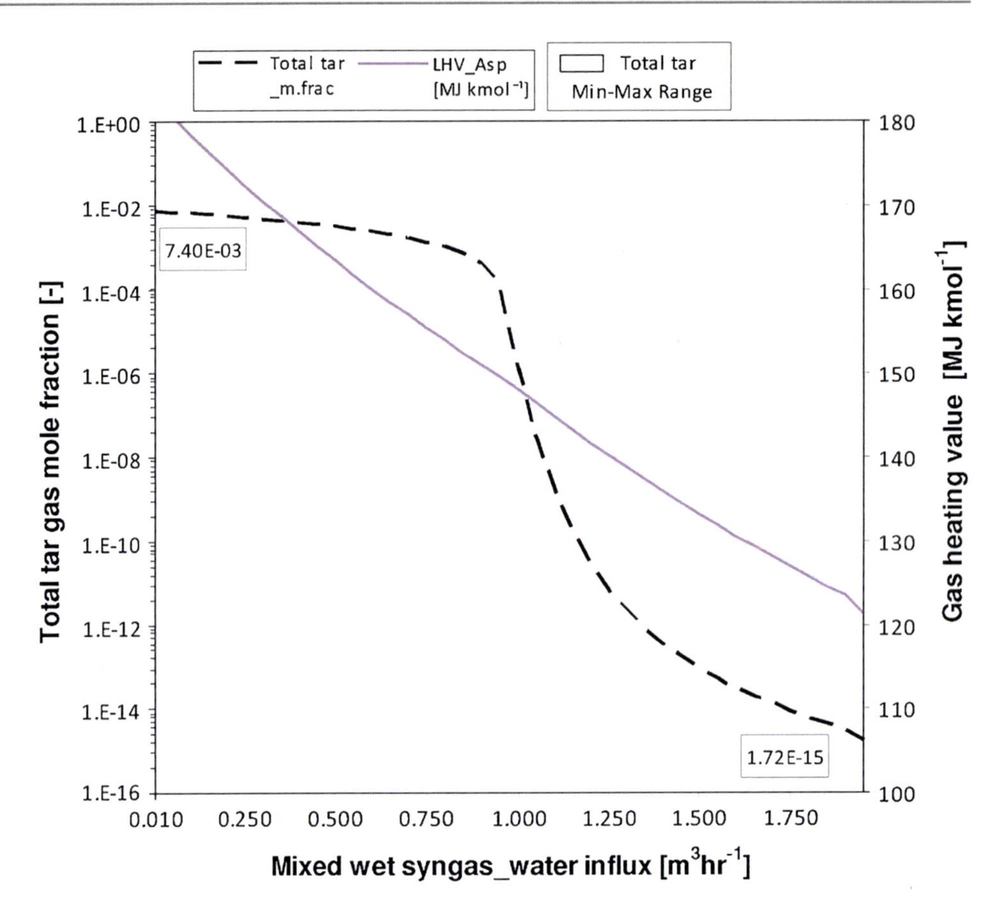

Fig. 193.5 Centralia trial simulation: total tar gas mole fractions and lower gas heating values (LHV) as a function of water influx: Mixed wet syngas

References

Bartke TC, Gunn RD (1983) The Hanna Wyoming underground coal gasification field test series. In: Krantz WB, Gunn RD (eds) Underground coal gasification—the state of the art. AIChE Symposium Series, vol 79(226), pp 4–14

Campbell GG, Brandenburg CF, Boyd RM (1974) Preliminary evaluation of underground coal gasification at Hanna. Technical progress report 82, edited by UW digital projects. Digitized collection western research institute: Laramie Wyoming, USA

Hill RW, Thorsness CB, Thompson DS (1983) The centralia partial-seam CRIP test. In: Proceedings of the 9th annual underground coal gasification symposium: Bloomingdale/Illinois, USA, (DOE/METC-84-7 CONF-830827), pp 10–22

Hill RW, Thorsness CB, Cena RJ, Stephens DR (1984) Results of the centralia underground coal gasification field test. In: Proceedings of the 10th annual underground coal gasification symposium: Williamsburg, VA, USA, (DOE/METC-85/5), pp 11–26

Klebingat S, Azzam R, Schulten M, Quicker P, Kempka T, Schlüter R, Fernandez-Steeger T (2013) Quantitative tar pollutant production and water solubility prognoses in the frame of Underground Coal Gasification (UCG) In: Thuro K (ed) Proceedings of the 19th conference on engineering geology and of the forum for young engineering geologists Munich, München, Techn. Univ., pp 589–594, 13–15th Mar 2013

Min W, Edgar TF (1987) The dynamic behaviour of a fixed-bed steam-oxygen coal gasifier disturbed by water influx II—model development and verification. Chem Eng Commun 52(4–6):195–213

Moskowtchuk W (1997) Mechanismen der In situ Vergasung unter Einbeziehen ukrainischer Kohlen. Labor- und Feldversuche. Dissertation. RWTH Aachen University: 164 S

Peng D, Robinson DB (1976) A new two-constant equation of state. Ind Eng Chem Fund 15(1):59–64

Considerations About the Integration of Geological and Geotechnical Studies Applied to Engineering Projects and to Environmental Impact Assessment in São Paulo State, Brazil

194

Bitar Omar Yazbek, Sofia Julia A.M. Campos, Amarilis Lucia C.F. Gallardo, Braga Tania de Oliveira, and Caio Pompeu Cavalhieri

Abstract

This paper brings to discussion the quality of the integration of geological and geotechnical studies that is achieved for the purpose of engineering projects and for environmental impact assessment. The integration of these studies in different phases of a development is analyzed, in order to aid, both at the same time and importance, the building of suitable constructions under the engineering and environmental sustainability points of view. We expect to contribute to a better understanding of the form and the current stage of this integration, especially in relation to the effective improvement from the geological and geotechnical knowledge acquired in both application fronts. Observations accomplished in infrastructure constructions were taken as reference, featuring roads, railways, pipelines, mines, development lands and landfills, accomplished in the last years in the state of São Paulo, Brazil, and submitted to the Environmental Impact Assessment (EIA) process, a tool that has helped enhance the integration of geological and geotechnical studies generated in both fronts. The results show that signs of integration exist but there still is a certain separation. Some technological and management challenges have been shown in a way to better improve the integration of the geological and geotechnical knowledge in new developments.

Keywords

Geological and geotechnical knowledge • Environmental impact assessment • Infrastructure constructions

194.1 Introduction

Studies involving geological and geotechnical knowledge are usually distinguished based on two application fronts: one aims at certifying the technical viability of engineering projects and at subsidizing construction and operation; and the other, a way of dealing with the environmental impacts and demonstrating the environmental viability.

In relation to engineering projects (which include a conceptual project, basic project and executive project), the required geological and geotechnical studies are linked to the challenge of predicting the interaction construction-physical environment behavior in order to ensure the execution of a safe, operational and technically suitable construction. Under the environmental viability point of view, and also trying to predict the interaction construction-physical

B.O. Yazbek · S.J.A.M. Campos (✉) · A.L.C.F. Gallardo · B.T. de Oliveira · C.P. Cavalhieri
Institute for Technological Research of São Paulo State, Sao Paulo, SP, Brazil
e-mail: scampos@ipt.br; labgeo@ipt.br

B.O. Yazbek
e-mail: omar@ipt.br

A.L.C.F. Gallardo
e-mail: amarilisgallardo@usp.br

A.L.C.F. Gallardo
Uninove, EPUSP, Sao Paulo, SP, Brazil

G. Lollino et al. (eds.), *Engineering Geology for Society and Territory – Volume 6*,
DOI: 10.1007/978-3-319-09060-3_194,

environment, the highest concern is to evaluate the adverse future consequences in relation to the environment.

Therefore, although there are different focuses, approaches and tools, the geological and geotechnical studies required in both application fronts include essentially the same object of analyses (the interaction construction-physical environment), a fact that highlights the importance of the integration accomplished in both fronts. Among the aspects that could be analyzed in relation to these studies, the technical cooperation between the involved teams in their preparation in each front and the effective integration of the geological and geotechnical knowledge developed in both applications are discussed. This is not only in favor of having a larger team and rationalization of resources, but also in the way of offering a refined and major comprehension about the construction-physical environment interaction.

194.2 Objectives, Materials and Methods

This paper aims at analyzing the relation established between geological and geotechnical studies elaborated for an engineering project and for environmental impact assessment and management, starting with the environmental impact assessment (EIA) concerning the same projects.

In order to reach the expected objectives, a bibliographic review was carried out about the issue. Studies performed for same project that approach geological and geotechnical knowledge in both application fronts were considered, as well as observations about the accomplished studies for engineering projects, and environmental studies related to cases in which the authors of the current work had some interaction, either supporting the environmental entity or the developer, in different stages of the development. Infrastructure construction cases were submitted to EIA procedures and the environmental license in the state of São Paulo (southeastern Brazil).

Only observations on large development cases and from different sectors were taken as reference, as well as the relatively recent ones back to the last 10 years, reaching a number of 41 developments (17 roads, 1 railway, 1 conveyor, 1 pipeline, 6 mines, 1 estuarine drainage channel, 4 development lands and 10 landfills).

In order to do the analyses on the basis of observations during each development and also by reviewing data and information, the AIA/EIR was considered in each case and the Basic Environmental Plan (BEP), as well as the denominated Construction Environmental Plan (CEP), the latter formalized only in some developments. Regarding the engineering projects, differences according to each case and casual checking of the conceptual project, basic project and executive project are pointed out in relation to specific features available.

We expect to identify elements that characterize the integration of the geological and geotechnical studies performed in both application fronts. One characteristic to be investigated is the context in which the integration occurs; in other words, the different stages of the development. Another characteristic is the various moments throughout the duration of a development in which the integration of the knowledge obtained in both application fronts tends to happen. Thus, it is essential to verify the relationship between the frequent stages in the preparation and execution of engineering projects and in EIA. A third characteristic is about the way which the cooperation and/or contribution between the technical teams occur, as well as a possible integration between the geological and geotechnical studies applied in both fronts.

Considering these characteristics, the level of integration reached between the geological and geotechnical studies for engineering projects and for environmental impact assessment and management were analyzed for each case. To have the result and the level, individual analyses were done, ranking the developments according to three levels (low, medium and high). The results were compiled in a chart, beginning with the predominant analyses. After making the observations for each case, the results are discussed and a summary of the most important conclusions is presented, expecting to cooperate to a first reflection about the theme.

194.3 Overview on the Integration of Geological and Geotechnical Studies

Although geological and geotechnical studies are found in the national and international literature, the integration of such studies in both application fronts is a topic that is hardly ever considered, especially when it comes to the analysis of the same development undergoing an EIA process.

Some works deal with the theme, which can be distinguished by physical environmental characteristics in engineering projects and in the environmental studies. Both are usually treated separately, the characteristics referring to geological and geotechnical studies performed on both fronts. Anderson (2006) states that many of the products that come from detailed geological and integrated mappings have been used as primary database to the future environmental planning of the North American state of North Dakota. Many unfavorable geological conditions are found, as clay deformation in soils of lacustrine origin, inappropriate supportability and presence of mass movements, which are among the most important causes of geological, geotechnical and environmental problems. There is also the difficulty in defining the railroad grade and the fact of recent and repeated seasonal floods in urban areas. The importance of the relationship between environmental geology and the geological processes in the physical environmental understanding, as well as the influence of geology in the engineering processes, especially regarding infrastructure constructions, are scenarios that must be taken into account together (Bell 2008). The author examines the probability of geological risk aspects, the importance of water and soil resources, environmental impact from mining, waste disposal and pollution in the environment, as well as many other aspects that result in environmental problems.

Concerning the integrating methods used, it is believed that the complex project elaboration implies the participation of several specialists, with different degrees and viewpoints. Therefore, the relationship among these specialists may become problematic in practice. Aiming at contributing to the effectiveness in the conception of large projects, Grebici (2007) developed a model that integrates different methods of cooperation, of organizing the process and of conception of other intermediate products and their adequacy to the development purposes, stressing out the importance of developing proper methods of cooperation. Indeed, many negative environmental impacts may be prevented or at least have its magnitude dramatically reduced by having effective cooperation between designers and environmental teams (Sánchez 2006). Sánchez and Hacking (2002) also emphasize the importance of linking the EIA to the Environmental Management System (EMS) of a development, using the studies conducted in order to elaborate the EIA/EIR for the construction management, operation and deactivation of developments, which in practice does not happen.

194.4 The Necessary Integration

The geological and geotechnical knowledge acquired or generated in engineering projects is also relevant in order to assess and manage environmental impacts and vice versa. However, it is possible to notice that physical field studies conducted in both application fronts are developed separately, commonly with distinct professionals and technical teams.

The demands associated with requirements that are specific to each context, such as the object and the scale of the generated cartographic products, contribute to such separation. While engineering project studies tend to focus on the knowledge of the underground physical environment and on scale detail, environmental geological and geotechnical studies aim at the whole understanding of the geodynamic surface, where possible negative consequences of a certain development tend to be highlighted by an environmental viewpoint. It is worth mentioning that it is regarded just as an aspect that is commonly observed rather than a rule. There are some studies that reveal exactly the opposite, as a result of specific demands. The scale is presented as a differential in some cases only. Also there are some situations where the impact is distinguished in details, as it happens in the cases of interferences in the groundwater related to mining, landfills and allotments. The situation where the connection between the studies conducted in both application fronts is noticeable refers to cases involving underground projects such as road or rail tunnels, in which the knowledge developed in each one is often useful to the other one. It is understood that the subsurface knowledge is as important as the surface one in projects with relevant interventions either on the ground or in the groundwater.

Table 194.1 Relative degree of integration of geological and geotechnical studies conducted for the purpose of engineering projects and assessment and management of environmental impacts, as overall prevalence observed in relation to all cases, according to the project phases proposed in IBAMA (2009)

Studies aimed at engineering projects		Studies aimed at assessment and management of environmental impacts			
		Environmental planning tools		Environmental management tools	
Project level	Engineering studies and projects	AIA/ EIR	BEP construction	CEP or EEP	EMP, EMS or BEP construction
Viability or pre-project	Conceptual project and feasibility studies	H	M	NA	NA
Project development	Basic project	M	M	L	L
	Executive project	L	L	M	L
Installing, building nd/or assembling	Executive project updated	L	L	M	L
Operating	Functional plan or operational project	L	NA	NA	M

Note EIS/EIR—Environmental Impact Study/Environmental Impact Report; BEP—Basic Environmental Plan; CEP—Construction Environmental Plan; EEP—Environmental Executive Project; EMP—Environmental Management Plan; EMS—Environmental Management System. L—Low; M—Medium H-High; NA—Not Applied

Table 194.1 was obtained using the cases discussed and the analysis of each one of them concerning one of the three ranking levels (High, Medium and Low), related to the integration of geological and geotechnical studies conducted with engineering projects, and assessment and environmental management purposes.

194.5 Conclusions

The results obtained with the completion of this study suggest the following conclusions: (**a**) although with different approaches and tools, both geological and geotechnical engineering projects as environmental studies include essentially the same object of analysis. This suggests the importance of the integration between geological and geotechnical studies carried out in both fronts of application; (**b**) geological and geotechnical studies conducted for the purpose of an engineering project, also when used in EIA, tend to facilitate the identification of significant environmental aspects and impacts. Likewise, geological and geotechnical knowledge acquired or generated during the EIA process, including environmental management to be carried out in phases of installation and operation of projects, has also been proven as useful to engineering projects. However, in view of the potential for integration, it is observed that much can proceed. There are signs of cooperation between the technical teams as well as integration between studies applied to engineering projects and the assessment and management of environmental impacts, but a certain separation still predominates, which is a possible situation influenced by the specific demands required in isolation by developers in each front; (**c**) the integration provided by the temporal and content matching in the achievement of engineering and geoenvironmental studies required for a project, due to the linking of EIA to environmental licensing, has been observed, but it is still well below the potential level. The fact that some new developments in the degree of integration shown indicate relatively higher situations can be better obtained with a larger number of cases; (**d**) both geological and geotechnical knowledge acquired during the feasibility stage and development of a large-scale engineering construction often fail to be fully used during the phases of installation and operation of such projects, a fact highlighted by the frequency degree of integration considered low and average these stages, which undermines the effectiveness of systematic planning of works and environmental impact assessment.

Finally, not only a greater synchronization among the activities of each context is required, but also actions and attitudes that facilitate the proximity of technological contents. It often seems that it depends more on the perception of professionals who lead the engineering projects, in order to foster the effective integration with environmental teams. The current challenge is to encourage cooperation of different perspectives on the same object and increase the exchange of knowledge acquired and generated in the stage of feasibility, development, installation and operation of a development, and these results could be achieved by approaching geologists and engineers involved in both fields.

References

Anderson FJ (2006) A highlight of environmental and engineering geology in Fargo, North Dakota, usa. Environ Geol 49:1034–1042
Bell FG (2008) Basic environmental and engineering geology. Whittles publishing, Caithness, Scotland, pp 342
Grebici K (2007) La maturite de l'information et le processus de conception collaborative. These pour obtenir le grade de docteur de l'institut national polytechnique de grenoble, specialite genie industriel, p 393
Ibama- instituto brasileiro do meio ambiente e dos recursos naturais renováveis – ibama. 2009. Disponível em: http://www.ibama.gov.br/. Acessado em 26 Oct 2009
Sánchez IE (2006) Avaliação de impacto ambiental: conceitos e métodos. Oficina de textos, são paulo, pp 496
Sánchez IE, Hacking T (2002) An approach to linking environmental impact assessment and environmental management systems. Impact assess proj appraisal, Guildford 20(1):25–38

Integrated Geological, Geotechnical and Hydrogeological Model Applied to Environmental Impact Assessment of Road Projects in Brazil

195

Sofia Julia A.M. Campos, Adalberto Aurelio Azevedo, Amarilis Lucia F.C. Gallardo, Pedro Refinetti Martins, Lauro Kazumi Dehira, and Alessandra Gonçalves Siqueira

Abstract

This paper discusses the use of an integrated geological, geotechnical and hydrogeological model as a tool for environmental impact assessment of new road projects in Brazil. The method consists of a geomechanical characterisation of rock masses and an identification of soil behavior along the proposed road alignment. It uses a morphostructural analysis integrated to a hydrogeological model in a Geographic Information System (GIS) platform aiming to identify environmental risks. Lineaments extracted from satellite imagery and digital terrain models (DTM) were analysed to identify patterns of orientation across the area of the project. Morphostructural domains were defined and grouped in areas of similar pattern. For each domain a statistical analysis of directions of lineaments was carried out to identify the main morphostructural directions. Structural field data was gathered for each domain. A systematic structural survey was performed in quarries and/or natural outcrops, identifying the rock type(s) and main discontinuities characteristics/parameters. The orientation data was statistically analysed in stereonets and sets of discontinuities were identified based on the maximum concentrations and the 85 % confidence interval. The integration of data was performed in a GIS environment. For each morphostructural domain, the lineament directions were compared to the field structural data and the main directions that control groundwater flow were identified. The hydrogeological model was also entered in the GIS and crossed with the product of the lineaments analysis to identify zones of higher hydraulic gradients and possible flow paths. The environmental impacts of the project were assessed, based on the integrated models developed.

Keywords

Engineering geology • Geotechnical • Road projects • Environmental impacts

S.J.A.M. Campos (✉) · A.A. Azevedo · A.L.F.C. Gallardo · L.K. Dehira · A.G. Siqueira
Institute for Technological Research of São Paulo State, São Paulo, SP, Brazil
e-mail: scampos@ipt.br; labgeo@ipt.br

A.A. Azevedo
e-mail: azevedoa@ipt.br

P.R. Martins
Beca Ltd, Auckland, New Zealand

A.L.F.C. Gallardo
Uninove, EPUSP, São Paulo, Brazil

195.1 Introduction

The importance of engineering geology to the understanding of the physical environment and the geological-geotechnical processes that operate at a site is paramount to the environmental assessment of infrastructure projects. Bell (2008) highlights several geological aspects that can influence infrastructure projects, potentially leading to environmental impacts. Many of these aspects can affect water resources and/or soils, in addition to other environmental problems.

G. Lollino et al. (eds.), *Engineering Geology for Society and Territory – Volume 6*,
DOI: 10.1007/978-3-319-09060-3_195, © Springer International Publishing Switzerland 2015

The development of complex projects necessarily imply the participation of several experts, from various backgrounds and distinct viewpoints, requiring different detailed degrees data integration, depending on the aspect to analyse. Where there is effective cooperation between the designers and the environmental team, many negative environmental impacts can be prevented or at least have their magnitude reduced significantly (Sánchez 2006).

Environmental assessments of road projects must ensure all available data and information are thoroughly examined to achieve more accurate results. Many products derived from detailed geologic mapping served as the primary database for future environmental planning, as outlined by Anderson (2006). The later author reported several unfavorable engineering geological conditions, such as deformation of clay soils of lacustrine origin, inadequate bearing capacity and the presence of regions favorable to mass movements, have been the cause of geotechnical and environmental issues.

This paper discusses the use of an integrated geological, geotechnical and hydrogeological model as a tool for environmental impact assessment of new road projects in Brazil.

Typically, road construction involves extensive earthworks such as cut and fill and excavation (bridges foundations and tunnels). The method proposed in this paper is best applied if used specifically to each of the different types of geotechnical works along the alignment. The aim of the method is to identify potential environmental risks associated with the implementation of the civil works. The effects of the earthworks on the physical environment are mainly change in topography and alteration of water level and/or quality. These effects, depending on their magnitude and the local geological conditions, can be associated, to the following impacts: landslides; erosion; siltation and lowering of the water table.

195.2 Objectives and Research Method

This paper aims to present a methodology for the assessment of environmental impacts of road projects on the physical environment during preliminary design. The method is based on integrated analyses of the geological, hydrogeological, geotechnical, geomechanical and environmental variables, compared against the engineering solutions proposed by designers. The primary contribution of the method is to provide results that assist in the development of an environmentally sound detailed design, optimizing engineering solutions while minimizing environmental impacts.

The methodology follows three major steps:

Step 1: *Definition of the geological and geotechnical model*: For each specific sector of the alignment the types and spatial distribution of soil and rock masses are identified. Focus is given to characterise the variability of the geotechnical properties of the materials present. The geological processes that may affect the site during construction and operation phases are identified. Uncertainties arising from the level of detail of the investigation are assessed and quantified.

Step 2: *Definition of the geomechanical behaviour:* For the rock and soil masses in which the different geotechnical works will be implemented the main characteristics are identified. Their geomechanical behaviour is recognised, with emphasis on the definition of the specific potential rupture mechanisms.

Step 3: *Identification of the risks and environmental impacts associated with the works*: Considering the particular requirement of the different geotechnical works, the anticipated geomechanical behaviour and the geotechnical uncertainties, the potential risks are identified. These risks are then compared to the land use characteristics at the specific site to assess the potential environmental impacts.

To follow these steps, for all road extension, the subsequent technical activities are carried out: (i) Geological and geotechnical data compilation from technical papers and various sources available for the project influence area. This includes any geotechnical investigation undertaken by the project owner or on their behalf at early stages of the project. (ii) Database organization in a GIS platform. (iii) Analysis of reports related to the technical requirements of the projects. (iv) Attendance of technical meetings organised by the project owner to follow the project progress and get the most updated geotechnical information. (v) Development of a Digital Terrain Model (DTM) for analysis of geological and geotechnical aspects in three dimensions. (vi) Representation of the proposed alignment in the GIS, highlighting the engineering solutions envisaged and the environmental constraints. (vii) Field surveys for geological, geotechnical and environmental aspects recognition in the proposed route.

195.3 Results

The main results achieved were the development of the DTM overlaid by the latest satellite image and the layout of the future road; morphostructural analysis, lineament density map, slope hill map, photo interpretation and potentiometric map. The DTM (Fig. 195.1) is an important three dimensions analysis tool for the proposed route as well as the geological and geotechnical aspects and the environmental constraints. The DTM contributed as significantly geomorphological component incorporation in geological and

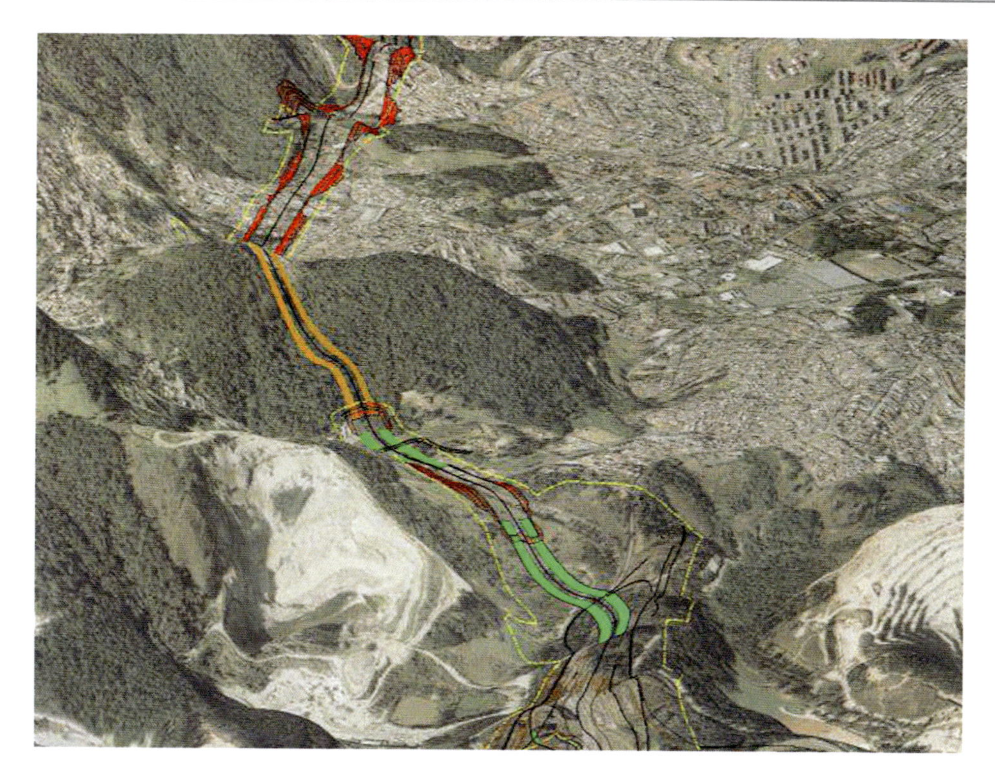

Fig. 195.1 Highway segment. Satellite image overlapped on the DTM. *Green* bridge or overpass; *orange* tunnel; *red* cut and fill; *yellow* right of way

geotechnical model of each specific work. The overlapped satellite image allows assessing risks considering also the use and occupancy. The morphostructural analysis was made to complement geological and geotechnical area characterization once boreholes conducted along the route construction provides the rock masses conditions only at specific sites. This morphostructural analysis comprised three distinct stages: extraction of lineaments, classification guidelines and preparation of lineament density map. The lineaments were extracted manually by shading relief image interpretation generated from the relief model. Using the ArcGIS tool resources, the total length of each lineament and its azimuth orientation were computed.

The lineament classification occurred relatively to water potential flow degrees in rock mass. This process took into account the orientation and the total length of the lineaments. Regarding orientation, should be considered to the following factors: field observations that suggest altered discontinuities orientations, suggesting greater water flow; reference works that indicate, on a regional scale, the orientation of the maximum horizontal stress compression since the structures parallel to these directions tend to be open, promoting water flow; in relation to the extension, it was considered that simply guidelines longer structures represent the most significant and therefore more water would lead. It is proposed to be created three extension classes that have statistical significance, compared to the universe of lineaments mapped. The density map of lineaments was made in order to produce continue information for the entire highway length. It was proposed to construct an appropriate size grid to cover the entire area covered by the project. The grid was divided into cells with one hectare (100 × 100 m) resolution, for each cell, calculated length of guidelines density per area (total length of guidelines on a single cell). The information was converted into points, corresponding to the center of each cell, and the cell values were interpolated by the Kriging method. As a final result, the map representing variations in density guidelines for the length of the stroke was obtained (Fig. 195.2). Areas with higher lineaments density could be interpreted as more massive fractured zones with poorer geomechanical characteristics. Two indirect investigation methods (slope hill and aerial photos analysis) were used to cover the entire track in order to identify possible deposits resulting from landslides, called talus bodies, which may interfere in the construction of the highway. The slope hill map was developed from the DTM and segmented into slopes classes, as the analytical region of the relief work. A sample output is shown in Fig. 195.3. The aerial photos analysis, was done in semi-detailed scale photos (e.g. 1:60,000 and 1:15,000 scales) and helped to confirm the talus bodies presence.

Field surveys must be programmed to confirm the presence of these masses that can undergo creep when saturated or by changing the water level or topography. The last stage

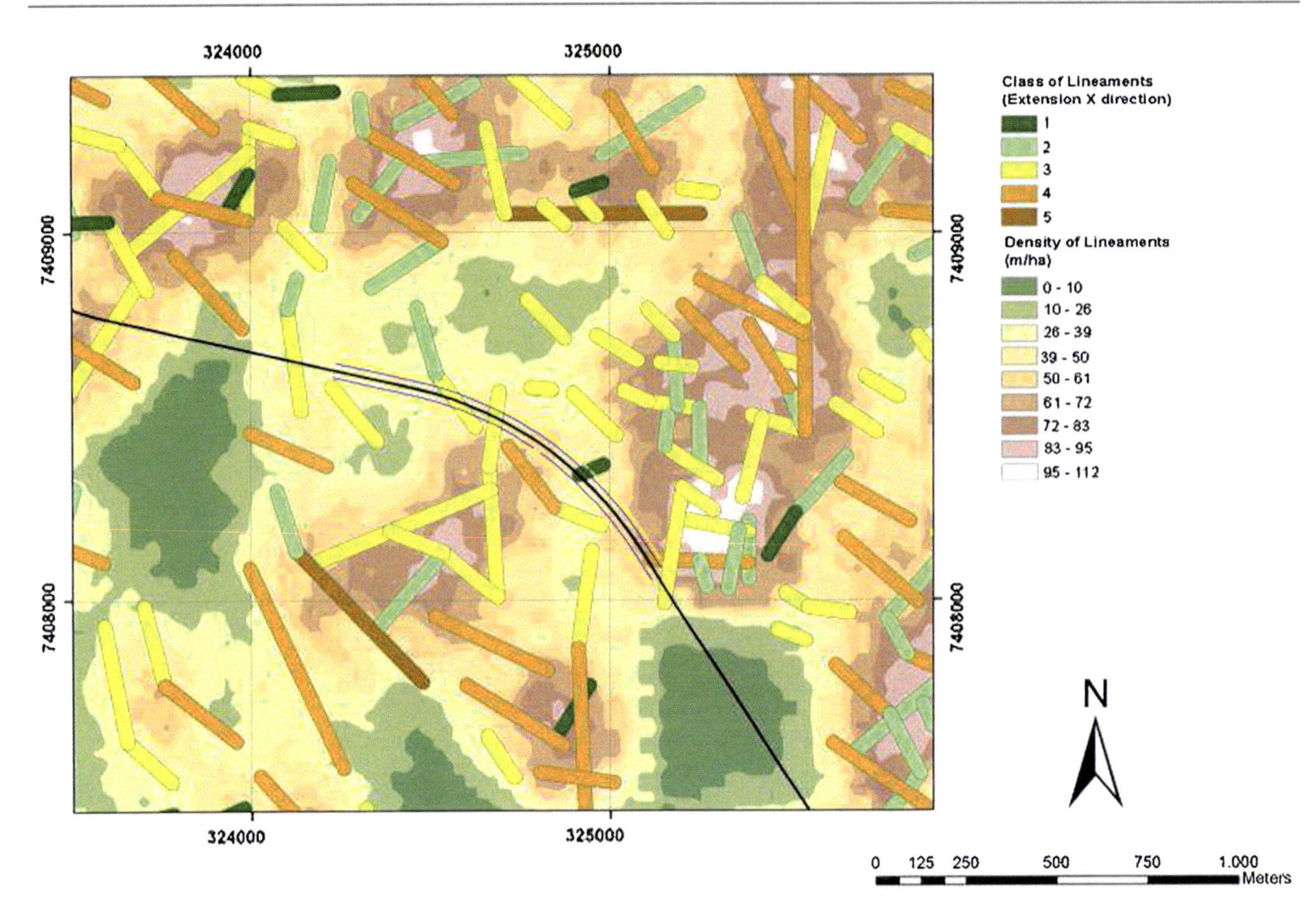

Fig. 195.2 Lineaments analysis in tunnel (in *Blue*) area construction

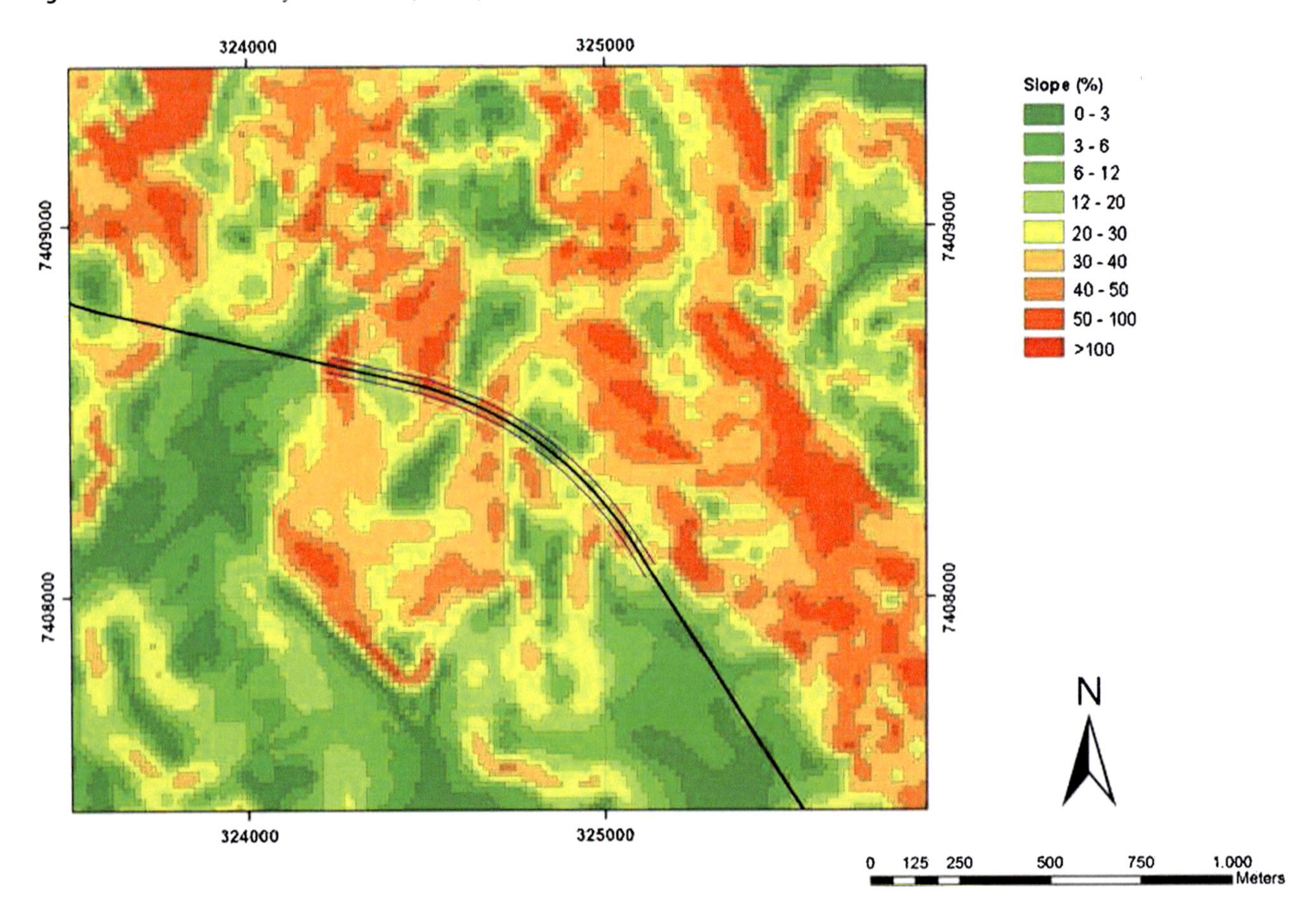

Fig. 195.3 Slope hill map for tunnel area. *Note* steep decline relief along the west entrance tunnel (in *blue*), favoring talus bodies occurrence

work was a potentiometric map development from the water level data obtained in boreholes, identifying sources (springs) and base level of the main drainages. This potentiometric map was crossed with the product of the lineaments analysis in GIS platform in order to identify zones of higher hydraulic gradients and possible flow paths. All models and maps were overlaid with the proposed road alignment (plan and profile). As main results from the method proposed, for possible environmental hazards, were identified: sections composed by rocks with varying degrees of alteration; stretches with geological constraints determining landslides; erosion in different horizons intercepted; deposits resulting from past landslides (talus bodies), including possible slope instabilization; stretches with interception of groundwater during excavation, including possible interception water table; Low, Medium or High structural conditioning water circulation around the potential for depletion of springs.

195.4 Conclusions

The model provides a morphostructural and hydrogeological analysis integration, still in the preliminary road design, for the prediction of the main environmental impacts associated with the implementation of road construction in tropical regions. The digital terrain model contributed significantly to the geomorphological component incorporation in geological and geotechnical model of each specific work area, as well as the type of intervention engineering proposal for each excerpt. The morphostructural analysis proved a valuable tool to complement geological and geotechnical information obtained through surveys (boreholes), in particular regarding the geological and structural constraints. These data were used for analyzes performed for tunnels and geotechnical solutions in hydrogeological characterization of the route presented. As main results from the method proposed, for possible environmental hazards, were identified sections composed by rocks with varying degrees of alteration and stretches with geological constraints determining potential landslides which was recommended alternative geotechnical solutions or adjust for their layout in plan; erosion in different horizons intercepted and recommended solution for each geotechnical differentiate horizon; identification of talus bodies or deposits resulting from past landslides, including possible instabilization, recommended adjust in layout plan; identify stretches with interception of groundwater during excavation, including possible interception water table, recommended specific drainage solution; For Low, Medium or High structural conditioning water circulation around the potential for depletion of springs was recommended further investigation and monitoring.

References

Anderson FJ (2006) A highlight of environmental and engineering geology in Fargo, North Dakota, USA. Environ Geol 49:1034–1042

Bell FG (2008) Basic environmental and engineering geology. Whittles Publishing, Caithness 342 pp

Sánchez LE (2006) Avaliação de Impacto Ambiental: Conceitos e Métodos. Oficina de Textos, São Paulo 496 pp

Author Index

A
Abbasi, Mehdi, 313
Abbasi, Mohsen, 313
Abdel-Ali, Chaouni, 95, 101
Accotto, Cristina, 591
Afonso, Maria José, 357, 819
Agui, Katsuhito, 793
Ahlam, Mounadel, 95, 101
Airoldi, Giulia, 657
Aldaihani, Humoud, 249
Almeida, Henrique, 819
Alzate, marta, 623
Anan, Shuji, 167
Antolini, Francesco, 705
Antonio, Dematteis, 963
Arnold, Patrick, 671
Arnold, Philippe, 671
Aryal, Sudar, 387
Asghari Kaljahi, Ebrahim, 295
Astolfi, Arianna, 689
Azam, Shahid, 381
Azevedo, Adalberto, 1071
Azzam, Rafig, 215, 1063

B
Babazadeh, Mahyar, 295
Bag, Ramakrishna, 537
Bahar, Ramdane, 1049
Baietto, Alessandro, 975
Baker, Paul, 299
Baldovin, Ezio, 825
Banks, Vanessa, 663
Barla, Marco, 705
Barykina, Olga, 91
Basaric, Irena, 285
Basu, Arindam, 865
Beaubien, Stanley Eugene, 1035
Bednarczyk, Zbigniew, 203
Benedetti, Gianluca, 967
Ben hassen, Mehdi, 113, 147
Berisavljevic, Dusan, 323
Berisavljevic, Zoran, 285, 323
Bernagozzi, Gabriele, 967
Bhatt, suresh, 195
Bianchi, Elio, 421
Bianchi, Gianpino Walter, 1007
Bianco, Fabrizio, 981
Bigarré, Pascal, 689
Bigi, Sabina, 1035
Bignall, Greg, 37
Bitar, Omar, 1067
Bitenc, Maja, 835
Bock, Helmut, 503
Bodnár, Nikolett, 851
Bonaga, Gilberto, 667
Bonetto, Sabrina, 709
Borbély, Dániel, 905
Borchardt, nicole, 627
Borgatti, Lisa, 421
Bornaz, Leandro, 721
Boronkay, Konstantinos, 839
Botu, Nicolae, 65
Bouffier, Christian, 689
Bo Wen, Ren, 875
Braga, Tania, 1067
Brandl, Johann, 937
Brenner, Olaf, 1039
Bricker, Stephanie, 663
Briganti, Renato, 967
Brink, George, 211
Brino, Lorenzo, 963
Brito, M. Graça, 415
Bugnano, Mauro, 657
Buocz, Ildikó, 901
Burger, Ulrich, 931, 975
Butscher, Christoph, 435
Buzzi, Olivier, 273

C
Cagna, Roberto, 657
Calista, Monia, 219
Campos, Sofia, 1067, 1071
Canora, Filomena, 757
Cao, Ling, 519
Cao, Liwen, 511
Capanema, Priscila, 1013
Castagna, Sara, 657
Castro, Laurenn, 1013
Cavalhieri, Caio, 1067
Cazacu, Gabriela Brandusa, 65

G. Lollino et al. (eds.), *Engineering Geology for Society and Territory – Volume 6*,
DOI: 10.1007/978-3-319-09060-3, © Springer International Publishing Switzerland 2015

Celestine, Okogbue, 27, 453
Chae, Byung-Gon, 883, 887
Chaminé, Helder I., 357, 819
Chen, Bao, 533, 541
Cheng, Qian-gong, 175, 185
Chen, Jianjie, 483
Chen, Rou-Fei, 153
Chen, Yonggui, 515, 541
Chikhaoui, Mohamed, 393
Choi, junghae, 883, 887
Chrysochoidis, Fragkiskos, 409
Ciufegni, Stefano, 981
Collins, Philip, 161, 243, 663, 779
Colombero, Chiara, 699
Comina, Cesare, 699
Continelli, Francesca, 967
Corsetti, marco, 731
Cui, Junwen, 87
Cui, Shuai, 1001
Cui, Yu-jun, 489, 499, 533
Culshaw, Martin, 31
Czap, Zoltan, 553

D
Daminato, antonio, 623
Daniela, Grigore, 65
Dao, Linh Quyen, 499
Davie, Colin, 453
Davis, gary, 337
De Angelis, Davide, 1035
Deffontaines, Benoit, 95, 101, 113, 139, 147, 153
De Freitas, Mike, 31
Dehira, Lauro, 1071
Delage, Pierre, 489, 499
Delgado Alonso-Martirena, Carlos, 3
Dematteis, Antonio, 985, 963
De Rosny, Julien, 689
Deva, Yogrndra, 947
Diederichs, Mark S., 41
Di Nunzio, Giuseppe, 731
Di Pasquale, Andrea, 31
Dippenaar, Matthys, 73
Djuric, Uros, 285
Dong, Qinghong, 813
Donnelly, Laurance, 31
Dorren, Luuk, 671
Durmeková, Tatiana, 927

E
Eduard, Kalustyan, 571
Edwards, Tom, 223
Eftekhari, Abbas, 599
Egal, Emmanuel, 1007
Eguchi, Takahiro, 793
Ene, Ezekwesili, 453
Ergul, Sibel, 761
Eusebio, Attilio, 591, 991
Evans, David, 31

F
Facello, Anna, 709
Fatima, El Hammichi, 95, 101
Favata, Giuseppe, 591
F. da Silva, ana paula, 399, 415
Feng, Fan, 1001
Fenton, Clark, 53, 567, 805
Fergason, Kenneth, 695
Fernandez-Steeger, Tomas, 1063
Ferranti, Flavia, 761
Ferreira, Fernando, 59
Ferrero, Anna Maria, 699, 709
Feyisa, Bayisa Regassa, 329
Figueira, Isabella, 1013
Filipello, Andrea, 715, 721
Fiorucci, Adriano, 981
Fityus, Stephen, 273
Fornari, Enrico, 591
Fortsakis, Petros, 409
Fortunato, Gerardo, 95, 101

G
Gallardo, Amarilis, 1067, 1071
Galperin, Anatoly, 1057
Gao, Weichao, 483
Gemander, Franziska, 465
Ghedhoui, rim, 139
Gherardi, Fabrizio, 971
Giles, David, 31, 249
Giraudo, Giorgio, 657
Gnavi, Loretta, 653
Görög, Peter, 905
Greenslade, Michael, 695
Gröger, Daniel, 65
Griffiths, James, 31
Grosse, Christian, 465
Guang Ming, REN, 875
Guerra, Cristiano, 967

H
Haerani, Evi, 107
Han, Bao, 17
Haneberg, William, 351
Han, Jimin, 483
Hassan, tabyaoui, 95, 101
Hedayatzadeh, mansour, 1019
Hesser, Jürgen, 551, 893
He, Yong, 229, 533
Höfer-Öllinger, Giorgio, 937
Hicks, Michael, 545
Hirnawan, 107
Homma, Shin-ichi, 167
Hong, Liu, 475
Hoxha, Dashnor, 393
Huang, Linchong, 649
Huo, Pan, 511
Hutchinson, D. Jean, 41

I
Igarashi, Toshifumi, 429

J
Jack, Christopher, 347
Jarek, Amanda, 1013
Jeon, cheolmin, 883
Jia, Mingyan, 483

Johnston, John, 273
Ju, Nengpan, 269
Junhai, An, 995
Józsa, Vendel, 443

K
Kagawa, Naoko, 107
Kaiser, Diethelm, 551
Kalinin, Ernest, 91
Kamera, Rita, 897
Kannan, Gopi, 921
Kasliwal, Vinod Kumar, 789
Kazilis, Nikolaos, 591
Keaton, Jeffrey, 363, 367
Kebede, Seifu, 329
Keegan, Tim, 223
Kempka, Thomas, 1063
Khademi Hamidi, Jafar, 1019
Khan, Fawad, 381
Köhler, Hans-Joachim, 215
Khmurchik, Vadim, 563
Khoshelham, Kourosh, 835
Kieffer, D. Scott, 835
Kingsland, Robert, 273, 387
Klebingat, Stefan, 1063
Kobayashi, Yoko, 167
Koungelis, Dimos, 799
Kovács, József, 851
Kovács, László, 897, 909, 915
Köppl, Florian, 753
Krastanov, Miroslav, 263
Käsling, Heiko, 465
Kulhawy, Fred H, 343

L
Lacerda, Luiz Alkimin, 1013
Langford, John Connor, 41
Larionov, Valery, 405
Laudo, Glarey, 963
Laureti, Roberto, 303, 611, 617, 633
Le Cor, Thomas, 809
Lege, Christian, 893
Leveinen, Jussi, 309
levin, mannie, 337
Li, Gen, 649
Lima, Celso, 59
Ling, Sixiang, 577
Lin, Weiren, 87
Li, Tianbin, 449
Liu, Char-Shine, 153
Liu, Fei, 813
Liu, Zhen, 583
Liu, Zhenghong, 229
LI, Xiang-Ling, 499
Lollino, Giorgio, 11
Lollino, Piernicola, 731
Lombardi, Salvatore, 1035
Lorenzo, Brino, 963
Lo Russo, Stefano, 255, 653
Lubo, Meng, 831
Lukas, Sven, 31

M
Machado, João, 59
Magalhaes, Samuel, 95, 101, 129
Makarycheva, Elizaveta, 405
Maksimovich, Nikolay, 563
Mancari, giuseppe, 623
Mandrone, Giuseppe, 699, 715, 721
Maria Elena, Parisi, 963
Marini, mattia, 623
Marinos, Vassilis, 409, 859
Martin, Christopher, 31
Martins, Pedro, 1071
Marzani, Alessandro, 667
Matejcek, Antonín, 927
Medeiros, Bruno, 725
Megyeri, Tamás, 905
Meng, Lubo, 449
Meriam, Lahsaini, 95, 101
Merrien-Soukatchoff, Véronique, 809
Mühling, Harry, 291
Miao, Jinli, 17
Michel, Stra, 623
Mielke, Philipp, 37
Milenkovic, Svetozar, 323
Millen, Bernard, 937
Mishra, akhila nath, 955
Moitre, Barbara, 981
Moncef, chabi, 739
Moreira, Patrícia, 819
Morelli, Gian Luca, 825
Morley, Anna, 31
Moser, Dorothee, 465
Moussa, Kacem, 393
Mo, XU, 475
Mészáros, Eszter, 915
M. Tóth, Tivadar, 897
Mukoyama, Sakae, 167
Munro, Rosalind, 367
Murgese, Davide, 657
Murton, Julian, 31
Muslim, Dicky, 107
Mutschler, Thomas, 679

N
Nadim, Charles-Edouard, 689
Nakaya, Masashi, 1031
Nanda, A, 921
Nankin, Rosen, 263
Nassim, HALLAL, 237
Nathalie, Monin, 963
Nawani, PC, 941
Nechnech, Ammar, 393
Nehler, Mathias, 37
Neves, Jorge, 59

Nguyen, Hieu Trung, 215
Nico, Giovanni, 731
Niitsuma, Hiroaki, 1031
Niu, Lihui, 515
Norbury, David, 31
Novakova, Lucie, 119
Novikov, Pavel, 405

O
Occhiena, Cristina, 689
Och, David, 387
Oke, Jeffrey, 869
Olivença, pedro, 47
Ondrášik, Rudolf, 927
Onuma, Kazuhiro, 1031
Oppizzio, Massimiliano, 981

P
Page, Michael, 299
Pal, Saikat, 921
Panasyan, Leili, 91
Panda, Bibhuti, 695
Parry, steve, 347
Pasculli, Antonio, 219
Patias, Josiele, 725
Patzelt, Johanna, 459
Paul, Benjamin, 893
Pejon, Osni Jose, 855
Pellicani, Roberta, 757
Pereira, jose henrique, 627
Perleros, Vassilis, 409
Pinheiro, Rogério, 357
Piraud, Jean, 1007
Pirulli, Marina, 689
Pitullo, Alfredo, 731
Polimeni, Santo, 967
Pombo, Joaquim, 399
Prountzopoulos, George, 409
Puglisi, Giuseppina Emma, 689

Q
Qin, Liaomao, 577
Qiu, Xin, 567
Quinta-Ferreira, Mário, 685

R
Rabia, mohamed chedly, 139
Rachid, BOUGDAL, 237
Raghuvanshi, Tarun, 329
Raith, Manuel, 465
Rakic, Dragoslav, 285
Ramos, Luís, 357
Ranfagni, Luca, 971
Rangeard, Damien, 809
Ren, Yong, 577
Ribeiro, Liliana, 279
Riccardo, Enrione, 1045
Riccardo, Torri, 963
Richter, Eva, 679
Richter, Ronald, 465
Riella, Alessandro, 991
Riemer, Wynfrith, 773
Riggi, Giuseppe, 967
Rivkin, Felix, 373
Robert, Alain-Alexandre, 1007
Robl, Klaus, 291
Rodgers, Chris, 273
Rodrigues, Aurora, 399
Rolando, Marco, 591
Romano, Fabio, 967
Rossi, Stefano, 971
Royse, katherine, 663
Rozgonyi-Boissinot, Nikoletta, 901
Rucker, Michael, 695
Ruggiero, Livio, 1035
Russell, Geoff, 387

S
Sacco, Pietro, 1035
Sadaoui, Omar, 1049
Sadaoui, Samir, 1049
Salih, Nihad, 779
Santoro, Federica, 219
Santos-Ferreira, Alexandre, 279
Santos, Vitor, 47, 415
Sappa, Giuseppe, 761
Sarigiannis, Dimitrios, 409
Sari, Mehmet, 843
Sarsembayeva, Assel, 243
Sasaki, Yasuhito, 167, 793
Sass, Ingo, 37
Scarano, Serena, 303, 611, 617, 633
Schlack, Frans, 299
Schlüter, Ralf, 1063
Schmitz, Heinz, 551
Schulten, Marc, 1063
Schwarz, Ludwig, 291
Sciarra, Nicola, 219
Sedighi, Majid, 525
Seferoglou, Konstantinos, 409
Seo, Yong-Seok, 883
Serangeli, Stefano, 303, 611, 617, 633, 1035
Serratrice, Jean François, 785
Shahri, Vijay, 921
Shao, Hua, 893
Sharda, Yash Pal, 947
Shen, Miao, 529
Shibayama, Motohiko, 107
Shinagawa, Shunsuke, 167
Shtrepi, Louena, 689
Shuqiang, Lu, 475
Sillen, Xavier, 499
Silva, Rui S., 819
Silveira, Rodrigo, 1013
Simon, Jérôme, 809
Singh, A K, 79
Siqueira, Alessandra, 1071
Soldo, Luca, 991
Somodi, Gábor, 915

Souza, Rafaela, 855
Spies, Thomas, 551
Spilotro, Giuseppe, 757
Spreafico, Margherita Cecilia, 667
Stead, Doug, 223
Steinacher, Reinhold, 557
Stoumpos, Georgios, 839, 859
Strom, Alexander, 645
Sturzenegger, Matthieu, 223
Sui, Wanghua, 749
Sun, Dongsheng, 87
Sun, Zhao, 511
Susic, Nenad, 323
Su, Wei, 529
Symeonidis, Konstantinos, 53
Szynkiewicz, Adam, 203

T
Taddia, Glenda, 255, 653
Takahashi, Manabu, 87
Tang, Anh Minh, 499
Tang, Chao-Sheng, 489
Taromi, majid, 599
Tartarello, Maria Chiara, 1035
Teixeira, José, 819
Testa, Daniela, 657
Thanh, Nguyen, 377
Thewes, Markus, 753
Thomas, Hywel, 525
Thuro, Kurosch, 459, 469, 753
Tochukwu Anthony, Ugwoke, 27
Toivanen, Tiina-Liisa, 309
Trigo, J. Filinto, 819
Tripathy, Snehasis, 537
Trizzino, Rosamaria, 317
Török, Ákos, 851, 901
Turki, Mohamed Moncef, 113

U
Ubertini, Francesco, 667
Ueshima, Masaaki, 107
Ulusay, Resat, 769
Umili, Gessica, 699, 709
Ündül, Ömer, 69
Utsuki, Shinji, 1031

V
Vardon, Philip, 545
Vasarhelyi, Balazs, 909
Vavilova, Vera, 571
Veiga, Anabela, 685
Vendramini, Mirko, 991
Verda, Vittorio, 255
Verzani, lorenzo, 1023
Vezocnik, Rok, 835
Villeneuve, Marlene, 439
Vinciguerra, Sergio, 699
Vitelli, Francesco, 757
Vittuari, Luca, 421, 667
Vlachopoulos, Nicholas, 41, 869
Vásárhelyi, Balázs, 443, 897
Vurbanov, Radoslav, 263

W
Wakolbinger, Walter, 291
Wang, Lianjie, 87
Wang, Qiong, 541
Wang, Yong, 511
Wang, Yu-feng, 175, 185
Watson, Paul, 249
Wen, Ann, 223
Wieser, Carola, 465
Wilfing, Lisa, 469
Wilkinson, Stephen, 805
Willms, David, 223
Winter, Mike, 641
Woods, Jonathan, 299
Woo, Ik, 887
Wu, Faquan, 17
Wu, Jie, 17
Wu, Jiu-jiang, 175, 185
Wu, Xi-yong, 577
Wu, Zhaoyang, 749

X
Xie, Lin, 1001
Xin lei, MA, 875
Xiong, Yonglin, 493
Xu, Yongfu, 519

Y
Yamamoto, Hiroyuki, 1031
Yan, Ming, 507
Yanna, YANG, 475
Ye, Weimin, 519, 529, 533, 541
Yokobori, Nohara, 429
Yoneda, Tetsuro, 429
Youcef, bouhadad, 125
Yousefibavil, Karim, 295
Yuan, Gexin, 483
Yu, Yongtang, 229

Z
Zaradkiewicz, Pawel, 291
Zhang, Bin, 1001
Zhang, Dingyang, 749
Zhang, Feng, 493
Zhang, Gailing, 749
Zhang, Henry, 387
Zhang, Huyuan, 507
Zhang, Qiang, 813
Zhang, Sumin, 229
Zhang, Wei, 229
Zhang, Xuezhe, 511
Zhang, Yawei, 541

Zhao, Jian, 901
Zhao, Jianjun, 269
Zhao, Siyuan, 577
Zhao, Weihua, 269
Zhao, zhenhua, 483
Zheng, Jianguo, 229
Zhou, Cuiying, 583
Zuo, Qiankun, 449
Zuquette, Lazaro Valentin, 725

Printed by Books on Demand, Germany